Resistive Circuit Analysis

Ohm's Law:

$v = Ri$

$i = Gv$

$G = 1/R$

Resistors in Series and Voltage Division:

$$R_{eq} = R_1 + R_2 + ... + R_n$$

$$v_n = \frac{R_n}{R_1 + R_2 + ... + R_n} V$$

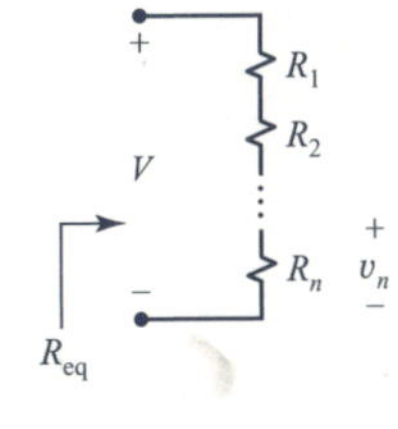

Resistors in Parallel and Current Division:

$$G_{eq} = G_1 + G_2 + ... + G_n$$

$$i_n = \frac{G_n}{G_1 + G_2 + ... + G_n} I$$

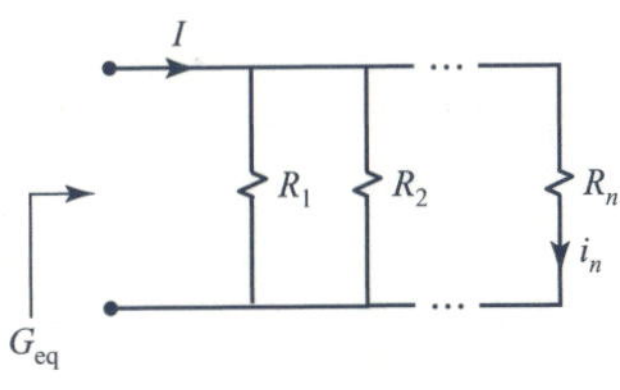

Special Case, Two Resistors in Parallel:

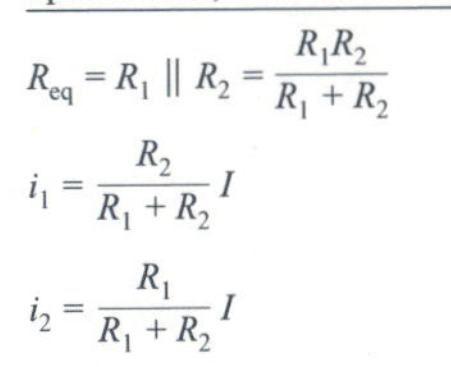

$$R_{eq} = R_1 \parallel R_2 = \frac{R_1 R_2}{R_1 + R_2}$$

$$i_1 = \frac{R_2}{R_1 + R_2} I$$

$$i_2 = \frac{R_1}{R_1 + R_2} I$$

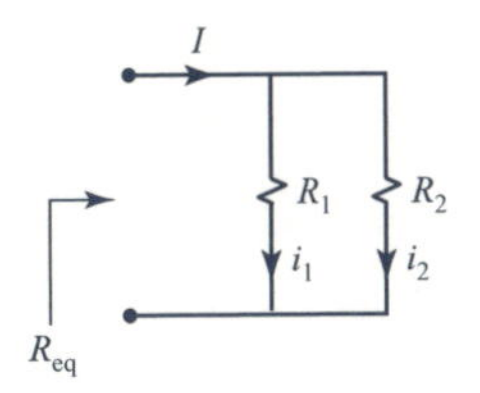

Source Transformations:

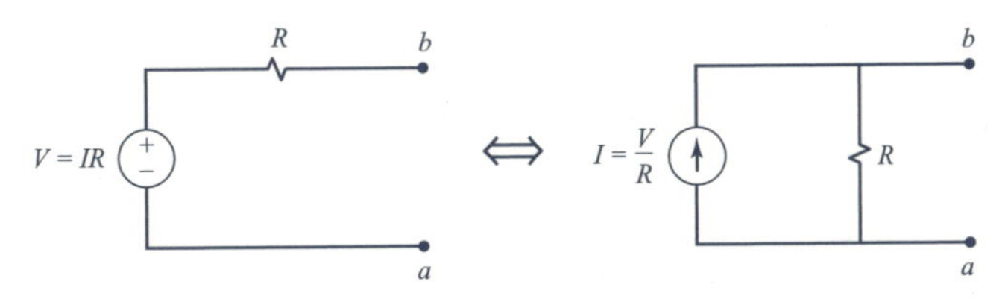

FUNDAMENTALS OF ELECTRIC CIRCUIT ANALYSIS

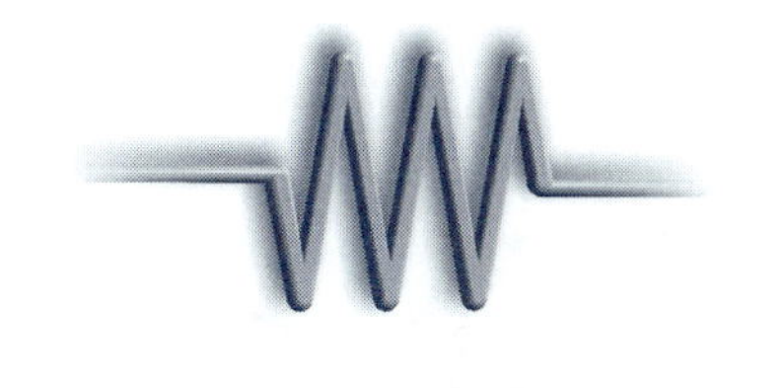

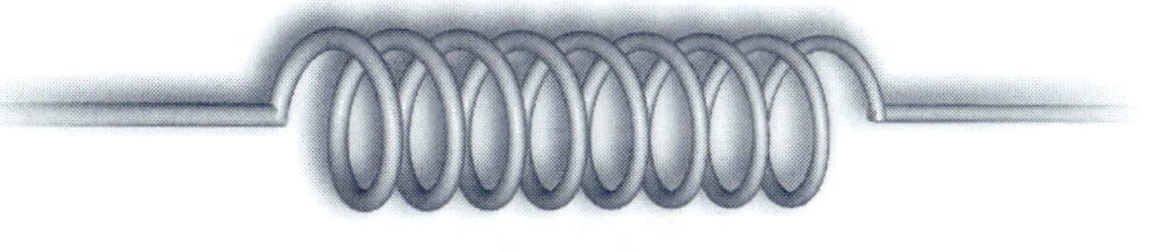

FUNDAMENTALS OF ELECTRIC CIRCUIT ANALYSIS

by

Clayton R. Paul

Department of Electrical and Computer Engineering
Mercer University

and

Emeritus Professor of Electrical Engineering
University of Kentucky

Acquistions Editor *Bill Zobrist*
Marketing Manager *Katherine Hepburn*
Senior Production Editor *Patricia McFadden*
Illustration Editor *Gene Aiello*
Senior Designer *Harry Nolan*
Interior Design *Michael Jung*
Cover Design *Sue Noli Design*
Cover Photo *Dr. Jeremy Burgess/Science Photo Library/Photo Researchers*

This book was set in *Times New Roman* by *TechBooks* and printed and bound by *Hamilton Printing.*
The cover was printed by *Phoenix Color.*

The book is printed on acid-free paper.

This book is free of errors. All text material and solutions have been meticulously reviewed numerous times. In addition, the author has worked all Example Problems, Exercise Problems and End-of-Chapter problems using several different solution methods to ensure that the answers are correct. The publisher and the author want to ensure that the presentation is of the highest quality. The reader should be focused on learning the content, not distracted by errors. You shouldn't have to learn from our mistakes. We welcome your comments on how the text can be further improved: Editor, Electrical Engineering, College Division, John Wiley & Sons, 605 Third Avenue, NY, NY 10158.

Library of Congress Cataloging-in-Publication Data:

Paul, Clayton R.
Fundamentals of circuit analysis / by Clayton R. Paul.
p. cm.
Includes bibliographical references.
ISBN 0-471-37195-5 (cloth : alk. paper)
1. Electric circuit analysis. I. Title.

TK454 .P285 2000
621.319'2--dc21 00-020738

ISBN 0-471-37195-5
Printed in the United States of America

10 9 8 7 6 5 4 3 2 1

To the humane and compassionate treatment of animals

Everything should be made as simple as possible, but no simpler.
Albert Einstein

Preface

This text is intended for a one-semester course in the analysis of linear electric circuits. The goal of this text is to insure that the reader has firmly mastered those skills that are necessary for analyzing the majority of electric circuits encountered in later courses of an electrical engineering curriculum as well as other disciplines in engineering. Only those topics that are necessary for achieving this goal are included and discussed. Hence this text is not intended to be a complete or theoretical exposition of all aspects of electric circuit theory. More thorough expositions are found in many traditional circuit analysis texts such as C.R. Paul, *Analysis of Linear Circuits* (McGraw-Hill, 1989). Such an exposition would not be possible in a one-semester course, nor is it necessary in order to insure a mastery of the essential circuit analysis skills. Later courses in an electrical engineering curriculum delve into those aspects in more depth. Without the ability to analyze relatively simple circuits reliably, the reader cannot comprehend the material in those later courses. Hence this text concentrates on a mastery of basic skills. It is this author's opinion that unnecessary detail tends to obscure a student's understanding of the basic concepts. Therefore there has been a deliberate attempt to minimize this detail to assist the reader in distinguishing the trees from the forest while at the same time not sacrificing the student's ability to master the essential circuit analysis skills.

It is also this author's belief that organization of the material is one of the important aspects of a student's learning and long-term retention of the basic skills. The organization of this text is somewhat different from the more traditional texts and is chosen for these pedagogical reasons.

Electrical engineering (EE) curricula have been contained in a four-year degree program since virtually the beginning of the discipline. A large amount of material has been inserted into that standard curriculum, and few topics have been deleted. There is little indication that this severely crowded four-year EE curriculum will be expanded to five years in the near future in order to relieve that pressure. Perhaps the introduction of the digital computer has had the most dramatic impact on that material. EE programs now offer combined degrees in electrical and computer engineering or separate tracks leading to a degree in either. In some programs up to one-fourth of the required EE courses concern computer topics and have been included in the curriculum in the last 10 years. The inclusion of a large amount of new material in the curriculum has resulted in severe course compression in EE curricula. Several courses have been shortened to make room for it. For example, the theory of electromagnetic fields, arguably the most fundamental EE topic, has been reduced from a two-semester sequence to a one-semester sequence in a large number of schools. The next most fundamental topic for an EE, analysis of electric circuits, remains the lone holdout as a two-semester sequence. There will be increasing pressure to reduce this to a one-semester sequence to accommodate the growing number of topics resulting from technological advances.

Many of the EE programs offer a Computer Engineering degree in which a one-semester course in electric circuit analysis is standard. This text would also be suitable for that use. In addition, many EE programs offer an electric circuits course as a service course to non-EEs such as mechanical engineers and civil engineers. These tend to also be one-semester courses for which this text would be suitable.

The skills for the analysis of resistive circuits are the most fundamental ones in circuit analysis and are thoroughly covered in the early chapters. The reader must be able to analyze, rapidly and without error, all such circuits in later courses, and this is one of the primary goals of the text. The second most fundamental circuit analysis skill and the second primary objective is the ability to analyze circuits whose sources are sinusoids (the phasor method). This skill once again uses the fundamental resistive circuit analysis skills with only the added burden of complex arithmetic. Once the student sees this important connection and realizes that all the resistive circuit analysis skills can be used in the phasor circuit, the solution of such circuits can be dealt with in a rapid and efficient manner. Understanding the concept of transient response, time constants in first-order circuits, and the overdamped, underdamped, and critically damped responses of second-order circuits is the third primary objective of the text. These are covered in a single, integrated chapter by the classical solution and the Laplace transform solution. Only the minimum of necessary detail of the Laplace transform needs to be covered in order to provide the reader with sufficient skill. Here again, all the resistive circuit analysis skills are used once again to solve these circuits.

Chapter 1 covers the basic concepts of voltage and current as well as the fundamental laws (KVL and KCL) governing those quantities. The fundamental circuit elements such as the independent sources and the resistor are not covered until Chapter 2. In Chapter 1, the elements are denoted as boxes, and hence the student can learn to apply KVL and KCL correctly without being distracted by the detail of what is in the boxes. Most of the mistakes made by students are in the application of KVL and KCL. Hence this chapter is a crucial part of their success in mastering the remaining topics. Certain other fundamental concepts, such as series and parallel elements and equivalent circuits, are also covered in this chapter without regard to the content of the element boxes. This also serves to emphasize that other quantities, such as the power delivered to or delivered by an element, are not directly related to the type of element (except in the case of a resistor, which is discussed in Chapter 2). Hence the ability of an independent source to deliver or absorb power in a multisource circuit is readily accepted by the student, and he/she has no preconceived notions that sources (independent or controlled) must always deliver power.

Chapter 2 is the most important chapter of the text, since it lays down the fundamental resistive circuit analysis skills common to all later analyses (phasor circuit analysis, classical time-domain analysis, and Laplace transform analysis). The fundamental circuit elements—the independent sources and the resistor—are discussed, and the fundamental circuit analysis skills of resistors in series and in parallel, voltage and current division, and source transformations are covered in considerable detail. The method of circuit reduction for single-source circuits is covered, as are the important single-loop and single-node-pair circuits. Most of the circuit reduction techniques seek to reduce the circuit to a single-loop or single-node-pair form. Hence the ability to rapidly and flawlessly analyze such circuits is crucial to a mastery of later skills. The solution of circuits containing more than one source is also demonstrated by the direct application of KVL, KCL, and Ohm's law. The key to the direct solution of simple circuits using this method is to express all other circuit variables in terms of the unknown, using KVL, KCL, and Ohm's law, and to resist the temptation to define any new unknown variables. Students find this powerful technique simple to apply once they have learned to resist the temptation to define other, extraneous unknown variables. The majority of the circuits that the students will be required to analyze in later courses are very simple and can be analyzed by these simple techniques. In addition, all the circuit analysis skills of the later chapters (superposition, Thevenin and Norton equivalents, phasor circuit analysis, and Laplace transform circuit analysis) rely heavily on the student having mastered these fundamental skills. Any deficiencies in these skills will show up and handicap the student. The controlled sources are also introduced, and the previous analysis methods are applied to circuits containing these sources. The concept of equivalent resistance at two terminals of a circuit containing controlled sources is also discussed. This is of considerable use in later electronics courses.

Additional techniques such as superposition, Thevenin, and Norton equivalent circuits as well as the node-voltage and mesh-current methods are discussed in Chapter 3. Superposition as well as the Thevenin and Norton equivalent circuit techniques rely heavily on the basic circuit analysis principles discussed in Chapter 2. Again, a deficiency in the student's understanding of and ability to apply the techniques of Chapter 2 will readily show up here. Finally we discuss the node-voltage and mesh-current analysis techniques.

Typical circuit analysis texts, and consequently the courses using them, cover the node-voltage and mesh-current techniques before most of the more fundamental skills of Chapter 2 are introduced. We have chosen to delay their discussion until the more fundamental resistive circuit analysis skills are firmly mastered, for several reasons. The first reason is that virtually all circuits (with a few minor exceptions) that the student will be required to analyze (by hand) in later courses can be analyzed using any or all of the previously discussed simpler techniques in Chapter 2 and do not require use of the node-voltage or mesh-current methods. The second reason is that students tend to rely on the first analysis method to which they are introduced to the exclusion of later methods. Hence we find that students will routinely apply node or mesh analysis to simple circuits if the node-voltage and mesh-current analysis methods are introduced first. This will often require the simultaneous solution of two or more equations in that many unknowns even though the circuit can be analyzed much faster and without the necessity of solving simultaneous equations by using the previously discussed simpler methods. In addition, the node-voltage and mesh-current methods, unlike the simpler methods of Chapter 2, provide virtually no insight into the factors affecting the circuit behavior. The primary objective in studying circuit analysis is in developing skills useful in circuit *design*. The simple analysis methods of Chapter 2 provide the insight into the development of those design skills, whereas the node-voltage and mesh-current methods do not. The author realizes that this decision to delay discussion of the node and mesh analysis methods is not universally accepted, but the above logic is sound and the author's experience in teaching circuits for over 30 years bears out its effectiveness.

Chapter 4 introduces the operational amplifier (op amp), which has become a standard circuit element. The properties of the actual op amp are discussed and simplified to those of the ideal op amp. These are used to analyze the standard op-amp circuits.

In Chapter 5 we discuss the energy storage elements. The discussion is traditional and mutual inductance as well as the ideal transformer are included.

The analysis of the steady-state response of circuits to sinusoidal sources (phasor circuit analysis) is covered in Chapter 6. Once the student understands that the original sinusoidal source (sine or cosine form) can be replaced with the complex exponential source using Euler's identity, it becomes obvious that the solution for the original source (sine or cosine form) can be obtained by taking the real or imaginary part of that response, using superposition. In this way we avoid having to convert all sources to cosines or all to sines—an error-prone process that is unnecessary. Once the idea of transforming the time-domain circuit to the frequency (phasor) domain is mastered, it becomes clear to the student that all of the previous resistive circuit analysis skills can now be utilized in solving that phasor circuit with only the added burden of complex arithmetic. Hence the student clearly sees that very few new concepts need be learned, and all the previously learned resistive circuit analysis skills can be reapplied with only the added complication of complex arithmetic. The notion of average power is discussed along with the concepts of resonance and elementary filters. Three-phase power distribution for balanced loads is also briefly discussed.

The time-domain (transient) response of circuits as a result of switching is treated in the classical fashion through the direct solution of the circuit differential equation in Chapter 7 for the simple first-order circuits and the series and parallel *RLC* circuits. The corresponding treatment of the time-domain response using the Laplace transform is also contained in Chapter 7. This coverage is sufficient to allow the analysis of most time-domain circuits that the student will encounter. The unit step and unit impulse functions are considered and the responses to them are obtained. Convolution is briefly introduced, as are the concepts of linearity and time invariance. Only the essential details of the Laplace transform required for the analysis of circuits are discussed in order for the student to master this technique in analyzing time-domain circuits.

Contrary to the organization of traditional circuit analysis texts, we have chosen to discuss phasors (Chapter 6) before transients (Chapter 7), for the following reason. The phasor circuit analysis skills are the most frequently required skills in later EE courses. Transient analysis skills are used much less frequently. Hence we chose this order of presentation for this hierarchical reason and to provide an earlier discussion of this fundamentally important skill. It is, however, feasible for the instructor to interchange this order of presentation if desired.

The appendix covers the essential techniques for the solution of systems of linear algebraic equations—an essential skill. Cramer's rule, Gauss elimination, and the essentials of matrix algebra are covered. Each chapter of the text concludes with examples of the use of PSPICE and/or MATLAB in solving the circuits of that chapter. Hence the use of PSPICE and MATLAB is thoroughly discussed throughout the text. The placement of the PSPICE and MATLAB material at the end of each chapter allows the instructor to omit this material if desired.

Included with each chapter are an abundance of *exercise problems*. These are relatively simple and allow the student to self-test whether he/she has mastered the previous material. Each chapter includes a large number of worked out *example problems* that illustrate the application of the material. Throughout the text and these example problems we have tried to point out the common mistakes that the author has observed over the years. It is important to do so in order to head off misunderstandings. A large number of *end-of-chapter problems* are also included. These are categorized according to the relevant sections of the text in order to assist the instructor. These range from relatively simple to moderately difficult. The answers to the majority of the exercise problems, example problems, and end-of-chapter problems are checked using PSPICE and/or MATLAB. The answers to the exercise problems and the end-of-chapter problems are provided at the end of the question for ready access by the reader. Examples of practical applications of the material are also provided at the end of each chapter. The important equations are highlighted by placing them in boxes.

A detailed solution manual for the end-of-chapter problems is available from the publisher for instructors who adopt the book. Additional instructor resources are available at the web site www.wiley.com/college/paul.

This format has been class tested in a one-semester course that all engineering students are required to complete in the School of Engineering at Mercer University. The course is 4 credit-hours and consists of three 50-minute lecture periods and one two-hour recitation period per week. The recitation is used for problem solving and the administration of exams. The recitation is an essential part of the success of this course. There is sufficient diversity of topics in the text to allow the instructor to tailor his/her choice to fit within a one-semester time frame and to fit his/her particular program. For example, instructors may choose to cover only the classical solution from Chapter 7 and omit the Laplace transform solution or vice versa. The Laplace transform presentation is compact but complete enough to demonstrate all aspects of the time-domain solution that are obtained with the classical solution. Some instructors may wish to defer discussion of op amps (Chapter 4) to a later electronics course.

This one-semester course represents the totality of the circuit analysis skills required of those students who choose to continue as EEs at Mercer University. Student acceptance of the format is virtually universal, and our experience has shown that they possess the necessary fundamental circuit analysis skills required by the later EE courses.

Clayton R. Paul
Macon, GA
October 1999

Table of Contents

Chapter 1 Basic Definitions and Laws **1**

1.1 Charge and Electric Forces *2*

1.2 Voltage *3*

1.3 Current and Magnetic Forces *7*

1.4 Lumped Circuit Elements and Power *10*

1.5 Kirchhoff's Current Law (KCL) *13*

1.6 Kirchhoff's Voltage Law (KVL) *18*

1.7 Conservation of Power *28*

1.8 Series and Parallel Connections of Elements *30*

1.9 Equivalent Circuits *32*

1.10 Redrawing Circuits in Equivalent Forms *34*

1.11 Application Examples *36*

1.11.1 Residential Power Distribution *36*

1.11.2 Automobile Storage Batteries *39*

Chapter 2 Basic Circuit Elements and Analysis Techniques **51**

2.1 The Independent Voltage and Current Sources *51*

2.2 The Linear Resistor and Ohm's Law *55*

2.3 Single-Loop and Single-Node-Pair Circuits *59*

2.4 Resistors in Series and in Parallel *67*

2.4.1 Circuit Solution by Circuit Reduction *73*

2.5 Voltage and Current Division *77*

2.6 Solutions for Circuits Containing More than One Source *85*

2.7 Source Transformations *90*

2.8 The Controlled (Dependent) Voltage and Current Sources *94*

2.8.1 Analysis of Circuits Containing Controlled Sources *97*

2.8.2 Equivalent Resistances of Circuits Containing Controlled Sources *100*

2.9 PSPICE Applications *105*

2.10 Application Examples *116*

2.10.1 Ammeters, Voltmeters, and Ohmmeters *116*

2.10.2 An Audio Amplifier *118*

Chapter 3 Additional Circuit Analysis Techniques 135

3.1 The Principle of Superposition *135*

3.2 The Thevenin Equivalent Circuit *142*

3.3 The Norton Equivalent Circuit *147*

3.4 Maximum Power Transfer *153*

3.5 The Node-Voltage Method *154*

3.5.1 Circuits Containing Voltage Sources *160*

3.6 The Mesh-Current Method *164*

3.6.1 Circuits Containing Current Sources *169*

3.7 PSPICE Applications *173*

3.8 MATLAB Applications *175*

Chapter 4 The Operational Amplifier (Op Amp). 191

4.1 The Actual Op Amp versus the Ideal Op Amp *191*

4.1.1 The Inverting Amplifier *194*

4.1.2 Negative Feedback and Saturation *195*

4.2 Other Useful Op-Amp Circuits *198*

4.2.1 The Noninverting Amplifier *198*

4.2.2 The Difference Amplifier *199*

4.2.3 The Summer *201*

4.2.4 The Buffer *203*

4.2.5 The Comparator *204*

4.3 Applications *204*

4.3.1 A Strain-Gauge Instrumentation Circuit *205*

4.3.2 A Photocell Instrumentation Circuit *206*

4.4 PSPICE Applications *206*

Chapter 5 The Energy Storage Elements. 213

5.1 The Capacitor *213*

5.1.1 Capacitors in Series and in Parallel *218*

5.1.2 Continuity of Capacitor Voltages *220*

5.2 The Inductor *220*

5.2.1 Inductors in Series and in Parallel *224*

5.2.2 Continuity of Inductor Currents *225*

5.3 Mutual Inductance *226*

5.3.1 The Ideal Transformer *231*

5.4 Response of the Energy Storage Elements to DC Sources *234*

5.5 The Differential Equations of a Circuit *235*

5.6 The Op-Amp Differentiator and Integrator *239*

5.7 PSPICE Applications *241*

5.8 Application Examples *250*

5.8.1 An Electronic Timer *250*

5.8.2 DC Power Distribution Sag in Digital Logic Circuits *251*

Chapter 6 Sinusoidal Excitation of Circuits 261

6.1 The Sinusoidal Source *261*

6.1.1 Representation of General Waveforms Via the Fourier Series *263*

6.1.2 Response of Circuits to Sinusoidal Sources *266*

6.2 Complex Numbers, Complex Algebra, and Euler's Identity *267*

6.3 The Phasor (Frequency-Domain) Circuit *274*

6.3.1 Representation of Sinusoidal Sources With Euler's Identity *274*

6.3.2 The Phasor Circuit *277*

6.4 Applications of Resistive-Circuit Analysis Techniques in the Phasor Circuit *280*

6.5 Circuits Containing More than One Sinusoidal Source *287*

6.5.1 Sources of the Same Frequency *288*

6.6 Power *292*

6.6.1 Power Relations for the Elements *296*

6.6.2 Power Factor *299*

6.6.3 Maximum Power Transfer *303*

6.6.4 Superposition of Average Power *304*

6.6.5 Effective (RMS) values of Periodic Waveforms *310*

6.7 Phasor Diagrams *314*

6.8 Frequency Response of Circuits *316*

6.8.1 Transfer Functions *316*

6.8.2 Resonance *320*

6.8.3 Elementary Electrical Filters *322*

6.8.4 Active (Op-Amp) Filters *326*

6.9 Commercial Power Distribution *327*

6.9.1 Wye-Connected Loads *329*

6.9.2 Delta-Connected Loads *332*

6.10 PSPICE Applications *333*

6.11 MATLAB Applications *337*

6.12 Application Examples *341*

6.12.1 AM Radio Tuner *341*

6.12.2 Crosstalk in Transmission Lines *342*

Chapter 7 General Excitation of Circuits.................. 361

7.1 First-Order Circuit Response *361*

7.1.1 The *RL* Circuit *362*

7.1.2 The *RC* Circuit *369*

7.2 Second-Order Circuit Response *373*

7.2.1 The Series *RLC* Circuit *373*

7.2.2 The Parallel *RLC* Circuit *383*

7.3 The Laplace Transform *387*

7.3.1 Important Properties of the Laplace Transform *388*

7.3.2 Transforms of Important Time Functions *389*

7.3.3 The Laplace Transforms of the *R*, *L*, and *C* Elements *394*

7.3.4 Determining Initial Conditions *401*

7.4 The Inverse Laplace Transform by Partial-Fraction Expansion *403*

7.4.1 Real and Distinct Poles *405*

7.4.2 Real and Repeated Poles *407*

7.4.3 Complex Conjugate Poles *409*

7.5 First-Order Circuit Response *413*

7.5.1 The *RL* Circuit *413*

7.5.2 The *RC* Circuit *417*

7.5.3 Additional Examples of First-Order Circuit Response *421*

7.6 Second-Order Circuit Response *424*

7.7 The Step and Impulse Responses *435*

7.8 Convolution *443*

7.9 Application of Linearity and Time Invariance *446*

7.10 PSPICE Applications *451*

7.11 MATLAB Applications *456*

7.12 Application Examples *464*

7.12.1 A Light Flasher *464*

7.12.2 Ringing in Digital Circuits *465*

7.12.3 Crosstalk in Digital Circuits *466*

APPENDIX Solution of Linear Algebraic Equations 485

A.1 Gauss Elimination *486*

A.2 Cramer's Rule *487*

A.3 Matrix Algebra *490*

Index 497

CHAPTER 1

Basic Definitions and Laws

This text concerns the analysis of linear electric circuits. No other skill is more basic for electrical engineers (EEs) than *circuit analysis*. Virtually all other EE courses rely on the mastery of the circuit analysis skills of this text. Deficiencies in understanding and being able to apply these skills will therefore handicap the reader throughout all later EE courses. Hence it is vitally important to master the application of these skills. It can be said that an engineer's function is characterized by the following:

> *Engineers develop mathematical models of physical systems for the purpose of design of those physical systems.*

This definition emphasizes that the mathematical analysis of models of systems is the heart of an engineer's daily work. Hence the development of skills for the analysis of those models, which are presented in this text, is crucial to the success of an engineer's daily work. The alternative to this development of a mathematical model and its subsequent analysis would be to construct a physical prototype of the system and adjust it in the laboratory until it satisfies the desired design objectives. This trial-and-error method is unsatisfactory, since endless adjustment of the system parameters would be made with no clear end in sight and no clear direction toward an optimum solution. On the other hand, a mathematical model of the system can be analyzed and designed either by hand or with a computer in a relatively straightforward, clear, and optimum manner. Once the performance of the mathematical model matches the design goals, we have reasonable certainty that the physical prototype will match those goals also. Developing the skills for analyzing the mathematical model also yields insight into the behavior of the physical system, which can be used to direct the design process.

As a simple example of this mathematical modeling process, consider a compact-disc sound system illustrated in Fig. 1.1. A simple circuit model of the physical system is shown and is composed of the typical electric circuit elements that we will consider. These elements are independent sources, resistors, capacitors, inductors, and dependent (controlled) sources. Upon completion of this text, the reader will be able to analyze electric circuit models similar to this for the important variables of interest (voltages and currents) and in addition will develop sufficient insight into the circuit behavior to allow the choice of element values to achieve a desired design goal. In essence, this is the objective of the text.

In order to refine our circuit analysis skills without being distracted by unnecessary detail, we will use simple values for the circuit elements (1 Ω, 12 A, etc.) rather than the more realistic

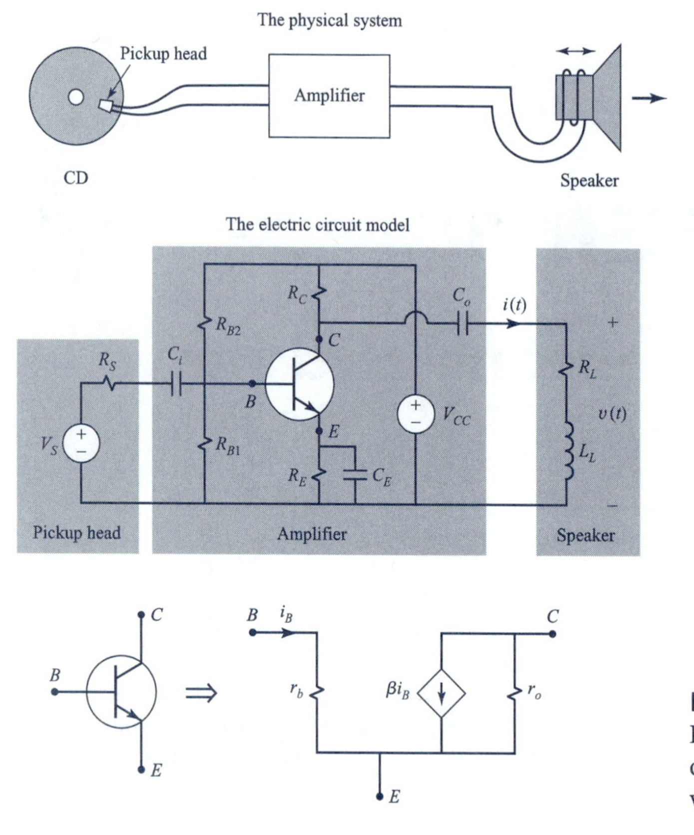

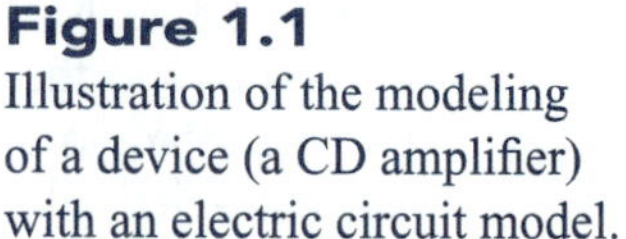

Figure 1.1
Illustration of the modeling of a device (a CD amplifier) with an electric circuit model.

values (1 kΩ, 12 mA, etc.). Once the analysis skills are mastered, it is a simple matter to handle circuits having the more realistic element values. Throughout the text we will use the internationally accepted SI system of units: meters (m), kilograms (kg), and seconds (s). Other units are derived as combinations of the basic SI units, e.g., the unit of force is the newton (N), where $1\ \text{N} = 1\ \text{kg-m/s}^2$. The important power-of-ten multipliers are pico (p, 10^{-12}), nano (n, 10^{-9}), micro (μ, 10^{-6}), milli (m, 10^{-3}), centi (c, 10^{-2}), kilo (k, 10^3), mega (M, 10^6), giga (G, 10^9).

Specific types of circuit elements will be considered in later chapters. In this chapter we will learn to apply the basic laws governing the voltages and currents of the elements which are interconnected to construct an electric circuit model. These laws dictate how the voltages and currents of elements are related for a particular interconnection without regard to the specific type of element.

1.1 Charge and Electric Forces

Forces produced by electric charges and their movement are the most basic items of interest in electrical engineering. An electron carries a negative charge of 1.602×10^{-19} C, where the unit of charge is the coulomb (C). Charges of the same sign repel each other, while charges of opposite sign attract. For example, consider two point charges, Q_1 and Q_2, separated a distance r as shown in Fig. 1.2. The force exerted on one charge by the other varies directly as the product of the charges (the sign of the charge is included in its value) and inversely as the square of the distance between them according to Coulomb's law:

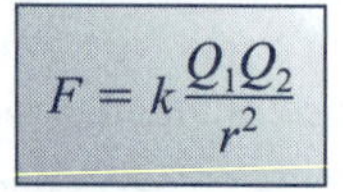

$$F = k\frac{Q_1 Q_2}{r^2} \tag{1.1}$$

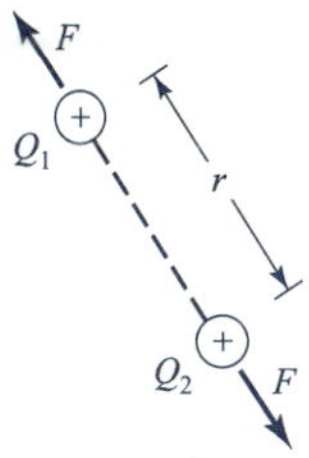

Figure 1.2
Illustration of Coulomb's law and the attraction/repulsion of charges.

The constant of proportionality, k, is approximately 9×10^9 when the other parameters are given in SI units and the surrounding medium is free space.

Exercise Problem 1.1

Determine the force exerted between the following charges: (a) $Q_1 = 1$ C, $Q_2 = 1$ C, $r = 1$ m, (b) $Q_1 = 3 \times 10^{-9}$ C, $Q_2 = 5 \times 10^{-6}$ C, $r = 2$ cm, (c) $Q_1 = -2$ nC, $Q_2 = 8$ pC, $r = 1$ μm.

Answers: 144 N, 9×10^9 N (1 million tons), 0.338 N.

Exercise Problem 1.2

Two equal charges, Q, are separated a distance d. Determine the distance to produce a force of 1 N for the following charges, (a) $Q = 1$ μC, (b) $Q = 1$ pC, (c) $Q = 5$ mC.

Answers: 9.49 cm, 94.9 nm, 474.34 m.

1.2 Voltage

Since charges exert forces on other charges, energy must be expended in moving a charge in the vicinity of other charges. Thus charges produce a type of force field.

The unit of energy is the joule (J), defined as the energy expended in the exertion of a force of one newton in moving an object through a distance of one meter (1 J = 1 N-m). For example, consider moving a charge q from point a to point b along a chosen path in the presence of another charge Q as illustrated in Fig. 1.3. Over certain portions of the path, the force exerted by Q may oppose the movement of q, while over other portions the force may be in a direction such as to aid the movement of q. Hence the net energy expended in moving the charge q from a to b, w_{ba}, may be positive or negative. This is similar to gravitational potential. For example, lifting a mass M to a height h above the earth as illustrated in Fig. 1.4 requires work on our part, and we say that

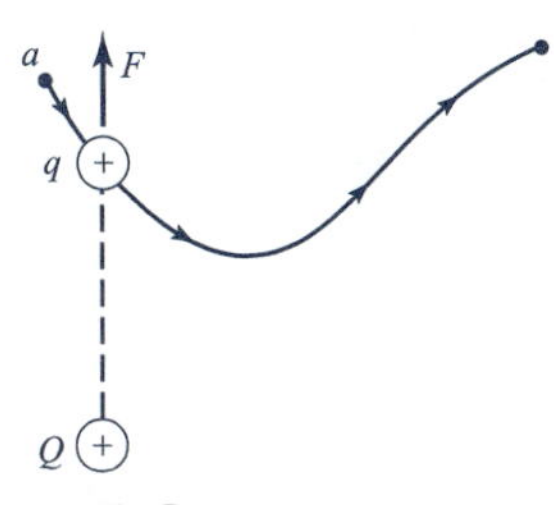

Figure 1.3
Illustration of the work required to move a charge along a path.

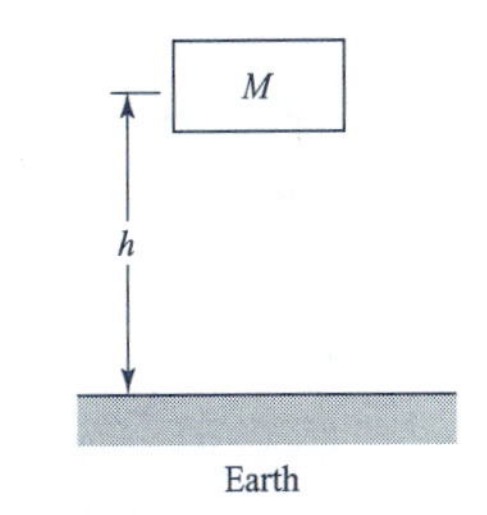

Figure 1.4
The similarly of gravitational potential to electric potential.

points above the earth are at a higher gravitational potential *with respect to the earth as a reference*. Similarly, if the movement of q from a to b as in Fig. 1.3 requires work w_{ba} on our part then we say that *the voltage of point b with respect to point a is the work per unit charge required to move the charge from point a to point b*:

$$v_{ba} = \frac{w_{ba}}{q} \tag{1.2}$$

Alternatively we say that there is a *potential difference* between points a and b with point b at the *assumed* higher potential. We use the words *assumed higher potential* because if this turns out to be a negative number then point a is actually at the higher potential. This will be an important concept throughout our work: *Assume a direction for the voltage and solve the required equations for its value*. If this value turns out to be a negative number, we can simply reverse the direction of the assumed voltage and remove the negative sign if we choose. However, it would be equally correct to retain the original assumed polarity and the negative sign in its value.

Determining the voltage between two points that is established by a point charge Q is a simple matter, as shown by the following example.

Example 1.1

A charge Q establishes a voltage between points a and b as shown in Fig. 1.5*a*. The two points are at radial distances R_a and R_b from the point charge. Determine that voltage v_{ba} between the two points (point b at the assumed higher potential).

Solution The direct method for determining this voltage is to move a test charge q (assume it is positive, but the value doesn't matter) along a straight path from a to b. If energy required to do so is positive, then the voltage of point b with respect to a is given by (1.2). Any other path can be taken between the two points. In order to simplify the computation of the energy, consider the path composed of an arc of constant radius R_a from point a to point c. Along this portion of the path, the force exerted on q is perpendicular to the path and hence no work is required to move q along this path. Along the remaining portion from c to b, the force exerted on q by Q is opposite the direction of movement, so that the net work required to move q from a to b is given by

$$\begin{aligned} w_{ba} &= -\int_{R_a}^{R_b} \underbrace{9 \times 10^9 \frac{qQ}{r^2}}_{F} \, dR \\ &= 9 \times 10^9 \, Qq\left(\frac{1}{R_b} - \frac{1}{R_a}\right) \end{aligned}$$

Hence, the voltage of point b with respect to point a is

$$\begin{aligned} v_{ba} &= \frac{w_{ba}}{q} \\ &= 9 \times 10^9 Q\left(\frac{1}{R_b} - \frac{1}{R_a}\right) \quad \text{V} \\ &= 9 \times 10^9 Q\left(\frac{1}{d^+} - \frac{1}{d^-}\right) \quad \text{V} \end{aligned} \tag{1.3}$$

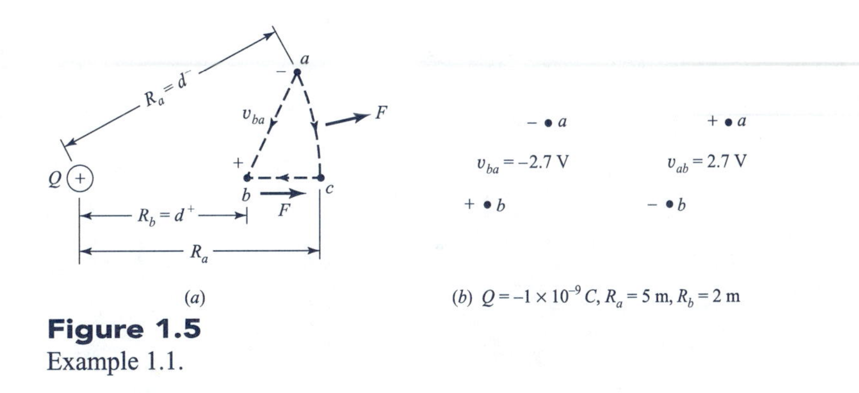

Figure 1.5
Example 1.1.

It is very important to observe that the distance from the charge to the *assumed* positive or higher potential point is denoted as R_b and the distance from the charge to the *assumed* negative or lower potential point is denoted as R_a. The sign of the voltage is as important as its magnitude, so in (1.3) we have shown this explicitly by naming R_b and R_a as d^+ and d^-, respectively, to highlight the proper distances. Observe that if point b is closer to Q than point a, i.e., $R_b < R_a$, then v_{ba} is a positive number, indicating that point b is at the higher potential. This is a sensible result, because work would be required to move q to points closer to Q. As an example, suppose that Q is a negative charge, $Q = -1 \times 10^{-9}$ C, and $R_a = 5$ m and $R_b = 2$ m. The resulting voltage of point b with respect to point a is $v_{ba} = -2.7$ V. This indicates that point a is at the higher potential than point b by 2.7 V. This is a sensible result in that Q being negative actually provides an attractive force for the movement of q from a to b. Hence we could denote the voltage between the two points in either of two equivalent ways as shown in Fig. 1.5*b*.

Example 1.2

Two charges are separated a certain distance as shown in Fig. 1.6*a*. Determine the voltage between the indicated points with assumed polarity shown.

Solution We may determine the voltage by superimposing the contributions due to each charge acting separately as shown in Fig. 1.6*b* and 1.6*c*. Applying (1.3) to the contribution from the 1-μC charge yields

$$\begin{aligned} v'_{xy} &= 9 \times 10^9 (1\ \mu\text{C}) \left(\frac{1}{1} - \frac{1}{1+1} \right) \\ &= 4500 \text{ V} \end{aligned}$$

Observe in applying (1.3) to this case, the point of assumed higher potential, point x, is closer to the charge, and hence $d^+ = 1$ m and $d^- = 1$ m + 1 m. Applying (1.3) to the contribution from the 5-μC charge yields

$$\begin{aligned} v''_{xy} &= 9 \times 10^9 (5\ \mu\text{C}) \left(\frac{1}{1+2} - \frac{1}{2} \right) \\ &= -7500 \text{ V} \end{aligned}$$

Observe in this case that the point of assumed higher potential, point x, is further from the charge than point y, and so in (1.3) $d^+ = 1$ m + 2 m and $d^- = 2$ m.

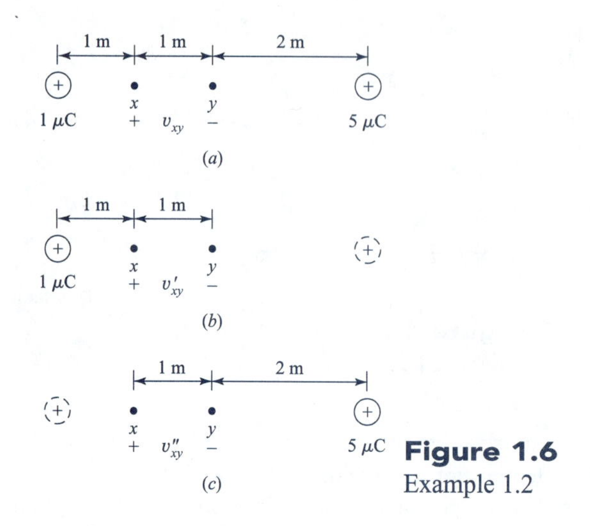

Figure 1.6
Example 1.2

It is always important to determine whether a result is reasonable to expect. For example, the voltage due to the 1-μC charge should be positive, since work will be required to move a positive test charge q from point y to point x because the 1-μC charge is positive and will exert an opposing force on our positive test charge and we are moving q from a point of *assumed* lower potential, point y, to the point of *assumed* higher potential, point x. Similarly, the voltage due to the 5-μC charge should be negative, because this positive charge would exert a force on the positive test charge q tending to move it in the direction from the point of *assumed* lower potential, point y, to the point of *assumed* higher potential, point x. Hence negative work would be required on our part; the field of the 5 μC provides the force to move the charge from y to x. Adding the two contributions gives the total voltage as

$$\begin{aligned} v_{xy} &= v'_{xy} + v''_{xy} \\ &= 4500 \text{ V} - 7500 \text{ V} \\ &= -3000 \text{ V} \end{aligned}$$

If we had asked for the voltage of point y with respect to point x, i.e., assumed point y at the higher potential, or positive, then

$$v_{yx} = 3000 \text{ V}$$

Exercise Problem 1.3

A 10-μC charge is placed at the origin of a rectangular coordinate system having x and y axes. Determine the voltage v_{ba} between (a) a ($x = 1$ cm, $y = 0$) and b ($x = 3$ cm, $y = 0$); (b) a ($x = 10$ cm, $y = 0$) and b ($x = 0$, $y = 10$ cm); and (c) a ($x = 0$, $y = 5$ m) and b ($x = 0$, $y = 1$ m).

Answers: 72 kV, −6 MV, 0 V.

Exercise Problem 1.4

Two charges are placed along the x axis of a rectangular coordinate system. $Q_1 = 2\ \mu\text{C}$ is located at the origin, while $Q_2 = -3\ \mu\text{C}$ is located at $x = 4$ m. Determine the voltage between two points on the y axis: point a at the assumed higher potential and located at $y = 3$ m, and point b at the assumed lower potential and located at $y = 2$ m.

Answer: $v_{ab} = -3000\ \text{V} + 637.38\ \text{V} = -2362.62\ \text{V}$.

1.3 Current and Magnetic Forces

Current is the rate of movement of electric charge. For example, consider Fig. 1.7a in which we show charge moving along a cylinder. Certain amounts of positive charge Q_R^+ and negative charge Q_R^- are moving to the right across a cross section of the cylinder, and similar quantities, Q_L^+ and Q_L^-, are moving to the left. The *net positive charge moving to the right* is

$$Q = Q_R^+ - Q_R^- - Q_L^+ + Q_L^-$$

since *negative charge moving in one direction is equivalent to positive charge moving in the opposite direction*. If we observe the charge crossing an area of the cylinder in a certain time interval, Δt, then *the current directed to the right is defined as the net positive charge transferred to the right per unit of time:*

$$i = \frac{\Delta Q}{\Delta t}$$

The unit of current is the ampere (A), defined as the movement of one coulomb of net positive charge past a point in one second (1 A = 1 C/s).

This rate of movement may not be constant but may vary with time. Hence the general definition of current is

$$i(t) = \frac{dQ(t)}{dt} \tag{1.4}$$

For example, consider the plot of net positive charge moving past a point shown in Fig. 1.8. Over the time interval $1\ \text{s} \leq t \leq 3\ \text{s}$ the net positive charge passing to the right is increasing at a rate of 1 C/s, and 1 A of current results. Over $3\ \text{s} \leq t \leq 5\ \text{s}$ the net positive charge passing to the right is decreasing at a rate of -1.5 C/s and hence the current at any point in this time interval is -1.5 A.

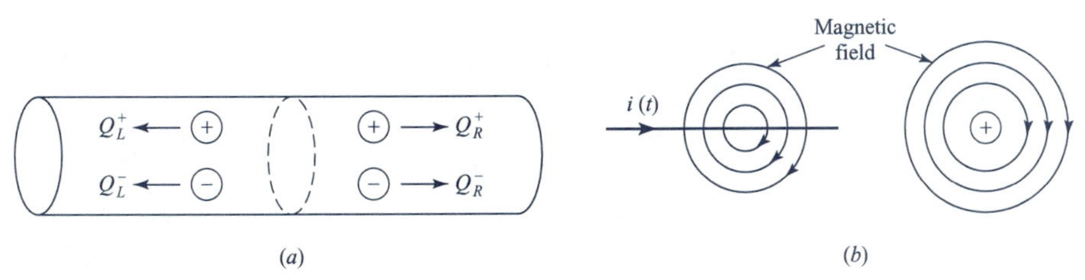

Figure 1.7
Illustration of (a) current as the rate of movement of electric charge, and (b) the magnetic field associated with that charge movement.

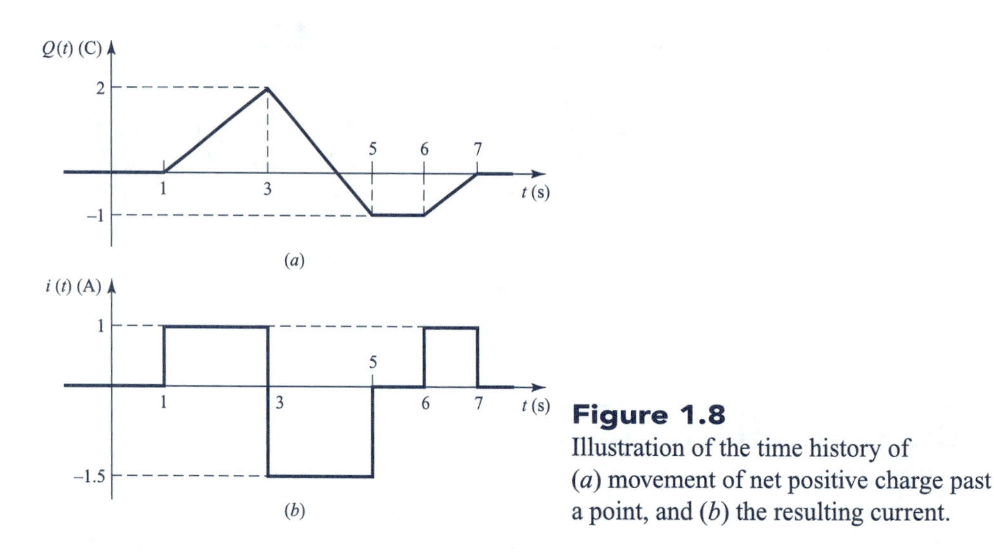

Figure 1.8
Illustration of the time history of (*a*) movement of net positive charge past a point, and (*b*) the resulting current.

Similarly, over 5 s $\leq t \leq$ 6 s the net positive charge passing to the right is constant at −1 C, and hence the current at any point in this time interval is zero. Finally, over 6 s $\leq t \leq$ 7 s the net positive charge transferred to the right is increasing at a rate of 1 C/s, and hence the current at any point in this time interval is 1A. At this point the reader should be able to interpret the derivative operation in (1.4) and hence the instantaneous current as the instantaneous *slope* of the charge transfer curve.

Conversely, the net positive charge passing a point is

$$Q(t) = \int_{-\infty}^{t} i(\tau)\, d\tau \tag{1.5}$$

so that the net positive charge passing the point between two times is

$$Q(t_2) - Q(t_1) = \int_{t_1}^{t_2} i(\tau)\, d\tau \tag{1.6}$$

Equation (1.5) can be interpreted as the *total area* under the current-versus-time curve from the beginning of time to the present time. For example, in Fig. 1.8, the net positive charge transferred to the right at t = 6 s is the sum of the areas under the $i(t)$ curve up to that time. These areas are 1 A × 2 s = 2 C over 1 s $\leq t \leq$ 3 s and −1.5 A × 2 s = −3 C over 3 s $\leq t \leq$ 5 s, yielding a total of −1 C at t = 6 s. Over the interval 6 s $\leq t \leq$ 7 s, 1 C = 1 A × 1 s is transferred, yielding a net total charge transfer of 0 for $t >$ 7 s.

In most metallic conductors such as wires, current is exclusively the movement of negative free electrons. Thus the charges are moving opposite to the direction of the current designation, yet the *net* positive charge movement is nevertheless in the direction we designate for $i(t)$.

Electric currents produce forces as do stationary electric charges. The forces produced by currents (movement of charge) are called magnetic forces, and they possess many of the properties of those produced by ordinary bar magnets. Figure 1.7*b* shows that the magnetic field lines about a current are cylindrical, as opposed to the electric field lines of stationary charges, which are directed radially away from the charge. The direction of the magnetic field lines is given by the right-hand rule: if the thumb of the right hand is pointed in the direction of the current, the fingers will point in the direction of the magnetic field. The needle of a compass that is placed in this field will align with the field lines.

Exercise Problem 1.5

The current is measured at a point in a conductor is found to be given by $i(t) = 0$ for $t \leq 0$, $i(t) = 5(1 - e^{-2t})$ A for $0 \text{ s} \leq t \leq 1$ s, and $i(t) = 31.945e^{-2t}$ A for $t \geq 1$ s. Determine the net positive charge transferred past this point in the direction of the current at (a) $t = 0.5$ s, (b) $t = 1$ s, (c) $t = 3$ s.

Answers: 4.96 C, 0.92 C, 2.838 C.

Exercise Problem 1.6

The positive charge passing a point to the right is sketched in Fig. E1.6. Determine the current directed to the right over the various time intervals.

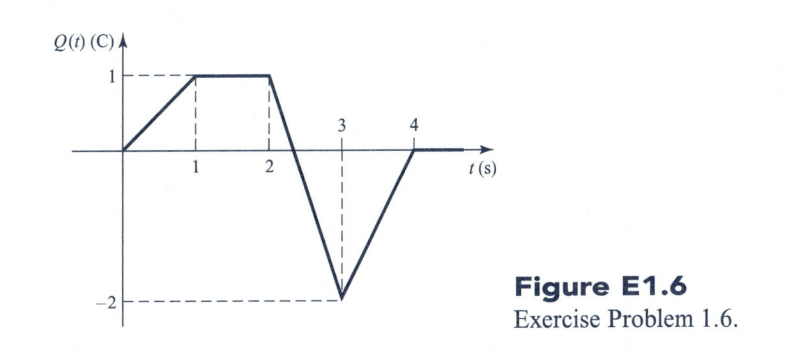

Figure E1.6
Exercise Problem 1.6.

Answers: 1 A for $0 < t < 1$ s, 0 A for $1 \text{ s} < t < 2$ s, -3 A for $2 \text{ s} < t < 3$ s, 2 A for $3 \text{ s} < t < 4$ s, and 0 for $t > 4$ s.

Exercise Problem 1.7

The current directed to the right is sketched in Fig. E1.7. Determine the net positive charge passing the point to the right at $t = 0.5, 1, 2, 3, 4, 5$ s.

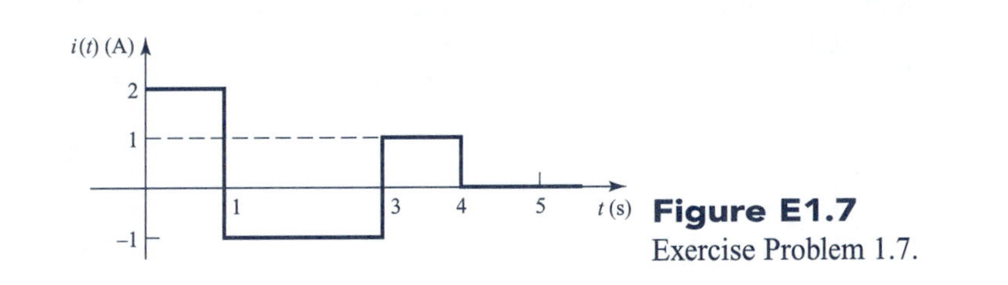

Figure E1.7
Exercise Problem 1.7.

Answers: 1, 2, 1, 0, 1, 1 C.

1.4 Lumped Circuit Elements and Power

The electric field of stationary charges extends throughout space, as does the magnetic field due to the movement of charge. The circuit elements that we will use to model electric circuits utilize these fields. Although these fields are distributed throughout space, we will consider their effects to be confined primarily to some small region about the element. Thus we will lump them into boxes as shown in Fig. 1.9. We refer to these symbols as *lumped elements*. Each element has two terminals, a and b, associated with it, which we refer to as *nodes*. The element also has a voltage, $v(t)$, between the two terminals, and a current, $i(t)$, entering one end of the element and exiting the other. The determination of these element voltages and currents for a particular interconnection of elements will be our goal in circuit analysis. In Chapters 2 and 5 we will discuss the specific types of electric circuit elements the boxes may contain. For the moment we will leave the element types undesignated, since the various interconnection laws that govern these voltages and currents for a particular interconnection of elements are not dependent on the types of elements within the boxes.

If we label the element voltage and current as shown in Fig. 1.9, where the current is *assumed* to enter the terminal of *assumed* higher voltage, then the element voltage and current are said to be labeled with the *passive sign convention*. The origin of this terminology has to do with the power delivered to or delivered by the element. To determine this we write the instantaneous power *delivered to the element* as the time derivative of the energy stored in the element, w_{ba}:

$$\begin{aligned} p(t) &= \frac{dw_{ba}(t)}{dt} \\ &= \frac{dw_{ba}(t)}{dq(t)} \frac{dq(t)}{dt} \\ &= v(t)\, i(t) \end{aligned} \tag{1.7}$$

The unit of power is the watt (W), where 1 W = 1 J/s. Thus *the instantaneous power delivered to (absorbed by) an element at some time t is the product of the voltage and current associated with that element at time t when the voltage and current are labeled with the passive sign convention.*

In order to illustrate this concept of power delivered to an element, consider charging a discharged automobile battery by connecting it to a fully charged one shown in Fig. 1.10. Connecting the + terminals and connecting the − terminals, the fully charged battery supplies current out its positive terminal into the positive terminal of the discharged battery. Since the current leaves the positive terminal of the fully charged battery, it is said to be *delivering* $v(t)i(t)$ watts of power. The current enters the positive terminal of the discharged battery, and hence this battery is said to be *absorbing* $v(t)i(t)$ watts of power. These important concepts are summarized in Fig. 1.11. The reader is advised to study and commit to memory all aspects of Fig. 1.11.

Figure 1.12 shows several examples of power calculation that should be studied carefully. The safest course of action is to formulate the power expression based on the voltage and current assumed polarities and then substitute the numerical values into that expression. For example, consider Fig. 1.12*d*. Since the assumed current direction enters the negative terminal of the

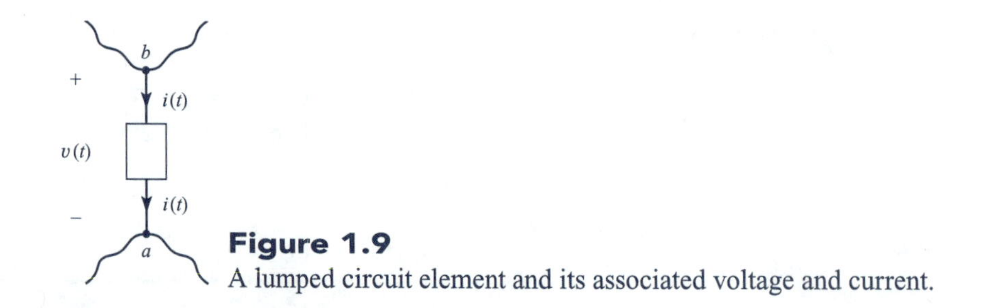

Figure 1.9
A lumped circuit element and its associated voltage and current.

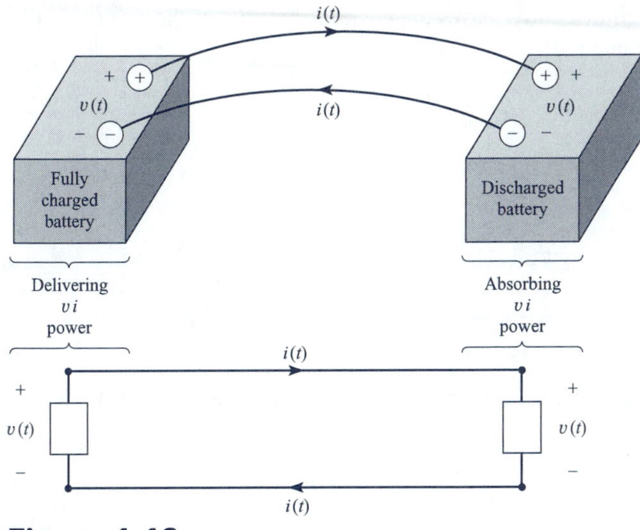

Figure 1.10
Charging a discharged automobile battery to illustrate the concept of power delivered to or absorbed by an element and the passive sign convention.

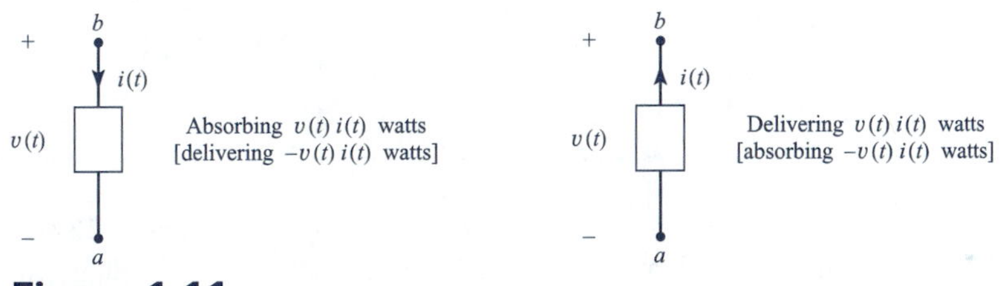

Figure 1.11
Illustration of the power delivered to (absorbed by) an element and the power delivered by the element.

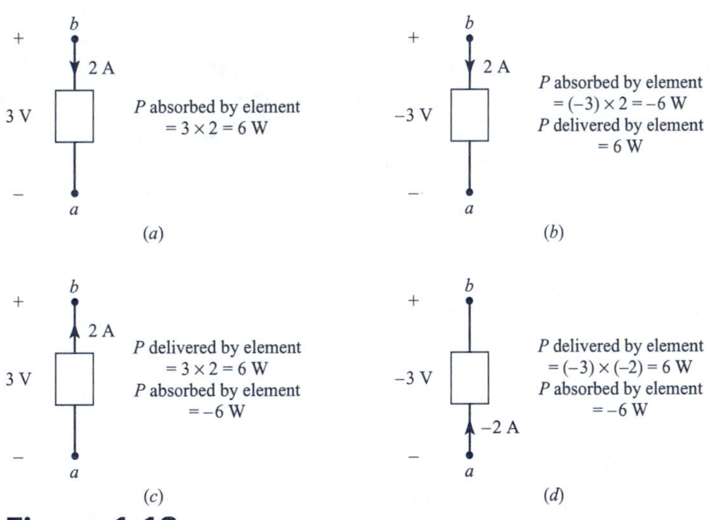

Figure 1.12
Examples of the computation of power delivered to or by an element.

assumed voltage (or equivalently, leaves the positive terminal of the assumed voltage), the power delivered by this element is

$$
\begin{aligned}
p_{\text{delivered by}} &= v(t)i(t) \\
&= (-3\ \text{V}) \times (-2\ \text{A}) \\
&= 6\ \text{W}
\end{aligned}
$$

Equivalently, we may say that this element is absorbing -6 W of power.

Exercise Problem 1.8

Determine the powers delivered to (absorbed by) the elements of Fig. E1.8.

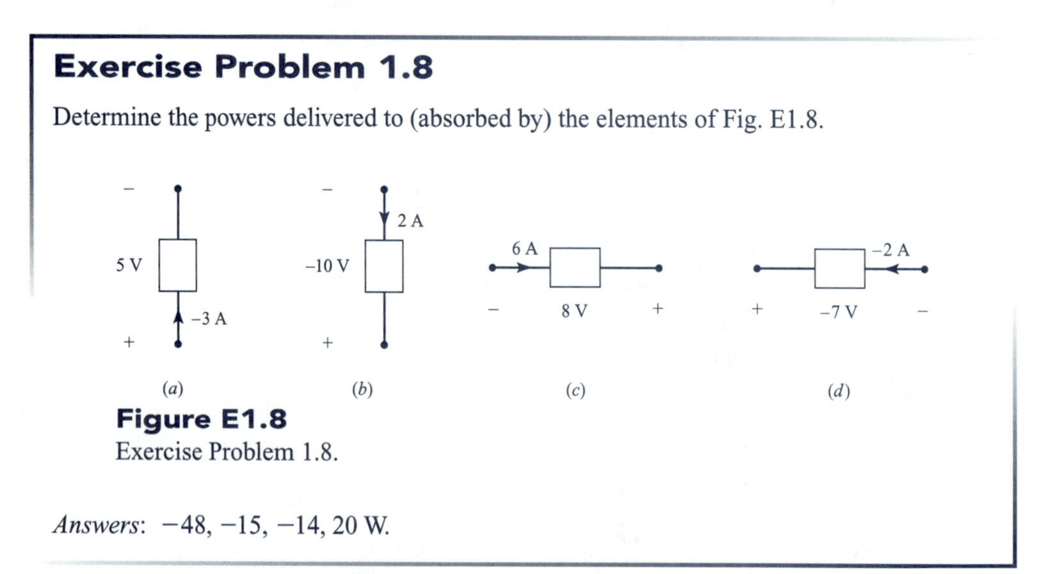

Figure E1.8
Exercise Problem 1.8.

Answers: $-48, -15, -14, 20$ W.

An *electric circuit is an interconnection of circuit elements*. Figure 1.13 shows an example of an electric circuit. Each element is labeled with a current and a voltage, which are given *assumed polarities*. Our task will be to determine the numerical values of those element voltages and currents such that they satisfy certain laws *for this particular interconnection*. In the following sections we will study the laws that govern the voltages and the currents of a particular interconnection. Kirchhoff's current law (KCL) dictates how the currents of a particular interconnection are related in order to satisfy certain physical laws (conservation of charge). Kirchhoff's voltage law (KVL) dictates how the voltages of a particular interconnection are related in order to also satisfy certain physical laws (conservation of energy). In later chapters we will study the various types of circuit elements that the boxes may contain. These element relations either relate the element voltage and current to each other (e.g., a resistor) or dictate the value of the element voltage or current (a source). For the moment we study the interconnection laws KVL and KCL, which are independent of the types of elements.

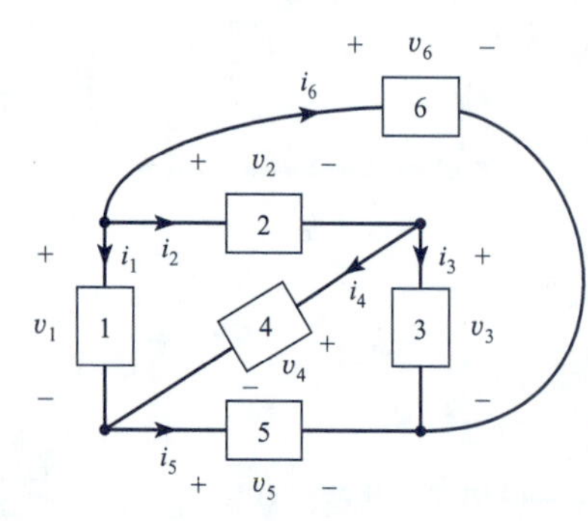

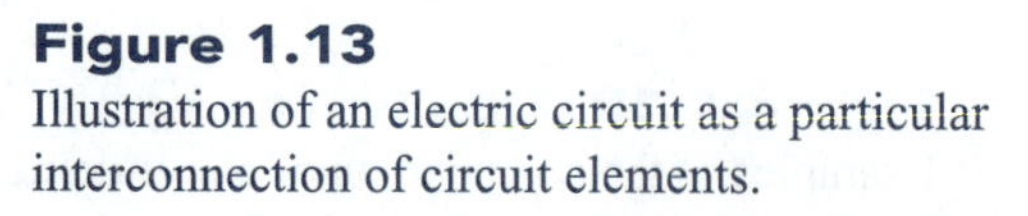

Figure 1.13
Illustration of an electric circuit as a particular interconnection of circuit elements.

1.5 Kirchhoff's Current Law (KCL)

Kirchhoff's current law, referred to simply as KCL, relates the currents in a particular circuit to each other. For example, consider the connection of five elements at a node shown in Fig. 1.14. KCL can be stated in any of three ways:

$$\text{The sum of the currents entering a node must equal zero} \tag{1.8a}$$

For Fig. 1.14, this translates to

$$i_1 + (-i_2) + i_3 + (-i_4) + i_5 = 0$$

since a current leaving a node is equivalent to the negative of that current entering that node and vice versa. Alternatively, we may write KCL as

$$\text{The sum of the currents leaving a node must equal zero} \tag{1.8b}$$

For Fig. 1.14, this translates to

$$(-i_1) + i_2 + (-i_3) + i_4 + (-i_5) = 0$$

Since both these results sum to zero, the third alternative (which the author prefers) is to write KCL as

$$\text{The sum of the currents entering a node must equal the sum of the currents leaving that node} \tag{1.8c}$$

For Fig. 1.14, this translates to

$$i_1 + i_3 + i_5 = i_2 + i_4$$

KCL must be satisfied at all nodes of a circuit. For example, consider the circuit of Fig. 1.15. Applying KCL at all nodes yields (as the reader should verify)

node a: $0 = i_1 + i_2$
node b: $i_2 + i_3 + i_{12} = 0$
node c: $0 = i_4 + i_{12}$
node d: $i_4 = i_5 + i_6$
node e: $i_6 + i_7 = 0$
node f: $i_8 + i_9 = i_7$
node g: $i_{11} = i_8$
node h: $i_1 + i_{10} = i_{11}$
node i: $i_5 = i_3 + i_9 + i_{10}$

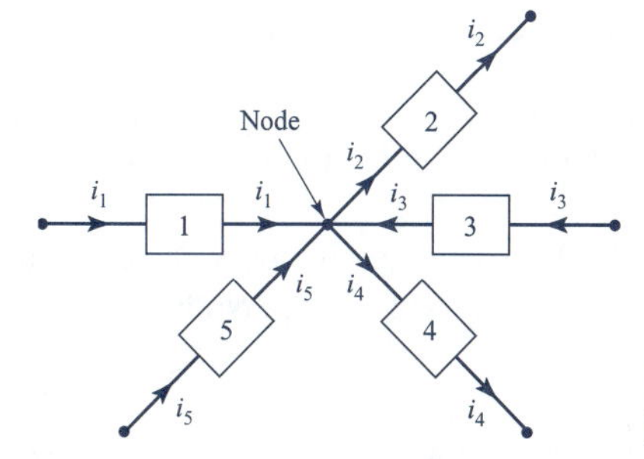

Figure 1.14
Illustration of Kirchhoff's current law (KCL).

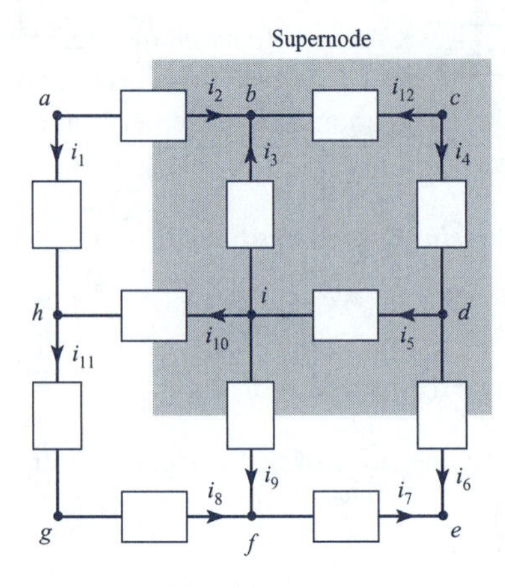

Figure 1.15
Illustration of Kirchhoff's current law (KCL).

KCL also applies to larger, closed regions of a circuit called *supernodes. A supernode is any closed region whose boundary intersects a circuit element only once.* Alternatively, *a supernode is a set of elements which when removed separates the circuit into two parts.* A supernode is shown in Fig. 1.15. A supernode encloses a set of nodes of the circuit. Applying KCL to this supernode yields

$$i_2 = i_6 + i_9 + i_{10}$$

But KCL at this supernode is simply the sum of the KCL equations for the nodes within the supernode:

$$\begin{array}{ll} \text{node } b: & i_2 + i_3 + i_{12} = 0 \\ \text{node } c: & -i_{12} - i_4 = 0 \\ \text{node } d: & i_4 = i_5 + i_6 \\ \text{node } i: & -i_3 = -i_5 + i_9 + i_{10} \\ \hline \Sigma: & i_2 = i_6 + i_9 + i_{10} \end{array}$$

KCL is equivalent to the statement that nodes (or supernodes) cannot accumulate or store charge—whatever charge enters a node must immediately leave that node. Hence KCL is a statement of conservation of charge. For example, if the total net positive charge entering a node is denoted by $Q_{\text{in}} = Q_1 + Q_2 - Q_3$ and $Q_{\text{in}} = 0$, then it follows that

$$\begin{aligned} i_{\text{in}} &= i_1 + i_2 - i_3 \\ &= \frac{dQ_1}{dt} + \frac{dQ_2}{dt} - \frac{dQ_3}{dt} \\ &= \frac{dQ_{\text{in}}}{dt} \\ &= 0 \end{aligned}$$

Example 1.3

Determine the currents i_x, i_y, i_z in the circuit of Fig. 1.16 by applying KCL.

Solution The key to applying KCL to determine an unknown current is to look for a node such that only one current entering or leaving it is unknown. Node *d* satisfies this, so we write KCL there to yield

$$i_x + 3\text{ A} = 2\text{ A}$$

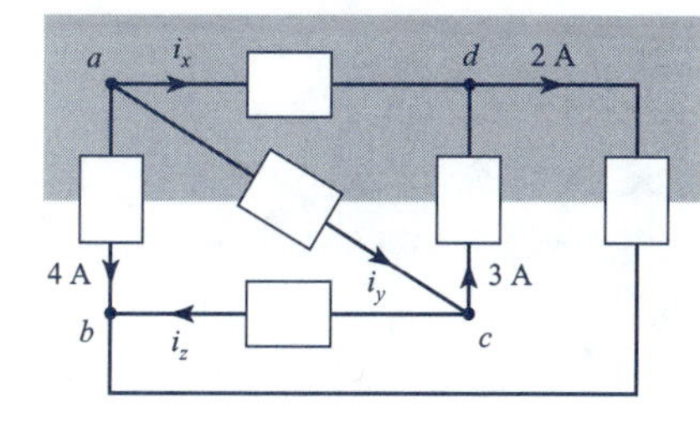

Figure 1.16
Example 1.3.

giving

$$i_x = -1\text{ A}$$

Knowing i_x, we can move to node a and obtain

$$i_x + i_y + 4\text{ A} = 0$$

Substituting the value of $i_x = -1$ A that was just determined gives

$$i_y = -3\text{ A}$$

Finally, we can apply KCL at node b to obtain

$$4\text{ A} + i_z + 2\text{ A} = 0$$

giving

$$i_z = -6\text{ A}$$

We could have also applied KCL at node c to obtain i_z from

$$i_y = i_z + 3\text{ A}$$

Substituting the previously determined value of $i_y = -3$ A into this again gives $i_z = -6$ A. We could also have applied KCL to the supernode containing nodes a and d as shown in Fig. 1.16 to obtain

$$i_y + 4\text{ A} + 2\text{ A} = 3\text{ A}$$

giving $i_y = -3$ A.

Example 1.4

Determine the currents i_x, i_y, i_z, i_w, i_v in the circuit of Fig. 1.17 by applying KCL.

Solution Again, the key to using KCL to determine unknown currents is to look for nodes where only one current in an attached branch is unknown. Nodes a and f qualify, so writing KCL at these gives

$$\text{node } a\text{:} \qquad 2\text{ A} = i_y + 1\text{ A}$$
$$\text{node } f\text{:} \qquad i_z + 1\text{ A} = 4\text{ A}$$

Solving these gives $i_y = 1$ A and $i_z = 3$ A. Applying KCL at node b gives

$$i_y + 2\text{ A} = i_x$$

Substituting the previously determined value of $i_y = 1$ A gives $i_x = 3$ A. Applying KCL at node e gives

$$i_w = 3\text{ A} + i_z$$

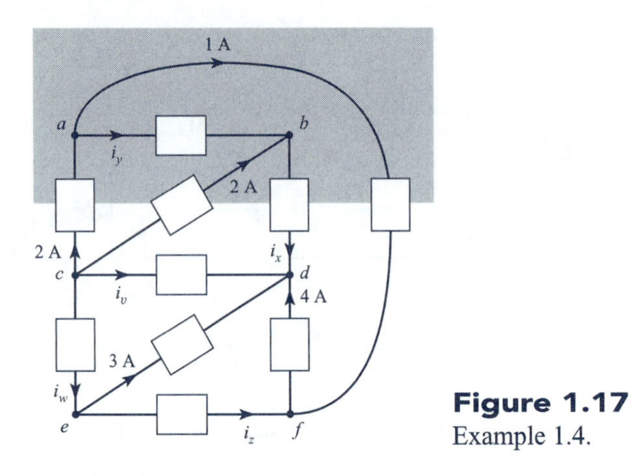

Figure 1.17
Example 1.4.

Substituting the previously determined value of $i_z = 3$ A yields $i_w = 6$ A. And now we are able to determine the remaining current, i_v, by applying KCL to node c to obtain

$$i_v + i_w + 2\text{ A} + 2\text{ A} = 0$$

Substituting the previously determined value of $i_w = 6$ A yields $i_v = -10$ A. There are several supernodes. One is drawn in Fig. 1.17. Applying KCL at that supernode yields

$$2\text{ A} + 2\text{ A} = i_x + 1\text{ A}$$

again yielding $i_x = 3$ A.

Example 1.5

Determine currents i_x, i_y, i_z, i_w in the circuit of Fig. 1.18 by applying KCL.

Solution Again we look for nodes such that there is only one unknown current in the attached branches. Nodes d and e are the only nodes that qualify, so we apply KCL to node d to obtain

$$i_y + 1\text{ A} = 3\text{ A}$$

giving $i_y = 2$ A. Knowing this, we may apply KCL at node c to obtain

$$i_x + 2\text{ A} = i_y + 4\text{ A}$$

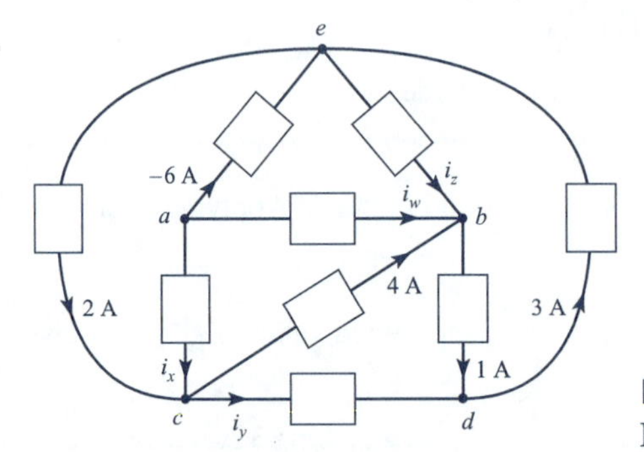

Figure 1.18
Example 1.5.

Substituting the previously determined value of $i_y = 2$ A gives $i_x = 4$ A. Next we apply KCL at node a to obtain

$$i_x + i_w + (-6\text{ A}) = 0$$

Substituting the previously determined value of $i_x = 4$ A gives $i_w = 2$ A. And finally we apply KCL at node b to give

$$i_z + i_w + 4\text{ A} = 1\text{ A}$$

Substituting the previously determined value of $i_w = 2$ A gives $i_z = -5$ A. KCL written at node e confirms these results:

$$(-6\text{ A}) + 3\text{ A} = 2\text{ A} + i_z$$

again giving $i_z = -5$ A.

Exercise Problem 1.9

In the circuit of Fig. E1.9, determine i_1, i_2, i_3.

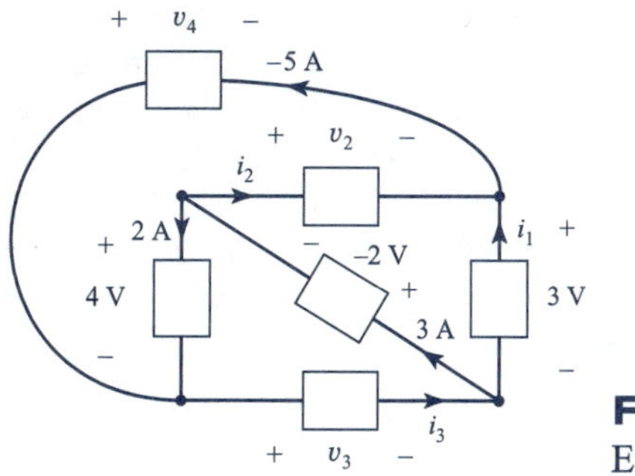

Figure E1.9
Exercise Problem 1.9.

Answers: 1 A, −6 A, −3 A.

Exercise Problem 1.10

In the circuit of Fig. E1.10, determine i_x, i_y, i_z.

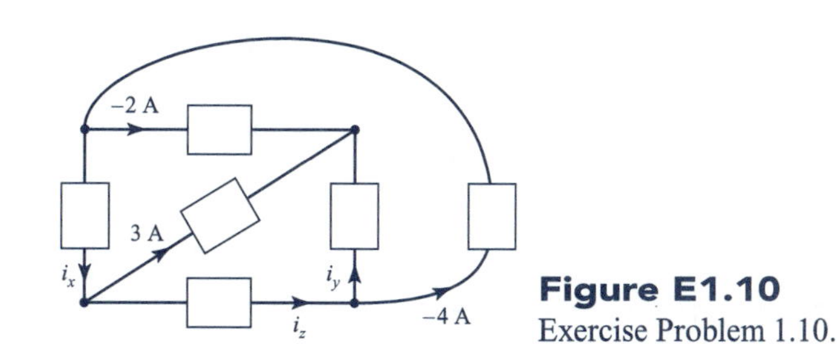

Figure E1.10
Exercise Problem 1.10.

Answers: −2 A, −1 A, −5A.

Exercise Problem 1.11

In the circuit of Fig. E1.11, determine i_x, i_y, i_z.

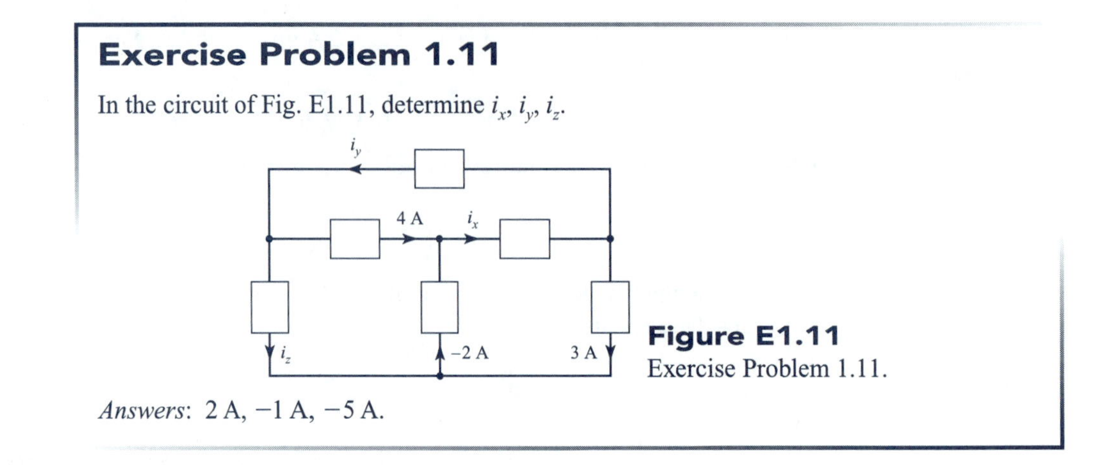

Figure E1.11
Exercise Problem 1.11.

Answers: 2 A, −1 A, −5 A.

1.6 Kirchhoff's Voltage Law (KVL)

Kirchhoff's voltage law (KVL) provides relationships between the element voltages of a circuit. Before stating KVL we need to define the terms *voltage drop, voltage rise,* and *circuit loop*.

Consider Fig. 1.19. *A circuit loop is a closed path formed by selecting a starting node, proceeding through the circuit elements in that path, and returning to the starting node going through an element only once.* The loop shown starts at node a (an arbitrarily chosen starting node), proceeds through element 1 to node b, then through element 2 to node c, then through element 3 to node d, and returns to the starting node by proceeding through element 4. Observe that not only does a circuit loop have a set of elements through which it passes, but there is also a direction of movement associated with that chosen loop. We arbitrarily select that direction of movement once we select the path. We could choose either a clockwise or a counterclockwise direction of movement, and either direction of traversal would be acceptable.

Next we define the terms *voltage rise* and *voltage drop*. Consider Fig. 1.20. An element has been labeled with an *assumed* voltage and an *assumed* polarity of that voltage. If the movement around the loop proceeds through an element in the direction toward the terminal of assumed higher voltage as shown in Fig. 1.20a, we say that we have traversed the element in the direction of a *voltage rise* of value v. On the other hand, if the direction of traversal has been toward the terminal of lower voltage as shown in Fig. 1.20b, we say that we have traversed the element in the direction of a *voltage drop* of value v. These notions stem from the idea that if we move a positive charge around the loop through the elements, its movement will require energy expenditure on our part if we proceed in the direction of a voltage rise and vice versa for a voltage drop. This is very similar to lifting a mass above the surface of the earth, in which case we say that we have moved the object in the direction of a gravitational potential rise.

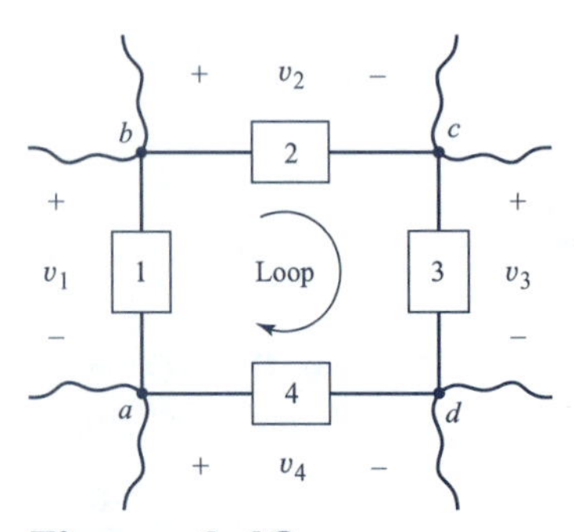

Figure 1.19
Illustration of a circuit loop.

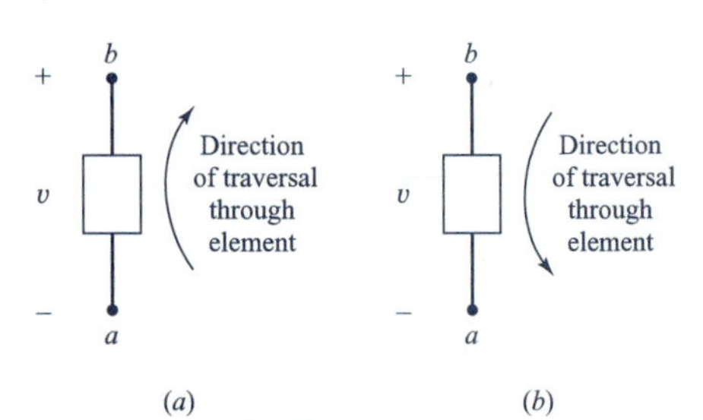

Figure 1.20
Illustration of the concepts of (a) voltage rise and (b) voltage drop.

Kirchhoff's voltage law can be stated in any of three ways:

The sum of the voltage rises encountered in traversing a loop must equal zero (1.9a)

For Fig. 1.19, this translates to

$$v_1 + (-v_2) + (-v_3) + v_4 = 0$$

since a voltage drop is equivalent to the negative of the corresponding voltage rise and vice versa. Alternatively, we may write KVL as

The sum of the voltage drops encountered in traversing a loop must equal zero (1.9b)

For Fig. 1.19, this translates to

$$(-v_1) + v_2 + v_3 + (-v_4) = 0$$

Since both these results sum to zero, the third alternative (which the author prefers) is to write KVL as

The sum of the voltage rises encountered in traversing a loop must equal the sum of the voltage drops (1.9c)

For Fig. 1.19, this translates to

$$v_1 + v_4 = v_2 + v_3$$

KVL must be satisfied around all loops of a circuit. The reader is cautioned that one sign error in any of these KVL equations renders the entire set worthless, so we must be very careful to apply KVL correctly. The same applies to KCL, but KVL seems to give more difficulty than KCL. The author prefers the following method. Consider Fig. 1.19 once again. Draw an imaginary line bisecting the loop so that some of the circuit elements are on one side of the line and the remaining elements of the loop are on the other side. The location chosen for this line is arbitrary. Next arbitrarily label one end of this line as positive (+) and the other end as negative (−). This is shown for the loop of Fig. 1.19 in Fig. 1.21. Then write KVL by equating the sum of the voltages on the left side of the line that have their positive values on the positive end of the line to the sum of the voltages on the right side of the line that have their positive values on the positive end of the line. Voltages that have their negative end on the positive end of the line are entered as negative quantities. For example, from Fig. 1.21 we have

$$v_1 - v_2 = v_3 - v_4$$

This seemingly trivial method has proven to minimize sign errors. As another example, consider the circuit loop shown in Fig. 1.22. Writing KVL in this fashion yields

$$-(-4) + 2 + (-3) + (-1) = -1 - (-2) + (-2) + 3$$

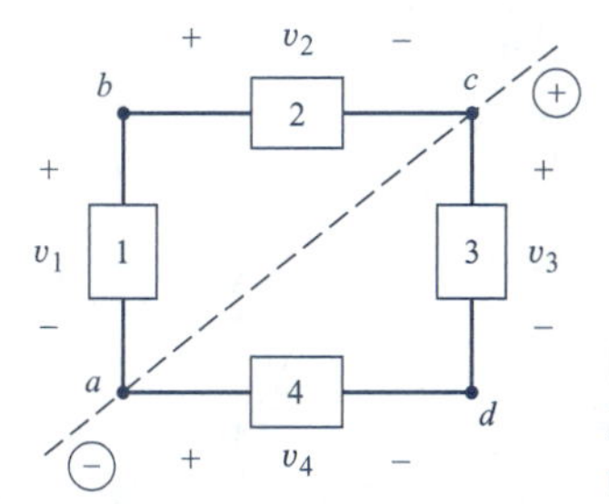

Figure 1.21
Illustration of a method for correctly writing KVL with reference to the circuit of Fig. 1.19.

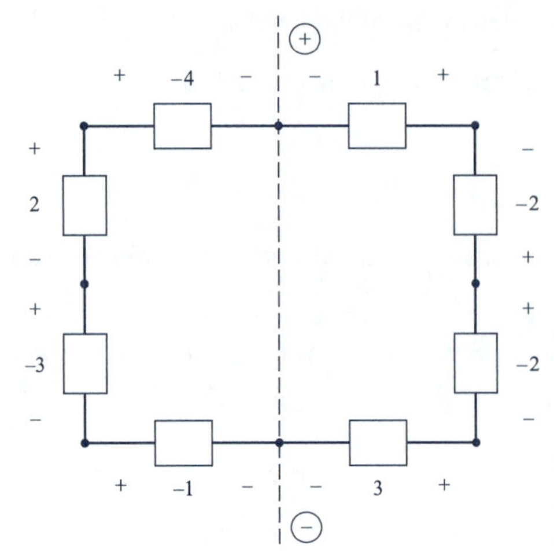

Figure 1.22
Another example illustrating a method for correctly writing KVL.

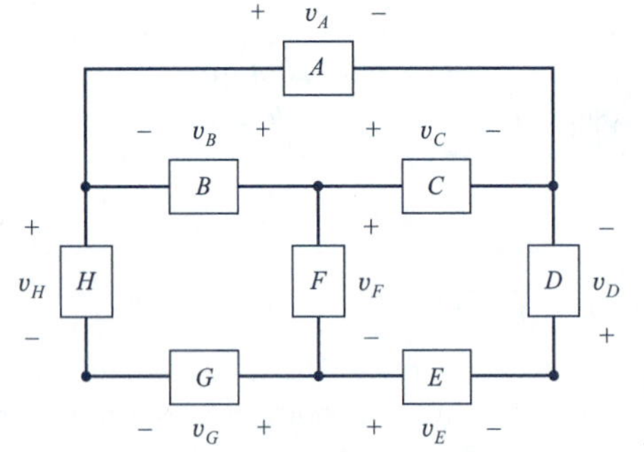

Figure 1.23
Another example of the application of KVL.

which is seen to be satisfied. Observe in this last example that the safest course of action is to apply KVL according to the polarity of the voltage, independent of its value, and then substitute the actual values of the voltages.

KVL must be satisfied around all loops of a circuit. Consider the circuit shown in Fig. 1.23. There are seven circuit loops:

$$
\begin{aligned}
&L_1: && ADEGH \\
&L_2: && HBFG \\
&L_3: && FCDE \\
&L_4: && ABC \\
&L_5: && HACFG \\
&L_6: && ADEFB \\
&L_7: && HBCDEG
\end{aligned}
$$

The seven loops are shown in Fig. 1.24. Each loop has the imaginary line drawn through it with the + and − ends labeled. Writing KVL in the above fashion yields (as the reader should verify)

$$
\begin{aligned}
L_1:&\quad -v_A + v_H = -v_D - v_E + v_G \\
L_2:&\quad v_B + v_H = v_F + v_G \\
L_3:&\quad -v_C + v_F = -v_D - v_E \\
L_4:&\quad -v_A = -v_C + v_B \\
L_5:&\quad -v_A + v_H = -v_C + v_F + v_G \\
L_6:&\quad -v_A - v_B = -v_D - v_E - v_F \\
L_7:&\quad -v_G + v_B + v_H = v_C - v_D - v_E
\end{aligned}
$$

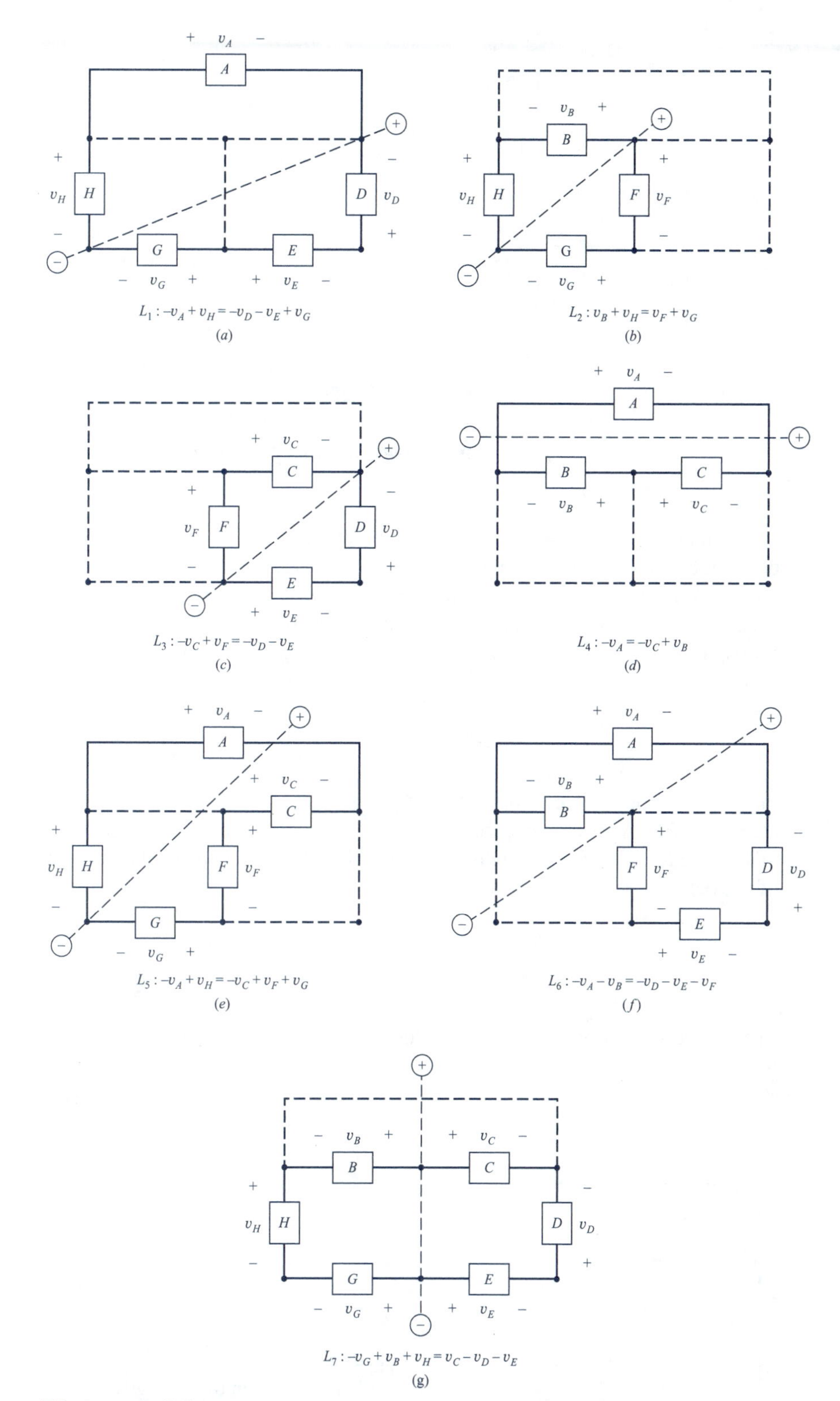

Figure 1.24
The seven loops in the circuit of Fig. 1.23, with KVL written for each.

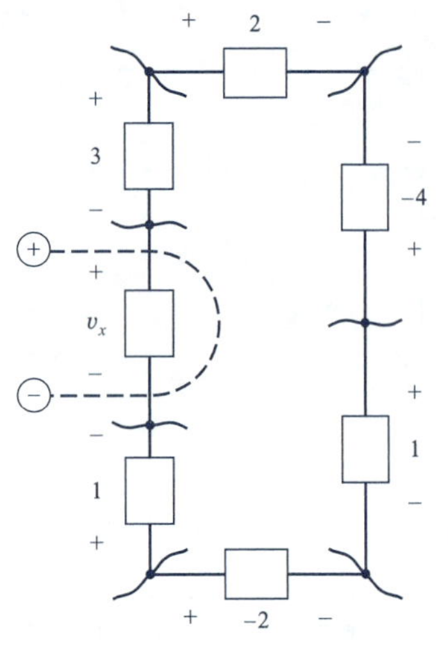

Figure 1.25
A useful variant of the method for writing KVL when only one voltage in the loop is unknown.

There is a useful variant of this method of writing KVL that is used when one wishes to determine an unknown voltage in a loop. For example, consider Fig. 1.25. The only unknown in the loop is the voltage v_x. The bisecting line is drawn so that *only* the element associated with v_x is on one side and the remaining elements are on the other side. The + sign of the bisecting line is assigned to the same end as the + sign of v_x. Writing KVL for this gives

$$\begin{aligned} v_x &= -3 + 2 - (-4) + 1 - (-2) + 1 \\ &= 7\text{ V} \end{aligned}$$

Figure 1.26 shows another application of this variant. All the voltages in the loop consisting of elements *ABCDE* are known except the voltage v across element *A*. Placing the bisecting line so that element *A* is on one side with the + sign of the bisecting line on the same end as the + sign of v gives

$$\begin{aligned} v &= -2 + (-1) - 3 - (-2) \\ &= -4\text{ V} \end{aligned}$$

The reader is well advised to study this simple but effective method for flawlessly writing KVL.

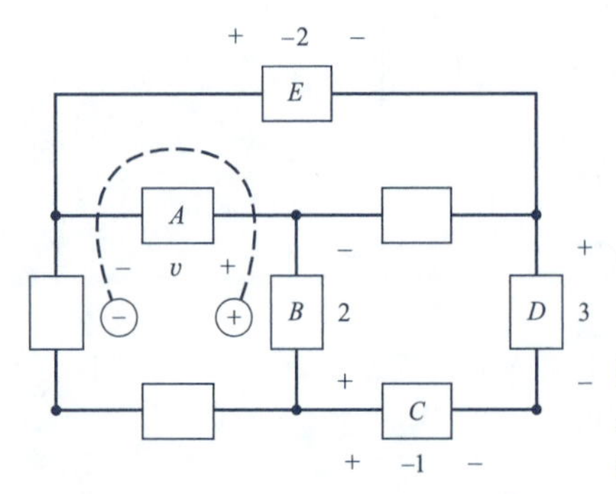

Figure 1.26
Another illustration of writing KVL for a loop where only one voltage is unknown.

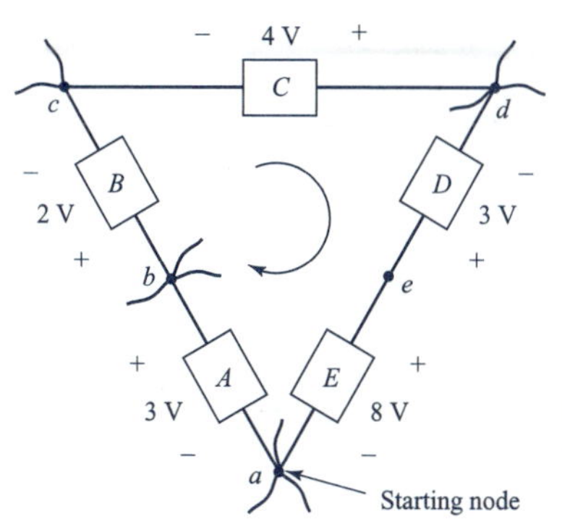

Figure 1.27
Illustration that KVL is a result of the law of conservation of energy.

KVL is essentially a statement of conservation of energy. If we move a charge q around a circuit loop, the sum of the energies expended in proceeding through a voltage rise and the sum of the energies expended in proceeding through a voltage drop should be equal, since we return to the starting point. For example, consider the loop shown in Fig. 1.27. Moving a charge q around the loop in the indicated direction results in the following energy expenditures:

Element	Energy Expenditure (J)
A	$3q$
B	$-2q$
C	$4q$
D	$3q$
E	$-8q$
	$\sum = 0$

If we factor out q from this sum we arrive at KVL:

$$(3\text{ V} - 2\text{ V} + 4\text{ V} + 3\text{ V} - 8\text{ V})\, q = 0$$

KVL must hold for *all* loops in a circuit, just as KCL must hold for *all* nodes and supernodes of that circuit. Also, each KVL equation applies to the voltages of that equation at the same instant of time, in the same way that each KCL equation applies to the currents of that equation at the same instant of time.

Example 1.6

Determine voltages v_x, v_y, v_z in the circuit of Fig. 1.28*a* by applying KVL.

Solution As with the application of KCL, the key to applying KVL in determining an unknown voltage is to select a loop that has only one unknown voltage. One loop satisfying this is the loop consisting of elements A, E, C, F as shown in Fig. 1.28*b*. Drawing the bisector around the unknown voltage v_x gives

$$v_x = -(-4\text{ V}) + (-3\text{ V}) + 2\text{ V}$$

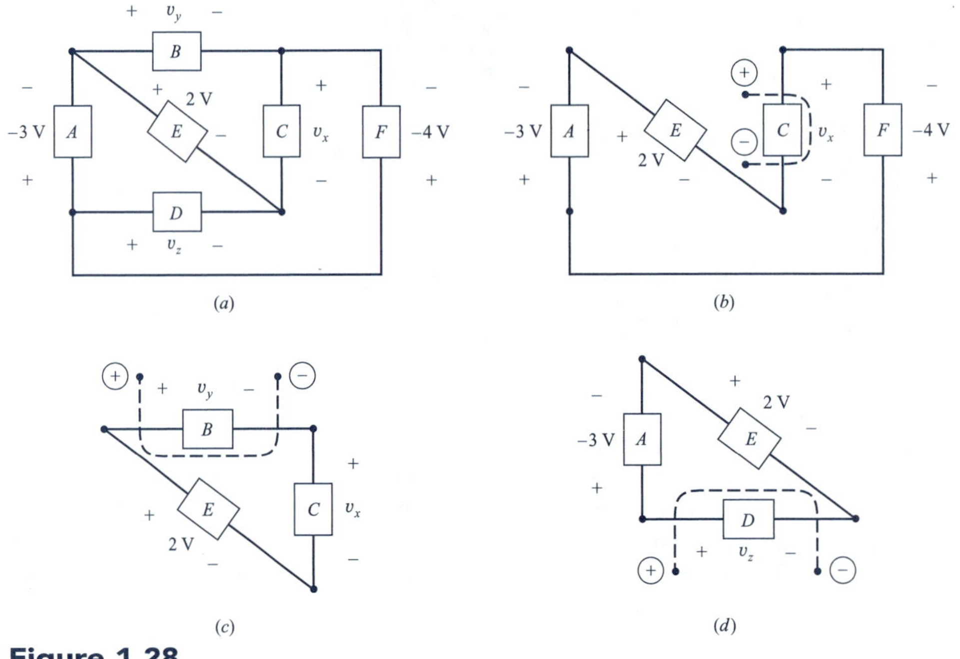

Figure 1.28
Example 1.6.

giving $v_x = 3$ V. Next we may select the loop containing elements B, C, E as shown in Fig. 1.28*c*, since this contains the unknown v_y and the other unknown in that loop, v_x, was previously determined. Drawing the bisector around the unknown v_y and writing KCL gives

$$v_y = 2\text{ V} - v_x$$

Substituting the previously determined $v_x = 3$ V yields $v_y = -1$ V. Finally we select a loop containing the final unknown voltage, v_z, containing elements A, D, E as shown in Fig. 1.28*d*. Drawing the bisector around v_z and writing KVL around this loop gives

$$v_z = (-3\text{ V}) + 2\text{ V}$$

giving $v_z = -1$ V.

Example 1.7

Determine voltages v_x, v_y, v_z, v_w in the circuit of Fig. 1.29*a* by applying KVL.

Solution Again we look for loops containing only one unknown voltage. There are four such loops shown in Fig. 1.29*b*, *c*, *d*, and *e*. Drawing the bisector around the unknown voltage and writing KVL give for these

$$\begin{aligned} v_x &= -(-6\text{ V}) - 3\text{ V} \\ v_y &= -(-6\text{ V}) - 5\text{ V} \\ v_z &= -2\text{ V} + 5\text{ V} \\ v_w &= -5\text{ V} + 2\text{ V} + (-4\text{ V}) \end{aligned}$$

from which we obtain $v_x = 3$ V, $v_y = 1$ V, $v_z = 3$ V, $v_w = -7$ V.

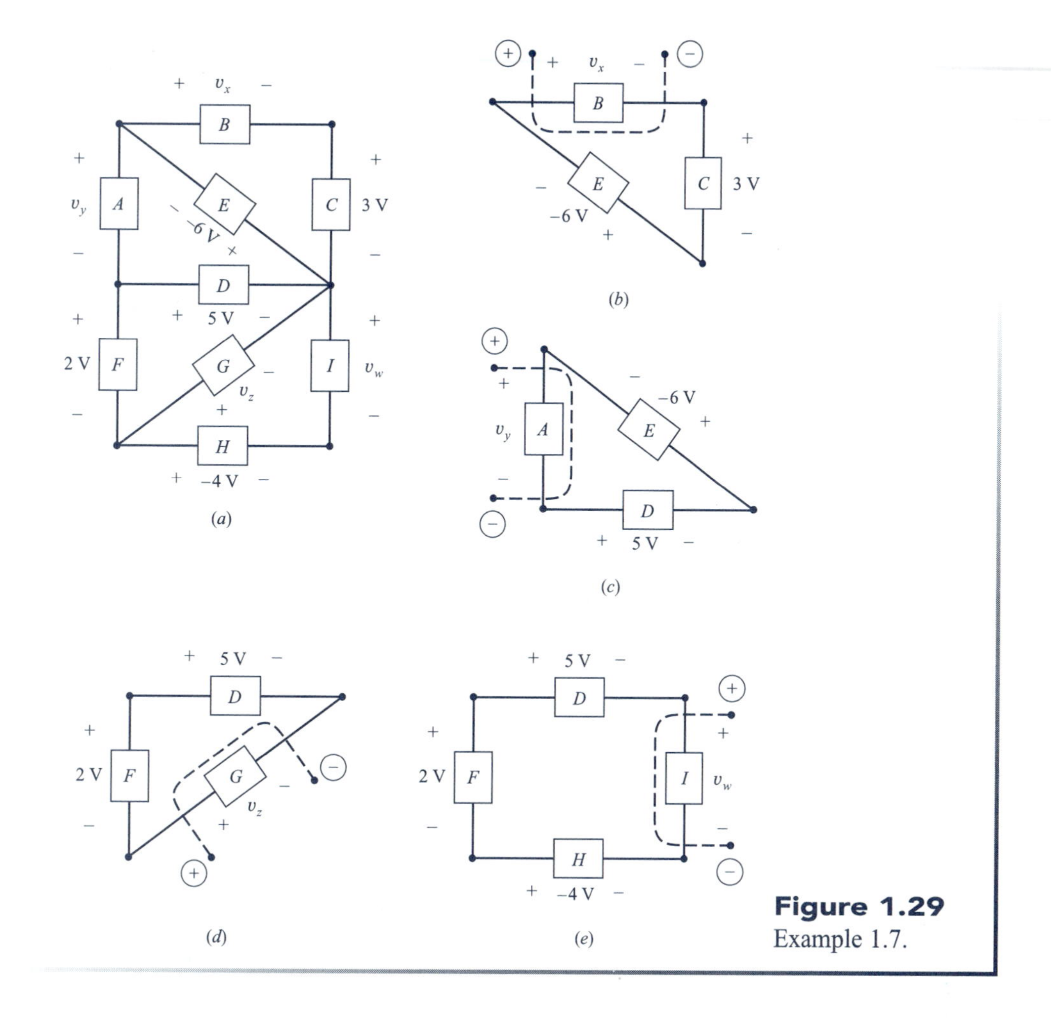

Figure 1.29
Example 1.7.

Example 1.8

Determine voltages v_x, v_y, v_z in the circuit of Fig. 1.30*a* by applying KVL.

Solution Again we look for loops containing only one unknown voltage. The loop containing elements *F*, *E*, *D*, *H* shown in Fig. 1.30*b* contains only one unknown voltage, v_z. Drawing the bisector and applying KVL gives

$$\begin{aligned} v_z &= -(-4\text{ V}) + 5\text{ V} + 2\text{ V} \\ &= 11\text{ V} \end{aligned}$$

The loop containing elements *C*, *D*, *E*, *G* shown in Fig. 1.30*c* contains two unknowns, v_x and v_z. But we just solved for v_z, so we draw the bisector and write KVL, obtaining

$$v_x = -6\text{ V} + 2\text{ V} - v_z$$

Substituting the previously determined value of $v_z = 11$ V yields $v_x = -15$ V. And finally, the loop containing elements *A*, *B*, *C*, *H* shown in Fig. 1.30*d* contains only the last unknown, v_y, from which we obtain

$$\begin{aligned} v_y &= 5\text{ V} + 6\text{ V} - 3\text{ V} \\ &= 8\text{ V} \end{aligned}$$

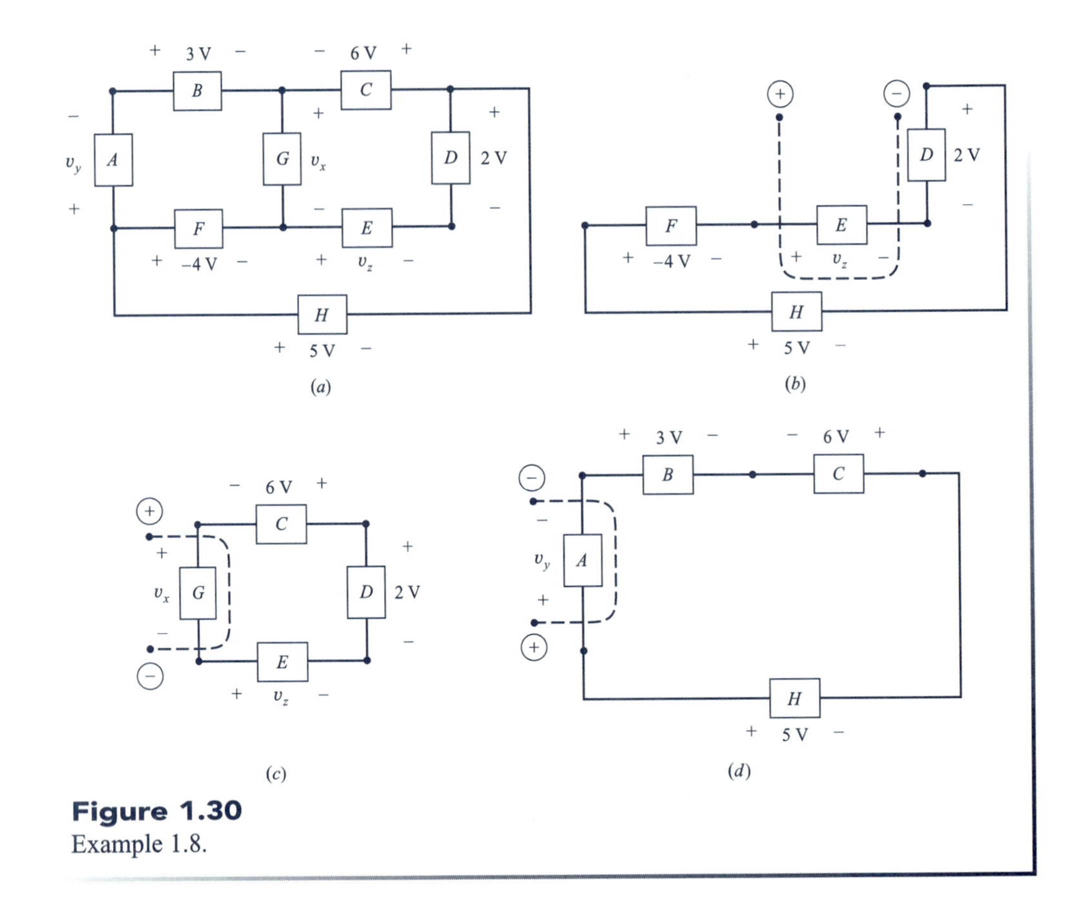

Figure 1.30
Example 1.8.

Exercise Problem 1.12

Determine voltages v_x, v_y, v_z in the circuit of Fig. E1.12 by applying KVL.

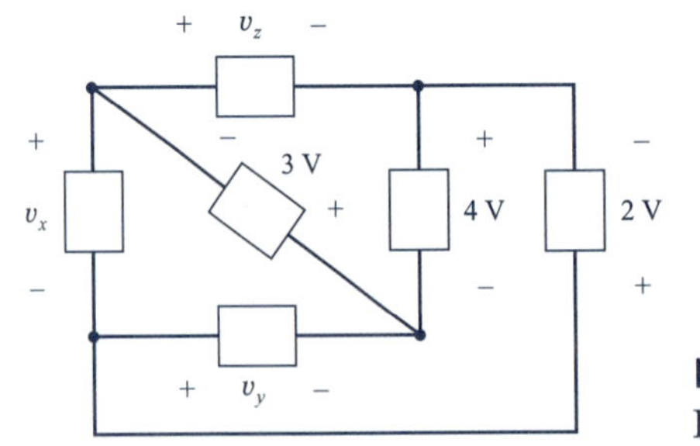

Figure E1.12
Exercise Problem 1.12.

Answers: −9 V, 6 V, −7 V.

Exercise Problem 1.13

Determine voltages v_x, v_y, v_z in the circuit of Fig. E1.13 by applying KVL.

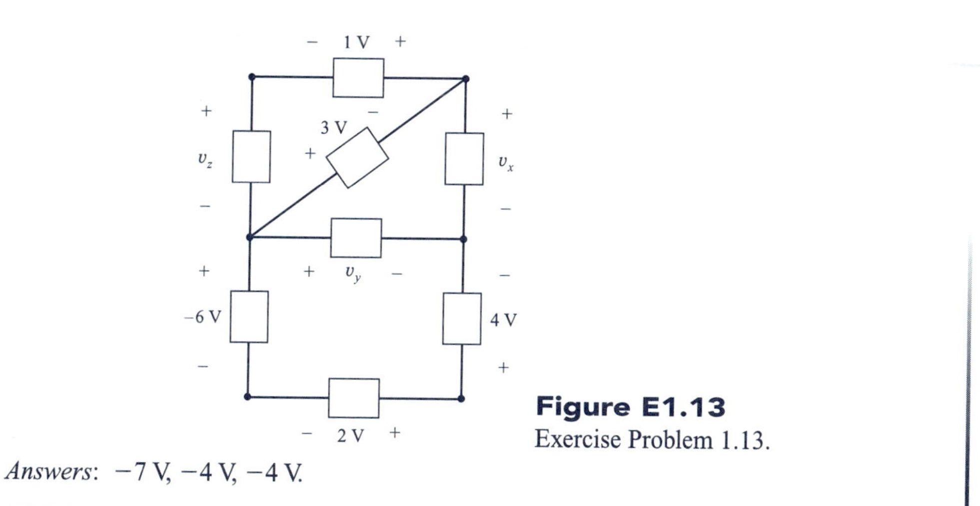

Figure E1.13
Exercise Problem 1.13.

Answers: −7 V, −4 V, −4 V.

Exercise Problem 1.14

Determine voltages v_x, v_y, v_z in the circuit of Fig. E1.14 by applying KVL.

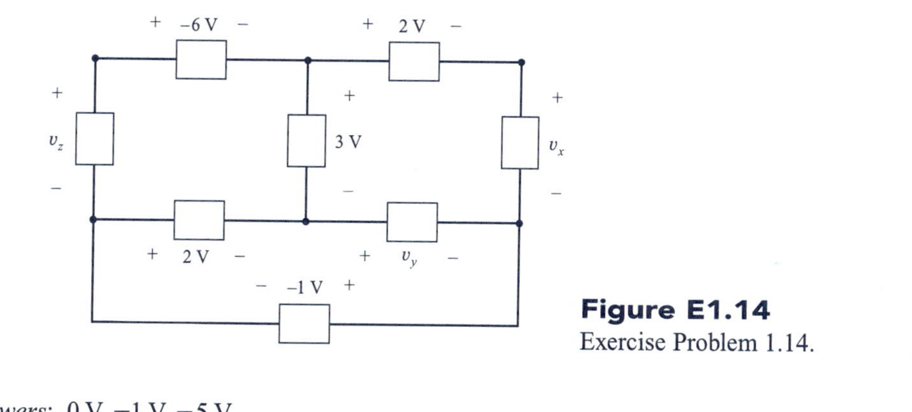

Figure E1.14
Exercise Problem 1.14.

Answers: 0 V, −1 V, −5 V.

Let us now combine KVL and KCL in one circuit solution.

Example 1.9

For the circuit of Fig. 1.31, determine voltage v and current i.

Solution Writing KVL around the loop containing elements E, H, B, A, and G gives

$$v = -3 + 2 + 1 - 4$$
$$= -4 \text{ V}$$

Applying KCL at the supernode cutting elements A, B, H, and E gives

$$1 + 4 + i_x = 3$$

or

$$i_x = -2 \text{ A}$$

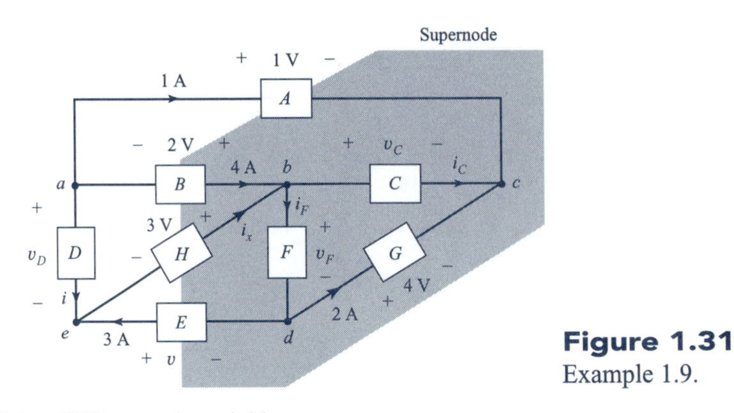

Figure 1.31
Example 1.9.

Now applying KCL at node e yields

$$i + 3 = i_x$$

or

$$i = -5\text{ A}$$

The remaining voltages and currents are obtained similarly as

$$v_D = 3 - 2 = 1\text{ V}$$
$$v_F = 3 + v = -1\text{ V}$$
$$v_C = 2 + 1 = 3\text{ V}$$
$$i_C = -2 - 1 = -3\text{ A}$$
$$i_F = 2 + 3 = 5\text{ A}$$

as the reader should verify.

Exercise Problem 1.15

In the circuit of Fig. E1.9, determine the voltages v_2, v_3, and v_4.

Answers: −1 V, −5 V, −2 V.

1.7 Conservation of Power

An electric circuit is considered to be a closed system with regard to energy, that is, the sum of the energies stored in all elements of the circuit at any time equals zero:

$$\sum_{\text{all elements}} w_i = 0 \tag{1.10}$$

If we differentiate this expression with respect to time we obtain the sum of the powers delivered to each element and arrive at the following important principle:

$$\begin{aligned}\sum_{\text{all elements}} p_i &= \sum_{\text{all elements}} v_i i_i \\ &= \frac{d}{dt} \sum_{\text{all elements}} w_i \\ &= 0\end{aligned} \tag{1.11}$$

where p_i is the power delivered to element i, and v_i and i_i are labeled on the elements with the passive sign convention. This is referred to as *the principle of conservation of power,* or Tellegen's theorem, and is stated as requiring that *the sum of the powers delivered to all the elements of a circuit at any time must equal zero.*

In a particular circuit, some circuit elements will be delivering power while others will be absorbing power at any instant of time. KVL and KCL and the principle of conservation of power are not all independent. In fact, any two imply the third.

Example 1.10

Verify conservation of power for the circuit of Fig. 1.31.

Solution The element currents and voltages were determined in Example 1.9. Computing the powers delivered to the elements gives

Element	Power Delivered to Element
A	$1\text{ A} \times 1\text{ V} = 1\text{ W}$
B	$-4\text{ A} \times 2\text{ V} = -8\text{ W}$
C	$-3\text{ A} \times 3\text{ V} = -9\text{ W}$
D	$-5\text{ A} \times 1\text{ V} = -5\text{ W}$
E	$-3\text{ A} \times (-4\text{ V}) = 12\text{ W}$
F	$5\text{ A} \times (-1\text{ V}) = -5\text{ W}$
G	$2\text{ A} \times 4\text{ V} = 8\text{ W}$
H	$-(-2\text{ A}) \times 3\text{ V} = 6\text{ W}$
	$\sum = 0\text{ W}$

Exercise Problem 1.16

Verify that the voltages and currents in the circuit of Fig. E1.9 that were determined in Exercise Problems 1.9 and 1.15 satisfy conservation of power.

Exercise Problem 1.17

Determine the unknown voltages and currents in the circuit of Fig. E1.17, and verify conservation of power.

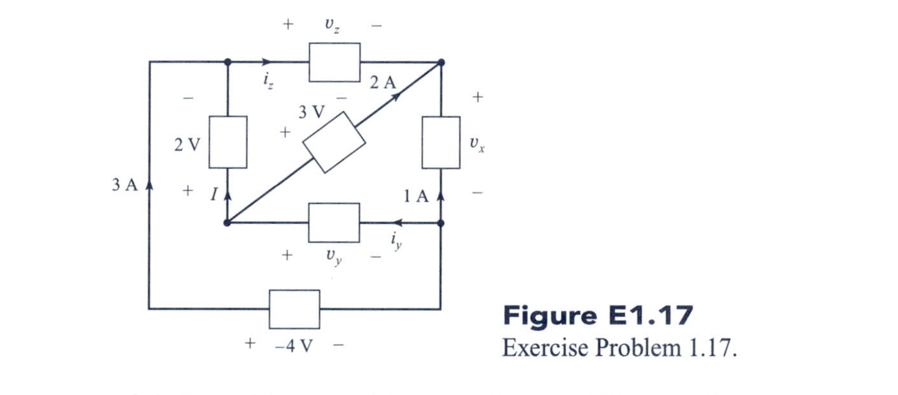

Figure E1.17
Exercise Problem 1.17.

Answers: $i_z = -3\text{ A}, I = -6\text{ A}, i_y = -4\text{ A}, v_z = 1\text{ V}, v_y = -2\text{ V}, v_x = -5\text{ V}.$

Exercise Problem 1.18

Determine the unknown voltages and currents in the circuit of Fig. E1.18, and verify conservation of power.

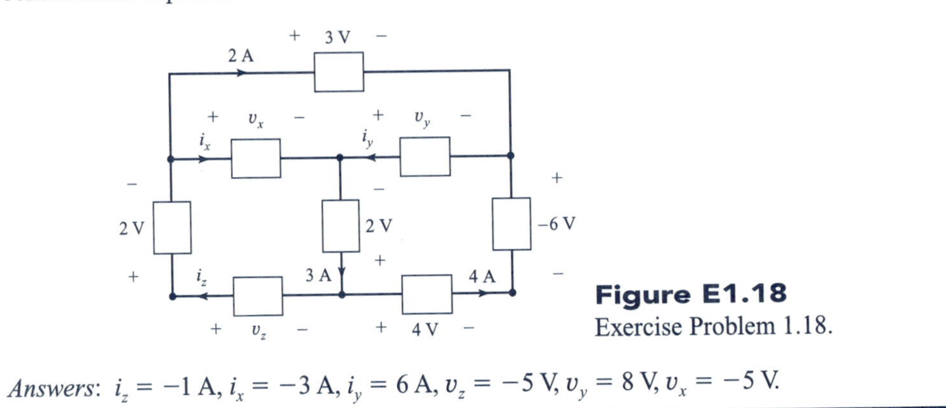

Figure E1.18
Exercise Problem 1.18.

Answers: $i_z = -1$ A, $i_x = -3$ A, $i_y = 6$ A, $v_z = -5$ V, $v_y = 8$ V, $v_x = -5$ V.

1.8 Series and Parallel Connections of Elements

We frequently have occasion to consider what are known as *series* and *parallel* connections of circuit elements. It is important that the reader understand the meanings of these terms, since many of the errors made in the analysis of circuits are a result of not doing so.

Two or more elements are said to be connected in *series* if *the currents through them must be equal* (because of KCL). A *series* connection of three elements is shown in Fig. 1.32*a*. The three elements are joined at nodes b and c. Elements A and B are said to be connected in series because the current i_A leaving element A and entering node b must, by KCL, equal the current i_B that enters element B and leaves node b, and therefore $i_A = i_B$. Similarly, elements B and C are said to be connected in series because KCL shows that $i_B = i_C$. Alternatively we may say that *two elements are connected in series if they are joined at a common node at which no other elements are attached.*

Two or more elements are said to be connected in *parallel* if *the voltages across them must be equal* (because of KVL). A *parallel* connection of three elements is shown in Fig. 1.32*b*. The

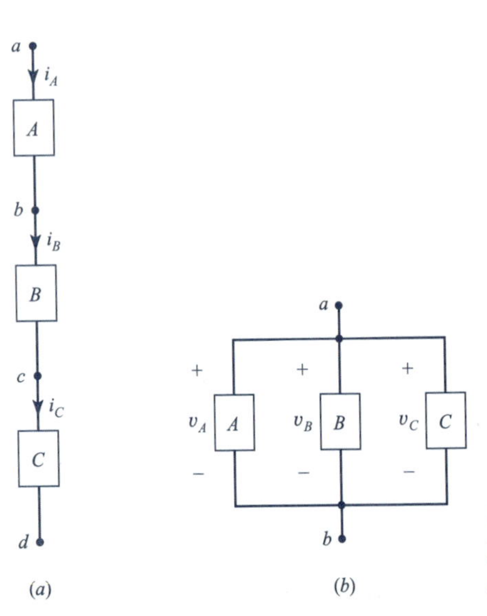

Figure 1.32
Illustration of (*a*) the series connection of elements, and (*b*) the parallel connection of elements.

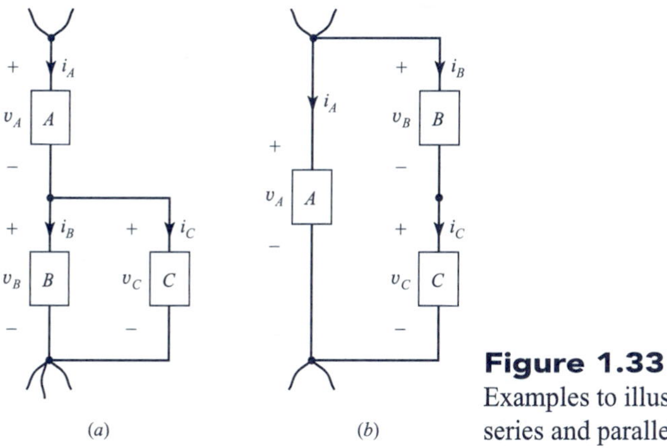

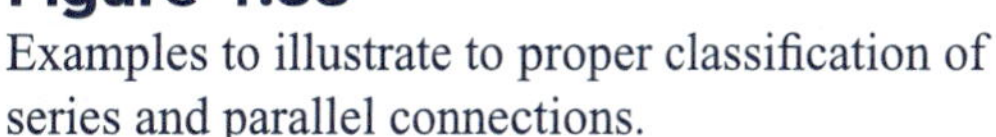

Figure 1.33
Examples to illustrate to proper classification of series and parallel connections.

three elements are connected at node a and at node b. The elements are said to be connected in parallel because KVL shows that the voltages across all three must be equal: $v_A = v_B = v_C$. Alternatively we may say that *two elements are connected in parallel if the elements are connected at both sets of terminals.*

Some common misconceptions and improper classification of typical connections are illustrated in Fig. 1.33. In Fig. 1.33*a* elements B and C are connected in parallel because, by KVL, $v_B = v_C$. However, elements A and B are not in series, nor are elements A and C, because their currents are not necessarily equal. For example, in order for A and B to be in series would require that $i_C = 0$, which is not necessarily true. In Fig. 1.33*b* elements B and C are in series because, by KCL, $i_B = i_C$. Elements A and B are not in parallel, nor are elements A and C, since their voltages are not necessarily equal. For example, in order for A and B to be in parallel would require that $v_C = 0$, which is not necessarily true.

Exercise Problem 1.19

Determine which elements in Fig. E1.19 are connected in series and which are connected in parallel.

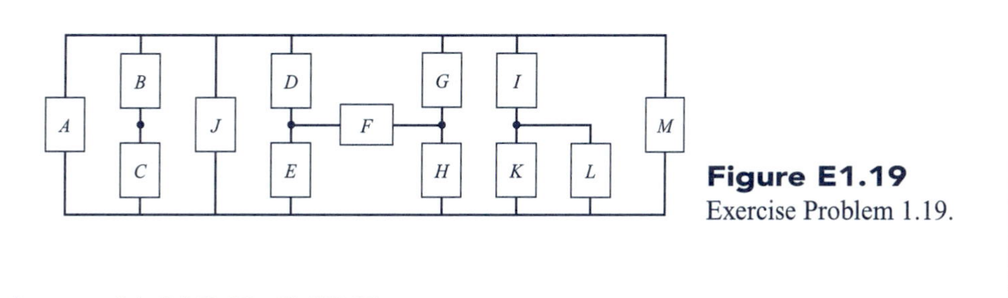

Figure E1.19
Exercise Problem 1.19.

Answers: $[A, J, M]$, $[B, C]$, $[K, L]$.

Exercise Problem 1.20

Determine which elements in Fig. E1.20 are connected in series and which are connected in parallel.

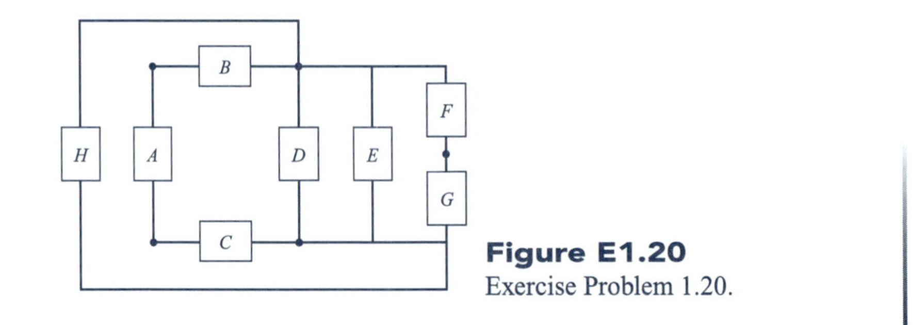

Figure E1.20
Exercise Problem 1.20.

Answers: [A, B, C], [F, G], [H, D, E].

Exercise Problem 1.21

Determine which elements in Fig. E1.21 are connected in series and which are connected in parallel.

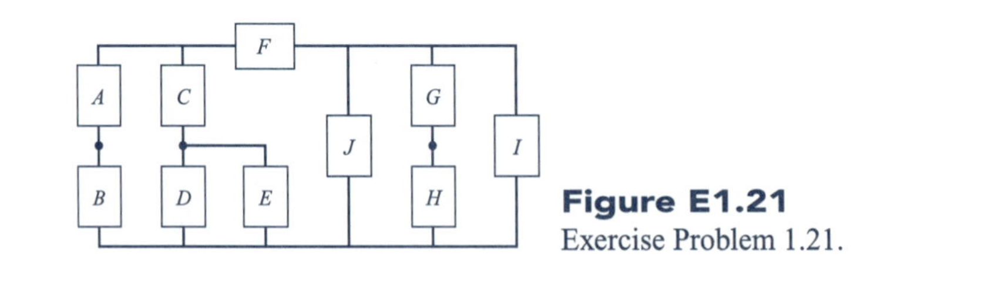

Figure E1.21
Exercise Problem 1.21.

Answers: [A, B], [G, H], [D, E], [I, J].

1.9 Equivalent Circuits

The concept of *equivalent circuits* is another important and powerful circuit analysis tool that we will frequently employ. But like the notions of series and parallel connections of elements, it is frequently misunderstood and misapplied. In order to illustrate this important principle, consider two circuits, circuit A and circuit B, that are to be joined at two nodes as shown in Fig. 1.34*a*. Suppose we replace the circuit to the right, circuit B, with another circuit, circuit C, as shown in Fig. 1.34*b*. *Circuit B and circuit C are said to be equivalent at their external terminals if they cause the same currents and voltages to occur in circuit A when either is attached to circuit A*. In order for this to be true, the voltage–current relations at the two external terminals must be identical, i.e., the relation between v_B and i_B for circuit B must be identical to the relation between v_C and i_C for circuit C. For example, if $v_B = 10i_B - 15$, then we must have $v_C = 10i_C - 15$ in order for circuits B and C to be equivalent at these two external terminals.

It is important to note that equivalence does not mean that the internal structure of circuits B and C must look the same. Generally they will not. But with respect to the currents and voltages that will be produced in circuit A when either B or C is attached to its terminals, there will be no

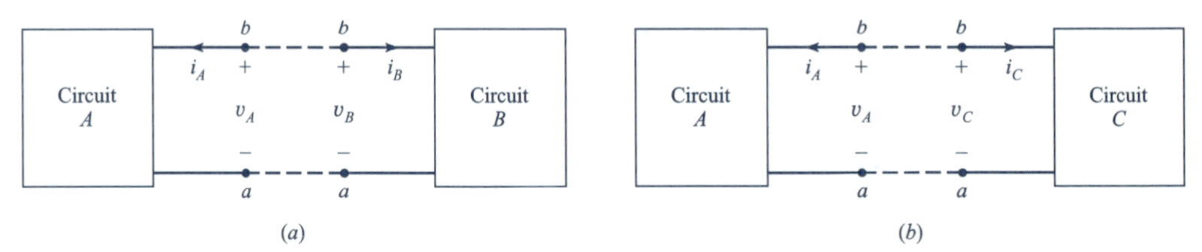

Figure 1.34
Illustration of the concept of equivalent circuits.

difference. So equivalence simply means that circuit A cannot determine which circuit is attached to its terminals, B or C.

We use this notion of equivalence on many occasions in order to replace a complicated circuit B with an equivalent circuit C whose internal structure is much simpler than B, thereby facilitating the determination of the voltages and currents in A. The true test of whether two two-terminal circuits are equivalent at their terminals is if their v–i relationships at those terminals are *identical*.

Example 1.11

Determine the value for I in circuit C of Fig. 1.35 in order for circuit C and circuit B to be equivalent at their two external terminals a and b.

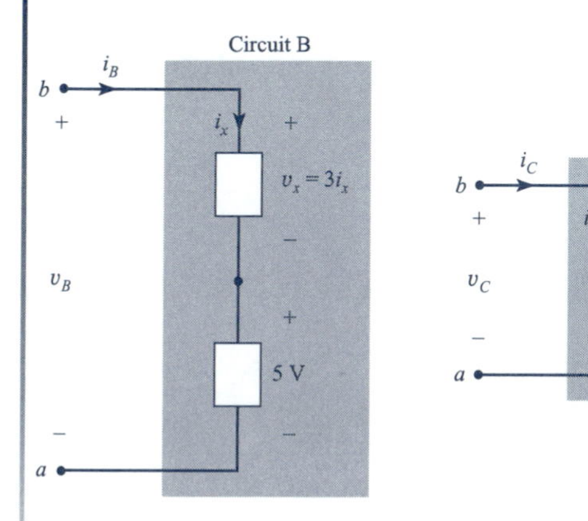

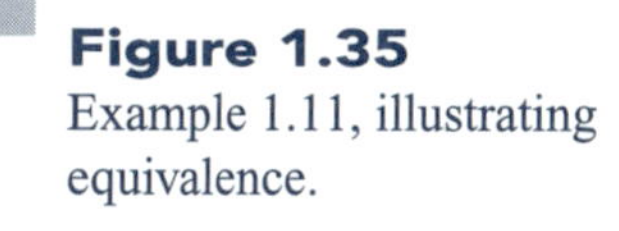

Figure 1.35 Example 1.11, illustrating equivalence.

Solution First determine the terminal voltage and current relationships. For circuit B we obtain using KVL

$$v_B = 3i_x + 5$$

and

$$i_x = i_B$$

Substituting gives

$$v_B = 3i_B + 5 \tag{a}$$

Applying KCL in circuit C gives

$$i_C = \frac{v_y}{3} + I$$

and

$$v_y = v_C$$

Substituting gives

$$i_C = \frac{v_C}{3} + I$$

Rewriting gives

$$v_C = 3i_C - 3I \tag{b}$$

Comparing the two results (a) and (b) shows that in order for circuits C and B to be equivalent at these two terminals, circuit C must have

$$I = -\tfrac{5}{3}\,\text{A}$$

Exercise Problem 1.22

Determine V and I in order that the two circuits in Fig. E1.22 will be equivalent at terminals ab.

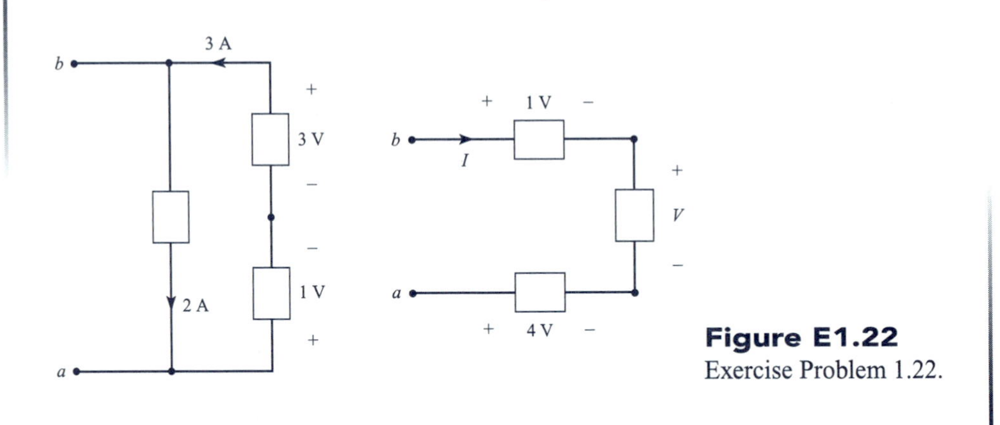

Figure E1.22
Exercise Problem 1.22.

Answers: $V = 5$ V, $I = -1$ A.

1.10 Redrawing Circuits in Equivalent Forms

The length and shape of the connection leads attached to the boxes (the elements) are not important. They can be treated as though they were rubber bands as shown in Fig. 1.36. This is referred to as the "*stretch-and-bend principle.*" It can frequently be used to redraw a circuit into an equivalent form that makes certain properties of the circuit much clearer. For example, consider Fig. 1.37. The circuit on the left can be redrawn in an equivalent form on the right using the stretch-and-bend principle. Note that this makes it clearer that elements C, B, and E are in series and elements A and D are in parallel.

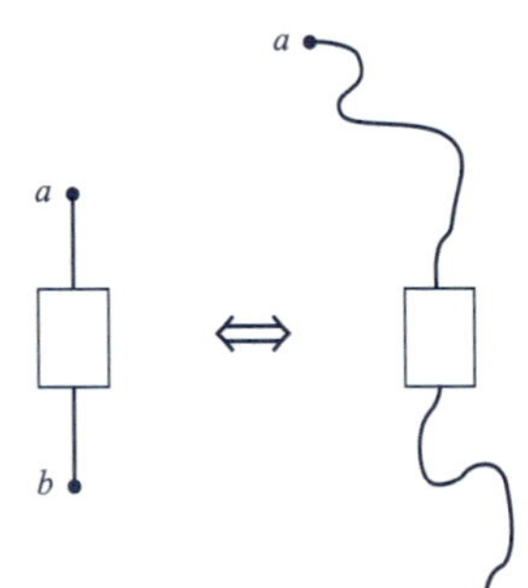

Figure 1.36
Illustration of the fact that the length and shape of the connection leads attached to an element are not important.

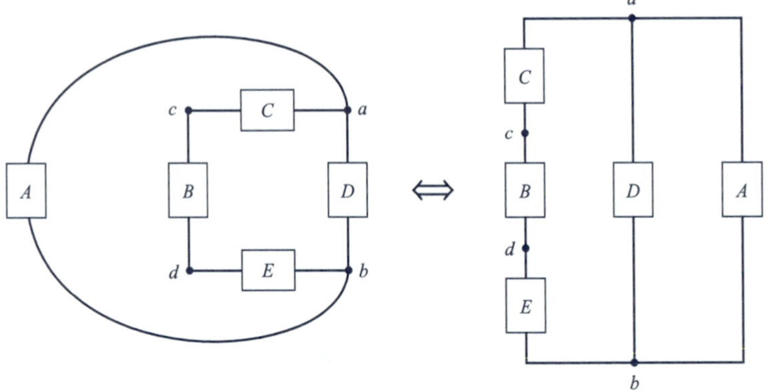

Figure 1.37
Redrawing a circuit into an equivalent form using the stretch-and-bend principle.

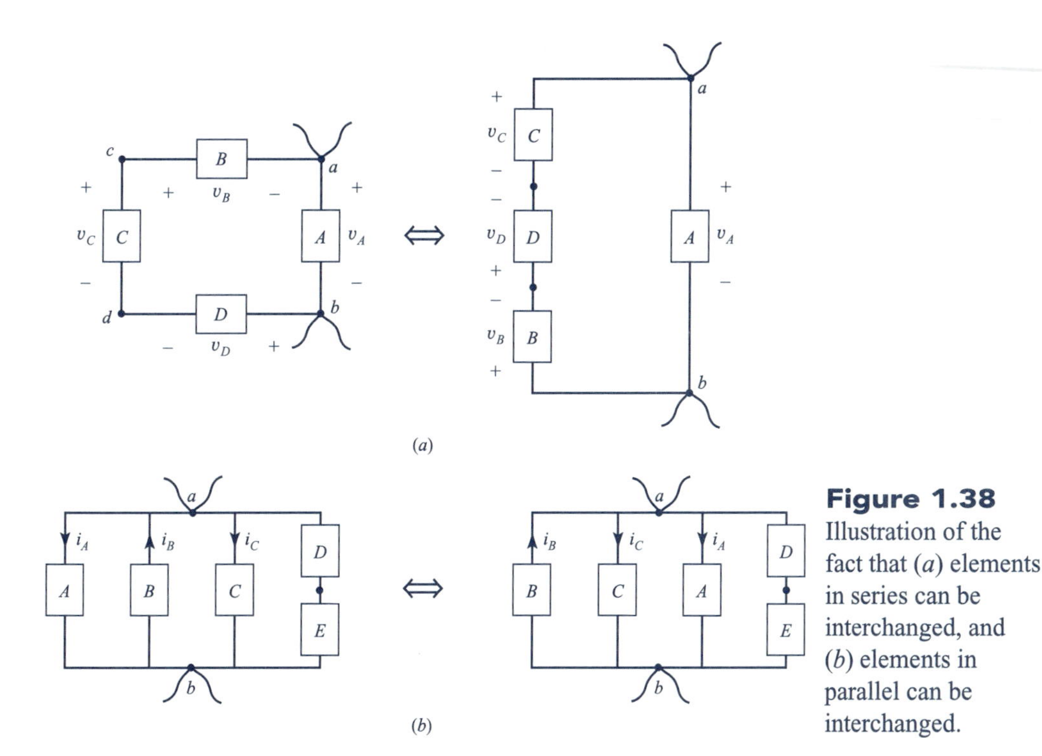

Figure 1.38
Illustration of the fact that (*a*) elements in series can be interchanged, and (*b*) elements in parallel can be interchanged.

Elements that are in series can be interchanged, and elements in parallel can be interchanged. This is illustrated in Fig. 1.38. In Fig. 1.38*a* elements B, C, and D are in series and their locations in the series connection can be interchanged as shown. Note that element A is not in series with element B, C, or D, since there are external connections at nodes a and b. Observe, however, that the *voltage* across an element remains with that element and *the polarity of that voltage with respect to nodes a and b must not be changed.* For example, note that the negative terminal of voltage v_B is at the node a end and the positive terminal is at the node b end. This must also be preserved in order that the redrawn circuit be equivalent to the original circuit. In Fig. 1.38*b* elements A, B, and C are in parallel and their locations in the parallel connection can be interchanged as shown. Note that elements D and E are in series but neither one is in parallel with element A, B, or C. Observe, however, that the *current* associated with an element remains with that element and the *direction of that current with respect to nodes a and b must not be changed.* For example, current i_B is directed towards node a and current i_A is directed toward node b. In the redrawn circuit this must be preserved in order that the redrawn circuit be equivalent to the original circuit.

Figure 1.39 shows what appears to be two nodes connected by only a connection lead. Actually, these are only one node, since there is no element between them and hence the voltages

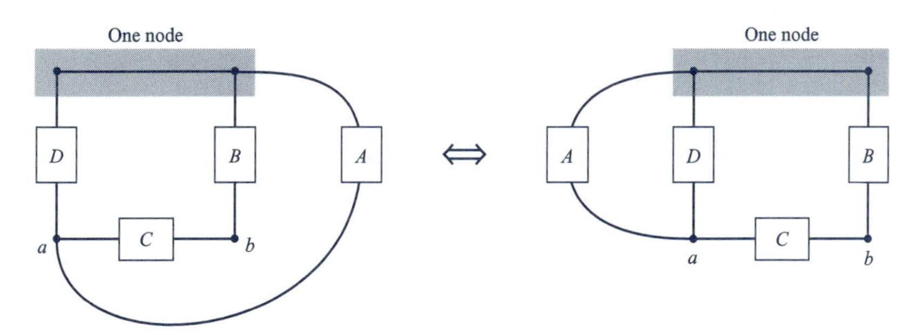

Figure 1.39
An element's connection point may be moved along a connection lead. Observe that there are only three nodes in this circuit.

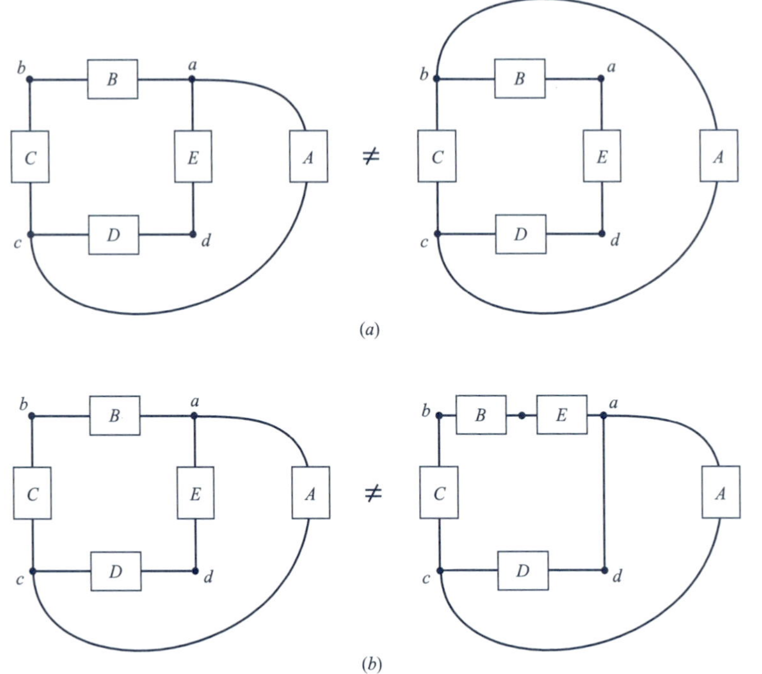

Figure 1.40 Illustration of common errors in redrawing a circuit: (*a*) cutting a connection lead and moving it to another node, and (*b*) cutting the connection leads of an element and inserting the element in another place.

of the two ends of this connection lead are the same. We can "slide" the connection point of element *A* to the other end of this connection lead and use the stretch-and-bend principle to draw an equivalent circuit as shown. This then shows that the elements *A* and *D* are in parallel.

Figure 1.40 shows some of the common errors in redrawing circuits. These errors fall into the categories of "cut and reattach," wherein a connection lead is cut and then reattached at some other point, or "cut and move," wherein the connection leads of an element are cut and the element is moved to (inserted at) another location. In the left circuit of Fig. 1.40*a*, element *A* is attached between nodes *a* and *c*. In the right circuit, it has been cut at node *a* and reattached to node *b*. These two circuits are not equivalent. In the left circuit of Fig. 1.40*b*, element *B* is in series with element *C*, and element *D* is in series with element *E*. In the right circuit the connection leads of element *E* have been cut and element *E* has been moved in series with element *B*. These two circuits are not equivalent.

The author has observed that redrawing circuits into forms that are *not equivalent* constitutes one of the most common and serious errors beginning students make in the analysis of a circuit. Redrawing a circuit into an equivalent form is often very useful is showing certain properties of the circuit and in making the analysis easier. However, if the circuit is redrawn into another form that is not equivalent to the original circuit, then a new and completely different circuit from the one intended has been formed. Analysis of this new circuit is a waste of time, because it is not the circuit that was to be analyzed.

1.11 Application Examples

Throughout this text we will give some examples of the application of the results of the chapter. As we proceed through the text, our ability to understand electric circuits will increase, and the types of application examples we can discuss will increase in sophistication. For the present we must be content with some rather simple ones that illustrate the important concepts of voltage, current, and power.

1.11.1 Residential Power Distribution Electric power is generated at some distant power plant where mechanical energy is converted to electrical energy by, for example, burning coal, which creates steam to drive electric turbines, which generate electrical energy. Nuclear energy and falling water at a dam are alternative means of turning the electric turbines. The electric energy is transmitted to a city via high-voltage power transmission lines. The power is transmitted in a three-phase form consisting of three power lines. This will be discussed in more detail in Chapter 6. The voltage is on the order of 365 kV. (The numbers mentioned here are typical for electric power distribution in the U.S.) This high voltage allows the corresponding current to be very small yet the product of voltage and current, the power transmitted, to be the same as for a low voltage and a high current. This high-voltage transmission is used in order to reduce the power losses due to the current passing through the resistance of the wires. (Resistance is discussed in the next chapter.) Once the power line has reached the point of distribution, such as a city, it is reduced in voltage in several stages with the use of transformers (discussed in Chapter 5).

Figure 1.41*a* shows how power is provided to the residence. The power-line voltage is a 7200-V waveform that varies sinusoidally with time at a rate of 60 Hz (60 cycles per second).

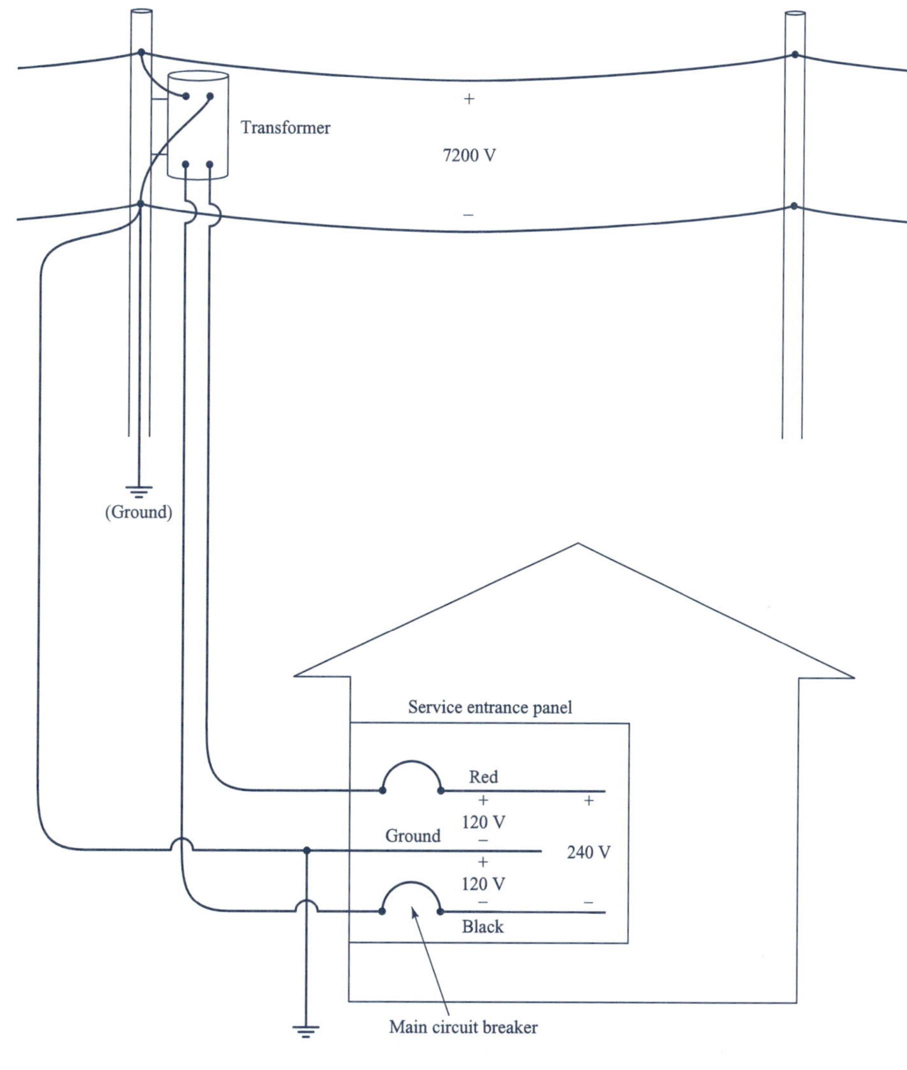

Figure 1.41
(continued)

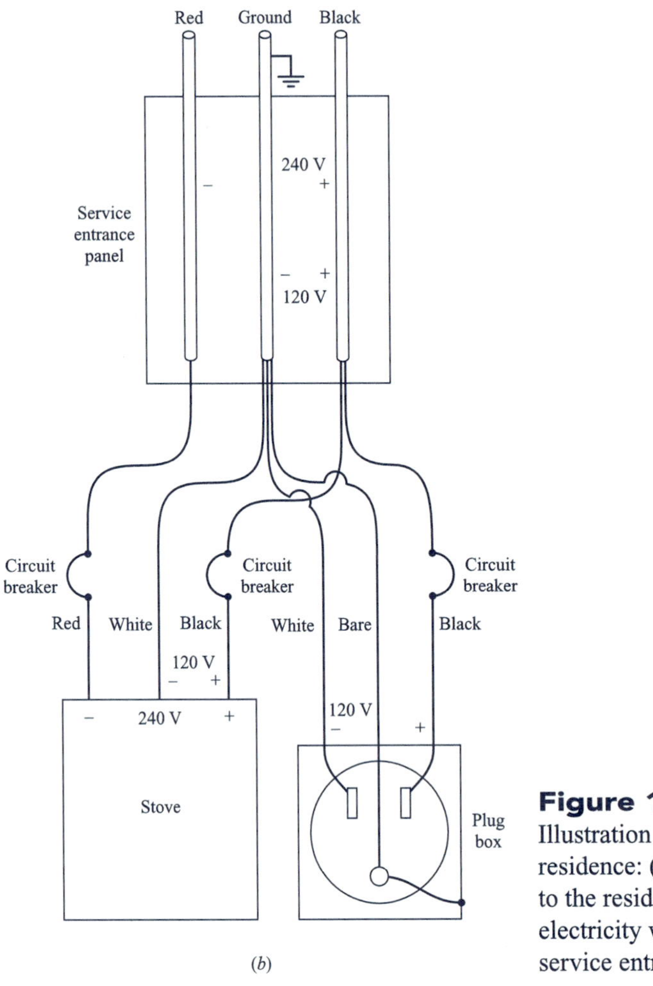

Figure 1.41
Illustration of electricity supply to residence: (*a*) supply from the power line to the residence, and (*b*) distribution of electricity within the residence via the service entrance panel.

This will be discussed in Chapter 6. Power poles at the residence carry two wires, of which one is connected to earth (grounded) and the other is at a voltage of 7200 V with respect to the lower wire (and ground). This is fed to a transformer, which steps the voltage down to 240 V. Three wires enter the home through the *service entrance panel*. Two of the wires are color-coded as black and red, while the third is a bare wire that is at ground potential. The voltage between the red and black wires is a 240-V sinusoid. Within the service entrance panel, these three wires are connected to three *bus bars* for providing the required power as shown in Fig. 1.41*b*.

Stoves, clothes dryers, and other high-power-consuming devices require both 240 and 120 V. The 240 V is provided between the red and black wires, while the 120 V is provided between either the red or the black and the neutral (ground) wire. Other, more numerous appliances such as lights, TVs, and general-purpose wall sockets are provided 120 V from the service entrance panel via three wires in a plastic-wrapped cable that is routed through the house as shown in Fig. 1.41*b*. One wire, whose insulation is color-coded black, is *hot* and is at 120 V with respect to the other wire, whose insulation is color-coded white and is connected to earth (ground) at the service entrance panel. In addition, a third, bare wire is carried in the cable. This is also connected to ground at the service entrance panel. This ground wire is also connected to the metallic housing of the socket. The current goes down the black wire and returns along the white wire. The purpose of the third (bare) ground wire is safety. Should the hot black wire become disconnected within the socket housing and come in contact with that metallic box, we want that to trip the circuit breaker, preventing the metallic housing from becoming hot at 120 V with respect to ground. If the ground wire were not connected to the metallic housing, the occurrence of the black or hot wire coming in contact with the metallic housing would make the housing at 120 V with respect to ground, causing a shock hazard if someone touched the housing, say, by touching the screw that holds the cover plate on. Once the black wire comes in contact with the metallic housing, current (a very large amount) flows back to the service entrance panel, tripping the circuit breaker for that circuit. Typical such circuit breakers are rated at 20 A, so that when the current going down the black wire exceeds 20 A (as it will with a short circuit in the housing), the circuit breaker trips, disconnecting power to the socket and returning it to a safe condition. Similar comments apply to the 240-V connection for the dryer or stove.

The power that the residence consumes is measured by a watt-hour meter installed where the power enters the service entrance panel. A typical reading lamp consumes 100 W at 120 V, giving a current draw of some 0.83 A. The watt-hour meter measures the product of power and time, giving a measure of the energy consumed by the residence. A typical electric power cost is about 7 cents/kW-h. Although the power consumption of a residence varies quite a bit from residence to residence, a typical residence consumes on the order of 30 kW-h per day. Hence the cost is on the order of $2.10/day, or some $60/month. A residence that consumes 30 kW-h in one day draws an average of some 1250 W (that would be the actual power consumption if it were constant over the 24-h day, which it obviously isn't). At a voltage of 240 V this gives a current of a little over 5 A. That is on the low side, because it is an average. On the other hand, suppose that power consumption occurred over 16 hours. In that case, the average power consumed would be 1875 W and at a voltage of 240 V, the average current draw would be about 8 A. Obviously, much higher peak power consumption occurs over shorter time periods (such as when a clothes dryer is running). A typical service entrance panel is rated at 200 A, so that at 240 V, we could consume 240 V $\times$ 200 A = 48 kW and be within the limits of the service-entrance panel fuse and circuit breaker capacity. But this would result in a very large bill from the power company.

1.11.2 Automobile Storage Batteries Automobile storage batteries typically supply 12 V for starting the vehicle. Once the vehicle is started, an alternator driven by the engine supplies power to recharge the battery. Suppose a battery has a 200-ampere-hour (A-h) capacity. Although this notion is idealized, in essence, this rating indicates that the battery can supply 200 A for 1 h or 1 A for 200 h. In fact, of course, as the battery runs down, the current supplied diminishes. Nevertheless, for illustrating the concepts, let us suppose that a 200-A-h battery is used to start the vehicle and during starting, the starter draws 40 A. It turns out that during staring the voltage of the battery drops from 12 V to around 9 V due to the internal resistance of the battery (to be discussed in the next chapter). If vehicle does not start, but we continue to turn the starter, how long will the battery last? Again, we idealize and suppose that the battery can supply 40 A until it is completely dead. If so, then the battery will last for some 200 A-h/40 A = 5 h. This will not actually happen, because the supplied current of the battery and its voltage will very soon diminish below what is required to turn the starter.

The total charge stored in the battery is the product of current and time (in seconds). Hence the 200-A-h battery stores 200 A $\times$ 3600 s = 720 kC of charge. Multiplying this by the voltage (assume a constant 12 V) gives the energy stored in the battery as 12 V $\times$ 200 A $\times$ 3600 s = 8.64 MJ. If the headlights of the vehicle are inadvertently left on after the engine is shut down and they consume 50 W, how long will the battery last? The current drawn by the lights is 50 W/12 V = 4.17 A. Hence the battery should last 200 A-h/4.17 A = 48 h—an unrealistic answer in that the battery would not be able to supply 4.17 A right up to the point that it is completely dead.

Problems

Section 1.1 Charge and Electric Forces

1.1-1 Two charges of equal magnitude 5×10^{-9} C but opposite sign are separated by a distance of 10 m. Determine the net force exerted on a positive charge, $q = 2 \times 10^{-9}$ C, that is placed midway between the two charges. *Answer*: 7.2×10^{-9} N.

1.1-2 Two positive charges, $Q_1 = 5 \times 10^{-9}$ C and $Q_2 = 2 \times 10^{-9}$ C, are held fixed and separated by a certain distance. A third, negative charge, $Q_3 = -2 \times 10^{-9}$ C, is introduced between them. Determine the ratio of the distance between Q_3 and Q_2 and the distance between Q_3 and Q_1 such that Q_3 will not move. *Answer*: 0.6325.

1.1-3 Two charges are held fixed in a two-dimensional coordinate system. One charge is negative, $Q_1 = -2 \times 10^{-9}$ C, and is placed at $x = 0$ m, $y = 2$ m. The other charge is positive, $Q_2 = 3 \times 10^{-9}$ C, and is placed at $x = 3$ m, $y = 0$ m. A positive charge, $q = 10^{-9}$ C, is placed at the origin, $x = 0$ m, $y = 0$ m. Determine the magnitude of the total force exerted on q and the direction of this resultant in terms of the angle measured from the x axis.
Answer: 5.408×10^{-9} N, 123.69°.

Section 1.2 Voltage

1.2-1 Two positive charges, $Q_1 = 2 \times 10^{-9}$ C and $Q_2 = 5 \times 10^{-9}$ C, are separated by a distance of 10 m. Determine the work required to move a positive charge $q = 1 \times 10^{-6}$ C along a straight line between Q_1 and Q_2 from a distance of 1 m away from Q_1 to a distance of 1 m away from Q_2.
Answer: 24×10^{-6} J.

1.2-2 Two charges, $Q_1 = 3 \times 10^{-6}$ C and $Q_2 = -2 \times 10^{-6}$ C, are held fixed and separated by a distance of 6 m. Determine the voltage v_{ba} between two points along a line between the two charges. Point a is 1 m from Q_1, and point b is 2 m from Q_2. *Answer*: −25.65 kV.

1.2-3 Two charges are held fixed in a rectangular coordinate system. Charge $Q_1 = 2 \times 10^{-9}$ C is located at $x = 0$ m and $y = 0$ m, and charge $Q_2 = 3 \times 10^{-9}$ C is located at $x = 5$ m and $y = 0$ m. Determine the voltage v_{ba} between point a at $x = 1$ m and $y = 0$ m and point b at $x = 3$ m and $y = 0$ m. *Answer*: −5.25 V.

1.2-4 Two charges are held fixed in a rectangular coordinate system. Charge $Q_1 = -1 \times 10^{-9}$ C is located at $x = 0$ m, $y = 0$ m, and charge $Q_2 = 3 \times 10^{-9}$ C is located at $x = 0$ m, $y = 2$ m. Determine the voltage v_{ba} between point b located at $x = 3$ m, $y = 0$ m and point a located at $x = 1$ m, $y = 0$ m. *Answer*: 1.41 V.

1.2-5 Determine the work required to move a positive charge of 2×10^{-7} C through a voltage increase of 100 V. *Answer*: 20 μJ.

1.2-6 Movement of a charge through a voltage increase of 1 kV requires 1 mJ of energy expenditure. Determine the charge moved. *Answer*: 1 μC.

1.2-7 A light bulb has 12 V applied across it. If 100 μC of charge is passed through it, determine the energy delivered to the bulb. *Answer*: 1.2 mJ.

Section 1.3 Current and Magnetic Forces

1.3-1 The current passing through an element is sketched in Fig. P1.3-1. Determine the net positive charge transferred in the direction of the current at $t = -1, -0.5, 0, 1, 2, 3, 4$ s.
Answers: $-0.5, -1, -1.5, -2, -1.5, -1, -1$ C.

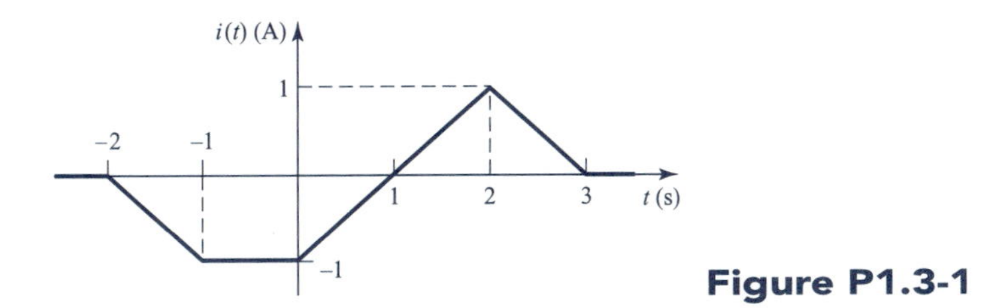

Figure P1.3-1

1.3-2 The charge passing a point to the right is sketched in Fig. P1.3-2. Sketch the current directed to the right as a function of time. *Answers*: -3 A for -2 s $< t < -1$ s, 0.5 A for -1 s $< t < 1$ s, 0 A for 1 s $< t <$ 2 s, 2 A for 2 s $< t <$ 3 s, -2 A for 3 s $< t <$ 4 s, 0 A for $t >$ 4 s.

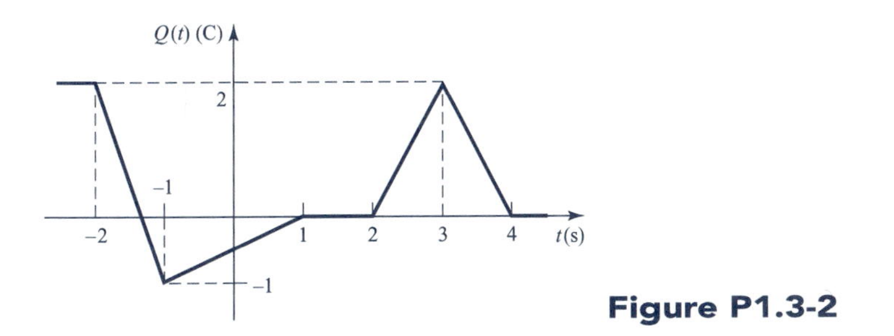

Figure P1.3-2

1.3-3 Automobile storage batteries are rated in terms of their terminal voltage (12 V) and their ampere-hour capacities. For a typical 12-V battery having a 115-A-h capacity, determine the length of time this battery will light a 6-W bulb. (Assume the battery voltage is constant at 12 V throughout the test, even though that is not true.) Determine the total energy stored in the battery before it is connected to the bulb. Determine the total amount of charge that has passed through the wires that connect the battery and the bulb. *Answers*: 230 h, 4.968 MJ, 414 kC.

1.3-4 The current through an element is sketched as a function of time in Fig. P1.3-4. Determine the net positive charge transferred through the element at $t = -1, 0, 1, 2, 3$, and 4 s in the direction of the current. *Answers*: $-1, -1.5, -1, 1, 1, 1$ C.

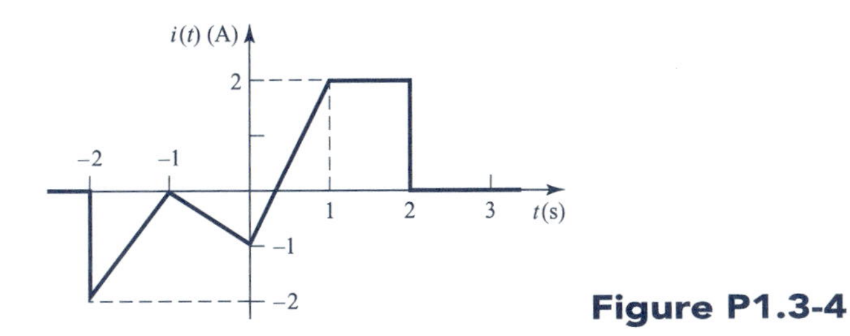

Figure P1.3-4

1.3-5 The net positive charge passing a point to the right is sketched in Fig. P1.3-5. Determine the current directed to the right at $t = -1.5, -0.5, 0.5, 1.5, 2.5, 3.5$ s.
Answers: $-2, 2, 3, 0, -3, 0$ A.

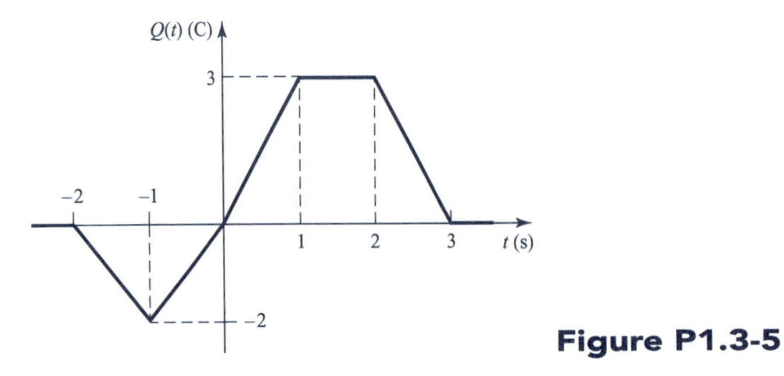

Figure P1.3-5

1.3-6 A current has the waveform described by $i(t) = 0$ for $t < 0$, $i(t) = 10e^{-3t}$ μA for $t \geq 0$. Determine the net positive charge transferred at $t = 0.5, 1, 5$ s. *Answers*: 2.59, 3.17, 3.33 μC.

1.3-7 The net positive charge transferred past a point is described by $q(t) = 5(1 - e^{-0.5t})$ μC for $t \geq 0$ and $q(t) = 0$ for $t \leq 0$. Determine the current at $t = 0.5, 1, 5$ s. *Answers*: 1.95, 1.52, 0.205 μA.

Section 1.4 Lumped Circuit Elements and Power

1.4-1 Determine the power delivered to (absorbed by) the elements shown in Fig. P1.4-1. *Answers*: (a) −6 W, (b) −12 W, (c) −6 W, (d) −4 W, (e) −6 W, (f) 12 W.

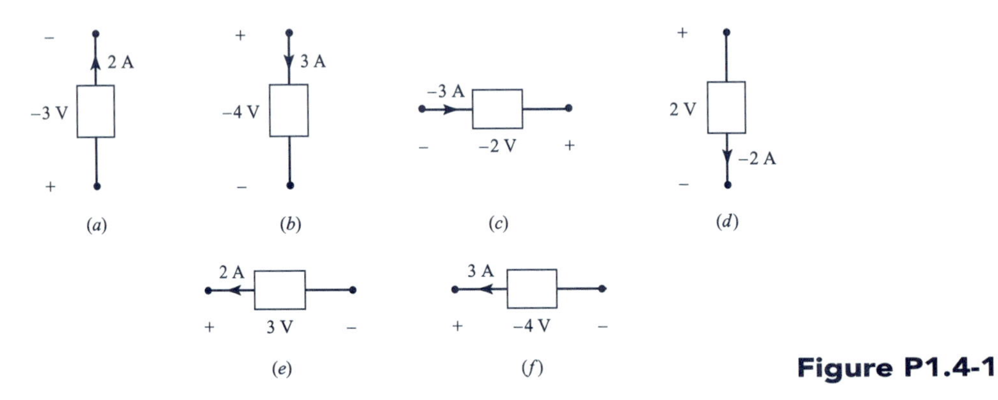

Figure P1.4-1

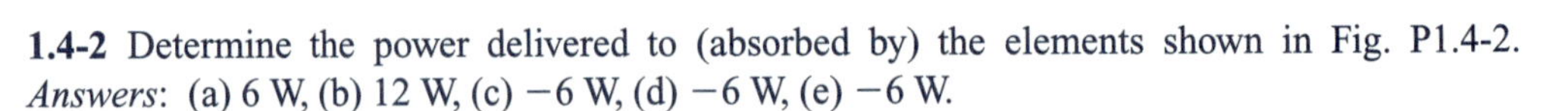

1.4-2 Determine the power delivered to (absorbed by) the elements shown in Fig. P1.4-2. *Answers*: (a) 6 W, (b) 12 W, (c) −6 W, (d) −6 W, (e) −6 W.

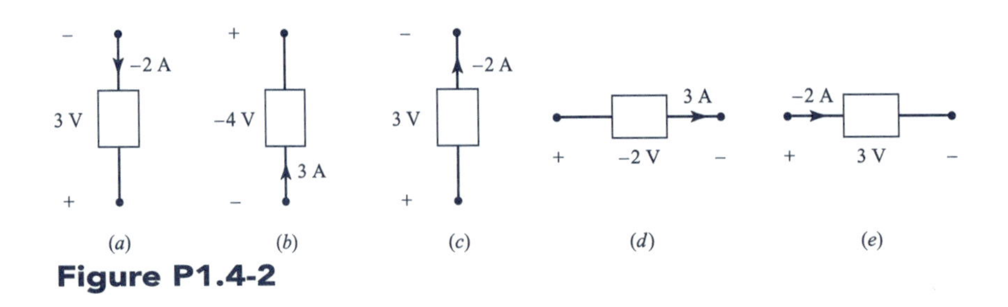

Figure P1.4-2

1.4-3 Determine the power delivered to (absorbed by) the elements shown in Fig. P1.4-3. *Answers*: (a) −6 W, (b) −6 W, (c) 6 W, (d) 20 W, (e) −18 W, (f) 30 W.

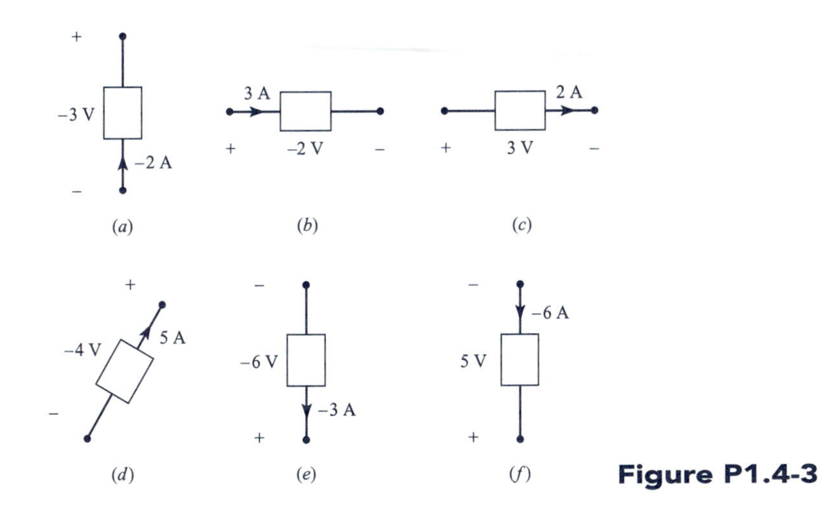

Figure P1.4-3

1.4-4 Determine the power delivered to (absorbed by) the elements shown in Fig. P1.4-4. *Answers*: (a) −16 W, (b) −30 W, (c) 18 W, (d) 8 W, (e) −21 W, (f) 40 W.

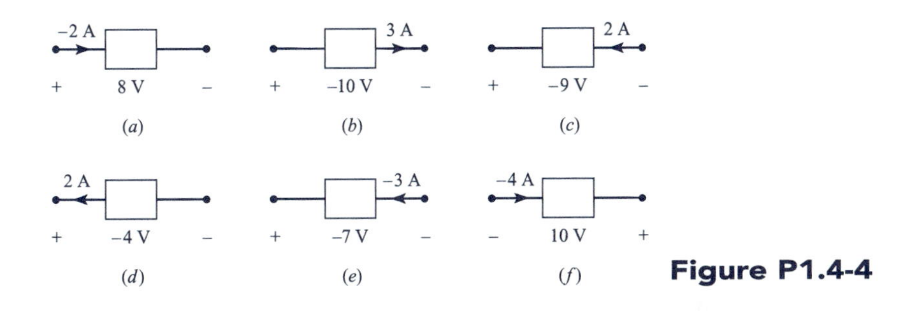

Figure P1.4-4

1.4-5 Determine the power delivered to (absorbed by) the elements shown in Fig. P1.4-5. *Answers*: (a) 20 W, (b) −12 W, (c) −28 W, (d) −6 W, (e) −15 W, (f) −10 W.

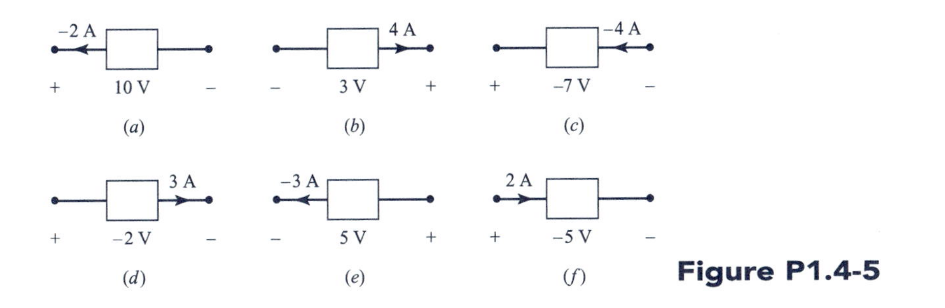

Figure P1.4-5

Section 1.5 Kirchhoff's Current Law (KCL)

1.5-1 Determine the currents i_x, i_y, i_z, i_w in the circuit of Fig. P1.5-1. *Answers*: $i_x = 1$ A, $i_y = 0$ A, $i_z = -3$ A, $i_w = 1$ A.

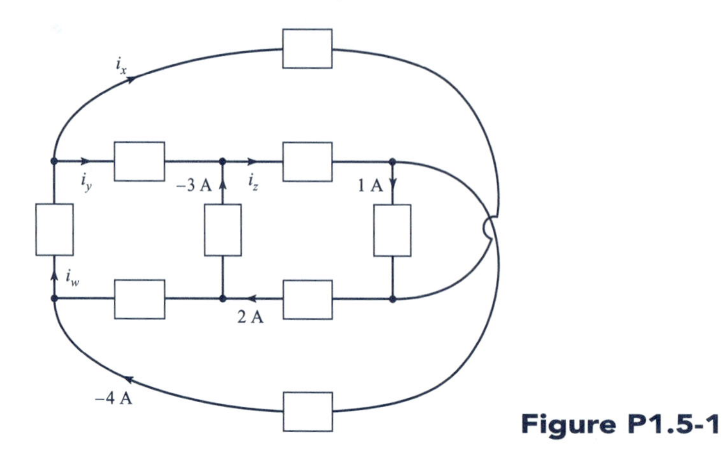

Figure P1.5-1

1.5-2 Determine the currents i_x, i_y, i_z in the circuit of Fig. P1.5-2.
Answers: $i_x = 4$ A, $i_y = 5$ A, $i_z = 2$ A.

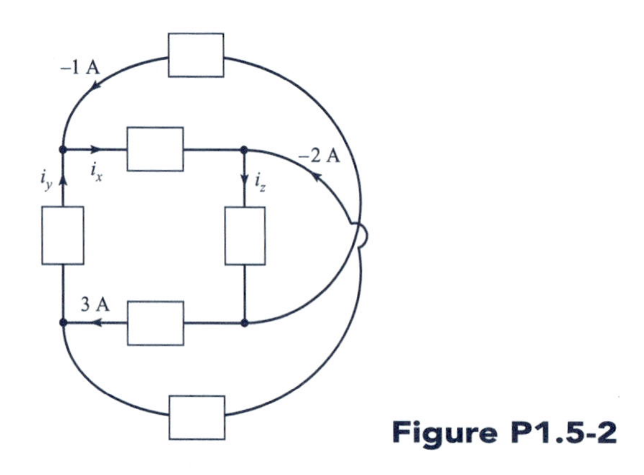

Figure P1.5-2

1.5-3 Determine the currents i_1, i_2, i_3, i_4 in the circuit of Fig. P1.5-3.
Answers: $i_1 = 4$ A, $i_2 = 5$ A, $i_3 = -\frac{11}{3}$ A, $i_4 = -\frac{14}{3}$ A.

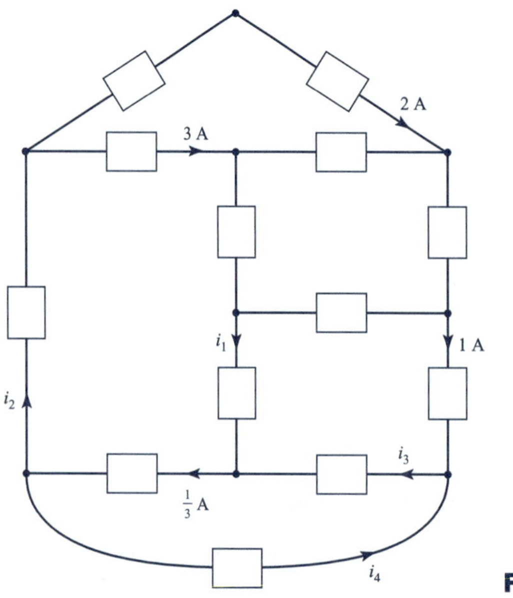

Figure P1.5-3

1.5-4 Determine the currents i_1, i_2 in the circuit of Fig. P1.5-4.
Answers: $i_1 = -3$ A, $i_2 = -1$ A.

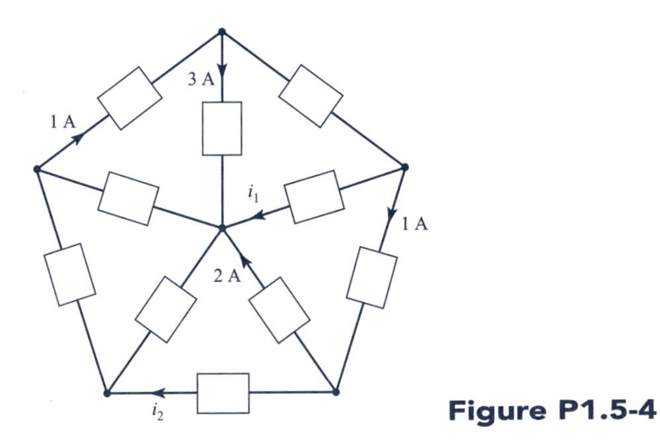

Figure P1.5-4

1.5-5 Determine the currents i_x, i_y in the circuit of Fig. P1.5-5.
Answers: $i_x = 10$ A, $i_y = -7$ A.

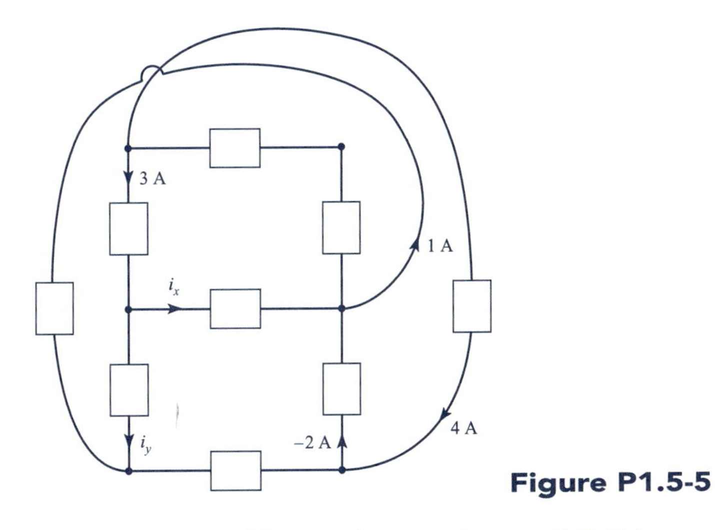

Figure P1.5-5

Section 1.6 Kirchhoff's Voltage Law (KVL)

1.6-1 Determine the voltages v_x, v_y, and v_{ba} in the circuit of Fig. P1.6-1.
Answers: $v_x = -6$ V, $v_y = -4$ V, $v_{ba} = 2$ V.

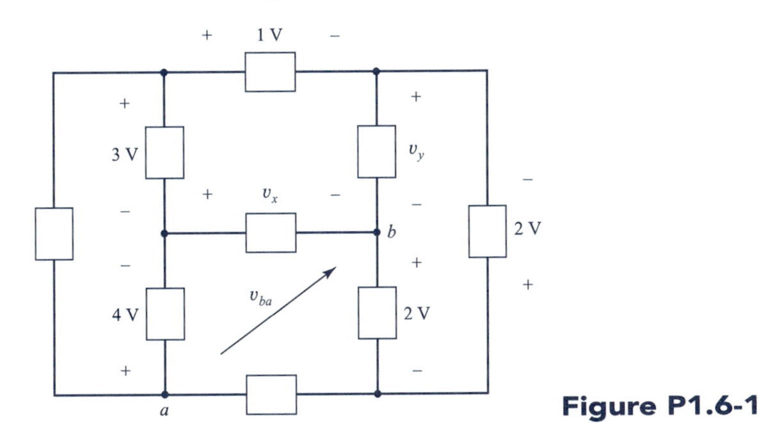

Figure P1.6-1

1.6-2 Determine the voltages v_x, v_y, v_z, and v_w in the circuit of Fig. P1.6-2.
Answers: $v_x = -8$ V, $v_y = 4$ V, $v_z = -2$ V, $v_w = 2$ V.

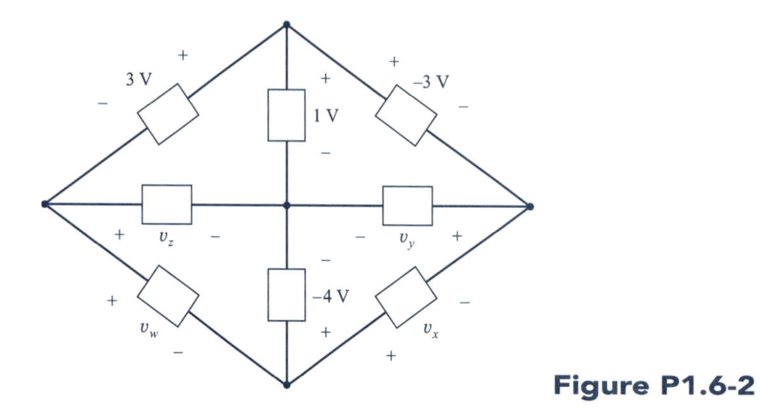

Figure P1.6-2

1.6-3 Determine the voltages v_x, v_y, and v_z in the circuit of Fig. P1.6-3. Can you determine the voltages v_w and v_q? *Answers:* $v_x = -1$ V, $v_y = 4$ V, $v_z = -4$ V; no, $v_w + v_q = -3$ V.

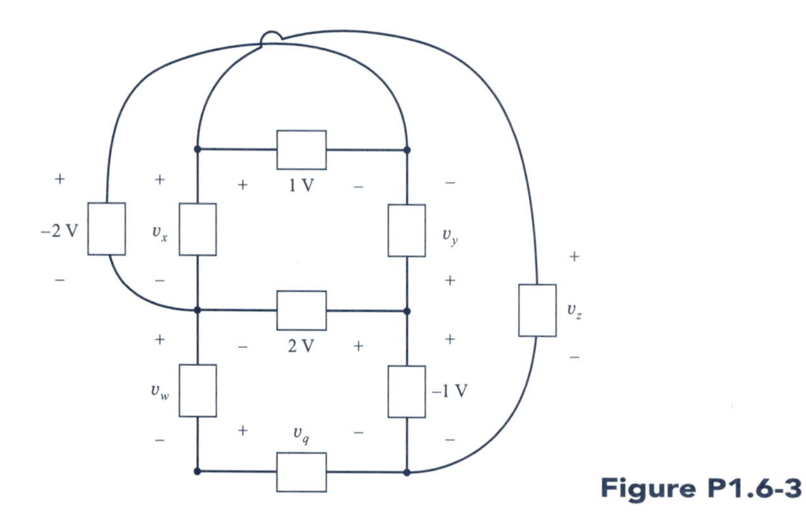

Figure P1.6-3

1.6-4 Determine the voltages v_x, v_y, and v_z in the circuit of Fig. P1.6-4. *Answers:* $v_x = -3$ V, $v_y = 5$ V, $v_z = 6$ V.

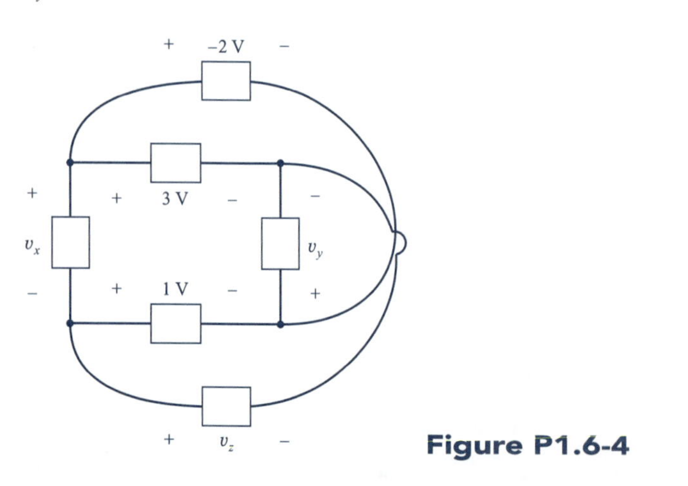

Figure P1.6-4

1.6-5 Determine the voltages v_x, v_y, and v_z in the circuit of Fig. P1.6-5. *Answers*: $v_x = -2$ V, $v_y = -1$ V, $v_z = -4$ V.

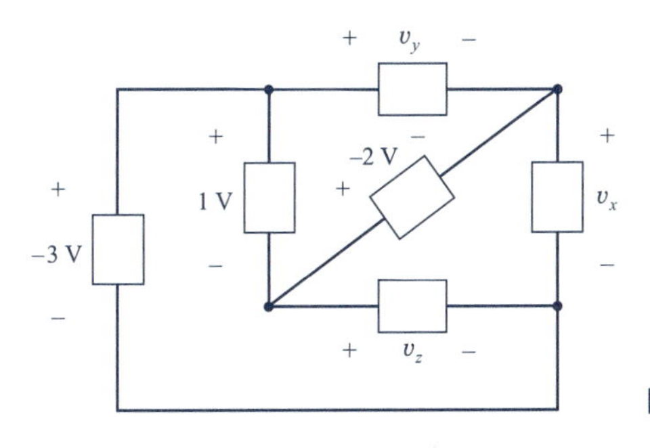

Figure P1.6-5

1.6-6 Determine the voltages v_x, v_y, v_z, and v_w in the circuit of Fig. P1.6-6.
Answers: $v_x = -1$ V, $v_y = -2$ V, $v_z = 4$ V, $v_w = 0$ V.

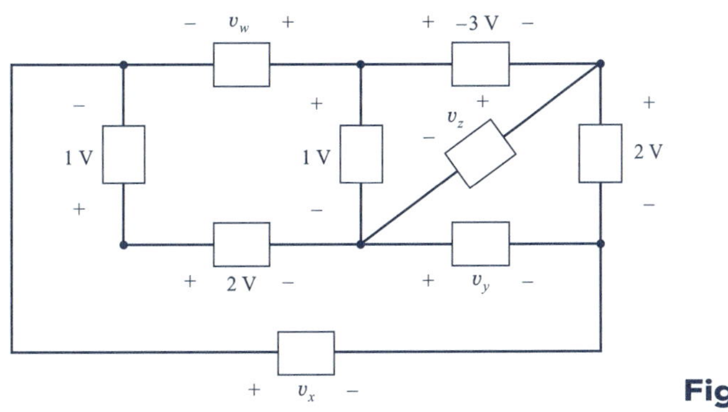

Figure P1.6-6

1.6-7 Determine the voltage v in the circuit of Fig. P1.6-7. *Answer*: -6 V.

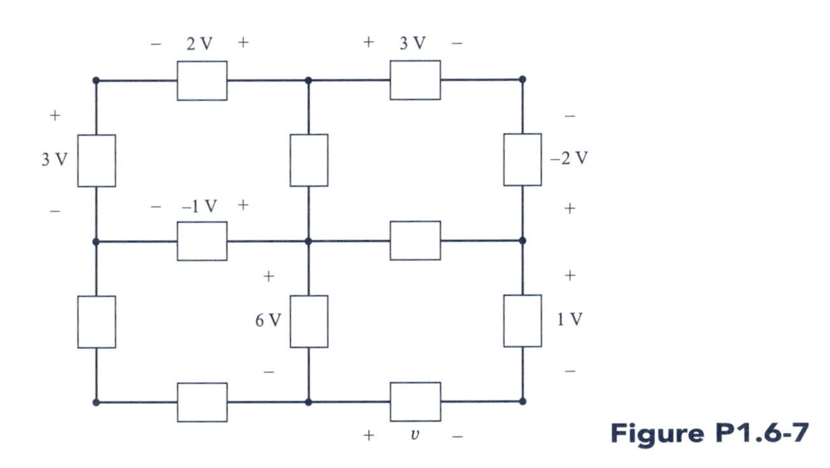

Figure P1.6-7

Section 1.7 Conservation of Power

1.7-1 Determine voltage v and current i in the circuit of Fig. P1.7-1. Determine the power delivered to element A. Solve for other voltages and currents, and check conservation of power. *Answers*: -5 V, -3 A, 42 W.

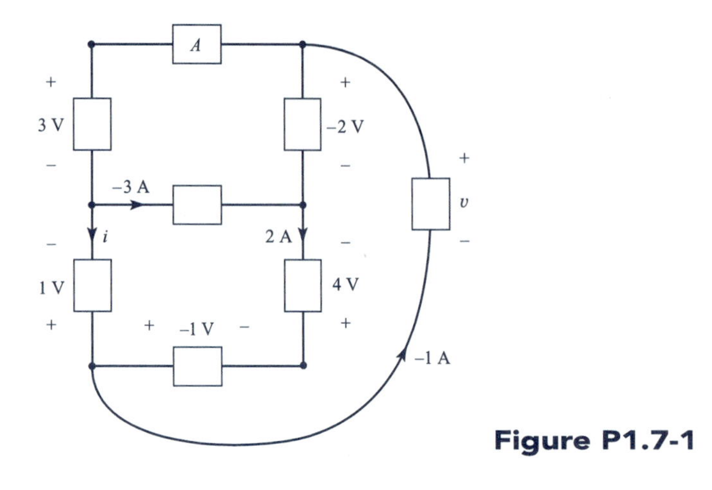

Figure P1.7-1

1.7-2 For the circuit in Fig. P1.7-2, determine the powers delivered to all four elements. Check conservation of power for this circuit. *Answers*: $P_1 = -4$ W, $P_2 = -4$ W, $P_3 = 12$ W, $P_4 = -4$ W.

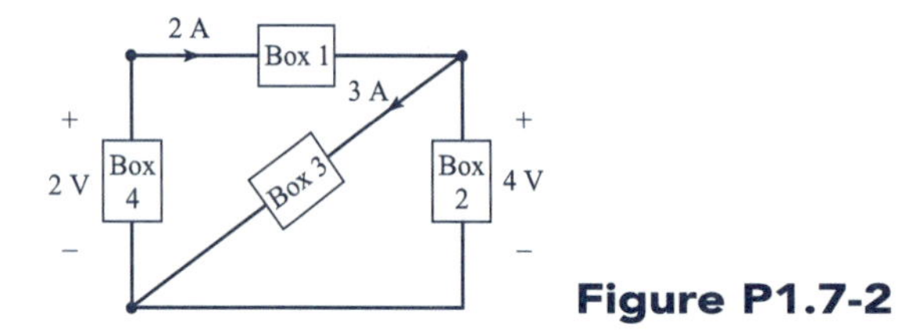

Figure P1.7-2

1.7-3 For the circuit in Fig. P1.7-3, determine the powers delivered to all elements. Check conservation of power for this circuit. *Answers*: $P_A = 8$ W, $P_B = 3$ W, $P_C = -1$ W, $P_D = 12$ W, $P_E = 2$ W, $P_F = -20$ W, $P_G = 4$ W, $P_H = -2$ W, $P_I = -6$ W.

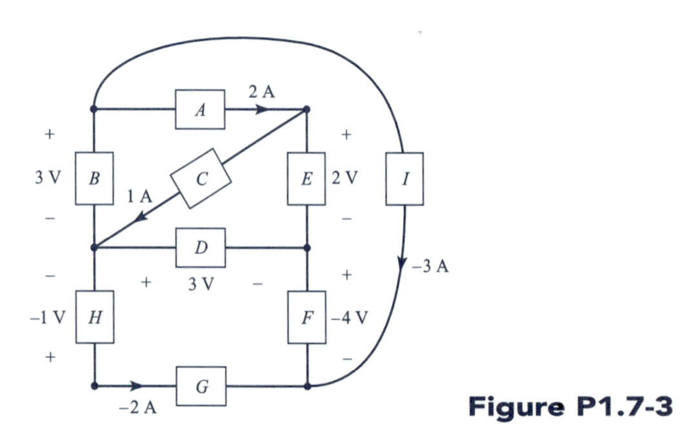

Figure P1.7-3

1.7-4 For the circuit in Fig. P1.7-4, determine the powers delivered to all elements. Check conservation of power for this circuit. *Answers*: $P_A = -8$ W, $P_B = -2$ W, $P_C = 6$ W, $P_D = 3$ W, $P_E = -7$ W, $P_F = 18$ W, $P_G = -6$ W, $P_H = -4$ W.

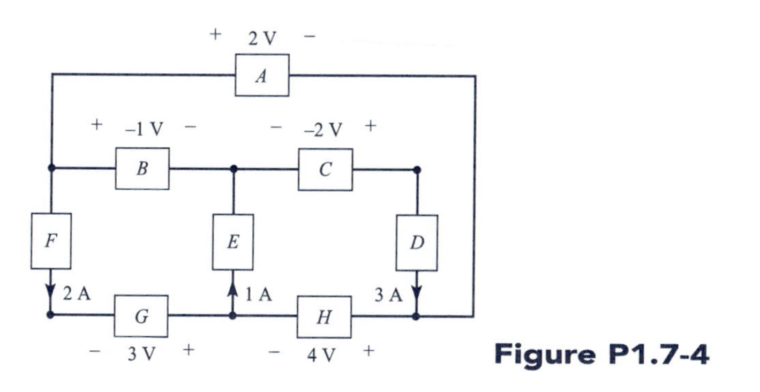

Figure P1.7-4

Section 1.8 Series and Parallel Connections of Elements

1.8-1 For the circuit shown in Fig. P1.8-1 determine which elements are in series and which are in parallel. *Answers*: series, [10,11], [1,2], [6,7]; parallel, [4,9].

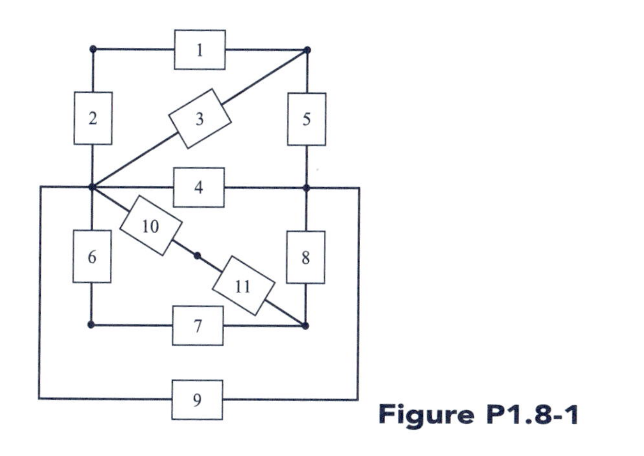

Figure P1.8-1

1.8-2 For the circuit shown in Fig. P1.8-2 determine which elements are in series and which are in parallel. *Answers*: series, [5,6,7]; parallel, [2,9], [1,3], [4,8].

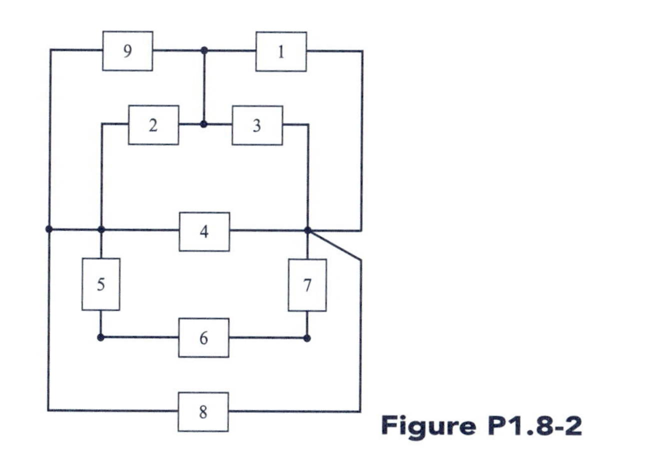

Figure P1.8-2

Section 1.9 Equivalent Circuits

1.9-1 Determine whether circuits A and B in Fig. P1.9-1 are equivalent at terminals a and b.
Answer: Yes.

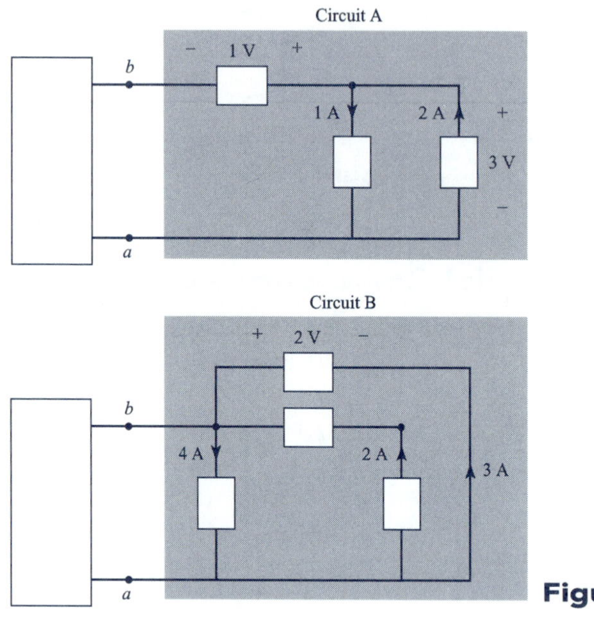

Figure P1.9-1

CHAPTER 2

Basic Circuit Elements and Analysis Techniques

Kirchhoff's voltage law (KVL), relates the element voltages in a circuit to each other, and Kirchhoff's current law (KCL), relates the element currents to each other in that circuit. So far we have not needed to describe the content of the element boxes, since KCL and KVL are independent of those contents. In this chapter we will examine the types of elements that the element boxes may contain. Some elements, the independent voltage and current sources, constrain the element voltage or current to be a specified value, while another element, the resistor, simply relates an element's voltage to its current without specifying either one. Still other elements, the dependent or controlled sources, relate the voltage or current of the element to the voltage or current associated with another element. In Chapter 5, we introduce the remaining elements needed for modeling an electric circuit: the inductor and the capacitor. The totality of these elements will allow us to devise accurate mathematical models of the majority of physical electrical devices. Our essential task is then to learn how to analyze a circuit model that is composed of these elements such that the element voltages and currents satisfy KVL, KCL, and the element relations for the particular interconnection of those elements (the circuit).

2.1 The Independent Voltage and Current Sources

The *independent voltage source,* whose symbol is shown in Fig. 2.1, constrains the element voltage to be a known (given) function of time t. The value of the element voltage is given as $v_S(t)$. The source is denoted as a circle having + and − signs in its interior to designate the polarity of the voltage $v_S(t)$. The element voltage $v(t)$ equals $v_S(t)$ when the + terminals of the two are aligned; otherwise $v(t) = -v_S(t)$. The current through the source, $i(t)$, is not determined as yet. Once the source is incorporated into a circuit, that circuit will then determine the current through the source. The specification of $v_S(t)$ can take many forms. A dc (direct current) waveform is shown in Fig. 2.2*a*. For this waveform, $v_S(t) = 3$ V and the voltage across the element does not change with time, much like that of a battery. An ac (alternating current) waveform is shown in Fig. 2.2*b*. For this waveform, $v_S(t) = 5 \sin 3t$ V and the voltage across the element changes sinusoidally. Observe that the complete specification of these sources is (a) the specification of the + and − signs in the circle and (b) the specification of $v_S(t)$.

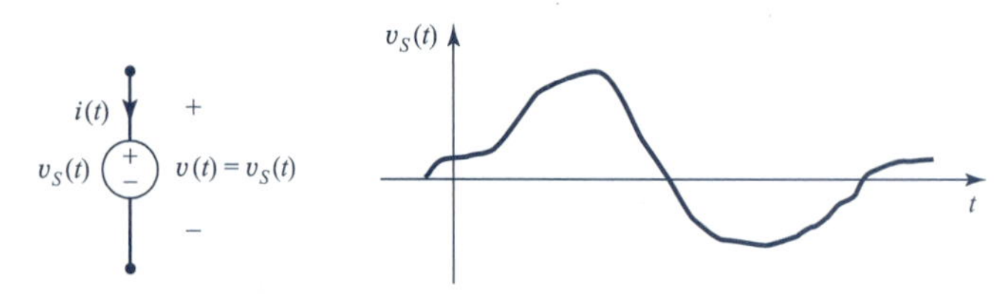

Figure 2.1
The independent voltage source.

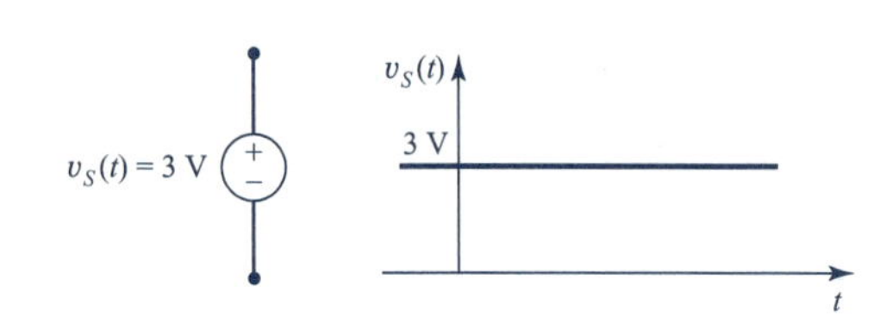

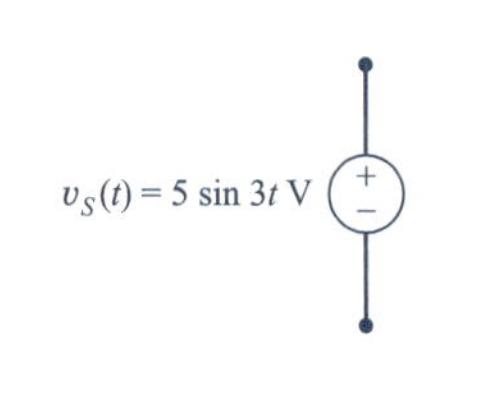

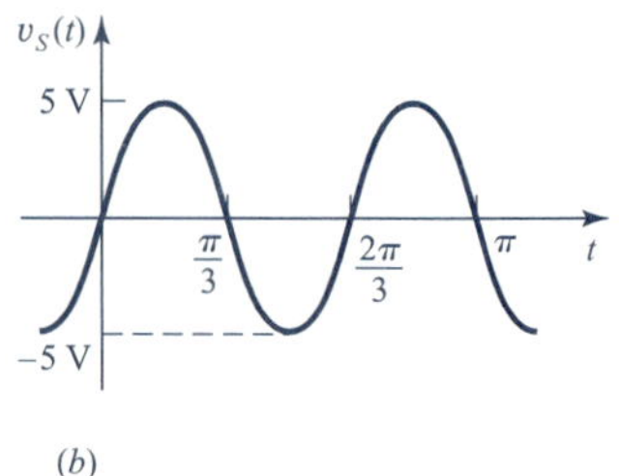

Figure 2.2
Types of voltage source waveforms: (*a*) dc, and (*b*) ac.

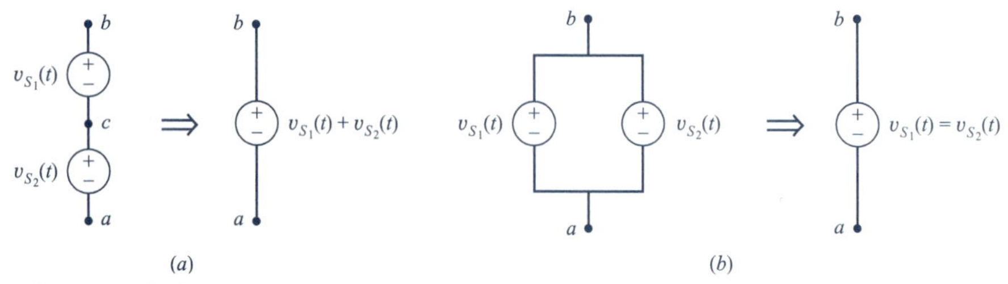

Figure 2.3
Combining voltage sources (*a*) in series and (*b*) in parallel.

Two independent voltage sources in series can be replaced, using KVL, with an equivalent voltage source whose value is the sum of the individual source voltages as shown in Fig. 2.3*a*. Two independent voltage sources in parallel shown in Fig. 2.3*b* must have identical values or else KVL would be violated.

Example 2.1

Residential power distribution was discussed in Section 1.11.1 of the previous chapter. Model the service entrance panel using independent voltage sources.

Solution Three wires are supplied to the residence from the power line (see Fig. 1.41 and Fig. 2.4*a*). One of the wires is bare and is connected to earth (ground). The other two wires have their insulations color coded as red and black. The voltage between the red and black wires is 240-V and is independent of the amount of current drawn by the residence. Hence we may model this source as a single voltage source as shown in Fig. 2.4*b*. The supplied voltage is a sinusoid at 60 Hz (cycles per second). The 240-V specification is rms. The peak voltage of the sinusoid is the rms value multiplied by $\sqrt{2}$, as we will see in Chapter 6. Hence the equation of the waveform of the source is $240\sqrt{2} \sin (2\pi \times 60 \times t)$ V. Within the service entrance panel a ground bus is provided, and the voltages between the red wire and ground and between the black wire and ground are $\frac{1}{2} \times 240$, 120 V rms. Hence we may

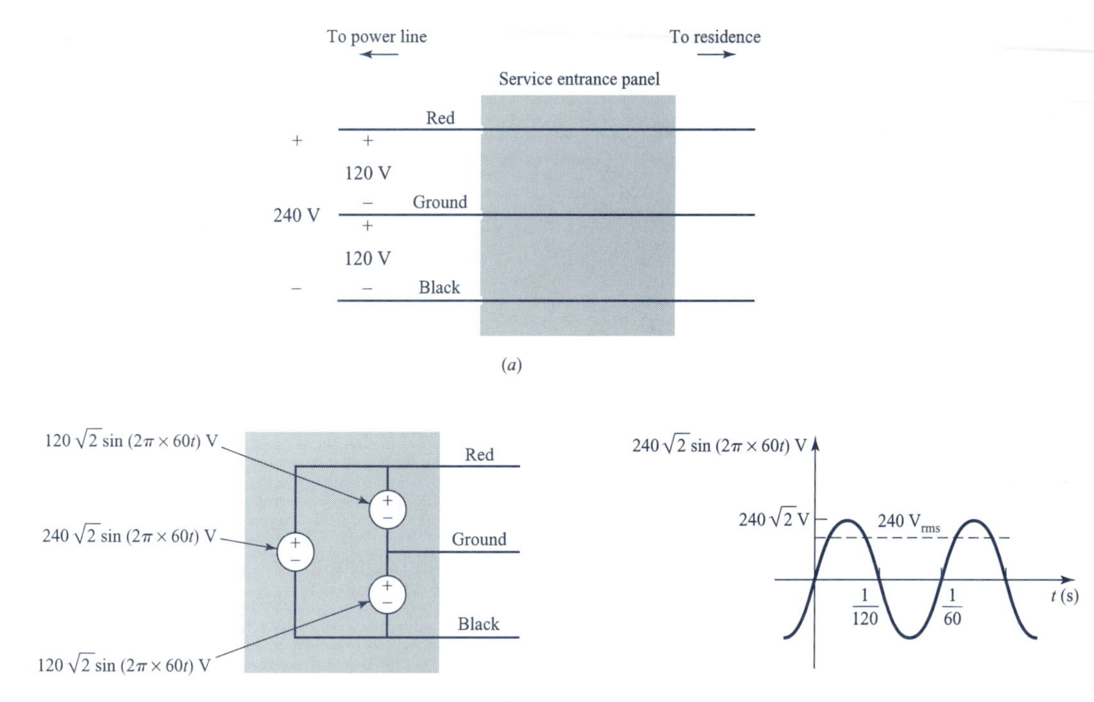

Figure 2.4
Example 2.1. Modeling residential power distribution as independent sources: (*a*) the service entrance panel of a residence, (*b*) the model, and (*c*) the voltage waveform.

model each of these as sinusoids having the equation of their waveforms as $120\sqrt{2}\sin(2\pi \times 60 \times t)$ V. Observe the polarities of these voltage sources. They must be such that KVL yields $240 = 120 + 120$.

Exercise Problem 2.1

Replace the combination of voltage sources shown in Fig. E2.1 with a single source that is equivalent at the two terminals.

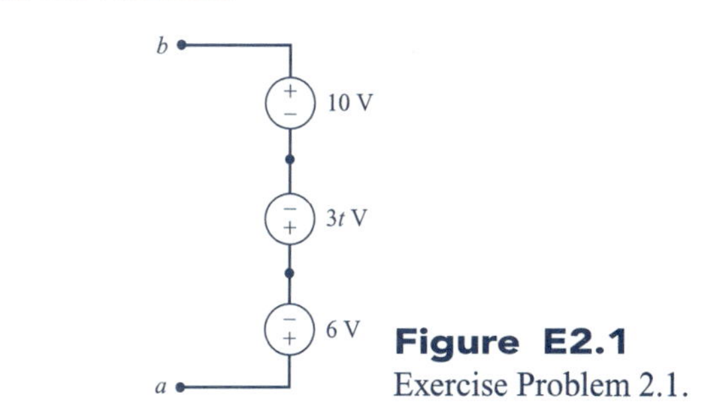

Figure E2.1
Exercise Problem 2.1.

Answer: The equivalent source has a value of $4 - 3t$ V and has its positive terminal at terminal b.

The *independent current source,* whose symbol is shown in Fig. 2.5, constrains the element current to be a known (given) function of time. The value of the element current is given as $i_S(t)$. The source is denoted by a circle having an arrow in its interior to designate the direction of its current. The element current $i(t)$ equals $i_S(t)$ when the current directions of the two are aligned;

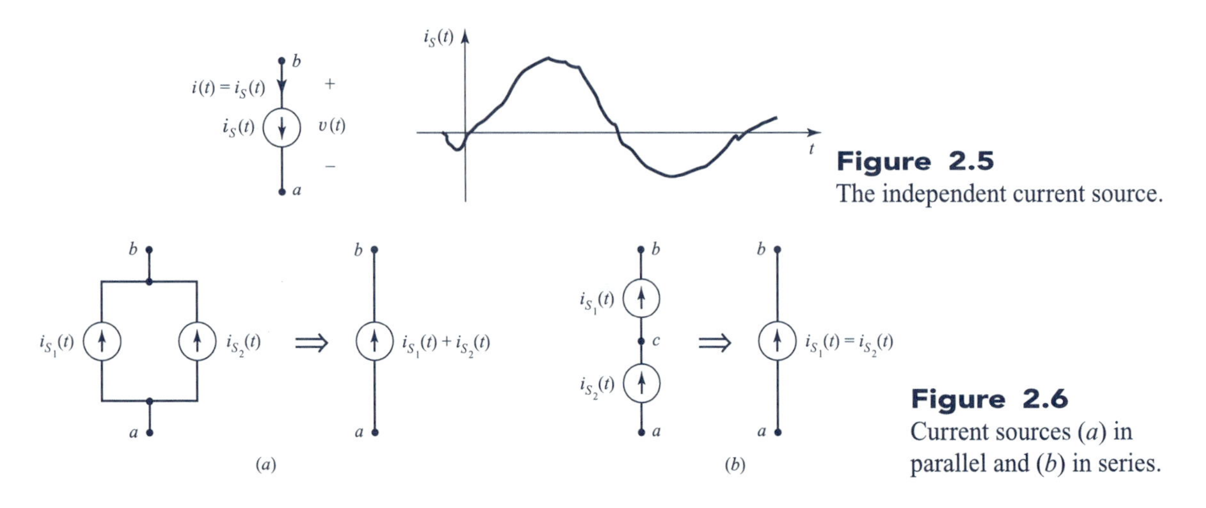

Figure 2.5
The independent current source.

Figure 2.6
Current sources (*a*) in parallel and (*b*) in series.

otherwise $i(t) = -i_S(t)$. The voltage across the source, $v(t)$, is not determined as yet. Once the source is incorporated into a circuit, that circuit will determine the voltage. The specification of $i_S(t)$ can also take many forms: dc, ac, etc. Observe that, as with the voltage source, the complete specification of this source is (a) the specification of the direction of the arrow inside the circle and (b) the specification of $i_S(t)$.

Two independent current sources in parallel can be replaced, using KCL, with an equivalent current source whose value is the sum of the individual source currents as shown in Fig. 2.6*a*. Two independent current sources in series as shown in Fig. 2.6*b* must have identical values or else KCL would be violated.

Example 2.2

Model a lightning stroke as a current source.

Solution A lightning stroke originates because of the separation of charge between a cloud and earth as shown in Fig. 2.7*a*. Negative charge accumulates on the cloud and positive charge (or an absence of negative charge) accumulates on the earth below the cloud. When that charge separation becomes large enough the air breaks down and a current flows from the earth to the cloud, neutralizing the charge separation. The waveform of the lightning current is sketched in Fig. 2.7*a* and rises to its peak of around 50 kA in around 1 $\mu s = 10^{-6}$ s and decays to zero rapidly thereafter. Since the lightning *channel* acts like a current source in that it is independent of objects in its path (such as airplanes), we may model it as an independent current source as shown in Fig. 2.7*b*. Observe the polarity of that current source: directed from the earth to the cloud.

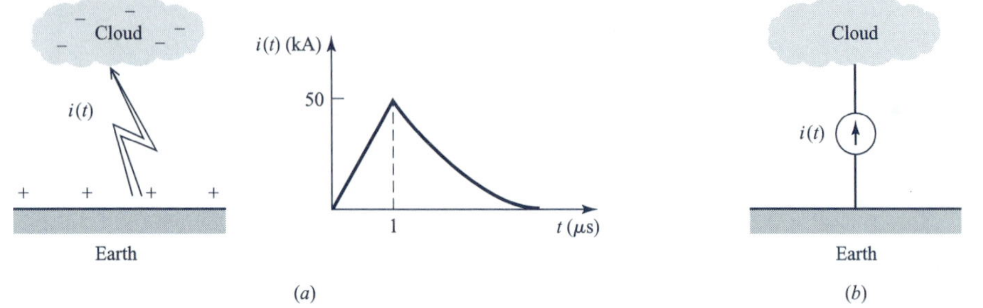

Figure 2.7
Example 2.2. Modeling a lightning stroke as a current source.

Exercise Problem 2.2

Replace the combination of current sources shown in Fig. E2.2 with a single source that is equivalent at the two terminals.

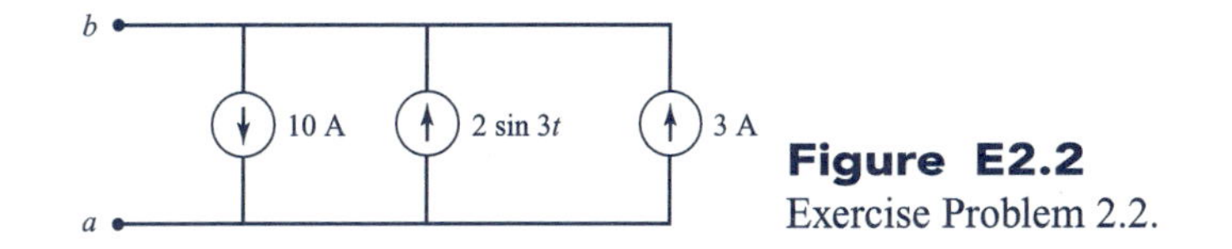

Figure E2.2
Exercise Problem 2.2.

Answer: The equivalent source has a value of $(-7 + 2 \sin 3t)$ A, which is directed from terminal a to terminal b.

Exercise Problem 2.3

Determine $v_S(t)$ and $i_S(t)$ for the circuit in Fig. E2.3 such that the two circuits are equivalent at terminals a and b.

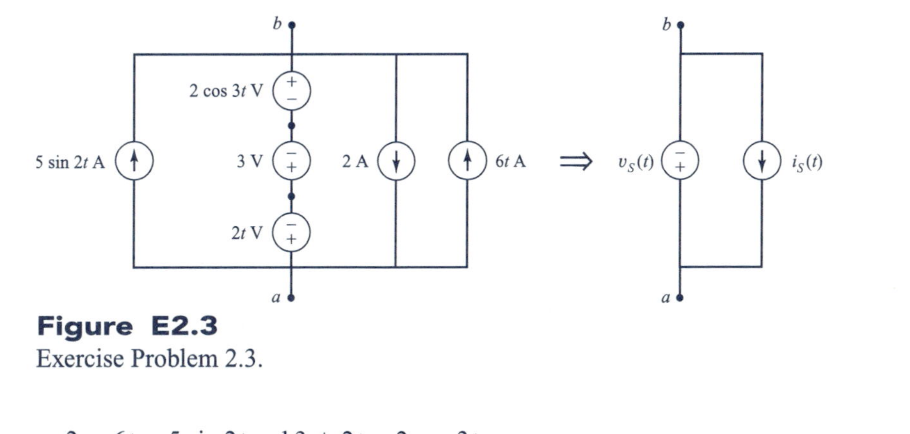

Figure E2.3
Exercise Problem 2.3.

Answers: $2 - 6t - 5 \sin 2t$ and $3 + 2t - 2 \cos 3t$.

2.2 The Linear Resistor and Ohm's Law

The independent sources considered in the previous section constrain (dictate) the element voltage or current, whereas the value of the other element variable remains unknown until the element is attached to a circuit. The resistor, on the other hand, does not constrain either the element voltage or the element current, but instead simply relates the two. The values of the resistor current and voltage are determined by the circuit that is attached to its two terminals.

The symbol for the linear resistor is shown in Fig. 2.8 along with the graph of the relationship between the resistor voltage and its current. For many materials, the voltage across a block of the material is simply related to the current through it with a straight-line relation, and this discovery is credited to a German physicist Georg Simon Ohm. Since the relation between the resistor voltage and its current is a straight line, we can dispense with the graph and simply write

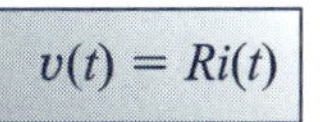

$$v(t) = Ri(t) \tag{2.1}$$

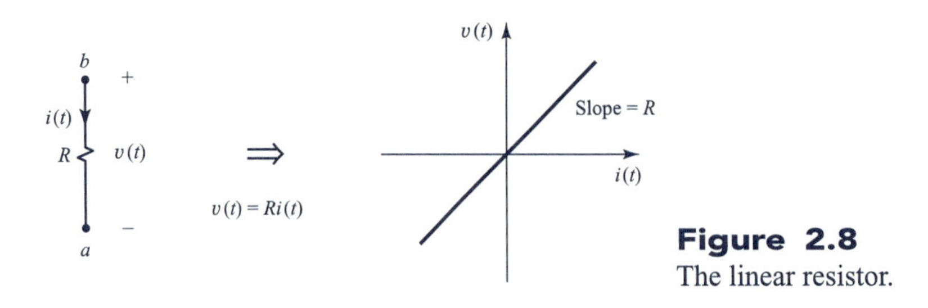

Figure 2.8
The linear resistor.

where R is the slope of the characteristic and is called the *resistance.* Its unit is the ohm (Ω), named in honor of Georg Simon Ohm. The reciprocal of resistance is *conductance,*

$$\boxed{G = \frac{1}{R}} \tag{2.2}$$

whose unit is the siemen (S), also named for a pioneer in the field. In terms of conductance, Ohm's law is written

$$\boxed{i(t) = G\,v(t)} \tag{2.3}$$

The power delivered to (absorbed by) a resistor is obtained as

$$\boxed{\begin{aligned} p(t) &= v(t)i(t) \\ &= i^2(t)\,R \\ &= Gv^2(t) \\ &= \frac{v^2(t)}{R} \end{aligned}} \tag{2.4}$$

Observe that the power delivered to a resistor is always positive, regardless of the sign of the voltage or current. Hence, *a resistor always absorbs power and never delivers power.* In a physical resistor, this absorbed power is dissipated in the form of heat.

Ohm's law in (2.1) and its reciprocal in (2.3) *presume* that the element voltage and current are *labeled with the passive sign convention* as shown in Fig. 2.8, i.e., *the current is assumed to enter the assumed positive voltage terminal.* If they are not, then a negative sign must be inserted into Ohm's law. Failure to observe this is a frequent mistake. For example, consider a true negative resistor shown in Fig. 2.9. Here the element voltage and current are again labeled with the passive sign convention, but

$$v(t) = -Ri(t)$$

and the power absorbed by a negative resistor is

$$\begin{aligned} p(t) &= v(t)i(t) \\ &= -i^2(t)\,R \\ &= -\frac{v^2(t)}{R} \end{aligned}$$

b
+
i(t)
−R
v(t)
v(t) = −Ri(t)
−
a

Figure 2.9
A negative resistor.

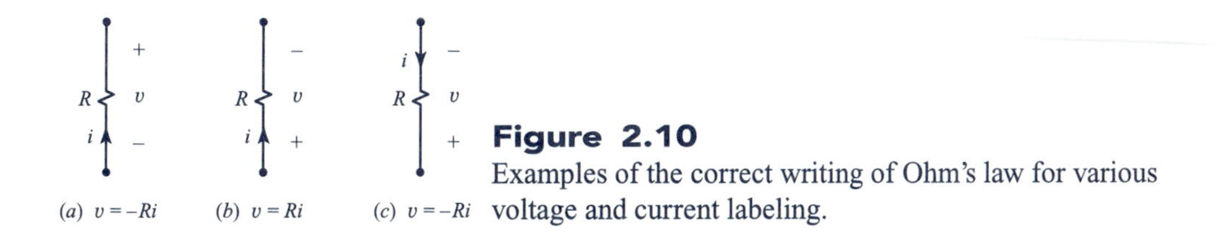

Figure 2.10
Examples of the correct writing of Ohm's law for various voltage and current labeling.

Observe that *a true negative resistor delivers power*. It is possible to construct negative resistors using elements such as tunnel diodes. But these require additional sources of energy such as batteries for operation, and the delivered power comes from these sources. We will only consider the positive resistor in this text.

Therefore the sign in Ohm's law is critically important. A frequent mistake is to write Ohm's law as in (2.1) without regard to the labeling of the resistor voltage and current. If the element voltage and current are *not* labeled with the passive sign convention, a minus sign must be inserted into Ohm's law in (2.1), i.e., $v(t) = -Ri(t)$ which does not represent a negative resistor but is simply due to labeling the resistor voltage and current contrary to the passive sign convention. The safest course of action is, when given the choice, to label the resistor voltage and current according to the passive sign convention. Some important examples are illustrated in Fig. 2.10 and should be studied sufficiently so that the reader will not make this critical sign error in applying Ohm's law.

Example 2.3

Write Ohm's law for the four resistors shown in Fig. 2.11.

Solution In Fig. 2.11*a*, the current is assumed to enter the negative voltage terminal. Since this is contrary to the passive sign convention, Ohm's law becomes

$$v = -3i$$

Observe that it can also be said that the current leaves the positive voltage terminal. In Fig. 2.11*b*, the current is again assumed to enter the negative voltage terminal, and since this is contrary to the passive sign convention, Ohm's law becomes

$$v = -5i$$

In Fig. 2.11*c*, the current is assumed to enter the positive voltage terminal. Since this conforms to the passive sign convention, Ohm's law becomes

$$v = 2i$$

In Fig. 2.11*d*, the current is assumed to leave (enter) the positive (negative) voltage terminal. Since this is contrary to the passive sign convention, Ohm's law becomes

$$v = -10i$$

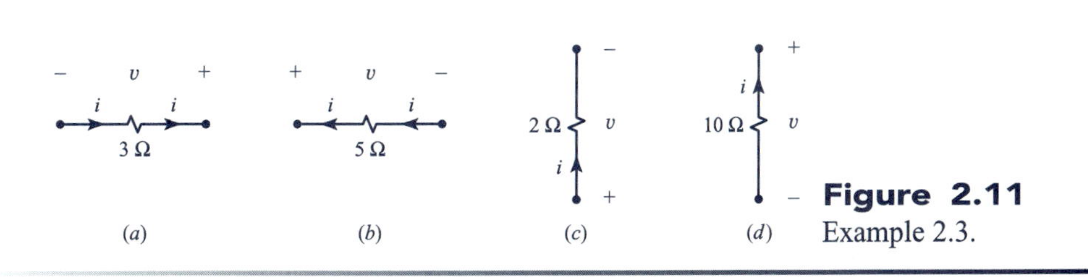

Figure 2.11
Example 2.3.

Example 2.4

Determine the unknown voltage or current for the four resistors in Fig. 2.12.

Solution In Fig. 2.12a the voltage and current are not labeled with the passive sign convention and hence the unknown current is

$$\begin{aligned} i &= -\frac{10\ \text{V}}{5\ \Omega} \\ &= -2\text{A} \end{aligned}$$

In Fig. 2.12b the voltage and current are again not labeled with the passive sign convention. Hence the unknown voltage is

$$\begin{aligned} v &= -3\ \Omega \times 2\ \text{A} \\ &= -6\ \text{V} \end{aligned}$$

In Fig. 2.12c the current enters the positive voltage terminal (or leaves the negative voltage terminal) and the labeling conforms to the passive sign convention, so that

$$\begin{aligned} v &= 6\ \Omega \times 3\ \text{A} \\ &= 18\ \text{V} \end{aligned}$$

And finally, the voltage and current in Fig. 2.12d also conform to the passive sign convention, and hence Ohm's law yields

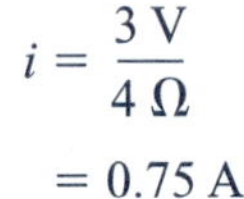

$$\begin{aligned} i &= \frac{3\ \text{V}}{4\ \Omega} \\ &= 0.75\ \text{A} \end{aligned}$$

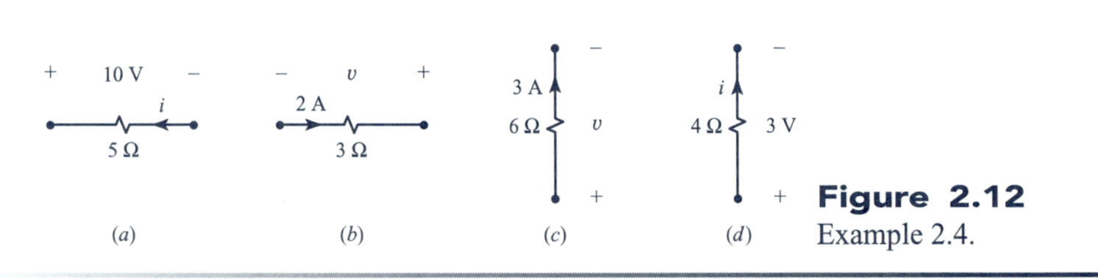

Figure 2.12
Example 2.4.

Exercise Problem 2.4

Determine the indicated unknown voltage or current for the resistors of Fig. E2.4.

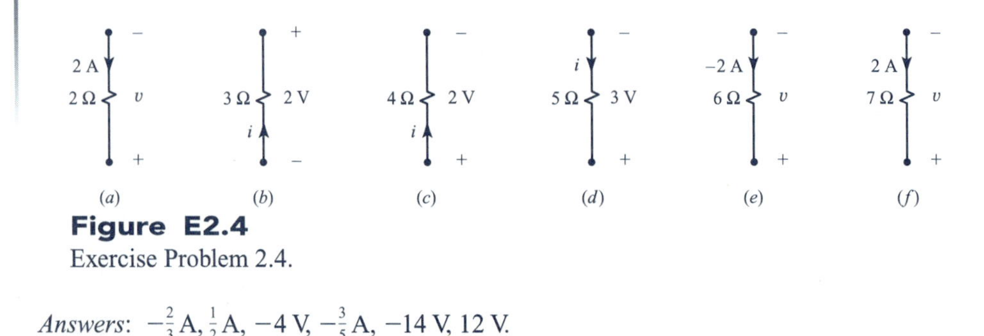

Figure E2.4
Exercise Problem 2.4.

Answers: $-\frac{2}{3}$ A, $\frac{1}{2}$ A, −4 V, $-\frac{3}{5}$ A, −14 V, 12 V.

2.3 Single-Loop and Single-Node-Pair Circuits

We now consider the two most fundamental circuits that will be encountered: the single-loop circuit and the single-node-pair circuit. First consider the *single-loop circuit* shown in Fig. 2.13. KCL shows that there will be one current common to all elements. So let us designate it as I and *arbitrarily* assume a direction for it, in this case, clockwise around the loop. By KCL, each element in the loop will have current I through it, and therefore we have essentially completely applied KCL to this circuit. Next we apply Ohm's law to the resistors to express their voltages in terms of this current. To aid in correctly applying Ohm's law, denote the direction of current I through each resistor and label the resulting voltage according to the passive sign convention. We have now completely applied Ohm's law and are left with only KVL to apply. Drawing an imaginary line through the circuit and labeling its ends with + and − signs, we write KVL as

$$-v_2 - R_2 I + v_1 - R_1 I = R_3 I + v_3 + R_4 I - v_4$$

Rewriting gives

$$I = \frac{v_1 - v_2 - v_3 + v_4}{R_1 + R_2 + R_3 + R_4}$$

This shows an important and rapid way of analyzing a single-loop circuit:

1. *Assume a direction for I flowing around the loop.*
2. *The resulting current is then*

$$I = \frac{\sum \text{(voltage sources } pushing \text{ in the direction of } I)}{\sum \text{(resistors in the loop)}} \tag{2.5}$$

3. *The voltages across the individual resistors can be obtained as the product of I and the resistance with the appropriate sign to conform to Ohm's law.*

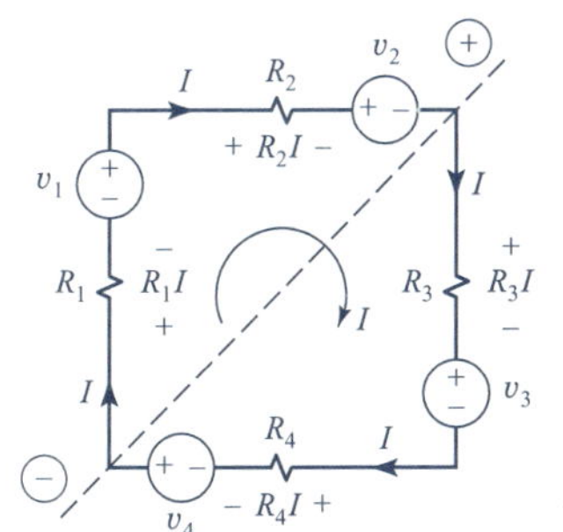

Figure 2.13
Illustration of a single-loop circuit.

Example 2.5

Determine voltages v_1 and v_2 in the circuit of Fig. 2.14*a*.

Solution Arbitrarily assuming a current I in the clockwise direction around the circuit as shown in Fig. 2.14*b*, we write

$$I = \frac{4\text{ V} - 3\text{ V} + 5\text{ V} - 10\text{ V}}{1\,\Omega + 2\,\Omega + 4\,\Omega + 6\,\Omega}$$

$$= \frac{-4\text{ V}}{13\,\Omega}$$

$$= -0.308\text{ A}$$

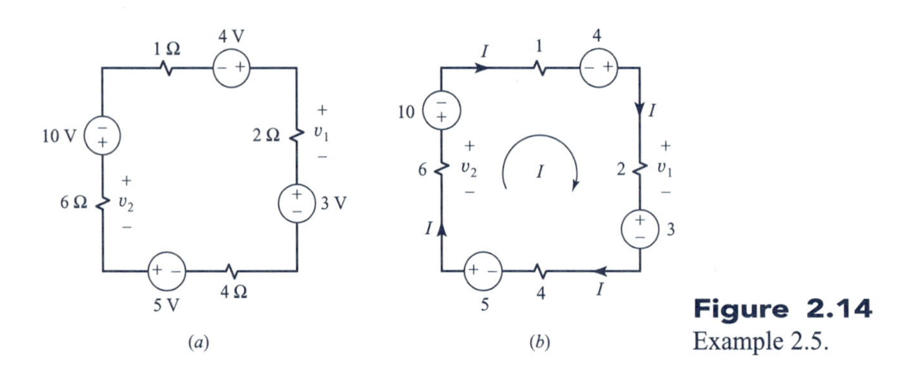

Figure 2.14
Example 2.5.

In determining the voltages, we observe that v_1 is labeled with the passive sign convention with respect to current I, so that

$$\begin{aligned} v_1 &= 2\,\Omega \times I \\ &= -0.615 \text{ V} \end{aligned}$$

Voltage v_2 is *not* labeled with the passive sign convention with respect to current I, so that

$$\begin{aligned} v_2 &= -6\,\Omega \times I \\ &= 1.846 \text{ V} \end{aligned}$$

Example 2.6

Determine voltages v_1, v_2, and v_3 in the circuit of Fig. 2.15*a*.

Solution Arbitrarily assuming a current I in the counterclockwise direction (for variety) around the circuit as shown in Fig. 2.15*b*, we write

$$\begin{aligned} I &= \frac{3\text{ V} - 5\text{ V} + 8\text{ V} - 2\text{ V}}{1\,\Omega + 2\,\Omega + 3\,\Omega + 4\,\Omega + 5\,\Omega} \\ &= \frac{4\text{ V}}{15\,\Omega} \\ &= 0.267\text{ A} \end{aligned}$$

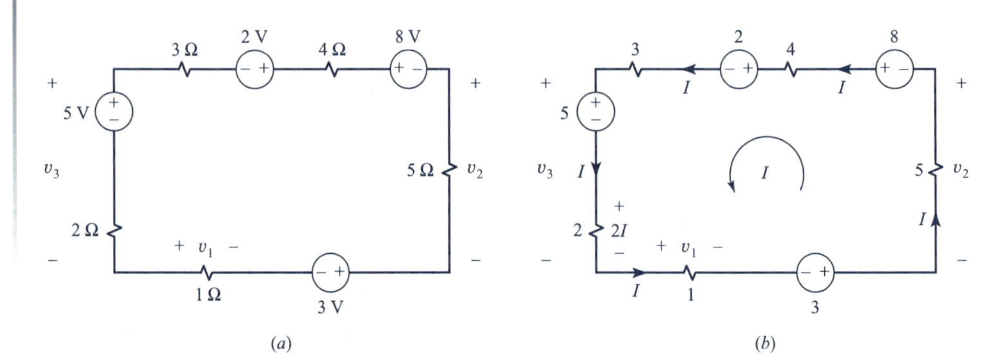

Figure 2.15
Example 2.6.

In determining the voltages, we observe that v_1 is labeled with the passive sign convention with respect to current I, so that

$$\begin{aligned} v_1 &= 1\,\Omega \times I \\ &= 0.267\text{ V} \end{aligned}$$

Voltage v_2 is *not* labeled with the passive sign convention with respect to current I, so that

$$\begin{aligned} v_2 &= -5\,\Omega \times I \\ &= -1.333\text{ V} \end{aligned}$$

Note that voltage v_3 is across the 5-V voltage source and the 2-Ω resistor. Hence the voltage v_3 is, by KVL, the sum of the voltages across these two elements. First we label the voltage across the 2-Ω resistor, $2I$, with the passive sign convention. Hence the voltage v_3 is

$$\begin{aligned} v_3 &= 5\text{ V} + 2\,\Omega \times I \\ &= 5\text{ V} + 2\,\Omega \times (0.267\text{ A}) \\ &= 5\text{ V} + 0.533\text{ V} \\ &= 5.533\text{ V} \end{aligned}$$

Example 2.7

Determine voltages v_1 and v_2 in the circuit of Fig. 2.16*a*.

Solution Arbitrarily assuming a current I in the clockwise direction around the circuit as shown in Fig. 2.16*b*, we write

$$\begin{aligned} I &= \frac{3\text{ V} - 5\text{ V} - 6\text{ V} - 10\text{ V}}{2\,\Omega + 3\,\Omega + 4\,\Omega + 3\,\Omega} \\ &= \frac{-18\text{ V}}{12\,\Omega} \\ &= -1.5\text{ A} \end{aligned}$$

In determining the voltages, we observe that v_1 is labeled with the passive sign convention with respect to current I, so that

$$\begin{aligned} v_1 &= 3\,\Omega \times I \\ &= 3\,\Omega \times (-1.5\text{ A}) \\ &= -4.5\text{ V} \end{aligned}$$

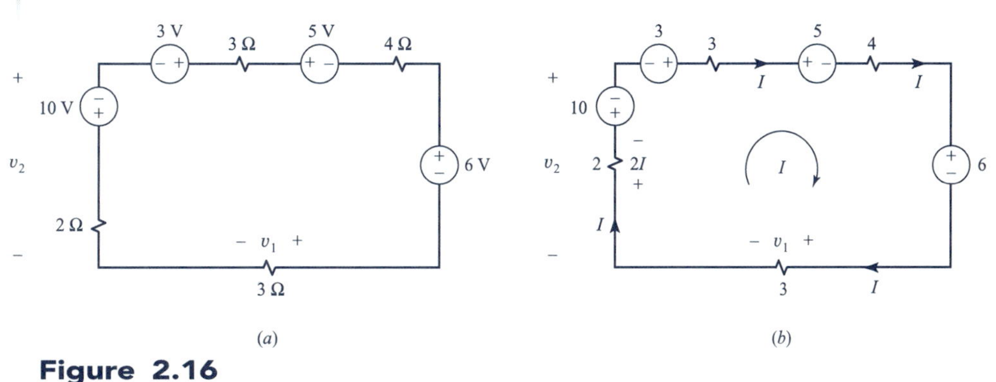

Figure 2.16
Example 2.7.

The voltage v_2 is across both the 10-V source and the 2-Ω resistor. Hence v_2 is, according to KVL, the sum of the voltages across the two elements. First express the voltage across the 2-Ω resistor as $2I$ volts, and label this with the passive sign convention. Next write KVL as

$$\begin{aligned} v_2 &= -10\ \text{V} - 2I \\ &= -10\ \text{V} - 2\ \Omega \times (-1.5\ \text{A}) \\ &= -7\ \text{V} \end{aligned}$$

Exercise Problem 2.5

Determine the voltages v_x and v_y in the circuit of Fig. E2.5.

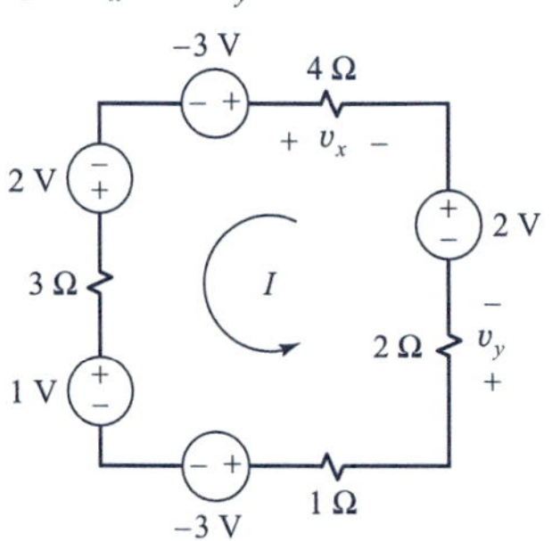

Figure E2.5
Exercise Problem 2.5.

Answers: −1.2 V, 0.6 V.

Exercise Problem 2.6

Determine the voltages v_1 and v_2 in the circuit of Fig. E2.6.

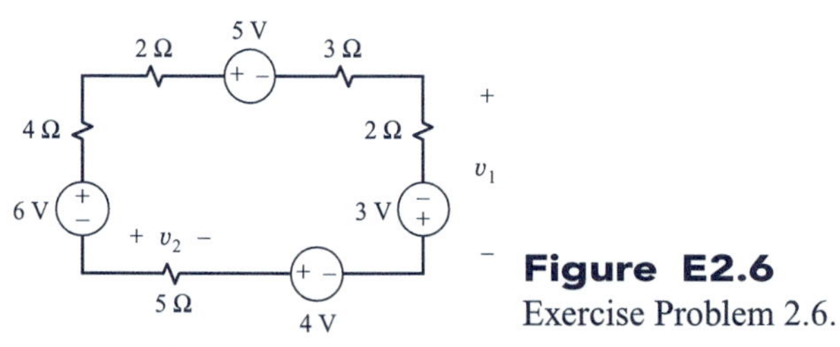

Figure E2.6
Exercise Problem 2.6.

Answers: −2 V, −2.5 V.

Exercise Problem 2.7

Determine the voltages v_1 and v_2 in the circuit of Fig. E2.7.

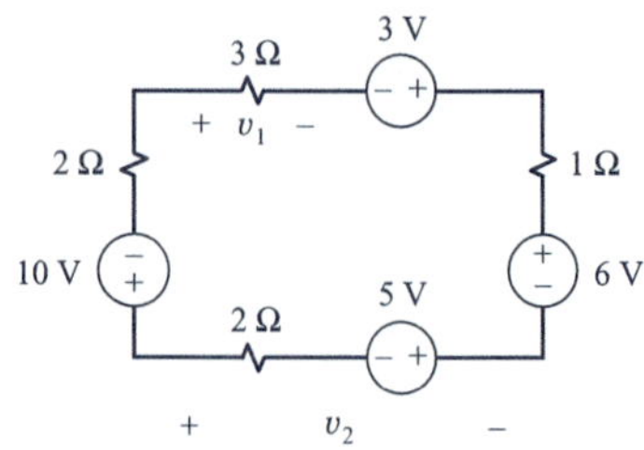

Figure E2.7
Exercise Problem 2.7.

Answers: −6.75 V, −0.5 V.

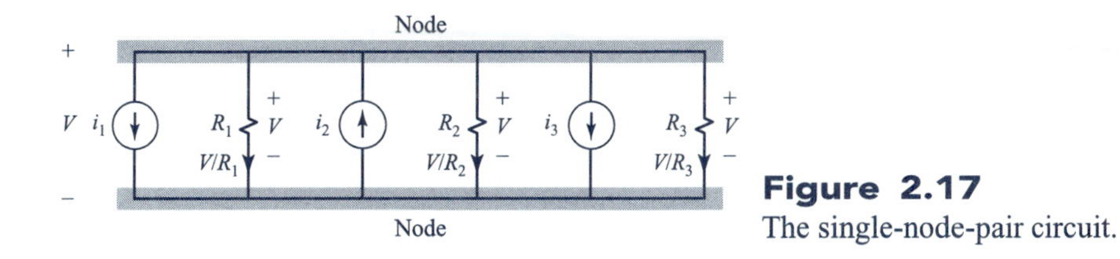

Figure 2.17
The single-node-pair circuit.

Next consider the *single-node-pair circuit* shown in Fig. 2.17. KVL shows that there will be one common voltage across all elements. So let us designate it as V and assume a direction for it, in this case, positive at the top. Each element will, according to KVL, have voltage V across it, positive at the top of the element. We have essentially completely applied KVL for this circuit. Next we apply Ohm's law to the resistors to express their currents in terms of this voltage. To aid in correctly applying Ohm's law, label the voltage V across each resistor and label the resulting current according to the passive sign convention. We have now completely applied Ohm's law and are left with only KCL to apply which we apply to the top node to yield:

$$i_2 = i_1 + \frac{V}{R_1} + \frac{V}{R_2} + i_3 + \frac{V}{R_3}$$

Rewriting gives

$$V = \frac{i_2 - i_1 - i_3}{\dfrac{1}{R_1} + \dfrac{1}{R_2} + \dfrac{1}{R_3}}$$

This shows an important and rapid way of analyzing a single-node-pair circuit:

1. *Assume a direction for V across all elements.*
2. *The resulting voltage is then*

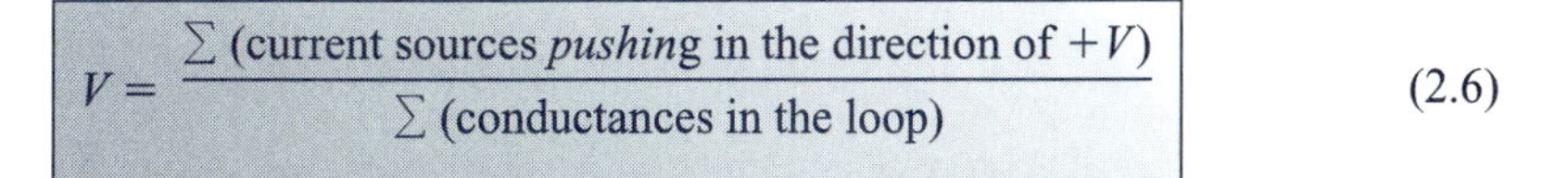

$$V = \frac{\sum \text{(current sources } \textit{pushing} \text{ in the direction of } +V)}{\sum \text{(conductances in the loop)}} \quad (2.6)$$

3. *The currents through the individual resistors can be obtained as the ratio of V to the resistance with the appropriate sign to conform to Ohm's law.*

Example 2.8

Determine currents i_1 and i_2 in the circuit of Fig. 2.18*a*.

Solution First label the voltage V as shown in Fig. 2.18*b*. Then write

$$\begin{aligned} V &= \frac{-2\text{ A} + 5\text{ A} - 4\text{ A}}{\frac{1}{4}\text{ S} + \frac{1}{3}\text{ S} + \frac{1}{1}\text{ S}} \\ &= \frac{-1\text{ A}}{\frac{19}{12}\text{ S}} \\ &= -0.632\text{ V} \end{aligned}$$

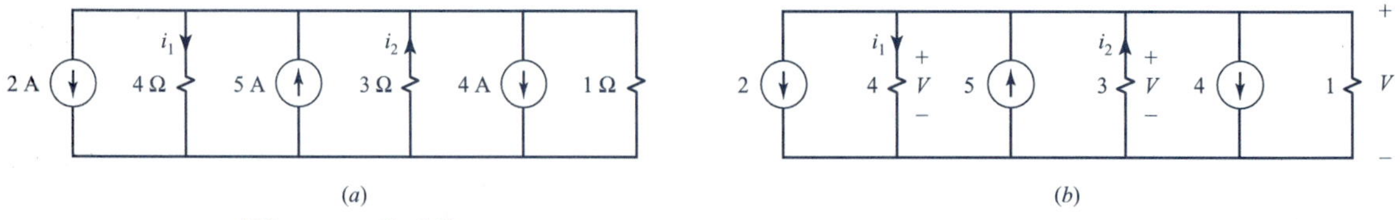

Figure 2.18
Example 2.8.

Ohm's law gives the currents through the resistors. Current i_1 is labeled with the passive sign convention with respect to voltage V. Hence

$$
\begin{aligned}
i_1 &= \frac{V}{4\ \Omega} \\
&= \frac{-0.632\ \text{V}}{4\ \Omega} \\
&= -0.158\ \text{A}
\end{aligned}
$$

Current i_2 is *not* labeled with the passive sign convention with respect to voltage V. Hence

$$
\begin{aligned}
i_2 &= -\frac{V}{3\ \Omega} \\
&= -\frac{-0.632\ \text{V}}{3\ \Omega} \\
&= 0.211\ \text{A}
\end{aligned}
$$

Example 2.9

Determine currents i_1 and i_2 in the circuit of Fig. 2.19*a*.

Solution First label the voltage V as shown in Fig. 2.19*b*. Then write

$$
\begin{aligned}
V &= \frac{4\ \text{A} - 5\ \text{A} + 6\ \text{A}}{\frac{1}{2}\ \text{S} + \frac{1}{3}\ \text{S}} \\
&= \frac{5\ \text{A}}{\frac{5}{6}\ \text{S}} \\
&= 6\ \text{V}
\end{aligned}
$$

Ohm's law gives the currents through the resistors. Current i_1 is *not* labeled with the passive sign convention with respect to voltage V. Hence

$$
\begin{aligned}
i_1 &= -\frac{V}{2\ \Omega} \\
&= -\frac{6\ \text{V}}{2\ \Omega} \\
&= -3\ \text{A}
\end{aligned}
$$

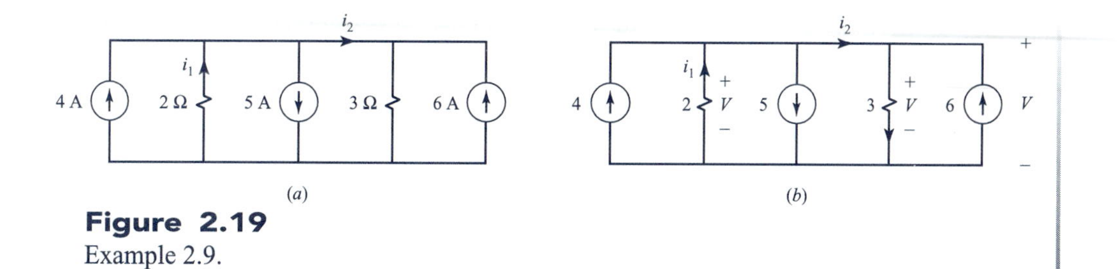

Figure 2.19
Example 2.9.

Current i_2 is the sum of the currents through the 3-Ω resistor and the 6-A current source. First label the current through the 3-Ω resistor according to the direction of V and the passive sign convention (down). Hence

$$\begin{aligned} i_2 &= \frac{V}{3\ \Omega} - 6\text{ A} \\ &= 2\text{ A} - 6\text{ A} \\ &= -4\text{ A} \end{aligned}$$

Example 2.10

Determine currents i_1 and i_2 in the circuit of Fig. 2.20*a*.

Solution First label the voltage V as shown in Fig. 2.20*b*. Then write

$$\begin{aligned} V &= \frac{-10\text{ A} + 5\text{ A} - 2\text{ A}}{\frac{1}{4}\text{ S} + \frac{1}{4}\text{ S} + \frac{1}{3}\text{ S}} \\ &= \frac{-7\text{ A}}{\frac{5}{6}\text{ S}} \\ &= -8.4\text{ V} \end{aligned}$$

Ohm's law gives the currents through the resistors. Current i_1 is labeled with the passive sign convention with respect to voltage V. Hence

$$\begin{aligned} i_1 &= \frac{V}{3\ \Omega} \\ &= \frac{-8.4\text{ V}}{3\ \Omega} \\ &= -2.8\text{ A} \end{aligned}$$

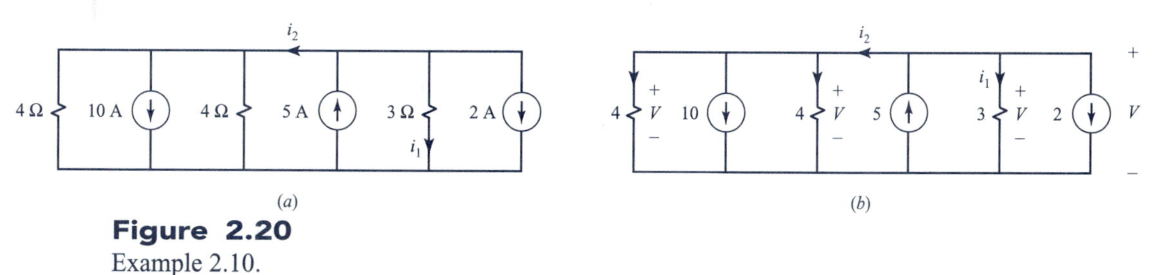

Figure 2.20
Example 2.10.

Current i_2 is the sum of the currents through the 3-Ω resistor, the 5-A current source, and the 2-A current source. Applying KCL yields

$$i_2 = 5\text{ A} - i_1 - 2\text{ A}$$
$$= 5.8\text{ A}$$

This can also be calculated as the sum of the currents through the 4-Ω resistors and the 10-A source:

$$i_2 = \frac{V}{4\,\Omega} + \frac{V}{4\,\Omega} + 10\text{ A}$$
$$= 5.8\text{ A}$$

Exercise Problem 2.8

Determine the currents i_x and i_y in the circuit of Fig. E2.8.

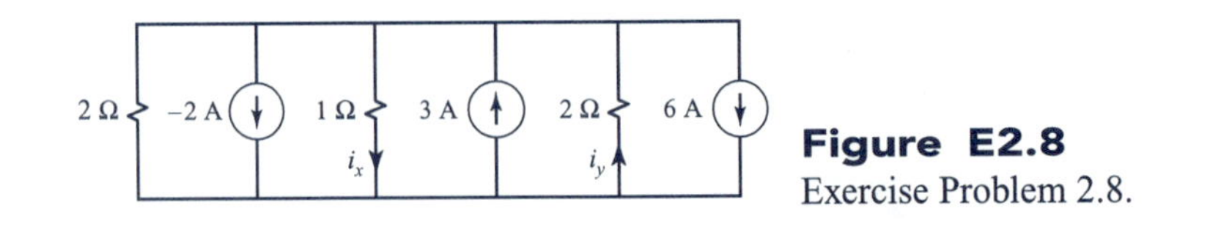

Figure E2.8
Exercise Problem 2.8.

Answers: −0.5 A, 0.25 A.

Exercise Problem 2.9

Determine the currents i_1 and i_2 in the circuit of Fig. E2.9.

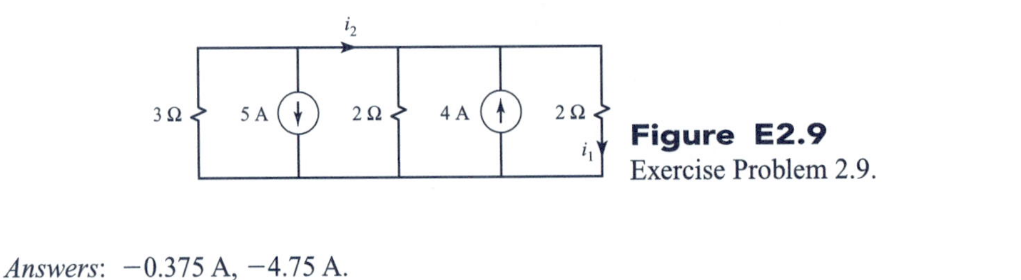

Figure E2.9
Exercise Problem 2.9.

Answers: −0.375 A, −4.75 A.

Exercise Problem 2.10

Determine the currents i_1 and i_2 in the circuit of Fig. E2.10.

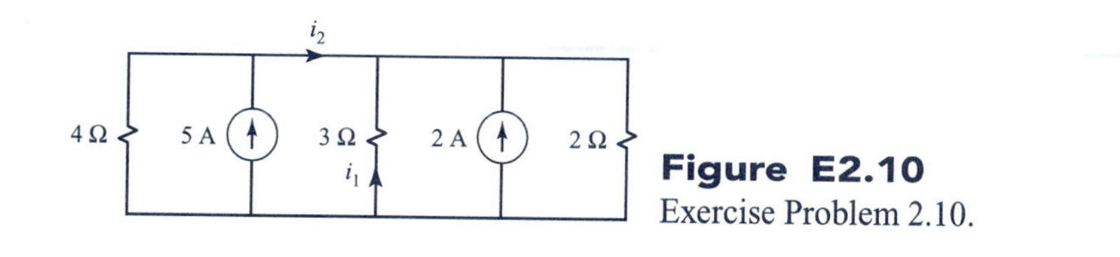

Figure E2.10
Exercise Problem 2.10.

Answers: −2.154 A, 3.385 A.

2.4 Resistors in Series and in Parallel

Resistors that are in series or in parallel can be replaced with equivalent resistances such that the resulting currents and voltages in any circuit attached to the terminals will remain unchanged. First consider the case of three resistors in *series* as shown in Fig. 2.21. KCL shows that, because they are in series, the currents through them are identical. In order to replace them with an equivalent resistance, the v–i relationship at the external terminals of the series connection and at the terminals of the equivalent resistance must be the same. Writing KVL for the series connection yields

$$\begin{aligned} v &= v_1 + v_2 + v_3 \\ &= R_1 i_1 + R_2 i_2 + R_3 i_3 \end{aligned}$$

Observing that, because they are in series,

$$i = i_1 = i_2 = i_3$$

and substituting gives

$$v = \underbrace{(R_1 + R_2 + R_3)}_{R_{eq}}\, i$$

Therefore n resistors in series can be replaced with an equivalent resistance at the external terminals of value

$$\boxed{\begin{aligned} R_{eq} &= \sum \text{(resistors in series)} \\ &= R_1 + R_2 + \cdots + R_n \end{aligned}} \tag{2.7}$$

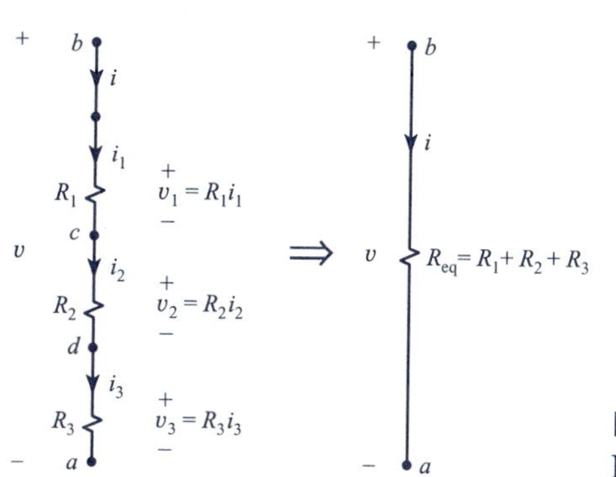

Figure 2.21
Resistors in series and their equivalent resistance.

Example 2.11

Determine the equivalent resistance seen at terminals *ab* in Fig. 2.22.

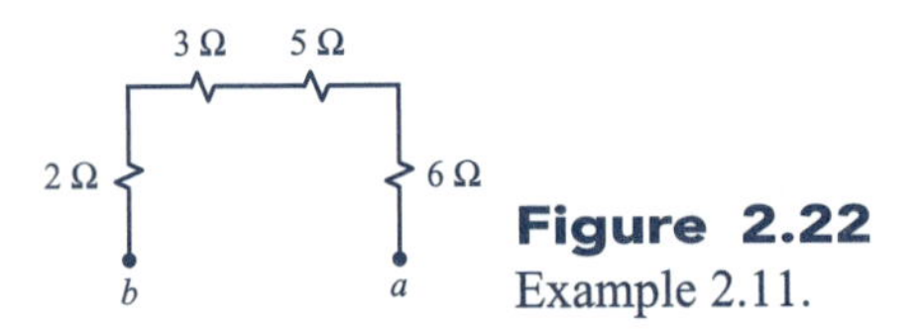

Figure 2.22
Example 2.11.

Solution Adding the resistors, since they are in series, gives

$$R_{eq} = 2\ \Omega + 3\ \Omega + 5\ \Omega + 6\ \Omega = 16\ \Omega.$$

Exercise Problem 2.11

Determine the equivalent resistance seen at terminals *ab* in Fig. E2.11.

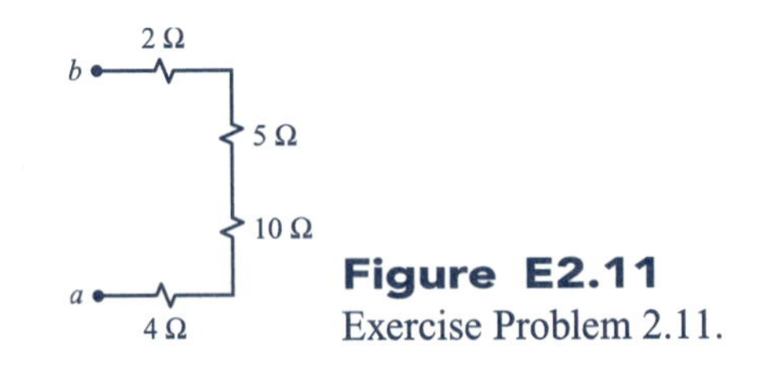

Figure E2.11
Exercise Problem 2.11.

Answer: 21 Ω.

Next, consider the case of three resistors in *parallel* as shown in Fig. 2.23. KVL shows that, because they are in parallel, the voltages across them are identical. In order to replace them with an equivalent conductance, the v–i relationship at the external terminals of the parallel connection and at the terminals of the equivalent conductance must be the same. Writing KCL yields

$$\begin{aligned} i &= i_1 + i_2 + i_3 \\ &= G_1 v_1 + G_2 v_2 + G_3 v_3 \end{aligned}$$

where we have written Ohm's law in terms of conductances. Observing that since they are in parallel,

$$v = v_1 = v_2 = v_3$$

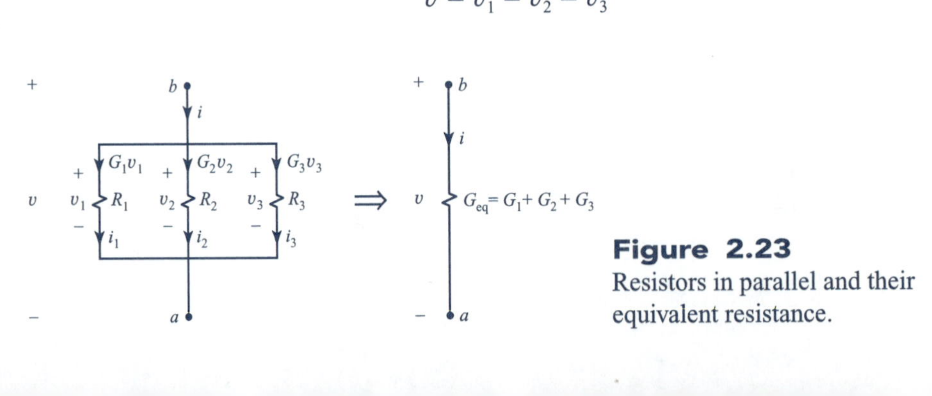

Figure 2.23
Resistors in parallel and their equivalent resistance.

and substituting gives

$$i = \underbrace{(G_1 + G_2 + G_3)}_{G_{eq}} v$$

Therefore n resistors in parallel can be replaced with an equivalent resistance at the external terminals whose *conductance* is

$$\begin{aligned} G_{eq} &= \sum (\text{conductances in parallel}) \\ &= G_1 + G_2 + \cdots + G_n \end{aligned} \tag{2.8}$$

We tend to deal with the resistance of a resistor more frequently than its conductance, so (2.8) can be more usefully written for the parallel combination as

$$\begin{aligned} \frac{1}{R_{eq}} &= \sum (\textit{reciprocals} \text{ of the resistances in parallel}) \\ &= \frac{1}{R_1} + \frac{1}{R_2} + \cdots + \frac{1}{R_n} \end{aligned} \tag{2.9}$$

Example 2.12

Determine the equivalent resistance seen at the terminals *ab* in Fig. 2.24.

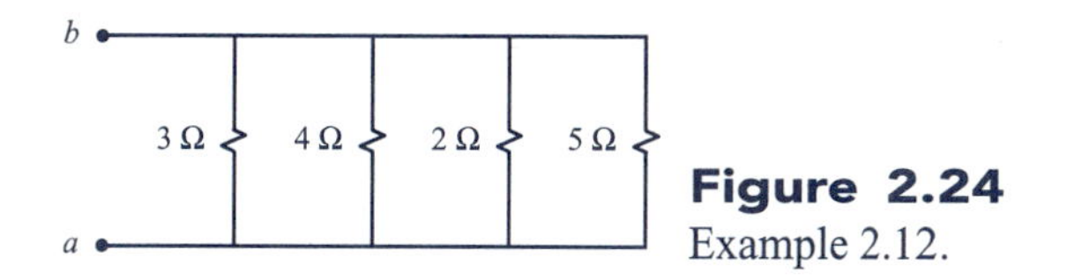

Figure 2.24
Example 2.12.

Solution Since the resistors are in parallel, their equivalent conductance is

$$\begin{aligned} \frac{1}{R_{eq}} &= \frac{1}{3\ \Omega} + \frac{1}{4\ \Omega} + \frac{1}{2\ \Omega} + \frac{1}{5\ \Omega} \\ &= 1.283\ \Omega \end{aligned}$$

Inverting this gives

$$R_{eq} = 0.779\ \Omega$$

Exercise Problem 2.12

Determine the equivalent resistance seen at terminals *ab* in Fig. E2.12.

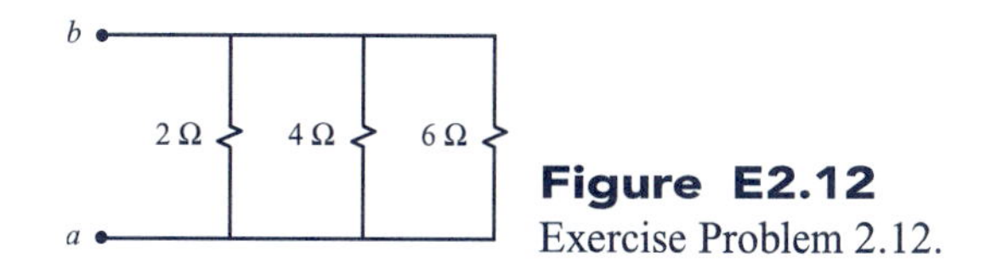

Figure E2.12
Exercise Problem 2.12.

Answer: 1.091 Ω.

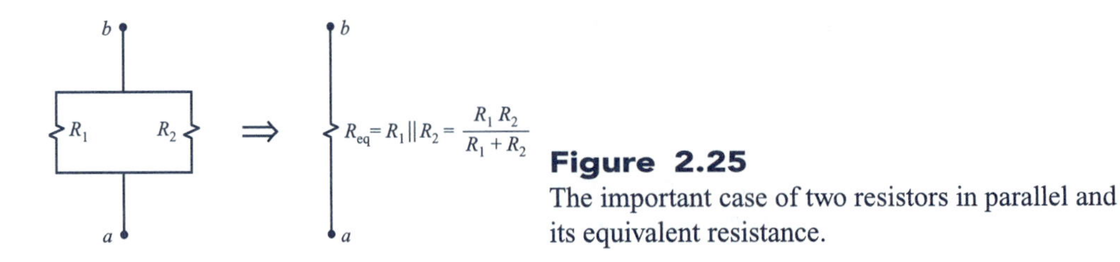

Figure 2.25
The important case of two resistors in parallel and its equivalent resistance.

There is an important case of *two resistors in parallel,* shown in Fig. 2.25, that will prove to be very useful in our future calculations. Using (2.9) for this case gives

$$R_{eq} = R_1 \| R_2 \equiv \frac{R_1 R_2}{R_1 + R_2} \tag{2.10}$$

The symbol of two vertical bars in this formula is read "in parallel with." Some important special cases of two resistors in parallel are illustrated in Fig. 2.26. Two *equal* resistances in parallel are equivalent to a single resistance whose value is one-half the value of each in the parallel combination:

$$R \| R = \frac{R}{2} \tag{2.11}$$

Also, the equivalent resistance of the parallel combination of two resistances can be no larger than the smaller of the two; the larger the disparity in the two values, the closer the result is to the value of the smaller resistance.

The result in (2.10) of course only applies to two resistors in parallel, but it can be applied to the computation of the equivalent resistance of more than two parallel resistances by repeatedly applying it as illustrated in Fig. 2.27. In doing so it is important to label the terminals where a replacement is made (in this case with ×'s) so that it will be clear where that replacement was made.

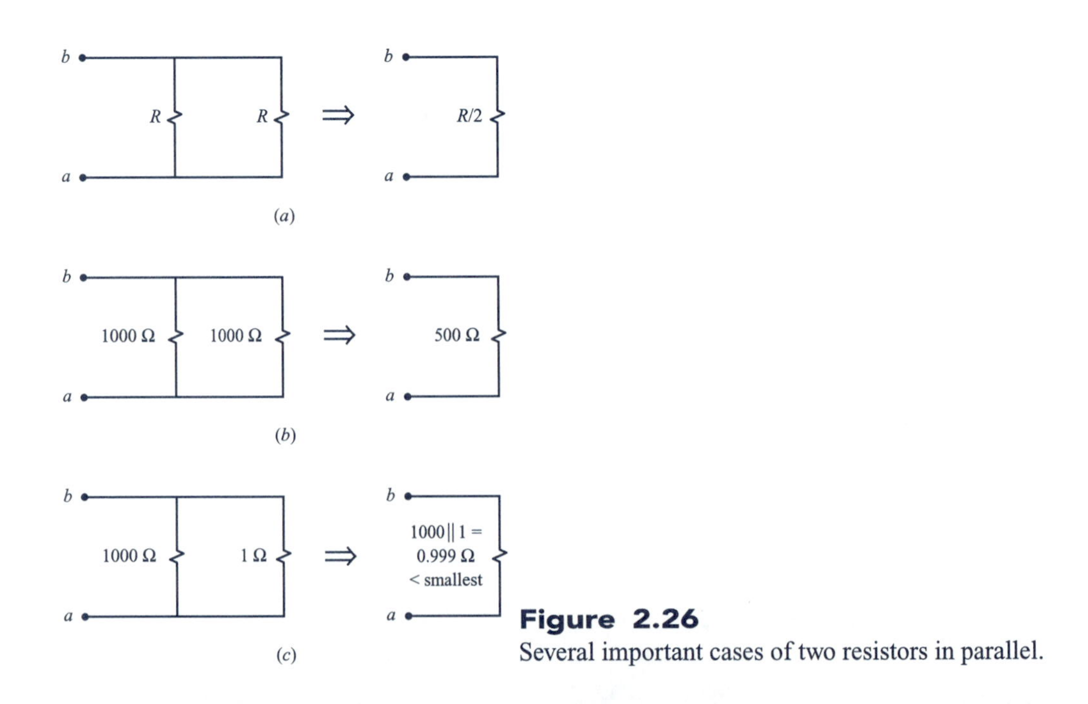

Figure 2.26
Several important cases of two resistors in parallel.

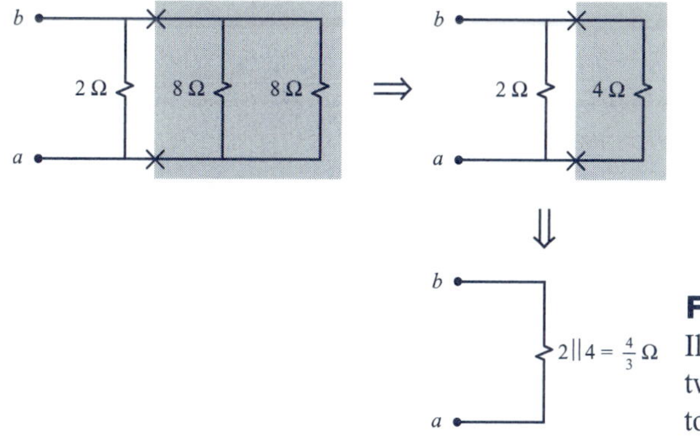

Figure 2.27
Illustration of the repeated use of the two-resistors-in-parallel reduction rule to obtain an equivalent resistance.

Example 2.13

Determine the equivalent resistance at terminals *ab* in the circuit of Fig. 2.28.

Solution Although we could combine any two resistors, we choose to combine the 4-Ω resistors, since it is easy to see that 4 Ω ‖ 4 Ω = 2 Ω, which is then in parallel with the 3-Ω resistor, giving the circuit of Fig. 2.28*b*. From that circuit we again use the two-resistors-in-parallel rule to obtain $R_{eq} = 2\ \Omega \,\|\, 3\ \Omega = \frac{6}{5}\ \Omega$.

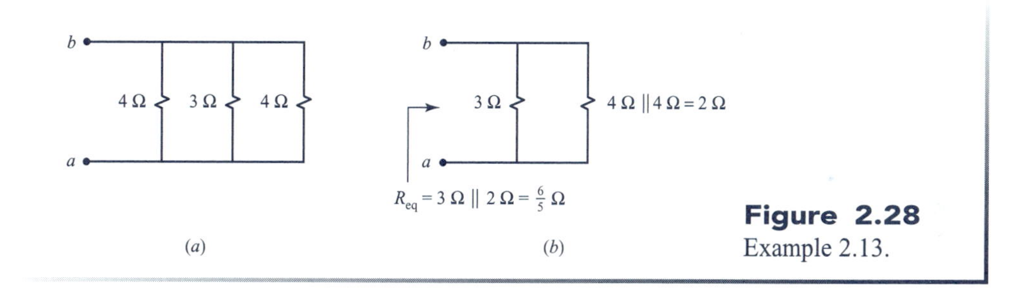

Figure 2.28
Example 2.13.

Exercise Problem 2.13

Determine the equivalent resistance seen at terminals *ab* in Fig. E2.13.

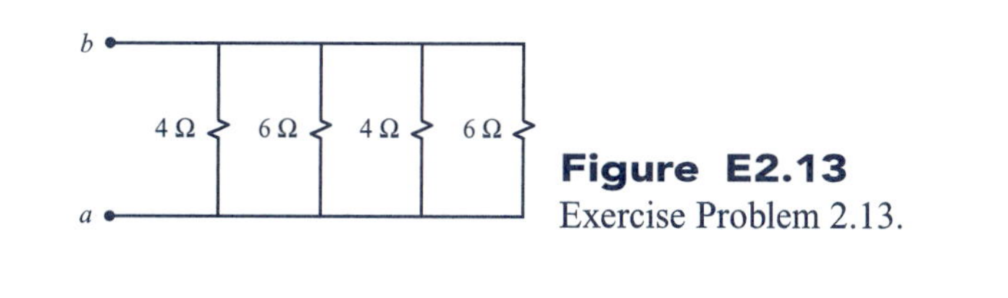

Figure E2.13
Exercise Problem 2.13.

Answer: 1.2 Ω.

Perhaps the most frequent mistake made in circuit analysis is *failing to correctly identify which voltage and current variables are lost in a reduction*. The following example illustrates the combination of series and parallel reductions and the potential for not correctly identifying the remaining voltages and currents in the reduced equivalent circuit.

Example 2.14

Determine the equivalent resistance at terminals *ab* in the circuit of Fig. 2.29.

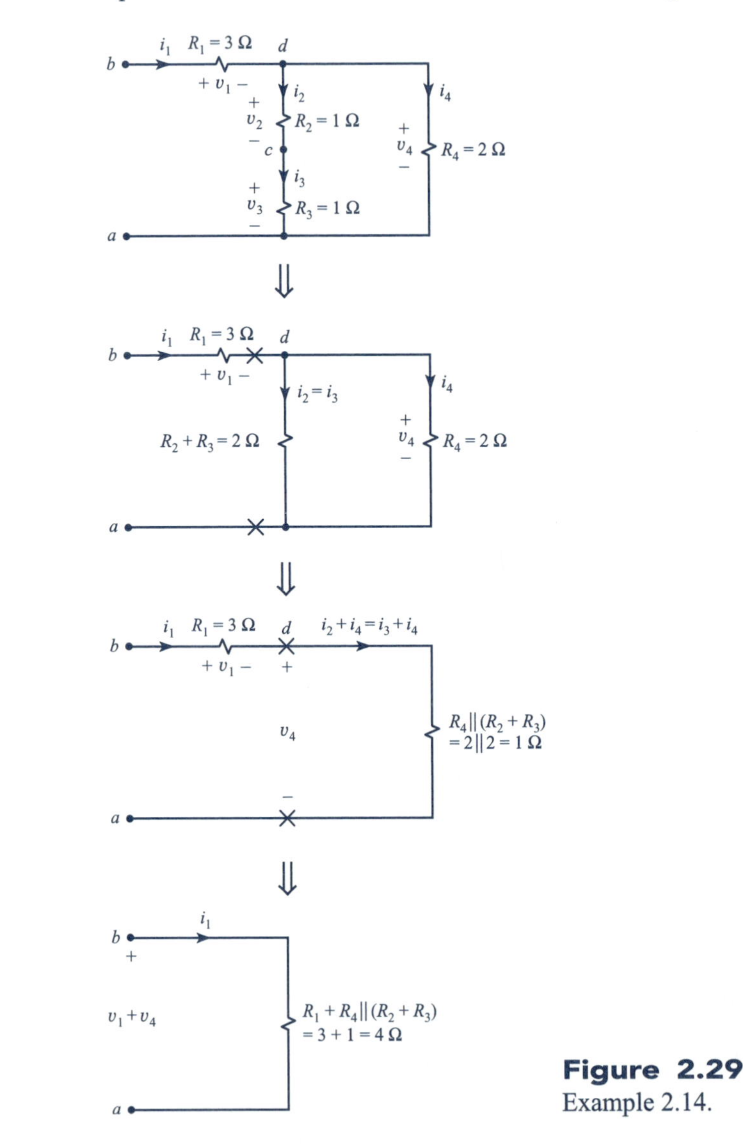

Figure 2.29
Example 2.14.

Solution First we add $R_2 = 1\ \Omega$ and $R_3 = 1\ \Omega$ in series to obtain an equivalent resistance of $R_2 + R_3 = 2\ \Omega$. Note that in doing so, node *c* has disappeared and voltages v_2 and v_3 have disappeared (although their sum is now across the equivalent 2-Ω resistor and the current through it is $i_2 = i_3$). Next we combine this equivalent resistance and $R_4 = 2\ \Omega$, which are in parallel, to yield an equivalent combination of $R_4 \parallel (R_2 + R_3) = 2 \parallel 2 = 1\ \Omega$. Note that here current i_4 is lost in the reduction but the voltage across this equivalent resistance is still v_4. Observe that the current through this equivalent resistor is, by KCL, the sum of the currents through the two resistors, $i_2 + i_4$ or $i_3 + i_4$. Finally we combine the previous equivalent resistance in series with R_1 to produce the final equivalent resistance at terminals *ab* of $R_1 + R_4 \parallel (R_2 + R_3) = 3 + 1 = 4\ \Omega$. Note in this final step that voltage v_1 has been lost, but the current through the final equivalent resistance is still i_1. The voltage across the terminals is the sum $v_1 + v_4$. The reader is well advised to study this example carefully.

Exercise Problem 2.14

Determine the equivalent resistance at terminals *ab* for the circuit of Fig. E2.14.

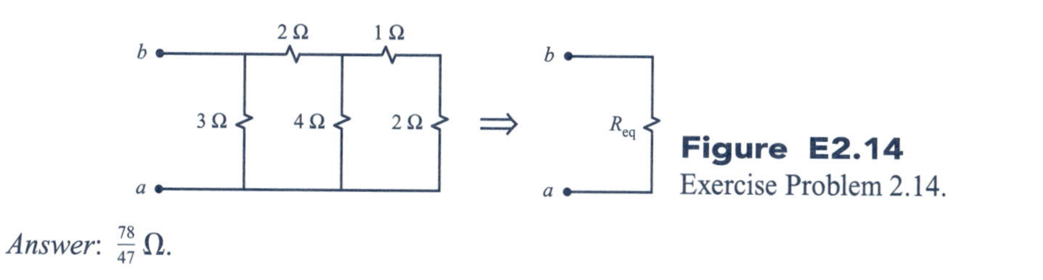

Figure E2.14
Exercise Problem 2.14.

Answer: $\frac{78}{47}$ Ω.

Example 2.15

Determine the equivalent resistance seen at terminals *ab* in Fig. 2.30*a*.

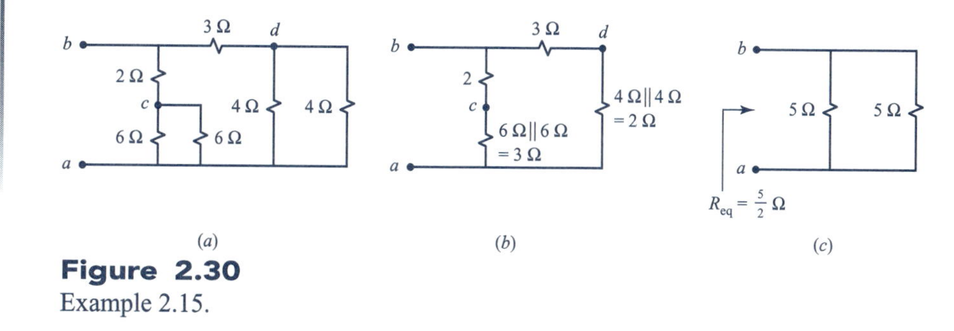

Figure 2.30
Example 2.15.

Solution First we combine the 4-Ω resistors that are in parallel to give 2 Ω, and then we combine the 6-Ω resistors that are in parallel to give 3 Ω, giving the circuit of Fig. 2.30*b*. Then we add the 2-Ω and 3-Ω resistors that are in series to give two 5-Ω resistors in parallel whose equivalent resistance is 2.5 Ω.

Exercise Problem 2.15

Determine the equivalent resistance seen at terminals *ab* in Fig. E2.15.

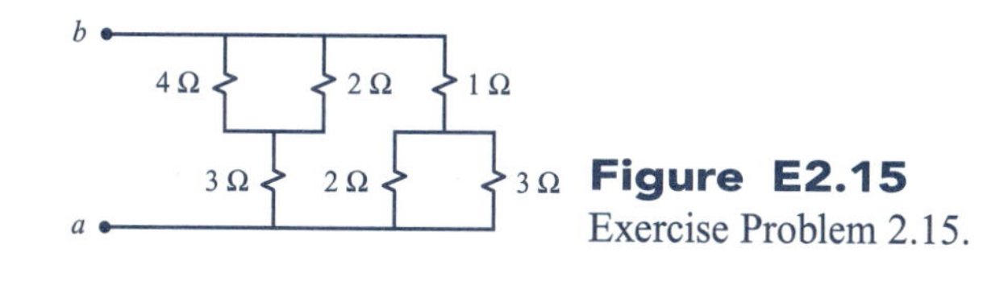

Figure E2.15
Exercise Problem 2.15.

Answer: 1.459 Ω.

2.4.1 Circuit Solution by Circuit Reduction The notion of equivalent resistances can be used to solve circuits that have only one independent source by successively reducing the circuit attached to the source to give an R_{eq} at its terminals. Using Ohm's law, we can easily determine the current leaving the terminals of the source and work back into the resistive circuit to determine any other voltages and currents of interest. Although this technique is very powerful and useful, there is a very common and serious mistake that is frequently made in its application: *failure to correctly identify which nodes, voltages, and currents are lost at each stage in the reduction*. Labeling nodes will greatly reduce the likelihood of such mistakes being made.

Example 2.16

Determine the currents i, i_x, i_y, and voltages v_x, v_y in the circuit of Fig. 2.31*a*.

Solution Combine the 1-Ω and 2-Ω resistors that are in series into an equivalent resistance of 3 Ω as shown in Step 1 in Fig. 2.31*b*. Note in this step that node *d* and voltage v_y have disappeared. Current i_y, however, remains, because it is the current through the series combination. Next combine this 3-Ω resistor and the 2-Ω resistor that is in parallel with it to yield an equivalent resistance of $\frac{6}{5}$ Ω as shown in Step 2 in Fig. 2.31*c*. Note in this step that currents i_x and i_y have disappeared, but voltage v_x remains, since it is the voltage across the parallel combination, or equivalently between nodes *c* and *a*. Finally, combine this $\frac{6}{5}$-Ω resistor with the 1-Ω resistor that is in series with it, giving an equivalent resistance of $\frac{11}{5}$ Ω at the terminals of the source, as shown in Step 3 in Fig. 2.31*d*. Note in this step that node *c*

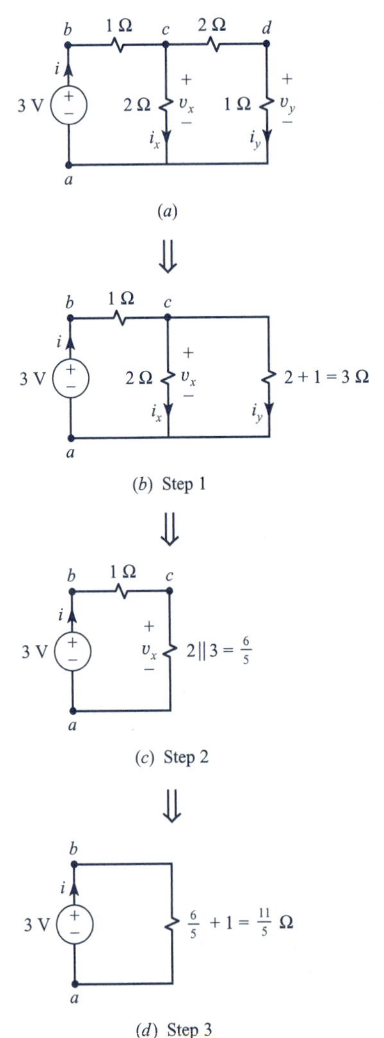

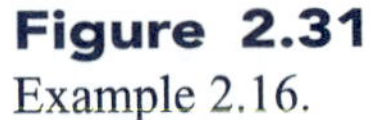

Figure 2.31
Example 2.16.

and voltage v_x have disappeared. In this final step we have a simple single-loop circuit and are able to solve for the current i:

$$\begin{aligned} i &= \frac{3\text{ V}}{\frac{11}{5}\ \Omega} \\ &= \tfrac{15}{11}\text{ A} \end{aligned}$$

Voltage v_x can be determined from Step 2 as

$$\begin{aligned} v_x &= \tfrac{6}{5}\ \Omega \times i \\ &= \tfrac{18}{11}\text{ V} \end{aligned}$$

Proceeding back to Step 1, we can determine

$$\begin{aligned} i_x &= \frac{v_x}{2\ \Omega} \\ &= \tfrac{9}{11}\text{ A} \end{aligned}$$

and

$$\begin{aligned} i_y &= \frac{v_x}{3\ \Omega} \\ &= \tfrac{6}{11}\text{ A} \end{aligned}$$

And, finally, from the original circuit we may determine

$$\begin{aligned} v_y &= 1\ \Omega \times i_y \\ &= \tfrac{6}{11}\text{ V} \end{aligned}$$

This example illustrates that one must be careful to identify which voltages and currents have been lost at each stage of a reduction. Labeling nodes will help avoid such misidentifications.

Example 2.17

Determine the currents i, I, and voltage V in the circuit of Fig. 2.32*a*.

Solution Combine the 1-Ω resistors that are in series to yield 2 Ω, as shown in Step 1 in Fig. 2.32*b*. Observe that node d has been lost in this reduction but current I remains, since it is the current through the series combination. Next combine this 2-Ω resistor that is in

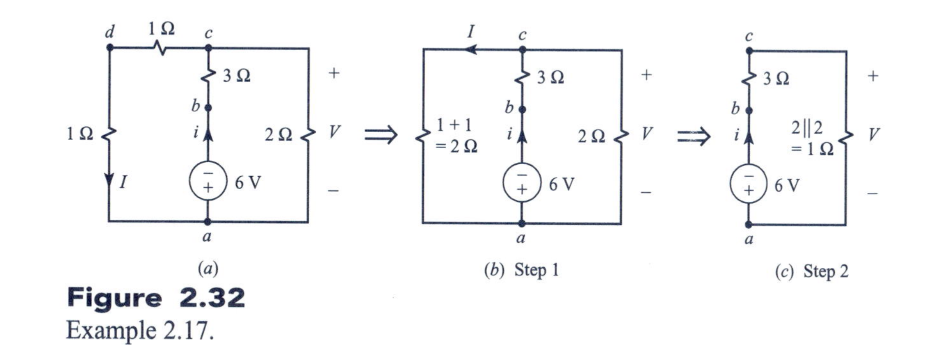

Figure 2.32
Example 2.17.

parallel with the other 2-Ω resistor to yield an equivalent 1-Ω resistor as shown in Step 2 in Fig. 2.32*c*. Note that current I has disappeared in this reduction but voltage V remains, since it is across the parallel combination. This final step gives a single-loop circuit, which can be solved for the current i as

$$i = \frac{-6\text{ V}}{3\ \Omega + 1\ \Omega} = -\tfrac{3}{2}\text{ A}$$

From this we obtain the voltage V as

$$V = i \times 1\ \Omega = -\tfrac{3}{2}\text{ V}$$

Moving back to Step 1, we obtain the current I as

$$I = \frac{V}{2\ \Omega} = -\tfrac{3}{4}\text{ V}$$

Example 2.18

Determine voltages V and v_x and currents I and i_x in the circuit of Fig. 2.33*a*.

Solution Combine the 2-Ω resistors that are in parallel to yield the circuit in Step 1 shown in Fig. 2.33*b*. Observe that current I has been lost in this reduction but voltage v_x remains, since it was across the parallel combination. Similarly, current i_x remains. Finally, combine this 1-Ω resistor and the 1-Ω resistor that is in series with it to yield the circuit of Step 2 shown in Fig. 2.33*c*. Note that node c and voltage v_x have disappeared in this reduction, but voltage V remains, since it was across the parallel combination. Also current i_x remains. At this stage we could reduce the two 2-Ω resistors to an equivalent 1-Ω resistor across the current source. However, this is not necessary, since we have a single-node-pair circuit and the voltage can be determined as

$$V = \frac{5\text{ A}}{\tfrac{1}{2}\ \Omega + \tfrac{1}{2}\ \Omega} = 5\text{ V}$$

The current i_x can also be determined using Ohm's law as

$$i_x = \frac{V}{2\ \Omega} = \tfrac{5}{2}\text{ A}$$

Moving back to Step 1, we determine the voltage v_x using Ohm's law as

$$v_x = i_x \times 1\ \Omega = \tfrac{5}{2}\text{ V}$$

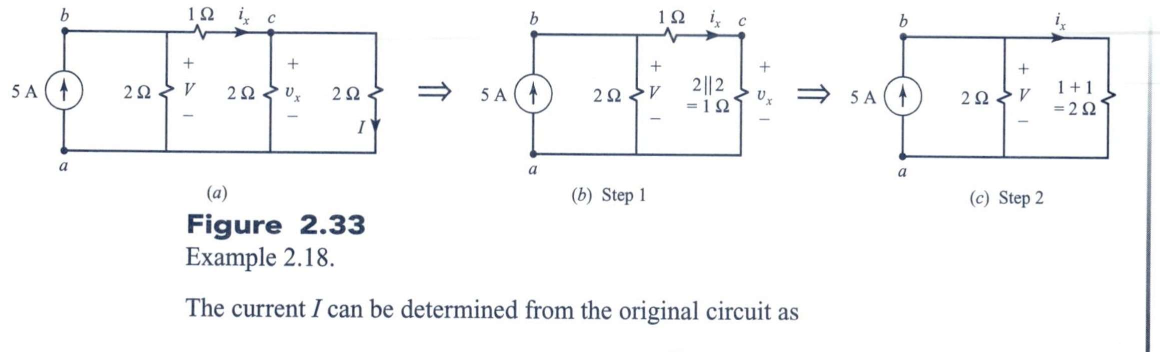

Figure 2.33
Example 2.18.

The current I can be determined from the original circuit as

$$I = \frac{v_x}{2\ \Omega}$$

$$= \tfrac{5}{4}\text{ A}$$

Exercise Problem 2.16

Determine v_x, v_y, and i_z in the circuit of Fig. E2.16.

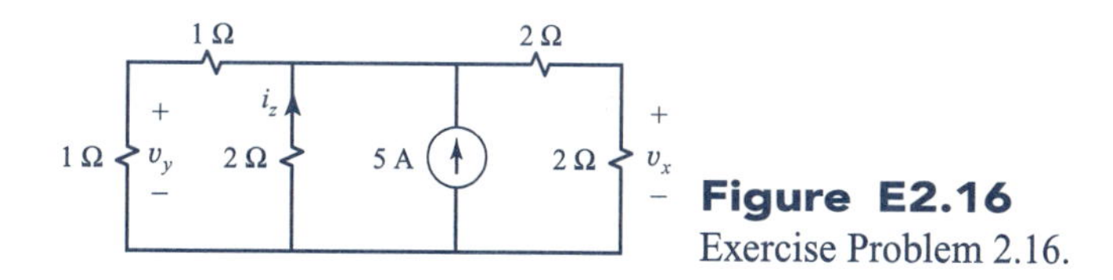

Figure E2.16
Exercise Problem 2.16.

Answers: 2 V, 2 V, and −2 A. Check conservation of power.

Exercise Problem 2.17

Determine v_y and i_x in the circuit of Fig. E2.17.

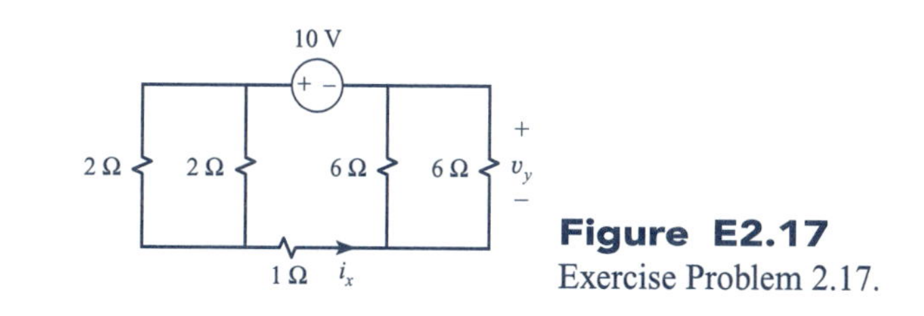

Figure E2.17
Exercise Problem 2.17.

Answers: 2 A and −6 V. Check conservation of power.

2.5 Voltage and Current Division

In this section we consider two other important and useful solution tools: voltage and current division. Consider the series connection of three resistors across which a voltage source is applied as shown in Fig. 2.34. Since this is a single-loop circuit, the current through all elements can be determined as

$$i = \frac{v_S(t)}{R_1 + R_2 + R_3}$$

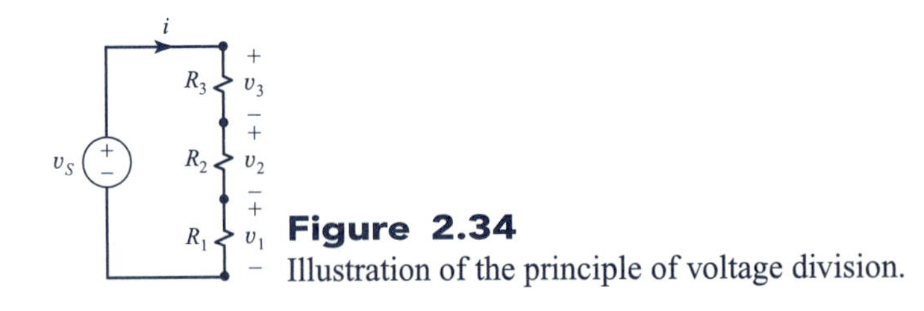

Figure 2.34
Illustration of the principle of voltage division.

The individual voltages across the resistors are obtained as the product of this current and the appropriate resistance:

$$v_1 = R_1 i = \frac{R_1}{R_1 + R_2 + R_3} v_S(t)$$

$$v_2 = R_2 i = \frac{R_2}{R_1 + R_2 + R_3} v_S(t)$$

$$v_3 = R_3 i = \frac{R_3}{R_1 + R_2 + R_3} v_S(t)$$

In general we see that for n resistors in series *the voltage across the series connection divides according to the ratio of the resistor associated with that voltage to the sum of the resistances in the series:*

$$v_i = \frac{R_i}{R_1 + R_2 + \cdots + R_n} v_S(t) \tag{2.12}$$

Example 2.19

Determine the voltage V in the circuit of Fig. 2.35.

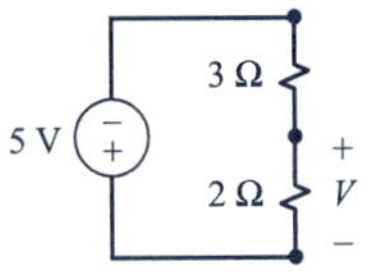

Figure 2.35
Example 2.19.

Solution By voltage division

$$V = -\frac{2\ \Omega}{2\ \Omega + 3\ \Omega} 5\ \text{V}$$
$$= -2\ \text{V}$$

This example shows that we must be careful of polarities in applying voltage division.

Example 2.20

Determine the voltages v_x and v_y in the circuit of Fig. 2.36.

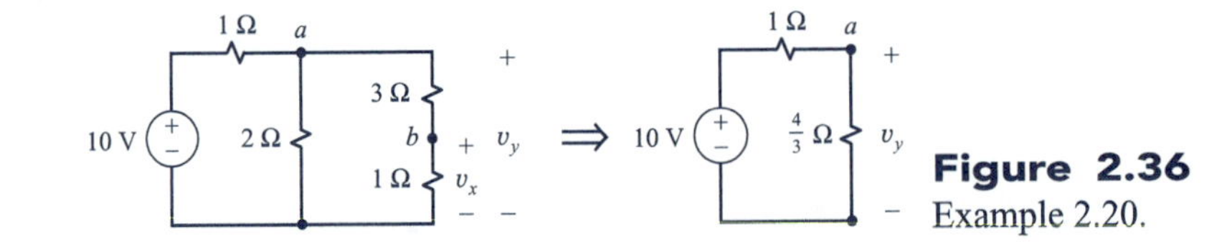

Figure 2.36
Example 2.20.

Solution First make the reduction as shown by combining the 3-Ω and 1-Ω resistors in series, which are in parallel with the 2-Ω resistor. Note that voltage v_x has been lost in this reduction. Nevertheless, voltage v_y may be determined in this reduced circuit by voltage division as

$$
\begin{aligned}
v_y &= \frac{\frac{4}{3}\,\Omega}{\frac{4}{3}\,\Omega + 1\,\Omega}\,10\text{ V} \\
&= \tfrac{40}{7}\text{ V}
\end{aligned}
$$

Moving back to the original circuit, we may determine v_x from v_y by using voltage division as

$$
\begin{aligned}
v_x &= \frac{1\,\Omega}{1\,\Omega + 3\,\Omega}\,v_y \\
&= \tfrac{10}{7}\text{ V}
\end{aligned}
$$

This example has shown that the voltage $v_S(t)$ in (2.12) need not be a voltage source but is simply the total voltage across the series connection.

Example 2.21

Determine the voltage V in the circuit of Fig. 2.37*a*.

Solution If we knew the voltage at node a with respect to the bottom of the circuit, we could use voltage division to obtain

$$
V = \frac{2\,\Omega}{2\,\Omega + 3\,\Omega + 4\,\Omega}\,V_a
$$

In order to determine V_a we use resistor combinations to give the circuit of Fig. 2.37*b*, and from that we determine

$$
\begin{aligned}
V_a &= (9\,\Omega \,\|\, 9\,\Omega) \times 10\text{ A} \\
&= \tfrac{9}{2}\,\Omega \times 10\text{ A} \\
&= 45\text{ V}
\end{aligned}
$$

Hence the desired voltage is, from Fig. 2.37*c*,

$$
\begin{aligned}
V &= \frac{2\,\Omega}{2\,\Omega + 3\,\Omega + 4\,\Omega}\,V_a \\
&= \tfrac{2}{9} \times 45\text{ V} \\
&= 10\text{ V}
\end{aligned}
$$

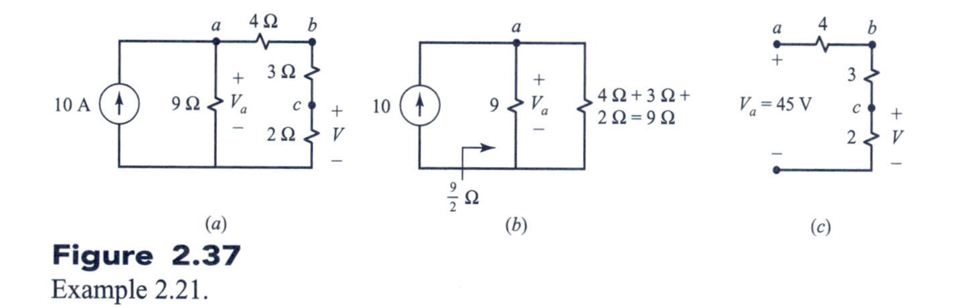

Figure 2.37
Example 2.21.

Exercise Problem 2.18

Determine the voltage V in the circuit of Fig. E2.18.

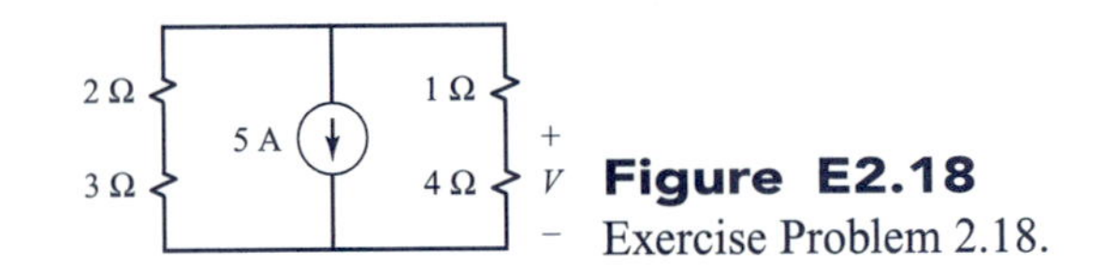

Figure E2.18
Exercise Problem 2.18.

Answer: −10 V.

Exercise Problem 2.19

Determine the voltage V in the circuit of Fig. E2.19.

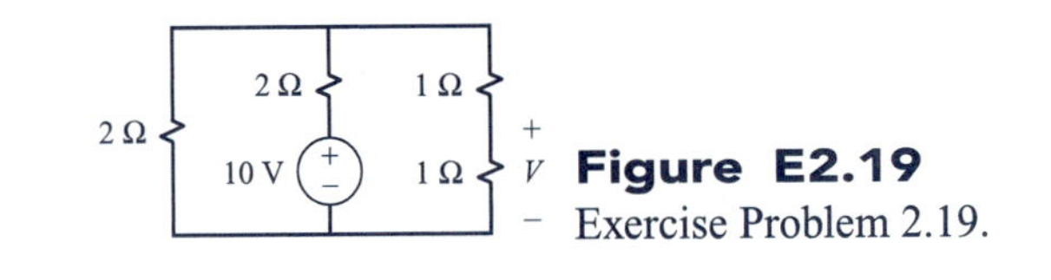

Figure E2.19
Exercise Problem 2.19.

Answer: $\frac{5}{3}$ V.

The next rule, *current division,* is similar to the voltage division rule. Consider the parallel connection of three resistors with a current source applied across the parallel connection as shown in Fig. 2.38. Since this is a single-node-pair circuit, the voltage across all elements can be determined as

$$v = \frac{i_S(t)}{G_1 + G_2 + G_3}$$

The individual currents through the resistors are obtained as the product of this voltage and the appropriate conductance:

$$i_1 = G_1 v = \frac{G_1}{G_1 + G_2 + G_3} i_S(t)$$

$$i_2 = G_2 v = \frac{G_2}{G_1 + G_2 + G_3} i_S(t)$$

$$i_3 = G_3 v = \frac{G_3}{G_1 + G_2 + G_3} i_S(t)$$

In general we see that for n resistors in parallel *the current applied to the parallel combination divides according to the ratio of the conductance associated with that current to the sum of the*

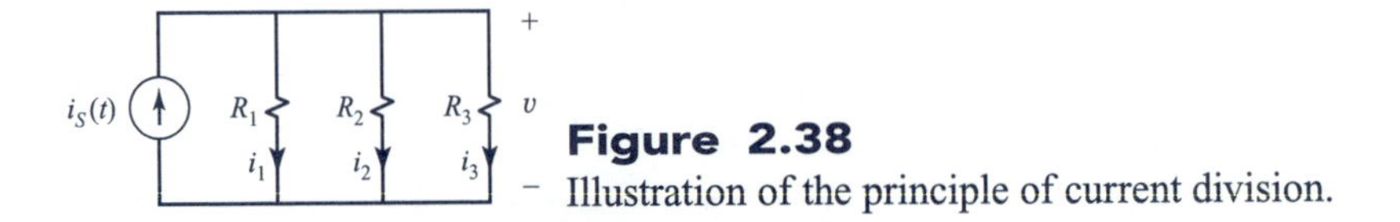

Figure 2.38
Illustration of the principle of current division.

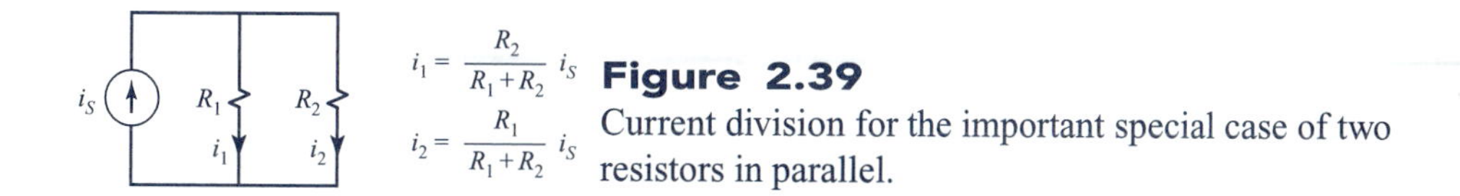

Figure 2.39
Current division for the important special case of two resistors in parallel.

conductances in the parallel combination:

$$i_1 = \frac{G_i}{G_1 + G_2 + \cdots + G_n} i_S(t) \tag{2.13}$$

We tend to deal in resistances rather than conductances. An important special case is shown in Fig. 2.39 where, by writing (2.13) in terms of resistances, we can write the current division formula in terms of resistances:

$$i_1 = \frac{R_2}{R_1 + R_2} i_S \tag{2.14a}$$

$$i_2 = \frac{R_1}{R_1 + R_2} i_S \tag{2.14b}$$

For this special case of two resistors in parallel, the current applied to the parallel combination divides according to the ratio of the *opposite resistor* and the sum of the two resistors in the parallel connection. We will have numerous occasions to apply this important special case of current division for two parallel resistors.

Example 2.22

Determine the currents I and i_x by current division in the circuit of Fig. 2.40*a*.

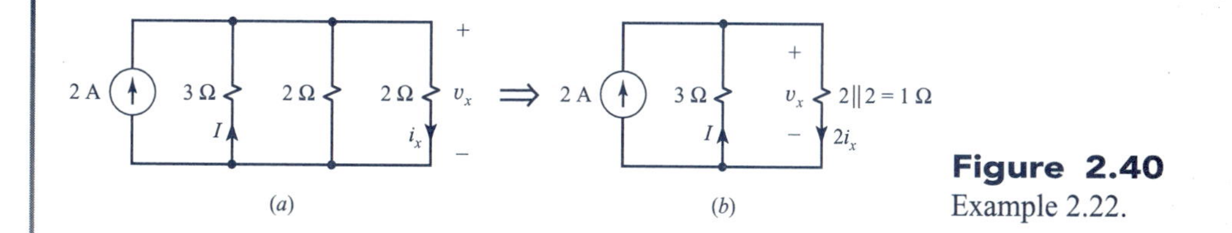

Figure 2.40
Example 2.22.

Solution Reduce the two 2-Ω resistors that are in parallel to an equivalent 1-Ω resistor as shown in Fig. 2.40*b*. Note that the current i_x has disappeared in this reduction. From this circuit we may use the two-resistor current division rule to obtain I as

$$I = -\frac{1\ \Omega}{1\ \Omega + 3\ \Omega} 2\ \text{A}$$

$$= -\tfrac{1}{2}\ \text{A}$$

In order to determine the current i_x we first determine the voltage v_x as

$$v_x = -3\ \Omega \times I$$

$$= \tfrac{3}{2}\ \text{V}$$

Note the minus sign required by the passive sign convention. The current i_x can now be found from the original circuit as

$$i_x = \frac{v_x}{2\ \Omega} = \tfrac{3}{4}\ \text{A}$$

Also observe in the original circuit that the current through the two parallel 2-Ω resistors must both be i_x, since the voltage across them is the same. Hence, in the reduced circuit, the current through the parallel combination is, by KCL, $2i_x$. Hence we could have applied current division to the reduced circuit to obtain

$$2i_x = \frac{3\ \Omega}{3\ \Omega + 1\ \Omega} 2\ \text{A} = \tfrac{3}{2}\ \text{A}$$

so that $i_x = \frac{3}{4}$ A as before.

The following examples illustrate some other useful but subtle applications of current division.

Example 2.23

Determine the voltage v_x in the circuit of Fig. 2.41*a*.

Solution First we label currents I and i_x for the purposes of using current division. Next we make a reduction shown in Fig. 2.41*b*. Note that current i_x and voltage v_x have disappeared in this reduction. This is a single-loop circuit, so that I may be determined as

$$I = \frac{10\ \text{V}}{1\ \Omega + \frac{4}{3}\ \Omega} = \tfrac{30}{7}\ \text{A}$$

Next we apply current division in the original circuit to yield

$$i_x = \frac{2\ \Omega}{2\ \Omega + 4\ \Omega} I = \tfrac{10}{7}\ \text{A}$$

From this we obtain v_x as

$$v_x = -1\ \Omega \times i_x = -\tfrac{10}{7}\ \text{V}$$

Again observe the minus sign in this last result, required by the passive sign convention.

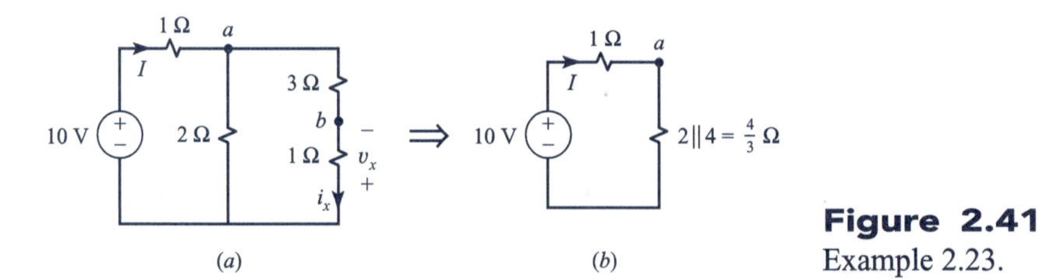

Figure 2.41 Example 2.23.

Several such current divisions can be used repeatedly, as the following example shows.

Example 2.24

Determine the currents I and i_x in the circuit of Fig. 2.42*a*.

Solution Making the reduction shown in Fig. 2.42*b* and applying current division yields

$$i_x = \frac{3\ \Omega}{3\ \Omega + 2\ \Omega} 5\ \text{A}$$
$$= 3\ \text{A}$$

Returning to the original circuit and applying current division gives

$$I = \tfrac{1}{2} i_x$$
$$= \tfrac{3}{2}\ \text{A}$$

Note the use of the equal-resistor current division rule.

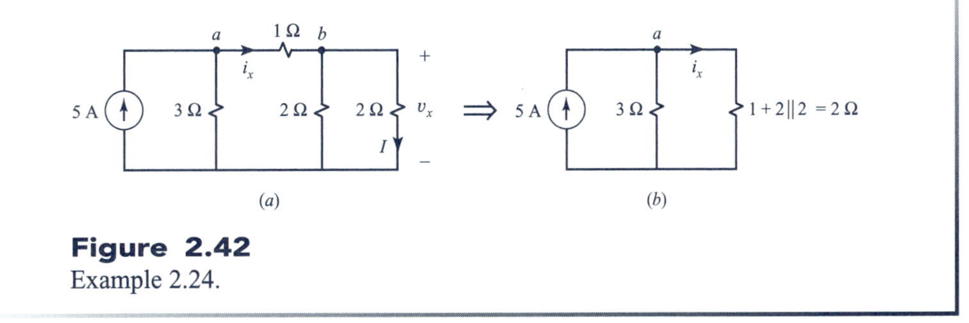

Figure 2.42
Example 2.24.

The next example illustrates the power in using all of previous results to solve some seemingly complicated circuits.

Example 2.25

Determine currents i_x, i_y, and I as well as voltage v_x in the circuit of Fig. 2.43*a*.

Solution First make the reduction shown in Fig. 2.43*b*. Note that current I has disappeared in this reduction. Next make the reduction shown in Fig. 2.43*c*. Note that the current i_y has disappeared in this final stage of the reduction but the voltage v_x remains. Nevertheless we may determine the current i_x for this single-loop circuit as

$$i_x = \frac{10\ \text{V}}{2\ \Omega + \frac{2}{3}\ \Omega}$$
$$= \tfrac{15}{4}\ \text{A}$$

From this result we determine v_x using Ohm's law as

$$v_x = \tfrac{2}{3}\ \Omega \times i_x$$
$$= \tfrac{5}{2}\ \text{V}$$

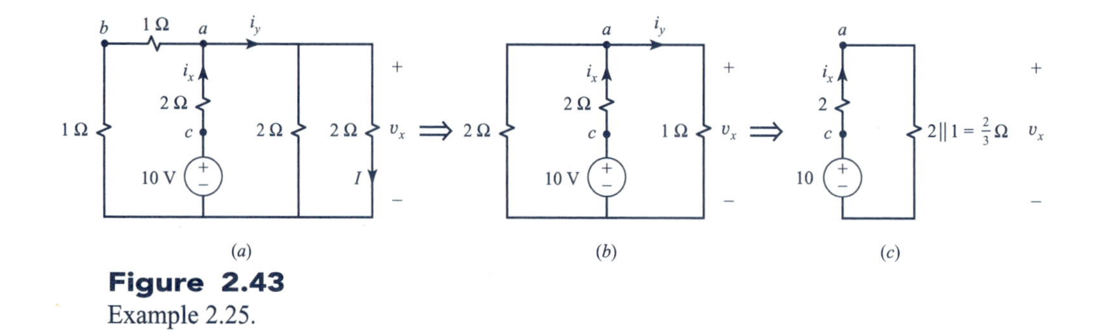

Figure 2.43
Example 2.25.

This could have been more easily determined using voltage division as

$$\begin{aligned} v_x &= \frac{\frac{2}{3}\,\Omega}{\frac{2}{3}\,\Omega + 2\,\Omega}\,10\text{ V} \\ &= \tfrac{5}{2}\text{ V} \end{aligned}$$

Proceeding back to Fig. 2.43*b*, we determine the current i_y by Ohm's law in terms of v_x as

$$\begin{aligned} i_y &= \frac{v_x}{1\,\Omega} \\ &= \tfrac{5}{2}\text{ A} \end{aligned}$$

Alternatively this could have been determined using current division as

$$\begin{aligned} i_y &= \frac{2\,\Omega}{2\,\Omega + 1\,\Omega}\,i_x \\ &= \tfrac{5}{2}\text{ A} \end{aligned}$$

And finally, proceeding back to the original circuit, we obtain the current I by Ohm's law:

$$\begin{aligned} I &= \frac{v_x}{2\,\Omega} \\ &= \tfrac{5}{4}\text{ A} \end{aligned}$$

or by current division:

$$\begin{aligned} I &= \tfrac{1}{2}\,i_y \\ &= \tfrac{5}{4}\text{ A} \end{aligned}$$

Exercise Problem 2.20

Determine current i and voltage v in the circuit of Fig. E2.20 using current and voltage division.

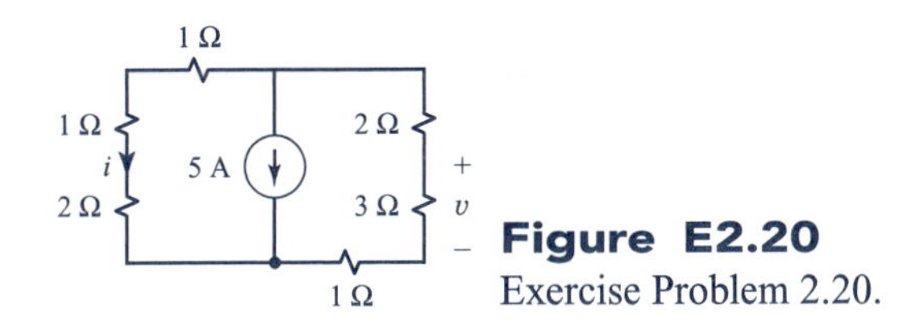

Figure E2.20
Exercise Problem 2.20.

Answers: −6 V and −3 A.

Exercise Problem 2.21

Determine v_x, v_y, and i_z in the circuit of Fig. E2.16 using voltage and current division.

Answers: 2 V, 2 V, and −2 A.

Exercise Problem 2.22

Determine v_y and i_x in the circuit of Fig. E2.17 using voltage and current division.

Answers: 2 A and −6 V.

2.6 Solutions for Circuits Containing More than One Source

We now have a number of methods for solving a circuit. But most of these assumed that there was only one independent source in the circuit, so that we could make a sequence of resistor reductions back to that source. In the case of circuits that contain more than one source, these methods cannot be used. (unless they are single-loop or single-node-pair circuits.) For example, in a circuit containing more than one source, if we attempt to use resistor reductions to reduce back to one of the sources, those reductions will contain the other source, and we cannot combine resistors in series or in parallel when they contain a source. In order to do so, we would have to ignore that source, which, of course, is illogical.

In this section we outline a very direct and effective way of solving circuits that can handle circuits containing more than one source. The method simply systematically applies KVL, KCL, and Ohm's law. Before doing so it is important to point out that for a simple series connection of two or more elements such as shown in Fig. 2.44*a* it is obvious by now that the currents through all elements are equal. This is called a *simple node*. Similarly, for a parallel connection of two or more elements such as shown in Fig. 2.44*b* it is equally obvious that the voltages across all elements are equal. This is called a *simple loop*. We need not waste time methodically reapplying KCL or KVL to rediscover this. The only times we need to directly apply KCL are for the connection of more than two elements at a node as shown in Fig. 2.44*c*, i.e., elements not in series. These will be referred to as *nonsimple nodes*. Similarly, the only time we need to directly apply KVL is for a loop whose elements are not connected at both pairs of nodes, i.e., elements not in parallel, such as shown in Fig. 2.44*d*. These will be referred to as *nonsimple loops*.

Our method will be referred to as the *direct method* and is the direct application of KCL to all nonsimple nodes, KVL to all nonsimple loops, and Ohm's law to every resistor. The key to the effective use of this method is to obtain all currents and voltages using these laws *in terms of the unknown. Resist the temptation to define any more extraneous unknowns unless there is no choice.*

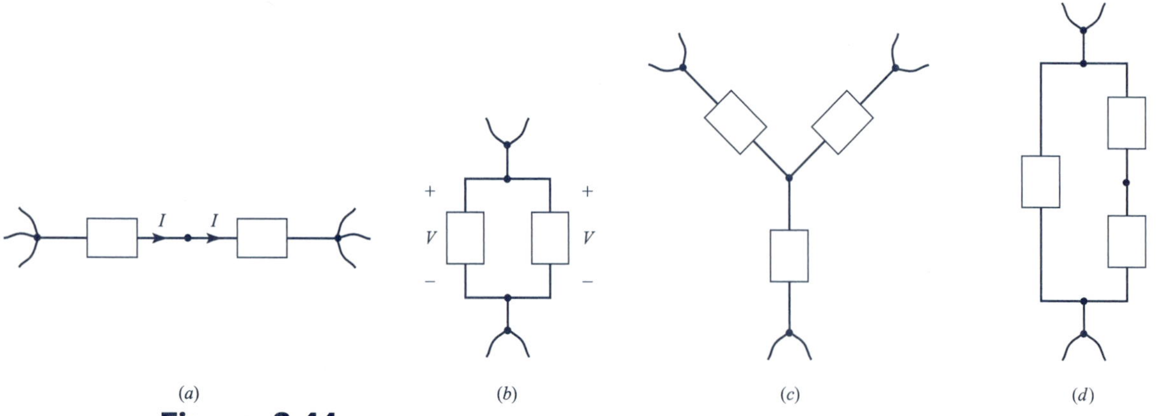

Figure 2.44
Illustration of the concepts of simple and nonsimple nodes and loops: (*a*) a simple node, (*b*) a simple loop, (*c*) a nonsimple node, (*d*) a nonsimple loop.

There are a few simple elements to the procedure that will greatly aid in completing it properly and avoiding confusion and mistakes. Once a few problems have been solved using this procedure, it will no longer be necessary to perform all the steps: the method will become straightforward. The steps are:

Step 1: Choose the unknown to be solved for. Choose only one if there is more than one unknown to be determined; the other unknown(s) can be easily determined once the selected unknown has been solved for.

Step 2: Place small checkoff boxes near the circuit diagram so that the three laws can be checked off methodically, i.e.,

- ☐ KVL (nonsimple loops)
- ☐ KCL (nonsimple nodes)
- ☐ Ohm's law (all resistors)

All steps must be checked off, or else we are not through with the solution. This is very important to do in the early stages of learning this procedure. After a few problems are completed by explicitly writing these checkoff boxes on the diagram, however, it can be done mentally. Systematically applying these laws and checking them off will immediately show what step is left to complete and will lead us directly to the solution.

Step 3: In applying Ohm's law, label the current direction through the resistor with an arrow *and* place + and − signs on that resistor to avoid making a sign mistake in applying Ohm's law. We certainly know by now that $V = RI$, but the polarity of the voltage and current must conform to the passive sign convention or else a negative sign is required in Ohm's law. A sign error in Ohm's law is a very common mistake.

Example 2.26

Determine the current i in the circuit of Fig. 2.45*a* using the direct method.

Solution There is only one nonsimple node: node a. (The lower node will result in the same KCL equation as at node a.) There is only one nonsimple loop: the loop consisting of the 10-V voltage source, the 2-Ω resistor, and the 3-Ω resistor. (Once again, the voltage

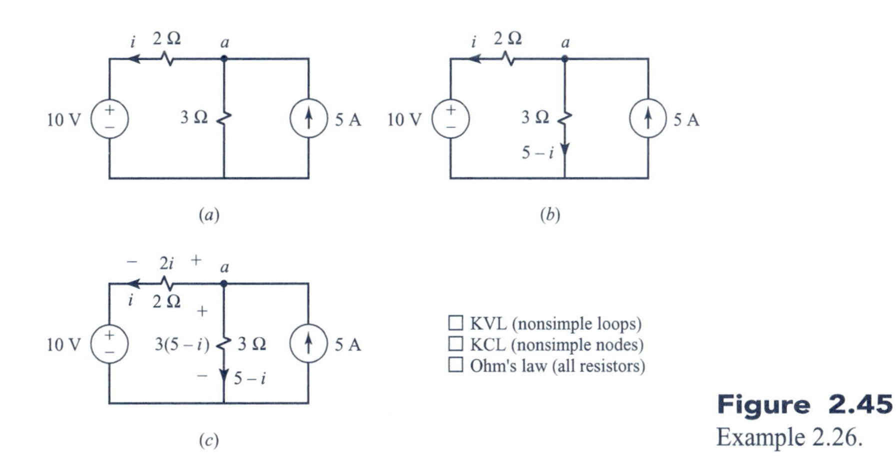

Figure 2.45
Example 2.26.

across the current source is not necessarily zero which is a frequent and incorrect assumption.) First we place the three checkoff boxes near the diagram, representing the three (and only) laws to apply. We cannot apply Ohm's law, because we do not yet know the voltage or the current associated with each resistor. Similarly, we cannot apply KVL, because we do not know the voltage across each element. Hence we are led to apply KCL first. We apply KCL at node a in order to write the current through the 3-Ω resistor in terms of the unknown current i. This gives the current down through the 3-Ω resistor as 5-i, as shown in Fig. 2.45b. We have exhausted the application of KCL for this circuit. Next apply Ohm's law to the two resistors to give the voltages across them as shown in Fig. 2.45c. We have now exhausted Ohm's law and are left with only KVL, which need only be applied around the nonsimple loop to yield

$$10\text{ V} = -2i + 3(5 - i)$$

Solving yields

$$i = 1\text{ A}$$

Observe that we have applied in this process all the laws that we have: KCL at the nonsimple nodes, KVL around the nonsimple loops, and Ohm's law for every resistor.

Example 2.27

Determine voltage V in the circuit of Fig. 2.46a using the direct method.

Solution First we write the checkoff boxes near the circuit diagram. There is only one nonsimple node: node a. Again we cannot apply Ohm's law, since we do not know the voltage and currents associated with the resistors. Applying KCL at node a will not be useful, because we would have to define three unknown currents: one through each of the 2-Ω resistors, and one through the 1-Ω resistor. We must resist the temptation to define these extraneous unknowns, which would complicate the solution. Hence we are led to apply KVL. Apply KVL around the nonsimple loop consisting of the 2-Ω resistor in series with the 3-V voltage source and the other 2-Ω resistor, and then around the nonsimple loop consisting of the 2-Ω resistor in series with the 3-V voltage source and the 1-Ω resistor, to determine the voltages across these resistors in terms of V as shown in Fig. 2.46b. This gives the voltage across the 2-Ω and 1-Ω resistors as V-3 and completes application of KVL for the circuit. Next apply Ohm's law to the three resistors to determine the currents through them

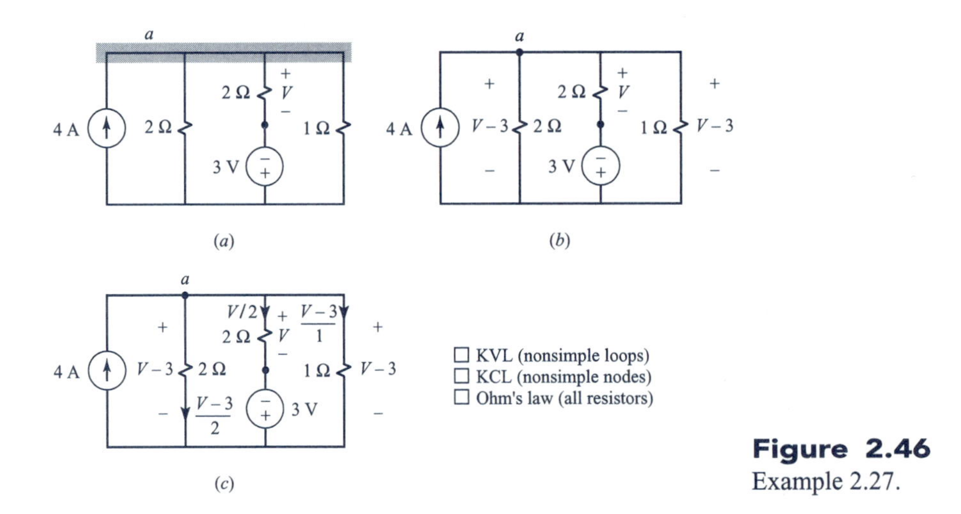

Figure 2.46
Example 2.27.

in terms of V as shown in Fig. 2.46c. This completes application of Ohm's law. Finally apply KCL to the only nonsimple node (the one at the top) to obtain

$$4\text{ A} = \frac{V-3}{2\ \Omega} + \frac{V}{2\ \Omega} + \frac{V-3}{1\ \Omega}$$

Solving this yields

$$V = \tfrac{17}{4}\text{ V}$$

Example 2.28

Determine the voltage V and current I in the circuit of Fig. 2.47a using the direct method.

Solution First we select the unknown to be solved for. We *arbitrarily* select current I. We cannot apply Ohm's law, because we do not yet know the voltage or the current associated with each resistor. Similarly, we cannot apply KVL, because we do not know the voltage across each element. Hence we are led to apply KCL first. First apply KCL at the only nonsimple node (node a at the top of the current source) to obtain the current through the 3-V voltage source in terms of I as I-2, as shown in Fig. 2.47b. This completes the

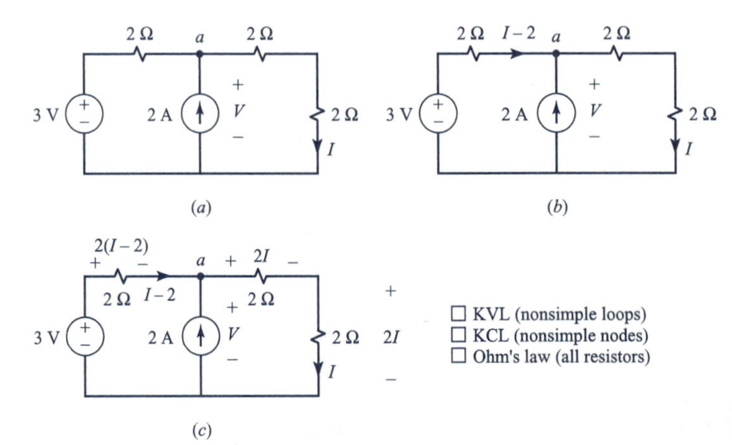

Figure 2.47
Example 2.28.

application of KCL. Next apply Ohm's law to all resistors in terms of I as shown in Fig. 2.47*c*. Finally apply KVL around the only nonsimple loop for which we know the voltages across all elements: the outside loop. (We do not know the voltage across the 2-A current source. Do not assume it is zero!) This yields

$$3\text{ V} = 2(I - 2) + 2I + 2I$$

Solving this yields

$$I = \tfrac{7}{6}\text{ A}$$

From this we obtain V by writing KVL around the loops as

$$\begin{aligned} V &= 2I + 2I \\ &= 3\text{ V} - 2\,(I - 2) \\ &= \tfrac{14}{3}\text{ V} \end{aligned}$$

The vast majority of the circuits one encounters and is willing to solve by hand (as opposed to using a digital computer) are directly solvable by this simple and effective method.

Exercise Problem 2.23

Determine current I in the circuit of Fig. E2.23 by the direct method.

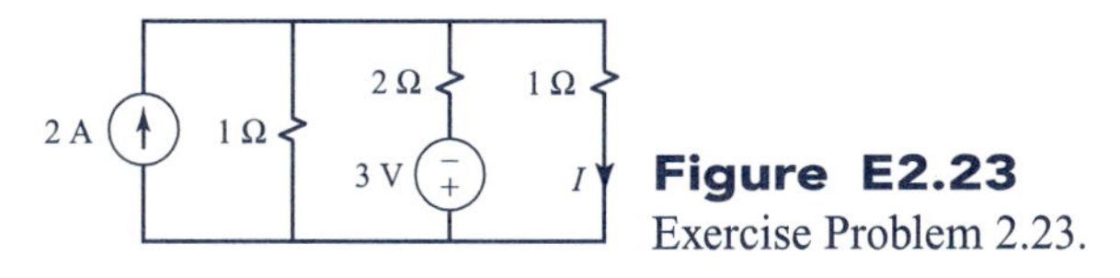

Figure E2.23
Exercise Problem 2.23.

Answer: $\frac{1}{5}$ A.

Exercise Problem 2.24

Determine voltage V in the circuit of Fig. E2.24 by the direct method.

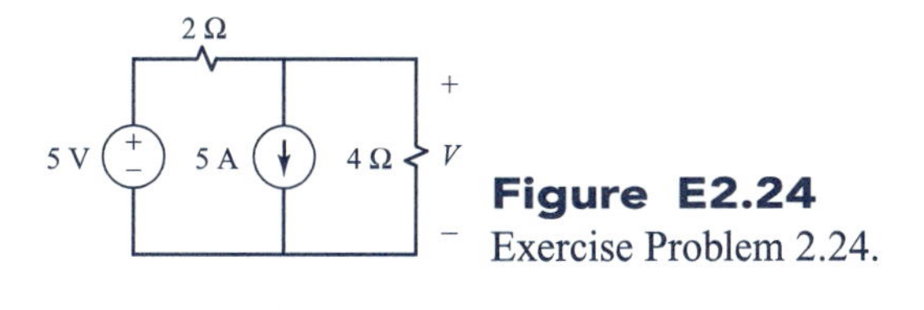

Figure E2.24
Exercise Problem 2.24.

Answer: $-\frac{10}{3}$ V.

Exercise Problem 2.25

Determine current I in the circuit of Fig. E2.25 by the direct method.

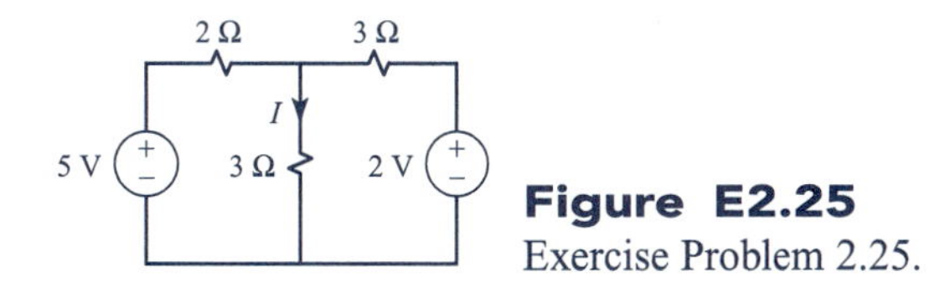

Figure E2.25
Exercise Problem 2.25.

Answer: $\frac{19}{21}$ A.

2.7 Source Transformations

We add one final solution technique to our arsenal: the method of *source transformation*. Figure 2.48 shows this important transformation. The two circuits are equivalent if their v–i relations are identical. Writing this relation for the left circuit using KVL gives

$$v = R_S i + V_S \tag{2.15}$$

Similarly, writing this relation for the right circuit using KCL gives

$$i = \frac{v}{R_S} - I_S \tag{2.16}$$

Rewriting (2.16) gives

$$v = R_S i + \underbrace{R_S I_S}_{V_S} \tag{2.17}$$

Hence the two are equivalent at terminals ab if V_S and I_S are related as

$$\boxed{V_S = R_S I_S} \tag{2.18}$$

We can convert from one structure to another so long as we convert the source value according to (2.18).

Perhaps the most common mistake made in applying this transformation is not observing the correct polarity of the two sources for equivalence: *the polarity of the voltage source V_S in one form must be such that it tends to push in the direction of the current I_S of the current source in the other form*. Two examples are shown in Fig. 2.49 and should be carefully studied.

The origin of the term "source" for these configurations is explained in Fig. 2.50. Consider an automobile storage battery. When no current is being drawn from its terminals, the terminal

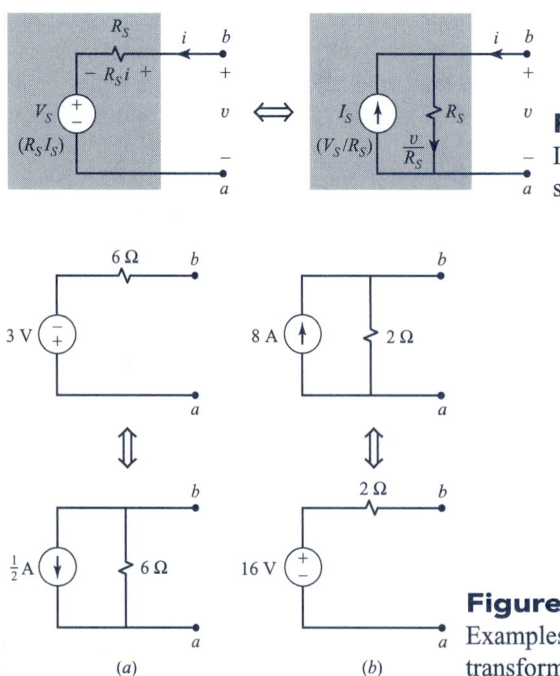

Figure 2.48 Illustration of the important principle of source transformations.

Figure 2.49 Examples of the correct use of source transformations.

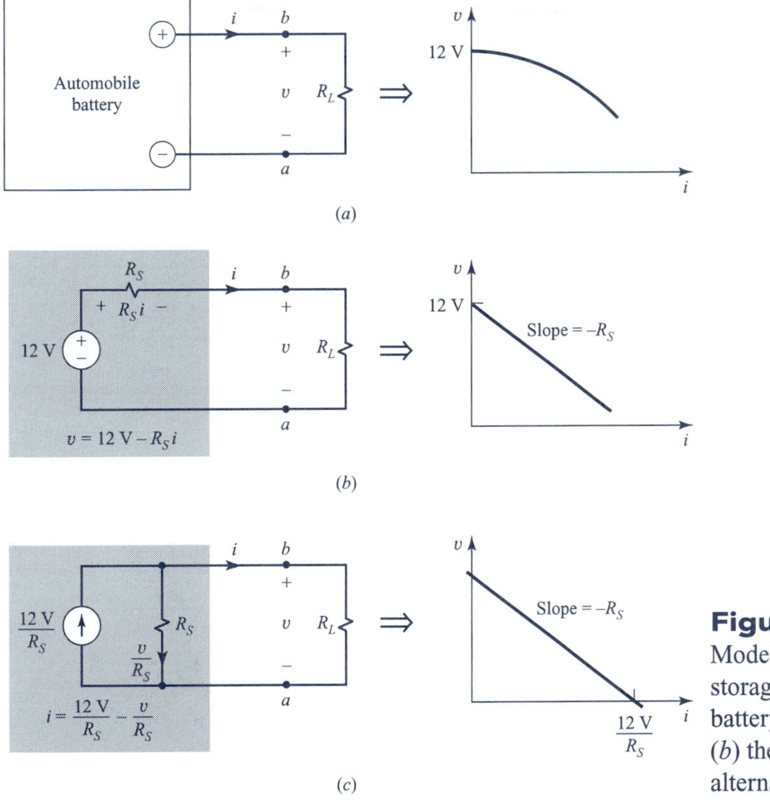

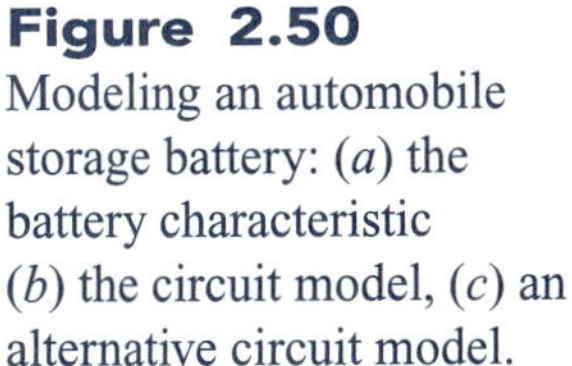

Figure 2.50 Modeling an automobile storage battery: (*a*) the battery characteristic (*b*) the circuit model, (*c*) an alternative circuit model.

voltage is 12 V. When current is being drawn from its terminals as in starting the vehicle, this terminal voltage drops to a value below 12 V. Modeling this battery as a 12-V independent voltage in series with a resistor R_S as shown in Fig. 2.50*b* simulates the terminal characteristic of the actual battery quite well. The output current i passing through the resistance R_S generates a voltage across that resistance of $R_S i$, which subtracts from the 12-V voltage of the source, resulting in lowering of the output voltage as current increases as is observed in the actual battery. Hence we say that this combination of an independent source in series with a resistor gives a reasonable *model* of the battery. Converting this to an equivalent representation of an independent current source of value $(12\text{ V})/R_S$ in parallel with the same resistance with a source transformation, as shown in Fig. 2.50*c*, gives an identical output characteristic to that of Fig. 2.50*b*. Therefore either configuration would be a good approximation of the actual source (the automobile battery) and hence the origin of the name *source transformation*.

The primary use of this transformation is to convert a circuit to a single-loop circuit or a single-node-pair circuit when no such circuit existed before the transformation. The following example illustrates this use.

Example 2.29

Determine the currents i and i_x in the circuit of Fig. 2.51*a* using a source transformation. Also solve this circuit using the direct method.

Solution Apply a source transformation to transform the 5-A current source that is in parallel with the 3-Ω resistor to a 15-V voltage source that is in series with the 3-Ω resistor as shown in Fig. 2.51*b*. Whenever making a source transformation it is highly recommended to mark the terminals where the transformation is made with small ×'s as is done

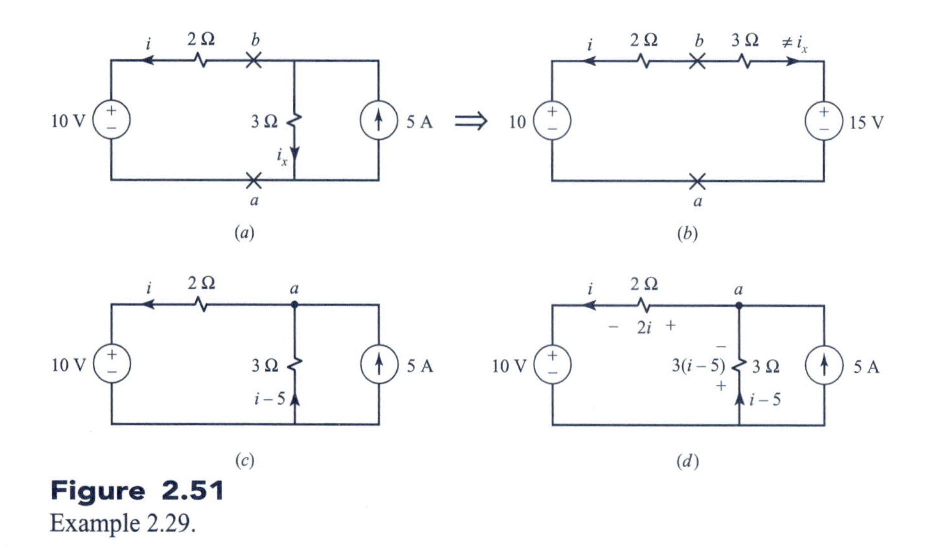

Figure 2.51
Example 2.29.

in the figure. Now the circuit of Fig. 2.51*b* is a single-loop circuit whose solution is easy to obtain as

$$i = \frac{15\text{ V} - 10\text{ V}}{2\ \Omega + 3\ \Omega}$$
$$= 1\text{ A}$$

A frequent and serious error is often made in applying source transformations and is illustrated in Fig. 2.51*b*: *the current through R_S does not remain with R_S in a source transformation*. Hence we have lost the current i_x in applying the transformation. But this is not a problem, since we can determine it from the original circuit in terms of i using KCL as

$$i_x = 5\text{ A} - i$$
$$= 4\text{ A}$$

An alternative method is the direct method. Applying KCL at the only nonsimple node, the top node, yields the current through the 3-Ω resistor as shown in Fig. 2.51*c*. Applying Ohm's law to express the voltages across the two resistors in terms of i yields the circuit of Fig. 2.51*d*. We have exhausted KCL and Ohm's law, and the only remaining law is KVL. Applying KVL around the only nonsimple loop of that circuit yields

$$10\text{ V} = -2i - 3(i - 5)$$

Solving this gives

$$i = 1\text{ A}$$

once again. The current i_x is

$$i_x = -(i - 5)$$
$$= 4\text{ A}$$

Example 2.30

Determine current I and voltage V in the circuit of Fig. 2.47 of Example 2.28 using source transformations.

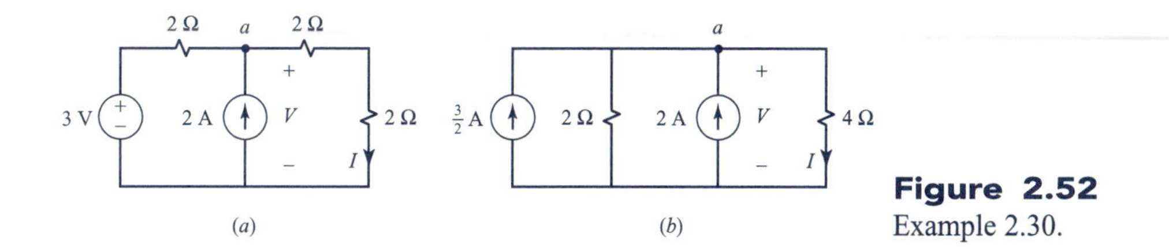

Figure 2.52
Example 2.30.

Solution Transform the 3-V source in series with the 2-Ω resistor using a source transformation as shown in Fig. 2.52*b*. This gives a single-node-pair circuit from which we determine

$$V = \left(\tfrac{3}{2}\,\text{A} + 2\,\text{A}\right) \times \underbrace{2\,\Omega \| 4\,\Omega}_{\frac{4}{3}\,\Omega}$$
$$= \tfrac{14}{3}\,\text{V}$$

as obtained in Example 2.28. From this we obtain, using Ohm's law,

$$I = \frac{\text{V}}{4\,\Omega}$$
$$= \tfrac{7}{6}\,\text{A}$$

Example 2.31

Determine voltage V in the circuit of Fig. 2.46 that was solved in Example 2.27.

Solution Convert the series connection of the 3-V source in series with the 2-Ω resistor using a source transformation as shown in Fig. 2.53*b*. Observe that the voltage V across the 2-Ω resistor in that combination is lost in the transformation. Finally add the current sources and combine the 2-Ω resistors to yield the circuit of Fig. 2.53*c*, from which we obtain, by current division,

$$i = \frac{1\,\Omega}{1\,\Omega + 1\,\Omega}\frac{5}{2}\text{A}$$
$$= \tfrac{5}{4}\,\text{A}$$

From this and the original circuit we obtain V using KVL as

$$V = 1\,\Omega \times i + 3\,\text{V}$$
$$= \tfrac{17}{4}\,\text{V}$$

as obtained in Example 2.27.

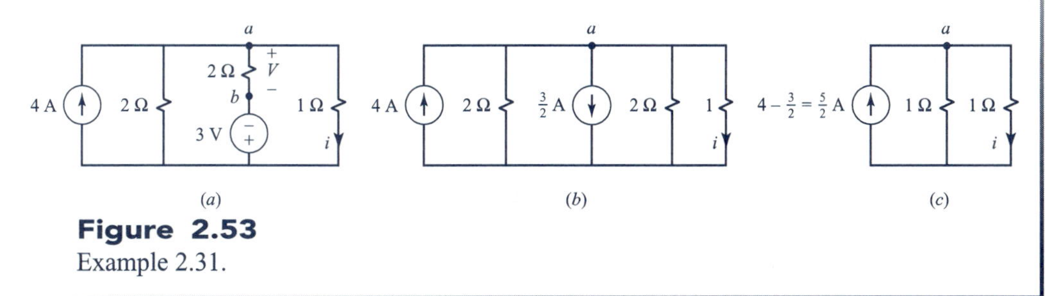

Figure 2.53
Example 2.31.

Exercise Problem 2.26

Determine V and I in the circuit of Fig. E2.26 by using source transformations.

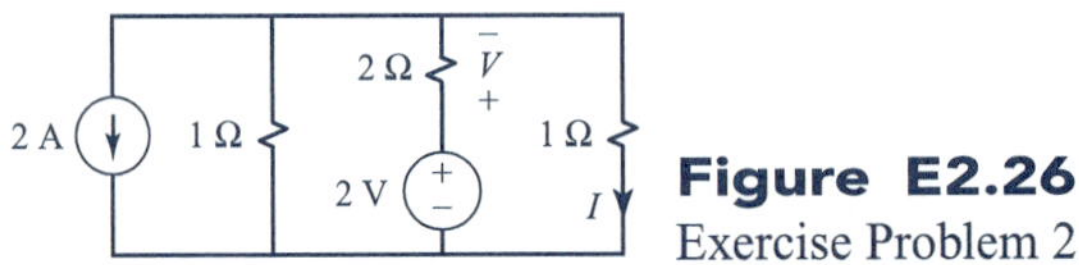

Figure E2.26
Exercise Problem 2.26.

Answers: $\frac{12}{5}$ V and $-\frac{2}{5}$ A.

Exercise Problem 2.27

Determine V and I in the circuit of Fig. E2.27 by using source transformations.

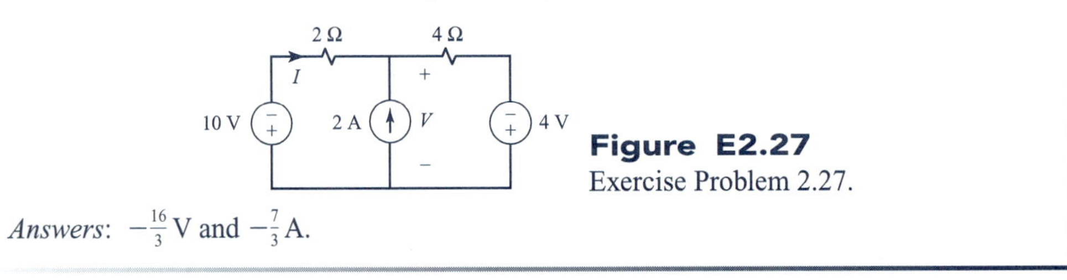

Figure E2.27
Exercise Problem 2.27.

Answers: $-\frac{16}{3}$ V and $-\frac{7}{3}$ A.

Exercise Problem 2.28

Determine V and I in the circuit of Fig. E2.28 by using source transformations.

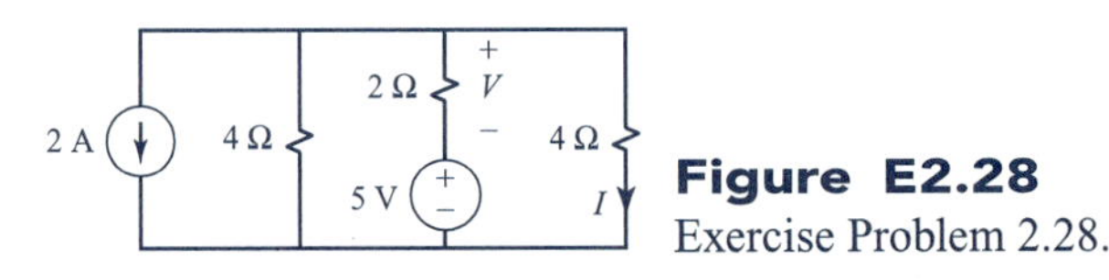

Figure E2.28
Exercise Problem 2.28.

Answers: $-\frac{9}{2}$ V and $\frac{1}{8}$ A.

2.8 The Controlled (Dependent) Voltage and Current Sources

The independent sources considered previously constrained either the element voltage or current to be a known value. The other element variable is, as yet, unknown, but will be determined once the source is attached to a circuit.

On the other hand, controlled or dependent sources, considered in this section, relate the element voltage or current to the voltage or current *associated with another element*. The four types of controlled sources are shown in Fig. 2.54. Observe that the symbols for controlled sources are diamonds as opposed to circles for independent sources. The type of source is once again shown by either + and − signs or an arrow within the diamond. The polarity of the source is determined by the symbols inside the diamonds. The values of the sources are given as a constant (whose value is given) multiplied by the controlling variable.

Figure 2.54*a* shows a voltage-controlled voltage source (VCVS). We will sometimes refer to the elements as *branches*. The branch containing the source is referred to as the *controlled branch,* and the element having the controlling variable is referred to as the *controlling branch*. For the VCVS in Fig. 2.54*a*, the controlling variable is the voltage, v_x, across the controlling

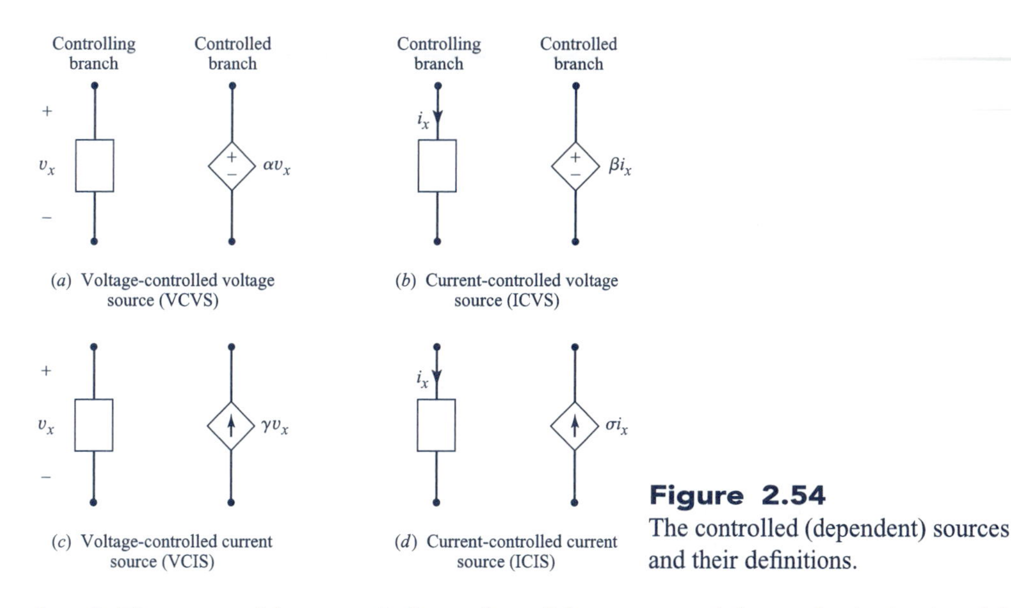

Figure 2.54
The controlled (dependent) sources and their definitions.

branch. The output of the source is the product of the constant α (whose value is given) and the controlling variable v_x. This output is known *only* when the controlling voltage v_x is known.

Figure 2.54*b* shows a current-controlled voltage source (ICVS). This is similar to the VCVS, but the controlling variable is the current through the controlling branch, i_x, and the constant β is given. A voltage-controlled current source (VCIS) is shown in Fig. 2.54*c*. The output of this source is a current, $\gamma\, v_x$, and the controlling variable is the voltage across the controlling branch. And finally, the current-controlled current source (ICIS) is shown in Fig. 2.54*d*.

There are three items that must be specified in order to completely specify any controlled source:

1. *the polarity of the source (+ and − signs or an arrow within the diamond),*
2. *the source constant α, β, γ, σ, and*
3. *the controlling variable, by completely specifying the controlling branch and the polarity of the controlling variable on that branch.*

Failure to specify any one of these three items leaves the source not completely specified.

Controlled sources are used primarily to model *electronic devices*. For example, Fig. 2.55*a* shows the model of a junction field-effect transistor (JFET), which is a very common element

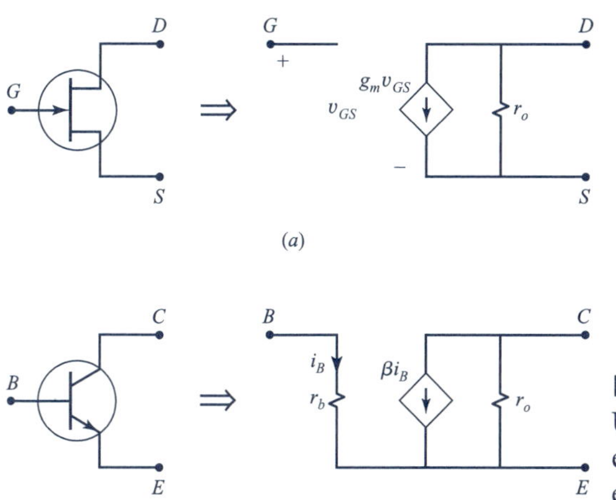

Figure 2.55
Use of controlled sources to model electronic devices: (*a*) the junction field effect transistor (JFET) and (*b*) the bipolar junction transistor (BJT).

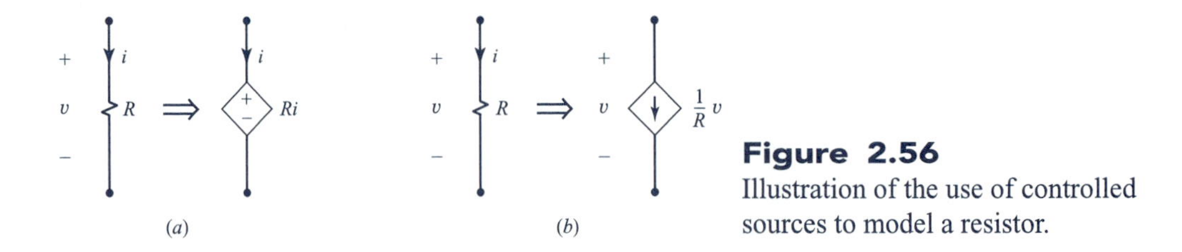

Figure 2.56
Illustration of the use of controlled sources to model a resistor.

used to construct electronic amplifiers and digital computers. Figure 2.55*b* shows another common electronic element, the bipolar junction transistor (BJT), and its model. This element is also used to construct electronic amplifiers and digital computers. Without controlled sources we would not be able to model these important electrical components. Figure 2.56 shows that we can also model resistors with controlled sources where the controlling variable is in the controlled branch. Hence we could construct all of our previous circuits using only independent and controlled sources. This is not done, however, since it would unnecessarily obscure the solution.

The location of the controlling variable *and* its polarity are crucial elements of its specification. The following example shows how this can affect the circuit solution.

Example 2.32

Determine the value of i_x in the circuit of Fig. 2.57.

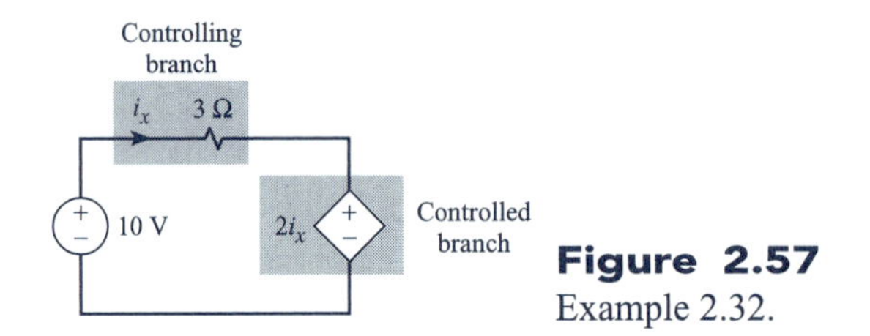

Figure 2.57
Example 2.32.

Solution Using Ohm's law to find the voltage across the 3-Ω resistor and applying KVL yields

$$10\text{ V} = 3\ \Omega \times i_x + 2i_x$$

Solving this gives $i_x = 2$ A. Suppose we reverse the direction of the controlling variable, i_x, perhaps through a labeling error. In that case we obtain $10\text{ V} = -3\ \Omega \times i_x + 2i_x$ with the result that $i_x = -10$ A. Clearly the polarity of the controlling variable is a critical part of the definition of the controlled source.

Example 2.33

Determine the current i_x in the circuit of Fig. 2.58.

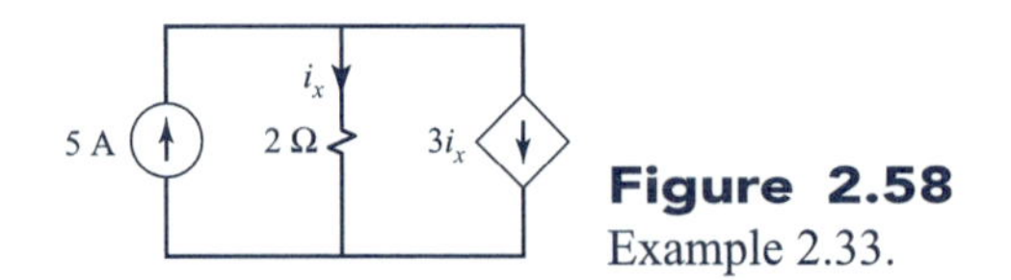

Figure 2.58
Example 2.33.

Solution Applying KCL at the upper node gives

$$5\text{ A} = i_x + 3i_x$$

Solving gives $i_x = \frac{5}{4}$ A.

Exercise Problem 2.29

Determine the voltage v_x in the circuit of Fig. E2.29.

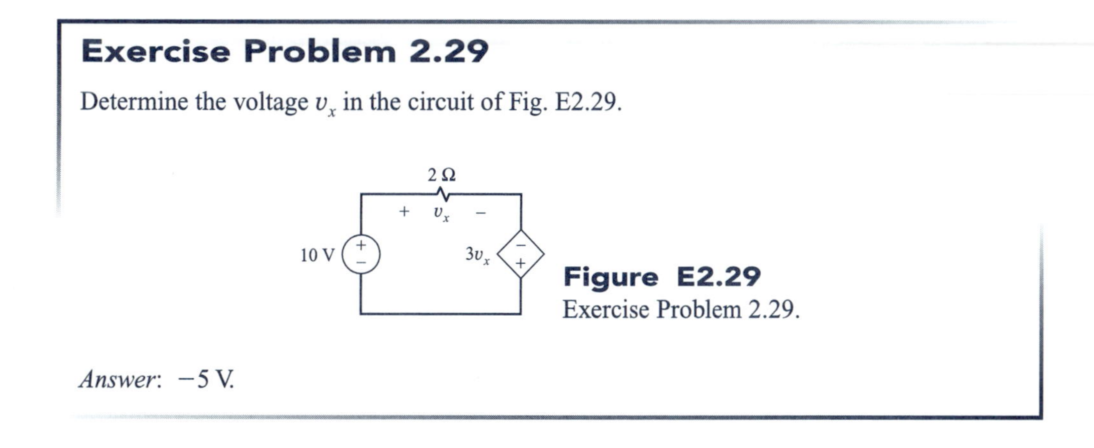

Figure E2.29
Exercise Problem 2.29.

Answer: -5 V.

Exercise Problem 2.30

Determine current i_x in the circuit of Fig. E2.30.

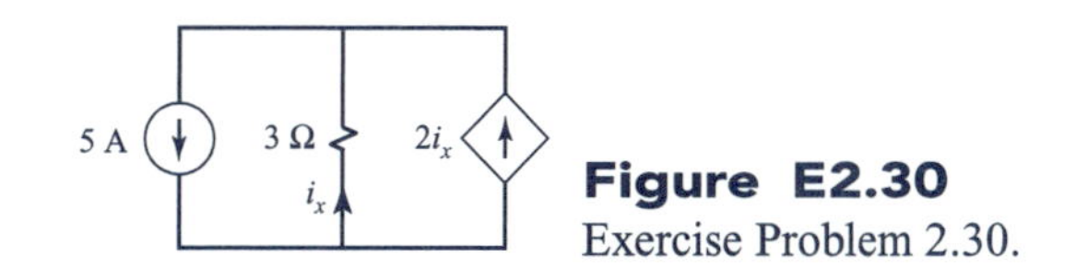

Figure E2.30
Exercise Problem 2.30.

Answer: $\frac{5}{3}$ A.

2.8.1 Analysis of Circuits Containing Controlled Sources

Many of our previous solution techniques can be used in circuits containing controlled sources, but we much exercise a bit more care with those circuits. The following two examples show the use of source transformations (and a common mistake) as well as the application of the direct method in solving circuits containing controlled sources.

Example 2.34

Determine the voltage v_x in the circuit of Fig. 2.59*a*.

Solution The 10-V source in series with the 2-Ω resistor can be converted using a source transformation as shown in Fig. 2.59*b*. This is a single-node-pair circuit, whose solution is

$$v_x = (2\,\Omega \,\|\, 2\,\Omega) \times (5\text{ A} + 3v_x)$$

Solving this gives

$$v_x = -\tfrac{5}{2}\text{ V}$$

Controlled sources can also be transformed using source transformations in the same way as independent sources. Figure 2.59*c* illustrates this. It is important to note that the controlling variable *does not remain with the resistor that it is associated with in the original circuit.* Note in Fig. 2.59*c* that v_x is still between node *a* and the lower node, as it should be, and does not go with the 2-Ω resistor that was in parallel with the controlled source. In order to solve

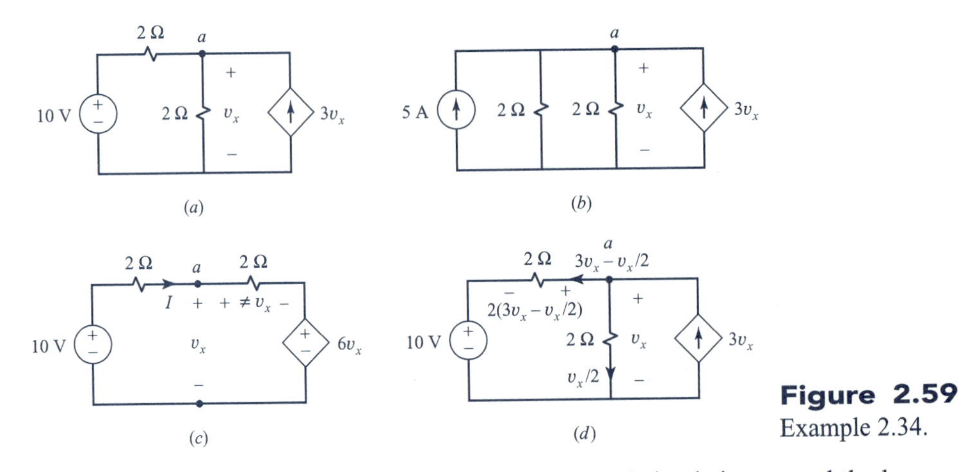

Figure 2.59 Example 2.34.

the single-loop circuit of Fig. 2.59*c* we define a current I circulating around the loop and determine

$$I = \frac{10\text{ V} - 6v_x}{2\ \Omega + 2\ \Omega}$$

But this gives only one equation in the two unknowns. The second equation can be obtained by using KVL to write v_x in terms of I:

$$v_x = 10\text{ V} - 2\ \Omega \times I$$

Substituting this into the previous equation gives

$$v_x = -\tfrac{5}{2}\text{ V}, \qquad I = \tfrac{25}{4}\text{ A}$$

as before. Figure 2.59*d* shows the application of the direct method. Ohm's law gives the current down through one 2-Ω resistor, and KCL at node *a* gives the current through the other 2-Ω resistor, with the subsequent application of Ohm's law giving its voltage. This completely uses Ohm's law and KCL. Applying KVL around the nonsimple loop gives

$$10\text{ V} = -2\ \Omega \times \left(3v_x - \frac{v_x}{2\ \Omega}\right) + v_x$$

Solving this once again gives

$$v_x = -\tfrac{5}{2}\text{ V}$$

Example 2.35

Determine the current i_x in the circuit of Fig. 2.60*a*.

Solution Figure 2.60*b* shows applying a source transformation to the 10-V source. Observe that the controlling variable, i_x, does not follow the 1-Ω resistor, since that current is directed into node *a*. However, since this is a single-node-pair circuit, we can define a voltage V and determine it as

$$V = (1\ \Omega \,\|\, 2\ \Omega) \times (10\text{ A} + 2i_x)$$

We need another equation in these two unknowns, V and i_x. The current down through the 2-Ω resistor is $V/2$, so that using KCL at node *a* we obtain

$$i_x = \frac{V}{2\ \Omega} - 2i_x$$

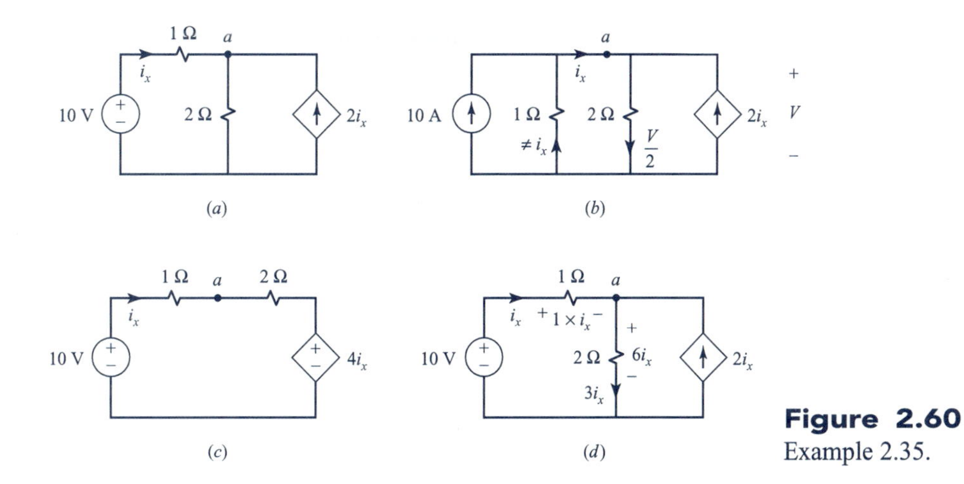

Figure 2.60
Example 2.35.

Solving this and the previous equation gives

$$i_x = \tfrac{10}{7}\text{ A}, \qquad V = \tfrac{60}{7}\text{ V}$$

As an alternative solution let us make a source transformation involving the controlled source as shown in Fig. 2.60*c*. This is a single-loop circuit, so that

$$i_x = \frac{10\text{ V} - 4i_x}{1\ \Omega + 2\ \Omega}$$

yielding $i_x = \frac{10}{7}$ A once again. Figure 2.60*d* shows the application of the direct method. Applying KCL at node *a* gives the current down through the 2-Ω resistor in terms of i_x. Using Ohm's law to determine the voltages across the 1-Ω and 2-Ω resistors and then using KVL around the nonsimple loop gives

$$10\text{ V} = 1\ \Omega \times i_x + 6i_x$$

Solving this gives, once again, $i_x = \frac{10}{7}$ A.

Example 2.36

Determine the current I in the circuit of Fig. 2.61*a*.

Solution We apply the direct method of solution. Label the voltage across the 3-Ω resistor using Ohm's law. Next apply KCL at the upper node to obtain the current I as shown in Fig. 2.61*b* in terms of the controlled-source controlling variable. Label the voltage across the 2-Ω resistor in terms of this current using Ohm's law. We have now exhausted Ohm's law

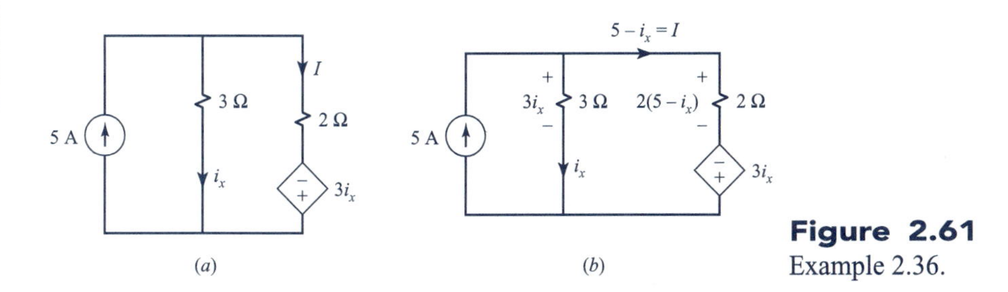

Figure 2.61
Example 2.36.

and KCL and are left with only KVL. Applying KVL around the loop containing the controlled source and the 2-Ω and 3-Ω resistors gives

$$3\ \Omega \times i_x = 2\ \Omega \times (5 - i_x) - 3i_x$$

Solving this gives $i_x = \frac{5}{4}$ A. Hence the current I is

$$I = 5 - i_x$$
$$= \frac{15}{4}\text{ A}$$

Exercise Problem 2.31

Determine voltage v in Fig. E2.31.

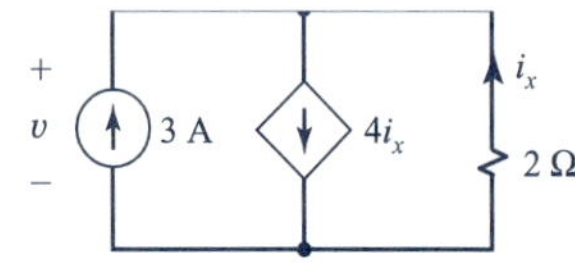

Figure E2.31
Exercise Problem 2.31.

Answer: −2 V.

Exercise Problem 2.32

Determine i_x and v_y in the circuit of Fig. E2.32.

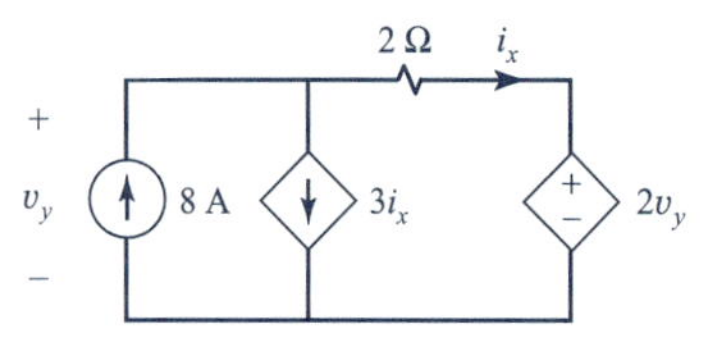

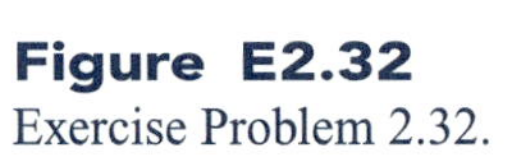
Figure E2.32
Exercise Problem 2.32.

Answers: 2 A and −4 V.

Exercise Problem 2.33

Determine v_x in the circuit of Fig. E2.33.

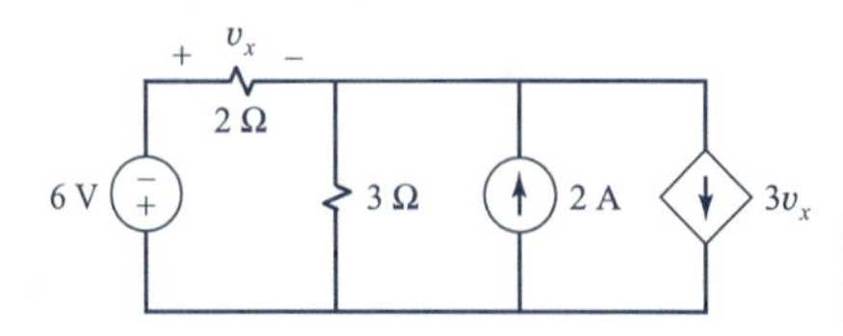

Figure E2.33
Exercise Problem 2.33.

Answer: $\frac{24}{13}$ V.

2.8.2 Equivalent Resistances of Circuits Containing Controlled Sources We will frequently need to determine the equivalent resistance at two terminals of a circuit that contains controlled sources. We cannot simply combine controlled sources using the series–parallel reduction rules developed for resistors. A simple, yet foolproof

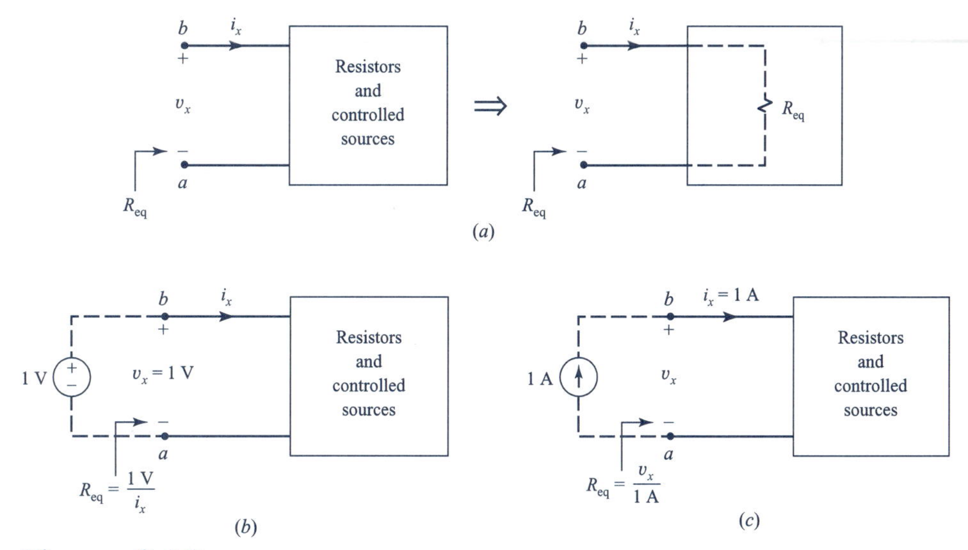

Figure 2.62
Illustration of the determination of equivalent resistances of circuits that contain controlled sources. (*b*) Apply a 1-V source, and compute the current i_x so that $R_{eq} = 1\text{ V}/i_x$. (*c*) Apply a 1-A source and compute the voltage v_x so that $R_{eq} = v_x/1\text{ A}$.

way of determining those equivalent resistances is illustrated in Fig. 2.62. The fundamental definition of equivalence is that the v–i relation at the two terminals of the two circuits must be the same in order for them to be equivalent. Hence define voltage v_x and current i_x at the two terminals *according to the passive sign convention*. It is important to understand that the passive sign convention applies to the terminals of the equivalent resistance within the box: *the current* i_x *enters the box at the terminal associated with the positive terminal of* v_x. (A frequent and *serious* mistake is to define the voltage and associated current contrary to the passive sign convention.) The equivalent resistance is then defined as the ratio of the voltage v_x and current i_x:

$$R_{eq} = \frac{v_x}{i_x} \tag{2.19}$$

Observe that the numerical values of v_x and i_x are unimportant: only their ratio matters. Figure 2.62*b* shows a simple way to facilitate this determination: constrain voltage v_x to be 1 V (or any other value) with an independent voltage source, and determine the resulting value of i_x. The equivalent resistance is then

$$R_{eq} = \frac{v_x = 1\text{V}}{i_x} \tag{2.20}$$

Figure 2.62*c* shows an alternative way: constrain current i_x to be 1 A (or any other value) with an independent current source, and determine the resulting value of v_x. The equivalent resistance is then

$$R_{eq} = \frac{v_x}{i_x = 1\text{A}} \tag{2.21}$$

The choice of whether to constrain v_x or i_x is somewhat arbitrary in that either will lead to the same result. However, if one constraint will pin down a controlling variable of a controlled source, that will be the best choice. The following examples illustrate that point.

Example 2.37

Determine the equivalent resistance at terminals *ab* for the circuit of Fig. 2.63*a*.

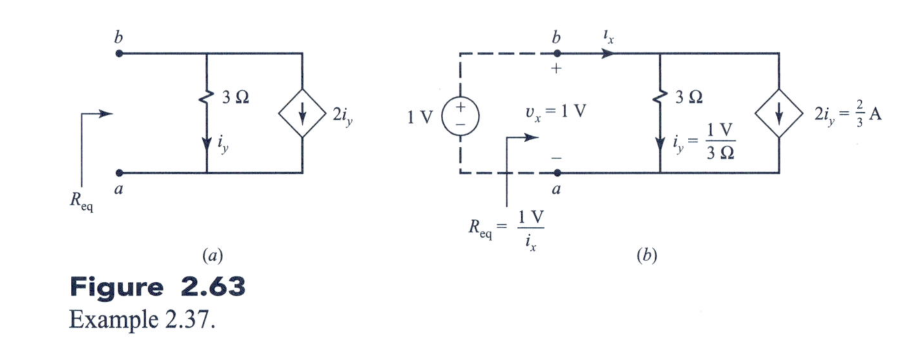

Figure 2.63
Example 2.37.

Solution If we apply a 1-V voltage source across terminals *ab*, that will constrain the current through the 3-Ω resistor, which is the controlling variable for the controlled source. Hence we apply a 1-V voltage source across the terminals as shown in Fig. 2.63*b*. This constrains the controlling variable to be

$$\begin{aligned} i_y &= \frac{1\text{ V}}{3\ \Omega} \\ &= \tfrac{1}{3}\text{ A} \end{aligned}$$

Hence, the value of the controlled source is

$$2i_y = \tfrac{2}{3}\text{ A}$$

The current i_x then becomes, with KCL applied at node *b*,

$$\begin{aligned} i_x &= i_y + 2i_y \\ &= 1\text{ A} \end{aligned}$$

Hence the equivalent resistance is

$$\begin{aligned} R_{\text{eq}} &= \frac{v_x = 1\text{ V}}{i_x} \\ &= 1\ \Omega \end{aligned}$$

Example 2.38

Determine the equivalent resistance at terminals *ab* for the circuit of Fig. 2.64*a*.

Solution Applying a 1-A current source to the terminals, we constrain the current i_y, which is the controlling variable for the controlled source. This is shown in Fig. 2.64*b*. Writing KVL around the loop gives

$$v_x = 1\text{ A} \times 1\ \Omega + 2i_y$$

But

$$i_x = -i_y = 1\text{ A}$$

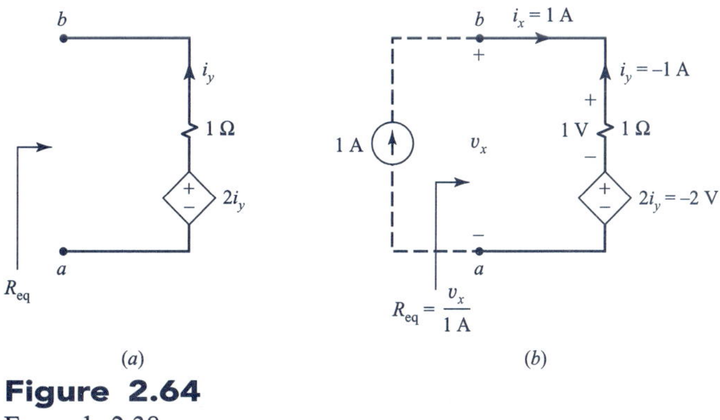

Figure 2.64
Example 2.38.

and hence

$$v_x = -1\text{ V}$$

Therefore the equivalent resistance is

$$\begin{aligned} R_{eq} &= \frac{v_x}{i_x = 1\text{ A}} \\ &= -1\ \Omega \end{aligned}$$

Since we have been careful to define v_x and i_x with the passive sign convention, we have a true negative resistance!

Example 2.39

Determine the equivalent resistance at terminals *ab* for the circuit of Fig. 2.65*a*.

Solution Applying a 1-A current source so that i_x will be constrained to $i_x = 1$ A as shown in Fig. 2.65*b* will constrain one of the controlling variables:

$$i_y = i_x = 1\text{ A}$$

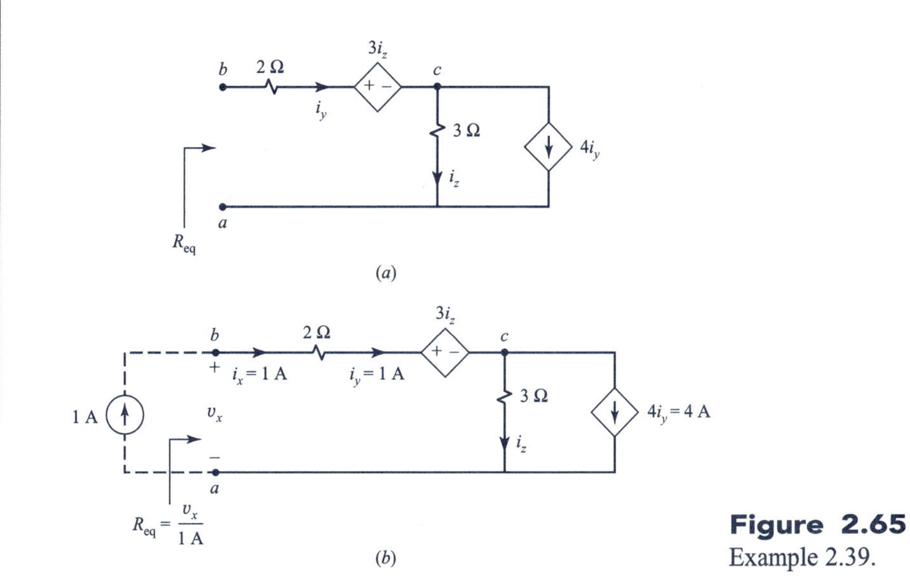

Figure 2.65
Example 2.39.

Applying KCL at node c gives

$$i_z + 4i_y = i_y = 1\text{ A}$$

Hence

$$i_z = -3\text{ A}$$

Applying KVL around the loop gives

$$\begin{aligned} v_x &= 2\ \Omega \times i_y + 3i_z + 3\ \Omega \times i_z \\ &= -16\text{ V} \end{aligned}$$

Thus the equivalent resistance becomes

$$\begin{aligned} R_{\text{eq}} &= \frac{v_x}{i_x = 1\text{ A}} \\ &= -16\ \Omega \end{aligned}$$

Once again we have a negative resistance.

Exercise Problem 2.34

Determine the equivalent resistance at terminals ab for the circuit shown in Fig. E2.34.

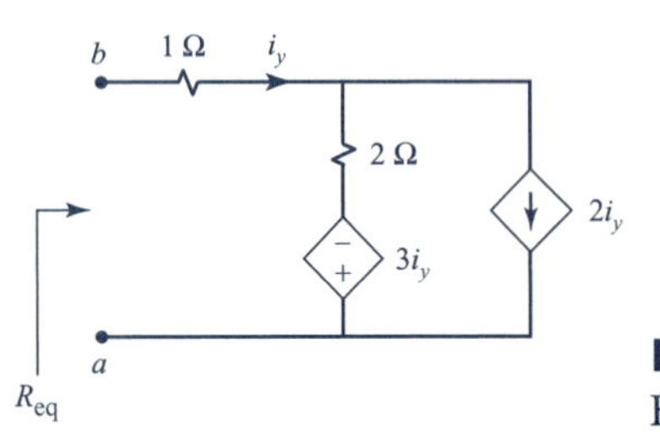

Figure E2.34
Exercise Problem 2.34.

Answer: $-4\ \Omega$.

Exercise Problem 2.35

Determine the equivalent resistance at terminals ab for the circuit shown in Fig. E2.35.

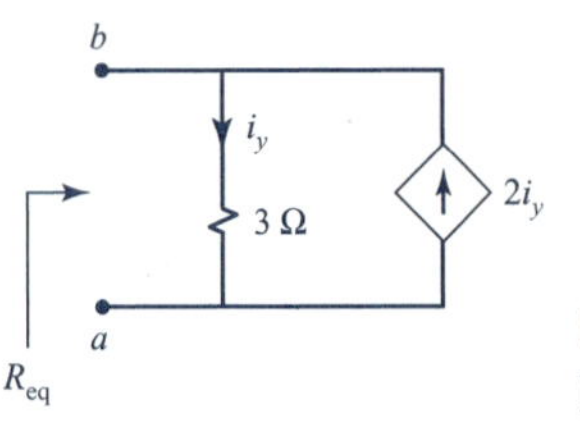

Figure E2.35
Exercise Problem 2.35.

Answer: $-3\ \Omega$.

Exercise Problem 2.36

Determine the equivalent resistance at terminals ab for the circuit shown in Fig. E2.36.

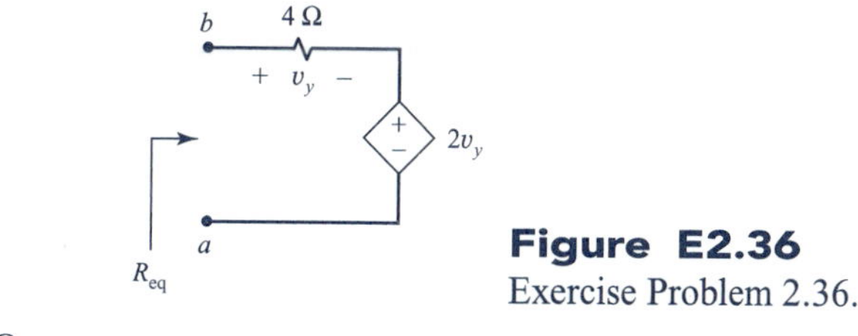

Figure E2.36
Exercise Problem 2.36.

Answer: 12 Ω.

Other analysis methods developed previously for circuits not containing controlled sources can be similarly used for circuits that contain these sources.

Example 2.40

Determine the voltages v_x and v_y in the circuit of Fig. 2.66, using voltage division where appropriate.

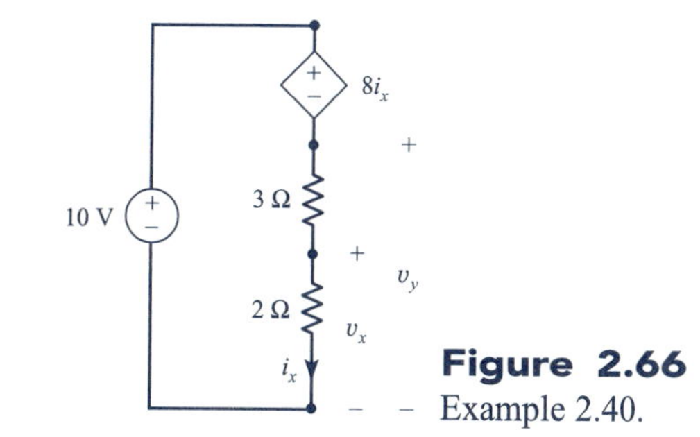

Figure 2.66
Example 2.40.

Solution This is a single loop circuit, so we can determine current i_x as

$$i_x = \frac{10\text{ V} - 8i_x}{3\ \Omega + 2\ \Omega}$$

Solving this for i_x yields

$$i_x = \tfrac{10}{13}\text{ A}$$

Hence the voltage v_y is

$$\begin{aligned} v_y &= 10\text{ V} - 8i_x \\ &= \tfrac{50}{13}\text{ V} \end{aligned}$$

Now by voltage division we can obtain

$$\begin{aligned} v_x &= \frac{2\ \Omega}{3\ \Omega + 2\ \Omega} v_y \\ &= \tfrac{20}{13}\text{ V} \end{aligned}$$

2.9 PSPICE Applications

Throughout this text we will introduce the reader to two computer programs, PSPICE and MATLAB, that are useful in providing circuit solutions as an alternative to and as a check of hand calculations. It is important that the reader have the proper perspective about these computer programs. They should not be viewed as a substitute for the hand calculation skills that we have developed and will continue to develop. Facility with these computer codes in the analysis of circuits can be developed in a very short time, but little or no insight into the circuit behavior will be obtained. Hence the reader will only have an ability to *analyze* a circuit. The more important role of an engineer is in the *design* of circuits to perform a desired task. The ability to use these computer-aided analysis programs will not provide that skill; hand calculations provide the insight that is crucial to developing circuit design.

The SPICE program was originally developed at the University of California at Berkeley in the mid 1970s for use on large mainframe computers. (SPICE is an acronym for Simulation Program with Integrated Circuit Emphasis.) The PSPICE program is the personal-computer version of SPICE and essentially has the SPICE program as its kernel. It was developed by the MicroSim Corporation, which was acquired by the OrCAD Corporation. A student version is available at no cost from the World Wide Web at http://www.orcad.com. A CD containing the student version of PSPICE can also be obtained from: OrCAD Inc. by down loading from the website or calling (800) 671-9505. All problems in this text can be solved with the student (educational) version of PSPICE. SPICE, together with its many variations such as PSPICE, has become the industry standard for the analysis of circuits and particularly electronic circuits.

The user writes a SPICE "program" that essentially describes, in the SPICE language, the types of elements in the circuit and their connection. The user is actually not using a programming language such as FORTRAN or BASIC, but is simply following rules and conventions that the developers of SPICE set up for describing the circuit to be analyzed. Since we will be using the PSPICE program in this text, we will describe its commands, which contain the original SPICE commands as a subset. Every PSPICE program consists of four parts: (a) a title statement (which must be present at the beginning of the program), (b) a set of circuit description statements, (c) a set of control statements, and (d) the .END statement (which must be at the end of the program):

Title statement

{Circuit description statements}

{Control statements}

.END

With the exception of the title statement and the .END statement, the statements can appear in any order.

PSPICE essentially writes the *node-voltage equations* from the user-supplied information and solves them to determine the element voltages and currents. We will discuss these node-voltage equations in Chapter 3, but for now it is only important to understand how they are defined so that we can interpret the output. Each element of the circuit has its nodes labeled by the user as shown in Fig. 2.67. In Fig. 2.67 the nodes at the ends of the element are labeled as N1 and N2. Usually these are given numerical values, but it is also acceptable to give them literal names, e.g., BOB. Every circuit must have a *reference node* denoted as 0. It is arbitrarily selected by the user. PSPICE solves for the node voltages, V(N1) and V(N2), that are positive with respect to the reference node. The voltage across the element can be obtained using V(N1,N2), which equals the difference V(N1)-V(N2).

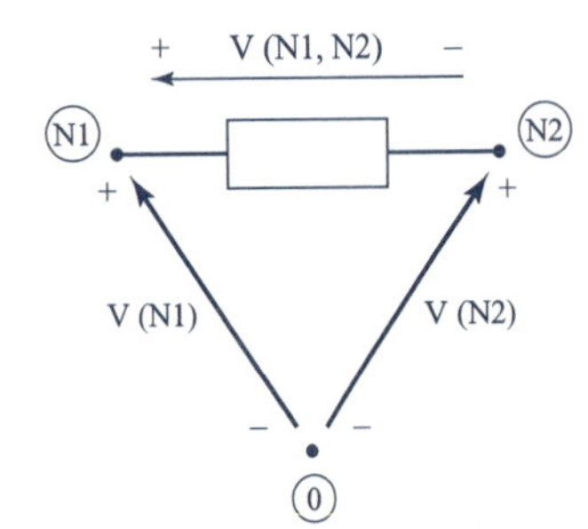

Figure 2.67
Definition of the node voltages used in SPICE and PSPICE.

Each branch or box in Fig. 2.67 contains one and only one of the following elements:

R	Resistor
V	Independent voltage source
I	Independent current source
G	Voltage-controlled current source
E	Voltage-controlled voltage source
F	Current-controlled current source
H	Current-controlled voltage source

The element description statement has three *fields*, which are separated by one or more blank spaces: (a) the element name which begins with one of the above letters, (b) the initial and final nodes at the ends of the element, and (c) the element value.

With reference to Fig. 2.68*a*, the resistor is described by

$$\underbrace{\text{RXXX}}_{\text{element name}}\quad\underbrace{\text{N1}}_{\text{initial node}}\quad\underbrace{\text{N2}}_{\text{final node}}\quad\underbrace{r}_{\text{element value}}$$

The element name must start with the letter R to inform PSPICE that this is a resistor, and the remainder XXX of the name can be anything the users wishes, e.g., RSILLY. The choice of which node is to be the initial (first-named) node is somewhat arbitrary. The current through the resistor can be requested by writing in the control statements I(RXXX), in which case this current is assumed to flow *from* the first-named (initial) node N1 *to* the last named (final) node N2.

With reference to Fig. 2.68*b*, the independent voltage source is described by

$$\underbrace{\text{VXXX}}_{\text{element name}}\quad\underbrace{\text{N1}}_{\text{initial node}}\quad\underbrace{\text{N2}}_{\text{final node}}\quad\text{DC}\quad\underbrace{v_S}_{\text{element value}}$$

The ordering of the nodes in this statement is critical; the first-named (initial) node, N1, must be on the positive end of the source; otherwise the value of the source must be entered as a negative number. Inclusion of the item DC tells PSPICE that this is a dc source. Other waveforms can be specified, and these will be discussed as needed.

With reference to Fig. 2.68*c*, the independent current source is described by

$$\underbrace{\text{IXXX}}_{\text{element name}}\quad\underbrace{\text{N1}}_{\text{initial node}}\quad\underbrace{\text{N2}}_{\text{final node}}\quad\text{DC}\quad\underbrace{i_S}_{\text{element value}}$$

Again, the ordering of the nodes is critical; the current is assumed to flow *from* the first-named (initial) node, N1, *to* the last-named (final) node, N2; otherwise the value of the source must be entered with a negative number. Examples of these are shown in Fig. 2.69.

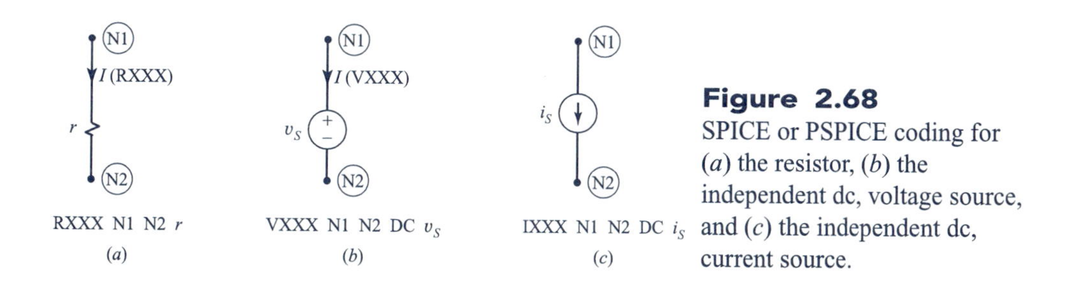

Figure 2.68 SPICE or PSPICE coding for (*a*) the resistor, (*b*) the independent dc, voltage source, and (*c*) the independent dc, current source.

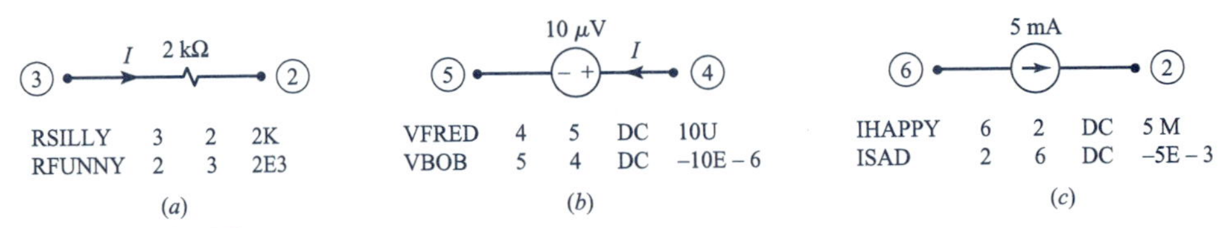

Figure 2.69
Examples of SPICE or PSPICE coding of elements.

Power-of-ten multipliers can be used in specifying the element values. These are

T = 1E12 (tera)	K = 1E3 (kilo)	N= 1E-9 (nano)
G = 1E9 (giga)	M = 1E-3 (milli)	P = 1E-12 (pico)
MEG = 1E6 (mega)	U = 1E-6 (micro)	F = 1E-15 (femto)

Specification of the resistor in Fig. 2.69*a* can be accomplished in several forms:

```
(1)   RSILLY   3   2   2K
(2)   RFUNNY   2   3   2E3
```

The sign of the solution for the current through this resistor will depend on the ordering of the nodes:

I = I(RSILLY)
I = -I(RFUNNY)

Similarly, specification of the independent voltage source can be of several forms as shown in Fig. 2.69*b*:

```
(1)   VFRED   4   5   DC   10U
(2)   VBOB    5   4   DC   -10E-6
```

Observe that the ordering of the nodes determines the sign of the value of the voltage source. It also determines the sign of the solution for the current through the source (which is assumed to flow through the source from the first-named node to the last-named node):

I = I(VFRED)
I = -I(VBOB)

And finally, specification of the independent current source can take on two forms as illustrated in Fig. 2.69*c*:

```
(1)   IHAPPY   6   2   DC   5M
(2)   ISAD     2   6   DC   -5E-3
```

Next we describe the control statements. The control statements accomplish two goals: (a) they specify how the solution is to be performed (for example, the values of a source that is to be varied) and (b) the desired output or solution variables. The first of these is the .DC, statement which has the form

```
.DC   SNAME   VSTART   VSTOP   VINCR
```

This allows us to sweep the values of an independent source in the circuit and have PSPICE recompute the solutions for each value without having to rerun the program for each of these values. The value of independent source (voltage or current) SNAME will be set at VSTART and the circuit solved, then the value will be set at VSTART+VINCR and the circuit re-solved, and so forth, until the value exceeds VSTOP, at which point the solution will be terminated. We can then request the solution for any circuit variable at each of these values of SNAME. This is a nice design feature, since we might want to know the value of a voltage source that will produce a certain solution. We can vary the source value and eventually home in on the value of this source that will give the desired solution.

However, in this text we will be interested in the circuit solution for fixed values of the independent sources. Hence we select one source in the circuit to be analyzed (any source) and write, for example,

```
.DC   VFRED   10U   10U   1
```

This statement will set the voltage source VFRED at 10 μV, and only one solution will be made, since the next value, 10 μV + 1 V, exceeds 10 μV.

The alternative to specifying the .DC statement is the .OP statement:

```
.OP
```

This will solve the circuit, but all node voltage solutions will be printed out. This can give a large amount of unnecessary output, and usually we are interested only in a few solution variables. Hence the .DC statement, which gives more flexibility in output, is the preferred form.

The second and final statement in the control group describes the desired output or solution variables. The form of this statement is

```
.PRINT   DC   OV1   OV2   OV3   ···
```

where OVX is the desired solution, e.g., V(3) (the voltage of node 3 with respect to the reference node), V(2,4) (the voltage of node 2 minus the voltage of node 4), or I(RSILLY) (the current through resistor RSILLY from the first-named node to the last-named node).

Comments can be placed anywhere in the program by preceeding them with a star (*):

```
*THIS IS NO FUN
```

These will not be processed by the program but will be printed in the output.

There are two methods of preparing and executing the PSPICE program. The first is the so-called DOS-based or direct-input method of directly preparing the program as described below, using a standard ASCII editor (which is provided with the PSPICE program) and storing it in a file, for example, XXXX.IN. Then we execute the file by typing

```
PSPICE   XXXX.IN
```

The output will be stored in the file XXXX.OUT for examination with an ASCII editor. The other method is by using the Windows-based version with the *schematic capture* feature in which the user picks and places elements, connects them, and gives them values directly on the screen. This feature avoids coding mistakes, such as elements connected incorrectly, which are difficult to see in the above program file. It nevertheless prepares a file from this visual information, such as we will directly prepare. We will concentrate on the direct-input mode discussed previously, since it is simple to learn and remember. For the simple circuits that we will consider, it is somewhat faster to prepare the input file using the direct method than with the schematic capture method.

Example 2.41

Determine the current I and voltage V in the circuit of Fig. 2.70*a* using hand calculations and with PSPICE.

Solution First we solve the circuit by hand by using a source transformation to convert the circuit to a single-loop circuit as shown in Fig. 2.70*b*. The current i_x in that circuit is

$$\begin{aligned} i_x &= \frac{10 \text{ mV} + 2.5 \text{ mV}}{1.5 \text{ k}\Omega + 500 \ \Omega} \\ &= 6.25 \ \mu\text{A} \end{aligned}$$

Hence the voltage V becomes

$$\begin{aligned} V &= 1.5 \text{ k} \times i_x \\ &= 9.375 \text{ mV} \end{aligned}$$

The current i_x is related to the current I by

$$\begin{aligned} I &= i_x - 5 \ \mu\text{A} \\ &= 1.25 \ \mu\text{A} \end{aligned}$$

Next we solve the circuit using PSPICE. First label the nodes as shown in Fig. 2.70*c*. The PSPICE program is

```
EXAMPLE 2.41
*THIS IS EXAMPLE 2.41
VFUN      1       0       DC      10M
RTOP      1       2       1.5K
RVERT     2       0       500
IHAPPY    2       0       DC      5U
.DC       IHAPPY  5U      5U      1
*THE VOLTAGE V IS V(1,2)
*THE CURRENT I IS I(RVERT)
.PRINT    DC      V(1,2)  I(RVERT)
.END
```

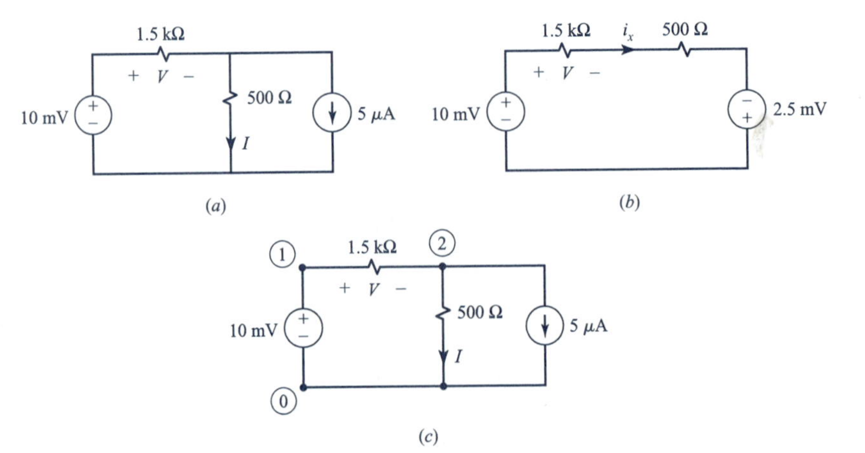

Figure 2.70
Example 2.41.

The PSPICE output gives the same values as computed by hand:

V(1,2) = 9.375E-3 V I(RVERT) = 1.25E -6 A

In order to demonstrate the criticality of the ordering of the nodes in the element specification statement we will change the node ordering in the above program to

```
EXAMPLE 2.41
*NODE ORDERING CHANGED
VFUN     0    1      DC     -10M
RTOP     2    1      1.5K
RVERT    0    2      500
IHAPPY   0    2      DC     -5U
.DC      IHAPPY     -5U     -5U    1
.PRINT   DC   V(1,2)        I(RVERT)
*THE VOLTAGE V IS V(1,2)
*THE CURRENT I IS NOW THE NEGATIVE OF I(RVERT)
.END
```

Note in the specification of the voltage source that the first-named node is on the negative end of the source, and hence the value of the source must be given as −10 mV. Similarly, in the specification of the current source the current direction is from the last-named node to the first-named node, and hence the value of the current source must be entered as −5 μA. This also necessitates entering −5 μA and −5 μA in the .DC statement that iterates IHAPPY. Finally observe that the sequence of node numbering for the resistors is reversed from the original. This does not affect the voltage across RTOP as V(1,2), but does affect the current through resistor RVERT, since, because the first-named node is now 0 and the last-named node is 2, when we request I(RVERT) we get the negative of current I:

V(1,2) = 9.375E-3 V I(RVERT) = −1.25E-6 A

The remaining elements are the controlled sources, whose specification statements are

1. *Voltage-controlled current source:*

   GXXX N1 N2 n m γ

2. *Voltage-controlled voltage source:*

   EXXX N1 N2 n m α

3. *Current-controlled current source:*

   FXXX N1 N2 VXXX σ

4. *Current-controlled voltage source:*

   HXXX N1 N2 VXXX β

These are illustrated in Fig. 2.71. The rules for ordering the nodes of the controlled source, N1 and N2, follow those of the independent sources: voltage sources have the first-named node on the positive end of the source, and the current of a current source flows from the first-named node to the second-named node. The controlling variables for these sources also follow the node convention as though they were output variables.

For example, for the voltage-controlled current source in Fig. 2.71*a* the controlling variable is the voltage v_x across the controlling branch. The nodes of this branch are called m and n. In the controlled-source specification statement the nodes of the controlling branch are ordered so that the controlling variable v_x is positive at the first-named node, n. The last item in the

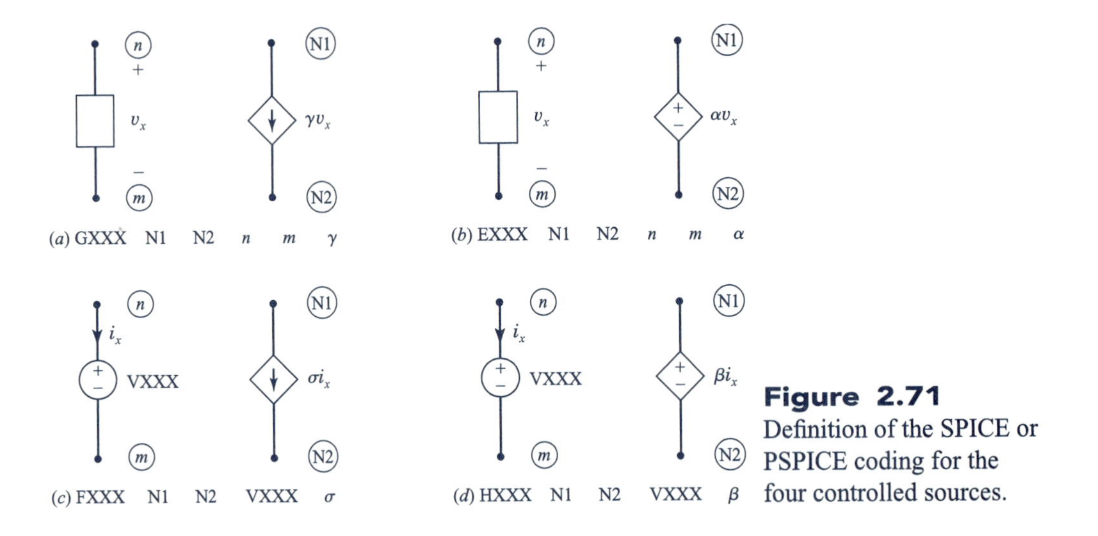

Figure 2.71 Definition of the SPICE or PSPICE coding for the four controlled sources.

specification statement is the value of the controlled source, γ (which multiplies v_x to give the source output).

In the case of a voltage-controlled voltage source shown in Fig. 2.71*b* there are no changes in the ordering of the controlling branch nodes, *n* and *m*. But the node ordering of the source branch follows the convention of a voltage source. In the case of the current-controlled sources in Fig. 2.71*c* and *d*, the controlling variable must be the current through a voltage source that constitutes the controlling branch. This fixes the ordering of the nodes in the *specification statement for that voltage source.* In other words, when we specify the voltage source in the controlling branch, we must insure that the controlling current i_x enters the positive terminal of that voltage source, which conforms to the usual convention as though this controlling current were an output variable. In the case that the controlling current is not through a voltage source in the original circuit, the usual technique is to insert a 0-V voltage source in that controlling branch and use its current as the controlling variable.

Example 2.42

Determine the current i_x and voltage v_x in the circuit of Fig. 2.72*a* with PSPICE and with hand calculations.

Solution To solve this by hand we use the direct method. Applying KCL at the top node in Fig. 2.72*b* gives the current directed to the right through the 2-kΩ resistor as $i_x - 2 \times 10^{-3} v_x$. Hence voltage v_x becomes, according to Ohm's law,

$$v_x = 2\ \text{k}\Omega \times (i_x - 2 \times 10^{-3}\, v_x)$$

or

$$v_x = 400 i_x$$

Applying KVL around the only nonsimple loop gives

$$10\ \text{mV} = v_x + 500 i_x + 1\ \text{k}\Omega \times i_x$$

Substituting the relation $v_x = 400 i_x$ gives

$$10\ \text{mV} = 1900 i_x$$

or

$$i_x = 5.263\ \mu\text{A}$$

and

$$\begin{aligned} v_x &= 400 i_x \\ &= 2.105\ \text{mV} \end{aligned}$$

The PSPICE circuit with nodes labeled is shown in Fig. 2.72*c*. Note that we have inserted a zero-volt voltage source in series with the 1-kΩ resistor to sample the controlling current i_x. The PSPICE program is

```
EXAMPLE 2.42
*AN EXAMPLE OF THE USE OF  CONTROLLED SOURCES
VS          1     0        DC       10M
R1          1     2        2K
HSOURCE           2        3        VTEST      500
R2          3     4        1K
GSOURCE           0        3        1          2     2M
VTEST       4     0        DC       0
.DC         VS    10M      10M      1
*THE VOLTAGE VX IS V(1,2)
*THE CURRENT IX IS I(VTEST) OR I(R2)
.PRINT      DC    V(1,2)        I(VTEST)       I(R2)
.END
```

The results are

V(1,2) = 2.105E-3   I(VTEST) = 5.263E-6   I(R2) = 5.263E-6

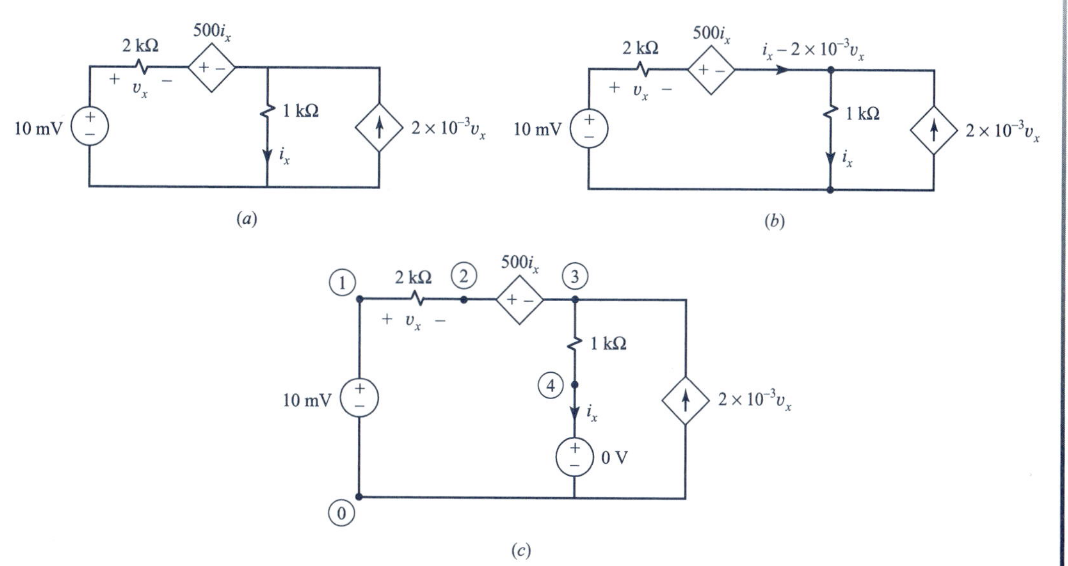

Figure 2.72
Example 2.42.

as was obtained with hand calculation. Again, in order to illustrate the importance of sequencing of nodes in the element specification statements we will modify the above program as

```
EXAMPLE 2.42
*NODE SEQUENCING MODIFIED
VS          1     0     DC     10M
R1          2     1     2K
HSOURCE           3     2      VTEST     500
R2          4     3     1K
GSOURCE           0     3      2         1      -2M
VTEST       0     4     DC     0
.DC         VS    10M   10M    1
*THE VOLTAGE VX IS V(1,2)
*THE CURRENT IX IS -I(VTEST) OR -I(R2)
.PRINT      DC    V(1,2)       I(VTEST)     I(R2)
.END
```

The results are

V(1,2) = 2.105E-3 I(VTEST) = −5.263E-6 I(R2) = −5.263E-6

Now let us discuss the changes. There is no change in the node sequencing in the voltage-source specification statement. Resistor R1 has the node sequencing reversed, but this is of no consequence, since we do not use the current through it. The voltage v_x is still V(1,2) regardless of the ordering of the nodes for R1. The ordering of the nodes for the source HSOURCE is reversed, necessitating a negative sign in the source value of 500. But note that in the specification of the zero-volt voltage source through which the controlling current i_x is taken, the nodes are reversed from the original specification, negating the other negative sign, so that the source value remains 500. Resistor R2 has its node sequencing reversed from the original, so that the current i_x is now -I(R2). Similarly, since the node sequencing of VTEST is reversed, the current i_x is now −I(VTEST).

As a final example we will use PSPICE to determine the equivalent resistance, R_{eq}.

Example 2.43

Determine the equivalent resistance seen at terminals *ab* of the circuit in Fig. 2.73*a* by using hand calculations and using PSPICE.

Solution In order to compute R_{eq} by hand, we apply a 1-V source to the terminals and compute the current I as shown in Fig. 2.73*b*. Applying KCL at the top node of the controlled source gives the current down through the 1-kΩ resistor as $3i_x$. Writing KVL gives

$$1\text{ V} = 500 i_x + 1\text{ k}\Omega \times 3 i_x$$

or

$$i_x = \frac{1}{3500} = 0.2857\text{ mA}$$

The total current I becomes

$$I = \frac{1\ \text{V}}{1\ \text{k}\Omega} + i_x$$
$$= 1.286\ \text{mA}$$

Hence the equivalent resistance becomes

$$R_{\text{eq}} = \frac{1\ \text{V}}{I}$$
$$= 777.8\ \Omega$$

Next we determine the equivalent resistance using PSPICE. We simply simulate what we did by hand, i.e., apply a 1-V source and compute the current as shown in Fig. 2.73*c*. Observe that we have inserted a 0-V voltage source to sample I and another to sample the controlling variable i_x. The nodes are numbered as shown, and the PSPICE program becomes

```
EXAMPLE 2.43
*DETERMINING EQUIVALENT RESISTANCES
VS          1     0     DC     1
VTEST1      1     2     DC     0
R1          2     0     1K
R2          2     3     500
VTEST2      3     4     DC     0
FSOURCE           0     4      VTEST2    2
R3          4     0     1K
.DC         VS    1     1      1
.TF         I(VTEST1)   VS
*THE EQUIVALENT RESISTANCE IS THE RATIO OF VS AND I(VTEST1)
.PRINT      DC    I(VTEST1)
.END
```

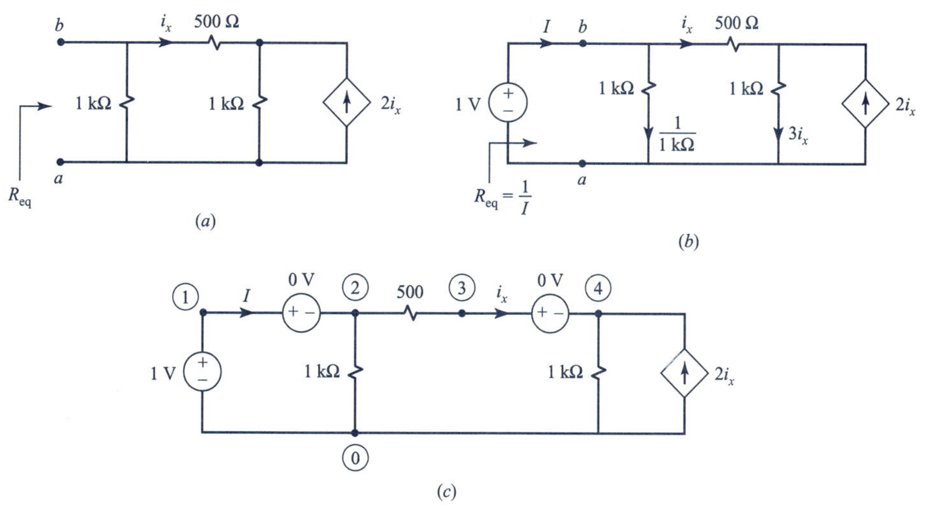

Figure 2.73
Example 2.43.

The result is I(VTEST1)=1.286E-3 as before. We have also used the *transfer function* option, .TF, to compute this ratio. The format of the .TF statement is

.TF output variable input source

This computes the ratio (output variable)/(input source), which in this case is the reciprocal of R_{eq}. But it also prints out the INPUT RESISTANCE AT VS=7.778E2, which was computed by hand.

PSPICE is a very simple program to use, but one must be careful to insure that it is analyzing the intended circuit. The most common error in using PSPICE is inattention to the proper node sequencing in the element specification statements. If the user makes an error in, for example, specifying a resistor as TRES, PSPICE will point this out and refuse to execute the program (since no element names begin with T). On the other hand, if the user makes an error (not an error as PSPICE sees it) in specifying the intended and correct sequence of nodes in an element specification statement, PSPICE will *not* point this out and the user will get an output. However, it will not be the solution for the intended circuit. After some practice one can become proficient in the correct use of PSPICE. It is advisable that one always check a hand calculation with PSPICE. If the two agree, there is a very high probability that the answer is correct. If they do not agree, then a mistake has been made; the hand calculation may be correct and the PSPICE result for the wrong circuit, or vice versa, or both may be incorrect. On getting the two calculations to agree, one achieves a sense of satisfaction.

2.10 Application Examples

In this final section we will discuss the measurement of voltage, current, and resistance as well as the construction of an audio amplifier. There are so-called multimeters that combine the measurement of all three variables into one instrument.

Today's electronic measurement equipment is very compact and accurate. The implementation of these functions in the electronic meter is slightly more involved than we will present, but the principles are the same.

Construction of amplifiers is also somewhat more complex than what we will present; their design is considered in more detail in electronics courses. Nevertheless, a simple amplifier provides the opportunity to show how controlled sources can be used for modeling.

2.10.1 Ammeters, Voltmeters, and Ohmmeters Ammeters measure current, voltmeters measure voltage, and ohmmeters measure resistance. All three meters utilize the *d'*Arsonval meter movement shown in Fig. 2.74*a*. A magnet produces a magnetic field, and a coil of wire is placed in that field. When the current I_m that is to be measured is passed through the coil, it too produces another magnetic field that interacts with the magnetic field of the magnet. This causes the coil to rotate. A needle is fixed to the coil and a scale is calibrated to show the current. The ideal ammeter would have zero resistance, but a real ammeter has a small resistance, denoted as R_m, due to the resistance of the wire of the coil. A suitable model of the meter is shown in Fig. 2.74*b* and consists of resistance R_m in series with an ideal meter.

These meters are rated in terms of their full-scale current and full-scale voltage. For example, a meter might be rated as 10 mA and 30 mV. This means that when 30 mV is applied across the coil, 10 mA will flow through it and will produce full-scale deflection of the needle. Hence for this case the resistance of the meter coil is

$$R_m = \frac{30\text{ mV}}{10\text{ mA}}$$
$$= 3\ \Omega$$

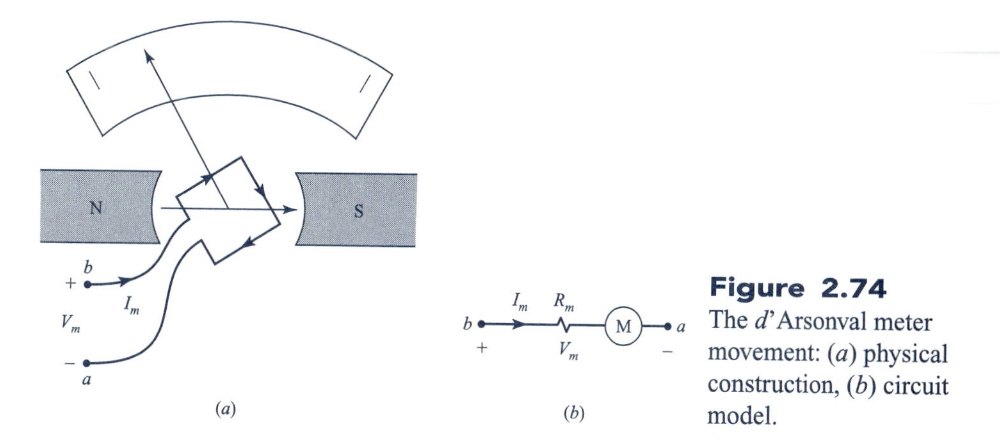

Figure 2.74
The *d'*Arsonval meter movement: (*a*) physical construction, (*b*) circuit model.

A particular meter has a specific maximum current that will produce full-scale deflection. If we wish to measure some larger current, then a shunt resistor, R_{shunt}, can be placed in parallel with the meter in order to accomplish this.

Example 2.44

A *d'*Arsonval meter having a full-scale current of 10 mA and voltage of 30 mV is to be used to measure a maximum current of 200 mA. Determine the value of shunt resistance that will allow the meter to produce full-scale deflection for 200 mA.

Solution With reference to Fig. 2.75*a* we want $I = 200$ mA when $I_m = 10$ mA. Hence we have a current divider:

$$I_m = \frac{R_{\text{shunt}}}{R_{\text{shunt}} + R_m} I$$

or

$$R_{\text{shunt}} = \frac{R_m}{\dfrac{I}{I_m} - 1}$$

Substituting $R_m = 3\ \Omega$ yields $R_{\text{shunt}} = 0.158\ \Omega$.

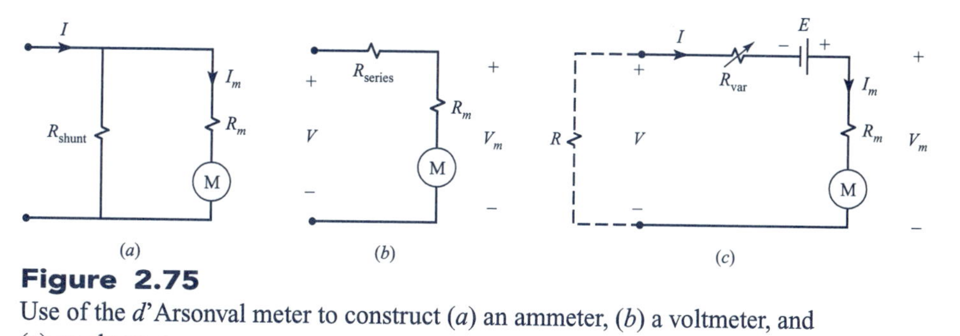

Figure 2.75
Use of the *d'*Arsonval meter to construct (*a*) an ammeter, (*b*) a voltmeter, and (*c*) an ohmmeter.

Next we discuss the construction of a voltmeter shown in Fig. 2.75*b*. Unlike the ammeter, a resistance, R_{series}, is inserted in series with the meter to allow for measurement of voltages larger than the maximum meter voltage.

Example 2.45

A *d'*Arsonval meter having a full-scale current of 10 mA and voltage of 30 mV is to be used to measure a maximum voltage of 500 mV. Determine the value of series resistance that will allow the meter to produce full-scale deflection for 500 mV.

Solution With reference to Fig. 2.75*b*, we want $V = 500$ mV when $V_m = 30$ mV. Hence we have a voltage divider:

$$V_m = \frac{R_m}{R_{\text{series}} + R_m} V$$

or

$$R_{\text{series}} = R_m \left(\frac{V}{V_m} - 1 \right)$$

Substituting $R_m = 3\ \Omega$ yields $R_{\text{series}} = 47\ \Omega$.

And finally we discuss the construction of a device for measuring resistance, an ohmmeter. The ohmmeter essentially consists of a battery, a *d'*Arsonval meter, and an adjustable resistance as shown in Fig. 2.75*c*. The battery serves to pass current through the unknown resistance, and the adjustable resistor, R_{var}, serves to zero the meter. In this case, as opposed to the previous two applications of the meter, we want

$$R = \infty, \quad I_m = 0 \quad \text{(zero meter deflection)}$$
$$R = 0, \quad I_m = 10 \text{ mA} \quad \text{(full-scale meter deflection)}$$

Example 2.46

Design an ohmmeter that uses a *d'*Arsonval meter having full-scale current of 10 mA and full-scale voltage of 30 mV. Determine the resistance measured when the meter indicates half-scale deflection.

Solution With reference to Fig. 2.75*c*, we want full-scale deflection for $R = 0\ \Omega$. In this case $V = 0$ V, $I = I_m = 10$ mA, $V_m = 30$ mV. Hence we obtain from the circuit

$$\begin{aligned} E &= (R_{\text{var}} + R_m) \times I_m \\ &= (R_{\text{var}} + 3\ \Omega) \times (10 \text{ mA}) \end{aligned}$$

Suppose we choose a value for the battery that is commonly available of $E = 9$ V. Solving gives $R_{\text{var}} = 897\ \Omega$. For half-scale deflection, $I_m = 5$ mA. From the circuit we obtain

$$E = (R + R_{\text{var}} + R_m) \times I_m$$

or

$$9 \text{ V} = (R + 897\ \Omega + 3\ \Omega) \times 5 \text{ mA}$$

Solving gives $R = 900\ \Omega$. Three-quarter deflection ($I_m = 7.5$ mA) would indicate a resistance of $R = 300\ \Omega$, and one-quarter deflection ($I_m = 2.5$ mA) would indicate a resistance of $R = 2700\ \Omega$.

2.10.2 An Audio Amplifier A simple amplifier designed to increase the magnitude of the voltage from a microphone and apply it to a speaker is shown in Fig. 2.76*a*. The microphone is modeled as having an internal resistance of 10 Ω and a sinusoidal voltage of 1 mV

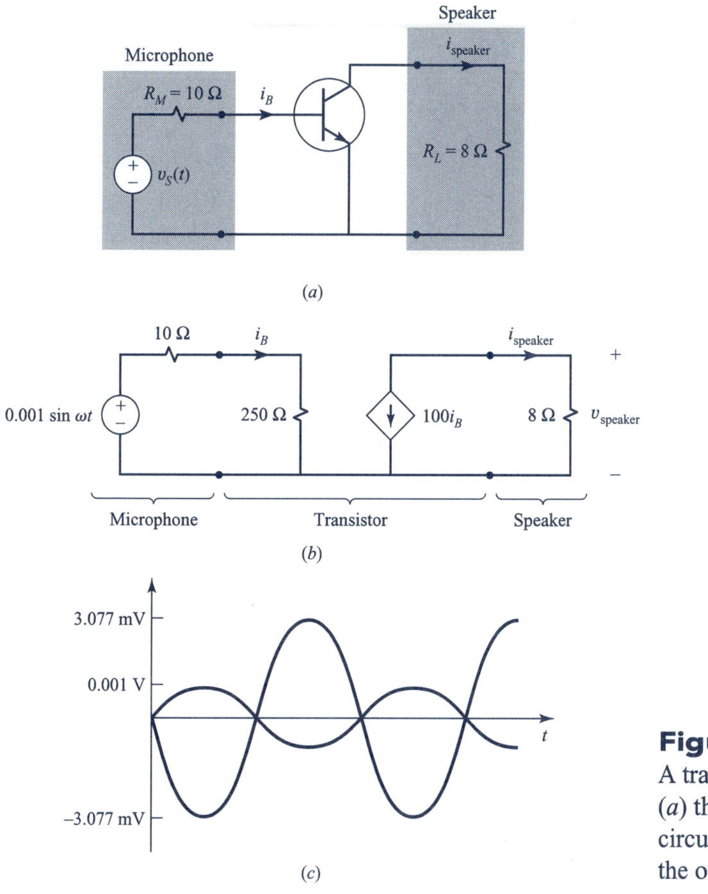

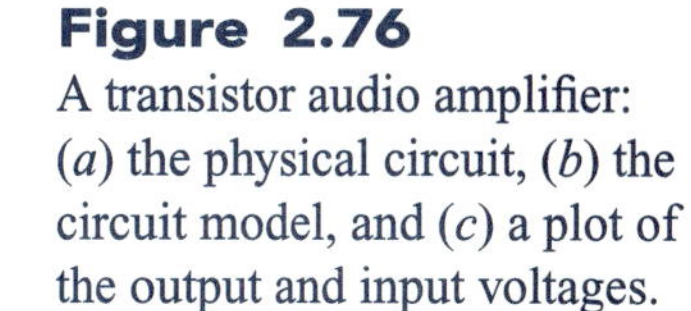

Figure 2.76
A transistor audio amplifier: (*a*) the physical circuit, (*b*) the circuit model, and (*c*) a plot of the output and input voltages.

peak. The speaker is modeled as a resistance of 8 Ω, representing the resistance of the coil of wire in that speaker. Replacing the bipolar transistor with its equivalent circuit given in Fig. 2.55*b* yields the *model* shown in Fig. 2.76*b*. From this circuit we obtain

$$\begin{aligned} i_B &= \frac{1 \sin \omega t \text{ mV}}{10\ \Omega + 250\ \Omega} \\ &= 3.846 \sin \omega t\ \mu\text{A} \end{aligned}$$

This is the controlling variable for the controlled source, and the speaker current is

$$\begin{aligned} i_{\text{speaker}} &= -100 i_B \\ &= -100 \times 3.846 \sin \omega t\ \mu\text{A} \\ &= -0.3846 \sin \omega t \text{ mA} \end{aligned}$$

Hence the voltage across the speaker terminals is

$$\begin{aligned} v_{\text{speaker}} &= 8\ \Omega \times i_{\text{speaker}} \\ &= -3.077 \sin \omega t \text{ mV} \end{aligned}$$

These voltages are plotted in Fig. 2.76*c*. In this case we say that the amplifier provides a voltage gain of (3.077 mV)/(1 mV) = 3.077. More elaborate amplifiers giving much larger voltage gains are designed in electronics courses, but they also use the basic controlled-source model of the transistor as well as the circuit analysis principles we have studied. Only the complexity of the circuit model changes.

Problems

Section 2.1 The Independent Voltage and Current Sources

2.1-1 Determine V and I such that the two circuits in Fig. P2.1-1 are equivalent at terminals ab. *Answers*: $V = -6 + 2t - 3\sin 4t$, $I = 2\cos t - 5 + 3t$.

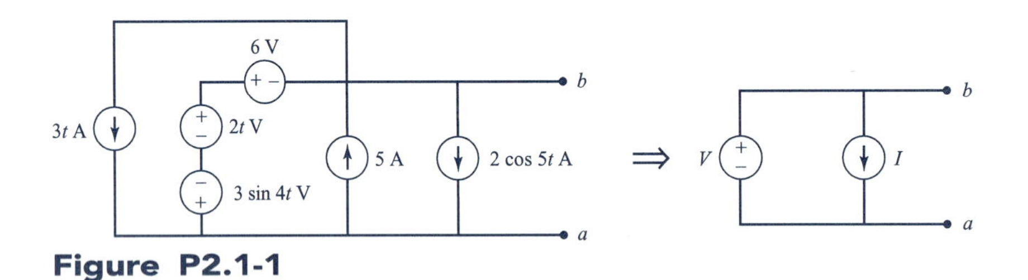

Figure P2.1-1

2.1-2 Determine the voltage v at $t = 2$ s in the circuit of Fig. P2.1-2. *Answer*: $v(2) = -20$ V.

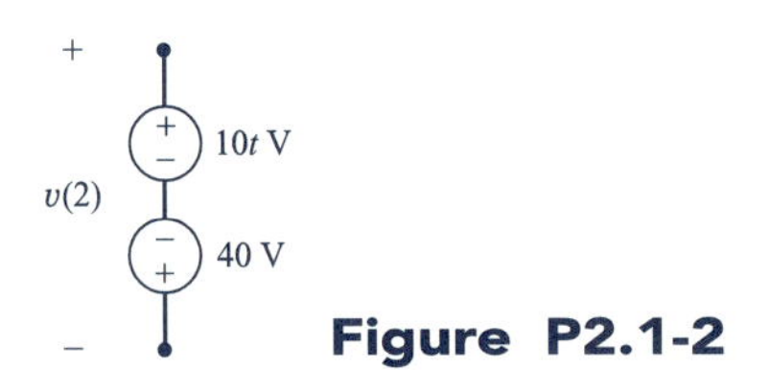

Figure P2.1-2

2.1-3 Determine the current i at $t = 4$ s in the circuit of Fig. P2.1-3. *Answer*: $i(4) = -20$ A.

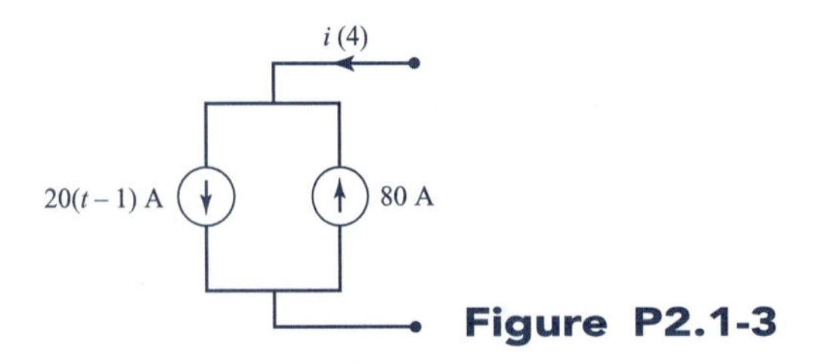

Figure P2.1-3

Section 2.2 The Linear Resistor and Ohm's Law

2.2-1 Determine the indicated v or i for the resistors of Fig. P2.2-1. *Answers*: (a) $v = -6$ V, (b) $v = 1$ V, (c) $i = -30$ A, (d) $v = -50$ V, (e) $v = -5$ V, (f) $i = -0.1$ A, (g) $v = -60$ μV, (h) $i = 100$ A, (i) $v = 0.67$ mV.

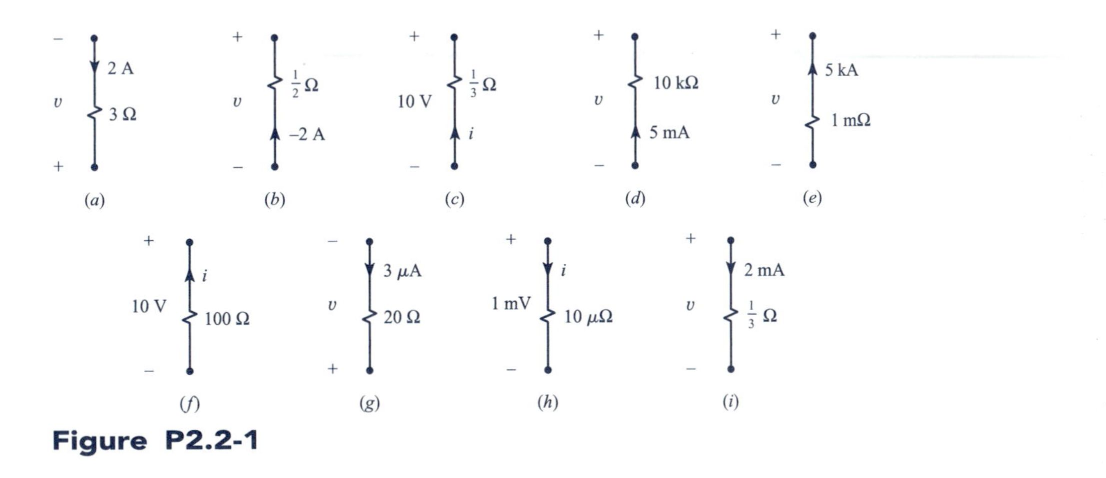

Figure P2.2-1

2.2-2 Determine the indicated v or i for the resistors of Fig. P2.2-2. *Answers*: (a) $v = 12$ V, (b) $v = -10$ V, (c) $v = -20$ V, (d) $i = -10$ A, (e) $i = 6$ A, (f) $i = -250$ μA.

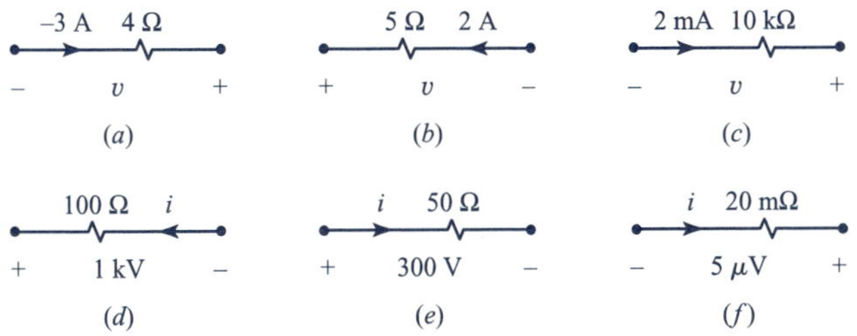

Figure P2.2-2

2.2-3 Determine the power dissipated in the resistors in Fig. P2.2-3.
Answers: (a) $p = 250$ μW, (b) $p = 0.25$ W, (c) $p = 1$pW, (d) $p = 10$ W, (e) $p = 12.5$ W, (f) $p = 2.5$ W.

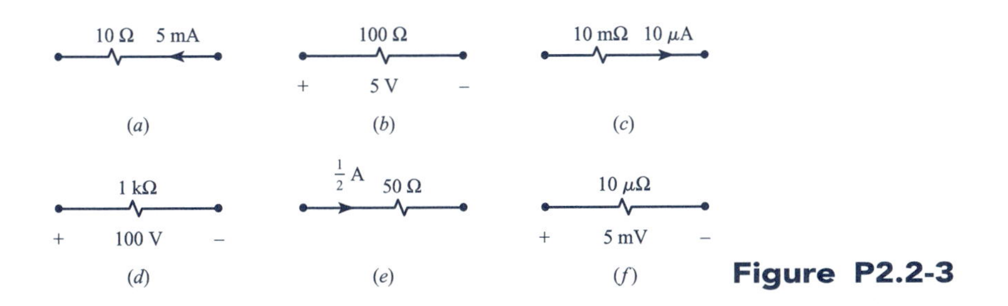

Figure P2.2-3

2.2-4 Determine the power dissipated in the resistors in Fig. P2.2-4.
Answers: (a) $p = 5$ W, (b) $p = 400$ W, (c) $p = 400$ μW, (d) $p = 450$ W, (e) $p = 25$ nW, (f) $p = 120$ μW.

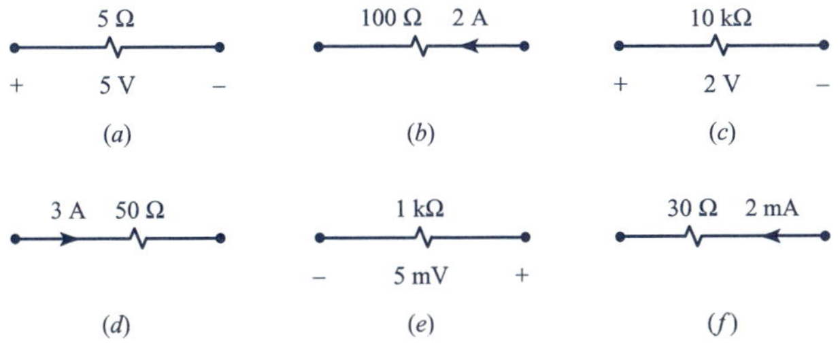

Figure P2.2-4

2.2-5 Determine the indicated v or i for the resistors of Fig. P2.2-5 as well as the power dissipated in those resistors. *Answers*: (a) $v = -16$ V, $p = 128$ W, (b) $i = -0.5$ mA, $p = 2.5$ nW, (c) $v = 10$ V, $p = 0.01$ W, (d) $R = 83.3$ mΩ, $p = 300$ μW, (e) $R = 4$ Ω, $p = 16$ W, (f) $R = 3$ mΩ, $p = 12$ nW.

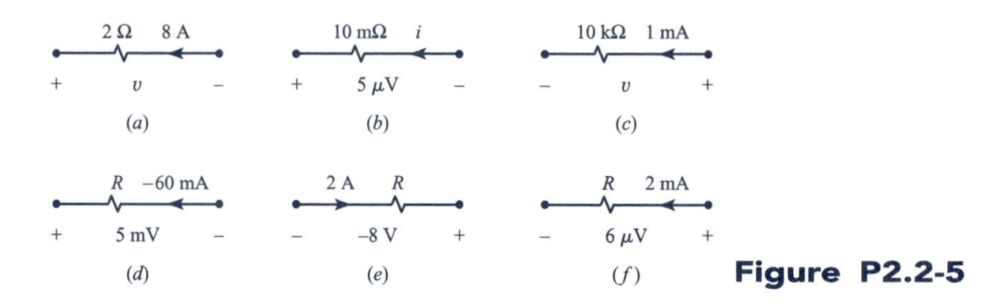

Figure P2.2-5

2.2-6 Determine the current i in Fig. P2.2-6. *Answers*: (a) $i = -2$ A, (b) $i = -0.5$ A, (c) $i = 2.5$ A, (d) $i = -1.5$ mA, (e) $i = 7.5$ A, (f) $i = -3.75$ μA.

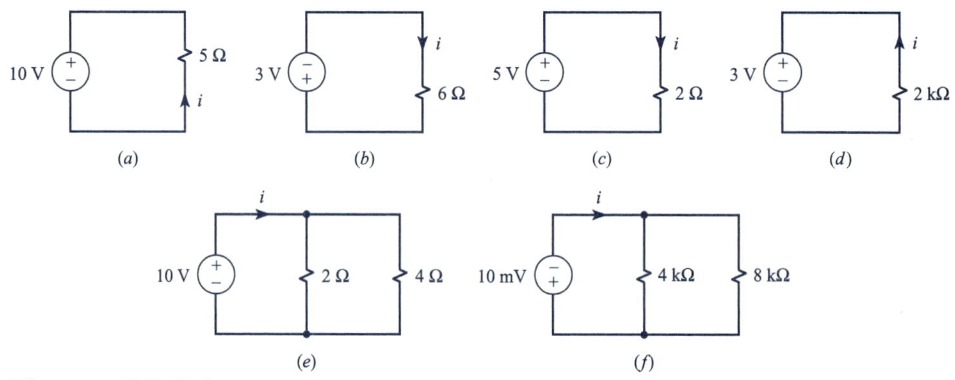

Figure P2.2-6

2.2-7 Determine the voltage v in Fig. P2.2-7. *Answers*: (a) $v = 6$ V, (b) $v = -32$ V, (c) $v = 20$m V, (d) $v = -50$ V, (e) $v = 20$ V, (f) $v = -60$ V.

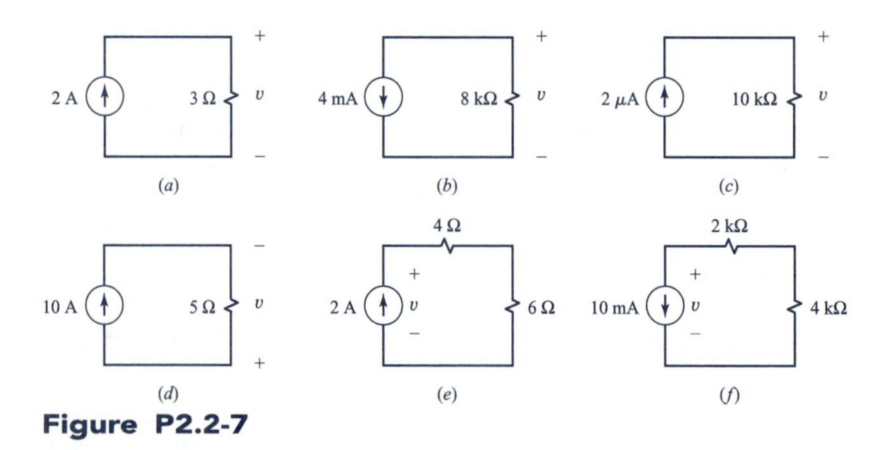

Figure P2.2-7

Section 2.3 Single-Loop and Single-Node-Pair Circuits

2.3-1 Determine i and v in the circuits of Fig. P2.3-1. *Answers*: (a) $\frac{5}{3} - t$ A, (b) $1 - \sin 2t$ V.

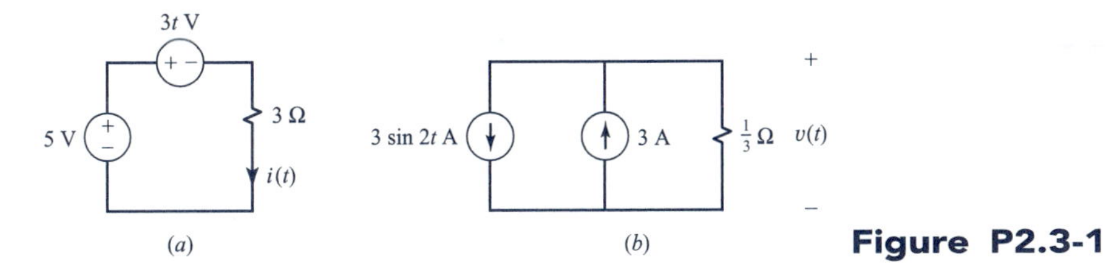

Figure P2.3-1

2.3-2 Determine I and V in the circuit of Fig. P2.3-2. *Answers*: $I = -0.3$ A, $V = -1.2$ V.

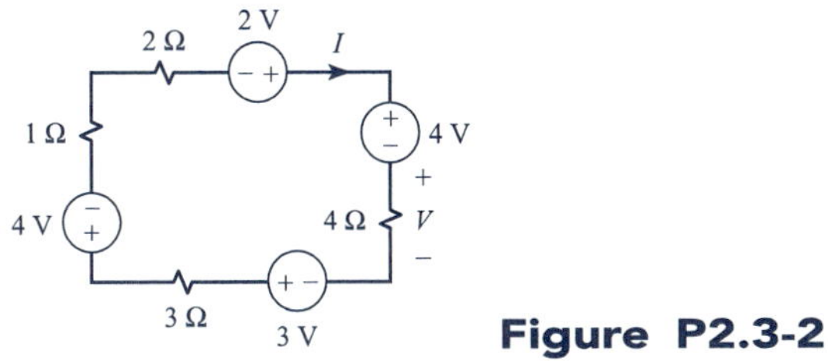

Figure P2.3-2

2.3-3 Determine I and V in the circuit of Fig. P2.3-3. *Answers*: $I = \frac{1}{3}$ A, $V = -\frac{1}{3}$ V.

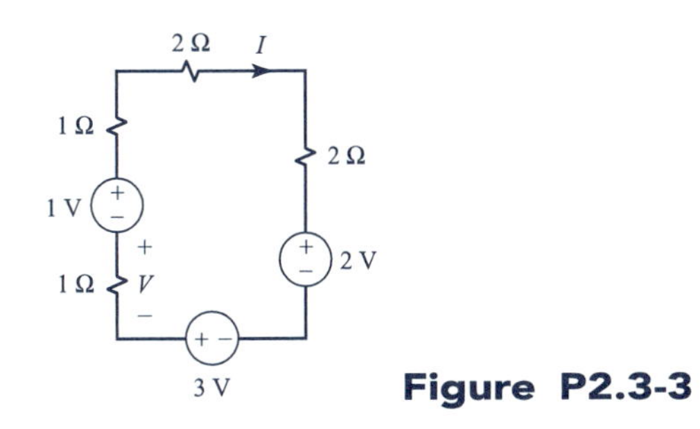

Figure P2.3-3

2.3-4 Determine I, v_x, and v_y in the circuit of Fig. P2.3-4. *Answers*: $I = \frac{1}{2}$ A, $v_x = 1$ V, $v_y = -\frac{1}{2}$ V.

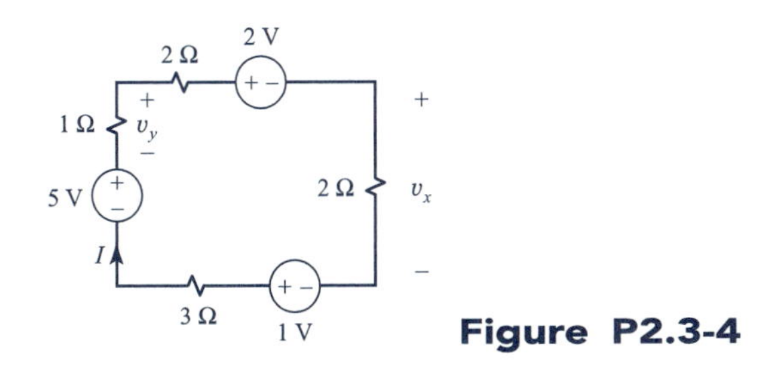

Figure P2.3-4

2.3-5 Determine I, v_x, and v_y in the circuit of Fig. P2.3-5. *Answers*: $I = \frac{5}{8}$ A, $v_x = -\frac{15}{8}$ V, $v_y = \frac{5}{8}$ V.

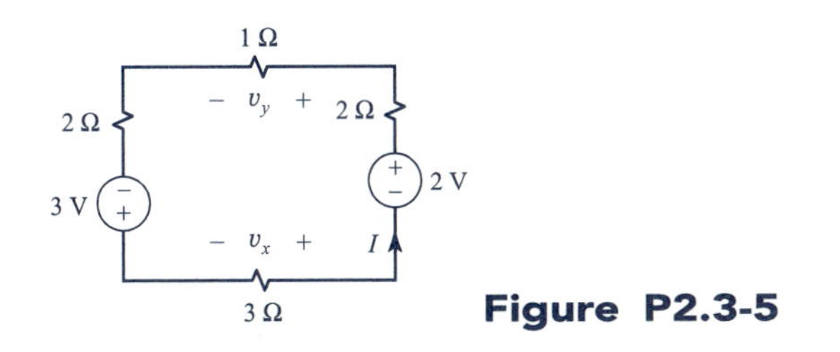

Figure P2.3-5

2.3-6 Determine I, v_x, and v_y in the circuit of Fig. P2.3-6. *Answers*: $I = -\frac{1}{2}$ A, $v_x = 2$ V, $v_y = -1$ V.

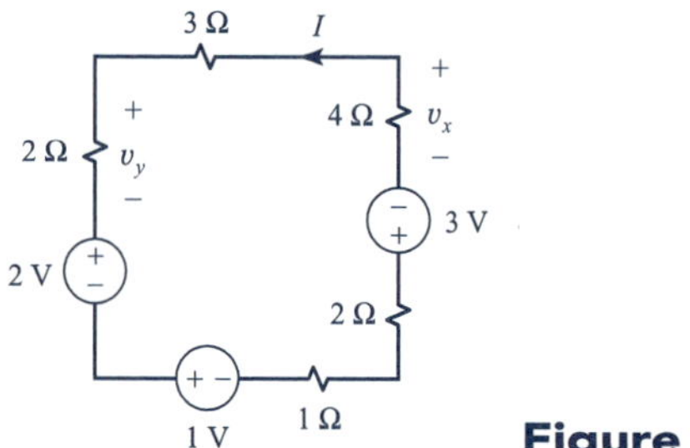

Figure P2.3-6

2.3-7 Determine I, v_x, and v_y in the circuit of Fig. P2.3-7. *Answers*: $I = -\frac{1}{3}$ A, $v_x = -\frac{1}{3}$ V, $v_y = \frac{2}{3}$ V.

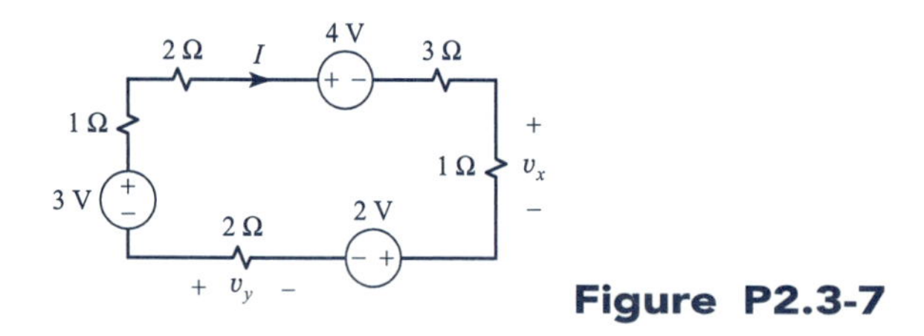

Figure P2.3-7

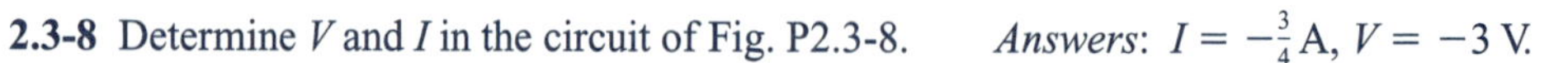

2.3-8 Determine V and I in the circuit of Fig. P2.3-8. *Answers*: $I = -\frac{3}{4}$ A, $V = -3$ V.

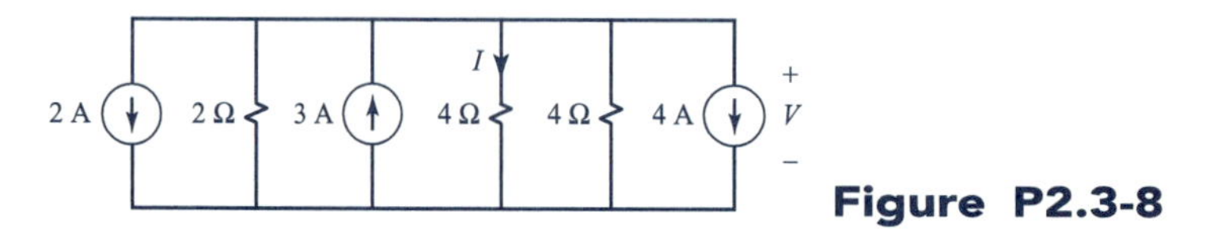

Figure P2.3-8

2.3-9 Determine V and I in the circuit of Fig. P2.3-9. *Answers*: $I = 1$ A, $V = 1$ V.

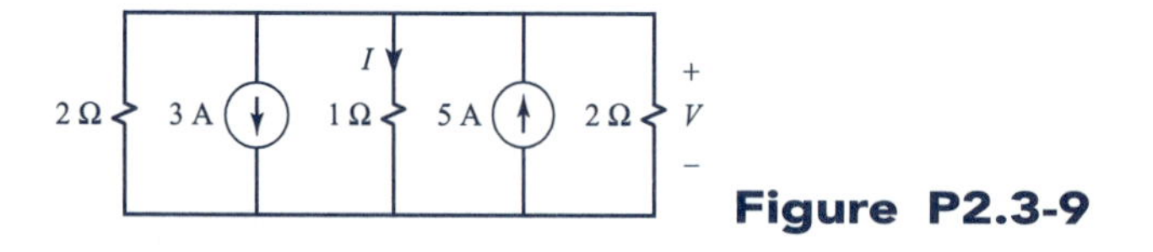

Figure P2.3-9

2.3-10 Determine V and I in the circuit of Fig. P2.3-10. *Answers*: $I = -1$ A, $V = -2$ V.

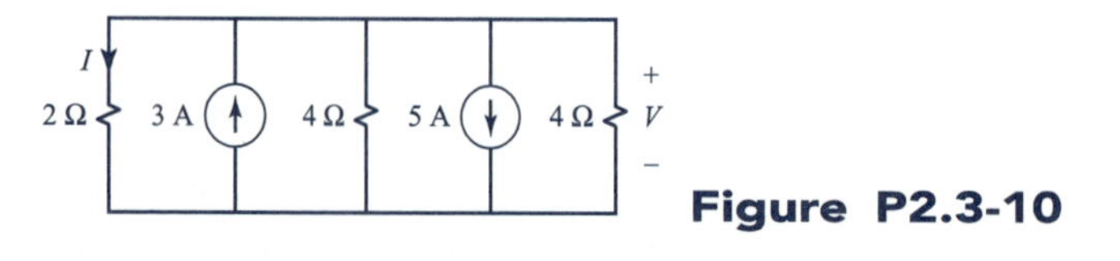

Figure P2.3-10

2.3-11 Determine V and I in the circuit of Fig. P2.3-11. *Answers*: $I = \frac{1}{3}$ A, $V = 3$ V.

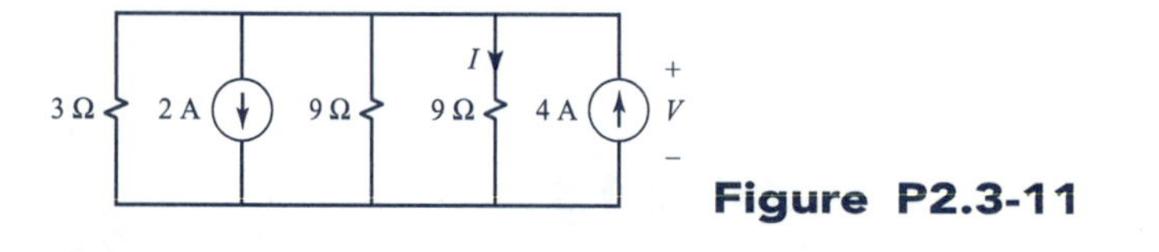

Figure P2.3-11

2.3-12 Determine V and I in the circuit of Fig. P2.3-12. *Answers*: $I = \frac{1}{2}$ A, $V = -1$ V.

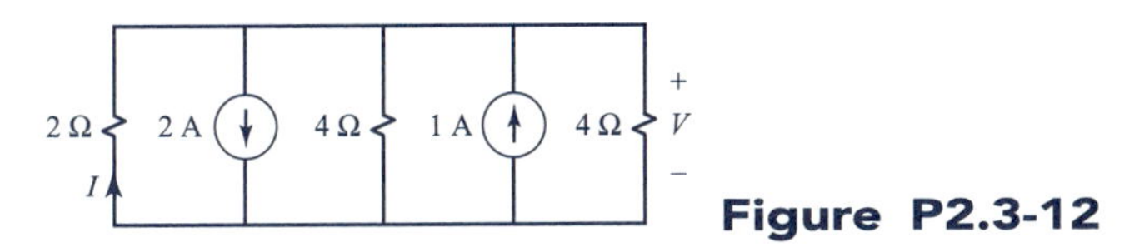

Figure P2.3-12

2.3-13 Determine V and I in the circuit of Fig. P2.3-13. *Answers*: $I = -\frac{1}{4}$ A, $V = \frac{3}{2}$ V.

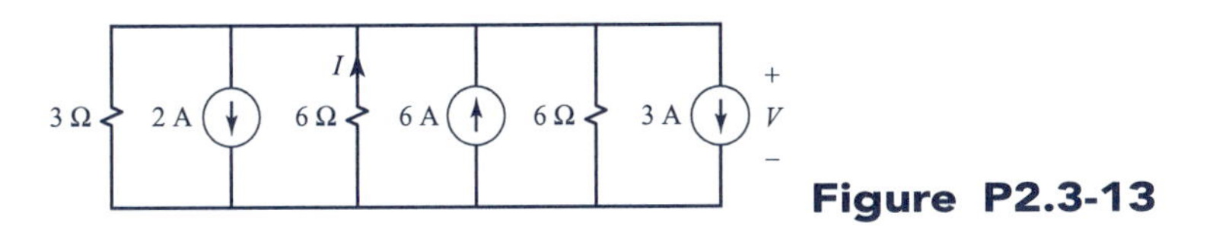

Figure P2.3-13

Section 2.4 Resistors in Series and in Parallel

2.4-1 Determine the equivalent resistance at terminals *ab* for the circuit shown in Fig. P2.4-1. *Answer*: $\frac{46}{5}$ Ω.

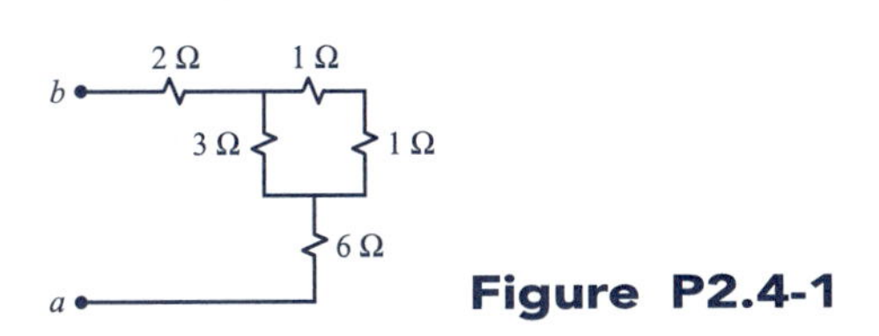

Figure P2.4-1

2.4-2 Determine the equivalent resistance at terminals *ab* for the circuit shown in Fig. P2.4-2. *Answer*: 5 Ω.

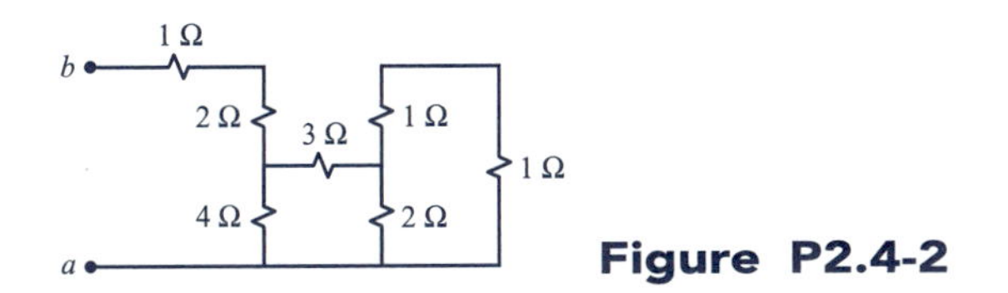

Figure P2.4-2

2.4-3 Determine the equivalent resistance at terminals *ab* for the circuit shown in Fig. P2.4-3. *Answer*: $\frac{26}{7}$ Ω.

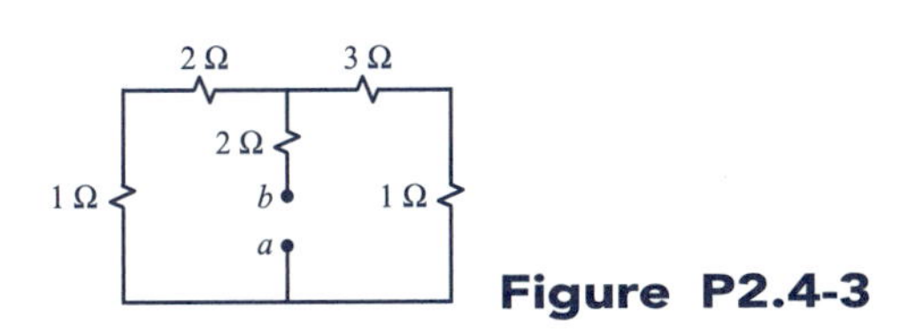

Figure P2.4-3

2.4-4 Determine the equivalent resistance at terminals *ab* for the circuit shown in Fig. P2.4-4. *Answer*: $\frac{163}{21}$ Ω.

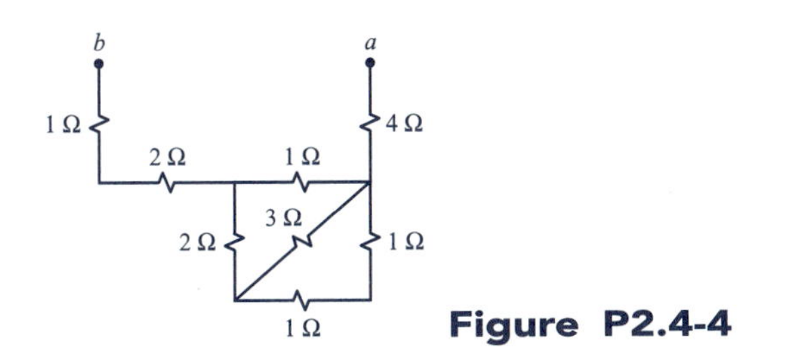

Figure P2.4-4

2.4-5 Determine the equivalent resistance at terminals *ab* for the circuit shown in Fig. P2.4-5.
Answer: $\frac{18}{7}\,\Omega$.

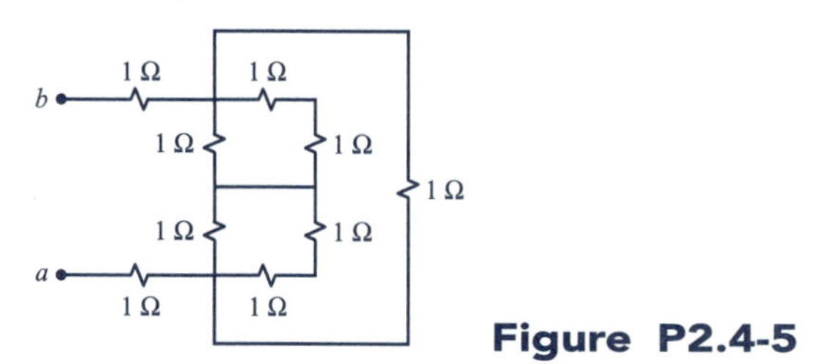

Figure P2.4-5

2.4-6 Determine v_x and i_x in the circuit of Fig. P2.4-6 by using the circuit reduction technique.
Answers: $v_x = 2.5$ V, $i_x = \frac{5}{4}$ A.

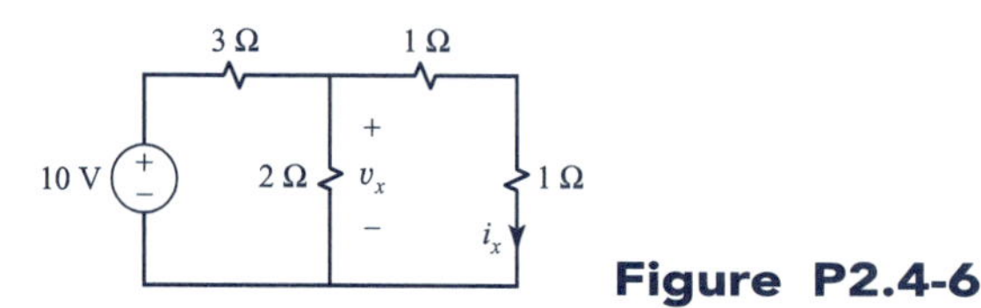

Figure P2.4-6

2.4-7 Determine v_x and i_x in the circuit of Fig. P2.4-7 by using the circuit reduction technique.
Answers: $v_x = \frac{5}{2}$ V, $i_x = \frac{5}{2}$ A.

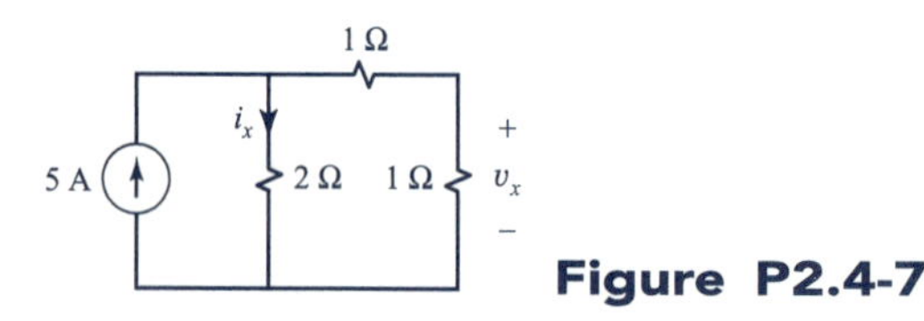

Figure P2.4-7

2.4-8 Determine v_x and i_x in the circuit of Fig. P2.4-8 by using the circuit reduction technique.
Answers: $v_x = -\frac{20}{3}$ V, $i_x = -\frac{10}{3}$ A.

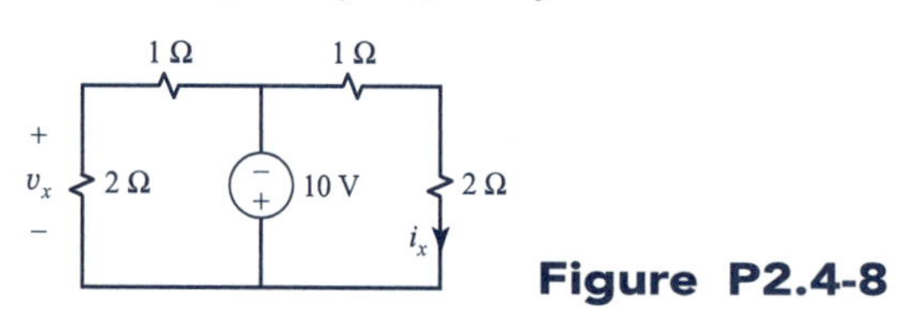

Figure P2.4-8

2.4-9 Determine v_x and i_x in the circuit of Fig. P2.4-9 by using the circuit reduction technique.
Answers: $v_x = \frac{30}{7}$ V, $i_x = \frac{10}{7}$ A.

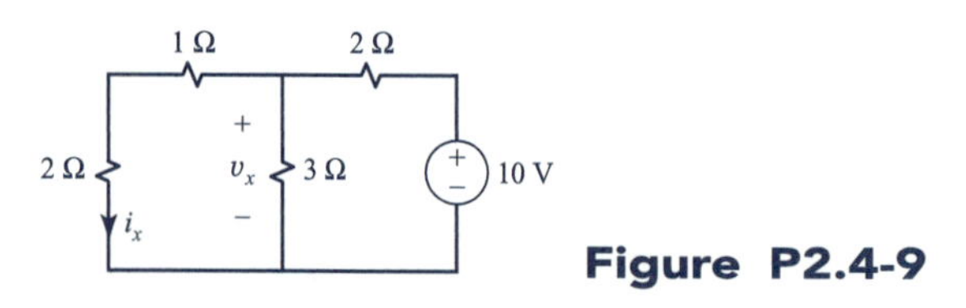

Figure P2.4-9

2.4-10 Determine v_x and i_x in the circuit of Fig. P2.4-10 by using the circuit reduction technique.
Answers: $v_x = \frac{8}{3}$ V, $i_x = \frac{4}{3}$ A.

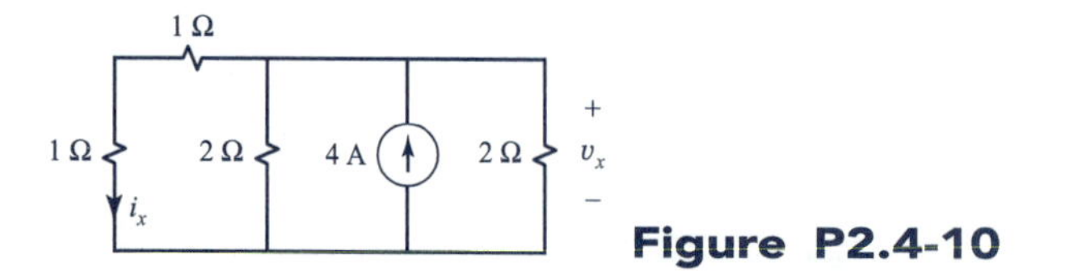

Figure P2.4-10

2.4-11 Determine v_x and i_x in the circuit of Fig. P2.4-11 by using the circuit reduction technique.
Answers: $v_x = \frac{5}{2}$ V, $i_x = \frac{5}{4}$ A.

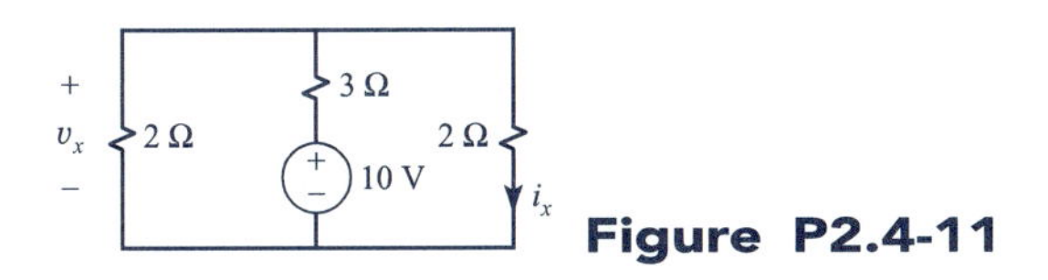

Figure P2.4-11

2.4-12 Determine v_x and i_x in the circuit of Fig. P2.4-12 by using the circuit reduction technique.
Answers: $v_x = \frac{12}{5}$ V, $i_x = \frac{3}{5}$ A.

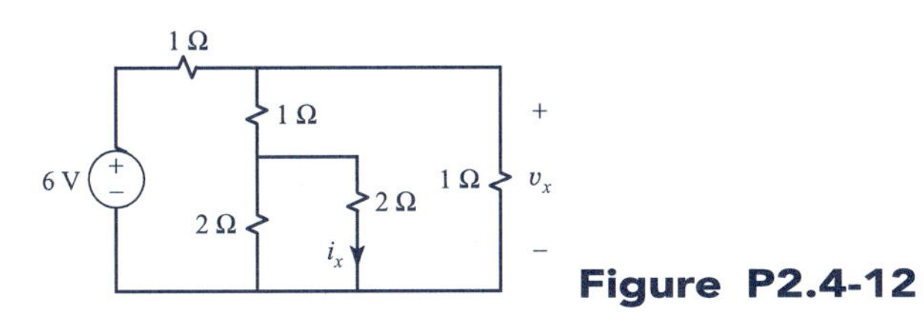

Figure P2.4-12

2.4-13 Determine v_x and i_x in the circuit of Fig. P2.4-13 by using the circuit reduction technique.
Answers: $v_x = \frac{5}{2}$ V, $i_x = -\frac{15}{4}$ A.

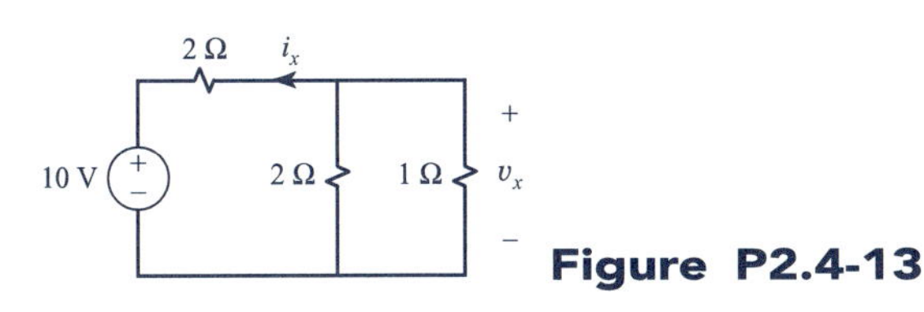

Figure P2.4-13

2.4-14 Determine v_x and i_x in the circuit of Fig. P2.4-14 by using the circuit reduction technique.
Answers: $v_x = -8$ V, $i_x = -6$ A.

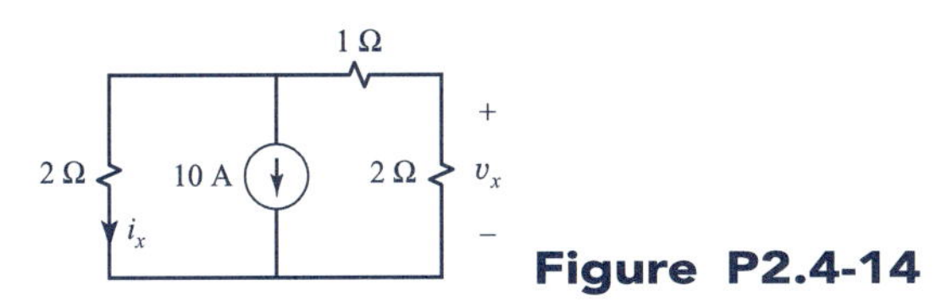

Figure P2.4-14

2.4-15 Determine v_x and i_x in the circuit of Fig. P2.4-15 by using the circuit reduction technique.
Answers: $v_x = -\frac{5}{4}$ V, $i_x = -\frac{5}{4}$ A.

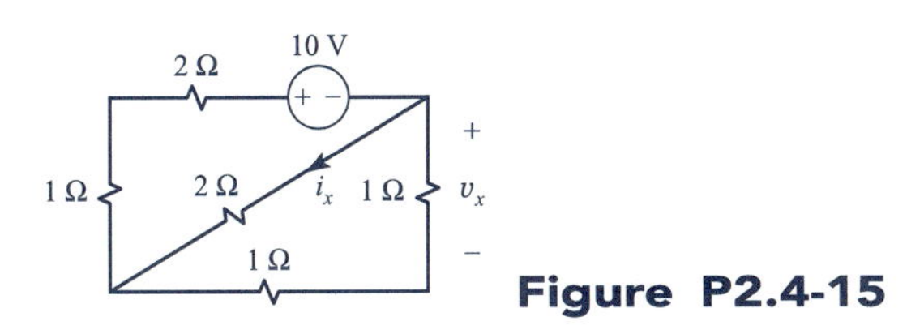

Figure P2.4-15

2.4-16 Determine v_x and i_x in the circuit of Fig. P2.4-16 by using the circuit reduction technique.
Answers: $v_x = -\frac{28}{3}$ V, $i_x = -4$ A.

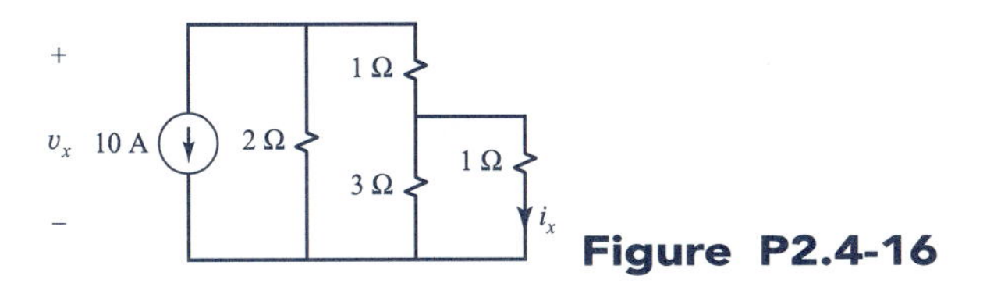

Figure P2.4-16

2.4-17 Determine v_x and i_x in the circuit of Fig. P2.4-17 by using the circuit reduction technique.
Answers: $v_x = -2$ V, $i_x = -\frac{1}{3}$ A.

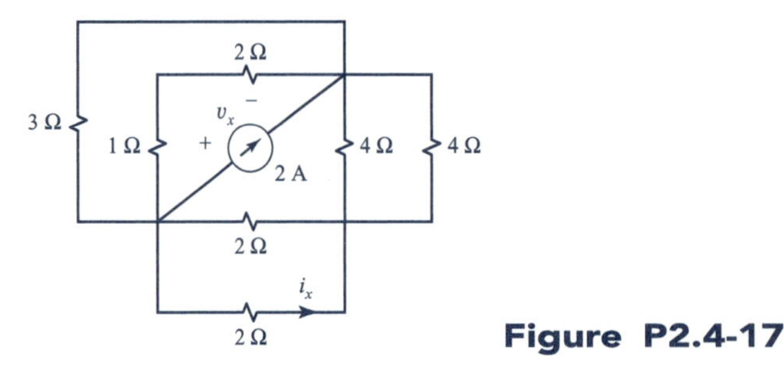

Figure P2.4-17

2.4-18 Determine v_x and i_x in the circuit of Fig. P2.4-18 by using the circuit reduction technique.
Answers: $v_x = 6$ V, $i_x = -1$ A.

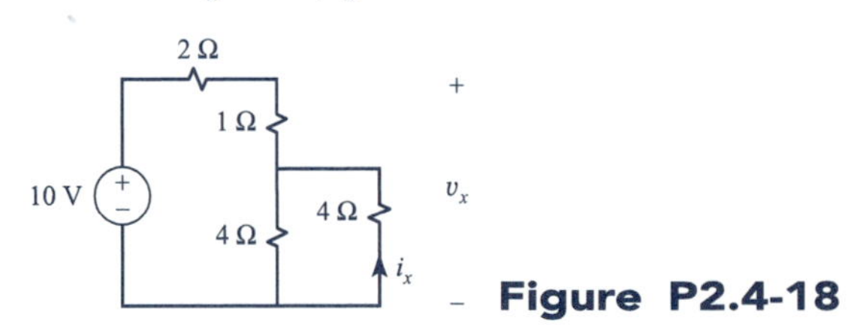

Figure P2.4-18

2.4-19 Determine v_x and i_x in the circuit of Fig. P2.4-19 by using the circuit reduction technique.
Answers: $v_x = -\frac{160}{31}$ V, $i_x = -\frac{120}{31}$ A.

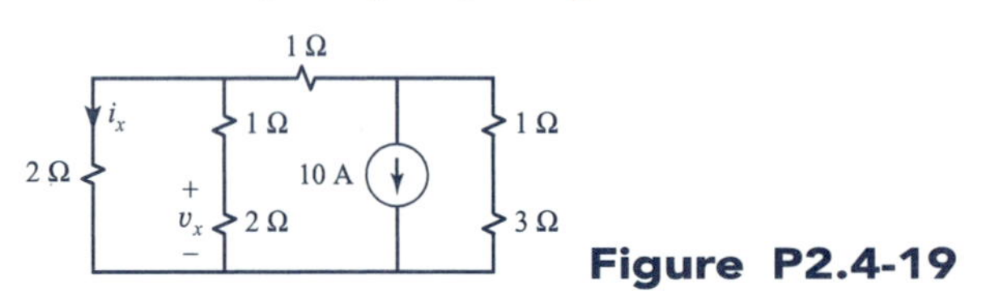

Figure P2.4-19

2.4-20 Determine v_x and i_x in the circuit of Fig. P2.4-20 by using the circuit reduction technique.
Answers: $v_x = -\frac{45}{7}$ V, $i_x = -\frac{15}{7}$ A.

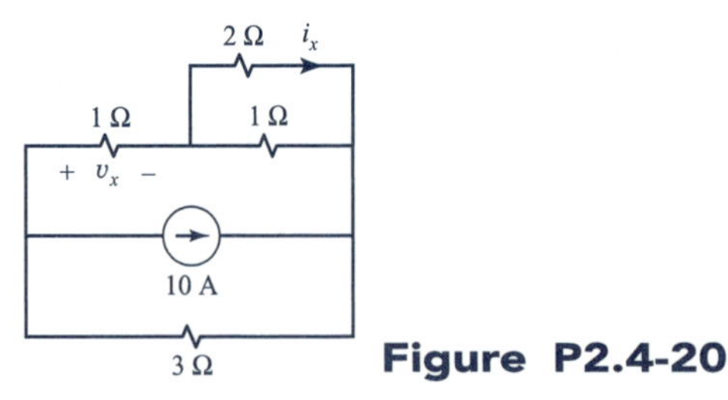

Figure P2.4-20

2.4-21 Determine v_x and i_x in the circuit of Fig. P2.4-21 by using the circuit reduction technique.
Answers: $v_x = \frac{30}{11}$ V, $i_x = -\frac{10}{11}$ A.

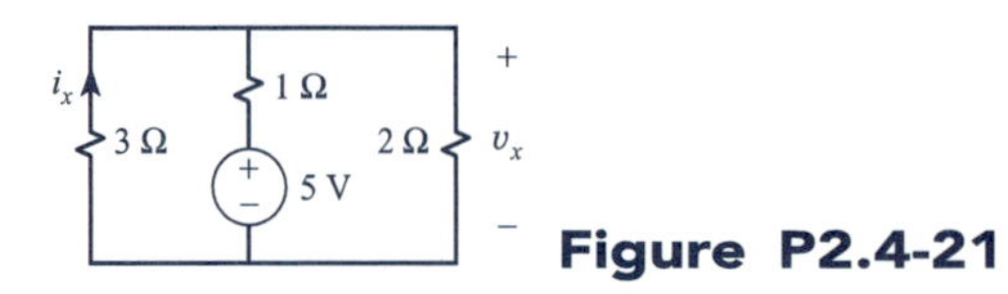

Figure P2.4-21

Section 2.5 Voltage and Current Division

2.5-1 Determine v_x and i_x in the circuit of Fig. P2.4-6 using voltage division and/or current division. *Answers*: $v_x = \frac{5}{2}$ V, $i_x = \frac{5}{4}$ A.

2.5-2 Determine v_x and i_x in the circuit of Fig. P2.4-7 using voltage division and/or current division. *Answers*: $v_x = \frac{5}{2}$ V, $i_x = \frac{5}{2}$ A.

2.5-3 Determine v_x and i_x in the circuit of Fig. P2.4-8 using voltage division and/or current division. *Answers*: $v_x = -\frac{20}{3}$ V, $i_x = -\frac{10}{3}$ A.

2.5-4 Determine v_x and i_x in the circuit of Fig. P2.4-9 using voltage division and/or current division. *Answers*: $v_x = \frac{30}{7}$ V, $i_x = \frac{10}{7}$ A.

2.5-5 Determine v_x and i_x in the circuit of Fig. P2.4-10 using voltage division and/or current division. *Answers*: $v_x = \frac{8}{3}$ V, $i_x = \frac{4}{3}$ A.

2.5-6 Determine v_x and i_x in the circuit of Fig. P2.4-11 using voltage division and/or current division. *Answers*: $v_x = \frac{5}{2}$ V, $i_x = \frac{5}{4}$ A.

2.5-7 Determine v_x and i_x in the circuit of Fig. P2.4-12 using voltage division and/or current division. *Answers*: $v_x = \frac{12}{5}$ V, $i_x = \frac{3}{5}$ A.

2.5-8 Determine v_x and i_x in the circuit of Fig. P2.4-13 using voltage division and/or current division. *Answers*: $v_x = \frac{5}{2}$ V, $i_x = -\frac{15}{4}$ A.

2.5-9 Determine v_x and i_x in the circuit of Fig. P2.4-14 using voltage division and/or current division. *Answers*: $v_x = -8$ V, $i_x = -6$ A.

2.5-10 Determine v_x and i_x in the circuit of Fig. P2.4-16 using voltage division and/or current division. *Answers*: $v_x = -\frac{28}{3}$ V, $i_x = -4$ A.

2.5-11 Determine v_x and i_x in the circuit of Fig. P2.4-18 using voltage division and/or current division. *Answers*: $v_x = 6$ V, $i_x = -1$ A.

2.5-12 Determine v_x and i_x in the circuit of Fig. P2.4-19 using voltage division and/or current division. *Answers*: $v_x = -\frac{160}{31}$ V, $i_x = -\frac{120}{31}$ A.

2.5-13 Determine v_x and i_x in the circuit of Fig. P2.4-20 using voltage division and/or current division. *Answers*: $v_x = -\frac{45}{7}$ V, $i_x = -\frac{15}{7}$ A.

2.5-14 Determine v_x and i_x in the circuit of Fig. P2.4-21 using voltage division and/or current division. *Answers*: $v_x = \frac{30}{11}$ V, $i_x = -\frac{10}{11}$ A.

Section 2.6 Solutions for Circuits Containing More Than One Source

2.6-1 Determine v_x and i_x in the circuit of Fig. P2.6-1 using the direct method.
Answers: $v_x = \frac{18}{5}$ V, $i_x = \frac{6}{5}$ A.

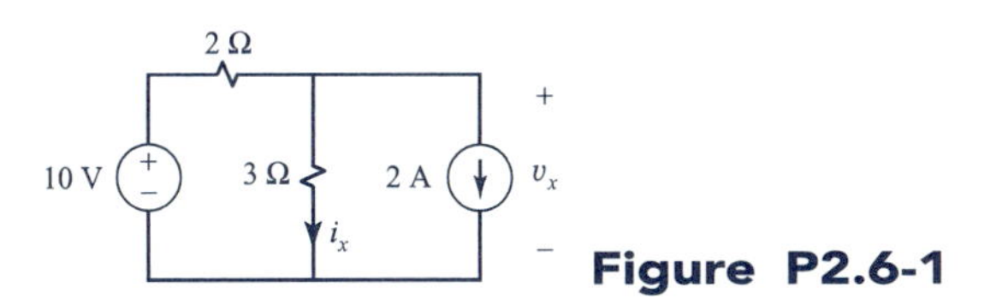

Figure P2.6-1

2.6-2 Determine v_x and i_x in the circuit of Fig. P2.6-2 using the direct method.
Answers: $v_x = \frac{18}{5}$ V, $i_x = \frac{6}{5}$ A.

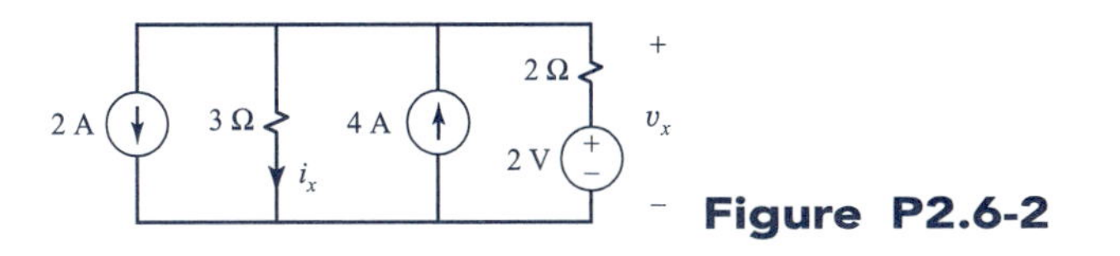

Figure P2.6-2

2.6-3 Determine v_x and i_x in the circuit of Fig. P2.6-3 using the direct method.
Answers: $v_x = \frac{28}{5}$ V, $i_x = \frac{7}{5}$ A.

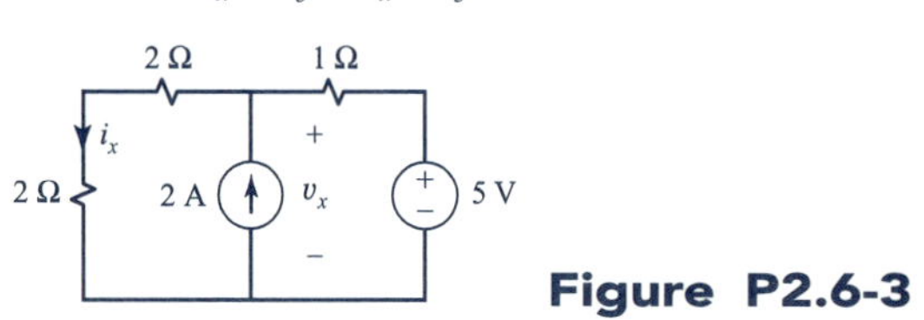

Figure P2.6-3

2.6-4 Determine v_x and i_x in the circuit of Fig. P2.6-4 using the direct method.
Answers: $v_x = -4$ V, $i_x = -2$ A.

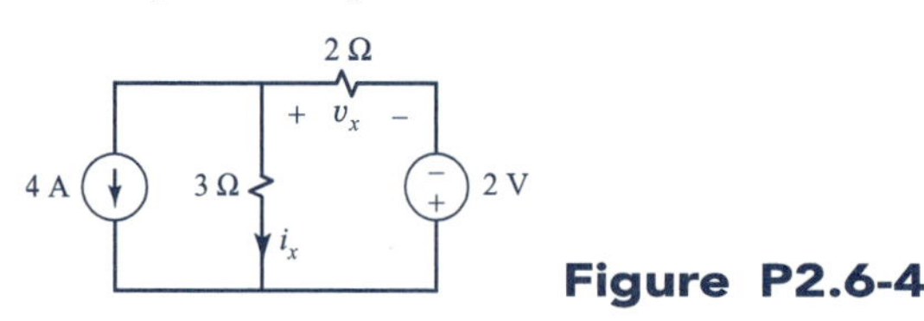

Figure P2.6-4

2.6-5 Determine v_x and i_x in the circuit of Fig. P2.6-5 using the direct method.
Answers: $v_x = -\frac{12}{5}$ V, $i_x = -\frac{3}{5}$ A.

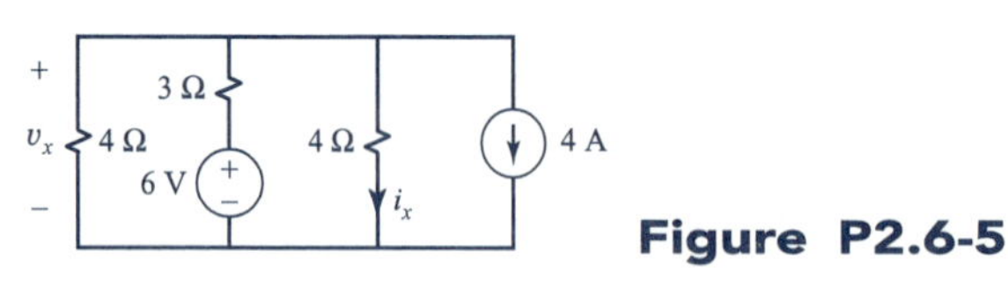

Figure P2.6-5

2.6-6 Determine v_x and i_x in the circuit of Fig. P2.6-6 using the direct method.
Answers: $v_x = -\frac{36}{7}$ V, $i_x = -\frac{2}{7}$ A.

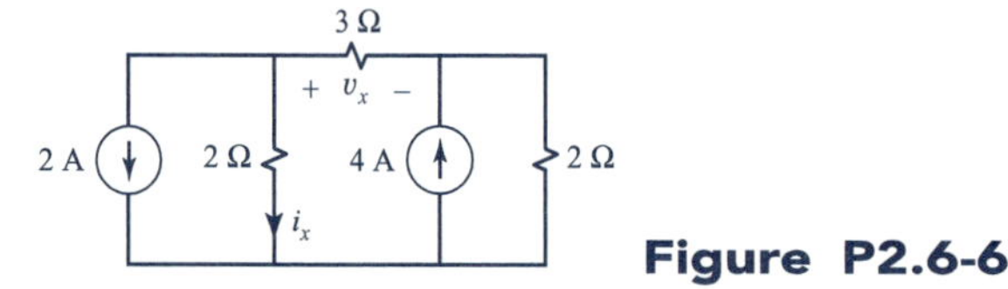

Figure P2.6-6

Section 2.7 Source Transformations

2.7-1 Determine v_x and i_x in the circuit of Fig. P2.6-1 using source transformation(s).
Answers: $v_x = \frac{18}{5}$ V, $i_x = \frac{6}{5}$ A.

2.7-2 Determine v_x and i_x in the circuit of Fig. P2.6-2 using source transformation(s).
Answers: $v_x = \frac{18}{5}$ V, $i_x = \frac{6}{5}$ A.

2.7-3 Determine v_x and i_x in the circuit of Fig. P2.6-3 using source transformation(s).
Answers: $v_x = \frac{28}{5}$ V, $i_x = \frac{7}{5}$ A.

2.7-4 Determine v_x and i_x in the circuit of Fig. P2.6-4 using source transformation(s).
Answers: $v_x = -4$ V, $i_x = -2$ A.

2.7-5 Determine v_x and i_x in the circuit of Fig. P2.6-5 using source transformation(s).
Answers: $v_x = -\frac{12}{5}$ V, $i_x = -\frac{3}{5}$ A.

2.7-6 Determine v_x and i_x in the circuit of Fig. P2.6-6 using source transformation(s).
Answers: $v_x = -\frac{36}{7}$ V, $i_x = -\frac{2}{7}$ A.

2.7-7 Determine v_x and i_x in the circuit of Fig. P2.7-7 using source transformation(s).
Answers: $v_x = \frac{12}{5}$ V, $i_x = -\frac{6}{5}$ A.

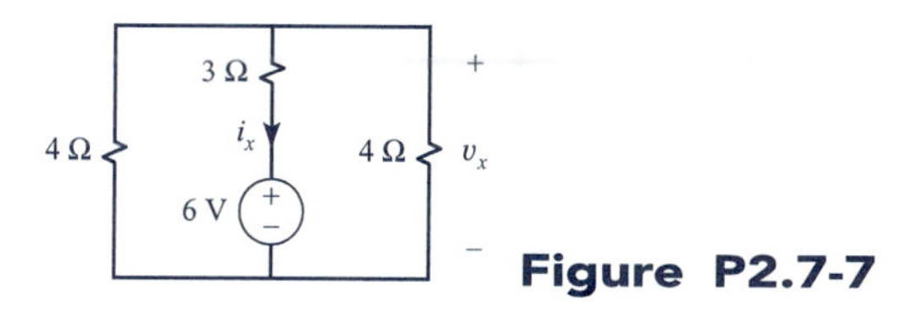

Figure P2.7-7

2.7-8 Determine v_x and i_x in the circuit of Fig. P2.7-8 using source transformation(s). *Answers*: $v_x = 2$ V, $i_x = 2$ A.

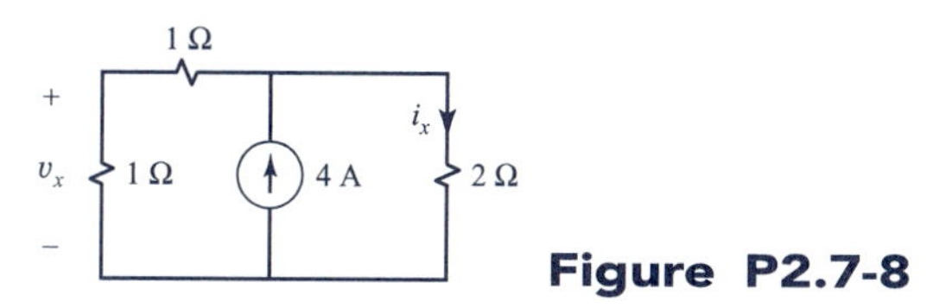

Figure P2.7-8

Section 2.8 The Controlled (Dependent) Voltage and Current Sources

2.8-1 Determine v_x and i_x in the circuit of Fig. P2.8-1. *Answers*: $v_x = 6$ V, $i_x = 2$ A.

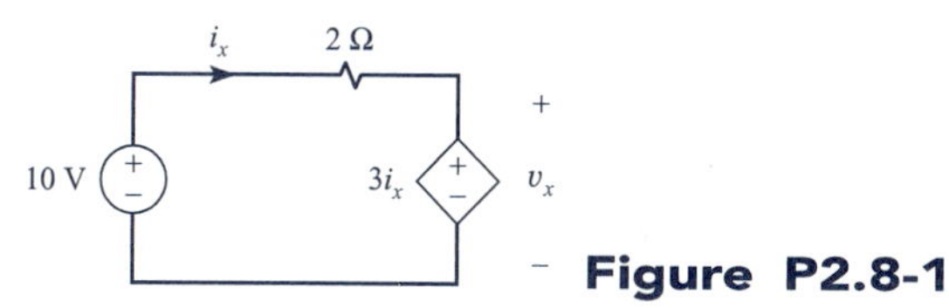

Figure P2.8-1

2.8-2 Determine v_x and i_x in the circuit of Fig. P2.8-2. *Answers*: $v_x = -40$ V, $i_x = 20$ A.

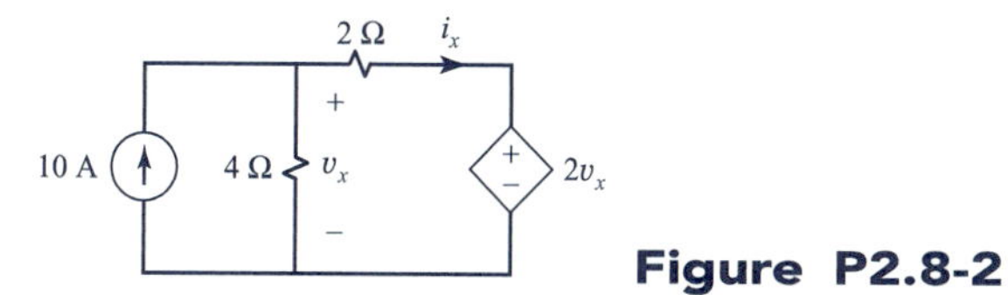

Figure P2.8-2

2.8-3 Determine v_x and i_x in the circuit of Fig. P2.8-3. *Answers*: $v_x = \frac{15}{8}$ V, $i_x = \frac{5}{8}$ A.

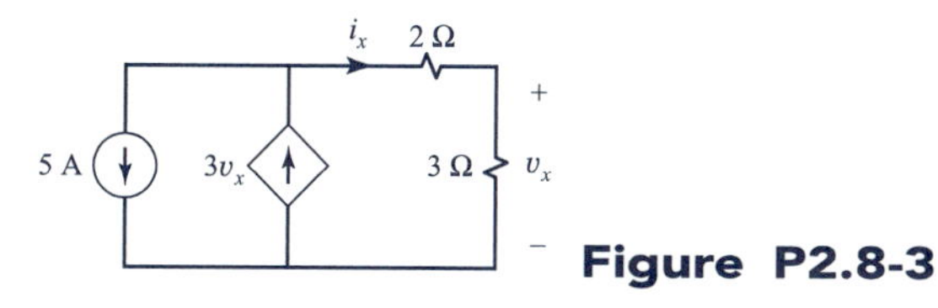

Figure P2.8-3

2.8-4 Determine v_x and i_x in the circuit of Fig. P2.8-4. *Answers*: $v_x = 10$ V, $i_x = 10$ A.

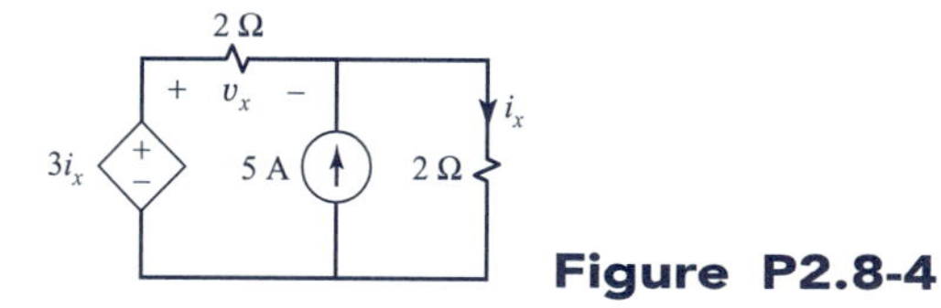

Figure P2.8-4

2.8-5 Determine v_x and i_x in the circuit of Fig. P2.8-5. *Answers*: $v_x = -10$ V, $i_x = 10$ A.

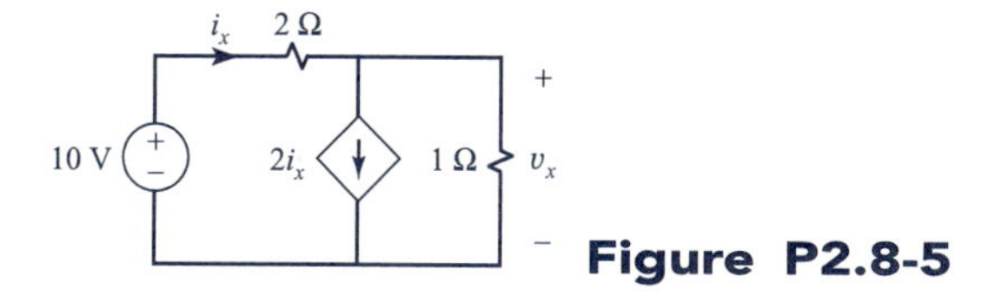

Figure P2.8-5

2.8-6 Determine v_x and i_x in the circuit of Fig. P2.8-6. *Answers*: $v_x = -20$ V, $i_x = 30$ A.

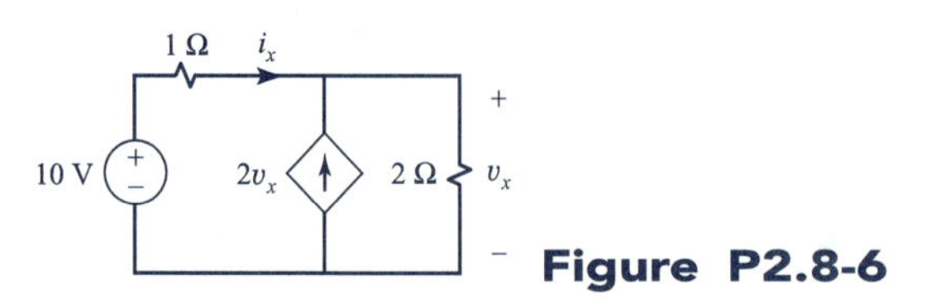

Figure P2.8-6

2.8-7 Determine v_x and i_x in the circuit of Fig. P2.8-7. *Answers*: $v_x = \frac{6}{7}$ V, $i_x = -\frac{6}{7}$ A.

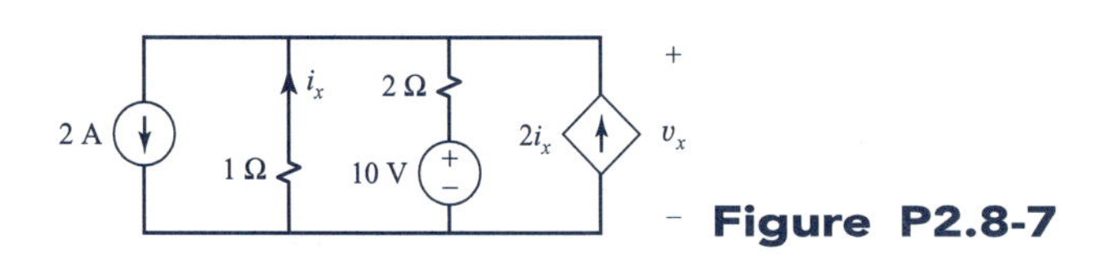

Figure P2.8-7

2.8-8 Determine v_x and i_x in the circuit of Fig. P2.8-8. *Answers*: $v_x = -\frac{3}{5}$ V, $i_x = \frac{9}{5}$ A.

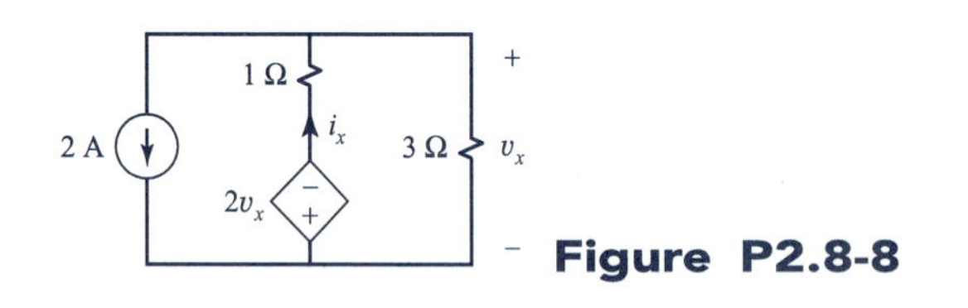

Figure P2.8-8

2.8-9 Determine v_x and i_x in the circuit of Fig. P2.8-9. *Answers*: $v_x = 1$ V, $i_x = -2$ A.

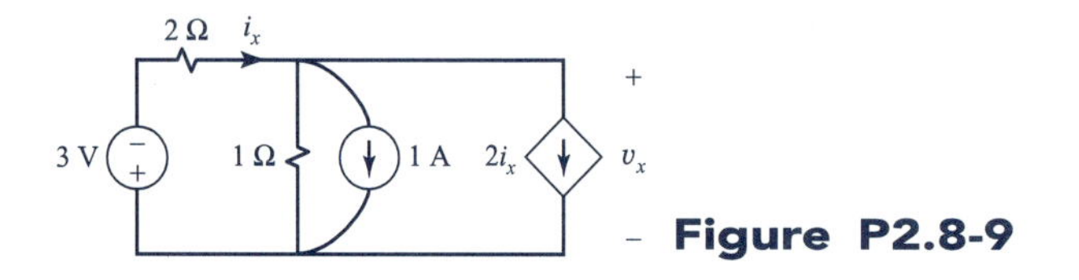

Figure P2.8-9

2.8-10 Determine v_x and i_x in the circuit of Fig. P2.8-10. *Answers*: $v_x = 5$ V, $i_x = -\frac{9}{2}$ A.

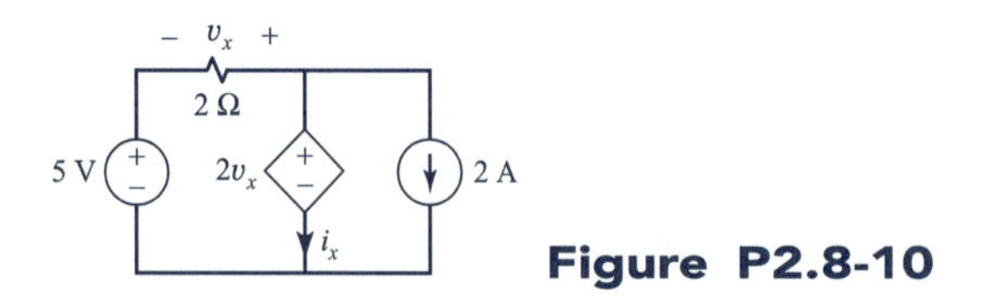

Figure P2.8-10

2.8-11 Determine the equivalent resistance at terminals *ab* for the circuit of Fig. P2.8-11. *Answer*: $R_{eq} = 0\ \Omega$.

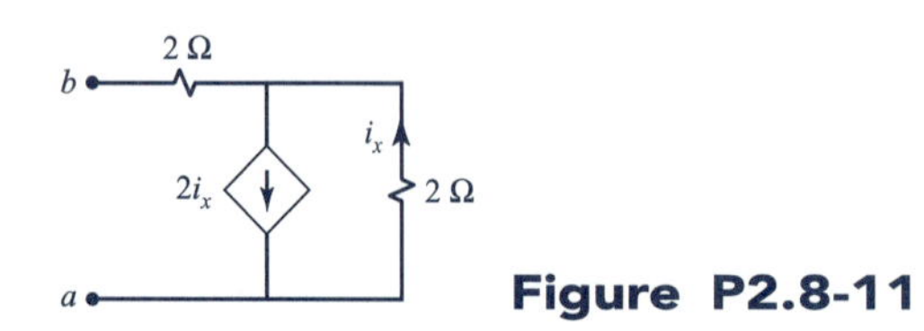

Figure P2.8-11

2.8-12 Determine the equivalent resistance at terminals *ab* for the circuit of Fig. P2.8-12. *Answer*: $R_{eq} = 13\ \Omega$.

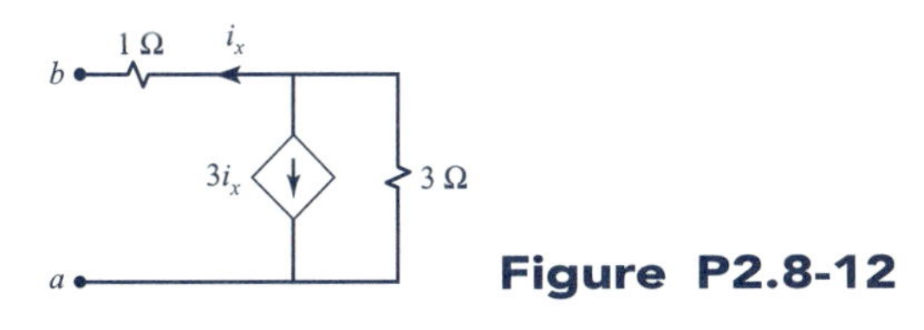

Figure P2.8-12

2.8-13 Determine the equivalent resistance at terminals *ab* for the circuit of Fig. P2.8-13.
Answer: $R_{eq} = \frac{5}{3}\ \Omega$.

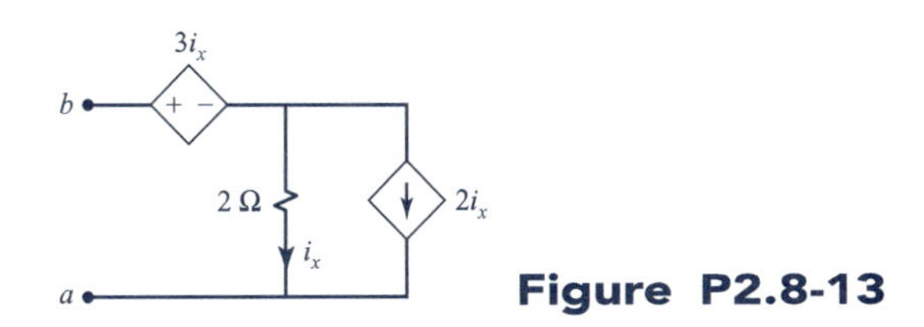

Figure P2.8-13

2.8-14 Determine the equivalent resistance at terminals *ab* for the circuit of Fig. P2.8-14.
Answer: $R_{eq} = -\frac{3}{2}\ \Omega$.

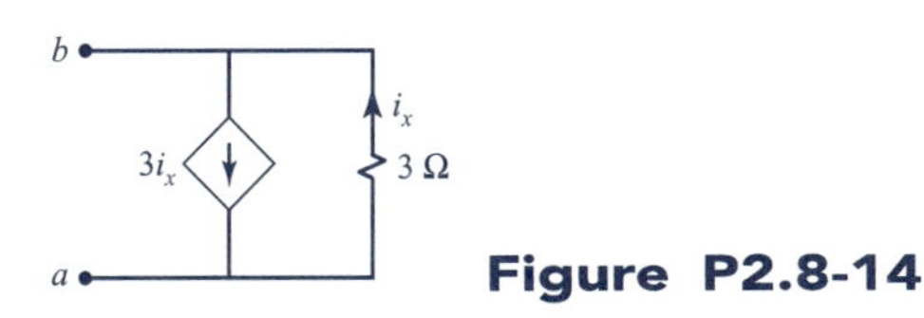

Figure P2.8-14

2.8-15 Determine the equivalent resistance at terminals *ab* for the circuit of Fig. P2.8-15.
Answer: $R_{eq} = \frac{3}{8}\ \Omega$.

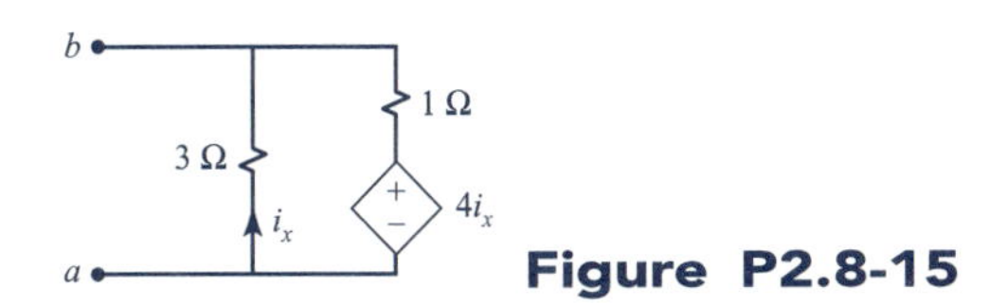

Figure P2.8-15

2.8-16 Determine the equivalent resistance at terminals *ab* for the circuit of Fig. P2.8-16.
Answer: $R_{eq} = -2\ \Omega$.

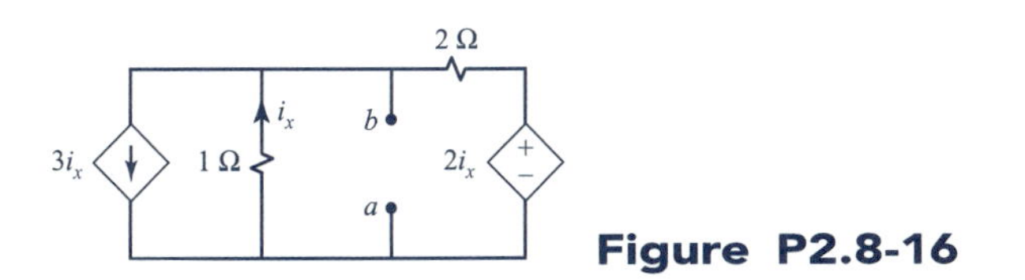

Figure P2.8-16

Section 2.9 PSPICE Applications

2.9-1 Use PSPICE to determine the voltage v_x and current i_x in the circuit of Fig. P2.6-1.
Answers: $v_x = 3.6$ V, $i_x = 1.2$ A.

2.9-2 Use PSPICE to determine the voltage v_x and current i_x in the circuit of Fig. P2.6-2.
Answers: $v_x = 3.6$ V, $i_x = 1.2$ A.

2.9-3 Use PSPICE to determine the voltage v_x and current i_x in the circuit of Fig. P2.6-3.
Answers: $v_x = 5.6$ V, $i_x = 1.4$ A.

2.9-4 Use PSPICE to determine the voltage v_x and current i_x in the circuit of Fig. P2.6-4.
Answers: $v_x = -4$ V, $i_x = -2$ A.

2.9-5 Use PSPICE to determine the voltage v_x and current i_x in the circuit of Fig. P2.6-5. *Answers*: $v_x = -2.4$ V, $i_x = -0.6$ A.

2.9-6 Use PSPICE to determine the voltage v_x and current i_x in the circuit of Fig. P2.6-6. *Answers*: $v_x = -5.143$ V, $i_x = -0.2857$ A.

2.9-7 Use PSPICE to determine the voltage v_x and current i_x in the circuit of Fig. P2.7-7. *Answers*: $v_x = 2.4$ V, $i_x = -1.2$ A.

2.9-8 Use PSPICE to determine the voltage v_x and current i_x in the circuit of Fig. P2.7-8. *Answers*: $v_x = 2$ V, $i_x = 2$ A.

2.9-9 Use PSPICE to determine the voltage v_x and current i_x in the circuit of Fig. P2.8-1. *Answers*: $v_x = 6$ V, $i_x = 2$ A.

2.9-10 Use PSPICE to determine the voltage v_x and current i_x in the circuit of Fig. P2.8-2. *Answers*: $v_x = -40$ V, $i_x = 20$ A.

2.9-11 Use PSPICE to determine the voltage v_x and current i_x in the circuit of Fig. P2.8-3. *Answers*: $v_x = 1.875$ V, $i_x = 0.625$ A.

2.9-12 Use PSPICE to determine the voltage v_x and current i_x in the circuit of Fig. P2.8-4. *Answers*: $v_x = 10$ V, $i_x = 10$ A.

2.9-13 Use PSPICE to determine the voltage v_x and current i_x in the circuit of Fig. P2.8-5. *Answers*: $v_x = -10$ V, $i_x = 10$ A.

2.9-14 Use PSPICE to determine the voltage v_x and current i_x in the circuit of Fig. P2.8-6. *Answers*: $v_x = -20$ V, $i_x = 30$ A.

2.9-15 Use PSPICE to determine the voltage v_x and current i_x in the circuit of Fig. P2.8-7. *Answers*: $v_x = 0.8571$ V, $i_x = -0.8571$ A.

2.9-16 Use PSPICE to determine the voltage v_x and current i_x in the circuit of Fig. P2.8-8. *Answers*: $v_x = -0.6$ V, $i_x = 1.8$ A.

2.9-17 Use PSPICE to determine the voltage v_x and current i_x in the circuit of Fig. P2.8-9. *Answers*: $v_x = 1$ V, $i_x = -2$ A.

2.9-18 Use PSPICE to determine the voltage v_x and current i_x in the circuit of Fig. P2.8-10. *Answers*: $v_x = 5$ V, $i_x = -4.5$ A.

2.9-19 Use PSPICE to determine the equivalent resistance seen at the terminals *ab* for the circuit of Fig. P2.8-11. *Answer*: $R_{eq} = 0\ \Omega$.

2.9-20 Use PSPICE to determine the equivalent resistance seen at the terminals *ab* for the circuit of Fig. P2.8-12. *Answer*: $R_{eq} = 13\ \Omega$.

2.9-21 Use PSPICE to determine the equivalent resistance seen at the terminals *ab* for the circuit of Fig. P2.8-13. *Answer*: $R_{eq} = 1.667\ \Omega$.

2.9-22 Use PSPICE to determine the equivalent resistance seen at the terminals *ab* for the circuit of Fig. P2.8-14. *Answer*: $R_{eq} = -1.5\ \Omega$.

2.9-23 Use PSPICE to determine the equivalent resistance seen at the terminals *ab* for the circuit of Fig. P2.8-15. *Answer*: $R_{eq} = 0.375\ \Omega$.

2.9-24 Use PSPICE to determine the equivalent resistance seen at the terminals *ab* for the circuit of Fig. P2.8-16. *Answer*: $R_{eq} = -2\ \Omega$.

CHAPTER 3

Additional Circuit Analysis Techniques

In the previous chapter we learned how to analyze a large class of relatively simple circuits. The majority of the circuits one encounters and is willing to solve by hand (as opposed to a computer solution) are solvable by those simple and straightforward methods. Those methods will form the cornerstone of virtually all our later solution methods. We will cast all our later circuit solutions into some variant of those methods. For example, in Chapters 6 and 7 we will investigate solving general circuits that contain, in addition to resistors and controlled sources, other nonresistive elements such as the inductor and capacitor. In Chapter 6 we will investigate a method of solving those circuits when they are driven by sinusoidal independent sources. This method will be referred to as the *phasor method*. In the final chapter, Chapter 7, we will investigate the solution of those circuits when they are driven by independent sources having arbitrary time variation. This method will be referred to as the *Laplace-transform method*. Both of those methods will essentially cast the solution into that of an equivalent resistive circuit, so that the resistive-circuit analysis skills we have already learned will again be brought to bear on those solutions. In this chapter we will add some additional solution techniques to our arsenal of circuit analysis techniques: the method of superposition, the use of Thevenin and Norton equivalent circuits, and the node-voltage and mesh-current methods. These additional methods allow us to handle more complicated circuits.

3.1 The Principle of Superposition

Circuits containing more than one independent source added some difficulty in the analysis, since we could not reduce back to a single source. The principle of superposition reduces the solution of a circuit containing more than one independent source to the sum of the solutions of circuits each of which contains only one independent source. Hence in each of those single-source circuits we can use our now familiar circuit reduction methods.

Superposition only applies to *linear circuits*. A linear circuit is one that contains only linear elements. All of our previously investigated elements—the resistor and the controlled sources—are linear elements, meaning that their v–i characteristics are straight lines through the origin of those characteristics. The remaining elements, the inductor and the capacitor considered in

Chapter 5, are also linear elements. The independent sources constrain an element voltage or current and do not relate an element's voltage to its own or another element's current. Hence the independent sources have no effect on a circuit being linear.

The *principle of superposition* can be stated as follows. *If a circuit composed of linear elements contains N independent sources, any element voltage or current in that circuit is composed of the sum of N contributions, each of which is due to one of the independent sources acting individually when all others are set equal to zero (deactivated).* A deactivated independent voltage source is replaced with a *short circuit* as shown in Fig. 3.1*a*. A deactivated independent current source is replaced with an *open circuit* as shown in Fig. 3.1*b*. In other words, *a short circuit is an element that has zero voltage across it.* Similarly, *an open circuit is an element that has zero current through it.*

To illustrate the application of the principle of superposition consider the circuit shown in Fig. 3.2*a*. The circuit contains two independent sources: an independent voltage source and an independent current source. We are interested in determining the voltage v across R_1 and the current i through R_2. The principle of superposition provides that v and i are composed of the sums of the solutions obtained with each source operating individually as shown in Fig. 3.2*b* and *c*. Each of these single-source circuits can be more readily solved using the methods of the previous chapter. For example, the voltage due to the independent voltage source shown in Fig. 3.2*b* can be determined by voltage division as

$$v' = \frac{R_1}{R_1 + R_2} v_S$$

and the current can be determined for this single-loop circuit as

$$i' = \frac{v_S}{R_1 + R_2}$$

Figure 3.1
Illustration of (*a*) replacing a deactivated voltage source with a short circuit, and (*b*) replacing a deactivated current source with an open circuit.

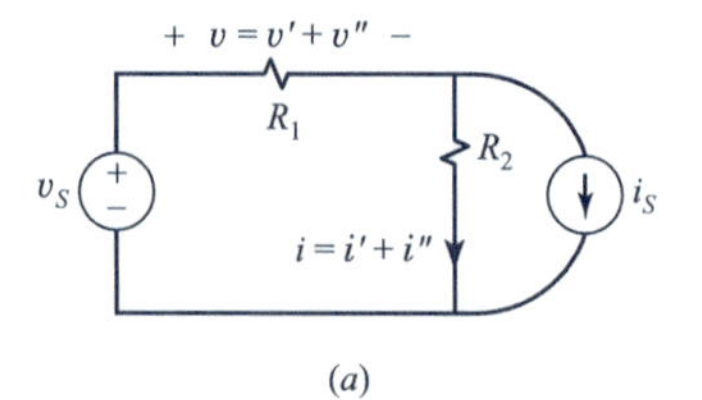

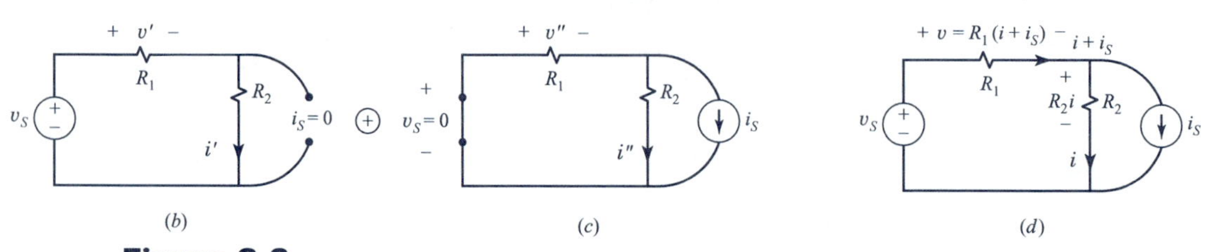

Figure 3.2
Illustration of the application of the principle of superposition.

Similarly, the portions of the solutions due to the independent current source can be obtained from the circuit of Fig. 3.2*c*. The voltage can be easily determined for this single-node-pair circuit as

$$v'' = \frac{R_1 R_2}{R_1 + R_2} i_S$$

and the current can be determined by current division as

$$i'' = -\frac{R_1}{R_1 + R_2} i_S$$

The total voltages and currents are the sums of these contributions:

$$v = \frac{R_1}{R_1 + R_2} v_S + \frac{R_1 R_2}{R_1 + R_2} i_S$$

$$i = \frac{v_S}{R_1 + R_2} - \frac{R_1}{R_1 + R_2} i_S$$

As a check, this can be determined using the direct method of the previous chapter as shown in Fig. 3.2*d*. Applying KCL at the upper node gives the voltage v as

$$v = R_1 \times (i + i_S)$$

Similarly the voltage across R_2 is obtained using Ohm's law as $R_2 \times i$. We have exhausted KCL and Ohm's law. Applying KVL around the only nonsimple loop gives

$$v_S = R_1(i + i_S) + R_2 i$$

Solving this gives

$$i = \frac{v_S}{R_1 + R_2} - \frac{R_1}{R_1 + R_2} i_S$$

as by superposition. The voltage v can then be determined as

$$\begin{aligned} v &= R_1 \times (i + i_S) \\ &= \frac{R_1}{R_1 + R_2} v_S + \frac{R_1 R_2}{R_1 + R_2} i_S \end{aligned}$$

again confirming the result obtained by superposition.

Example 3.1

Determine voltage V and current I in the circuit of Fig. 3.3*a* using superposition. Confirm this result using any other method.

Solution The circuit containing the 3-A current source with all other sources deactivated is shown in Fig. 3.3*b*. From that we obtain by current division

$$\begin{aligned} I' &= -\frac{3\,\Omega + 4\,\Omega}{1\,\Omega + 2\,\Omega + 3\,\Omega + 4\,\Omega} \times 3\text{ A} \\ &= -\tfrac{21}{10}\text{ A} \end{aligned}$$

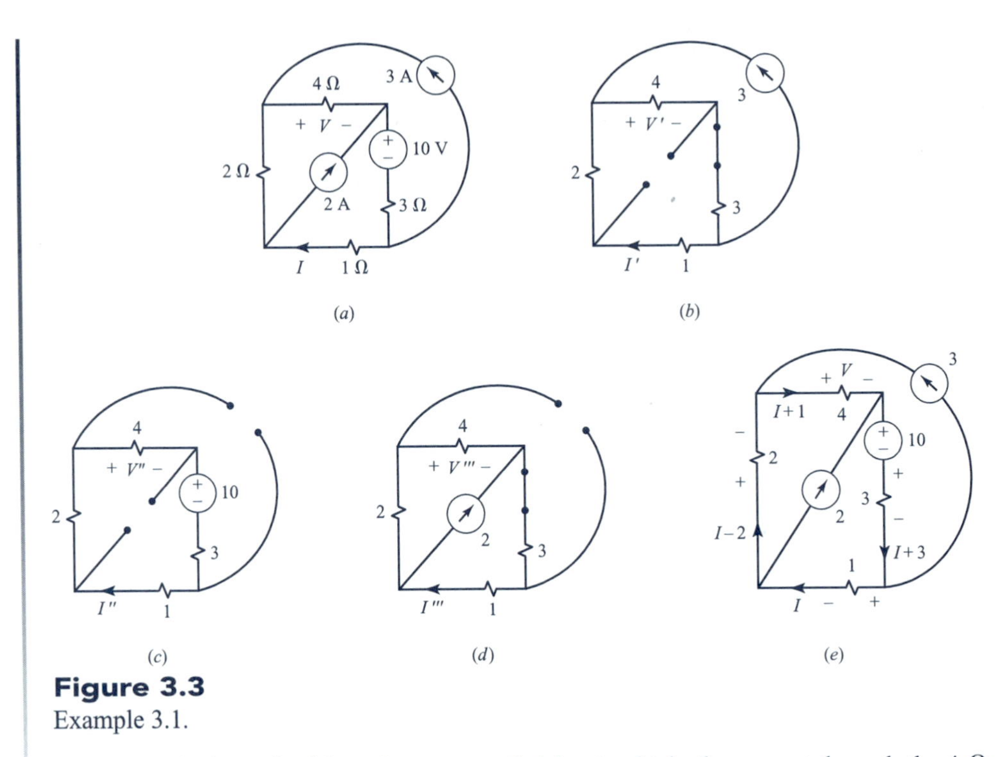

Figure 3.3
Example 3.1.

The voltage is determined by using current division to obtain the current through the 4-Ω resistor:

$$V' = 4\,\Omega \times \frac{1\,\Omega + 2\,\Omega}{1\,\Omega + 2\,\Omega + 3\,\Omega + 4\,\Omega} \times 3\,\text{A}$$

$$= \tfrac{36}{10}\,\text{V}$$

The circuit containing the 10-V voltage source with all other sources deactivated is shown in Fig. 3.3*c*. From that single-loop circuit we obtain

$$I'' = -\frac{10\,\text{V}}{1\,\Omega + 2\,\Omega + 3\,\Omega + 4\,\Omega}$$

$$= -1\,\text{A}$$

The voltage is determined using this current as

$$V'' = 4\,\Omega \times I''$$

$$= -4\,\text{V}$$

Alternatively this voltage could have been determined using voltage division:

$$V'' = -\frac{4\,\Omega}{1\,\Omega + 2\,\Omega + 3\,\Omega + 4\,\Omega} \times 10\,\text{V}$$

$$= -4\,\text{V}$$

The circuit containing the 2-A current source with all other sources deactivated is shown in Fig. 3.3*d*. From that we obtain by current division

$$I''' = \frac{2\,\Omega + 4\,\Omega}{1\,\Omega + 2\,\Omega + 3\,\Omega + 4\,\Omega} \times 2\,\text{A}$$

$$= \tfrac{12}{10}\,\text{A}$$

The voltage is determined by using current division to obtain the current through the 4-Ω resistor:

$$V''' = -4\ \Omega \times \frac{1\ \Omega + 3\ \Omega}{1\ \Omega + 2\ \Omega + 3\ \Omega + 4\ \Omega} \times 2\ \text{A}$$

$$= -\tfrac{32}{10}\ \text{V}$$

The total voltage and current are the sums of these:

$$I = I' + I'' + I'''$$

$$= -\tfrac{21}{10}\ \text{A} - 1\ \text{A} + \tfrac{12}{10}\ \text{A}$$

$$= -\tfrac{19}{10}\ \text{A}$$

and

$$V = V' + V'' + V'''$$

$$= \tfrac{36}{10}\ \text{V} - 4\ \text{V} - \tfrac{32}{10}\ \text{V}$$

$$= -\tfrac{36}{10}\ \text{V}$$

An alternative solution using the direct method is illustrated in Fig. 3.3*e*. Writing the currents through the resistors using KCL and using these currents in Ohm's law gives the voltages across them. Applying KVL around the loop containing the resistors gives

$$(I - 2) \times 2\ \Omega + (I + 1) \times 4\ \Omega + 10\ \text{V} + (I + 3) \times 3\ \Omega + I \times 1\ \Omega = 0$$

Solving this gives

$$I = -\tfrac{19}{10}\ \text{A}$$

and

$$V = (I + 1) \times 4\ \Omega$$

$$= -\tfrac{36}{10}\ \text{V}$$

as before.

Observe in this example that without superposition, the direct method is the only feasible method for solution, since in the original circuit (a) no resistors are in series or in parallel, (b) voltage and current division cannot be used, and (c) source transformations cannot be used. However, once we decompose the original circuit into the three circuits, each containing only one independent source, any of those methods become available for use. This facet of superposition is its greatest advantage: all single-source solution methods become available in each of the single-source circuits.

Exercise Problem 3.1

Determine the voltage v and current i in the circuit of Fig. E3.1. Check your result using any other method.

Answers: $\tfrac{4}{7}$ V and $\tfrac{6}{7}$ A.

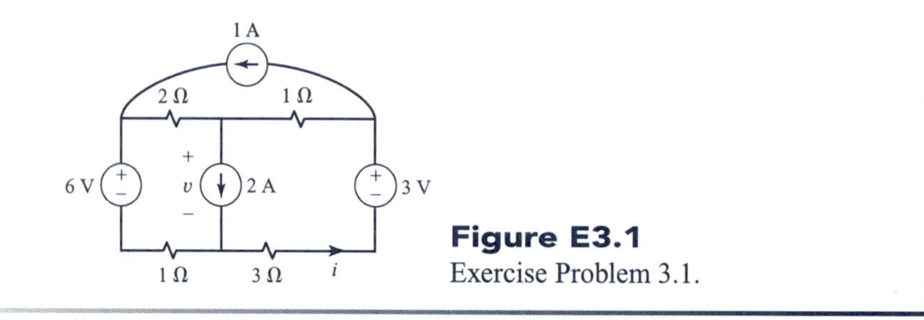

Figure E3.1
Exercise Problem 3.1.

The remaining illustration of superposition is for circuits containing dependent or *controlled* sources. It is important to note that superposition applies *only to independent sources*. In other words, only independent sources are to be deactivated in applying superposition.

Example 3.2

Determine, using superposition, the voltage v in the circuit of Fig. 3.4*a*.

Solution The three circuits resulting from superposition of the three independent sources are shown in Fig. 3.4*b*, *c*, and *d*. Figure 3.4*b* represents the effect of only the 10-V independent voltage source. Observe that the controlling variable of the controlled source, i', has one continuous path in the single-loop circuit. Applying KVL around the loop gives

$$10\text{ V} = 2\ \Omega \times i' - 4i' + 3\ \Omega \times i'$$

so that

$$i' = 10\text{ A}$$

Thus

$$\begin{aligned} v' &= -2\ \Omega \times i' \\ &= -20\text{ V} \end{aligned}$$

Figure 3.4*c* represents the effect of the 2-A independent current source. Applying KCL gives the current through the 2-Ω resistor as $2 - i''$. Writing KVL around the outside loop gives

$$3\ \Omega \times i'' = 4i'' + 2\ \Omega \times (2 - i'')$$

or

$$i'' = 4\text{ A}$$

Hence

$$\begin{aligned} v'' &= 2\ \Omega \times (2 - i'') \\ &= -4\text{ V} \end{aligned}$$

Observe that if we had attempted to solve this circuit by performing a source transformation on the 2-A independent current source and the 3-Ω resistor, we would have lost the controlled-source controlling variable i''.

Figure 3.4*d* represents the effect of the 8-A independent current source. Applying KCL gives the current through the 2-Ω resistor as $8 + i'''$. Writing KVL around the loop yields

$$3\ \Omega \times i''' = 4i''' - 2\ \Omega \times (8 + i''')$$

or

$$i''' = -16\text{ A}$$

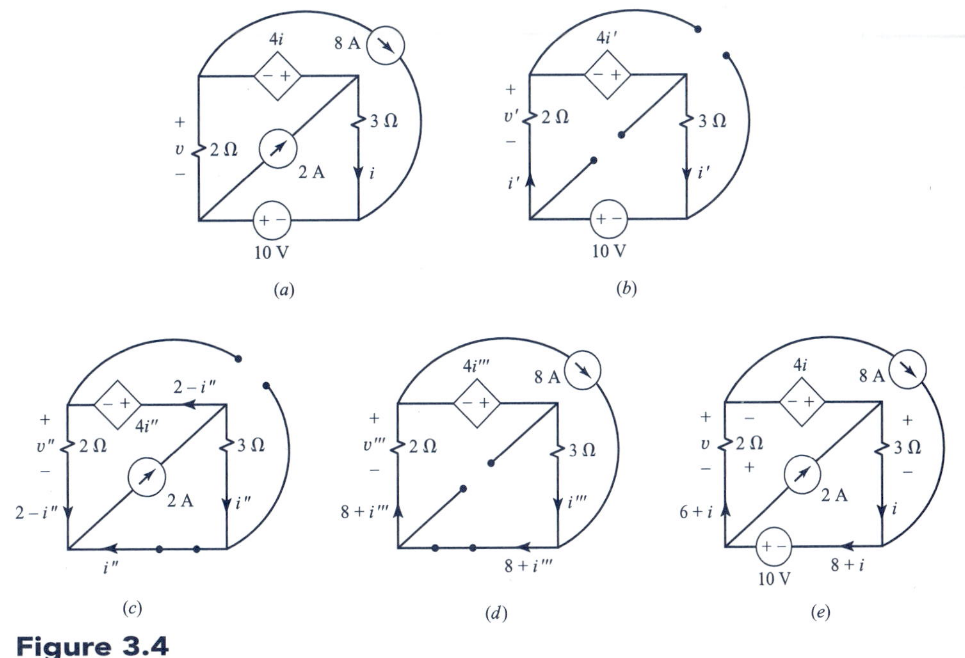

Figure 3.4
Example 3.2.

and

$$v''' = -2\,\Omega \times (8 + i''')$$
$$= 16\text{ V}$$

Thus, by superposition, the sum of these contributions gives the voltage v as

$$v = v' + v'' + v'''$$
$$= -20\text{ V} - 4\text{ V} + 16\text{ V}$$
$$= -8\text{ V}$$

Also we can obtain i as

$$i = i' + i'' + i'''$$
$$= 10\text{ A} + 4\text{ A} - 16\text{ A}$$
$$= -2\text{ A}$$

This can be confirmed using the direct method as shown in Fig. 3.4*e*. Applying KCL gives the current through the 2-Ω resistor as $6 + i$. Applying KVL around the loop gives

$$10\text{ V} = 2\,\Omega \times (6 + i) - 4i + 3\,\Omega \times i$$

or

$$i = -2\text{ A}$$

and

$$v = -2\,\Omega \times (6 + i)$$
$$= -8\text{ V}$$

as was determined by superposition.

This example has shown that controlled sources must not be deactivated in applying superposition. It may turn out that the controlling variable of a controlled source happens to be rendered zero in a particular superposition circuit, thereby deactivating that controlled source, but we did not deactivate it a priori.

Exercise Problem 3.2

Determine v and i in the circuit of Fig. E3.2 using superposition. Check your result using any other method.

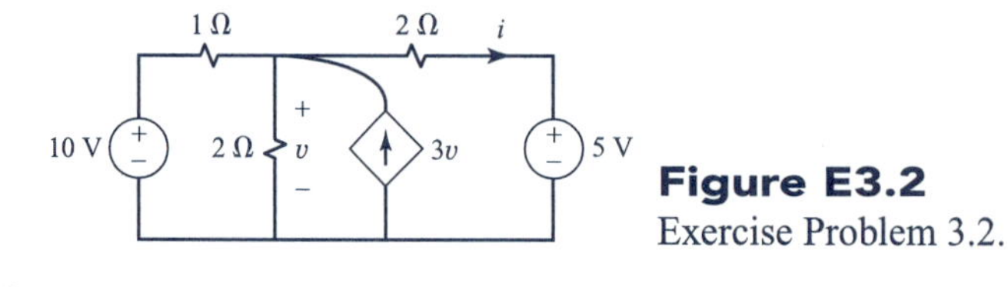

Figure E3.2
Exercise Problem 3.2.

Answers: $-\frac{25}{2}$ V and $-\frac{35}{4}$ A.

3.2 The Thevenin Equivalent Circuit

We next consider replacing a two-terminal linear circuit with a simpler, but equivalent, circuit at its two terminals. Consider separating a circuit into two portions that are attached at two common terminals as shown in Fig. 3.5*a*. The linear portion on the left may contain linear resistors, independent sources, and linear controlled sources. The remainder portion on the right may contain all these elements as well as nonlinear elements. We wish to replace the linear portion with an equivalent circuit such that its effect on the remainder portion

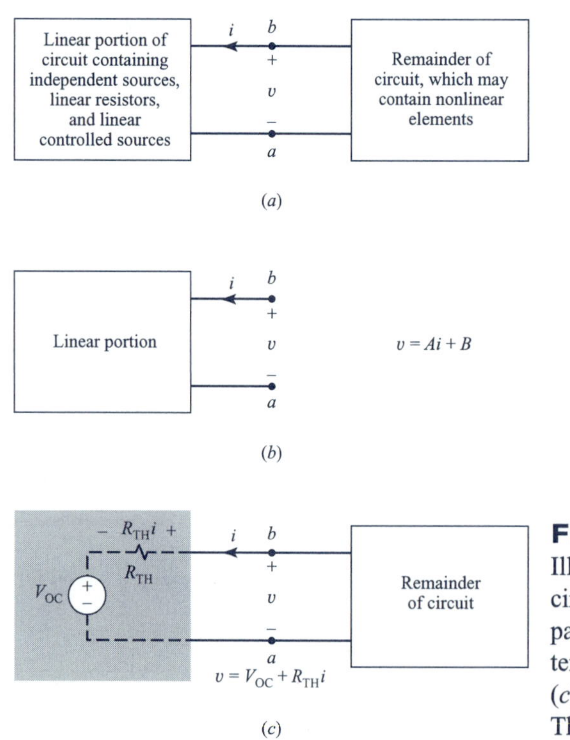

Figure 3.5
Illustration of the Thevenin equivalent circuit: (*a*) separating a circuit into two parts, (*b*) writing the v–i relation at the terminals of the linear part, and (*c*) representing the linear part with the Thevenin equivalent circuit.

is unchanged. In order for the replacement circuit to be equivalent to the original circuit at these two terminals, their v–i relations must be identical. How are the v and i of the linear circuit at terminals ab related? The only reasonable answer is as the equation of a straight line such as

$$v = Ai + B \tag{3.1}$$

as shown in Fig. 3.5*b*, since this portion of the circuit was stipulated to be linear. Now all we need to do is to obtain another circuit whose v–i relationship is the same as (3.1). Figure 3.5*c* shows a candidate, which is referred to as the *Thevenin equivalent circuit*. The circuit consists of the series combination of an independent voltage source, V_{OC}, and a resistor, R_{TH}. The meaning of the subscripts on these elements will become clear. To demonstrate that this circuit has the same relation as (3.1) and hence is equivalent, write KVL around the loop to yield

$$\boxed{v = R_{TH} i + V_{OC}} \tag{3.2}$$

Comparing (3.2) and (3.1) shows that $A = R_{TH}$ and $B = V_{OC}$. This equation shows that when the linear portion is separated from the remainder circuit and hence *open-circuited*, $i = 0$ and the voltage appearing at terminals ab is V_{OC}:

$$\boxed{V_{OC} = v|_{i=0}} \tag{3.3}$$

Hence V_{OC} is the *open-circuit voltage* appearing at the terminals. It should be clear that the *independent sources* contribute to V_{OC}. Hence if we *deactivate all independent sources* in the linear portion, leaving only resistors and controlled sources in it, R_{TH} is the equivalent resistance seen at terminals ab:

$$\boxed{R_{TH} = \frac{v}{i}\bigg|_{\text{all independent sources deactivated}}} \tag{3.4}$$

Note that v and i are labeled with the passive sign convention *with respect to the linear portion*. Hence if R_{TH} turns out to be a negative number, we have a true negative resistance. In essence, R_{TH} is simply the equivalent resistance seen looking into the terminals with the independent sources deactivated. But this was covered in Chapter 2, and hence there is nothing new in this second and final step of the procedure. Figure 3.6 summarizes the methods for computing V_{OC} and R_{TH}.

It is very important to label the terminals of the linear portion so that V_{OC} is inserted with the correct polarity, in this case positive at terminal b. A common mistake is to insert V_{OC} in such a way that the polarity is reversed. Another common mistake is not labeling v_x and i_x with the passive sign convention *with respect to the linear portion of the circuit* when computing R_{TH} as shown in Fig. 3.6*b*.

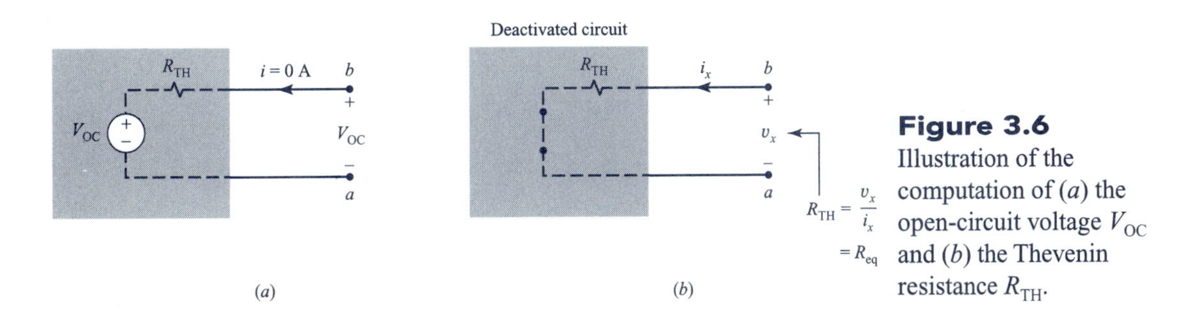

Figure 3.6 Illustration of the computation of (*a*) the open-circuit voltage V_{OC} and (*b*) the Thevenin resistance R_{TH}.

When computing the open-circuit voltage V_{OC}, it is important to label the current at the terminals as being zero (since the terminals are open-circuited, not allowing current flow through them) as illustrated in Fig. 3.6*a*. Many mistakes will be avoided if this is done. In addition, it will immediately show several important simplifications that are useful in solving for V_{OC}.

Example 3.3

Determine the current i in the circuit of Fig. 3.7 by reducing the circuit attached to the 5-Ω resistor to a Thevenin equivalent.

Solution Remove the remainder circuit (the 5-Ω resistor), and compute the open-circuit voltage as shown in Fig. 3.7*b*. Label the current at the terminals as being zero. This immediately shows that since $i = 0$ for the open circuit, the voltage across the 3-Ω resistor is zero. Hence the 3-Ω resistor does not enter into this computation, and V_{OC} appears across the 10-A current source. This illustrates the importance of labeling the zero-current condition at the terminals when computing V_{OC}. This open-circuit voltage can be determined by making

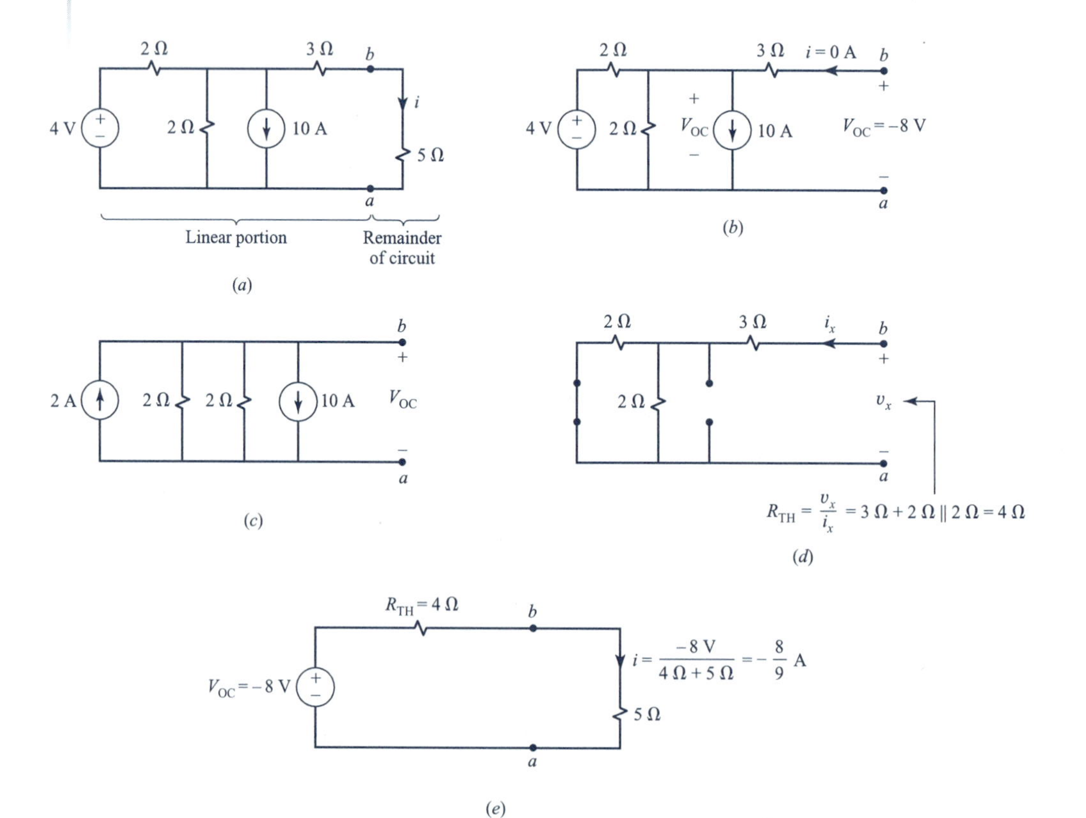

Figure 3.7
Example 3.3.

a source transformation with the 4-V voltage source and the 2-Ω resistor as shown in Fig. 3.7*c*. This is a single-node-pair circuit, and hence

$$\begin{aligned} V_{\mathrm{OC}} &= (2\,\Omega \,\|\, 2\,\Omega) \times (2\,\mathrm{A} - 10\,\mathrm{A}) \\ &= -8\,\mathrm{V} \end{aligned}$$

The computation of R_{TH} is shown in Fig. 3.7*d*. The two independent sources are deactivated, and we require the equivalent resistance seen at the terminals. Defining the terminal voltage v_x and current i_x gives

$$\begin{aligned} R_{\mathrm{TH}} &= \frac{v_x}{i_x} \\ &= 3\,\Omega + 2\,\Omega \,\|\, 2\,\Omega \\ &= 4\,\Omega \end{aligned}$$

Reconnecting the Thevenin equivalent circuit to the remainder circuit as shown in Fig. 3.7*e*, we have a single-loop circuit and hence

$$\begin{aligned} i &= \frac{-8\,\mathrm{V}}{4\,\Omega + 5\,\Omega} \\ &= -\tfrac{8}{9}\,\mathrm{A} \end{aligned}$$

The reader should verify this result directly from the original circuit in Fig. 3.7*a*, using superposition or any other method.

The presence of controlled sources presents no additional problems, as the following example shows.

Example 3.4

Determine current i in the circuit of Fig. 3.8*a* by use of a Thevenin equivalent.

Solution We *arbitrarily* choose to divide the circuit at the indicated terminals *ab*. Removing the remainder circuit and labeling the zero current at the terminals gives the circuit shown in Fig. 3.8*b*. Observe once again that the series 3-Ω resistor has no effect here, since its current is zero (a result of open-circuiting the terminals). We can obtain V_{OC} by using superposition. The two circuits each containing only one independent source are shown in Fig. 3.8*c*. Writing KVL around the loop containing the 10-V source, the 1-Ω resistor, and the controlled source in the first circuit gives

$$10\,\mathrm{V} = v' + 2v'$$

or

$$v' = \tfrac{10}{3}\,\mathrm{V}$$

Thus

$$\begin{aligned} V'_{\mathrm{OC}} &= 2v' \\ &= \tfrac{20}{3}\,\mathrm{V} \end{aligned}$$

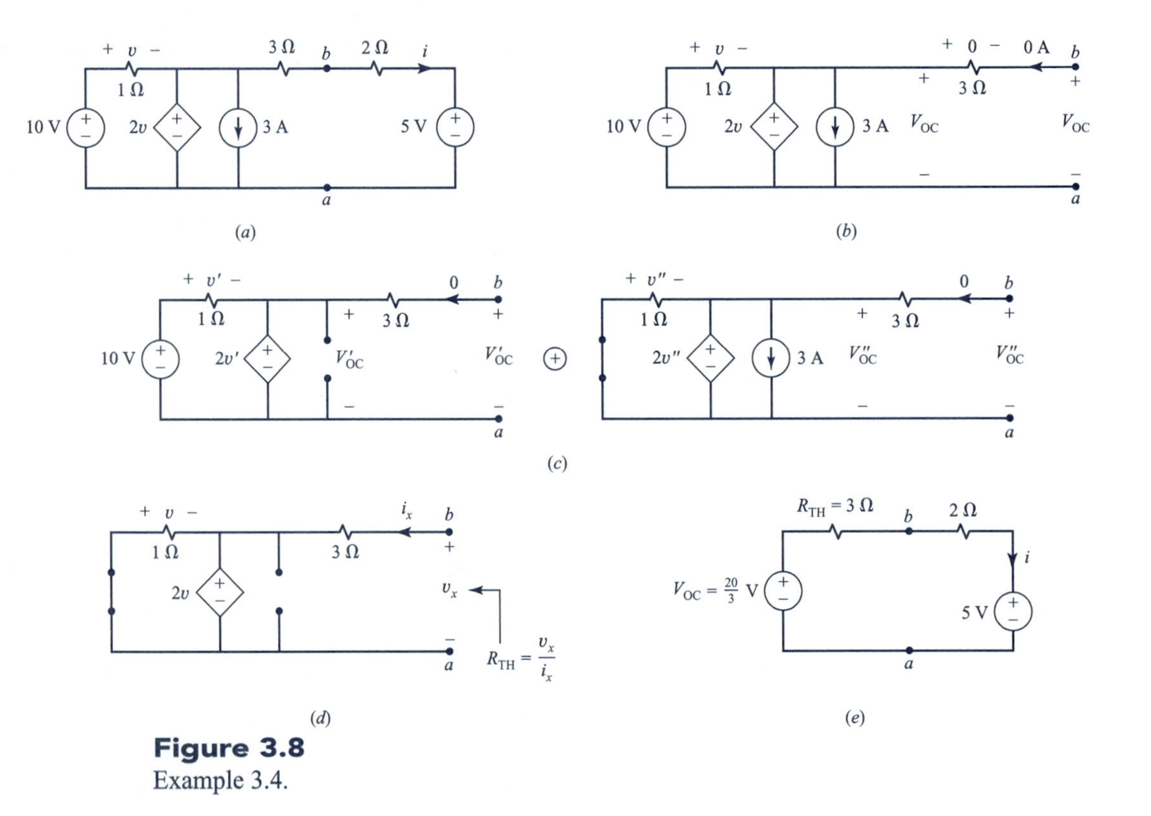

Figure 3.8
Example 3.4.

For the second circuit we obtain with KVL

$$v'' + 2v'' = 0$$

and hence

$$v'' = 0$$

and

$$\begin{aligned} V''_{OC} &= 2v'' \\ &= 0 \end{aligned}$$

Observe that in this second circuit, the 3-A current source has no effect on V_{OC}. This is due to the controlled source. The total open-circuit voltage becomes

$$\begin{aligned} V_{OC} &= V'_{OC} + V''_{OC} \\ &= \tfrac{20}{3}\,\text{V} + 0\,\text{V} \\ &= \tfrac{20}{3}\,\text{V} \end{aligned}$$

The circuit for determining R_{TH} has all *independent sources* deactivated and is shown in Fig. 3.8*d*. Defining the terminal voltage v_x and terminal current i_x with the passive sign convention, we obtain, using KVL,

$$v_x = 3\,\Omega \times i_x + 2v$$

But again

$$v + 2v = 0$$

and hence

$$v = 0$$

Thus

$$R_{TH} = \frac{v_x}{i_x} = 3\ \Omega$$

Reconnecting the Thevenin equivalent circuit to the remainder circuit gives the circuit of Fig. 3.8*e*, a single-loop circuit, from which we obtain

$$i = \frac{\frac{20}{3}\text{ V} - 5\text{ V}}{3\ \Omega + 2\ \Omega} = \frac{1}{3}\text{ A}$$

Exercise Problem 3.3

Determine the current i and voltage v in the circuit of Fig. E3.1 by use of a Thevenin equivalent circuit and placing the 3-Ω resistor in the remainder circuit.

Answers: $\frac{4}{7}$ V and $\frac{6}{7}$ A.

Exercise Problem 3.4

Determine the current i and voltage v in the circuit of Fig. E3.2 by use of a Thevenin equivalent circuit and placing the 5-V source in the remainder circuit.

Answers: $-\frac{35}{4}$ A and $-\frac{25}{2}$ V.

3.3 The Norton Equivalent Circuit

An alternative to the Thevenin equivalent circuit is the Norton equivalent circuit, shown in Fig. 3.9*b*. The validity of this equivalent circuit can be seen by making a source transformation on the Thevenin equivalent circuit. The Thevenin resistance remains the same and is calculated as before. The current source is the ratio of V_{OC} and R_{TH}. It can be determined by placing a *short circuit* across the terminals of the linear portion as shown in Fig. 3.9*c*. Hence the

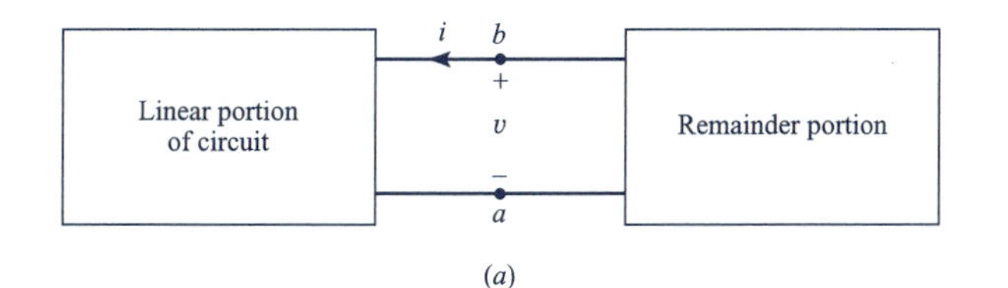

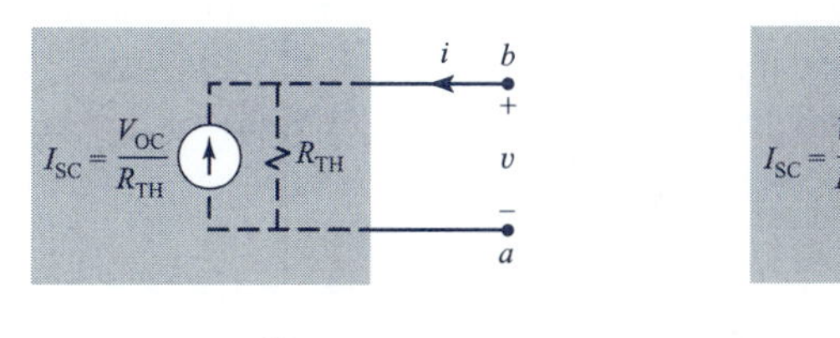

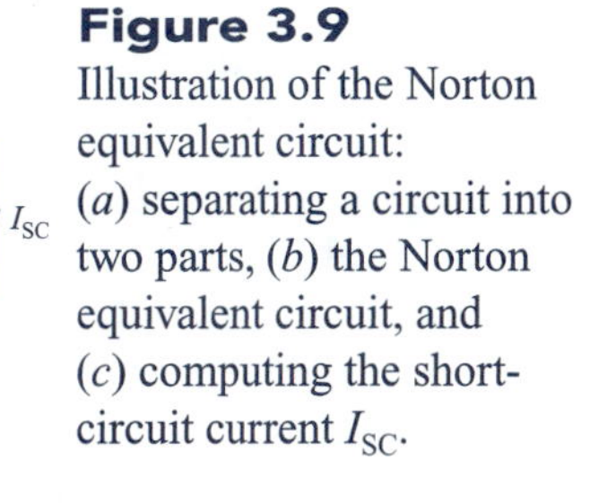

Figure 3.9
Illustration of the Norton equivalent circuit: (*a*) separating a circuit into two parts, (*b*) the Norton equivalent circuit, and (*c*) computing the short-circuit current I_{SC}.

current source of the Norton equivalent circuit, I_{SC}, represents the *short circuit current* through this short circuit placed across the terminals of the linear portion:

$$I_{SC} = -i|_{v=0} = \frac{V_{OC}}{R_{TH}} \tag{3.5}$$

It is vitally important when computing the short-circuit current I_{SC} to label the voltage across the terminals as zero, which is due to the short circuit across the terminals as illustrated in Fig. 3.9*c*. Doing so will immediately show various simplifications that can be used in solving for I_{SC}. This is essentially the dual to labeling the zero current at the terminals of the open circuit when computing V_{OC} for the Thevenin equivalent circuit.

Example 3.5

Determine the current *i* in the circuit of Fig. 3.7 that was considered in Example 3.3 by using a Norton equivalent circuit.

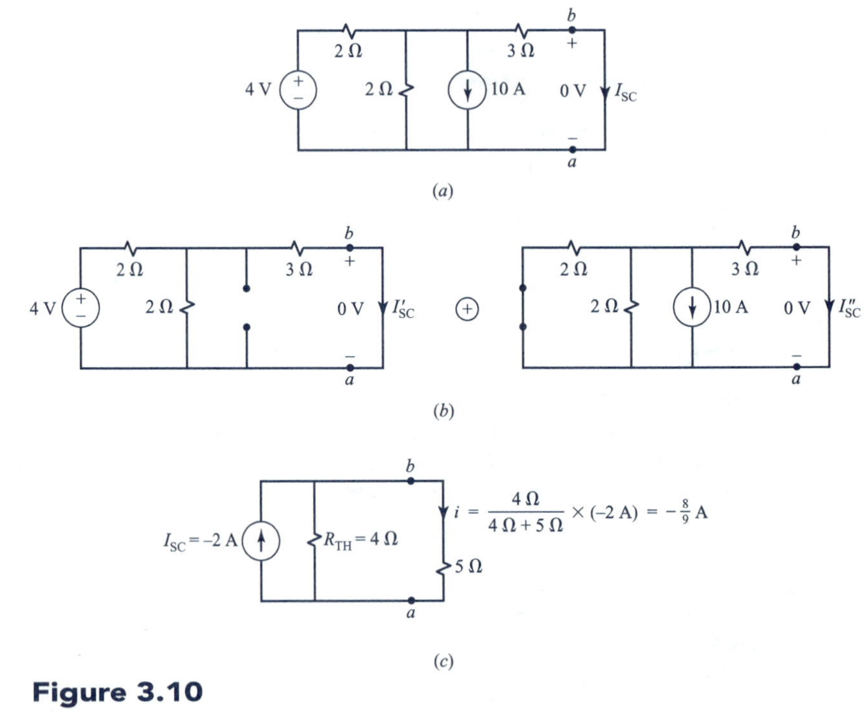

Figure 3.10
Example 3.5.

Solution The Thevenin resistance remains unchanged: $R_{TH} = 4\ \Omega$. The short-circuit current is determined from Fig. 3.10*a*. Observe that the zero-volt condition across the short circuit is inserted. Superposition is used to obtain this current, as shown in Fig. 3.10*b*, using current division in both cases:

$$I'_{SC} = \frac{2\ \Omega}{2\ \Omega + 3\ \Omega} \times \frac{4\ \text{V}}{2\ \Omega + 2\ \Omega \parallel 3\ \Omega}$$
$$= \tfrac{1}{2}\ \text{A}$$

and

$$I''_{SC} = -\frac{2\,\Omega \parallel 2\,\Omega}{2\,\Omega \parallel 2\,\Omega + 3\,\Omega} \times 10\,\text{A}$$
$$= -\tfrac{5}{2}\,\text{A}$$

Therefore

$$I_{SC} = I'_{SC} + I''_{SC}$$
$$= \tfrac{1}{2}\,\text{A} - \tfrac{5}{2}\,\text{A}$$
$$= -2\,\text{A}$$

Observe that

$$I_{SC} = \frac{V_{OC}}{R_{TH}}$$
$$= \frac{-8\,\text{V}}{4\,\Omega}$$
$$= -2\,\text{A}$$

Attaching the Norton equivalent circuit to the remainder circuit as shown in Fig. 3.10*c*, we obtain the current using current division as

$$i = \frac{4\,\Omega}{4\,\Omega + 5\,\Omega} \times (-2\,\text{A})$$
$$= -\tfrac{8}{9}\,\text{A}$$

as before.

Example 3.6

Determine the voltage V in the circuit of Fig. 3.11*a* using a Norton equivalent circuit.

Solution We arbitrarily partition the circuit into two parts as shown in Fig. 3.11*a*. We first determine the short-circuit current as shown in Fig. 3.11*b*. Observe again that the zero voltage across the short circuit is noted. Performing a source transformation gives the indicated circuit, and applying KCL at the upper node yields the current down through the $2\,\Omega \parallel 2\,\Omega = 1\text{-}\Omega$ resistor as $2i - \frac{5}{2}$. Applying KVL around the loop containing the two resistors gives

$$1\,\Omega \times \left(2i - \tfrac{5}{2}\right)\text{A} = 4\,\Omega \times i$$

which yields

$$I_{SC} = i$$
$$= -\tfrac{5}{4}\,\text{A}$$

The Thevenin equivalent resistance is found from the circuit of Fig. 3.11*c*. Since

$$i_x = -i$$

applying KVL gives

$$v_x = -4\,\Omega \times i + 2\,\Omega \parallel 2\,\Omega \times 2i$$
$$= 2i_x$$

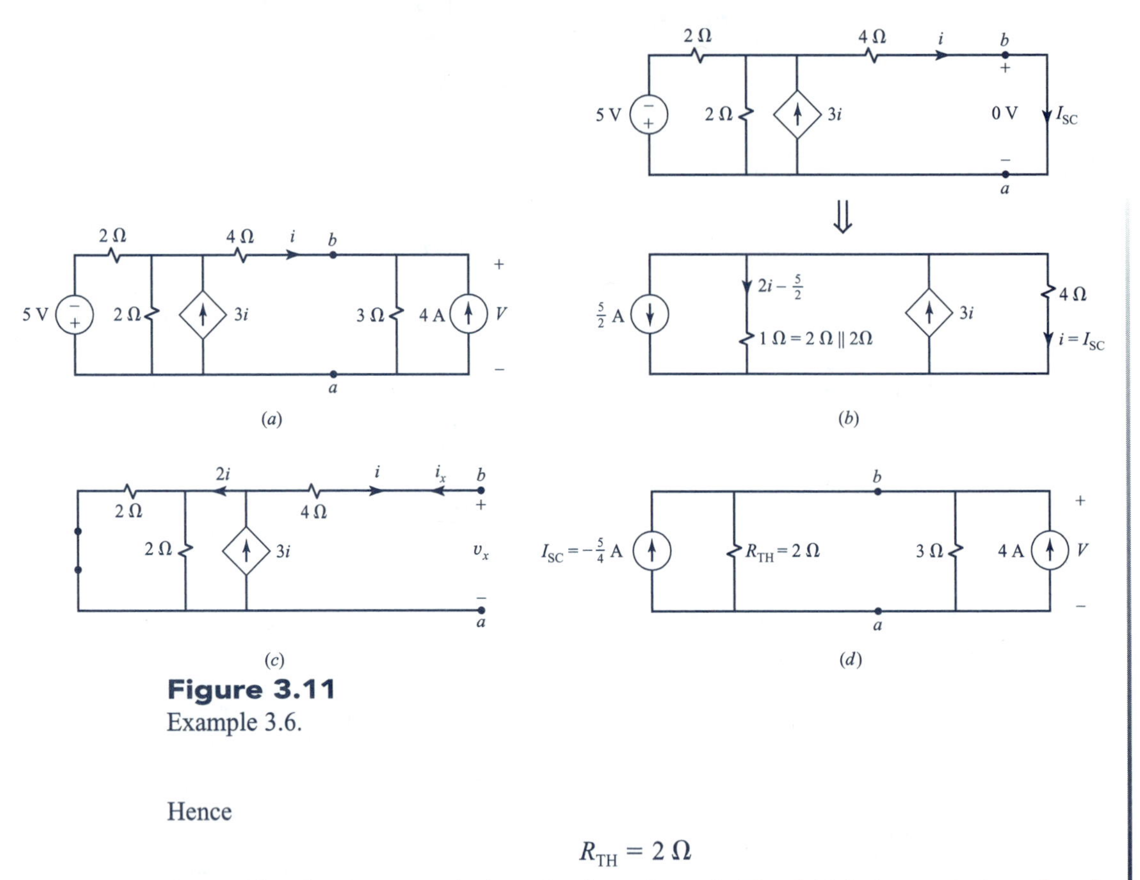

Figure 3.11
Example 3.6.

Hence

$$R_{TH} = 2\ \Omega$$

Reattaching the Norton equivalent circuit as shown in Fig. 3.11*d* gives a single-node-pair circuit, from which we obtain

$$V = (2\ \Omega \,\|\, 3\ \Omega) \times \left(-\tfrac{5}{4}\text{A} + 4\text{ A}\right)$$
$$= \tfrac{33}{10}\text{V}$$

Example 3.7

Figure 3.12*a* shows a circuit that is known as a *Wheatstone bridge*. It is useful in making precise measurements of resistances. The load resistor, R_L, represents the internal resistance of a meter that measures the current through it. It is possible to determine the value of an unknown resistor, R_4, by adjusting the (known) values of R_1, R_2, and R_3 so that the voltage across R_L is zero. Determine the Thevenin and Norton equivalent circuits for this device at terminals *ab*.

Solution First we determine the open-circuit voltage as shown in Fig. 3.12*b*. The zero current imposed by the open circuit is indicated. Because of this zero-current condition, the voltages across R_1 and R_2 can be determined by voltage division:

$$v_1 = \frac{R_1}{R_1 + R_3} V_S$$

$$v_2 = \frac{R_2}{R_2 + R_4} V_S$$

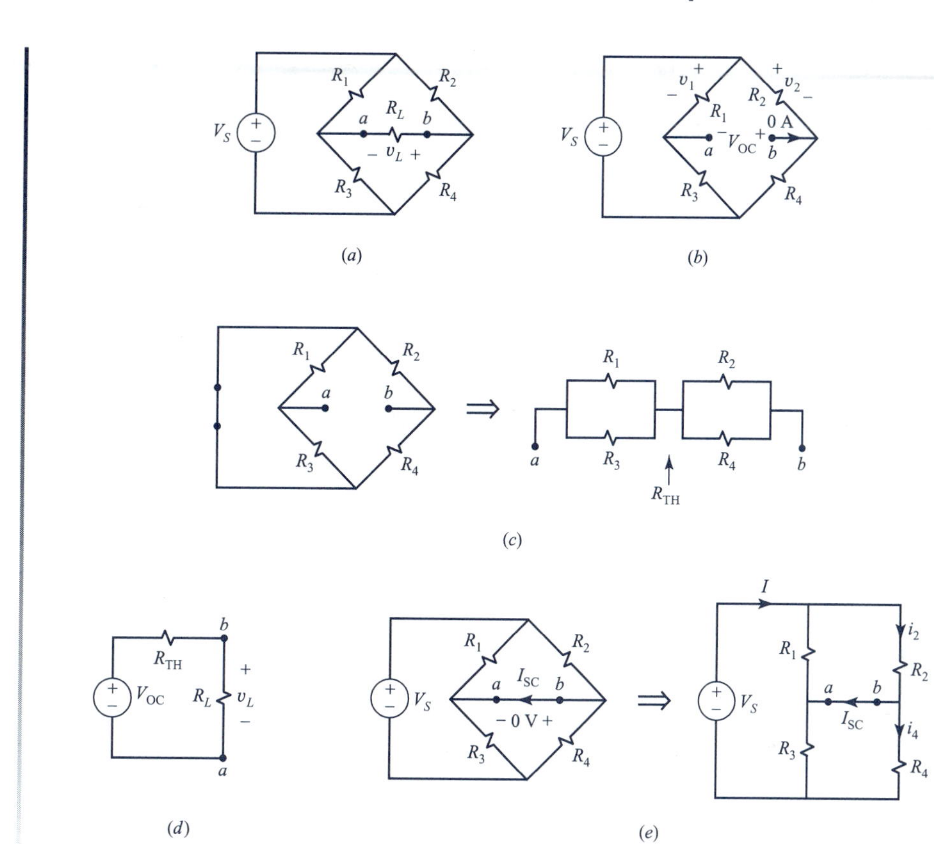

Figure 3.12

The Wheatstone bridge: Example 3.7.

The open-circuit voltage then becomes, using KVL,

$$
\begin{aligned}
V_{OC} &= v_1 - v_2 \\
&= \left(\frac{R_1}{R_1 + R_3} - \frac{R_2}{R_2 + R_4}\right) V_S \\
&= \frac{R_1 R_4 - R_2 R_3}{R_1 R_2 + R_1 R_4 + R_2 R_3 + R_3 R_4} V_S
\end{aligned}
$$

The Thevenin resistance is determined from Fig. 3.12*c* as

$$
\begin{aligned}
R_{TH} &= R_1 \,\|\, R_3 + R_2 \,\|\, R_4 \\
&= \frac{R_1 R_3}{R_1 + R_3} + \frac{R_2 R_4}{R_2 + R_4} \\
&= \frac{R_1 R_2 R_3 + R_1 R_2 R_4 + R_1 R_3 R_4 + R_2 R_3 R_4}{R_1 R_2 + R_1 R_4 + R_2 R_3 + R_3 R_4}
\end{aligned}
$$

Reattaching the Thevenin equivalent gives the circuit shown in Fig. 3.12*d*. From this we see that $v_L = 0$ only if $V_{OC} = 0$. But from our previous result we see that $V_{OC} = 0$ only if

$$R_1 R_4 = R_2 R_3$$

or

$$\frac{R_1}{R_2} = \frac{R_3}{R_4}$$

This is the condition for *balance* of the bridge.

The short-circuit current for the Norton equivalent circuit is determined from Fig. 3.12*e*. The zero-voltage condition imposed by the short circuit is indicated. To determine this we first determine current I as

$$\begin{aligned} I &= \frac{V_S}{R_1 \| R_2 + R_3 \| R_4} \\ &= \frac{V_S}{\dfrac{R_1R_2}{R_1 + R_2} + \dfrac{R_3R_4}{R_3 + R_4}} \\ &= \frac{R_1R_3 + R_1R_4 + R_2R_3 + R_2R_4}{R_1R_2R_3 + R_1R_2R_4 + R_1R_3R_4 + R_2R_3R_4} V_S \end{aligned}$$

Then currents i_2 and i_4 can be determined by current division as

$$i_2 = \frac{R_1}{R_1 + R_2} I$$

$$i_4 = \frac{R_3}{R_3 + R_4} I$$

Hence the short-circuit current becomes, by KCL, and substituting the result for I,

$$\begin{aligned} I_{SC} &= i_2 - i_4 \\ &= \left(\frac{R_1}{R_1 + R_2} - \frac{R_3}{R_3 + R_4}\right) I \\ &= \frac{R_1R_4 - R_2R_3}{R_1R_2R_3 + R_1R_2R_4 + R_1R_3R_4 + R_2R_3R_4} V_S \end{aligned}$$

Again we obtain the balance condition when $I_{SC} = 0$:

$$\frac{R_1}{R_2} = \frac{R_3}{R_4}$$

We can confirm our calculation of R_{TH} by combining these results as

$$\begin{aligned} R_{TH} &= \frac{V_{OC}}{I_{SC}} \\ &= \frac{R_1R_2R_3 + R_1R_2R_4 + R_1R_3R_4 + R_2R_3R_4}{R_1R_2 + R_1R_4 + R_2R_3 + R_3R_4} \end{aligned}$$

as before.

Exercise Problem 3.5

Determine the current i and voltage v in the circuit of Fig. E3.1 by use of a Norton equivalent circuit and placing the 3-Ω resistor in the remainder circuit.

Answers: $\frac{4}{7}$ V and $\frac{6}{7}$ A.

Exercise Problem 3.6

Determine the current i and voltage v in the circuit of Fig. E3.2 by use of a Norton equivalent circuit and placing the 5-V source in the remainder circuit.

Answers: $-\frac{35}{4}$ A and $-\frac{25}{2}$ V.

3.4 Maximum Power Transfer

One of the most important concepts in modeling is the basic *source–load* configuration shown in Fig. 3.13. All practical sources *effectively* have internal resistance represented by R_S. When the source is attached to a load, represented by R_L, all of the source open-circuit voltage, $v_S(t)$, will not appear across the terminals of the load. Some of $v_S(t)$ will be developed across the source resistance. The question arises as to what would be the optimum choice for R_L (if we have a choice) such that maximum power will be delivered from the source to the load. Ordinarily we do not have a choice as to the value of R_S, since it is the equivalent internal resistance of the source and as such it is not accessible.

Let us determine this optimum value of R_L to achieve maximum power transfer from the source to the load. The load current is

$$i_L = \frac{v_S}{R_S + R_L}$$

and the power delivered to the load is

$$\begin{aligned} p_L &= i_L^2 R_L \\ &= \frac{R_L}{(R_S + R_L)^2} v_S^2 \end{aligned}$$

The important question is: what value of R_L will maximize this expression? If $R_L = 0$ (a short circuit) then $p_L = 0$, and if $R_L = \infty$ (an open circuit) then $p_L = 0$. Hence the desired value is somewhere between these values: $0 < R_L < \infty$. To obtain the value of R_L that will maximize this expression we differentiate this expression with respect to R_L and set the result equal to zero, obtaining

$$\begin{aligned} \frac{dp_L}{dR_L} &= 0 \\ &= \frac{(R_S + R_L)^2 - 2R_L(R_S + R_L)}{(R_S + R_L)^4} v_S^2 \end{aligned}$$

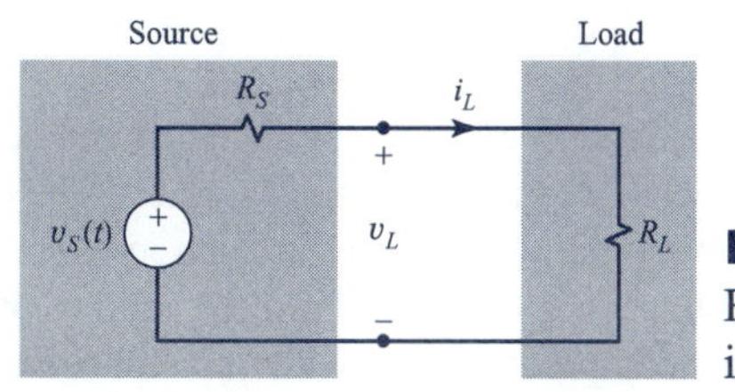

Figure 3.13
Representation of the general source–load configuration in simplified form.

whose solution is

$$R_L = R_S \tag{3.6}$$

Although this could give a minimum, one can show that the power expression has only a maximum. Thus *for maximum power transfer to a load, we should choose the load resistance to be equal to the internal resistance of the source.* Under these conditions, the maximum power delivered to the load is

$$p_L\Big|_{\max} = \frac{v_S^2}{4(R_S = R_L)} \tag{3.7}$$

Example 3.8

Determine the value of load resistance to achieve maximum power transfer to that load for the circuit of Fig. 3.14*a*.

Solution The circuit attached to the load can be reduced to a Thevenin equivalent circuit as shown in Fig. 3.14*b*. From that circuit we determine $R_L = R_{TH} = 4\ \Omega$, and the maximum power is $\frac{49}{16}$ W.

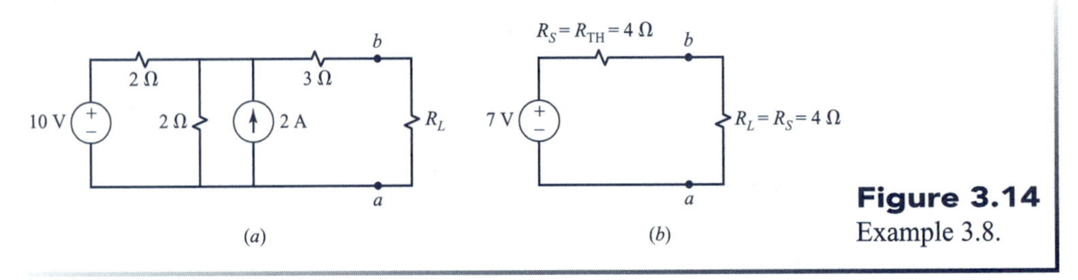

Figure 3.14
Example 3.8.

Exercise Problem 3.7

Determine the value of R_L for maximum power transfer in the circuit of Fig. E3.7 as well as that maximum power.

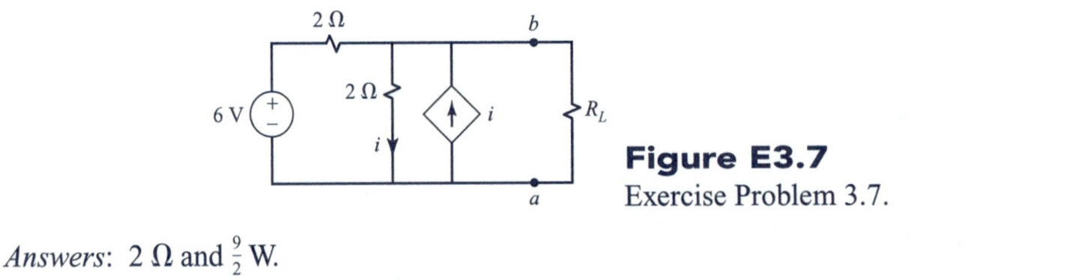

Figure E3.7
Exercise Problem 3.7.

Answers: 2 Ω and $\frac{9}{2}$ W.

3.5 The Node-Voltage Method

All of our previous analysis methods were entirely adequate for the analysis of relatively simple circuits. These simple circuits are important because they are amenable to hand calculations and hence provide considerable insight into the behavior of more general circuits. They also represent a large number of practical circuits. However, many other practical circuits consist of a large number of interconnected resistors, independent sources, and controlled sources. For these larger circuits, hand calculation is not feasible, and computer methods are used for their analysis.

In this and the following section we will investigate two techniques for the analysis of large circuits. These methods seek to systematically generate a set of simultaneous algebraic equations whose solution allows us to obtain all the other voltages and currents in the circuit. Most computer-aided circuit analysis routines generate the required equations using these methods or some variation of them. In fact, the most popular circuit analysis program, SPICE, uses a variation of the *node-voltage method,* which we will discuss, to systematically generate the required simultaneous equations that are to be solved. While these methods are quite straightforward, they have the drawback that they result in simultaneous equations to be solved. Solving simultaneous equations consisting of more than two equations in two unknowns by hand is very difficult and prone to error. Hence these larger sets of equations are solved using a digital computer.

In order to illustrate the node-voltage method, consider the circuit shown in Fig. 3.15*a*. There are a total of four nodes. We *arbitrarily* choose one node, node a, as a *reference node* to which the *node voltages* are referenced. Then we define the voltages of the other three nodes, b,c,d, with respect to this reference node as shown in Fig. 3.15*b*. The voltages V_b, V_c, V_d are said to be the node voltages and are defined between the respective node and the reference node, positive at the respective nodes. For example, node voltage V_b is defined between node b and the reference node, node a, and is positive at node b. Voltages between pairs of these nodes can be related to the node voltages using KVL. For example, the voltage between node c and node b positive at node c is $V_c - V_b$. Hence if we determine the solutions for the three node voltages, we can determine the voltages across every element in the circuit in terms of these node voltages.

The key to writing a set of three equations in these three node voltages is to *write KCL at the three nodes*. The first step is to *express the voltage across every resistor in terms of the node voltages and then use Ohm's law to write the current through that resistor in terms of those node voltages* as shown in Fig. 3.15*c*. It is vitally important to label the direction through the resistors with the passive sign convention with respect to the voltage across them so that a sign error will not occur in applying Ohm's law. For example, the current through R_2 and directed down is V_b/R_2. It is a good idea to label this current direction on the resistor with an arrow according to the passive sign convention so that Ohm's law for that resistor will be correctly written. Similarly, the current through R_1 and directed left to right is $(V_c - V_b)/R_1$ with the direction denoted by a small arrow. Once these resistor currents are determined, we *write KCL at the $n - 1$ nodes:*

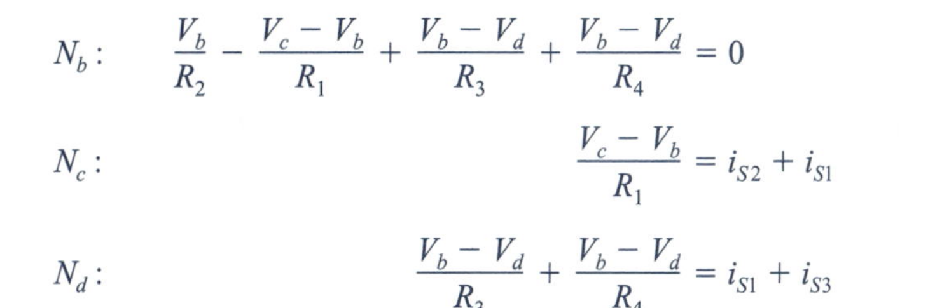

$$N_b: \quad \frac{V_b}{R_2} - \frac{V_c - V_b}{R_1} + \frac{V_b - V_d}{R_3} + \frac{V_b - V_d}{R_4} = 0$$

$$N_c: \quad \frac{V_c - V_b}{R_1} = i_{S2} + i_{S1}$$

$$N_d: \quad \frac{V_b - V_d}{R_3} + \frac{V_b - V_d}{R_4} = i_{S1} + i_{S3}$$

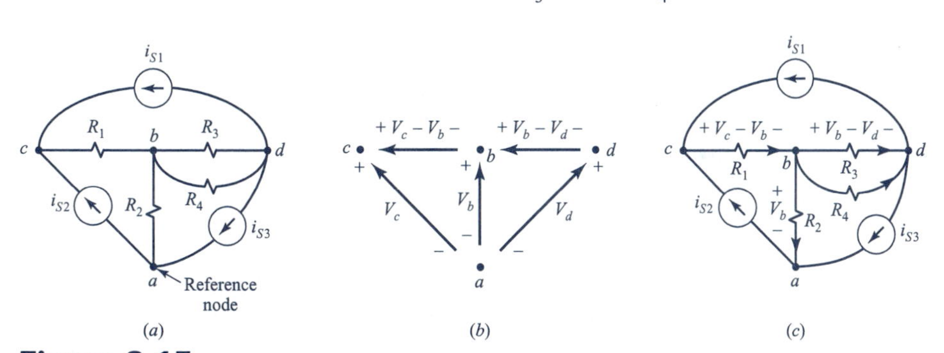

Figure 3.15
Illustration of the node-voltage method: (*a*) a general circuit, (*b*) definition of the node voltages, and (*c*) labeling of the original circuit with the node voltages.

Observe that we have placed the unknown quantities involving the node voltages on the left side and placed the known quantities (the current-source currents) on the right side. Collecting terms, we place these equations in *standard form* by organizing the coefficients of the node voltages and ordering them in the same fashion in each equation:

$$N_b: \quad \left(\frac{1}{R_1}+\frac{1}{R_2}+\frac{1}{R_3}+\frac{1}{R_4}\right)V_b - \frac{1}{R_1}V_c - \left(\frac{1}{R_3}+\frac{1}{R_4}\right)V_d = 0$$

$$N_c: \quad -\frac{1}{R_1}V_b + \frac{1}{R_1}V_c = i_{S2} + i_{S1}$$

$$N_d: \quad -\left(\frac{1}{R_3}+\frac{1}{R_4}\right)V_b + \left(\frac{1}{R_3}+\frac{1}{R_4}\right)V_d = -i_{S1} - i_{S3}$$

In matrix form (see the Appendix) these may be written compactly as

$$\mathbf{GV} = \mathbf{I}$$

where

$$\mathbf{G} = \begin{bmatrix} \frac{1}{R_1}+\frac{1}{R_2}+\frac{1}{R_3}+\frac{1}{R_4} & -\frac{1}{R_1} & -\left(\frac{1}{R_3}+\frac{1}{R_4}\right) \\ -\frac{1}{R_1} & \frac{1}{R_1} & 0 \\ -\left(\frac{1}{R_3}+\frac{1}{R_4}\right) & 0 & \frac{1}{R_3}+\frac{1}{R_4} \end{bmatrix}$$

$$\mathbf{V} = \begin{bmatrix} V_b \\ V_c \\ V_d \end{bmatrix}$$

$$\mathbf{I} = \begin{bmatrix} 0 \\ i_{S1} + i_{S2} \\ -i_{S1} - i_{S3} \end{bmatrix}$$

Once these are solved, all other element voltages and currents are easily determined.

Example 3.9

Write the node-voltage equations for the circuit of Fig. 3.16*a*, and solve for the current I. Use superposition to check the result.

Solution Node d is arbitrarily chosen as the reference node. The other node voltages are then designated positive at their respective nodes with respect to this reference node, as shown in Fig. 3.16*b*. Similarly the other element voltages are written in terms of these node voltages. The original circuit with the resistor voltages labeled in terms of these node voltages and arrows denoting the resulting currents according to Ohm's law is shown in Fig. 3.16*c*. Writing KCL at the three nodes a, b, c, yields

$$N_a: \quad \frac{V_a}{2\,\Omega} - \frac{V_b - V_a}{1\,\Omega} = 3\text{ A}$$

$$N_b: \quad \frac{V_b - V_a}{1\,\Omega} - \frac{V_c - V_b}{2\,\Omega} = -2\text{ A}$$

$$N_c: \quad \frac{V_c}{1\,\Omega} + \frac{V_c - V_b}{2\,\Omega} = -3\text{ A}$$

Combining terms and putting the equations in standard form yields

$$N_a: \quad \left(\frac{1}{1\,\Omega} + \frac{1}{2\,\Omega}\right)V_a - \frac{1}{1\,\Omega}V_b = 3\text{ A}$$

$$N_b: \quad -\frac{1}{1\,\Omega}V_a + \left(\frac{1}{1\,\Omega} + \frac{1}{2\,\Omega}\right)V_b - \frac{1}{2\,\Omega}V_c = -2\text{ A}$$

$$N_c: \quad -\frac{1}{2\,\Omega}V_b + \left(\frac{1}{1\,\Omega} + \frac{1}{2\,\Omega}\right)V_c = -3\text{ A}$$

or in matrix form

$$\begin{bmatrix} \frac{3}{2} & -1 & 0 \\ -1 & \frac{3}{2} & -\frac{1}{2} \\ 0 & -\frac{1}{2} & \frac{3}{2} \end{bmatrix} \begin{bmatrix} V_a \\ V_b \\ V_c \end{bmatrix} = \begin{bmatrix} 3 \\ -2 \\ -3 \end{bmatrix}$$

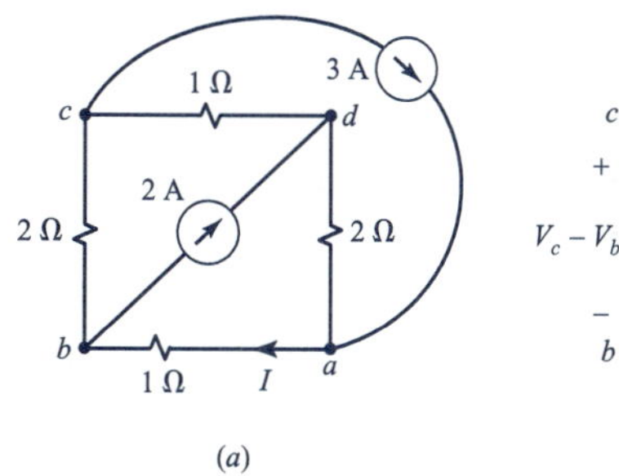

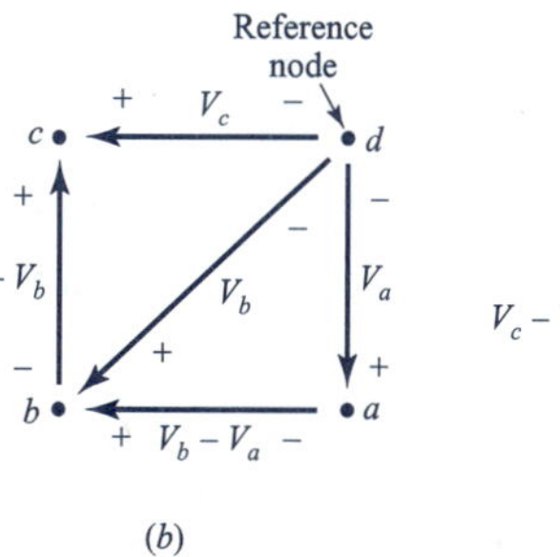

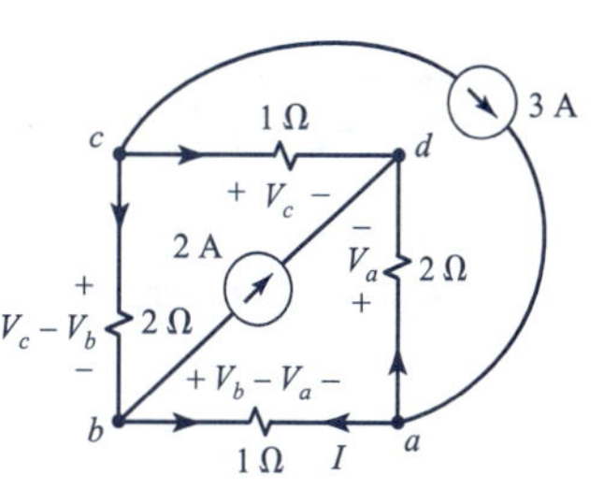

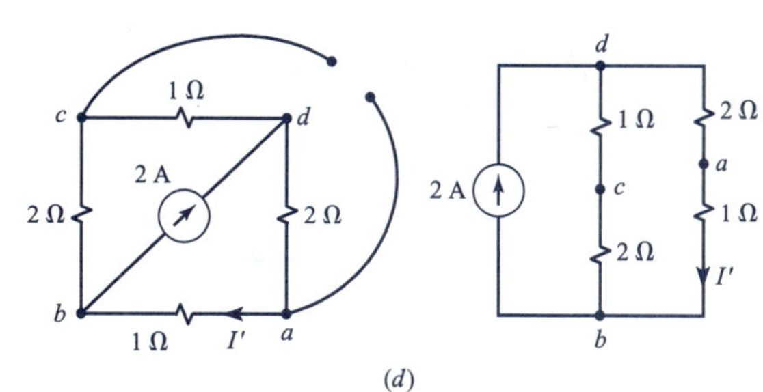

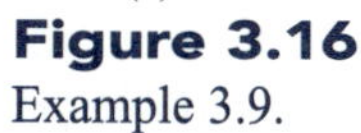

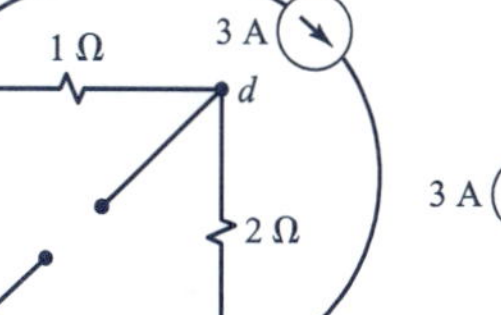

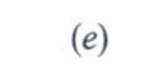

Figure 3.16
Example 3.9.

Solving these gives

$$V_a = 1\text{ V}$$
$$V_b = -\tfrac{3}{2}\text{ V}$$
$$V_c = -\tfrac{5}{2}\text{ V}$$

Current I becomes

$$I = -\frac{V_b - V_a}{1\ \Omega}$$
$$= \tfrac{5}{2}\text{ A}$$

Utilizing superposition, we obtain the two circuits shown in Fig. 3.16*d* and *e*. These are redrawn to facilitate the application of current division. The results are

$$I' = \frac{1\ \Omega + 2\ \Omega}{1\ \Omega + 2\ \Omega + 1\ \Omega + 2\ \Omega} \times 2\text{ A}$$
$$= 1\text{ A}$$
$$I'' = \frac{1\ \Omega + 2\ \Omega}{1\ \Omega + 2\ \Omega + 1\ \Omega + 2\ \Omega} \times 3\text{ A}$$
$$= \tfrac{3}{2}\text{ A}$$

Hence

$$I = I' + I''$$
$$= 1\text{ A} + \tfrac{3}{2}\text{ A}$$
$$= \tfrac{5}{2}\text{ A}$$

as was obtained with node-voltage equations.

Exercise Problem 3.8

Determine current I in the circuit of Fig. E3.8 by writing node-voltage equations. Check your result using superposition or source transformations.

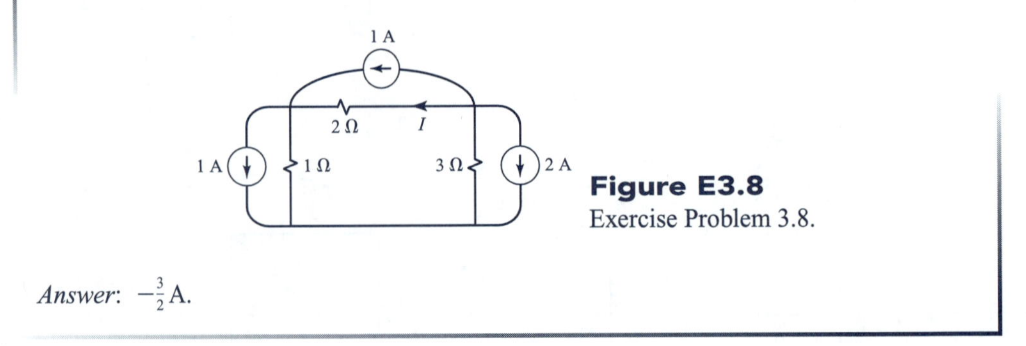

Figure E3.8
Exercise Problem 3.8.

Answer: $-\tfrac{3}{2}$ A.

The previous examples illustrated writing the node-voltage equations for circuits that contained independent current sources. Circuits containing *controlled* current sources are handled similarly with the added requirement that we write the controlling variable in terms of the node voltages.

Example 3.10

Write the node-voltage equations for the circuit of Fig. 3.17*a*.

Solution The node voltages with respect to the chosen reference node, node d, are shown in Fig. 3.17*b* and on the circuit diagram in Fig. 3.17*c*. Writing KCL at nodes a, b, and c gives

$$N_a: \quad \frac{V_a - V_b}{R_1} - \alpha i_x = i_S$$

$$N_b: \quad -\frac{V_a - V_b}{R_1} + \frac{V_b}{R_2} + \frac{V_b - V_c}{R_3} = 0$$

$$N_c: \quad -\frac{V_b - V_c}{R_3} + \frac{V_c}{R_4} = -i_S$$

The controlling variable is written in terms of the node voltages as

$$i_x = \frac{V_b - V_c}{R_3}$$

Substituting this and collecting terms yields the node-voltage equations as

$$N_a: \quad \frac{1}{R_1}V_a - \left(\frac{1}{R_1} + \frac{\alpha}{R_3}\right)V_b + \frac{\alpha}{R_3}V_c = i_S$$

$$N_b: \quad -\frac{1}{R_1}V_a + \left(\frac{1}{R_1} + \frac{1}{R_2} + \frac{1}{R_3}\right)V_b - \frac{1}{R_3}V_c = 0$$

$$N_c: \quad -\frac{1}{R_3}V_b + \left(\frac{1}{R_3} + \frac{1}{R_4}\right)V_c = -i_S$$

In matrix form these become

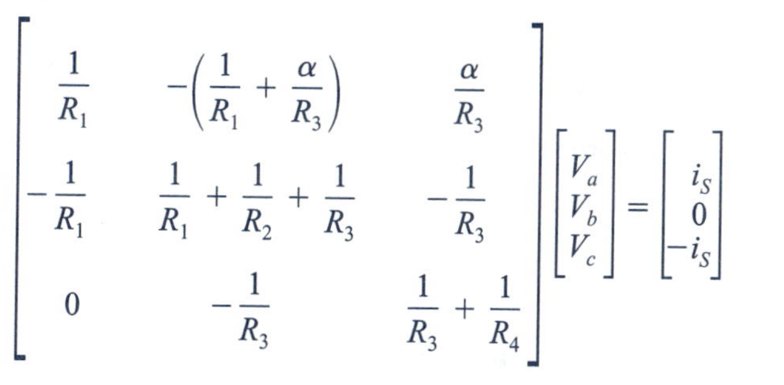

$$\begin{bmatrix} \frac{1}{R_1} & -\left(\frac{1}{R_1} + \frac{\alpha}{R_3}\right) & \frac{\alpha}{R_3} \\ -\frac{1}{R_1} & \frac{1}{R_1} + \frac{1}{R_2} + \frac{1}{R_3} & -\frac{1}{R_3} \\ 0 & -\frac{1}{R_3} & \frac{1}{R_3} + \frac{1}{R_4} \end{bmatrix} \begin{bmatrix} V_a \\ V_b \\ V_c \end{bmatrix} = \begin{bmatrix} i_S \\ 0 \\ -i_S \end{bmatrix}$$

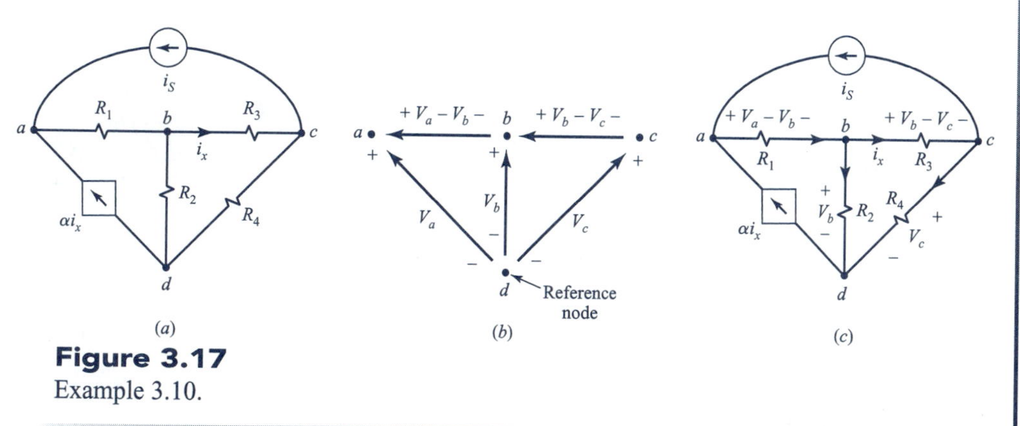

Figure 3.17
Example 3.10.

Exercise Problem 3.9

Determine voltage V in the circuit of Fig. E3.9.

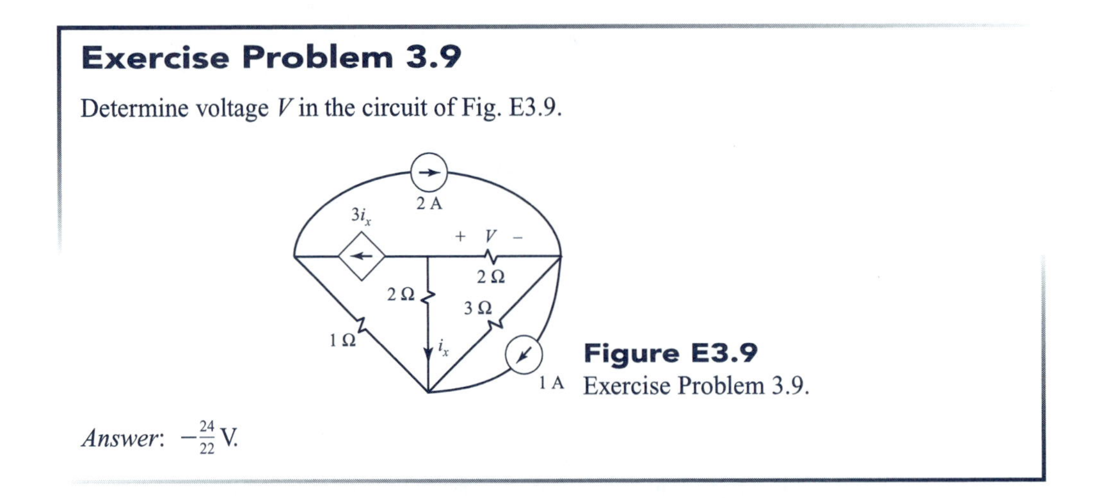

Figure E3.9
Exercise Problem 3.9.

Answer: $-\frac{24}{22}$ V.

3.5.1 Circuits Containing Voltage Sources The previous circuits contained current sources, which facilitated writing KCL. In this section we will consider writing node-voltage equations for circuits that contain *voltage sources* (independent or controlled). The basic problem with voltage sources is that we do not know the current through them, nor can we relate that current to the voltage of the element. Hence there appears to be a difficulty in writing KCL at the nodes on either side of the voltage source. In actuality this is not the problem it seems, since *voltage sources either constrain the node voltages or relate one node voltage to another.* The following example illustrates this situation.

Example 3.11

Write the node-voltage equations for the circuit in Fig. 3.18*a*.

Solution Node d is arbitrarily chosen as the reference node, and the node voltages are defined in the usual fashion as shown in Fig. 3.18*b*. Observe that the two voltage sources provide important constraints between the node voltages:

$$V_c = v_{S1}$$
$$V_a - V_b = v_{S2}$$

Observe that node voltage V_c is no longer an unknown, since it is directly related to v_{S1}. Also note that one of the remaining node voltages V_a and V_b is also no longer an unknown, since they are related by v_{S2}. Hence only one of the three node voltages is unknown, and therefore

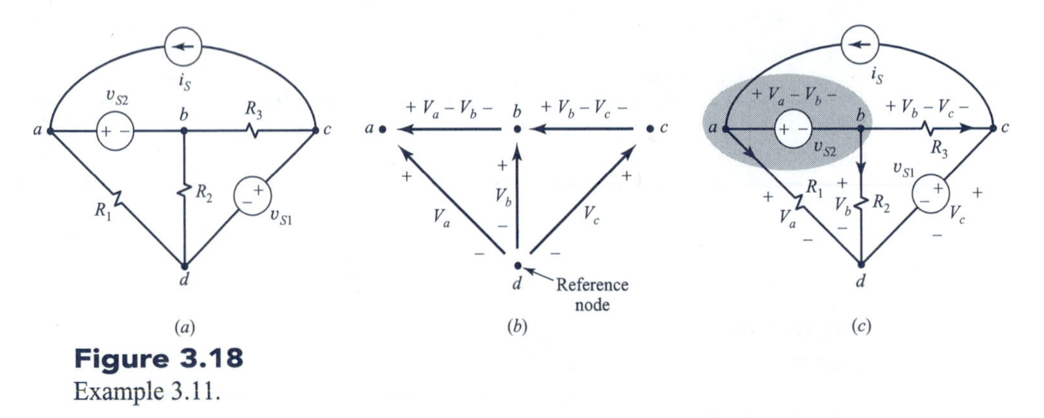

Figure 3.18
Example 3.11.

we only need one node-voltage equation in this unknown node voltage. It will do no good to write KCL at node c, since we do not know the current through voltage source v_{S1}. The remaining node-voltage equation can be obtained by placing a *supernode* around voltage source v_{S2} and writing KCL for that supernode as

$$\frac{V_a}{R_1} + \frac{V_b}{R_2} + \frac{V_b - V_c}{R_3} = i_S$$

Substituting the above two constraints generated by the two voltage sources (and arbitrarily choosing V_a instead of V_b to be the unknown) yields

$$\left(\frac{1}{R_1} + \frac{1}{R_2} + \frac{1}{R_3}\right)V_a = i_S + \frac{1}{R_3}v_{S1} + \left(\frac{1}{R_2} + \frac{1}{R_3}\right)v_{S2}$$

Solving this equation gives V_a, and the remaining node voltages, V_b and V_c, can be related to this through the constraints imposed by the two voltage sources.

This example has indicated the following general rule for writing the node-voltage equations for circuits that contain voltage sources:

1. *Choose a reference node, and define the node voltages in the usual fashion.*
2. *Write the constraints that are imposed on the node voltages by the voltage sources.*
3. *Draw supernodes around all voltage sources that have neither end connected to the reference node.*
4. *Write KCL for all supernodes and all remaining nodes except those nodes that have a voltage source connected between that node and the reference node.*
5. *Substitute the constraints imposed by the voltage sources into these equations, and place them in standard form.*

From these results it is easy to show that *in a circuit containing n nodes and N_{V_S} voltage sources (independent or controlled), the number of node-voltage equations equals*

$$n - 1 - N_{V_S}$$

Example 3.12

Determine current I in the circuit of Fig. 3.19*a* by writing node-voltage equations.

Solution The node voltages are defined in Fig. 3.19*b*, and the voltage across each resistor is written in terms of these node voltages. First observe the constraints imposed by the voltage sources:

$$V_c - V_a = 1\text{ V}$$
$$V_b = -3\text{ V}$$

Since the 3-V voltage source is connected between node b and the reference node, we will not write KCL at node b. Drawing a supernode around nodes a and c and writing KCL for that supernode gives

$$\frac{V_a}{2\,\Omega} + \frac{V_a - V_b}{2\,\Omega} + \frac{V_c}{3\,\Omega} = 2\text{ A}$$

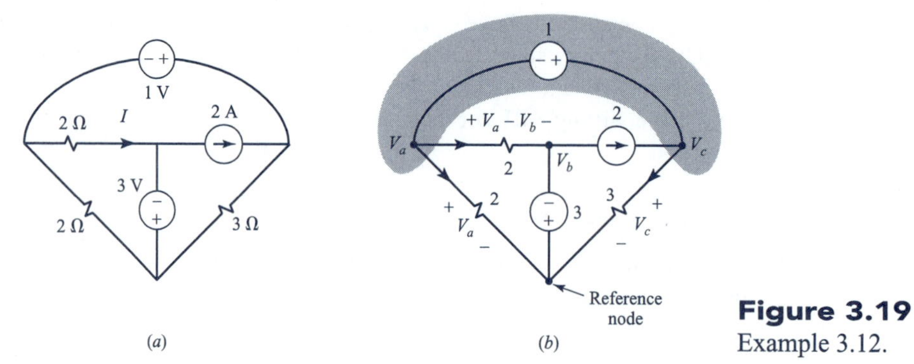

Figure 3.19
Example 3.12.

Rearranging these gives

$$V_a - \frac{V_b}{2} + \frac{V_c}{3} = 2$$

Substituting the voltage source constraints gives

$$V_a + \frac{3}{2} + \frac{V_a + 1}{3} = 2$$

Solving gives $V_a = \frac{1}{8}$ V. The current I is

$$\begin{aligned} I &= \frac{V_a - V_b}{2\ \Omega} \\ &= \frac{\frac{1}{8} + 3}{2} \\ &= \frac{25}{16}\text{ A} \end{aligned}$$

Exercise Problem 3.10

Determine current I in the circuit of Fig. E3.10 by writing node-voltage equations.

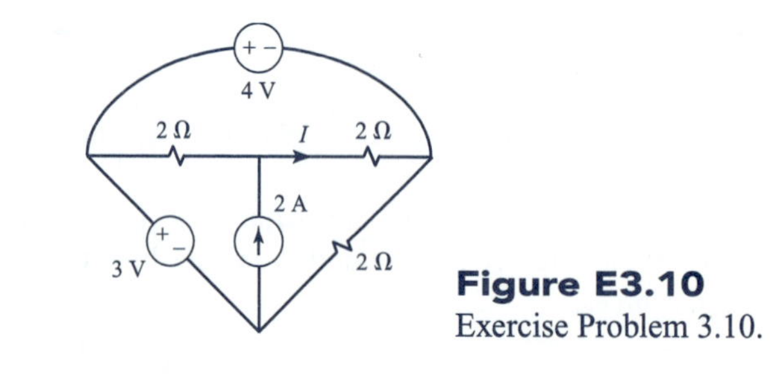

Figure E3.10
Exercise Problem 3.10.

Answer: 2 A.

Controlled voltages do not change this basic procedure: we only need to write the controlling variable in terms of the node voltages.

Example 3.13

Write the node-voltage equations for the circuit of Fig. 3.20*a*.

Solution The reference node is chosen as node *a*, and the node voltages are defined in Fig. 3.20*b*. The circuit with these node voltages labeled is shown in Fig. 3.20*c*. A supernode is drawn around voltage source v_S enclosing nodes *c* and *b*, since neither of these nodes is the reference node. The constraints imposed by the voltage sources are written as

$$V_d = \alpha i_x$$
$$V_b - V_c = -v_S$$

The controlling variable is

$$i_x = -\frac{1}{R_3} V_b$$

Only one node voltage is unknown. We arbitrarily choose that to be V_b. Hence these constraints become

$$V_d = -\frac{\alpha}{R_3} V_b$$
$$V_c = V_b + v_S$$

Node *d* has the controlled voltage source attached between it and the reference node so, we do not write KCL at this node. The only remaining node is the supernode that encloses v_S and nodes *c* and *d*, so that KCL yields

$$N_b + N_c: \qquad \frac{1}{R_1}(V_c - V_d) + \frac{1}{R_2} V_c + \frac{1}{R_3} V_b = -i_S$$

Substituting the above constraints and writing in standard form gives the node-voltage equation to be solved:

$$\left(\frac{1}{R_1} + \frac{\alpha}{R_1 R_3} + \frac{1}{R_2} + \frac{1}{R_3}\right) V_b = -i_S - \left(\frac{1}{R_1} + \frac{1}{R_2}\right) v_S$$

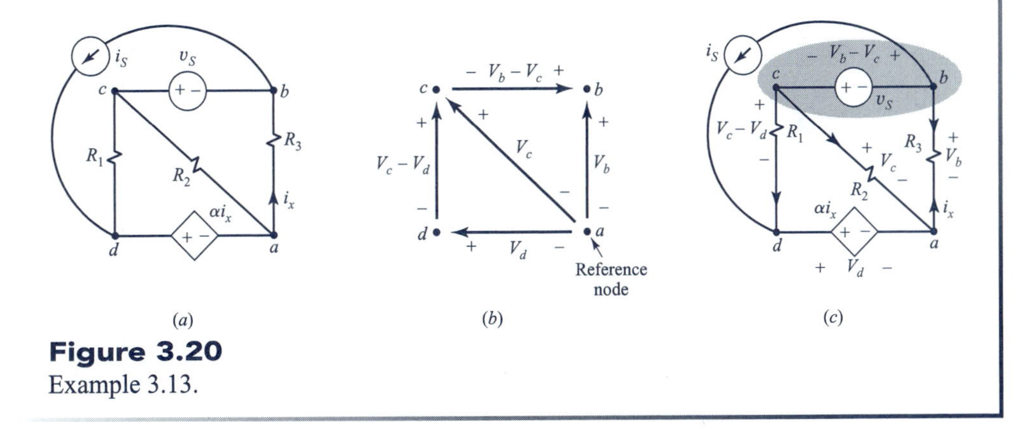

Figure 3.20
Example 3.13.

Exercise Problem 3.11

Determine the voltage v in the circuit of Fig. E3.11 by writing node-voltage equations. Confirm your result using superposition.

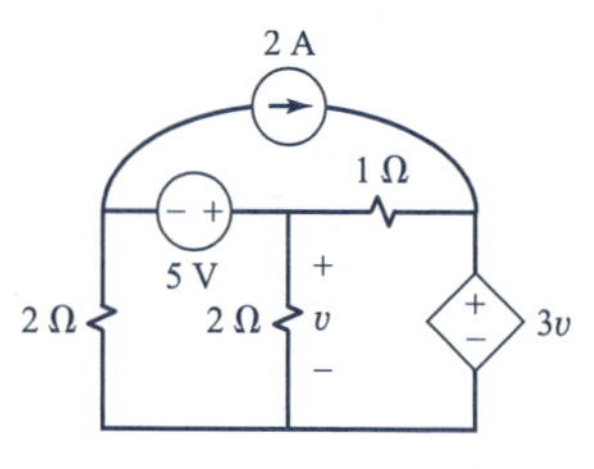

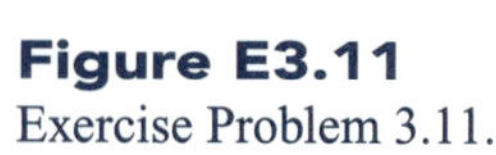

Figure E3.11
Exercise Problem 3.11.

Answer: $-\frac{1}{2}$ V.

3.6 The Mesh-Current Method

In a circuit containing n nodes and N_{V_S} voltage sources (independent or controlled), the number of node-voltage equations is $n - 1 - N_{V_S}$. The following alternative method of writing a set of simultaneous equations, the *mesh-current method,* generates $b - n + 1 - N_{I_S}$ equations, where b is the number of branches (elements) and N_{I_S} is the number of current sources (independent or controlled). Minimizing the number of simultaneous equations one is required to solve is an important consideration. Hence these considerations will dictate whether node-voltage equations or mesh-current equations should be written for a particular circuit.

The mesh-current method is, in a sense, the dual to the node-voltage method. A circuit mesh can be thought of as a pane in a window in that *a mesh is a circuit loop that does not enclose any elements.* Some examples are shown in Fig. 3.21. The *mesh currents* are fictitious currents that are defined to flow only around the mesh. Figure 3.22 shows a three-mesh circuit and the mesh currents. We arbitrarily define each mesh current as flowing clockwise or counterclockwise around the mesh. Which direction we choose is completely arbitrary, and all the mesh currents do not have to flow in the same direction. Figure 3.22 also shows that the currents through elements that are on an outside mesh are solely the mesh current of that mesh, whereas the currents flowing through elements that are shared between two meshes are combinations of those mesh currents according to KCL. It is vitally important that one label the total current through each element in terms of the mesh currents flowing through that element. For resistors, it is also vitally important to label with + and − signs the resulting voltage across the resistors with the passive sign convention, so that a sign error will not be made when we later apply Ohm's law.

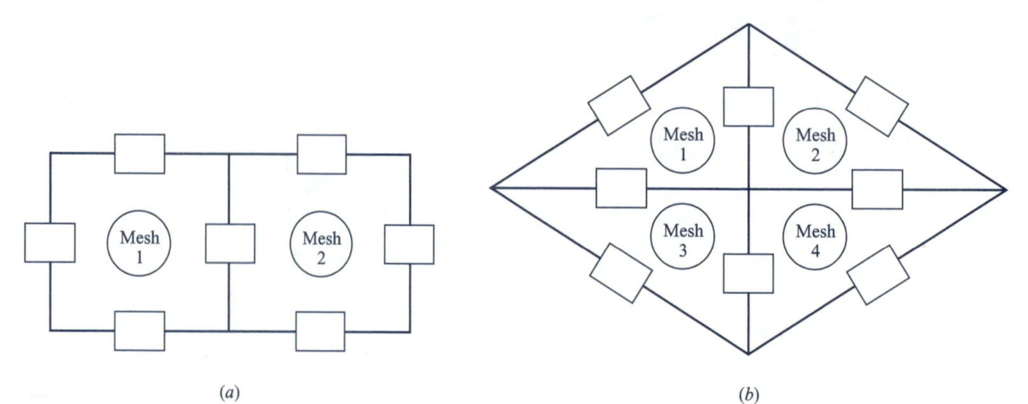

Figure 3.21
Illustrations of the concept of a mesh.

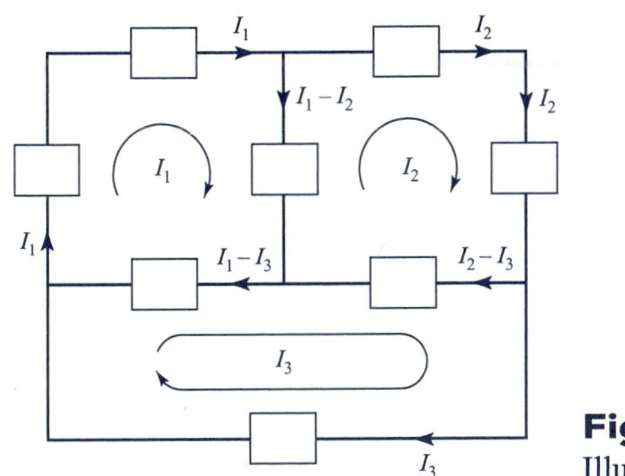

Figure 3.22
Illustration of the definition of mesh currents.

The mesh-current method generates a set of simultaneous equations in terms of the mesh currents, which when solved yields the currents through all elements. The following are the general rules for writing mesh-current equations:

1. *Define the mesh currents.*
2. *Write the total current through each element in terms of the mesh currents flowing through them.*
3. *Write KVL around each mesh.*
4. *Put these equations in standard form, and solve them for the mesh currents.*

Observe that these rules for writing mesh-current equations are essentially the dual to the rules for writing node-voltage equations.

For example, consider the circuit of Fig. 3.23. The mesh currents are defined (arbitrarily chosen to flow in the clockwise direction). The total current through each resistor is indicated, and a set of + and − voltage signs are placed on that resistor in accordance with the current direction to insure that Ohm's law will be correctly written for that resistor. Then KVL is written around each mesh to give

$$
\begin{aligned}
M_1&: \quad R_1(I_1) + R_3(I_1 - I_2) + R_4(I_1 - I_3) = v_{S1} - v_{S3} \\
M_2&: \quad R_5(I_2) - R_3(I_1 - I_2) - R_2(I_3 - I_2) = v_{S2} + v_{S3} \\
M_3&: \quad R_6(I_3) + R_2(I_3 - I_2) - R_4(I_1 - I_3) = -v_{S2}
\end{aligned}
$$

Grouping terms and placing in standard form yields

$$
\begin{aligned}
M_1&: \quad (R_1 + R_3 + R_4)I_1 - R_3 I_2 - R_4 I_3 = v_{S1} - v_{S3} \\
M_2&: \quad -R_3 I_1 + (R_2 + R_3 + R_5)I_2 - R_2 I_3 = v_{S2} + v_{S3} \\
M_3&: \quad -R_4 I_1 - R_2 I_2 + (R_2 + R_4 + R_6)I_3 = -v_{S2}
\end{aligned}
$$

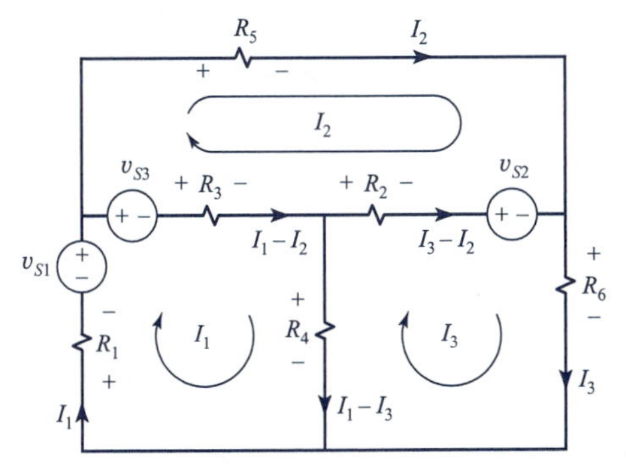

Figure 3.23
An example of writing the mesh-current equations.

These mesh-current equations can be placed in matrix form as

$$\mathbf{RI} = \mathbf{V}$$

where

$$\mathbf{R} = \begin{bmatrix} R_1 + R_3 + R_4 & -R_3 & -R_4 \\ -R_3 & R_2 + R_3 + R_5 & -R_2 \\ -R_4 & -R_2 & R_2 + R_4 + R_6 \end{bmatrix}$$

$$\mathbf{I} = \begin{bmatrix} I_1 \\ I_2 \\ I_3 \end{bmatrix}$$

$$\mathbf{V} = \begin{bmatrix} v_{S1} - v_{S3} \\ v_{S2} + v_{S3} \\ -v_{S2} \end{bmatrix}$$

Once these equations are solved, the current through each element can be written in terms of the mesh currents.

Example 3.14

Write the mesh-current equations for the circuit of Fig. 3.24*a*, and determine the current I.

Solution Mesh currents I_1 and I_3 are chosen to flow in the counterclockwise direction in Fig. 3.24*a*, and mesh current I_2 is chosen to flow in the clockwise direction. This is done to simply demonstrate that the choice of directions for the mesh currents does not affect the solution. The total currents through each resistor are labeled, and a set of + and − voltage signs are placed on that resistor in accordance with the current direction to insure that Ohm's law will be correctly written for that resistor. Then KVL is written around each mesh to give

$$\begin{aligned} M_1&: \quad 2(I_1 + I_2) + 1(I_1 - I_3) = 3 \\ M_2&: \quad 2(I_1 + I_2) + 3(I_2 + I_3) + 1(I_2) = 0 \\ M_3&: \quad 3(I_2 + I_3) - 1(I_1 - I_3) = 2 \end{aligned}$$

Rearranging in standard form yields

$$\begin{aligned} M_1&: \quad 3I_1 + 2I_2 - I_3 = 3 \\ M_2&: \quad 2I_1 + 6I_2 + 3I_3 = 0 \\ M_3&: \quad -I_1 + 3I_2 + 4I_3 = 2 \end{aligned}$$

or in matrix form

$$\begin{bmatrix} 3 & 2 & -1 \\ 2 & 6 & 3 \\ -1 & 3 & 4 \end{bmatrix} \begin{bmatrix} I_1 \\ I_2 \\ I_3 \end{bmatrix} = \begin{bmatrix} 3 \\ 0 \\ 2 \end{bmatrix}$$

Solving these give the mesh currents as

$$\begin{aligned} I_1 &= \tfrac{69}{11}\,\text{A} \\ I_2 &= -5\,\text{A} \\ I_3 &= \tfrac{64}{11}\,\text{A} \end{aligned}$$

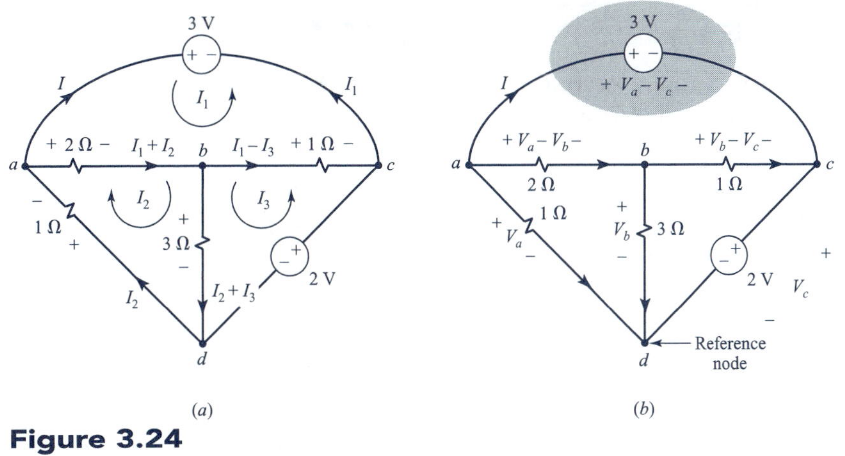

Figure 3.24
Example 3.14.

Hence, the current I becomes

$$
\begin{aligned}
I &= -I_1 \\
&= -\tfrac{69}{11}\,\text{A}
\end{aligned}
$$

In order to demonstrate the point that fewer simultaneous equations may be generated with another method, we will write the node-voltage equations. We choose node d as the reference node as shown in Fig. 3.24b and define the node voltages. These are labeled on the circuit diagram, and current directions through the resistors are also labeled. The two independent voltage sources provide the following constraints:

$$
\begin{aligned}
V_c &= 2\,\text{V} \\
V_a - V_c &= 3\,\text{V}
\end{aligned}
$$

or

$$V_a = 5\,\text{V}$$

Only one node voltage is unknown: V_b. Since the 2-V voltage source is connected between node c and the reference node, we will not write KCL at node c. It is sufficient to write KCL at node b, giving

$$N_b: \quad \tfrac{1}{3}V_b + \tfrac{1}{1}(V_b - V_c) - \tfrac{1}{2}(V_a - V_b) = 0$$

Substituting the above constraints into this equation yields

$$N_b: \quad \left(\tfrac{1}{3} + \tfrac{1}{1} + \tfrac{1}{2}\right)V_b = 2 + \tfrac{5}{2}$$

giving

$$V_b = \tfrac{27}{11}\,\text{V}$$

The current I becomes

$$
\begin{aligned}
I &= -\frac{V_a}{1} - \frac{V_a - V_b}{2} \\
&= -\tfrac{69}{11}\,\text{A}
\end{aligned}
$$

as before.

Exercise Problem 3.12

Determine the voltage v in the circuit of Fig. E3.12 by writing mesh-current equations. Confirm your result by any other method.

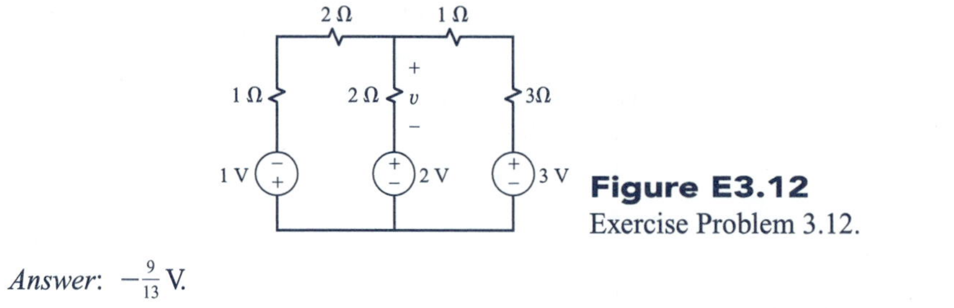

Figure E3.12
Exercise Problem 3.12.

Answer: $-\frac{9}{13}$ V.

Controlled voltage sources are treated in the same fashion as independent voltage sources: one only needs to write the controlling variable in terms of the mesh currents.

Example 3.15

Determine the voltage v_x in the circuit of Fig. 3.25*a* by writing mesh-current equations.

Solution The mesh currents for the two meshes are defined in Fig. 3.25*b*, and the total current through each resistor is indicated. The controlling variable for the controlled source is

$$v_x = 3(I_1 - I_2)$$

The mesh-current equations become

$$M_1: \quad (1 + 2)I_1 + 3(I_1 - I_2) = -6 - 3 + 4$$
$$M_2: \quad (2 + 4)I_2 - 3(I_1 - I_2) + 4v_x = -4 + 1$$

Substituting the constraint on v_x and arranging in standard form yields

$$M_1: \quad 6I_1 - 3I_2 = -5$$
$$M_2: \quad 9I_1 - 3I_2 = -3$$

Solving these gives

$$I_1 = \tfrac{2}{3}\text{ A}$$
$$I_2 = 3\text{ A}$$

Hence,

$$v_x = 3(I_1 - I_2)$$
$$= -7\text{ V}$$

As an alternative solution we use source transformations, being careful not to lose the controlling variable, as shown in Fig. 3.25*c*. Writing KVL around the final single-loop circuit yields

$$\frac{2}{3}v_x + v_x = 2\left(\frac{4v_x - 1}{6} - 3\right) + 4$$

Solving this gives $v_x = -7$ V, as was obtained with mesh currents.

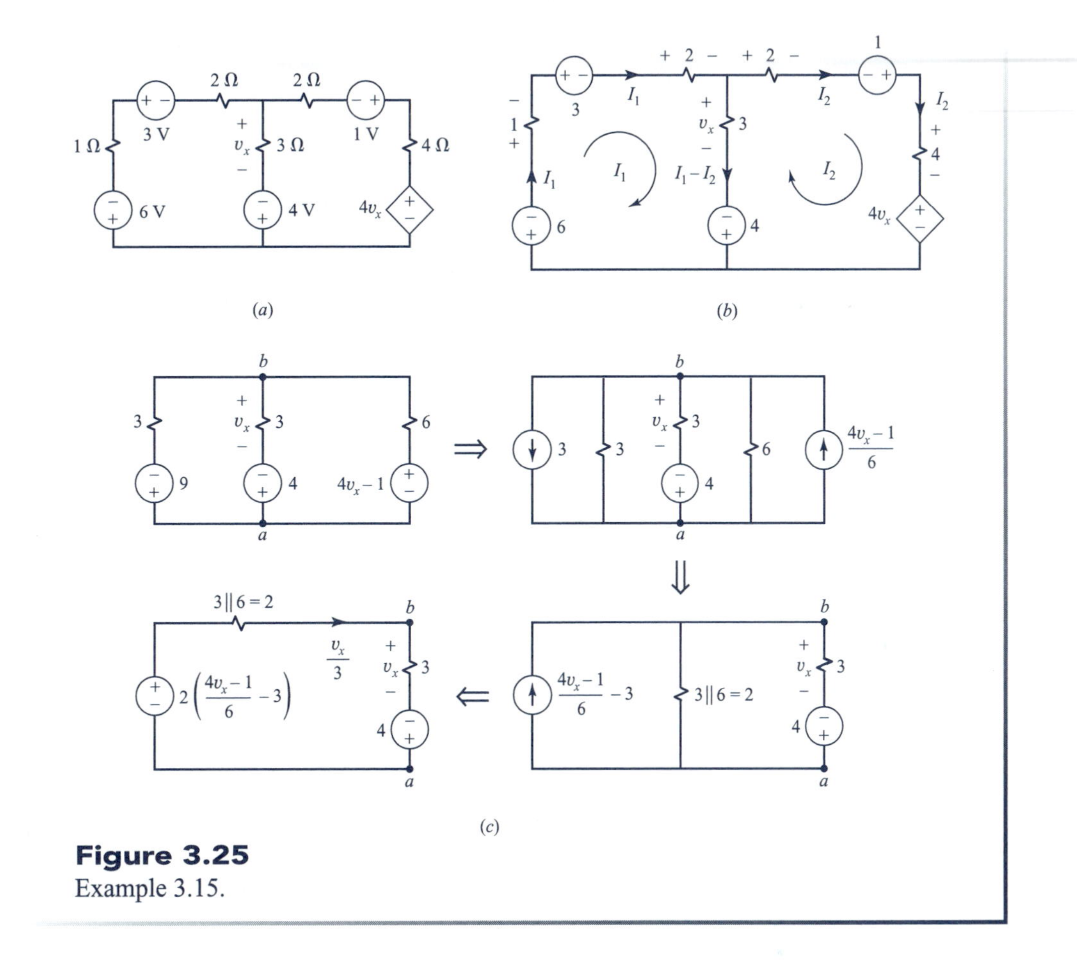

Figure 3.25
Example 3.15.

3.6.1 Circuits Containing Current Sources The previous circuits have demonstrated writing the mesh-current equations for circuits that contain voltage sources (independent or controlled). In this section we investigate writing the mesh-current equations for circuits that may also contain current sources (independent or controlled). This should be viewed as the dual to the method of writing the node-voltage equations for circuits that contain voltage sources. The total number of unknown mesh currents and resulting equations will be $b - n + 1 - N_{I_S}$ where N_{I_S} is the number of current sources (independent or controlled).

Consider the circuit shown in Fig. 3.26. We define the mesh currents in the usual fashion. Observe that the i_{S1} current source constrains mesh current I_3:

$$I_3 = -i_{S1}$$

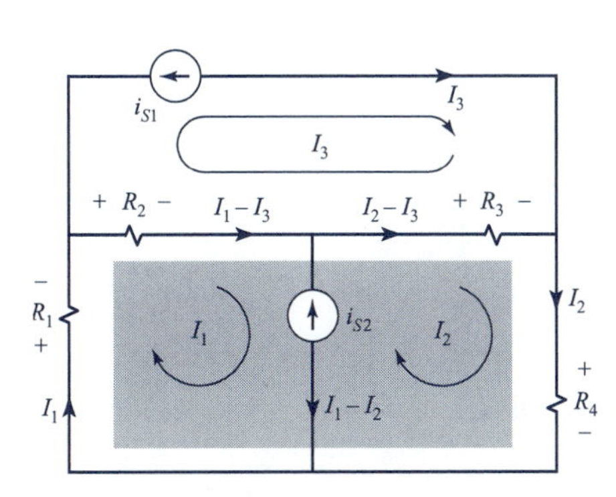

Figure 3.26
Example of writing mesh-current equations for circuits that contain current sources.

Similarly observe that the i_{S2} current source *relates* mesh currents I_1 and I_2, since the total current through that source is $I_1 - I_2$:

$$I_1 - I_2 = -i_{S2}$$

Hence only one of the three mesh currents is truly unknown. Thus we need only one equation in this unknown current. We should not write KVL around mesh 3, since the voltage across the current source is unknown. Similarly, the voltage across the i_{S2} current source is unknown, and we should not write KVL around mesh 2 or mesh 3. However, if we write KVL around the loop containing mesh 2 and mesh 3, we obtain

$$M_2 + M_3: \qquad R_1 I_1 + R_2(I_1 - I_3) + R_3(I_2 - I_3) + R_4 I_2 = 0$$

Substituting the above constraints gives

$$(R_1 + R_2 + R_3 + R_4)I_1 = -(R_2 + R_3)i_{S1} - (R_3 + R_4)i_{S2}$$

This has shown a general method for writing the mesh-current equations when a circuit contains current sources (independent or controlled):

1. *Define the mesh currents in the usual fashion.*
2. *Write the constraints that are imposed on the mesh currents by any current sources.*
3. *Draw loops around all pairs of meshes that share a current source.*
4. *Write KVL for all these loops and all other meshes except those meshes that have a current source in an outside branch.*
5. *Substitute the constraints imposed on the mesh currents by the current sources into these equations, and place them in standard form.*

From these results it is easy to show that *in a circuit containing b branches (elements), n nodes, and N_{I_S} current sources (independent or controlled), the number of mesh-current equations equals*

$$b - n + 1 - N_{I_S}$$

Observe that this is the dual procedure to writing node-voltage equations for circuits that contain voltage sources.

Example 3.16

Determine the voltage V in the circuit of Fig. 3.27a by writing mesh equations.

Solution The mesh currents are defined in Fig. 3.27b, and the total currents through the elements are labeled in terms of these mesh currents. First we write the constraints caused by the current sources:

$$I_1 = 1\text{ A}$$
$$I_2 - I_3 = 2\text{ A}$$

Observe that the voltages across the two current sources are not known, so we write KVL around the combination of mesh 3 and mesh 2:

$$3(I_1 - I_3) + 4(I_1 - I_2) - 2I_2 - 2I_3 = 0$$

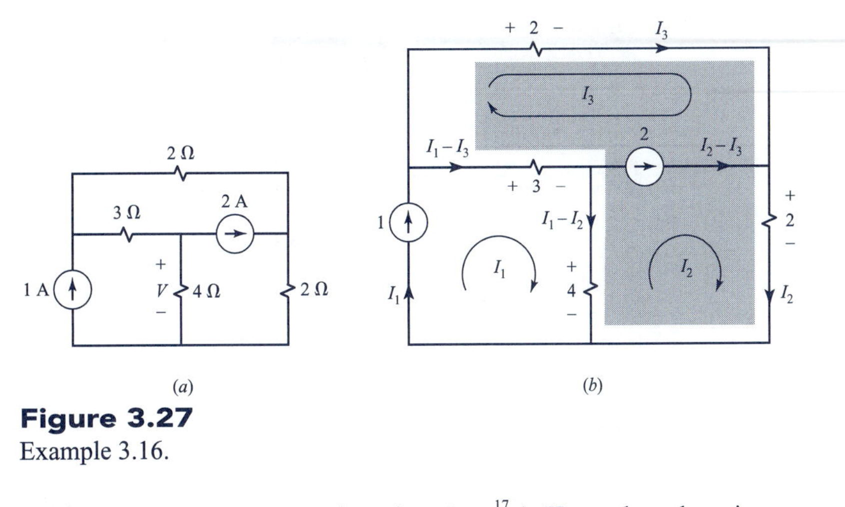

Figure 3.27
Example 3.16.

Substituting the current source constraints gives $I_2 = \frac{17}{11}$ A. Hence the voltage is

$$
\begin{aligned}
V &= 4(I_1 - I_2) \\
&= -\tfrac{24}{11}\text{ A}
\end{aligned}
$$

Exercise Problem 3.13

Determine the voltage V in the circuit of Fig. E3.13 by writing mesh-current equations.

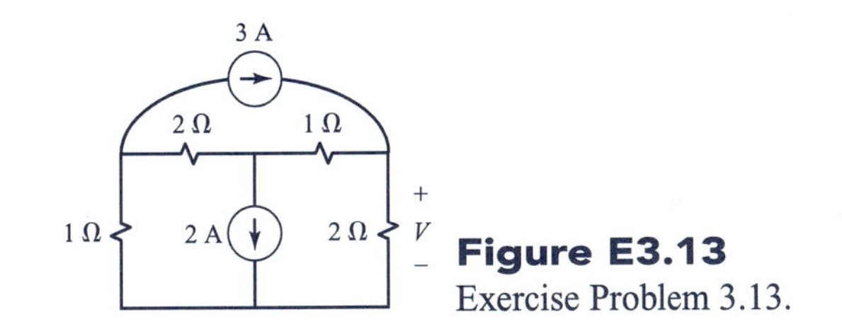

Figure E3.13
Exercise Problem 3.13.

Answer: 1 V.

Controlled current sources do not significantly alter this procedure. Simply treat those as independent current sources and write the controlling variable in terms of the mesh currents.

Example 3.17

Determine the voltage v_x in the circuit of Fig. 3.28*a* by writing mesh current equations. Check the result using any other method.

Solution Define the three mesh currents as shown, and write the total current through each element in terms of those mesh currents. The 3-A current source constrains the current of mesh 3:

$$I_3 = -3\text{ A}$$

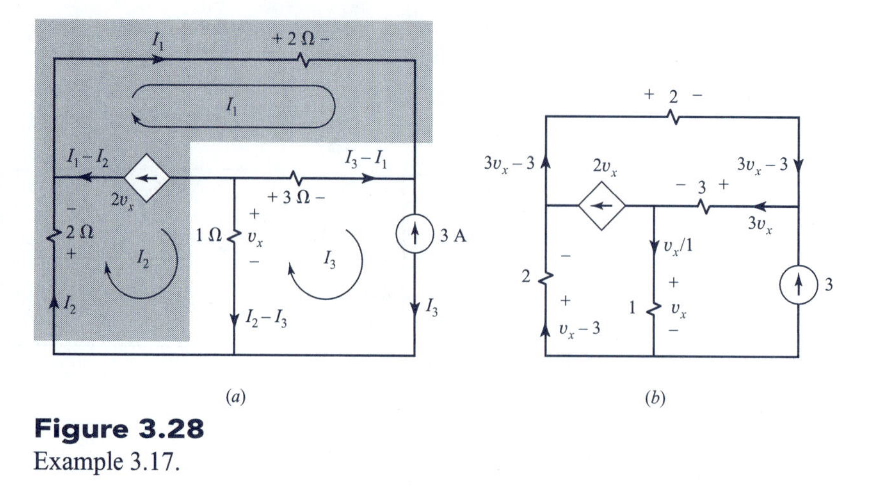

Figure 3.28
Example 3.17.

The controlling variable for the controlled current source is

$$v_x = 1\ \Omega \times (I_2 - I_3)$$

The controlled current source constrains $I_1 - I_2$ as

$$2v_x = (I_1 - I_2)$$

Substituting these gives

$$I_1 = 3I_2 + 6$$

Observe that mesh 3 contains a current source in the outside edge, so KVL should not be written around this mesh. Also observe that meshes 1 and 2 share the controlled current source, so a loop is drawn around these two meshes and KVL is written around that loop, giving

$$2I_2 + 2I_1 - 3(I_3 - I_1) + 1(I_2 - I_3) = 0$$

Substituting the above constraints gives

$$I_2 = -\tfrac{7}{3}\text{ A}$$

Hence

$$v_x = 1\ \Omega \times (I_2 - I_3)$$
$$= \tfrac{2}{3}\text{ V}$$

An alternate method is direct solution. Figure 3.28*b* shows the current through each element labeled using Ohm's law and KCL. Writing KVL yields

$$2(v_x - 3) + 2(3v_x - 3) + 3(3v_x) + v_x = 0$$

giving $v_x = \frac{2}{3}$ V once again.

Exercise Problem 3.14

Determine the current I in the circuit of Fig. E3.14 by writing mesh-current equations.

Answer: $\frac{3}{4}$ A.

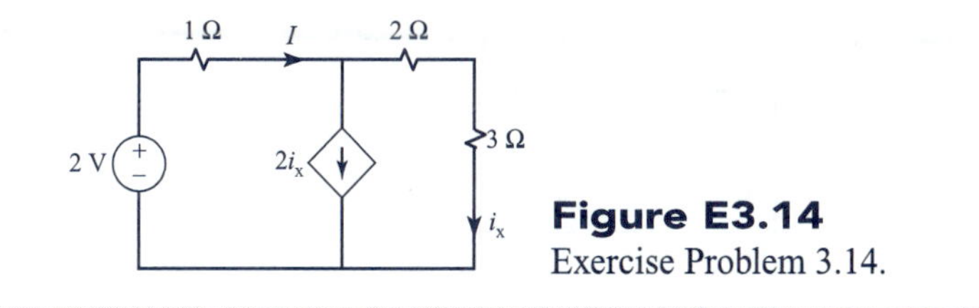

Figure E3.14
Exercise Problem 3.14.

3.7 PSPICE Applications

In this section we continue the illustration of the use of PSPICE to provide circuit solutions.

Example 3.18

Determine the current I in the circuit of Fig. 3.29*a* by using PSPICE and by using a Thevenin equivalent circuit.

Solution The circuit with nodes labeled is shown in Fig. 3.29*b*. We have inserted 0-V voltage sources to sample the controlling current i_x as well as the desired solution current I. We could have simply requested the current through the 500-Ω resistor to obtain the current I, but the controlled-source controlling current, i_x, must be associated with a voltage source.

```
EXAMPLE 3.18
VS        1     0     DC     5
R1        1     2     500
R2        2     3     1K
R3        3     4     2K
VTEST1    4     0     DC     0
HSOURCE         3     5      VTEST1     500
R4        5     6     500
VTEST2    6     0     DC     0
.DC       VS    5     5      1
*THE CURRENT I IS I(VTEST2)
.PRINT    DC    I(VTEST2)
.END
```

The result is I = I(VTEST2) = 1.875E-3.

Next we solve this by hand using a Thevenin equivalent reduction. The circuit with the 500-Ω resistor removed is shown in Fig. 3.29*c*, from which we obtain with KVL

$$5\text{ V} = 500\ \Omega \times i_x + 1\text{ k}\Omega \times i_x + 2\text{ k}\Omega \times i_x$$

giving

$$i_x = \frac{5\text{ V}}{3500\ \Omega}$$
$$= 1.429\text{ mA}$$

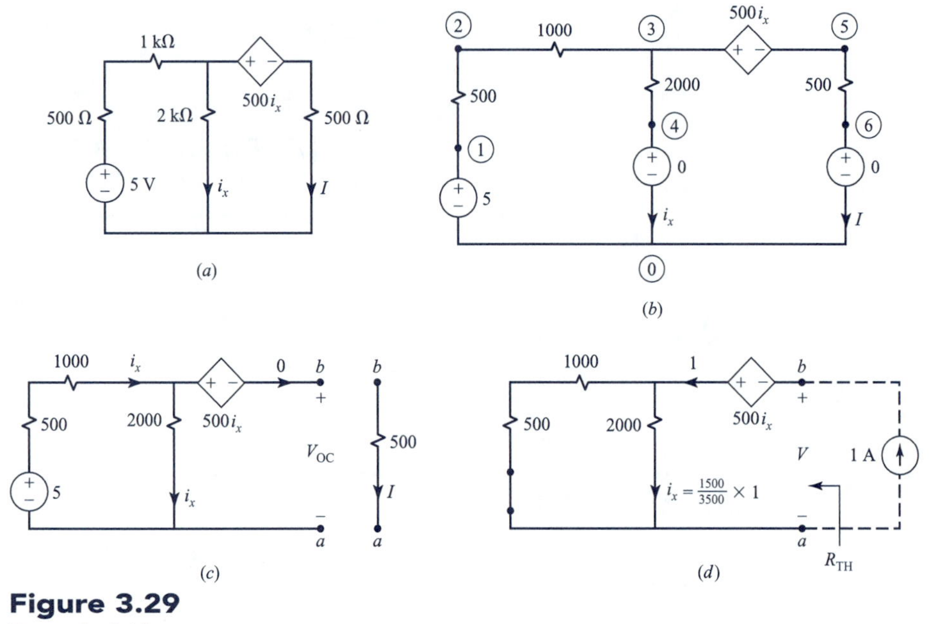

Figure 3.29
Example 3.18.

Note that because of the open circuit, the current i_x goes through the 500-Ω and 1-kΩ resistors. Hence

$$V_{OC} = 2\text{ k}\Omega \times i_x - 500i_x$$
$$= 2.143\text{ V}$$

Deactivating the independent voltage source gives the circuit shown in Fig. 3.29*d* for determining the Thevenin resistance. Applying a 1-A current source, the current i_x is determined by current division as

$$i_x = \frac{1500\ \Omega}{1500\ \Omega + 2000\ \Omega} \times 1\text{ A}$$
$$= 0.4286\text{ A}$$

We compute the voltage V as

$$V = 2\text{ k}\Omega \times i_x - 500i_x$$
$$= 1500 \times 0.4286$$
$$= 642.9\text{ V}$$

Hence the Thevenin resistance is

$$R_{TH} = \frac{V}{1\text{ A}}$$
$$= 642.9\ \Omega$$

Reattaching the 500-Ω resistor, we compute the current as

$$I = \frac{V_{OC}}{R_{TH} + 500\ \Omega}$$
$$= 1.875\text{ mA}$$

as was obtained with PSPICE.

3.8 MATLAB Applications

The next computer program that we will introduce and use as an aid in solving circuits is the MATLAB program, which is marketed by The Math Works, Inc., 24 Prime Park Way, Natick, MA 01760-1500. A reduced-cost student version is available from the Prentice-Hall Publishing Company. Information about the student version of MATLAB can be obtained from the web site www.mathworks.com. MATLAB was originally written to deal with matrix operations, and although that remains one of its important uses, it has many other facilities that can be used in circuit solution. We will use MATLAB throughout the text where appropriate, but to begin our discussion we use an example.

Example 3.19

Solve the circuit of the previous example, Example 3.18, for the current I by writing mesh-current equations and using MATLAB to solve those simultaneous equations.

Solution Labeling mesh currents as shown in Fig. 3.30, we write, using KVL,

$$3500I_1 - 2000I_2 = 5$$
$$-2000I_1 + 2500I_2 + 500i_x = 0$$

Substituting

$$i_x = I_1 - I_2$$

gives

$$3500I_1 - 2000I_2 = 5$$
$$-1500I_1 + 2000I_2 = 0$$

These must be written in matrix form (see the Appendix) for use in MATLAB:

$$\mathbf{RI} = \mathbf{V}$$

where

$$\mathbf{R} = \begin{bmatrix} 3500 & -2000 \\ -1500 & 2000 \end{bmatrix}$$

$$\mathbf{I} = \begin{bmatrix} I_1 \\ I_2 \end{bmatrix}$$

$$\mathbf{V} = \begin{bmatrix} 5 \\ 0 \end{bmatrix}$$

The matrices are entered and solved interactively on the screen; there is no program to be prepared and submitted as with PSPICE. The following gives a track record of the solution for this problem.

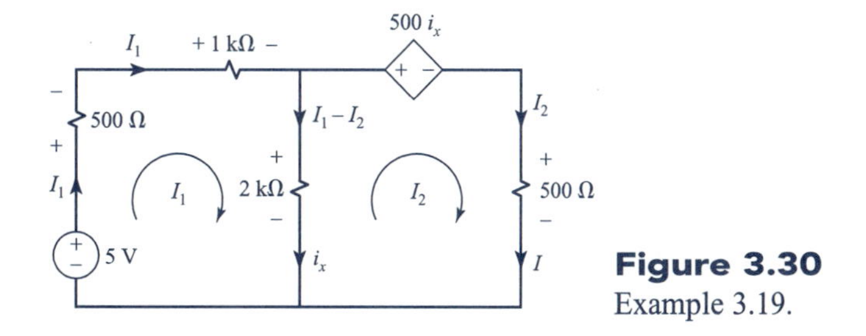

Figure 3.30
Example 3.19.

```
>>    R=[3500 -2000; -1500 2000]
R=
       3500       -2000
      -1500        2000
>>    V=[5;0]
V=
          5
          0
>> I=R\V
I=
2.5000e-003
1.8750e-003
>>I=inv(R)*V
I=
2.5000e-003
1.8750e-003
>>    quit
```

The symbols >> are the prompt for data entry. We first enter the matrix by rows. Each entry in a row is separated by at least one blank space or a comma. The rows of **R** are separated with a semicolon. The entire matrix is enclosed in brackets. Once this is entered, the program prints out the matrix as we would expect to see it in matrix form. If that is not desired, place a semicolon at the end of the line where the matrix is entered to suppress the printing of it. Next we input the vector **V** (two rows and one column). Observe that the first row is the lone entry 5 and the second row is the lone entry 0; the two are separated by a semicolon. The entire vector is enclosed in brackets. The next command, I=R\V, means that the vector **V** is to be multiplied on the left by the inverse of **R**. This gives the two mesh currents as $I_1 = 2.5\text{ E} - 3\text{ A}$ and $I_2 = I = 1.875\text{ E} - 3\text{ A}$, as was obtained in Example 3.18. The default format for printing numbers is the integer format (although the program stores the number with considerably more precision). We have changed the format to scientific notation by selecting the *short e* format from the OPTIONS menu. An alternative way of solving the matrix equations is to write I=inv(R)*V, which yields the same result. And finally we quit the program by simply typing *quit* at the prompt.

Although MATLAB does not directly solve circuits as PSPICE does, it has many functions that are useful in circuit analysis. Throughout the text we will be illustrating those applications.

Problems

Section 3.1 The Principle of Superposition

3.1-1 Determine current i_x in the circuit of Fig. P3.1-1 using superposition.
Answer: $i_x = -\frac{1}{5}$ A.

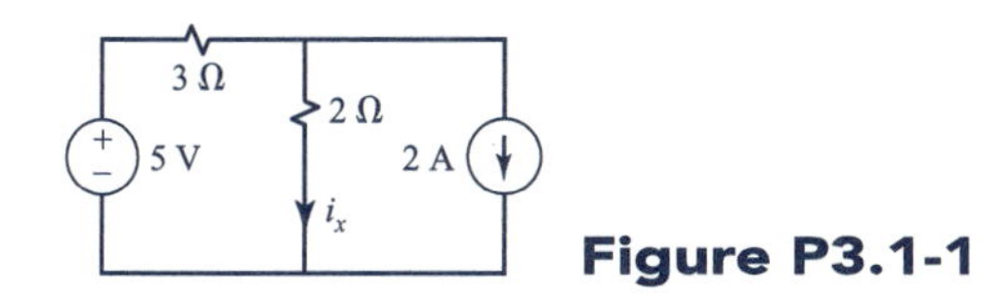

Figure P3.1-1

3.1-2 Determine current i_x in the circuit of Fig. P3.1-2 using superposition.
Answer: $i_x = -\frac{11}{3}$ A.

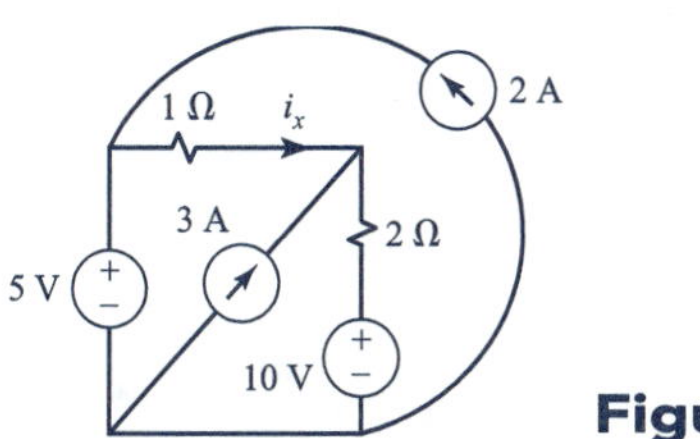

Figure P3.1-2

3.1-3 Determine current i_x in the circuit of Fig. P3.1-3 using superposition.
Answer: $i_x = -\frac{1}{2}$ A.

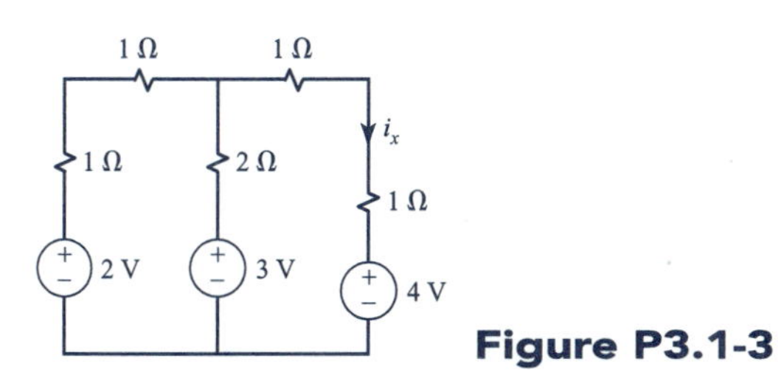

Figure P3.1-3

3.1-4 Determine current i_x in the circuit of Fig. P3.1-4 using superposition.
Answer: $i_x = \frac{2}{3}$ A.

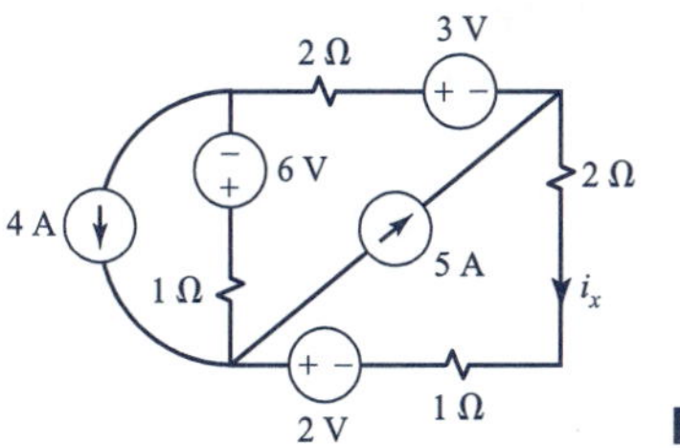

Figure P3.1-4

3.1-5 Determine current i_x in the circuit of Fig. P3.1-5 using superposition.
Answer: $i_x = -\frac{25}{6}$ A.

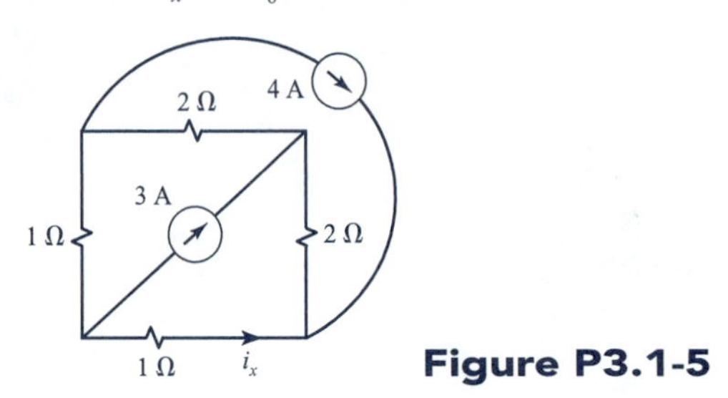

Figure P3.1-5

3.1-6 Determine current i_x and voltage v_x in the circuit of Fig. P3.1-6 using superposition.
Answers: $i_x = \frac{1}{3}$ A, $v_x = -\frac{13}{3}$ V.

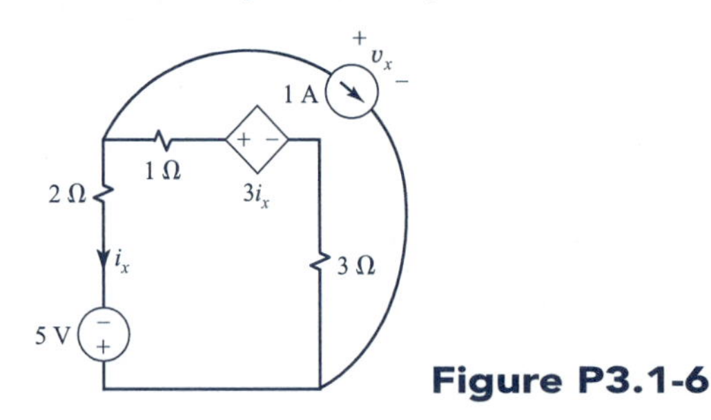

Figure P3.1-6

3.1-7 Determine current i_x and voltage v_x in the circuit of Fig. P3.1-7 using superposition.
Answers: $i_x = -\frac{1}{8}$ A, $v_x = \frac{19}{8}$ V.

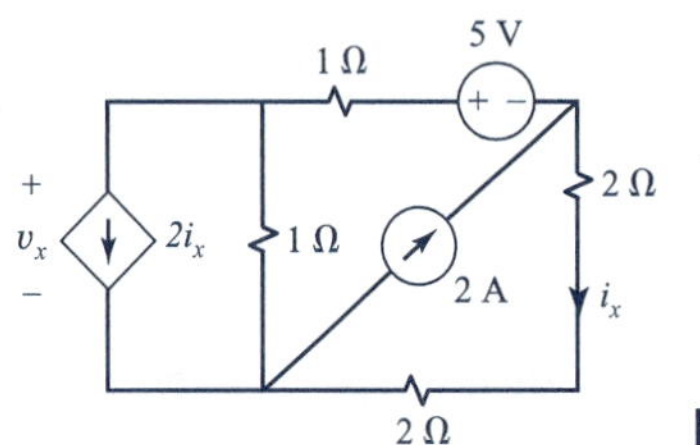

Figure P3.1-7

3.1-8 Determine current i_x and voltage v_x in the circuit of Fig. P3.1-8 using superposition.
Answers: $i_x = \frac{1}{2}$ A, $v_x = 1$ V.

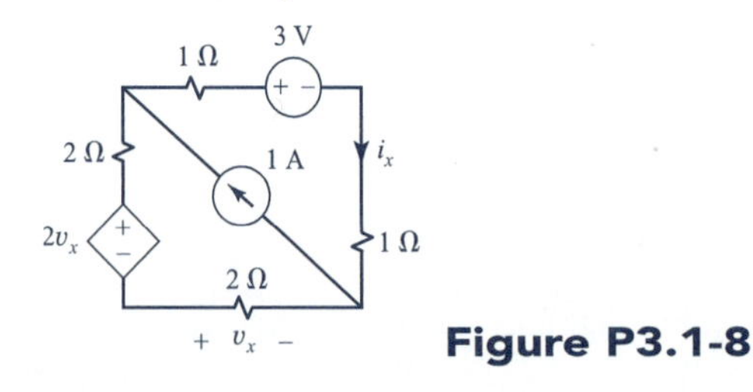

Figure P3.1-8

3.1-9 Determine current i_x and voltage v_x in the circuit of Fig. P3.1-9 using superposition.
Answers: $i_x = \frac{4}{15}$ A, $v_x = \frac{16}{15}$ V.

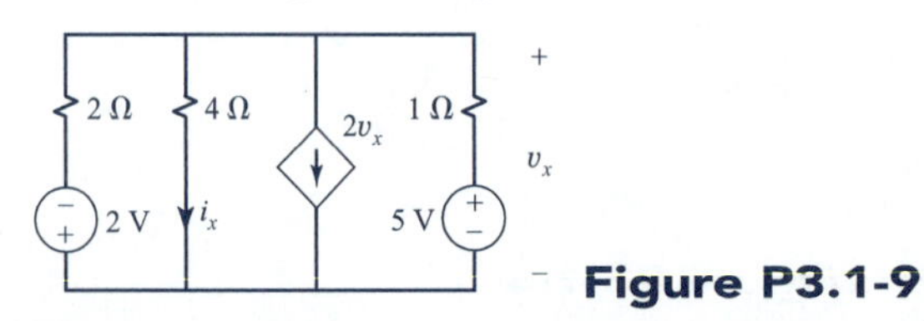

Figure P3.1-9

3.1-10 Determine current i_x and voltage v_x in the circuit of Fig. P3.1-10 using superposition.
Answers: $i_x = 1$ A, $v_x = 3$ V.

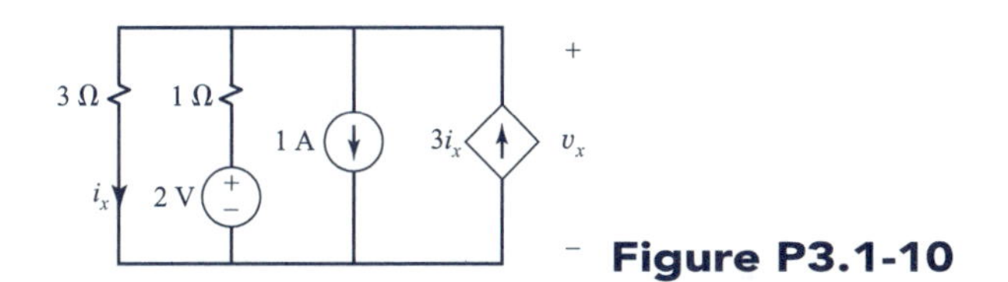

Figure P3.1-10

Section 3.2 The Thevenin Equivalent Circuit

3.2-1 Determine V_{OC} and R_{TH} such that the two circuits in Fig. P3.2-1 will be equivalent at terminals *ab*. *Answers:* $V_{OC} = \frac{11}{2}$ V, $R_{TH} = 3\ \Omega$.

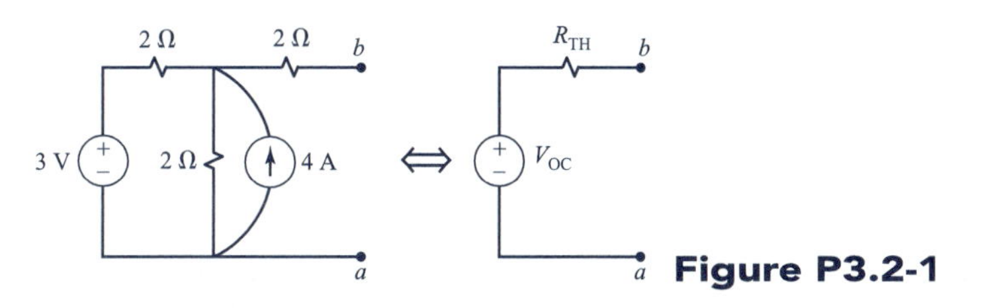

Figure P3.2-1

3.2-2 Determine V_{OC} and R_{TH} such that the two circuits in Fig. P3.2-2 will be equivalent at terminals *ab*. *Answers:* $V_{OC} = -\frac{11}{2}$ V, $R_{TH} = 5\ \Omega$.

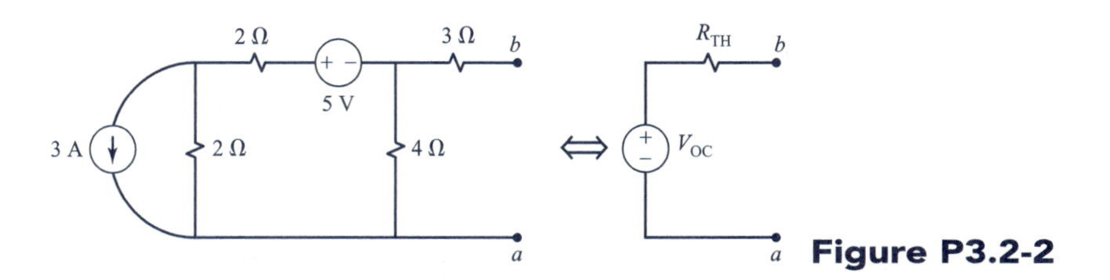

Figure P3.2-2

3.2-3 Determine V_{OC} and R_{TH} such that the two circuits in Fig. P3.2-3 will be equivalent at terminals *ab*. *Answers:* $V_{OC} = -6$ V, $R_{TH} = \frac{16}{3}\ \Omega$.

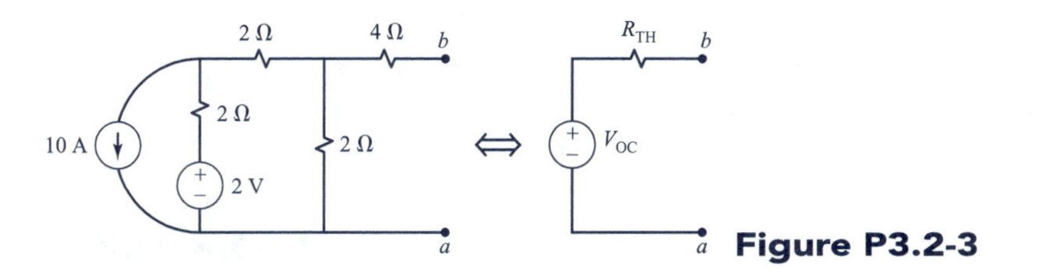

Figure P3.2-3

3.2-4 Determine V_{OC} and R_{TH} such that the two circuits in Fig. P3.2-4 will be equivalent at terminals *ab*. *Answers:* $V_{OC} = -7$ V, $R_{TH} = 5\ \Omega$.

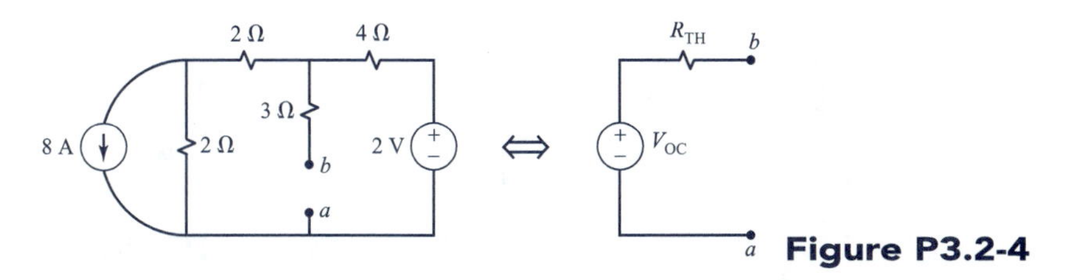

Figure P3.2-4

3.2-5 Determine V_{OC} and R_{TH} such that the two circuits in Fig. P3.2-5 will be equivalent at terminals *ab*. *Answers*: $V_{OC} = -9$ V, $R_{TH} = 4\ \Omega$.

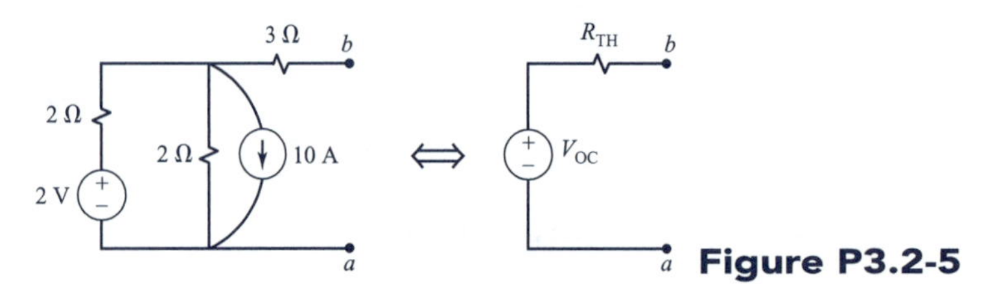

Figure P3.2-5

3.2-6 Determine V_{OC} and R_{TH} such that the two circuits in Fig. P3.2-6 will be equivalent at terminals *ab*. *Answers*: $V_{OC} = -1$ V, $R_{TH} = \frac{7}{3}\ \Omega$.

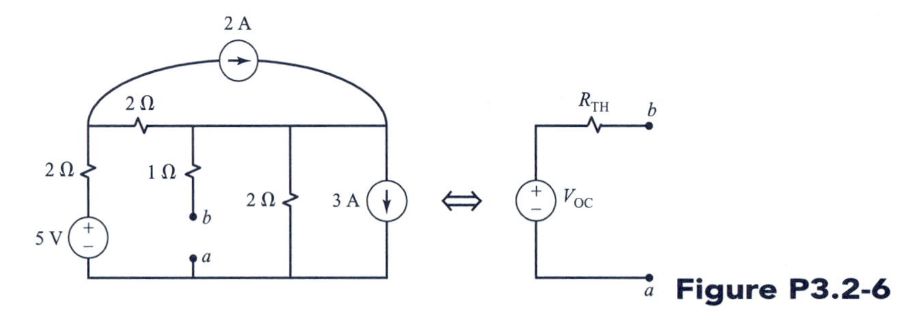

Figure P3.2-6

3.2-7 Determine V_{OC} and R_{TH} such that the two circuits in Fig. P3.2-7 will be equivalent at terminals *ab*. *Answers*: $V_{OC} = -7$ V, $R_{TH} = 3\ \Omega$.

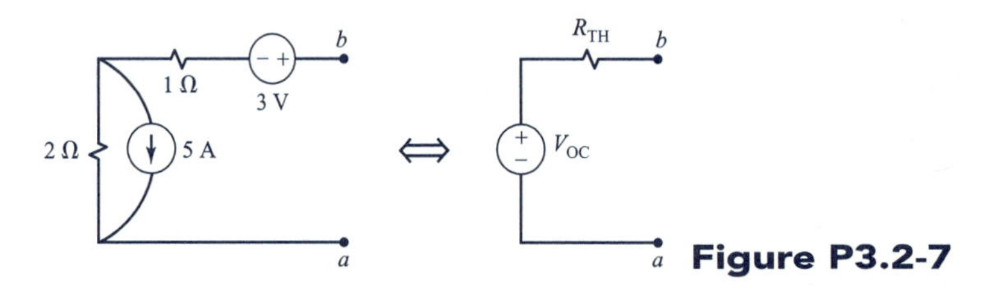

Figure P3.2-7

3.2-8 Determine V_{OC} and R_{TH} such that the two circuits in Fig. P3.2-8 will be equivalent at terminals *ab*. *Answers*: $V_{OC} = -6$ V, $R_{TH} = 2\ \Omega$.

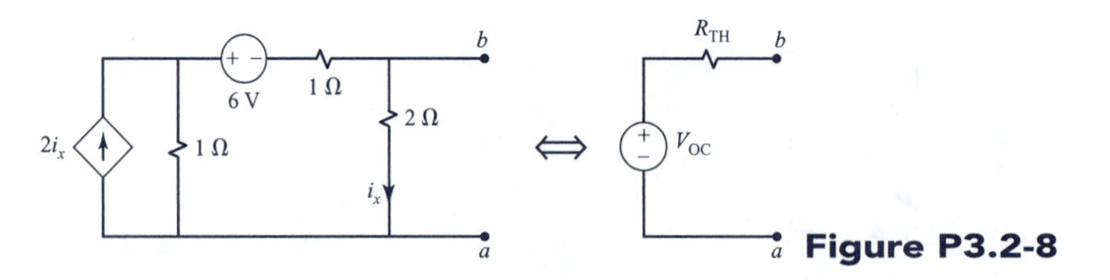

Figure P3.2-8

3.2-9 Determine V_{OC} and R_{TH} such that the two circuits in Fig. P3.2-9 will be equivalent at terminals *ab*. *Answers*: $V_{OC} = -2$ V, $R_{TH} = 5\ \Omega$.

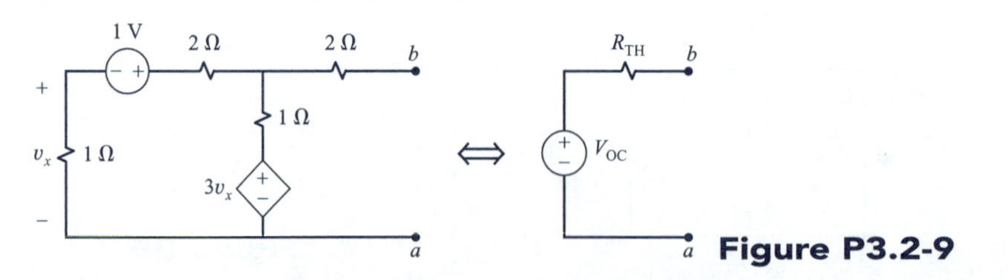

Figure P3.2-9

3.2-10 Determine V_{OC} and R_{TH} such that the two circuits in Fig. P3.2-10 will be equivalent at terminals *ab*. *Answers*: $V_{OC} = -6$ V, $R_{TH} = -1\ \Omega$.

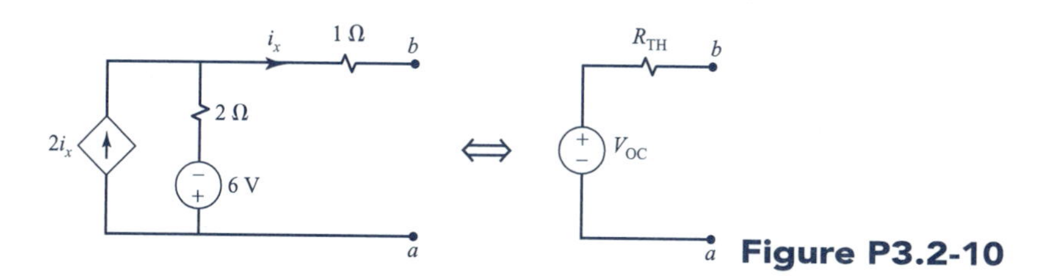

Figure P3.2-10

3.2-11 Determine V_{OC} and R_{TH} such that the two circuits in Fig. P3.2-11 will be equivalent at terminals *ab*. *Answers*: $V_{OC} = -2$ V, $R_{TH} = 1\ \Omega$.

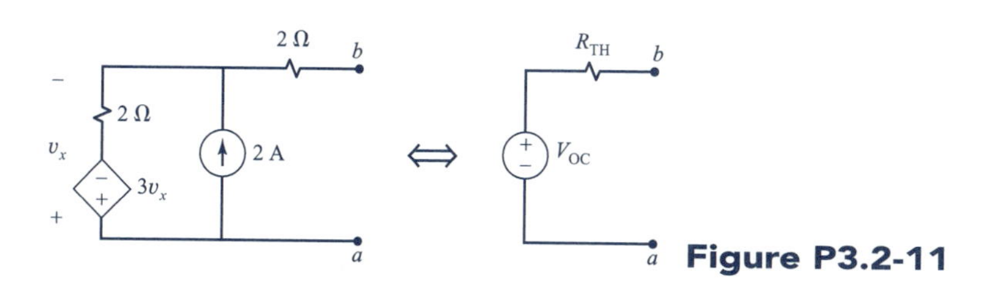

Figure P3.2-11

3.2-12 Solve for current I and voltage V in the circuit of Fig. P3.2-12 by reducing the indicated circuit to a Thevenin equivalent at terminals *ab*.
Answers: $V_{OC} = -\frac{5}{3}$ V, $R_{TH} = \frac{8}{3}\ \Omega$, $I = -\frac{35}{17}$ A, $V = -\frac{5}{17}$ V.

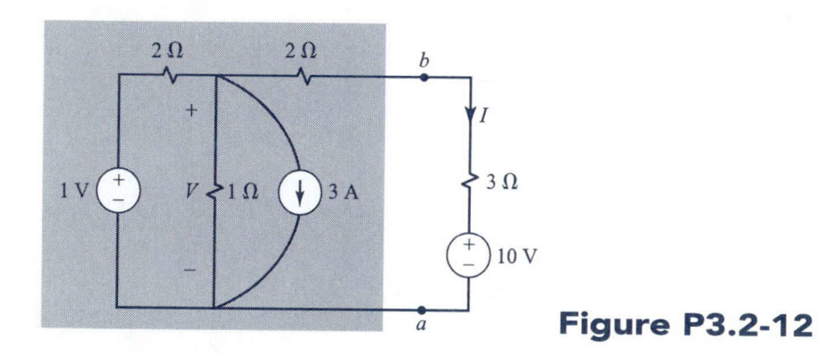

Figure P3.2-12

3.2-13 Solve for current I and voltage V in the circuit of Fig. P3.2-13 by reducing the indicated circuit to a Thevenin equivalent at terminals *ab*.
Answers: $V_{OC} = -12$ V, $R_{TH} = 3\ \Omega$, $I = -\frac{17}{5}$ A, $V = -\frac{4}{5}$ V.

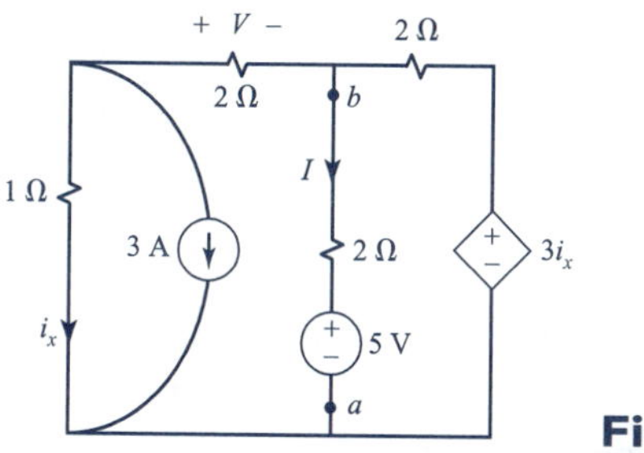

Figure P3.2-13

3.2-14 Solve for current I and voltage V in the circuit of Fig. P3.2-14 by reducing the indicated circuit to a Thevenin equivalent at terminals *ab*.
Answers: $V_{OC} = -3$ V, $R_{TH} = 5\ \Omega$, $I = -\frac{13}{5}$ A, $V = \frac{49}{5}$ V.

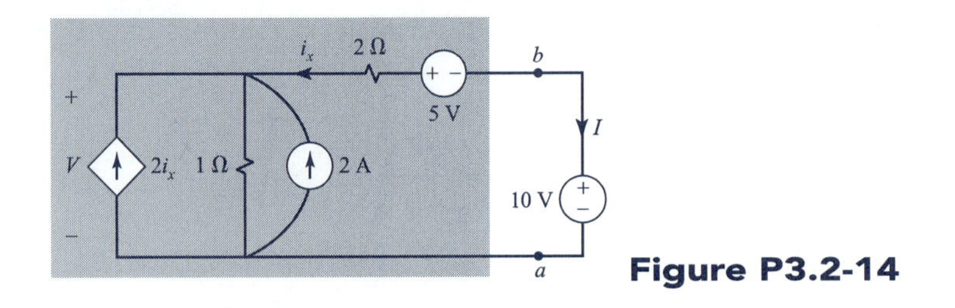

Figure P3.2-14

3.2-15 Solve for current I and voltage V in the circuit of Fig. P3.2-15 by reducing the indicated circuit to a Thevenin equivalent at terminals ab.
Answers: $V_{OC} = -1$ V, $R_{TH} = 3\ \Omega$, $I = -\frac{3}{4}$ A, $V = \frac{11}{4}$ V.

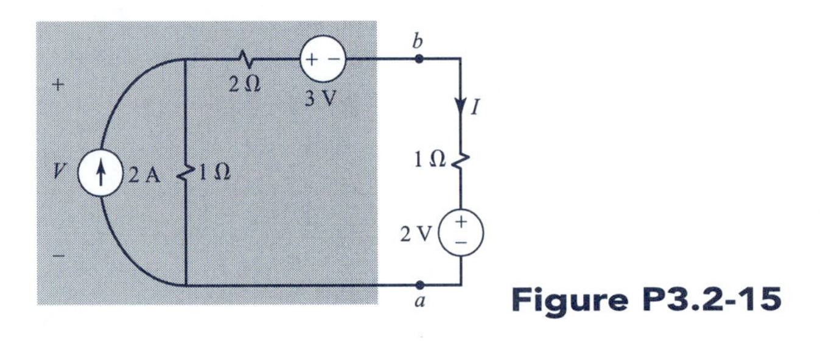

Figure P3.2-15

3.2-16 Solve for current I and voltage V in the circuit of Fig. P3.2-16 by reducing the indicated circuit to a Thevenin equivalent at terminals ab.
Answers: $V_{OC} = -1$ V, $R_{TH} = 3\ \Omega$, $I = \frac{1}{2}$ A, $V = -\frac{1}{2}$ V.

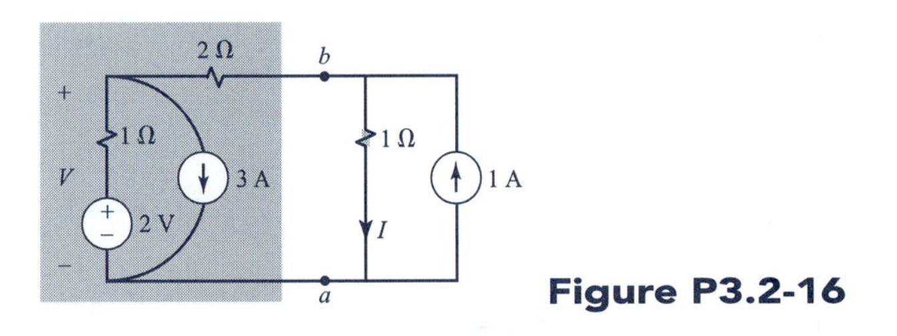

Figure P3.2-16

3.2-17 Solve for current I and voltage V in the circuit of Fig. P3.2-17 by reducing the indicated circuit to a Thevenin equivalent at terminals ab.
Answers: $V_{OC} = 2$ V, $R_{TH} = -1\ \Omega$, $I = 2$ A, $V = 6$ V.

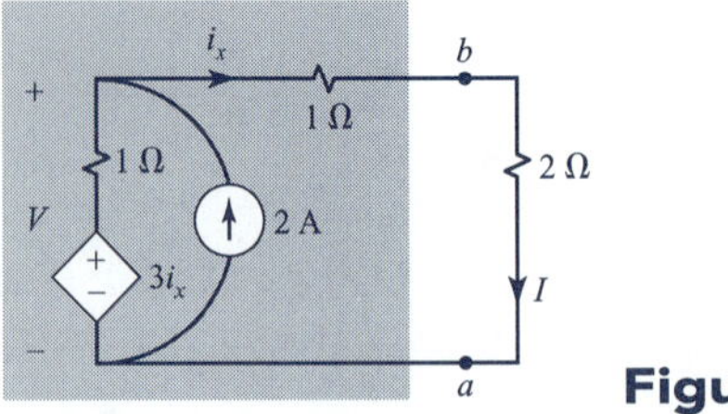

Figure P3.2-17

Section 3.3 The Norton Equivalent Circuit

3.3-1 Determine the Norton equivalent circuit at terminals ab in Fig. P3.2-1.
Answers: $I_{SC} = \frac{11}{6}$ A, $R_{TH} = 3\ \Omega$.

3.3-2 Determine the Norton equivalent circuit at terminals *ab* in Fig. P3.2-2.
Answers: $I_{SC} = -\frac{11}{10}$ A, $R_{TH} = 5\ \Omega$.

3.3-3 Determine the Norton equivalent circuit at terminals *ab* in Fig. P3.2-3.
Answers: $I_{SC} = -\frac{9}{8}$ A, $R_{TH} = \frac{16}{3}\ \Omega$.

3.3-4 Determine the Norton equivalent circuit at terminals *ab* in Fig. P3.2-4.
Answers: $I_{SC} = -\frac{7}{5}$ A, $R_{TH} = 5\ \Omega$.

3.3-5 Determine the Norton equivalent circuit at terminals *ab* in Fig. P3.2-5.
Answers: $I_{SC} = -\frac{9}{4}$ A, $R_{TH} = 4\ \Omega$.

3.3-6 Determine the Norton equivalent circuit at terminals *ab* in Fig. P3.2-6.
Answers: $I_{SC} = -\frac{3}{7}$ A, $R_{TH} = \frac{7}{3}\ \Omega$.

3.3-7 Determine the Norton equivalent circuit at terminals *ab* in Fig. P3.2-7.
Answers: $I_{SC} = -\frac{7}{3}$ A, $R_{TH} = 3\ \Omega$.

3.3-8 Determine the Norton equivalent circuit at terminals *ab* in Fig. P3.2-8.
Answers: $I_{SC} = -3$ A, $R_{TH} = 2\ \Omega$.

3.3-9 Determine the Norton equivalent circuit at terminals *ab* in Fig. P3.2-9.
Answers: $I_{SC} = -\frac{2}{5}$ A, $R_{TH} = 5\ \Omega$.

3.3-10 Determine the Norton equivalent circuit at terminals *ab* in Fig. P3.2-10.
Answers: $I_{SC} = 6$ A, $R_{TH} = -1\ \Omega$.

3.3-11 Determine the Norton equivalent circuit at terminals *ab* in Fig. P3.2-11.
Answers: $I_{SC} = -2$ A, $R_{TH} = 1\ \Omega$.

3.3-12 Solve for current I and voltage V in the circuit of Fig. P3.2-12 by reducing the indicated circuit to a Norton equivalent at terminals *ab*.
Answers: $I_{SC} = -\frac{5}{8}$ A, $R_{TH} = \frac{8}{3}\ \Omega$, $I = -\frac{35}{17}$ A, $V = -\frac{5}{17}$ V.

3.3-13 Solve for current I and voltage V in the circuit of Fig. P3.2-13 by reducing the indicated circuit to a Norton equivalent at terminals *ab*.
Answers: $I_{SC} = -4$ A, $R_{TH} = 3\ \Omega$, $I = -\frac{17}{5}$ A, $V = -\frac{4}{5}$ V.

3.3-14 Solve for current I and voltage V in the circuit of Fig. P3.2-14 by reducing the indicated circuit to a Norton equivalent at terminals *ab*.
Answers: $I_{SC} = -\frac{3}{5}$ A, $R_{TH} = 5\ \Omega$, $I = -\frac{13}{5}$ A, $V = \frac{49}{5}$ V.

3.3-15 Solve for current I and voltage V in the circuit of Fig. P3.2-15 by reducing the indicated circuit to a Norton equivalent at terminals *ab*.
Answers: $I_{SC} = -\frac{1}{3}$ A, $R_{TH} = 3\ \Omega$, $I = -\frac{3}{4}$ A, $V = \frac{11}{4}$ V.

3.3-16 Solve for current I and voltage V in the circuit of Fig. P3.2-16 by reducing the indicated circuit to a Norton equivalent at terminals *ab*.
Answers: $I_{SC} = -\frac{1}{3}$ A, $R_{TH} = 3\ \Omega$, $I = \frac{1}{2}$ A, $V = -\frac{1}{2}$ V.

3.3-17 Solve for current I and voltage V in the circuit of Fig. P3.2-17 by reducing the indicated circuit to a Norton equivalent at terminals *ab*.
Answers: $I_{SC} = -2$ A, $R_{TH} = -1\ \Omega$, $I = 2$ A, $V = 6$ V.

Section 3.4 Maximum Power Transfer

3.4-1 Determine the value of the resistor R_L in Fig. P3.4-1 such that maximum power will be delivered to it. Determine that maximum power.
Answers: $V_{OC} = -\frac{11}{2}$ V, $R_{TH} = 5\ \Omega$, $R_L = 5\ \Omega$, $p = \frac{121}{80}$ W.

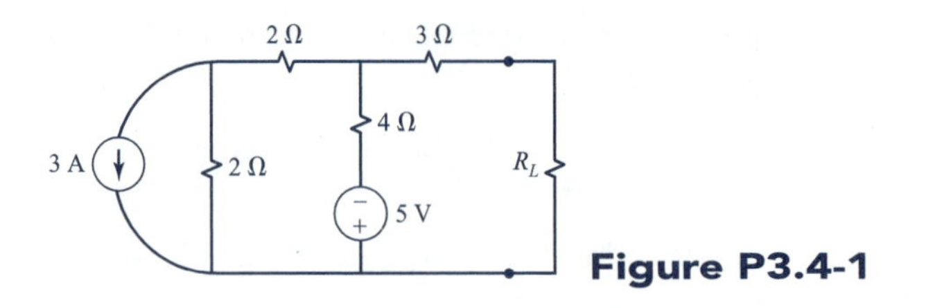

Figure P3.4-1

3.4-2 Determine the value of the resistor R_L in Fig. P3.4-2 such that maximum power will be delivered to it. Determine that maximum power.
Answers: $V_{OC} = 4$ V, $R_{TH} = \frac{2}{5}\ \Omega$, $R_L = \frac{2}{5}\ \Omega$, $p = 10$ W.

Figure P3.4-2

3.4-3 Determine the value of the resistor R_L in Fig. P3.4-3 such that maximum power will be delivered to it. Determine that maximum power.
Answers: $V_{OC} = -9$ V, $R_{TH} = 4\ \Omega$, $R_L = 4\ \Omega$, $p = \frac{81}{16}$ W.

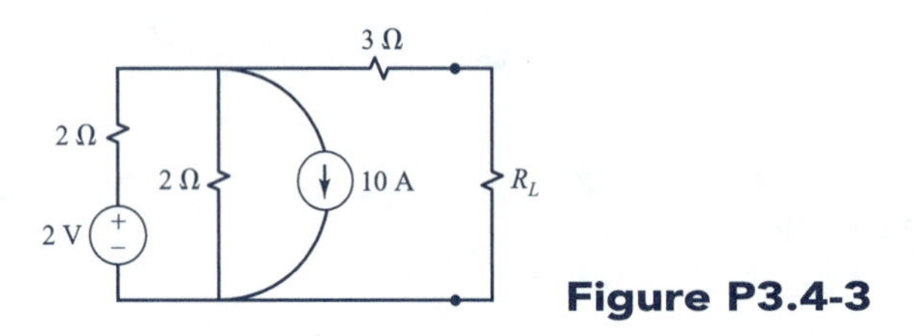

Figure P3.4-3

3.4-4 Determine the value of the resistor R_L in Fig. P3.4-4 such that maximum power will be delivered to it. Determine that maximum power.
Answers: $V_{OC} = -6$ V, $R_{TH} = 2\ \Omega$, $R_L = 2\ \Omega$, $p = \frac{9}{2}$ W.

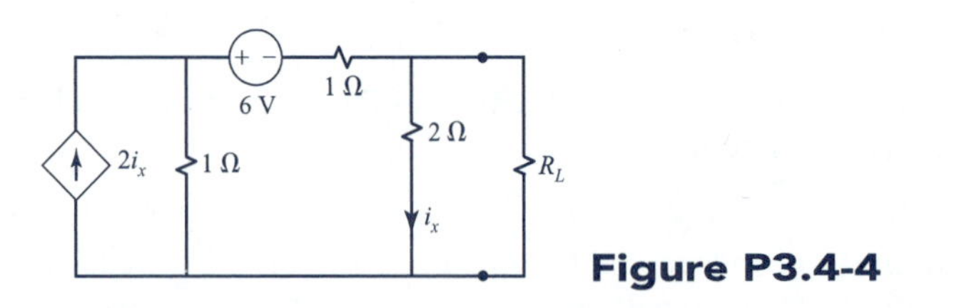

Figure P3.4-4

3.4-5 Determine the value of the resistor R_L in Fig. P3.4-5 such that maximum power will be delivered to it. Determine that maximum power.
Answers: $V_{OC} = 2$ V, $R_{TH} = 5\ \Omega$, $R_L = 5\ \Omega$, $p = \frac{1}{5}$ W.

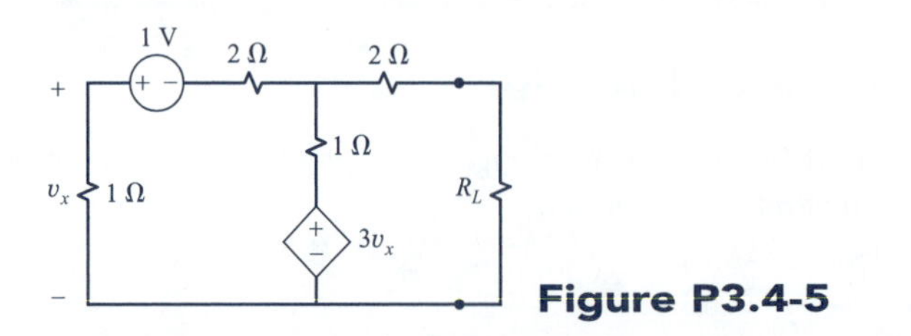

Figure P3.4-5

3.4-6 Determine the value of the resistor R_L in Fig. P3.4-6 such that maximum power will be delivered to it. Determine that maximum power.
Answers: $V_{OC} = 3$ V, $R_{TH} = 4\ \Omega$, $R_L = 4\ \Omega$, $p = \frac{9}{16}$ W.

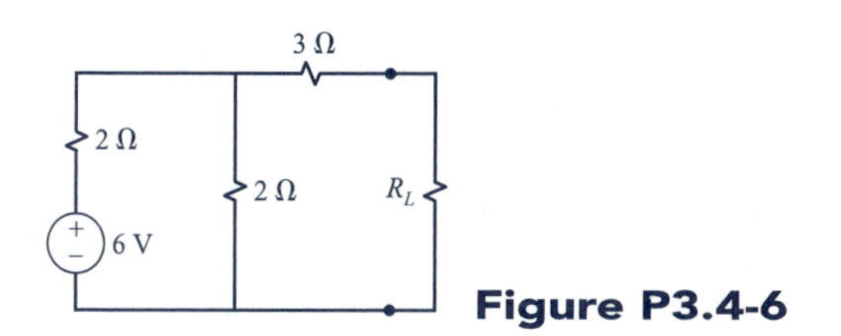

Figure P3.4-6

3.4-7 Determine the value of the resistor R_L in Fig. P3.4-7 such that maximum power will be delivered to it. Determine that maximum power.
Answers: $V_{OC} = -6$ V, $R_{TH} = 1\ \Omega$, $R_L = 1\ \Omega$, $p = 9$ W.

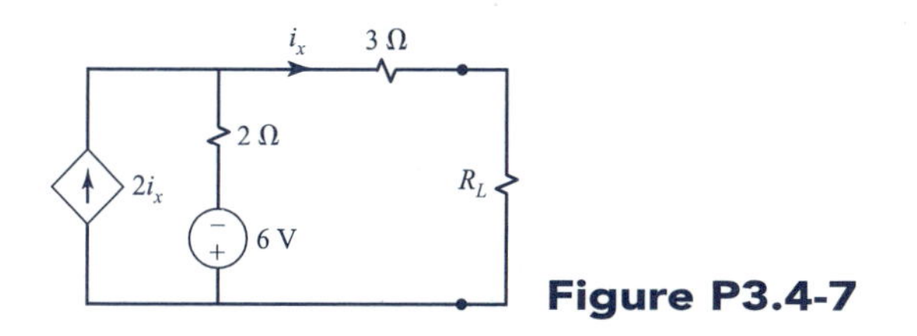

Figure P3.4-7

Section 3.5 The Node-Voltage Method

3.5-1 Determine current I in Fig. P3.5-1 by writing and solving node-voltage equations.
Answer: $I = 1.5$ A.

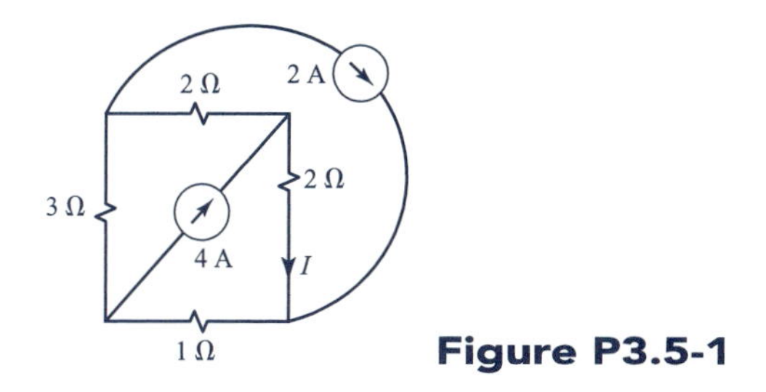

Figure P3.5-1

3.5-2 Determine current I in Fig. P3.5-2 by writing and solving node-voltage equations.
Answer: $I = -0.667$ A.

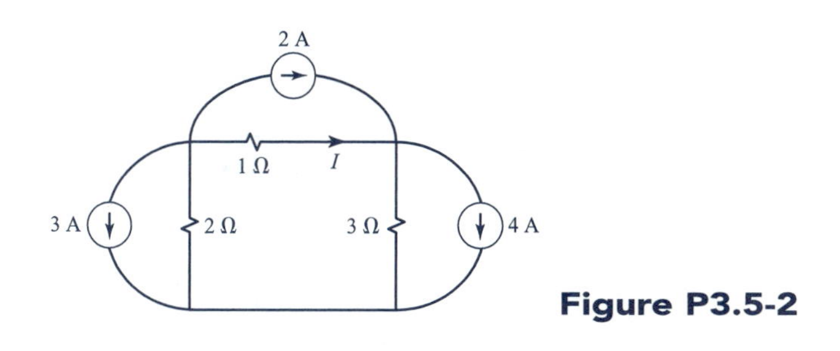

Figure P3.5-2

3.5-3 Determine current I in Fig. P3.5-3 by writing and solving node-voltage equations.
Answer: $I = -0.6$ A.

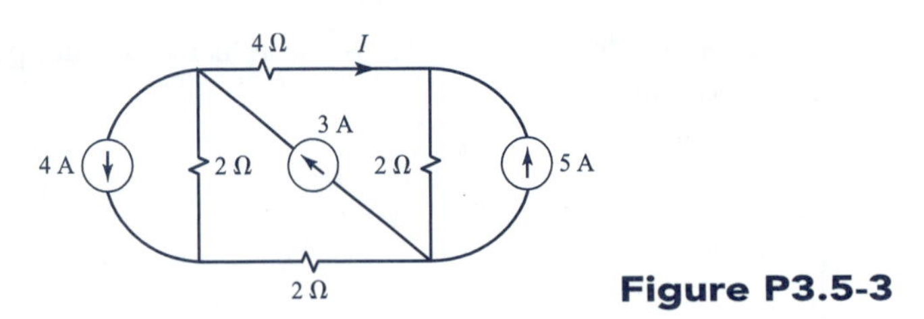

Figure P3.5-3

3.5-4 Determine current I in Fig. P3.5-4 by writing and solving node-voltage equations.
Answer: $I = 2.182$ A.

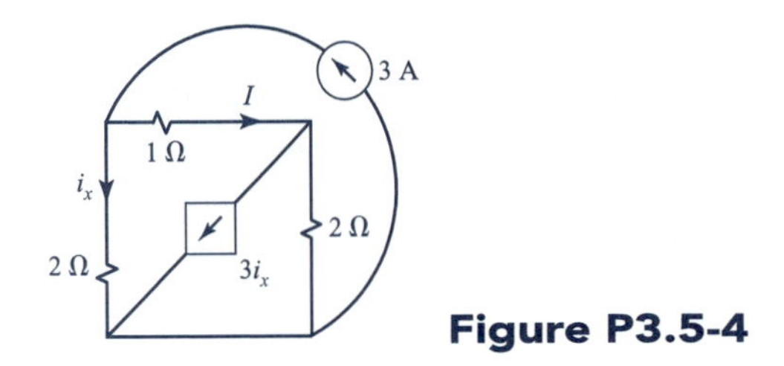

Figure P3.5-4

3.5-5 Determine current I in Fig. P3.5-5 by writing and solving node-voltage equations.
Answer: $I = 5.5$ A.

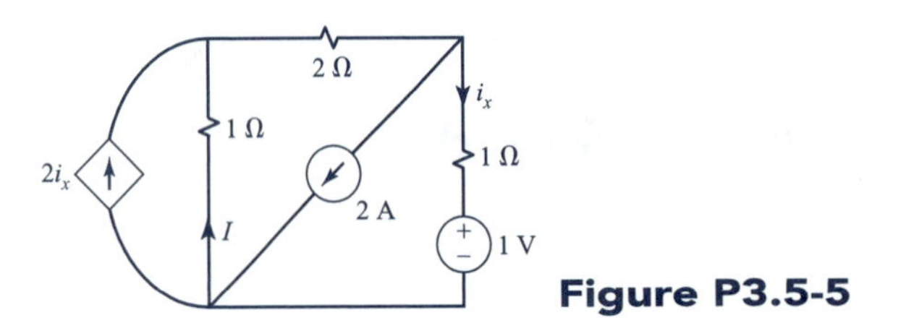

Figure P3.5-5

3.5-6 Determine current I in Fig. P3.5-6 by writing and solving node-voltage equations.
Answer: $I = -2.25$ A.

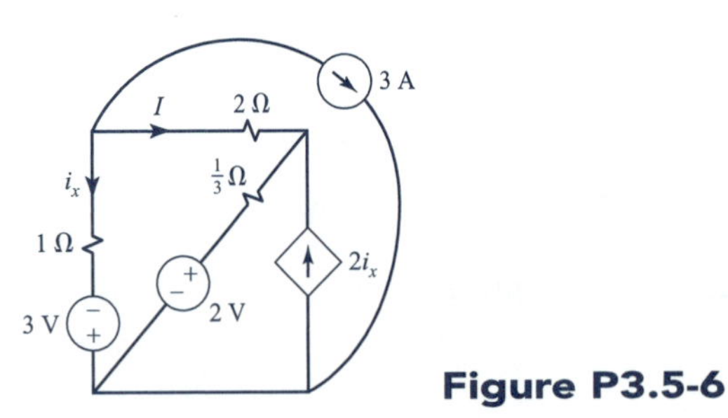

Figure P3.5-6

3.5-7 Determine current I in Fig. P3.5-7 by writing and solving node-voltage equations.
Answer: $I = 1.25$ A.

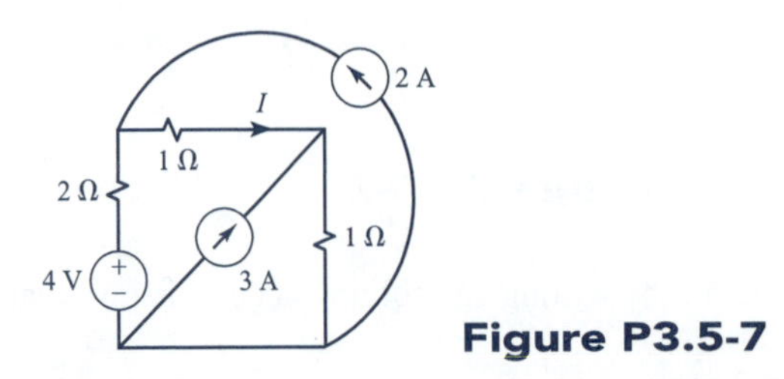

Figure P3.5-7

3.5-8 Determine current I in Fig. P3.5-8 by writing and solving node-voltage equations.
Answer: $I = 2.818$ A.

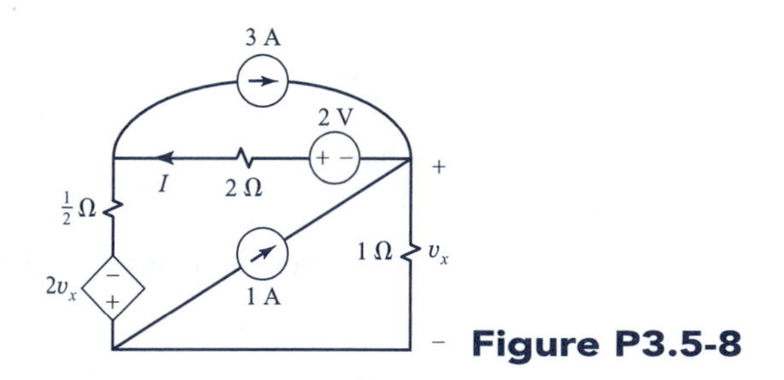

Figure P3.5-8

3.5-9 Determine current I in Fig. P3.5-9 by writing and solving node-voltage equations.
Answer: $I = 0.811$ A.

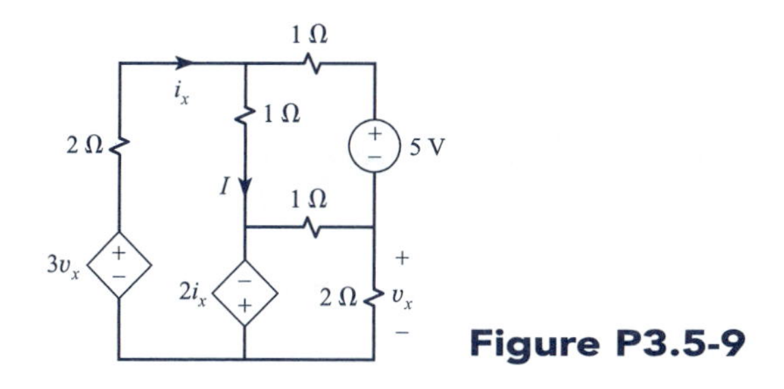

Figure P3.5-9

Section 3.6 The Mesh-Current Method

3.6-1 Determine voltage V in Fig. P3.6-1 by writing and solving mesh-current equations.
Answer: $V = 0.077$ V.

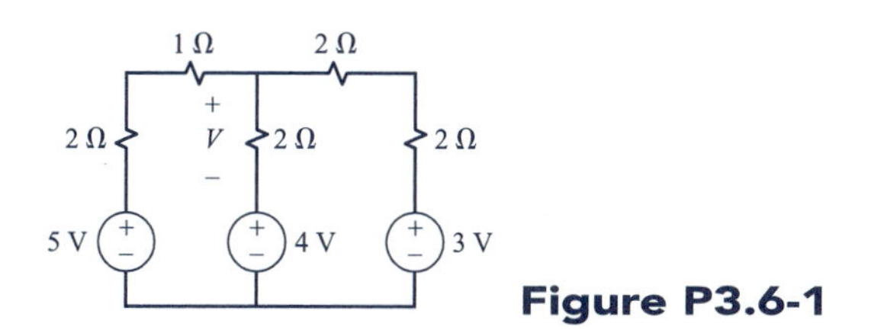

Figure P3.6-1

3.6-2 Determine voltage V in Fig. P3.6-2 by writing and solving mesh-current equations.
Answer: $V = -1.368$ V.

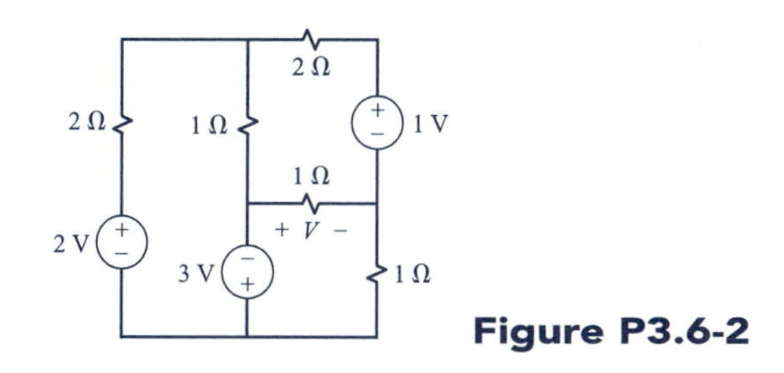

Figure P3.6-2

3.6-3 Determine voltage V in Fig. P3.6-3 by writing and solving mesh-current equations.
Answer: $V = 0.811$ V.

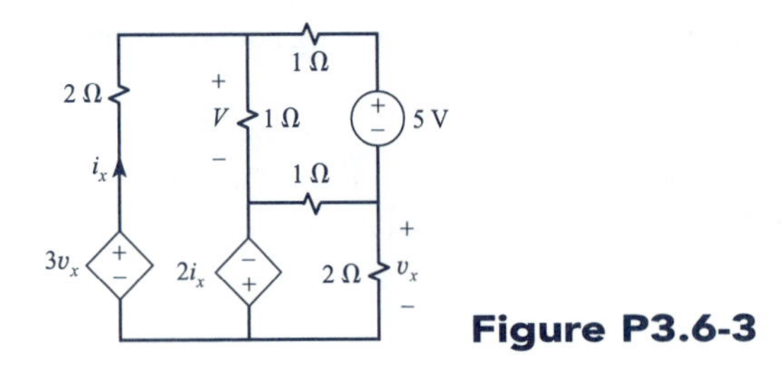

Figure P3.6-3

3.6-4 Determine voltage V in Fig. P3.6-4 by writing and solving mesh-current equations.
Answer: $V = -0.609$ V.

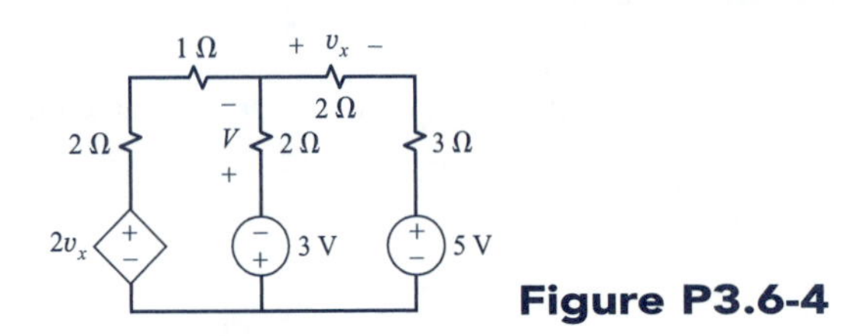

Figure P3.6-4

3.6-5 Determine voltage V in Fig. P3.6-5 by writing and solving mesh-current equations.
Answer: $V = -1.444$ V.

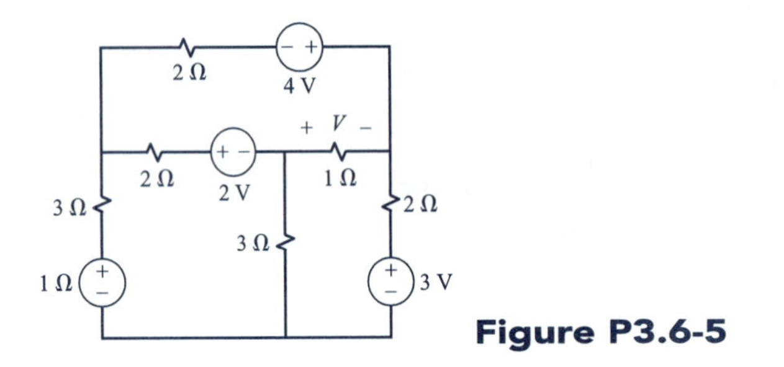

Figure P3.6-5

3.6-6 Determine voltage V in Fig. P3.6-6 by writing and solving mesh-current equations.
Answer: $V = 3.143$ V.

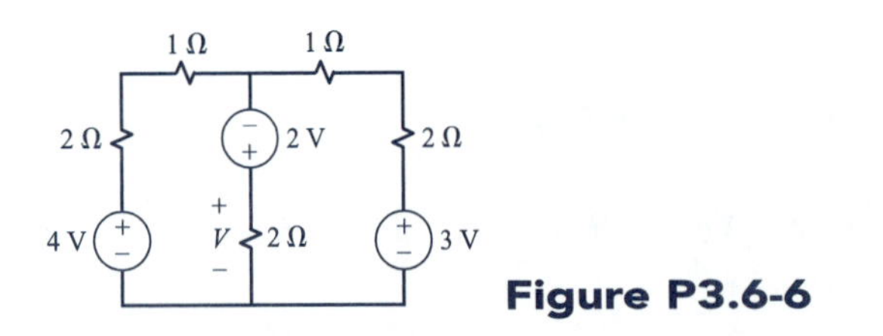

Figure P3.6-6

3.6-7 Determine voltage V in Fig. P3.6-7 by writing and solving mesh-current equations.
Answer: $V = -3.194$ V.

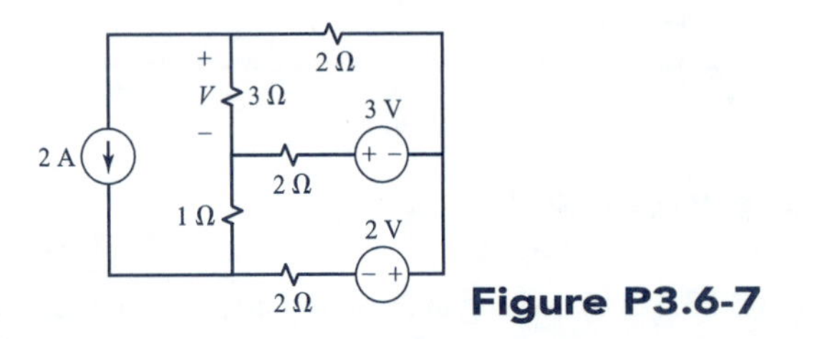

Figure P3.6-7

3.6-8 Determine voltage V in Fig. P3.6-8 by writing and solving mesh-current equations.
Answer: $V = 7.5$ V.

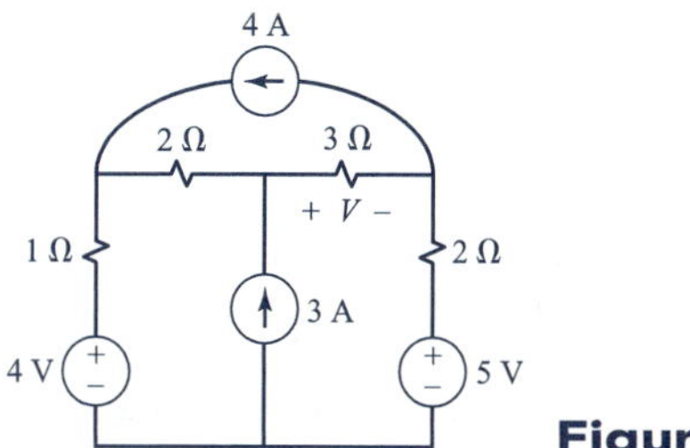

Figure P3.6-8

3.6-9 Determine voltage V in Fig. P3.6-9 by writing and solving mesh-current equations.
Answer: $V = -5$ V.

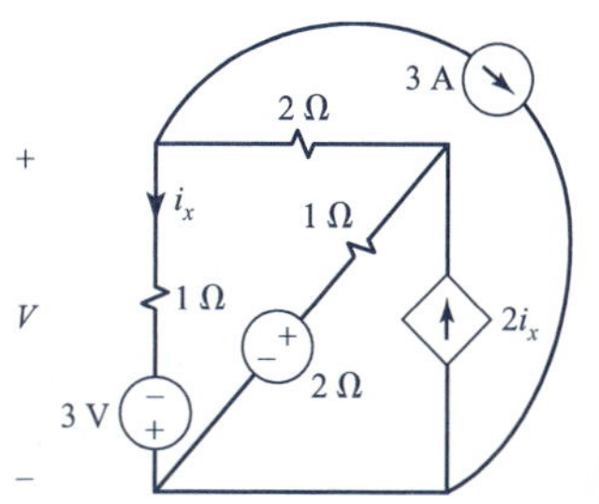

Figure P3.6-9

Section 3.7 PSPICE Applications

3.7-1 Determine current I in Fig. P3.5-1 using PSPICE. *Answer*: $I = 1.5$ A.

3.7-2 Determine current I in Fig. P3.5-2 using PSPICE. *Answer*: $I = -0.667$ A.

3.7-3 Determine current I in Fig. P3.5-3 using PSPICE. *Answer*: $I = -0.6$ A.

3.7-4 Determine current I in Fig. P3.5-4 using PSPICE. *Answer*: $I = 2.182$ A.

3.7-5 Determine current I in Fig. P3.5-5 using PSPICE. *Answer*: $I = 5.5$ A.

3.7-6 Determine current I in Fig. P3.5-6 using PSPICE. *Answer*: $I = -2.25$ A.

3.7-7 Determine current I in Fig. P3.5-7 using PSPICE. *Answer*: $I = 1.25$ A.

3.7-8 Determine current I in Fig. P3.5-8 using PSPICE. *Answer*: $I = 2.818$ A.

3.7-9 Determine current I in Fig. P3.5-9 using PSPICE. *Answer*: $I = 0.8108$ A.

3.7-10 Determine voltage V in Fig. P3.6-1 using PSPICE. *Answer*: $V = 0.07692$ V.

3.7-11 Determine voltage V in Fig. P3.6-2 using PSPICE. *Answer*: $V = -1.368$ V.

3.7-12 Determine voltage V in Fig. P3.6-3 using PSPICE. *Answer*: $V = 0.8108$ V.

3.7-13 Determine voltage V in Fig. P3.6-4 using PSPICE. *Answer*: $V = -0.6087$ V.

3.7-14 Determine voltage V in Fig. P3.6-5 using PSPICE. *Answer*: $V = -1.444$ V.

3.7-15 Determine voltage V in Fig. P3.6-6 using PSPICE. *Answer*: $V = 3.143$ V.

3.7-16 Determine voltage V in Fig. P3.6-7 using PSPICE. *Answer*: $V = -3.194$ V.

3.7-17 Determine voltage V in Fig. P3.6-8 using PSPICE. *Answer*: $V = 7.5$ V.

3.7-18 Determine voltage V in Fig. P3.6-9 using PSPICE. *Answer*: $V = -5.0$ V.

Section 3.8 MATLAB Applications

3.8-1 Use MATLAB to solve the node-voltage equations and determine I for the circuit of Fig. P3.5-2. *Answer*: $I = -0.667$ A.

3.8-2 Use MATLAB to solve the node-voltage equations and determine I for the circuit of Fig. P3.5-3. *Answer*: $I = -0.6$ A.

3.8-3 Use MATLAB to solve the node-voltage equations and determine I for the circuit of Fig. P3.5-7. *Answer*: $I = 1.25$ A.

3.8-4 Use MATLAB to solve the node-voltage equations and determine I for the circuit of Fig. P3.5-9. *Answer*: $I = 0.811$ A.

3.8-5 Use MATLAB to solve the mesh-current equations and determine V for the circuit of Fig. P3.6-1. *Answer*: $V = 0.077$ V.

3.8-6 Use MATLAB to solve the mesh-current equations and determine V for the circuit of Fig. P3.6-3. *Answer*: $V = 0.811$ V.

3.8-7 Use MATLAB to solve the mesh-current equations and determine V for the circuit of Fig. P3.6-5. *Answer*: $V = -1.444$ V.

3.8-8 Use MATLAB to solve the mesh-current equations and determine V for the circuit of Fig. P3.6-7. *Answer*: $V = -3.194$ V.

CHAPTER 4

The Operational Amplifier (Op Amp)

The operational amplifier, or op amp, which we will introduce in this chapter, is an electronic device that is composed of a large number of electronic elements such as the bipolar junction transistor (BJT) as well as numerous resistors. We will not be concerned about this internal structure, but will instead utilize its characteristics at the external terminals. Op amps have become as standard a circuit device as the resistor. They are inexpensive (less than $1 US) and are available in dual-inline packages (DIPs) suitable for insertion into printed circuit boards (PCBs).

4.1 The Actual Op Amp versus the Ideal Op Amp

Figure 4.1*a* shows the pin connections of a DIP containing a popular op amp, the μA741. Also shown on that diagram is the circuit symbol for the op amp. A dc power supply must be attached to pins 7 and 4 in order to power the internal electronics of the device. This is provided by a dual dc power supply as shown in Fig. 4.1*b* that consists of two dc voltage sources (batteries), V_{dc}, connected at a common, or *ground,* terminal. Typical dc voltages used to power the op amp lie between ± 12 V and ± 15 V. The negative terminal of this power supply is connected to pin 4, while the positive terminal is connected to pin 7. There are two inputs to the device: an inverting input v^- and a noninverting input v^+. These voltages are referenced to the ground of the dc power supply. The output is taken at pin 6, and this voltage is also referenced to the ground of the dc power supply. Two remaining terminals, pins 1 and 5, are used to *balance* the amplifier, the need for which will be discussed.

Figure 4.2*a* shows the circuit symbol of the op amp. Ordinarily the $+V_{dc}$ and $-V_{dc}$ dc power supply voltage terminals are omitted from the symbol but are understood to be present. Figure 4.2*b* shows the equivalent circuit of the op amp. The element R_i is the input resistance between the + and − input terminals. The element R_o is the output (Thevenin) resistance. The op amp is a difference amplifier in that the output voltage is proportional to the difference voltage, $v_d = v^+ - v^-$. This is provided by the voltage-controlled voltage source, $Av_d = A(v^+ - v^-)$. Values for these parameters for a typical op amp, the μA741, are

$$R_i = 2\ \text{M}\Omega$$
$$R_o = 75\ \Omega$$
$$A = 200{,}000$$

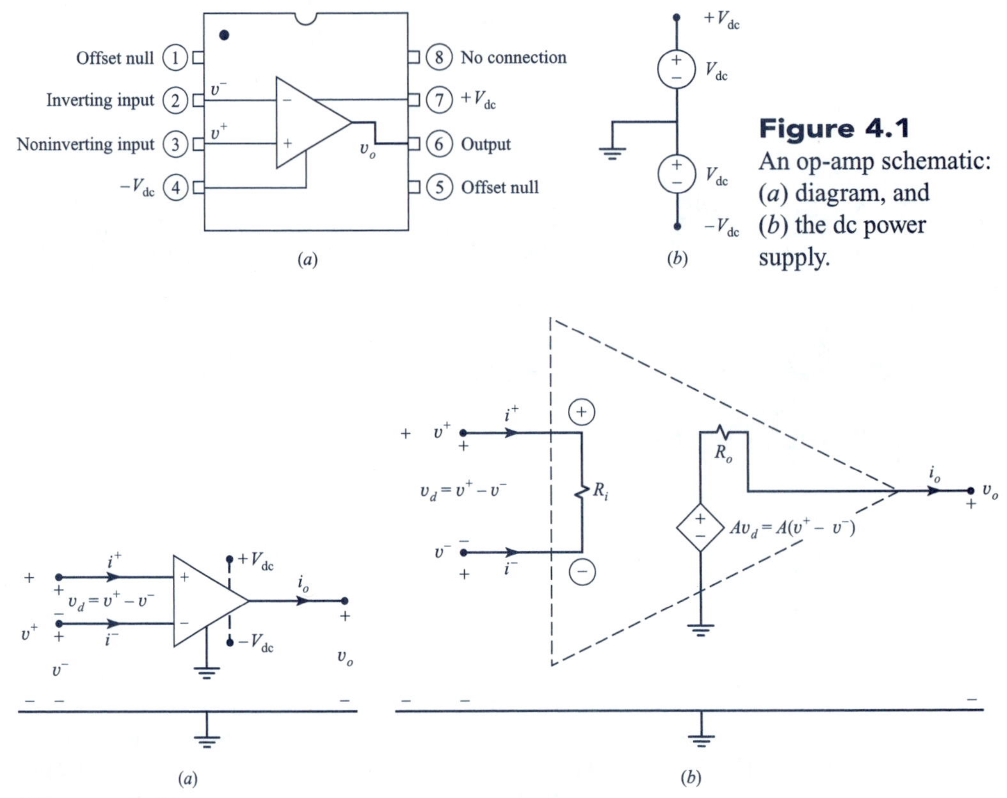

Figure 4.1
An op-amp schematic: (*a*) diagram, and (*b*) the dc power supply.

Figure 4.2
Op-amp models: (*a*) the external terminal definition, and (*b*) the internal model.

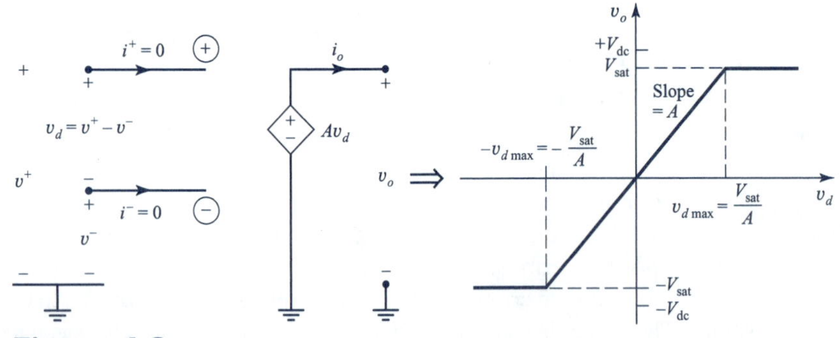

Figure 4.3
The op-amp input–output relation assuming infinite input resistance and zero output resistance.

Because of the large value of the input resistance, it will be assumed infinite in our later analyses. Similarly, because the output resistance is small, it is assumed zero. This leads to the approximate model of the op amp shown in Fig. 4.3, whose properties are

$$\begin{aligned} v_o &= Av_d \\ &= A\left(v^+ - v^-\right) \\ i^+ &= 0 \\ i^- &= 0 \end{aligned} \tag{4.1}$$

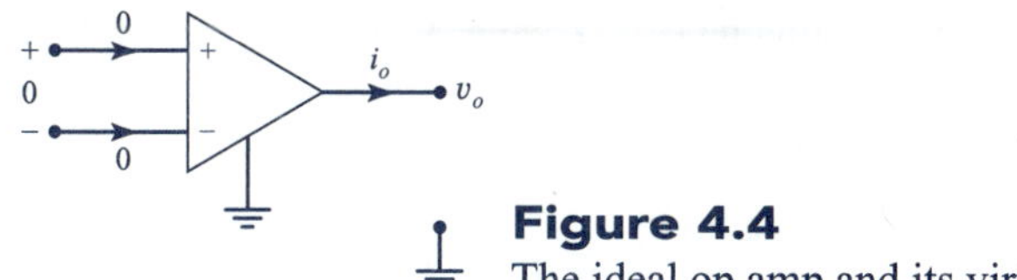

Figure 4.4
The ideal op amp and its virtual short-circuit properties.

The transfer characteristic is also shown in Fig. 4.3 and relates the output voltage to the difference input voltage, $v_d = v^+ - v^-$. Observe that there is a linear region where the output voltage is between the extremes $-V_{sat} \leq v_o \leq V_{sat}$. The level V_{sat} is referred to as the *saturation voltage* of the op amp. Typically, this is close to but slightly less than the dc power supply voltages: $V_{sat} \approx V_{dc}$. Hence, for linear operation, the output cannot exceed V_{sat}:

$$-V_{sat} \leq v_o \leq V_{sat} \tag{4.2}$$

The extreme values of the difference voltage at the ends of this linear region are $v_{d\max} = V_{sat}/A$. This is a very small range. For example, for the μA741, $v_{d\max} = V_{sat}/A = 14\text{ V}/200{,}000 = 70\ \mu\text{V}$. We will see that, for linear operation, this allows us to assume that the difference voltage is approximately zero. If fact, because the gain is so large, if we approximate it as infinity, $A \to \infty$, then $v_d \to 0$. This results in the properties of an *ideal op amp* shown in Fig. 4.4:

$$\left.\begin{aligned} v_d &= 0 \\ i^+ &= 0 \\ i^- &= 0 \end{aligned}\right\} \quad A \to \infty \tag{4.3}$$

Use of this ideal op-amp model rather than that of an actual op amp will greatly simplify the analysis of circuits containing it and will lead to results that closely approximate the true results, as we will see. The properties of the ideal op amp given in (4.3) are said to be the *virtual short-circuit* properties in that although the currents are zero, the voltage across the terminals is also zero. A true short circuit would have the voltage across the terminals zero but the current through it as yet undefined and not necessarily zero. Hence the word *virtual.*

The virtual short-circuit properties given in (4.3) are very powerful constraints. The analysis of a circuit containing an ideal op amp will become trivial if we always do the following:

Step 1: Label the ideal op amp with the virtual short-circuit conditions

$$v_d = 0, \qquad i^+ = 0, \qquad i^- = 0$$

Step 2: Apply the zero-input-current constraint at the noninverting terminal, $i^+ = 0$, by applying KCL to the circuit that is attached to that terminal.
Step 3: Apply the zero-input-current constraint at the inverting terminal, $i^- = 0$, by applying KCL to the circuit that is attached to that terminal.
Step 4: Apply the zero-difference-voltage constraint between the input terminals, $v_d = 0$, by applying KVL to all loops of the attached circuit that involve that voltage.

There are some additional requirements of practical op amps that must be considered in any application. The maximum current that can be drawn from the op-amp terminals is limited to $|i_o| \leq i_{sat}$, where i_{sat} is referred to as the *saturation current* of the op amp. Typical values of saturation current for low-power op amps are usually a few tens of milliamperes. Hence op amps cannot supply large currents. This means that load resistances as well as other resistors in the attached circuit are typically in the kilohm ($10^3\,\Omega$) range. Most of our applications of the op amp will involve operation in the linear region. It is necessary in all those applications to check the following two conditions to insure that operation is indeed in the linear region:

$$-V_{\text{sat}} \le v_o \le V_{\text{sat}} \\ -i_{\text{sat}} \le i_o \le i_{\text{sat}} \tag{4.4}$$

In addition, the output voltage of an op amp cannot change arbitrarily fast. This limitation is referred to as the *slew rate*. The slew rate is the maximum rate of change of the output voltage. Typical values are on the order of 1 V/μs. For example, suppose the output voltage of an op amp is $v_o = V \sin \omega t$. The maximum magnitude of the time derivative of this is $|dv_o/dt|_{\text{max}} = \omega V$. Suppose that $V = 10$ V and the slew rate is 1V/μs. The maximum frequency of a signal that can be correctly handled is $\omega = 2\pi f = (1\text{ V}/\mu\text{s})/10\text{ V} = 10^5$ rad/s, or 15.9 kHz. This means that the output will be distorted if we attempt to obtain the maximum output voltage swing. At lower levels of output voltage, higher frequencies can be faithfully processed.

The final nonideal property to be considered is offset voltage and current. If a short circuit is applied across the v^+ and v^- input terminals giving a difference voltage of zero ($v_d = v^+ - v^- = 0$), the output voltage will not be its ideal zero voltage, but will have a small (millivolt) component, which is referred to as *offset*. There are methods for determining this offset and also nulling it out using a circuit attached to pins 1 and 5 of Fig. 4.1*a*, but we will defer that discussion to later electronics courses.

For our purposes, we will assume the ideal op-amp properties given in (4.3), which provide reasonable approximations for most practical purposes.

4.1.1 The Inverting Amplifier A useful application of the op amp is the inverting amplifier shown in Fig. 4.5*a*. Substituting the finite-gain model of the op amp (with infinite input resistance and zero output resistance) yields the circuit of Fig. 4.5*b*. Because of the infinite input resistance, the current into the input terminals is zero, so we define, by applying KCL at the inverting input, a current I through resistors R_i and R_f. From that circuit we write

$$v_i + v_d = R_i I \\ v_o + v_d = -R_f I \\ v_o = A v_d \tag{4.5}$$

Solving these for the ratio v_o/v_i gives the *voltage gain* of the overall amplifier as

$$\frac{v_o}{v_i} = -\frac{A\dfrac{R_f}{R_i}}{A + 1 + \dfrac{R_f}{R_i}} \tag{4.6}$$

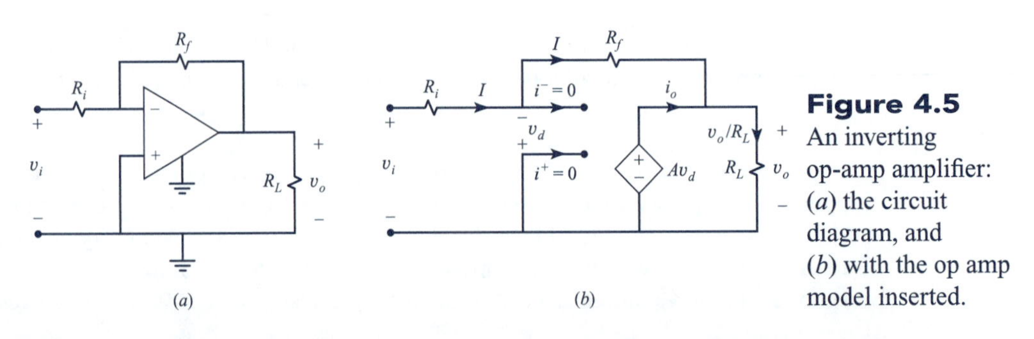

Figure 4.5
An inverting op-amp amplifier: (*a*) the circuit diagram, and (*b*) with the op amp model inserted.

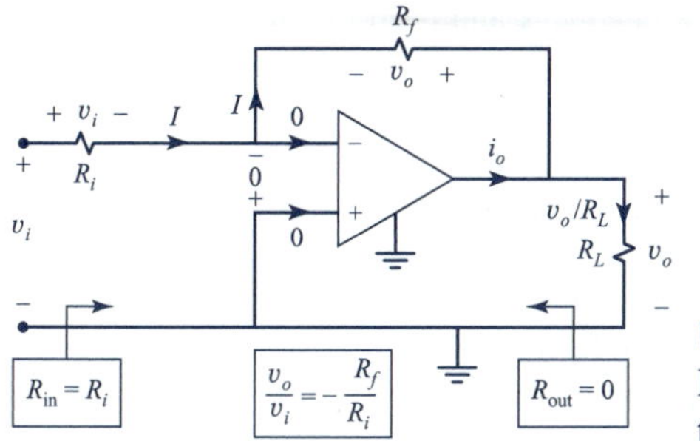

Figure 4.6
Properties of the inverting op-amp amplifier using the virtual short-circuit conditions.

If we let the gain of the op amp approach infinity, we obtain

$$\frac{v_o}{v_i} = -\frac{R_f}{R_i} \qquad A \to \infty \tag{4.7}$$

Hence, in order to construct an amplifer with a voltage gain of, for example, −10, we simply choose resistor values for R_f and R_i such that their ratio is 10. For example, if we choose $R_f = 100\ \text{k}\Omega$ and $R_i = 10\ \text{k}\Omega$, we will have constructed a voltage amplifier having a voltage gain of −10. This means that an input voltage of, for example, $v_i\,(t) = 1 \sin \omega t$ V yields an output voltage of $v_o\,(t) = -10 \sin \omega t$ V. The negative sign in the gain expression simply means that the output voltage is 180° out of phase with the input voltage. For example, as the input sinusoid increases, the output sinusoid decreases.

This result could have been more easily obtained by using the virtual short-circuit conditions given in (4.3). To show this we label the op amp's input terminals with those conditions, as shown in Fig. 4.6. Exercising the condition that the current into the noninverting terminal is zero, $i^+ = 0$, yields no significant result. Exercising the condition that the current into the inverting terminal is zero, $i^- = 0$, shows that, by KCL, the currents through R_f and R_i must be equal, so we label that current as I. Exercising the condition that the difference voltage is zero, $v_d = 0$, provides, by KVL, the following two conditions:

$$v_i = R_i I$$
$$v_o = -R_f I$$

Taking the ratio of these gives the voltage gain of the overall amplifier as obtained in (4.7), but much more simply. This is an indication of the considerable simplification that the virtual short-circuit conditions can provide. In addition, observe that the output voltage is independent of the load R_L. This means that the Thevenin resistance seen by R_L is *zero*. Hence we have an ideal voltage source at the output terminals. The input resistance to the overall amplifier is the ratio of the input voltage v_i and the input current I, which are labeled with the passive sign convention:

$$R_{in} = \frac{v_i}{I} = R_i \tag{4.8}$$

Therefore, in order to construct an amplifier having a voltage gain of −15 and an input resistance of 20 kΩ we choose $R_i = 20\ \text{k}\Omega$ and $R_f = 300\ \text{k}\Omega$. Hence the design of a voltage amplifier for use in, for example, amplifying the low-level output of a microphone is exceedingly simple.

4.1.2 Negative Feedback and Saturation It is important to observe that a portion of the output voltage is fed back through R_f to the negative, or inverting, input of the op amp. This is referred to as *negative feedback* and is an important property of all such uses of op amps. Negative feedback must be present for proper operation. It is an essential requirement for insuring that the difference voltage is zero ($v_d = 0$). For example, observe that in Fig. 4.5*b*, when v_o increases, a portion of the increase is added to the inverting input terminal, hence subtracting from v_d and thus driving v_d to zero. Suppose the feedback resistor, R_f, had been inadvertently connected to the positive, or noninverting, input terminal of the op amp. In that case, as v_o increased, a portion of the increase would be added to v_d, thus increasing v_o through the controlled source, Av_d, causing v_o to increase further, etc. Hence the output voltage would be driven into saturation (see Fig. 4.3), and the output would be stuck at $v_o = V_{sat}$. In this case no amplification would take place, because the output voltage would be dc and independent of changes in the input voltage v_i. When examining a potential op-amp circuit, one must always check to insure that negative feedback is present.

An additional property that must be verified for all op-amp circuits is to insure that, even though negative feedback is present, the output is not in saturation. Recall that the practical limits on the output voltage and current of the op amp are given in (4.4). If these are exceeded, the op amp is operating outside of the linear range, and again no amplification is taking place. (There are a few applications where we intentionally operate an op amp in saturation.) For example, consider the inverting amplifier shown in Fig. 4.6. Suppose that for this amplifier, $R_i = 5\ \text{k}\Omega$, $R_f = 50\ \text{k}\Omega$, $R_L = 100\ \Omega$, and $v_i(t) = 2 \sin \omega t$ V. The gain of this amplifier is -10, and hence the output voltage is $v_o(t) = -20 \sin \omega t$ V. For a typical μA741 op amp, $V_{sat} = 14$ V. The output voltage is clipped at V_{sat} for portions of the output voltage greater that V_{sat} as shown in Fig. 4.7. In this case *distortion* of the output waveform has occurred, rendering it useless. If we reduce the input to $v_i(t) = \sin \omega t$ V, then the output voltage is $v_o(t) = -10 \sin \omega t$ V and the op amp will not be driven into saturation.

In addition we must insure that the output current of the op amp does not exceed i_{sat}. At the peak of the output sinusoid, the total output current for the op amp is, by KCL, the sum of the currents in the load and the feedback resistor:

$$i_o = \frac{v_o}{R_L} + \frac{v_o}{R_f} \leq i_{sat}$$

For $R_i = 5\ \text{k}\Omega$, $R_f = 50\ \text{k}\Omega$, and $R_L = 100\ \Omega$ with a 1-V input sinusoid we obtain at the peak of the output sinusoid $i_o = -10\ \text{V}/100\ \Omega - 10\ \text{V}/50\ \text{k}\Omega = -100.2$ mA. For typical low-power op amps the output saturation current is typically a few tens of milliamperes. It is impossible for the op amp to provide the required current, and so this load cannot be successfully driven. Suppose we change the load resistance to $R_L = 10\ \text{k}\Omega$. This does not affect the overall gain of the amplifier, since that depends only on the ratio of R_f and R_i. In this case the output current at the peak of the sinusoid becomes $i_o = -1.2$ mA, which does not exceed typical values

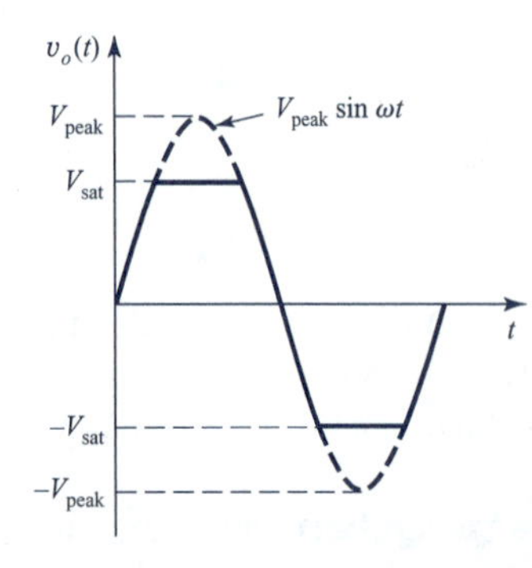

Figure 4.7
Illustration of the saturation of the output voltage.

of i_{sat} for low-power op amps. Hence small load resistances cannot be successfully driven by op-amp amplifier circuits because of the basic limitations of the op amp.

Example 4.1

Design an amplifer that has an input resistance of 10 kΩ, that has an overall gain of 20, and whose output is in phase with its input. The input voltage is a sinusoid with peak voltage of 50 mV, and the load resistance is 500 Ω. An op amp having $V_{sat} = 14$ V and $i_{sat} = 2$ mA is available for use.

Solution If we cascade two inverting amplifiers, the phase shift of one will cancel the phase shift of the other, giving an overall amplifier with a positive gain. The product of the two gains should equal 20, and the input resistance of the first stage should be 10 kΩ. Figure 4.8 shows one of many satisfactory designs. The first stage has a gain of -5 and the second stage has a gain of -4, giving an overall gain of 20. The input resistance to the first stage is 10 kΩ, and the input resistance to the second stage is 20 kΩ. For a peak input voltage of 50 mV, the peak output voltage of the first stage is 250 mV, and hence the output current of the first op amp is $-250 \text{ mV}/20 \text{ k}\Omega - 250 \text{ mV}/50 \text{ k}\Omega = -17.5\ \mu\text{A}$, which is well within the saturation current of the available op amp. The output voltage of the first stage is $-5 \times 50 \text{ mV} = -250$ mV, which does not exceed the saturation voltage of the op amp. Applying this to the second stage gives a final output voltage of $-4 \times (-250 \text{ mV}) = 1$ V, which is again within the saturation voltage of the available op amp. The output current of the second op amp is $1 \text{ V}/80 \text{ k}\Omega + 1 \text{ V}/500\ \Omega = 2.0125$ mA, which is approximately the saturation current of the available op amp, and hence a successful design has been achieved.

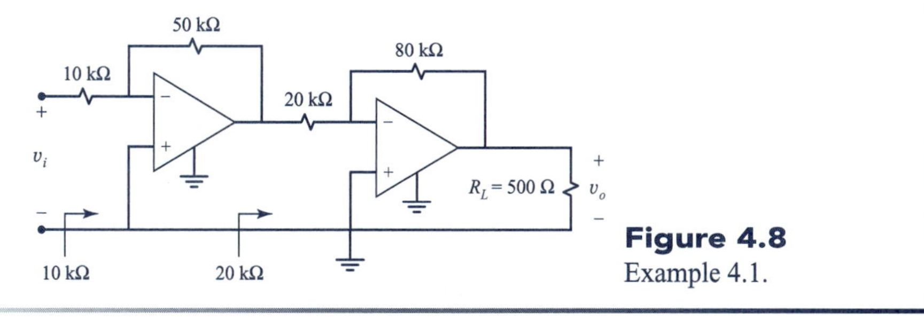

Figure 4.8
Example 4.1.

Exercise Problem 4.1

For the circuit of Fig. E4.1 determine the peak value of the output voltage and the maximum output current of the op amp.

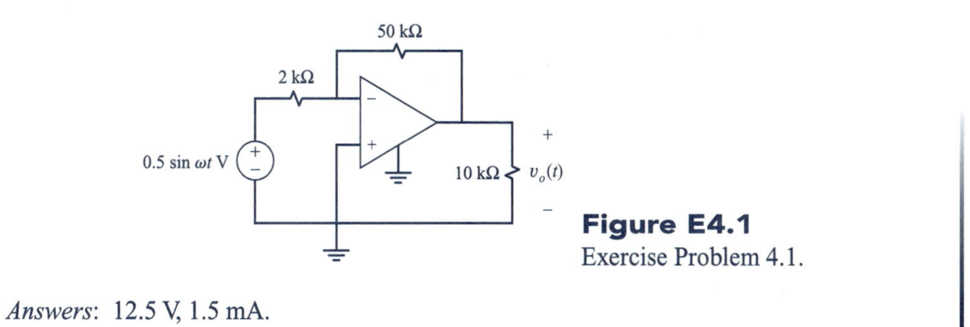

Figure E4.1
Exercise Problem 4.1.

Answers: 12.5 V, 1.5 mA.

4.2 Other Useful Op-Amp Circuits

There are a number of other useful op-amp circuits that can be used as building blocks for other applications.

4.2.1 The Noninverting Amplifier The amplifier previously considered and shown in Fig. 4.6 had a negative sign in the overall voltage gain expression. That signified that the output voltage is 180° out of phase with the input voltage. There are instances where this is undesirable. Figure 4.9 shows an amplifier whose output voltage is in phase with the input voltage. Examining the circuit shows that negative feedback is present, since the output is connected to the inverting input through R_f. To determine the overall voltage gain we label the virtual short-circuit conditions as shown. Examining the zero current of the noninverting terminal shows that the input current to the overall amplifier is zero and hence the input resistance of the amplifier is infinite, unlike the inverting amplifier:

$$R_{in} = \infty \tag{4.9}$$

Applying KCL at the inverting input shows that the currents through R and R_f are the same, so we label that current as shown. Applying KVL around loops that involve the zero difference voltage yields the following equations:

$$v_i = RI$$
$$v_o = RI + R_f I$$

Taking the ratio of these gives the overall voltage gain as

$$\frac{v_o}{v_i} = \frac{R + R_f}{R} = 1 + \frac{R_f}{R} \tag{4.10}$$

and we have created a noninverting amplifier. Observe that, as in the inverting amplifier, the output voltage is independent of the load resistor and hence the output resistance of the overall amplifier is zero: an ideal source.

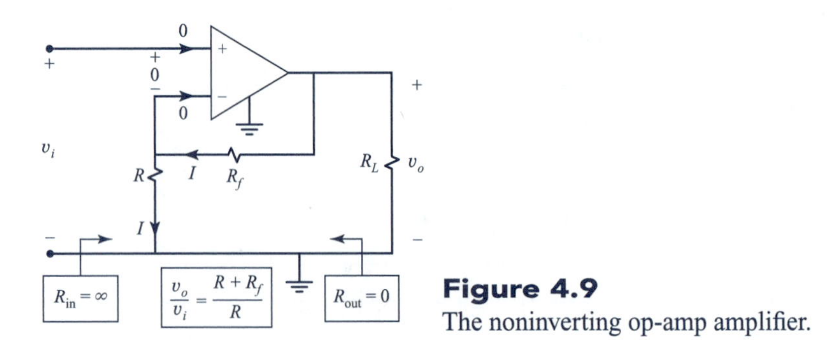

Figure 4.9
The noninverting op-amp amplifier.

Example 4.2

For the circuit of Fig. 4.10*a* determine the output voltage and determine whether the op amp is in saturation if $V_{sat} = 14$ V and $i_{sat} = 2$ mA.

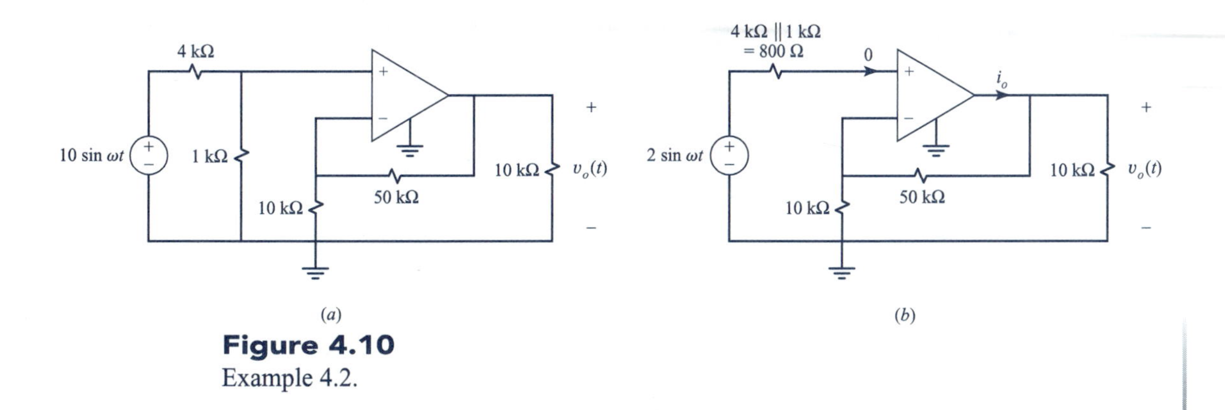

Figure 4.10
Example 4.2.

Solution The 4-kΩ and 1-kΩ resistors form a voltage divider (since the current into the noninverting terminal of the op amp is zero). Hence the circuit can be reduced to the form shown in Fig. 4.10*b*. From this circuit and (4.10) we determine that the gain is 6 and hence the output voltage is $v_o = 12 \sin \omega t$. The peak of this output voltage does not exceed the saturation voltage of the op amp. The peak output current of the op amp is 12 V/10 kΩ + 12 V/60 kΩ = 1.4 mA, which does not exceed the saturation current of the op amp.

Exercise Problem 4.2

Design a noninverting amplifier to have a gain of 10 and supply a load of 5 kΩ. Determine the maximum output voltage and current if the input voltage is a sinusoid of peak level of 1V.

Answers: $R = 5$ kΩ, $R_f = 45$ kΩ, 10 V, 2.2 mA.

4.2.2 The Difference Amplifier In instrumentation circuits, it is frequently necessary to subtract two voltages and to amplify the difference. Figure 4.11*a* shows a circuit that will accomplish this task. Since the current into the noninverting terminal is zero, we observe that the resistors R_3 and R_4 form a voltage divider with input voltage v_2. Hence we reduce the circuit as shown in Fig. 4.11*b*. The output voltage can be obtained by superposition from the two circuits shown in Fig. 4.11*c*. The circuit on the left is recognized as an inverting amplifier, so that

$$v_o' = -\frac{R_2}{R_1} v_1$$

The circuit on the right is recognized as a noninverting amplifier, so that

$$v_o'' = \frac{R_1 + R_2}{R_1} \left(\frac{R_4}{R_3 + R_4} v_2 \right)$$

Combining these gives the overall gain as

$$v_o = \frac{R_1 + R_2}{R_1} \left(\frac{R_4}{R_3 + R_4} v_2 \right) - \frac{R_2}{R_1} v_1$$

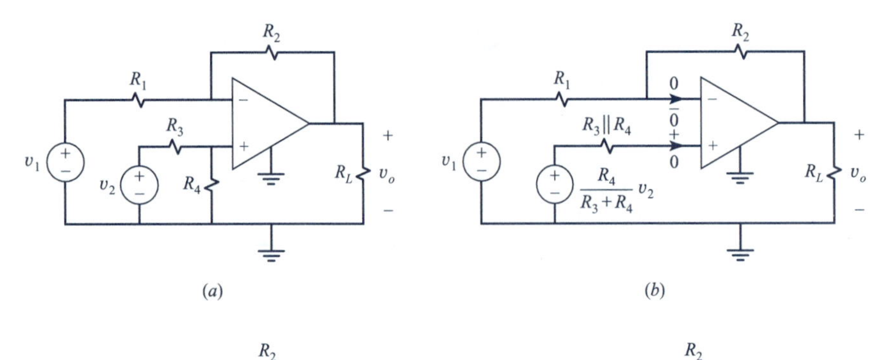

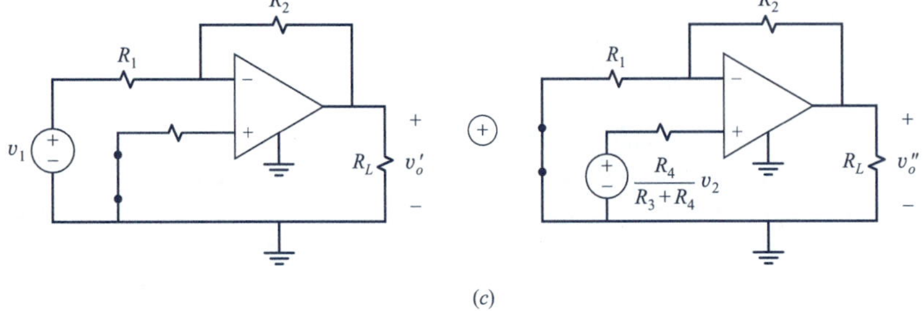

Figure 4.11
The op amp difference amplifier.

If we choose $R_3 = R_1$ and $R_4 = R_2$, this result becomes

$$v_o = \frac{R_2}{R_1}(v_2 - v_1) \tag{4.11}$$

and we have produced a circuit that amplifies the difference between two voltages.

Example 4.3

One of the difficulties with the difference amplifier of Fig. 4.11*a* is that if the voltages v_1 and v_2 have source resistances that are not zero (as assumed in the previous analysis), these will be added to R_1 and R_3 and hence affect the performance if their values are on the order of those of R_1 and R_3. Figure 4.12 shows an improved, instrumentation-quality difference amplifier. Determine the output of this amplifier, and discuss why it is independent of the source resistances of voltages v_1 and v_2.

Solution First observe that OA1 and OA2 are noninverting amplifiers. If we define a voltage at the central point between these as v_x, then OA2 can be analyzed as shown in Fig. 4.12*b*. Applying superposition, we see that the output is the sum of a noninverting amplifier response (due to v_2) and an inverting amplifier response (due to v_x):

$$v_{o2} = \frac{R_1 + R_2}{R_1} v_2 - \frac{R_2}{R_1} v_x$$

The output of OA1 is similar with v_2 replaced with v_1. The final stage is a difference amplifier with unity gain, so that

$$\begin{aligned} v_o &= v_{o2} - v_{o1} \\ &= \frac{R_2 + R_1}{R_1}(v_1 - v_2) \end{aligned} \tag{4.12}$$

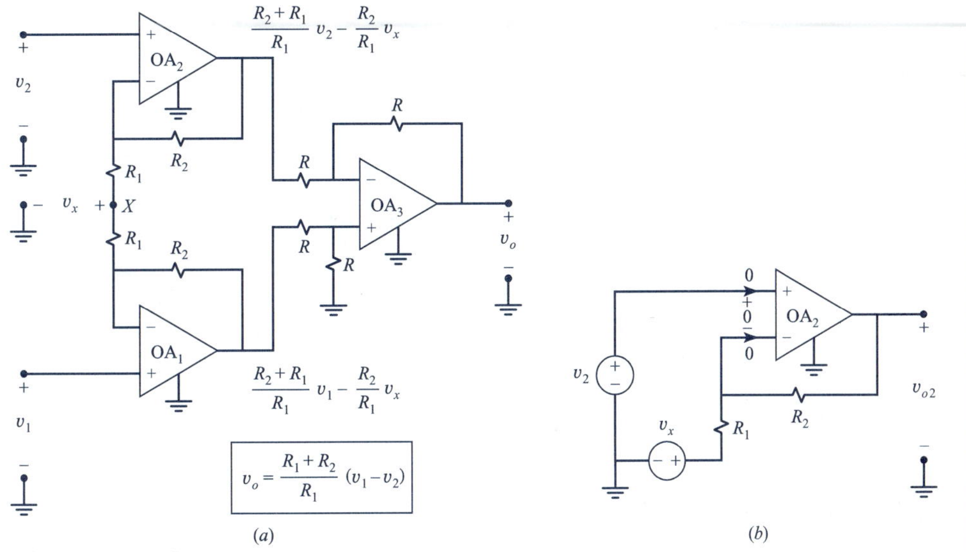

Figure 4.12
An instrumentation-quality op-amp difference amplifier.

and the v_x components cancel out. The input currents to the overall amplifier are zero, since these are the currents into the noninverting terminals of OA1 and OA2. Hence, voltages v_1 and v_2 may have nonzero source resistances, but this does not affect the performance of the amplifier.

Exercise Problem 4.3

Design a difference amplifier in the form of Fig. 4.11*a* to produce an output of $v_o = 5v_2 - 10v_1$.

Answers: $R_1 = 5\ \text{k}\Omega$, $R_2 = 50\ \text{k}\Omega$, $R_3 = 6\ \text{k}\Omega$, $R_4 = 5\ \text{k}\Omega$.

4.2.3 The Summer Figure 4.13 shows a circuit that will sum several voltages, weighting their contribution to the sum with arbitrary coefficients. Labeling currents as shown and applying the virtual short circuit conditions gives

$$I_i = \frac{v_i}{R_i}$$

Therefore

$$v_o = -R_f(I_1 + I_2 + \cdots + I_n)$$

Hence, the output is the sum of inverting amplifier responses to each individual input as

$$v_o = -\frac{R_f}{R_1}v_1 - \frac{R_f}{R_2}v_2 - \cdots - \frac{R_f}{R_n}v_n \tag{4.13}$$

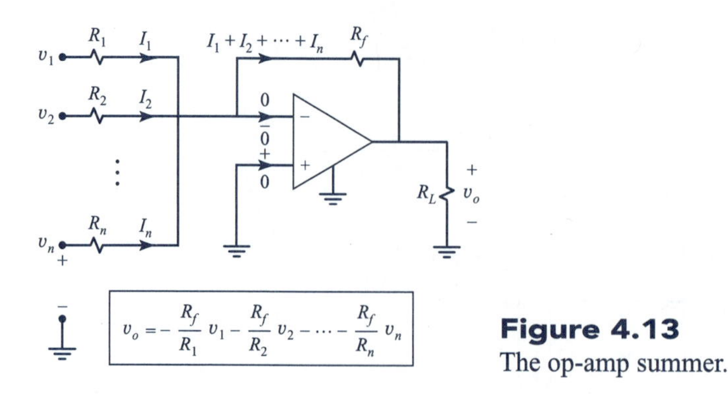

Figure 4.13
The op-amp summer.

The input resistance seen looking into terminal i is

$$R_{in,i} = R_i \tag{4.14}$$

Example 4.4

Design a circuit to sum three voltages in the proportion $v_o = 3v_1 + 2v_2 - 4v_3$.

Solution From (4.13) we could select $R_f = 60\ \text{k}\Omega$, $R_1 = 20\ \text{k}\Omega$, $R_2 = 30\ \text{k}\Omega$, $R_3 = 15\ \text{k}\Omega$. But since the summer inverts each input, this would give an output of $v_o = -3v_1 - 2v_2 - 4v_3$. Hence we need to invert v_1 and v_2 before applying them to the summer inputs. This can be done with unity-gain inverting amplifiers as shown in Fig. 4.14.

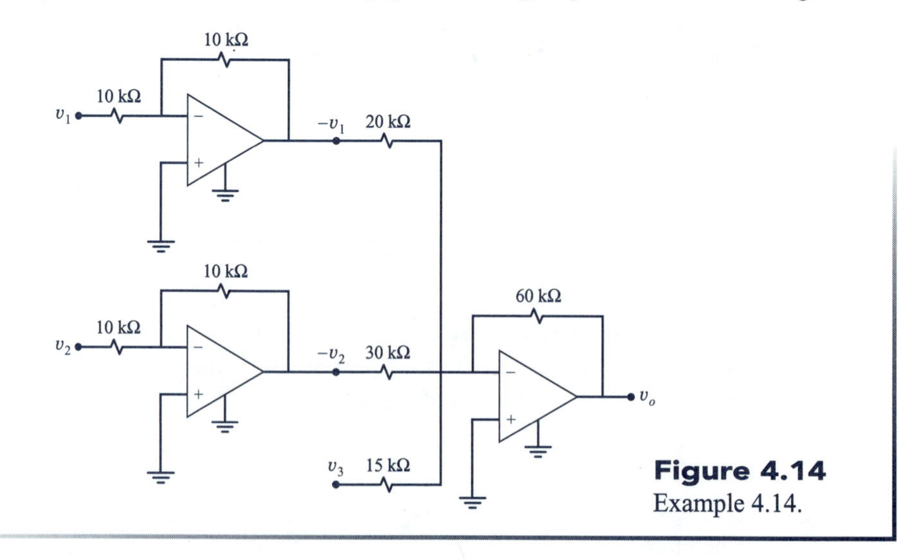

Figure 4.14
Example 4.14.

Exercise Problem 4.4

Design a circuit that will combine three voltages according to the ratio

$$v_o = -2v_1 + v_2 - 4v_3$$

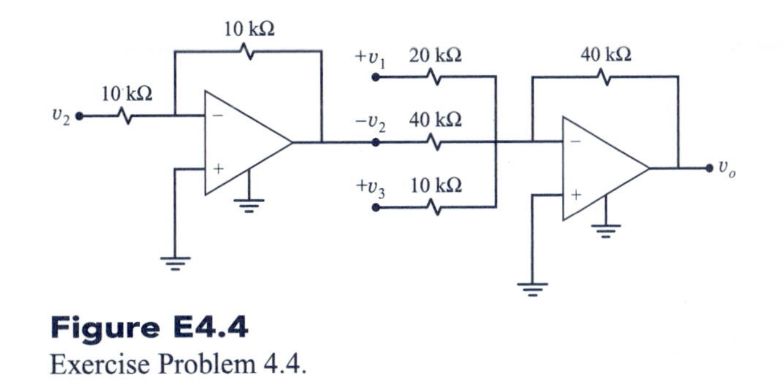

Figure E4.4
Exercise Problem 4.4.

Answer: Choose a summer with $R_f = 40$ kΩ, $R_1 = 20$ kΩ, $R_2 = 40$ kΩ, $R_3 = 10$ kΩ, and precede input 2 with a unity-gain inverting amplifier. The final design is shown in Fig. E4.4

4.2.4 The Buffer A load placed across the terminals of a source provides *loading* of that source in that as the load current increases, the voltage across the load decreases due to the internal voltage drop across the source resistance of the source. (See Fig. 2.50.) In order to prevent this loading, a *buffer* or *voltage follower* circuit, shown in Fig. 4.15, can be used. Observe that the output of the op amp is directly connected to the inverting terminal, providing the necessary negative feedback. Two important features of this circuit become immediately apparent once we label the virtual short-circuit conditions on the op amp. The first is that because the current into the noninverting terminal is zero, the buffer presents no loading to the source. Secondly, because of the zero difference voltage constraint, the output voltage equals the input voltage:

$$\boxed{v_o = v_i} \tag{4.15}$$

Hence, the device is also said to be a *voltage follower*, since the output voltage follows the input voltage.

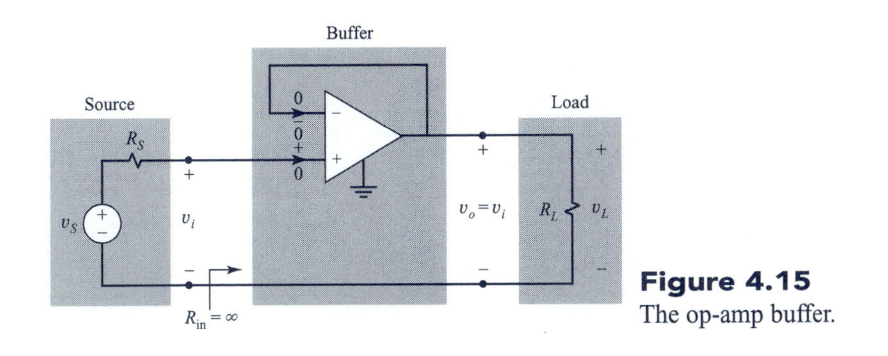

Figure 4.15
The op-amp buffer.

Exercise Problem 4.5

Suppose in the circuit of Fig. 4.15, we have $v_s = 5 \sin \omega t$, $R_S = 1$ kΩ, $R_L = 2500$ Ω. Determine the output voltage and output current of the op amp.

Answers: $v_L = 5 \sin \omega t$ V, $i_o = 2 \sin \omega t$ mA.

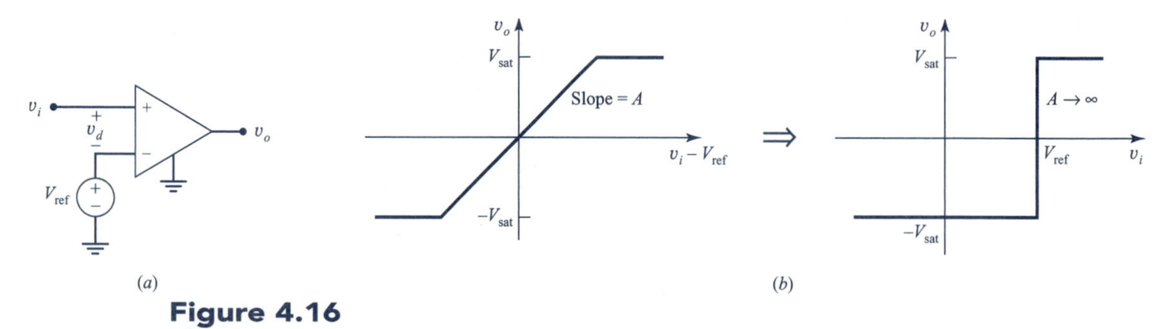

Figure 4.16
The op-amp comparator.

4.2.5 The Comparator The previous op-amp circuits have been operated in their linear regions. Care was taken to insure that they were not being driven into saturation or required to produce unreasonably large output currents. The comparator shown in Fig. 4.16*a* is intentionally driven into saturation in order to detect when a voltage exceeds a certain reference voltage, V_{ref}. In order to analyze this circuit, we first plot, assuming finite gain, the transfer characteristic of the op amp relating v_o to the difference voltage $v_d = v_i - V_{ref}$, as shown in Fig. 4.16*b*. Then we replot this with the x axis being v_i alone. This is done by simply shifting the previous graph to the right by V_{ref}. And finally we let the op-amp gain go to infinity. This produces the threshold curve shown in Fig. 4.16*b*, which has the following properties:

$$\begin{aligned} v_i < V_{ref} \quad &\Rightarrow \quad v_o = -V_{sat} \\ v_i > V_{ref} \quad &\Rightarrow \quad v_o = V_{sat} \end{aligned} \tag{4.16}$$

Hence, the device serves as a *threshold detector* for determining when a certain voltage has exceeded a reference voltage, i.e., the input voltage is *compared* with the reference voltage.

Exercise Problem 4.6

A circuit is to be designed such that the output will change value whenever the input exceeds 5 V. Only a 9-V battery is available for use.

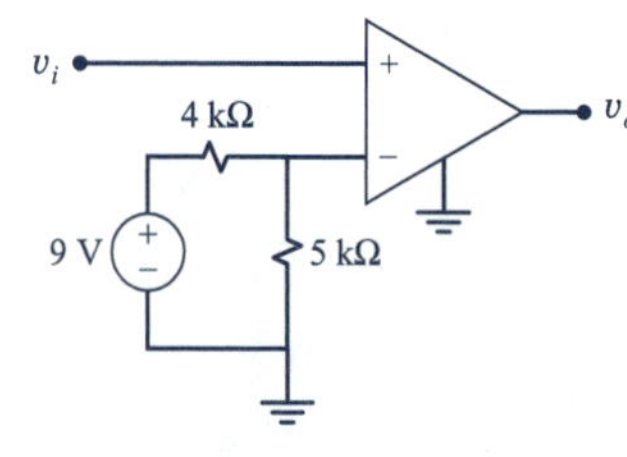

Figure E4.6
Exercise Problem 4.6.

Answer: The design is shown in Fig. E4.6.

4.3 Applications

Op amps are very useful in constructing instrumentation circuits. These instrumentation circuits consist of a transducer and an op-amp signal conditioner as shown in Fig. 4.17. The transducer is a device that converts a particular physical variable (light intensity, strain, temperature, pressure, etc.) into an analogous electrical signal (voltage or current). For example, a pressure transducer gives a voltage output that is proportional to the pressure in pounds per square inch. Usually this output is too small to be processed reliably or its range is not appropriate for the signal processor [such as a analog-to-digital converter (ADC)]. An op-amp signal conditioner can convert the transducer output to a level that is appropriate for signal processing so that the physical variable can be measured.

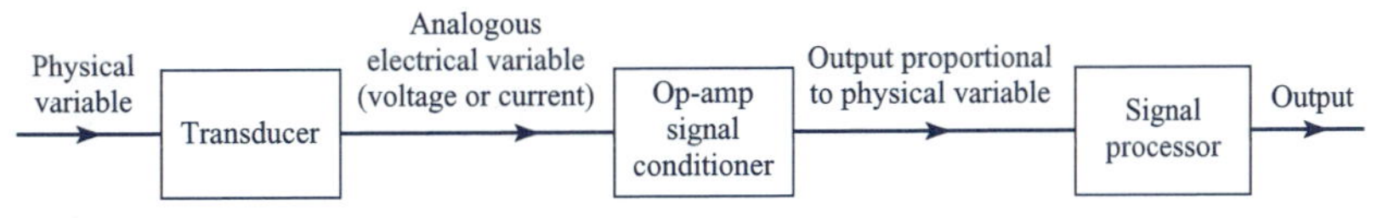

Figure 4.17
Illustration of the use of an op-amp signal conditioner with a transducer.

4.3.1 A Strain-Gauge Instrumentation Circuit A strain gauge is a variable resistor that, when applied to a structure such as a beam, measures the elongation (strain) of the structure in response to an applied force (stress). The resistance of a strain gauge is $R + \Delta R$. Typical values for the unstressed strain gauge are on the order of $R = 300\ \Omega$, and the maximum displacement resistance, ΔR, is on the order of a few ohms. This incremental change in resistance is too small to be reliably measured with a standard ohmmeter or to be used as the input to a signal processor (which may require voltage levels in the range of 0 to 10 V, for example).

Figure 4.18*a* shows the use of four matched strain gauges to measure the strain of a beam in response to an applied force. Two of the strain gauges are placed on the top of the beam, and two on the bottom. When the force is applied, the top strain gauges change their resistance to $R + \Delta R$ and the bottom strain gauges change their resistance to $R - \Delta R$. If we place these four strain gauges in the segments of a Wheatstone bridge discussed in Chapter 3 and apply a dc voltage V_S to the bridge as shown in Fig. 4.18*b*, the strain-gauge voltages are (by voltage division)

$$v_2 = \frac{R + \Delta R}{2R} V_S$$

$$v_1 = \frac{R - \Delta R}{2R} V_S$$

The difference voltage is

$$v_2 - v_1 = \frac{\Delta R}{R} V_S$$

A difference amplifier such as shown in Fig. 4.11*a* or Fig. 4.12*a* can be used to amplify the difference voltage, giving

$$v_o = \frac{R_2 + R_1}{R_1}(v_2 - v_1)$$

$$= \frac{R_2 + R_1}{R_1}\left(\frac{\Delta R}{R}\right)V_S$$

Appropriate choices of R_1, R_2, and V_S can give an output voltage level that is proportional to the strain and of a sufficient level to be processed by a signal processor such as a voltmeter.

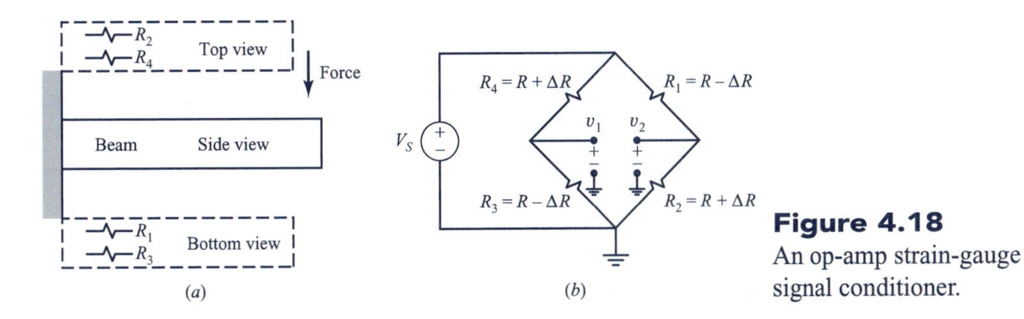

Figure 4.18
An op-amp strain-gauge signal conditioner.

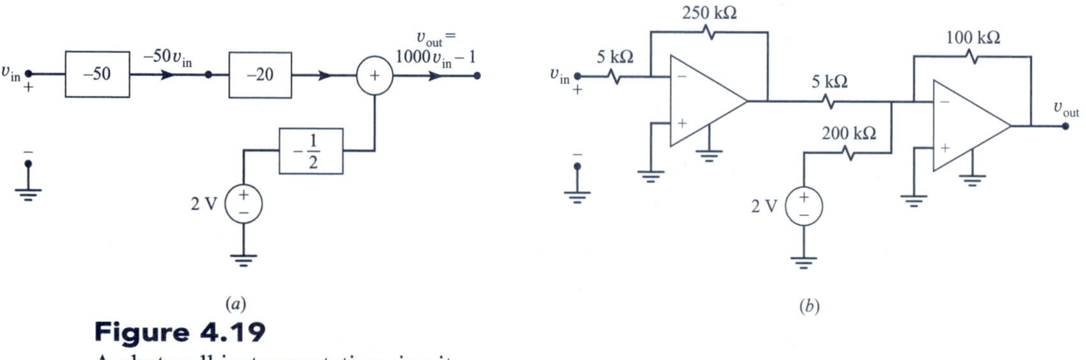

Figure 4.19
A photocell instrumentation circuit.

4.3.2 A Photocell Instrumentation Circuit A photocell is a device that converts light intensity (in lumens, abbreviated lm) to a voltage. For example, a typical photocell may have the following relation between light intensity and output voltage:

$$1 \text{ lm} \Rightarrow 1 \text{ mV}$$
$$10 \text{ lm} \Rightarrow 11 \text{ mV}$$

An ADC used to digitally process this signal may require voltage levels between 0 and 10 V. We will need a signal conditioner that amplifies this output voltage and also converts it to the desired range for the ADC. Hence we want a signal conditioner that provides 0 V when the input is 1 mV and provides 10 V when the input is 11 mV. This linear relation is

$$v_{out} = 1000v_{in} - 1$$

Figure 4.19*a* shows the block diagram of the proposed signal conditioner. It consists of an amplifier with a gain of −50, followed by a summer. The −50-gain input is fed through one input of the summer that has a gain of −20, giving a contribution to the output of $1000v_{in}$. An available dc voltage of 2 V is applied to the other input of the summer to give the remaining output contribution of −1 V. One possible implementation of this circuit using op amps is shown in Fig. 4.19*b*.

4.4 PSPICE Applications

PSPICE is very useful in modeling op amps and incorporating all the nonideal effects that would complicate an analysis. Figure 4.20*a* shows a nonideal model of an op amp that includes finite input resistance, nonzero output resistance, and finite gain. PSPICE has a nice feature that permits defining this as an element and using it in many places in a circuit without the need for redefining the circuit each time it is used. This is the *subcircuit* feature of PSPICE, and the subcircuit is defined using the following sequence of statements:

```
.SUBCKT <subcircuit name> <node list of subcircuit model>
circuit element and connection statements
.ENDS <subcircuit name>
```

The subcircuit is inserted into another circuit using the statement

```
Xname <node list of attachment in circuit> <subcircuit name>
```

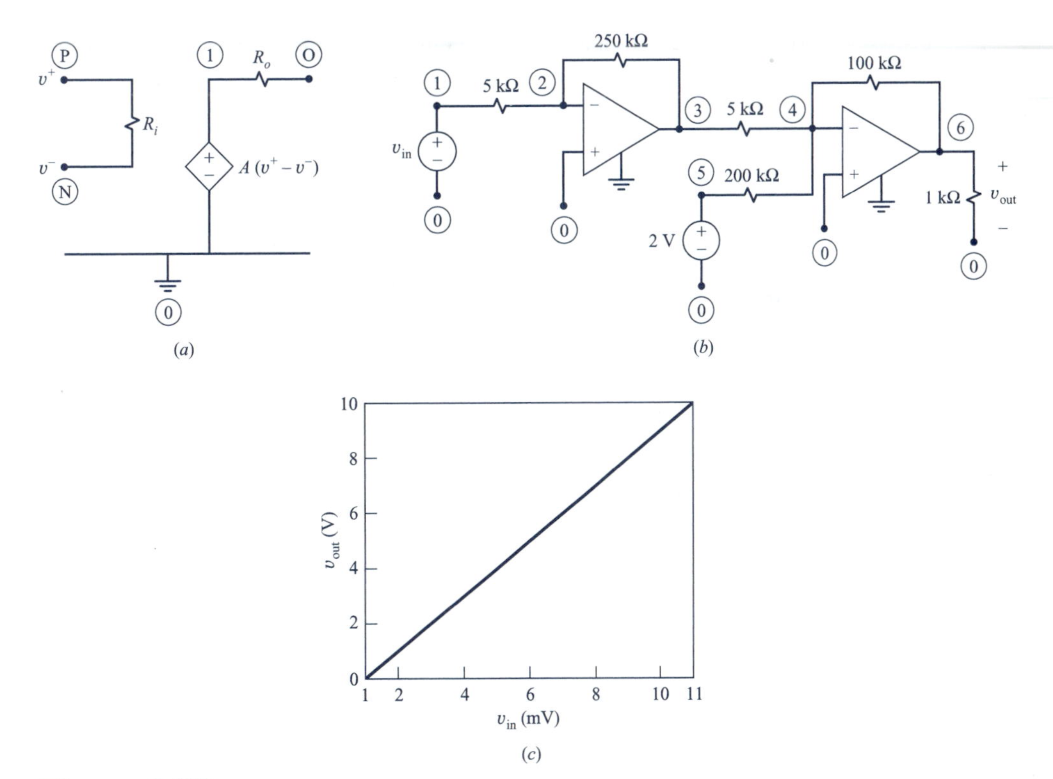

Figure 4.20
Use of the op amp in SPICE or PSPICE: (*a*) node labeling of the op amp, (*b*) node labeling of the op-amp photocell instrumentation circuit, and (*c*) the PSPICE results for the output voltage.

The node names in the subcircuit are distinct from the node names in the overall circuit. The node list of the subcircuit model and the node list of attachment in the circuit in the Xname statement are the only connection between the two circuits, and the sequence of the nodes in those statements is important. For example, suppose we have

```
.SUBCKT FUN 2 5 9
<statements>
.ENDS FUN
```

and in the circuit

```
XHAPPY 1 7 8 FUN
```

This indicates that the subcircuit FUN will be inserted into the circuit and that node 2 of FUN will be attached to node 1 in the circuit, node 5 of FUN will be attached to node 7 in the circuit, and node 9 of FUN will be attached to node 8 in the circuit. The node numbers of the subcircuit and those of the overall circuit are separate and distinct; only the zero (0) node of the two circuits is common.

Labeling the nodes of the op-amp circuit as shown in Fig. 4.20*a*, the subcircuit statements are

```
SUBCKT OPAMP P N O
RI P N 2MEG
E 1 0 P N 200K
RO 1 O 75
.ENDS OPAMP
```

The photocell transducer circuit of Fig. 4.19*b* will be used as an example of the simulation. The nodes are labeled in Fig. 4.20*b*, and the PSPICE program becomes

```
PHOTOCELL SIGNAL CONDITIONER
VIN 1 0 DC 1
R11 1 2 5K
R12 2 3 250K
XOA1 0 2 3 OPAMP
VTWO 5 0 DC 2
R21 3 4 5K
R22 5 4 200K
R23 4 6 100K
XOA2 0 4 6 OPAMP
RL 6 0 1K
.SUBCKT OPAMP P N O
<insert subcircuit model>
.ENDS OPAMP
.DC VIN 1M 11M .01M
.PROBE
.END
```

The input voltage is swept from 1 to 11 mV in steps of 0.01 mV, and the output is plotted in Fig. 4.20*c*. The statement .PROBE causes all node voltages to be stored for later plotting versus the incremented source values of VIN. Apparently, the nonideal parameters have only moderate effect. For example, for $v_{in} = 1$ mV, we have $v_{out} = -259.4$ μV, and for $v_{in} = 11$ mV, we have $v_{out} = 9.996$ V. Midway in the range, $v_{in} = 5$ mV, we obtain $v_{out} = 3.998$ V where the desired output is 4 V.

Problems

Section 4.1 The Actual Op Amp versus the Ideal Op Amp

4.1-1 Design an inverting op-amp amplifier for a voltage gain of −25, an input impedance of 5 kΩ, and a load of 2 kΩ. If the input voltage is $v_i(t) = 0.5 \sin \omega t$ V, determine the output voltage and the output current for the op amp.
Answers: $R_i = 5$ kΩ, $R_f = 125$ kΩ, $v_o = -12.5 \sin \omega t$ V, $i_o = -6.35 \sin \omega t$ mA.

4.1-2 Design an inverting op-amp amplifier for a voltage gain of −10, an input impedance of 10 kΩ, and a load of 5 kΩ. If the input voltage is $v_i(t) = \sin \omega t$ V, determine the output voltage and the output current for the op amp.
Answers: $R_i = 10$ kΩ, $R_f = 100$ kΩ, $v_o = -10 \sin \omega t$ V, $i_o = -2.1 \sin \omega t$ mA.

4.1-3 Design an inverting op-amp amplifier for a voltage gain of −5, an input impedance of 15 kΩ, and a load of 8 kΩ. If the input voltage is $v_i(t) = 2 \sin \omega t$ V, determine the output voltage and the output current for the op-amp.
Answers: $R_i = 15$ kΩ, $R_f = 75$ kΩ, $v_o = -10 \sin \omega t$ V, $i_o = -1.383 \sin \omega t$ mA.

4.1-4 Design an inverting op-amp amplifier for a voltage gain of −12, an input impedance of 8 kΩ, and a load of 2 kΩ. If the input voltage is $v_i(t) = 0.5 \sin \omega t$ V, determine the output voltage and the output current for the op amp.
Answers: $R_i = 8$ kΩ, $R_f = 96$ kΩ, $v_o = -6 \sin \omega t$ V, $i_o = -3.063 \sin \omega t$ mA.

4.1-5 Determine the voltage gain of the amplifier of Fig. P4.1-5.
Answer: $A_v = -(R_2R_3 + R_4R_3 + R_2R_4)/R_1R_3$.

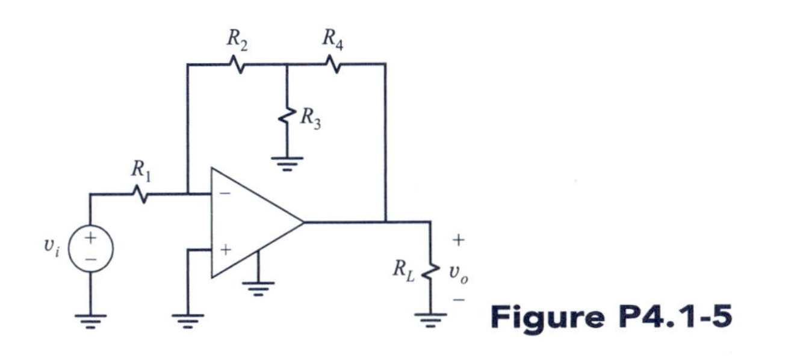

Figure P4.1-5

4.1-6 Determine the output voltage, v_o, in terms of the input voltages v_1 and v_2 in the circuit of Fig. P4.1-6. Determine the output current of each op amp if $v_1 = 0.1 \sin \omega t$ V and $v_2 = 0.5 \sin \omega t$ V. *Answers*: $v_o = 4v_1 - 2v_2$, $v_o = -0.6 \sin \omega t$ V, $i_o = -70 \sin \omega t$ μA.

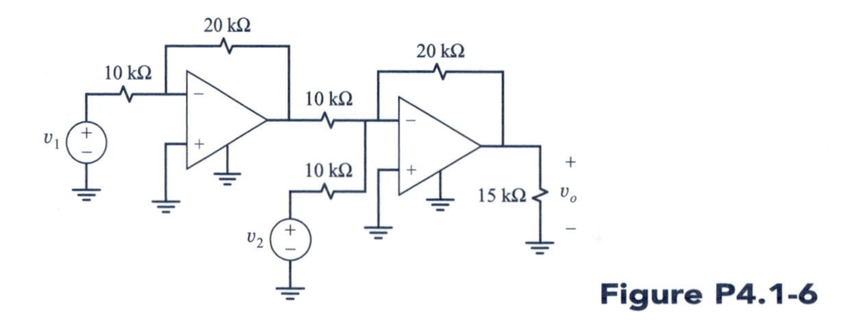

Figure P4.1-6

Section 4.2 Other Useful Op-Amp Circuits

4.2-1 Design a noninverting op-amp amplifier for a voltage gain of 20 and a load of 1 kΩ. Determine the output voltage and output op-amp current if $v_i = 0.1 \sin \omega t$ V.
Answers: $R = 5$ kΩ, $R_f = 95$ kΩ, $v_o = 2 \sin \omega t$ V, $i_o = 2.02 \sin \omega t$ mA.

4.2-2 Design a noninverting op-amp amplifier for a voltage gain of 15 and a load of 5 kΩ. Determine the output voltage and output op-amp current if $v_i = 0.5 \sin \omega t$ V.
Answers: $R = 5$ kΩ, $R_f = 70$ kΩ, $v_o = 7.5 \sin \omega t$ V, $i_o = 1.6 \sin \omega t$ mA.

4.2-3 Determine the ratio of the output current to input voltage for the circuit of Fig. 4.2-3.
Answer: $i_o/v_{\text{in}} = 10^{-4}$.

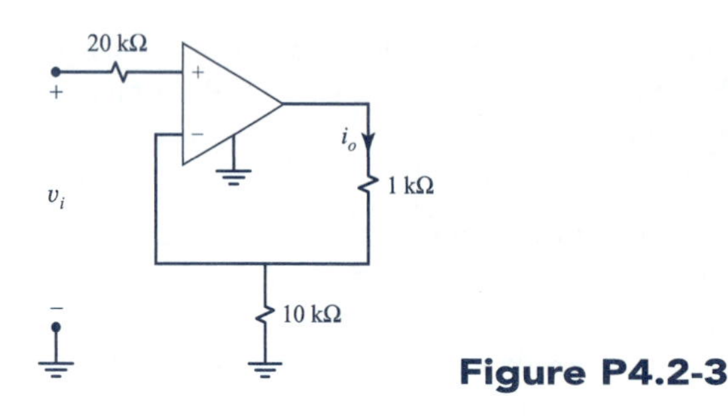

Figure P4.2-3

4.2-4 Determine the voltage gain of the circuit of Fig. P4.2-4. *Answer*: $A_v = 11$.

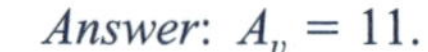

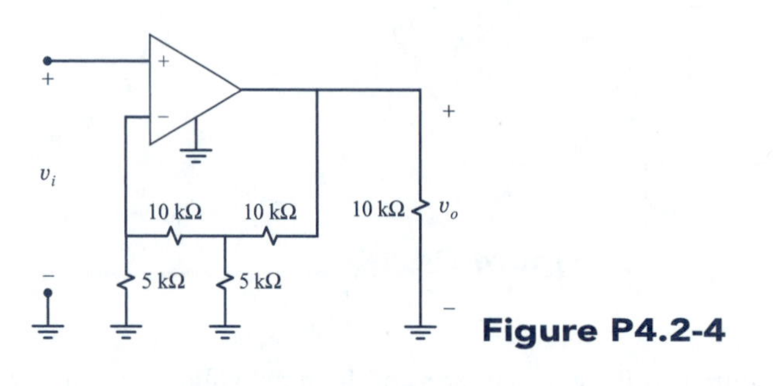

Figure P4.2-4

4.2-5 Determine the output voltage for the circuit of Fig. P4.2-5. *Answer*: $v_o = 12$ V.

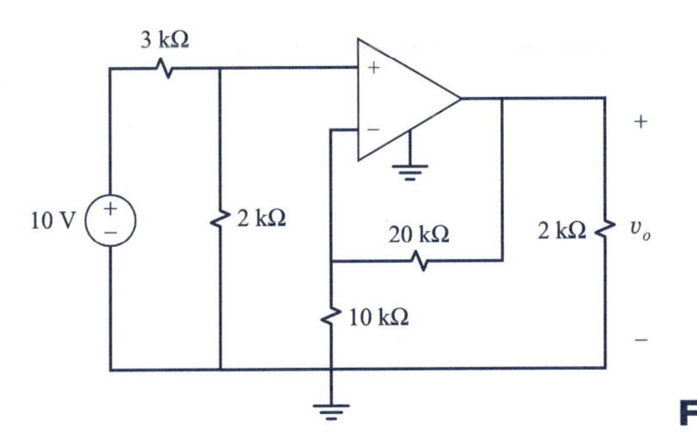

Figure P4.2-5

4.2-6 Determine the output voltage for the circuit of Fig. P4.2-6. *Answer*: $v_o = -11$ V.

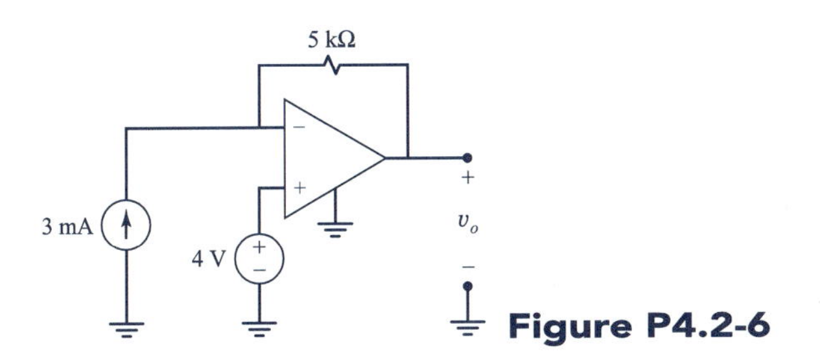

Figure P4.2-6

4.2-7 Determine the output voltage of the circuit of Fig. P4.2-7.
Answer: $v_o = 3$ V.

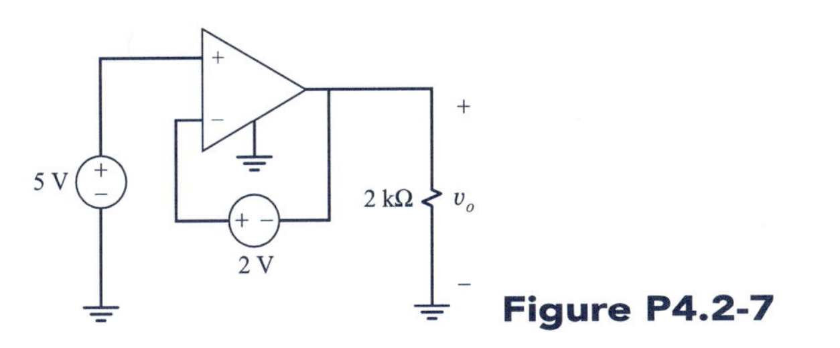

Figure P4.2-7

Section 4.3 Applications

4.3-1 Design a circuit that will provide a relationship between output voltage and input voltage of $v_o = 10v_i - 4$. Two op amps, a 9-V battery, and assorted resistors are available.
Answer: A summer with $v_o = -v_1 - \frac{4}{9}v_2$, an inverting amplifier with $v_1 = -10v_i$, and $v_2 = 9$ V.

4.3-2 Design a circuit that will provide a relationship between output voltage and input voltage of $v_o = -20v_i - 10$. Two op amps, a 5-V battery, and assorted resistors are available.
Answer: A summer with $v_o = -20v_i - 2v_2$, and $v_2 = 5$ V.

4.3-3 Design a circuit that will provide a relationship between output voltage and input voltage of $v_o = 5v_i + 2$. Two op amps, a 9-V battery, and assorted resistors are available.
Answer: A summer with $v_o' = -5v_i - 0.222v_2$, followed by an inverting amplifier with $v_o = -v_o'$ and $v_2 = 9$ V.

4.3-4 Design a circuit that will translate an input voltage range of $-20 \text{ mV} \le v_i \le 20$ mV to an output voltage range of $0 \le v_o \le 5$ V. Two op amps, a 1-V battery, and assorted resistors are available. *Answer*: A summer with $v_o' = -12.5v_i - 0.25v_2$, followed by an inverting amplifier with $v_o = -10v_o'$ and $v_2 = 1$ V.

4.3-5 Design a circuit that will convert an input voltage range of $-5\text{ mV} \leq v_i \leq 10\text{ mV}$ to an output current into a 1-kΩ load of $0 \leq i_o \leq 2\text{ mA}$. Two op amps, a 1.5-V battery, and assorted resistors are available. *Answer*: A summer with $v_o' = -13.3v_i - 0.04433v_2$, followed by an inverting amplifier with $v_o = -10v_o'$ and $v_2 = 1.5\text{ V}$.

Section 4.4 PSPICE Applications

4.4-1 Use PSPICE to verify the results of Problem 4.1-1.
Answers: $v_o = -12.5\text{ V}$, $i_o = -6.349\text{ mA}$.

4.4-2 Use PSPICE to verify the results of Problem 4.1-2.
Answers: $v_o = -9.999\text{ V}$, $i_o = -2.100\text{ mA}$.

4.4-3 Use PSPICE to verify the results of Problem 4.1-6.
Answers: $v_o = -0.6\text{ V}$, $i_{o1} = -30\ \mu\text{A}$, $i_{o2} = -70\ \mu\text{A}$.

4.4-4 Use PSPICE to verify the results of Problem 4.2-3.
Answer: $i_o = 100\ \mu\text{A}$ for $v_i = 1\text{V}$.

4.4-5 Use PSPICE to verify the results of Problem 4.2-4.
Answer: $v_o = 11\text{ V}$ for $v_i = 1\text{V}$.

CHAPTER 5

The Energy Storage Elements

In this chapter we will introduce the two remaining circuit elements: the capacitor and the inductor. Unlike the resistor, which dissipates power, the capacitor and the inductor *store energy* and do not dissipate power. Of course, these are *ideal elements,* like the resistor and the voltage and current sources considered previously. Physical inductors, for example, are typically composed of coils of wire as in an electric heater. These physical inductors dissipate power because of the resistance of the wire. Hence we might model a physical inductor as the series combination of an ideal inductor and an ideal resistor.

An additional significant change imposed by these energy storage elements is the following. All of the previous circuits that contained resistors and controlled and independent sources are governed by *algebraic equations*, i.e., the circuit equations resulting from the application of KVL, KCL, and the element relations contained no derivatives. Once we include a capacitor or an inductor in a circuit, the governing equations will become ordinary differential equations, which are somewhat more difficult to solve than algebraic equations. The remaining two chapters will investigate the solution of these differential equations. This added analysis difficulty is not without benefit. Circuits containing capacitors or inductors provide a greater variety of signal processing than those that only contain resistors. An example is the construction of electrical filters, which will be considered in the next chapter.

5.1 The Capacitor

The first energy storage element we will consider is the capacitor. Essentially the capacitor is a device that stores charge. Figure 5.1 shows a physical capacitor consisting of two metallic plates of area A separated by distance d. An independent voltage source is attached to the plates and transfers charge Q to the plate attached to the positive terminal and an equal but opposite charge to the other plate. This charge and the applied voltage are related by the *capacitance* of the capacitor as

$$Q = CV \tag{5.1}$$

The unit of capacitance is the coulomb per volt, or *farad* (F): where 1 F = 1 C/V. For a time-varying source $v(t)$ we have similarly

$$\boxed{q(t) = Cv(t)} \tag{5.2}$$

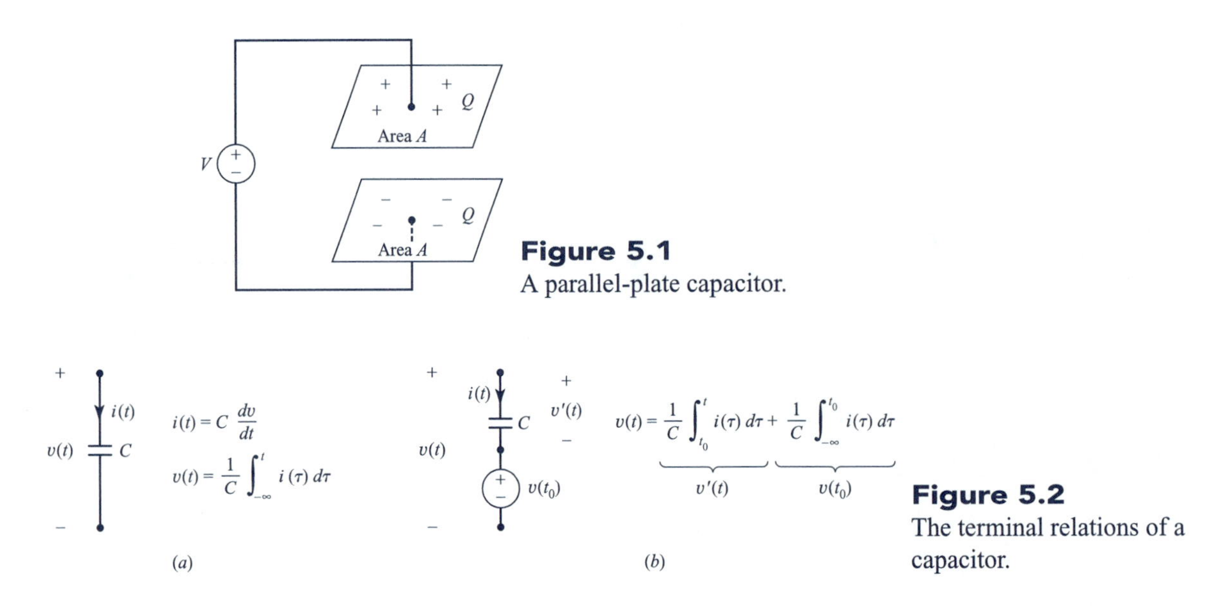

Figure 5.1
A parallel-plate capacitor.

Figure 5.2
The terminal relations of a capacitor.

For the parallel-plate capacitor shown in Fig. 5.1 with plates of area A (m^2) separated by a distance d (m) with air as the intervening medium, the capacitance is approximately*

$$C = 8.84 \times 10^{-12} \frac{A}{d} \qquad \text{F} \tag{5.3}$$

Usually the capacitance of practical capacitors is given in microfarads (1 μF = 10^{-6} F) or picofarads (1 pF = 10^{-12} F). For 1-F capacitor with a separation of 1 m between the plates, each would have to have an area of approximately 44,000 square miles.

The element symbol for the capacitor is shown in Fig. 5.2*a*. The current entering the terminals is equal to the rate of buildup of charge on the plates:

$$i(t) = \frac{dq(t)}{dt} \tag{5.4}$$

If we differentiate (5.2) with respect to time t and substitute into (5.4), we obtain the terminal voltage–current relation for the capacitor:

$$i(t) = C\frac{dv(t)}{dt} \tag{5.5}$$

Hence, the current through a capacitor is related to the rate-of-change of the voltage across it or, in other words, the *slope of the* $v(t)$ *characteristic*. Note that the voltage and current are labeled with the passive sign convention. If they were not so labeled, then a negative sign would be required in (5.5). This terminal relation can be inverted to yield

$$v(t) = \frac{1}{C}\int_{-\infty}^{t} i(\tau)\, d\tau \tag{5.6}$$

*C.R. Paul and S.A. Nasar, *Introduction to Electromagnetic Fields,* 2nd edition, McGraw-Hill, New York, 1987.

Hence, the voltage across the capacitor requires that we determine the *area under the i(t) curve.* This result shows that the voltage across a capacitor at some time t depends on all the past current through that capacitor. This makes sense because the integral of the current over all previous time yields the present value of charge on the capacitor plates. (See equation (5.2).) We can break the integral into two parts as

$$v(t) = \frac{1}{C}\int_{t_0}^{t} i(\tau)\, d\tau + \frac{1}{C}\int_{-\infty}^{t_0} i(\tau)\, d\tau \tag{5.7}$$

Recognizing that the second integral represents the capacitor voltage at time t_0:

$$v(t_0) = \frac{1}{C}\int_{-\infty}^{t_0} i(\tau)\, d\tau \tag{5.8}$$

the equation for capacitor voltage in (5.7) can be written in terms of the *initial voltage* as

$$v(t) = \frac{1}{C}\int_{t_0}^{t} i(\tau)\, d\tau + v(t_0) \tag{5.9}$$

Hence if we know the voltage across the capacitor at some time t_0, we do not have to determine the area under the current curve from the beginning of time up to that time. We only need to determine the area under the current curve from t_0 to the present time. The relation in (5.9) can be represented as an equivalent circuit shown in Fig. 5.2*b,* consisting of a capacitor whose voltage at $t = t_0$ is zero in series with a voltage source $v(t_0)$ representing the voltage of the capacitor at $t = t_0$. This voltage source represents the net accumulated area under the current curve up to t_0.

The energy stored in a capacitor at a particular time (stored in the electric field between the plates*) can be found by integrating the expression for the instantaneous power delivered to the capacitor:

$$\begin{aligned} p(t) &= v(t)i(t) \\ &= Cv(t)\,\frac{dv(t)}{dt} \end{aligned} \tag{5.10}$$

to give

$$\begin{aligned} w(t) &= \int_{-\infty}^{t} p(\tau)\, d\tau \\ &= C\int_{-\infty}^{t} v(\tau)\,\frac{dv(\tau)}{d\tau}\, d\tau \\ &= \tfrac{1}{2}Cv(t)^2 - \tfrac{1}{2}Cv(-\infty)^2 \end{aligned} \tag{5.11}$$

Assuming that the capacitor voltage was zero at $t = -\infty$, the energy stored in the capacitor at some time t depends only on the voltage of the capacitor at that time:

$$w(t) = \tfrac{1}{2}Cv^2(t) \tag{5.12}$$

Hence the energy stored in the capacitor is proportional to the square of the capacitor voltage at that time t.

---

*C.R. Paul and S.A. Nasar, *Introduction to Electromagnetic Fields,* 2nd edition, McGraw-Hill, New York, 1987.

Example 5.1

A voltage source is applied to a 5-F capacitor as shown in Fig. 5.3*a*. Sketch the capacitor current and the stored energy as a function of time.

Solution The capacitor voltage equals the source voltage, and hence the capacitor current is determined from (5.5) as

$$i(t) = 5\,\frac{dv_S(t)}{dt}$$

Therefore the capacitor current is the product of the capacitance and the *instantaneous slope* of the source voltage curve as shown in Fig. 5.3*b*. Since the curve of $v_S(t)$ is piecewise linear, this is a simple computation involving the slopes of the linear segments of that curve. For example, for $-1\text{ s} \le t \le 1\text{ s}$ the slope of the $v_S(t)$ curve is $10\text{ V}/2\text{ s} = 5\text{ V/s}$. Multiplying this by the capacitance of 5 F gives the current over this time interval of 25 A. The energy stored in the capacitor is related to the square of the source voltage curve:

$$w(t) = \tfrac{1}{2} \times 5v_S^2(t)$$

which is sketched in Fig. 5.3*c*. For example, the value of $v_S(t)$ at $t = 1$ s is 10 V. Squaring this and multiplying by $C/2 = \frac{5}{2}$ gives the stored energy at that time: 250 J. The stored energy at $t = 2$ s is also 250 J, and it is 0 J at $t = -1$ s and $t = 3$ s. Since $v_S(t)$ varies linearly with t for $-1\text{ s} \le t \le 1$ s, its square varies as t^2.

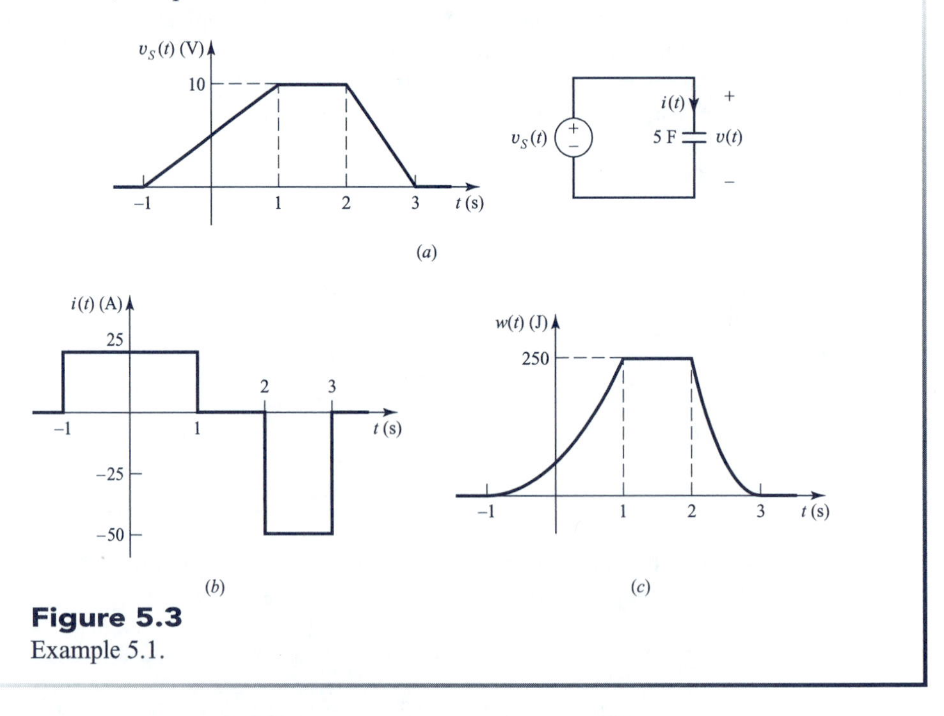

Figure 5.3
Example 5.1.

Exercise Problem 5.1

A 10-μF capacitor is driven by a voltage source as shown in Fig. E5.1. Determine the capacitor current at $t = 0.5$, 2, 3.5, and 4.5 ms.

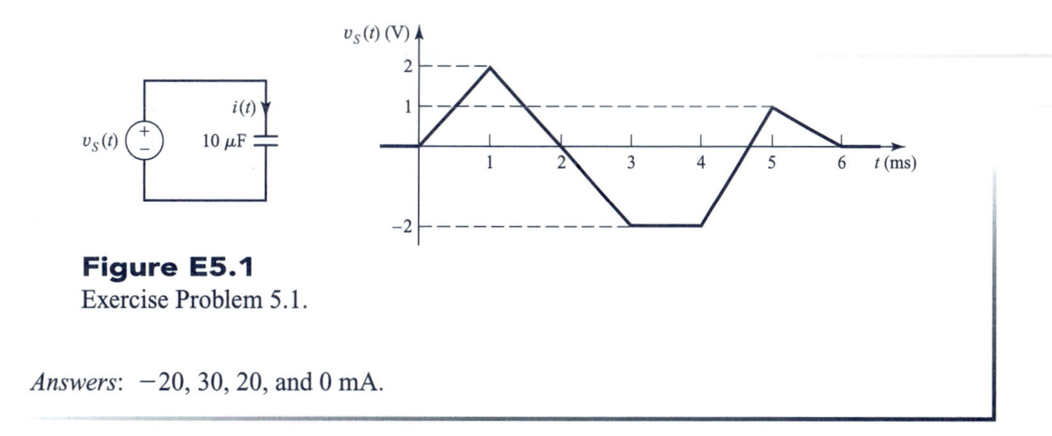

Figure E5.1
Exercise Problem 5.1.

Answers: −20, 30, 20, and 0 mA.

Example 5.2

A current source is applied to a 5-F capacitor as shown in Fig. 5.4*a*. Sketch the capacitor voltage as a function of time.

Solution The capacitor current is equal to the source current, $i_S(t)$, and hence the capacitor voltage is determined from (5.6) as

$$v(t) = \frac{1}{5}\int_{-\infty}^{t} i_S(\tau)\, d\tau$$

Hence the capacitor voltage is related to the area under the current source waveform. For example, at $t = 3$ s, the total area under the $i_S(t)$ curve is

$$\text{Area}\big|_{t=3\text{ s}} = 10 + 10 + 5 = 25$$

Hence $v(3\text{ s}) = 25/5 = 5$ V. In sketching the curve we simply determine the voltages at the points where the curve changes, $t = 1, 2, 3$ s, and then sketch in the shape between those points. From $t = -1$ s to $t = 1$ s the curve increases linearly with t and hence the area increases as t^2. From $t = 1$ s to $t = 2$ s the curve is independent of t and hence the area increases linearly with t. Similarly, from $t = 2$ s to $t = 3$ s the curve decreases linearly with t and hence the area increases as t^2.

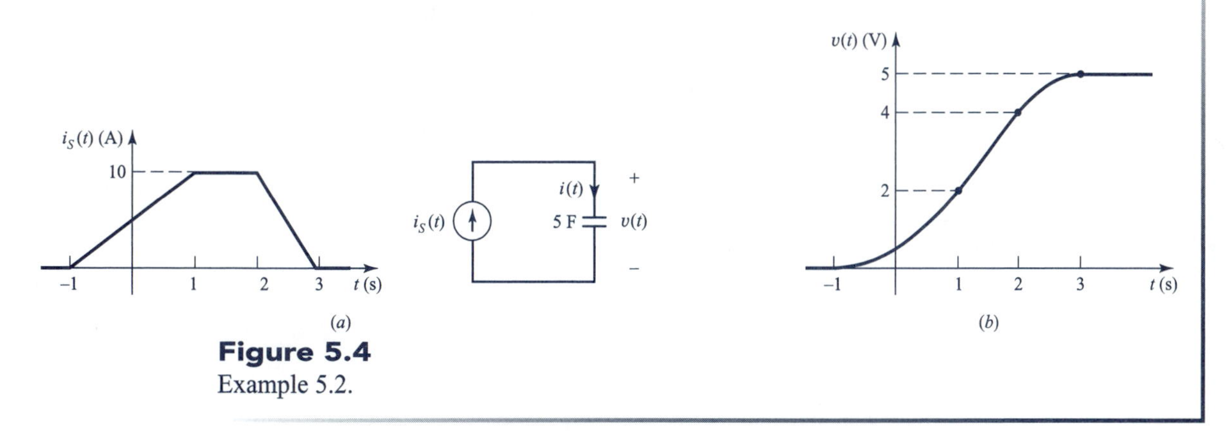

Figure 5.4
Example 5.2.

Exercise Problem 5.2

A 100-pF capacitor is driven by a current source as shown in Fig. E5.2. Determine the capacitor voltage at $t = 1, 1.5, 3$, and $5\ \mu s$.

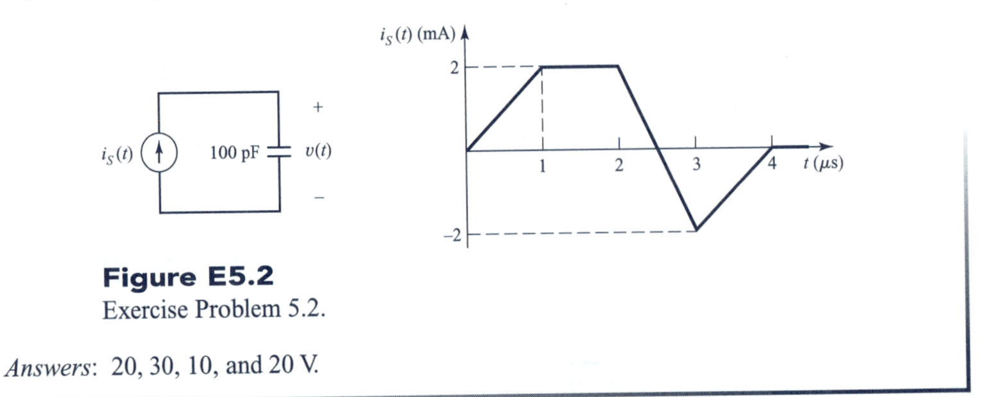

Figure E5.2
Exercise Problem 5.2.

Answers: 20, 30, 10, and 20 V.

5.1.1 Capacitors in Series and in Parallel Several capacitors in series or in parallel may be combined to yield equivalent capacitances at the terminals of the combination. For example, consider n capacitors in parallel as shown in Fig. 5.5. The terminal voltage is equal to the voltage of each of the capacitors:

$$v(t) = v_1(t) = v_2(t) = \cdots = v_n(t) \tag{5.13}$$

and the current entering the combination is the sum of the currents of each capacitor:

$$i(t) = i_1(t) + i_2(t) + \cdots + i_n(t) \tag{5.14}$$

Substituting the terminal relations for each of the capacitors given in (5.5) into (5.14), utilizing (5.13) in that expression, and rearranging gives

$$\begin{aligned} i(t) &= C_1 \frac{dv(t)}{dt} + C_2 \frac{dv(t)}{dt} + \cdots + C_n \frac{dv(t)}{dt} \\ &= \underbrace{(C_1 + C_2 + \cdots + C_n)}_{C_{\text{eq}}} \frac{dv(t)}{dt} \end{aligned} \tag{5.15}$$

Hence the equivalent capacitance of n capacitors in parallel is the sum of their capacitances:

$$\boxed{C_{\text{eq}} = C_1 + C_2 + \cdots + C_n} \tag{5.16}$$

and *capacitors in parallel combine like resistors in series*.

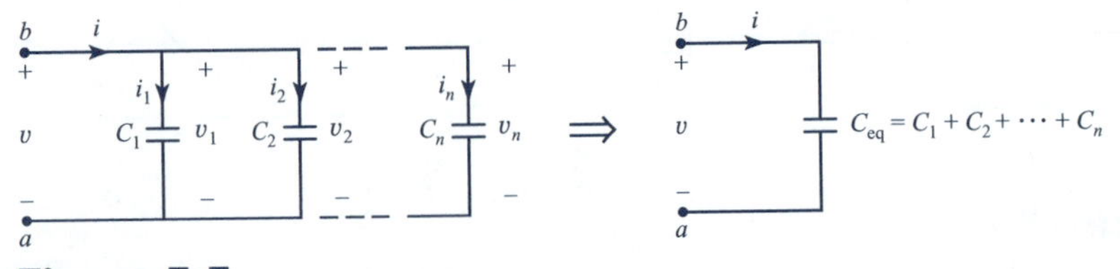

Figure 5.5
Capacitors in parallel and their equivalent capacitance.

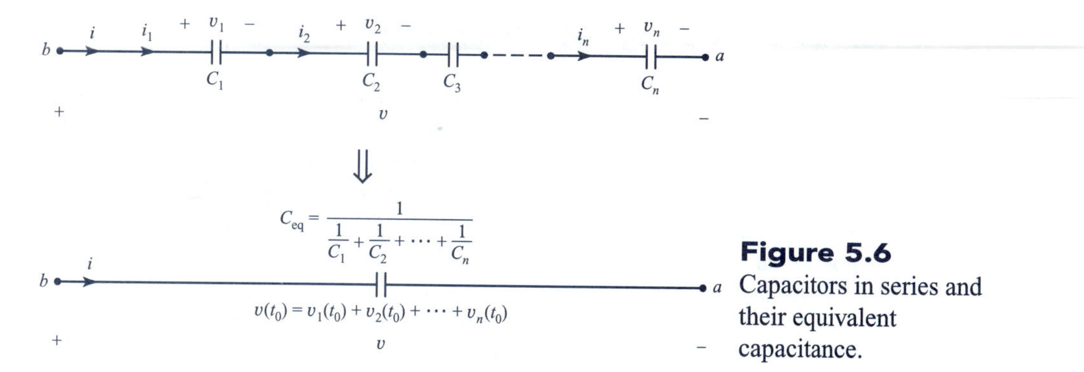

Figure 5.6 Capacitors in series and their equivalent capacitance.

In contrast, consider n capacitors in series shown in Fig. 5.6. The terminal voltage is the sum of the capacitor voltages:

$$v(t) = v_1(t) + v_2(t) + \cdots + v_n(t) \tag{5.17}$$

and the terminal current is equal to the current of each capacitor:

$$i(t) = i_1(t) = i_2(t) = \cdots = i_n(t) \tag{5.18}$$

Substituting the terminal relations of the capacitors given in (5.9) into (5.17), utilizing (5.18), and rearranging gives

$$\begin{aligned} v(t) &= \frac{1}{C_1}\int_{t_0}^{t} i_1(\tau)\,d\tau + v_1(t_0) + \frac{1}{C_2}\int_{t_0}^{t} i_2(\tau)\,d\tau + v_2(t_0) + \cdots \\ &\quad + \frac{1}{C_n}\int_{t_0}^{t} i_n(\tau)\,d\tau + v_n(t_0) \\ &= \underbrace{\left(\frac{1}{C_1} + \frac{1}{C_2} + \cdots + \frac{1}{C_n}\right)}_{\frac{1}{C_{eq}}}\int_{t_0}^{t} i(\tau)\,d\tau + v_1(t_0) + v_2(t_0) + \cdots + v_n(t_0) \end{aligned} \tag{5.19}$$

Hence the equivalent capacitance C_{eq} of the n capacitors in series is given by

$$\frac{1}{C_{eq}} = \frac{1}{C_1} + \frac{1}{C_2} + \cdots + \frac{1}{C_n} \tag{5.20}$$

having an initial voltage that is the sum of the initial voltages of each capacitor:

$$v(t_0) = v_1(t_0) + v_2(t_0) + \cdots + v_n(t_0) \tag{5.21}$$

Therefore, *capacitors in series combine like resistors in parallel.*

Exercise Problem 5.3

Determine the equivalent capacitance of the series connection of a 10-μF capacitor, a 6-μF capacitor, and a 2-μF capacitor.

Answer: 1.304 μF.

Determine the equivalent capacitance of a parallel connection of these capacitors.

Answer: 18 μF.

5.1.2 Continuity of Capacitor Voltages The voltage of a capacitor cannot change instantaneously. This important fact may be shown in a number of ways. First consider the terminal relation in (5.5) relating the capacitor current to the *derivative* of the capacitor voltage. If the capacitor voltage were to change instantaneously, the capacitor current would become infinite: an unrealistic situation. This may also be shown in the following alternative manner. Consider the terminal relation in (5.6) relating the terminal voltage to the *integral* of the capacitor current. Any realistic current waveform when integrated will produce a continuous function (the capacitor voltage). And finally consider the result in (5.12) relating the stored energy to the square of the capacitor voltage. This result shows that if the capacitor voltage were to change instantaneously, so would the stored energy: again an unrealistic situation. Hence *the voltage of a capacitor must be continuous in time and may not abruptly change value.*

The above arguments assumed physically realistic voltages and currents. In Chapter 7 we will examine the sole exception to this result. We will examine a mathematically useful but physically unreasonable function: the *impulse function*. If the capacitor current is represented by an impulse, then the capacitor voltage will change instantaneously. But this is the sole exception (a nonphysical one) to the above continuity of capacitor voltages.

5.2 The Inductor

The second and final energy storage element we will consider is the inductor. An inductor is a device that stores magnetic flux and hence stores energy in the magnetic field. Consider forming a length of wire into a loop and passing a current through it as shown in Fig. 5.7*a*. As discussed in Chapter 1, the current produces a magnetic field around the wire that circulates about the wire in a direction according to the right-hand rule. The magnetic field produced by this current penetrates the area of the loop formed by the wire. The total *magnetic flux* penetrating the area of the loop is a function of the current and the area of the loop and is denoted by $\psi(t)$; its units are webers (Wb). This flux is related to the current producing it by the inductance of the loop, L, as

$$\boxed{\psi(t) = Li(t)} \tag{5.22}$$

The units of inductance L are webers per ampere (Wb/A), or henrys (H): 1 H = 1 Wb/A.

An important law of physics, Faraday's law, provides that whenever a time-changing magnetic flux penetrates an area of a loop, a voltage will be induced in that loop (as though a voltage source had been inserted in it) whose value is equal to the time rate of change of the

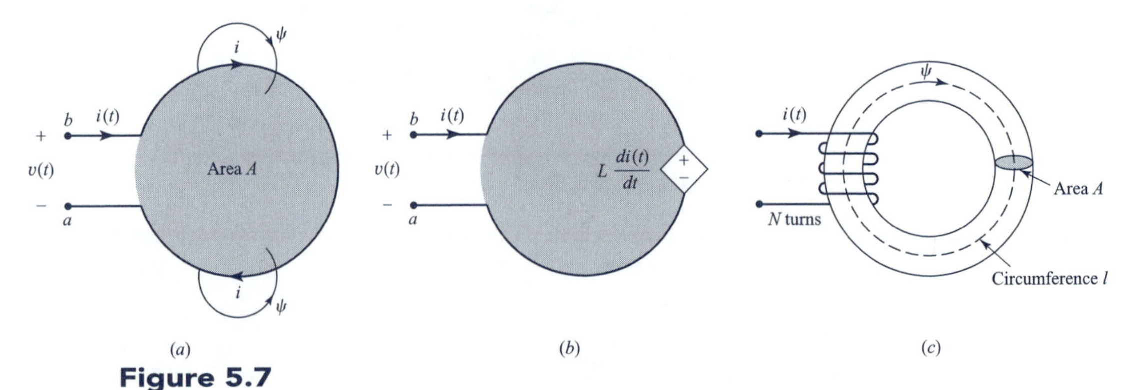

Figure 5.7
The inductor: (*a*) relationship of magnetic flux to inductance, (*b*) an equivalent circuit, and (*c*) a toroidal inductor.

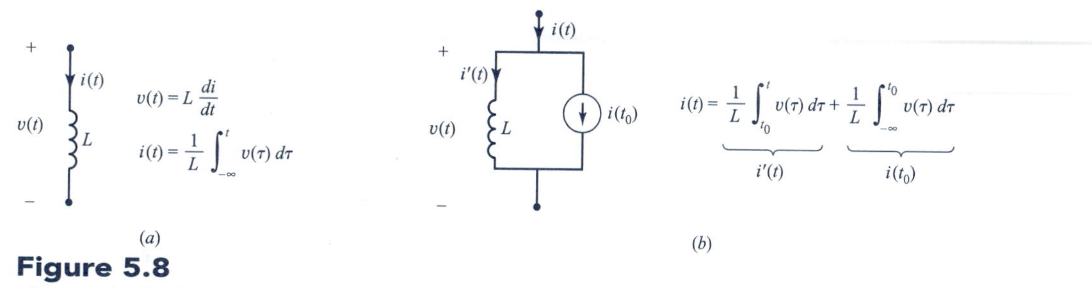

Figure 5.8
The terminal relations of an inductor.

magnetic flux penetrating the loop. Hence we can view this as a controlled source whose voltage is equal to

$$\begin{aligned} v(t) &= \frac{d\psi(t)}{dt} \\ &= L\frac{di(t)}{dt} \end{aligned} \tag{5.23}$$

and whose polarity is such that it tends to induce another current that would oppose the *change* in the original magnetic flux as shown in Fig. 5.7*b*.* If N turns are wound around a toroid having a cross-sectional area A, relative permeability μ_r (relative to air), and mean circumference l as shown in Fig. 5.7*c*, the inductance is approximately*

$$L \cong 4\pi \times 10^{-7}\,\mu_r \frac{N^2 A}{l} \qquad \text{H} \tag{5.24}$$

The inductances of practical inductors are usually given in millihenrys (1 mH = 10^{-3} H), microhenrys (1 μH = 10^{-6} H), or nanohenrys (1 nH = 10^{-9} H). The terminal relation between the voltage and current of the inductor is given in (5.23):

$$\boxed{v(t) = L\frac{di(t)}{dt}} \tag{5.25}$$

Hence, the inductor voltage is related to the rate-of-change of the inductor current or the *slope of the curve of the inductor current*. This relation may be inverted to yield

$$\boxed{i(t) = \frac{1}{L}\int_{-\infty}^{t} v(\tau)\,d\tau} \tag{5.26}$$

Thus the inductor current is related to the *net area under the curve of the the inductor voltage* up to the present time. The symbol of the inductor is shown in Fig. 5.8*a*. Observe that, as with the capacitor, the voltage and current are labeled with the passive sign convention. If they were not, a negative sign would be required in (5.25) and (5.26). In terms of the current at some time t_0 the relation in (5.26) can be written as

$$\boxed{i(t) = \frac{1}{L}\int_{t_0}^{t} v(\tau)\,d\tau + i(t_0)} \tag{5.27}$$

*C.R. Paul and S.A. Nasar, *Introduction to Electromagnetic Fields,* 2nd edition, McGraw-Hill, New York, 1987.

where the *initial current* at time t_0 is

$$i(t_0) = \frac{1}{L}\int_{-\infty}^{t_0} v(\tau)\, d\tau \tag{5.28}$$

This shows that the inductor can be represented as shown in Fig. 5.8*b*, as the parallel combination of an inductor with zero initial current and a current source representing the initial current $i(t_0)$ at $t = t_0$. Hence knowing the inductor at $t = t_0$ obviates the need for determining the area under the voltage curve from $-\infty$ up to t_0.

The energy stored in an inductor (in the magnetic field) is the integral of the instantaneous power delivered to it:

$$\begin{aligned} p(t) &= v(t)i(t) \\ &= Li(t)\,\frac{di(t)}{dt} \end{aligned} \tag{5.29}$$

giving

$$\begin{aligned} w(t) &= \int_{-\infty}^{t} p(\tau)\, d\tau \\ &= \tfrac{1}{2}Li^2(t) - \tfrac{1}{2}Li^2(-\infty) \\ &= \tfrac{1}{2}Li^2(t) \end{aligned} \tag{5.30}$$

Again we have assumed that the current through the inductor was zero at $t = -\infty$. Hence the energy stored in an inductor at some time t is related to the square of the current through the inductor at t.

Example 5.3

A current source is applied to a 5-H inductor as shown in Fig. 5.9*a*. Sketch the voltage across the inductor versus time.

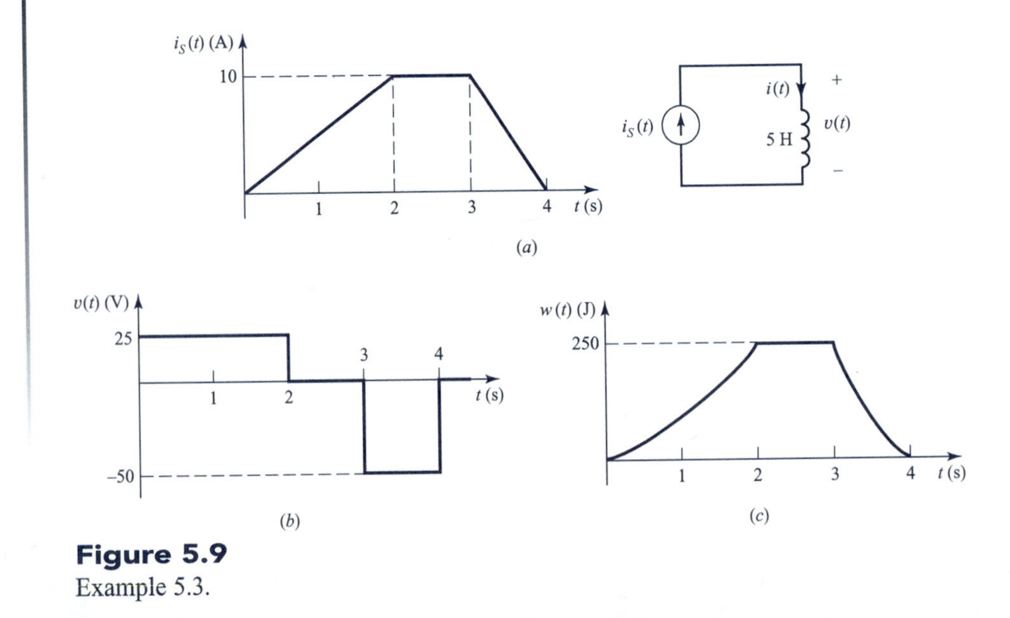

Figure 5.9
Example 5.3.

Solution The inductor current is numerically equal to the current source current, so that the inductor voltage is obtained from (5.25) as

$$v(t) = 5\frac{di_S(t)}{dt}$$

Hence the inductor voltage is related to the instantaneous slope of the current source current as shown in Fig. 5.9*b*. Similarly, the energy stored in the inductor is related to the square of the current-source current as

$$w(t) = \tfrac{1}{2} \times 5i_S^2(t)$$

which is sketched in Fig. 5.9*c*.

Exercise Problem 5.4

Replace the 100-pF capacitor in Fig. E5.2 with a 1-mH inductor and determine the inductor voltage at $t = 0.5$, 1.5, 2.5, and 3.5 μs.

Answers: 0, 2, 2, −4 V.

Example 5.4

A voltage source is applied to a 5-H inductor as shown in Fig. 5.10*a*. Sketch the inductor current versus time.

Solution The inductor voltage is equal to the voltage-source voltage, and hence the inductor current is given by (5.26) as

$$i(t) = \frac{1}{5}\int_{-\infty}^{t} v_S(\tau)\, d\tau$$

Thus the inductor current is related to the area under the voltage source curve. The result is given in Fig. 5.10*b*. The simplest way of computing this is to determine the net area accumulated at 1, 2, and 3 s and sketch the curve between these values. Since the voltage-source curve is constant over those intervals, the resulting current curve varies linearly with t.

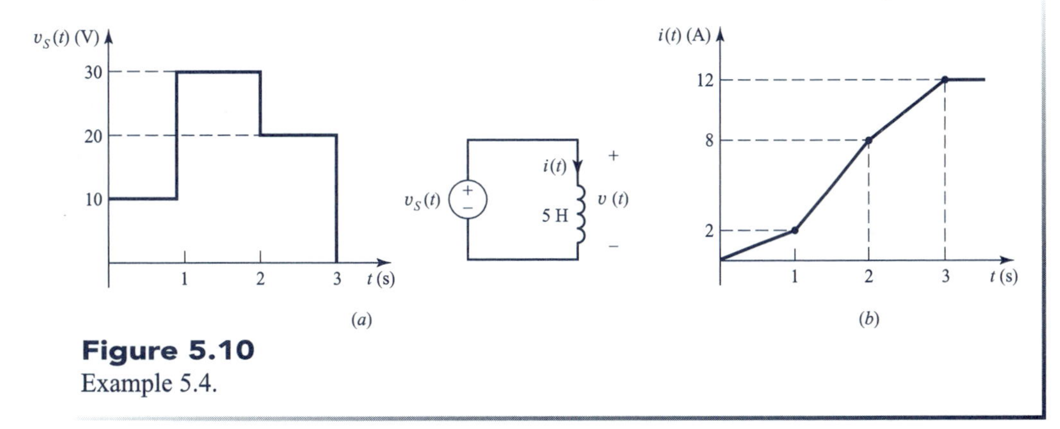

Figure 5.10
Example 5.4.

Exercise Problem 5.5

Replace the 10-μF capacitor in Fig E5.1 with a 10-μH inductor, and determine the inductor current at t = 1, 3, 4, and 6 ms.

Answers: 100, −100, 100, and −100 A.

5.2.1 Inductors in Series and in Parallel The remaining properties of inductors are duals of the corresponding properties for capacitors. Consider the series connection of n inductors shown in Fig. 5.11. The terminal voltage across the series combination is the sum of the voltages across each inductor:

$$v(t) = v_1(t) + v_2(t) + \cdots + v_n(t) \tag{5.31}$$

and the terminal current is equal to the current of each inductor:

$$i(t) = i_1(t) = i_2(t) = \cdots = i_n(t) \tag{5.32}$$

Substituting these results into the inductor relations given in (5.25) yields

$$\begin{aligned} v(t) &= L_1 \frac{di(t)}{dt} + L_2 \frac{di(t)}{dt} + \cdots + L_n \frac{di(t)}{dt} \\ &= \underbrace{(L_1 + L_2 + \cdots + L_n)}_{L_{\text{eq}}} \frac{di(t)}{dt} \end{aligned} \tag{5.33}$$

Hence the equivalent inductance of n inductances in series is the sum of their inductances:

$$L_{\text{eq}} = L_1 + L_2 + \cdots + L_n \tag{5.34}$$

and *inductors in series combine like resistors in series.*

In contrast, consider n inductors in parallel as shown in Fig. 5.12. The terminal voltage is equal to the voltage of each of the inductors:

$$v(t) = v_1(t) = v_2(t) = \cdots = v_n(t) \tag{5.35}$$

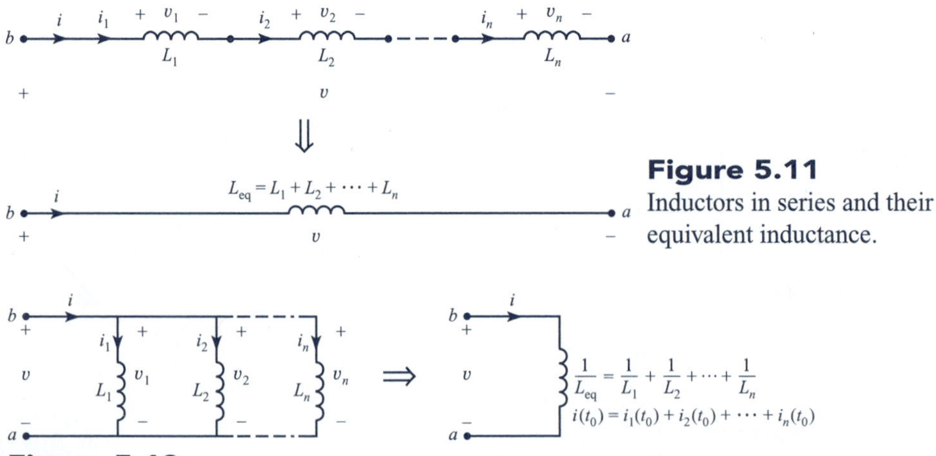

Figure 5.11
Inductors in series and their equivalent inductance.

Figure 5.12
Inductors in parallel and their equivalent inductance.

and the current entering the combination is the sum of the currents of each inductor:

$$i(t) = i_1(t) + i_2(t) + \cdots + i_n(t) \tag{5.36}$$

Substituting the terminal relations for each of the inductors given in (5.26) into (5.36), utilizing (5.35) and rearranging gives

$$\begin{aligned} i(t) &= \frac{1}{L_1}\int_{t_0}^{t} v_1(\tau)\,d\tau + i_1(t_0) + \frac{1}{L_2}\int_{t_0}^{t} v_2(\tau)\,d\tau + i_2(t_0) + \cdots + \frac{1}{L_n}\int_{t_0}^{t} v_n(\tau)\,d\tau + i_n(t_0) \\ &= \underbrace{\left(\frac{1}{L_1} + \frac{1}{L_2} + \cdots + \frac{1}{L_n}\right)}_{\frac{1}{L_{eq}}}\int_{t_0}^{t} v(\tau)\,d\tau + i_1(t_0) + i_2(t_0) + \cdots + i_n(t_0) \end{aligned} \tag{5.37}$$

Hence the equivalent inductance L_{eq} of n inductors in parallel is given by

$$\frac{1}{L_{eq}} = \frac{1}{L_1} + \frac{1}{L_2} + \cdots + \frac{1}{L_n} \tag{5.38}$$

having an initial current that is the sum of the initial currents of each inductor:

$$i(t_0) = i_1(t_0) + i_2(t_0) + \cdots + i_n(t_0) \tag{5.39}$$

Therefore, *inductors in parallel combine like resistors in parallel.*

Exercise Problem 5.6

Determine the equivalent inductance of the series connection of a 10-μH inductor, a 6-μH inductor, and a 2-μH inductor.

Answer: 18 μH.

Determine the equivalent inductance of a parallel connection of these inductors.

Answer: 1.304 μH.

5.2.2 Continuity of Inductor Currents The current of an inductor cannot change instantaneously. This important fact is the dual to the previous observation that capacitor voltages may not change instantaneously and may be demonstrated in a similar fashion. From the terminal relation in (5.25) relating the inductor voltage to the *derivative* of the inductor current we see that if the inductor current were to change instantaneously, the inductor voltage would become infinite: an unrealistic situation. Similarly, the terminal relation in (5.26) relating the terminal current to the *integral* of the inductor voltage shows that any realistic voltage waveform when integrated will produce a continuous function (the inductor current). And finally, the result in (5.30) relating the stored energy to the square of the inductor current shows that if the inductor current were to change instantaneously, so would the stored energy: again an unrealistic situation. Hence *the current of an inductor must be continuous in time and may not abruptly change value*. The sole exception to this result will be an impulse of voltage across an inductor, in which case the inductor current may change instantaneously. But the impulse function is not a physically realistic function, so that the result does apply to physical currents and voltages.

5.3 Mutual Inductance

We have seen how a time-varying current passing through the windings of an inductor can produce a time-varying magnetic flux, which induces a voltage via Faraday's law in those windings. If this magnetic flux happens to also link (pass through) some other windings, it will also induce a voltage in those windings. The two sets of windings are then said to be *coupled* via a *mutual inductance*. For example, consider two coils shown in Fig. 5.13*a*. Current $i_1(t)$ in coil 1 produces a magnetic flux, $\psi_{21}(t)$, that also passes through the area of the second coil and is called the *mutual flux*. Not all of the magnetic flux generated by $i_1(t)$ will pass through the second coil. The portion that does not couple with the second coil is denoted as $\psi_{11}(t)$ and is called the *leakage flux*. The total magnetic flux produced in coil 1 is $\psi_1 = \psi_{11} + \psi_{21}$. If the number of turns of coil 1 is N_1 and the number of turns of coil 2 is N_2, then the total flux passing through coil 1 is $\Phi_1 = N_1(\psi_{11} + \psi_{21})$ and the total flux passing through coil 2 is $\Phi_{21} = N_2\psi_{21}$. The self-inductance L_1 of coil 1 is the ratio of the total flux passing through all N_1 turns, $\Phi_1 = N_1(\psi_{11} + \psi_{21})$, to the current producing it, $i_1(t)$:

$$\Phi_1 = L_1 i_1 \tag{5.40}$$

The mutual inductance M_{21} between coil 1 and coil 2 is the ratio of the total flux passing through coil 2, $\Phi_{21} = N_2\,\psi_{21}$, to the current in coil 1 that produces it, $i_1(t)$:

$$\Phi_{21} = M_{21} i_1 \tag{5.41}$$

According to Faraday's law, a voltage will be induced in coil 1 that is proportional to the time rate of change of the magnetic flux that passes through that coil:

$$\begin{aligned} v_1(t) &= \frac{d\Phi_1}{dt} \\ &= L_1 \frac{di_1(t)}{dt} \end{aligned} \tag{5.42}$$

Similarly, a voltage will be induced in coil 2 that is proportional to the time rate of change of the magnetic flux that passes through that coil:

$$\begin{aligned} v_2(t) &= \frac{d\Phi_{21}}{dt} \\ &= M_{21} \frac{di_1(t)}{dt} \end{aligned} \tag{5.43}$$

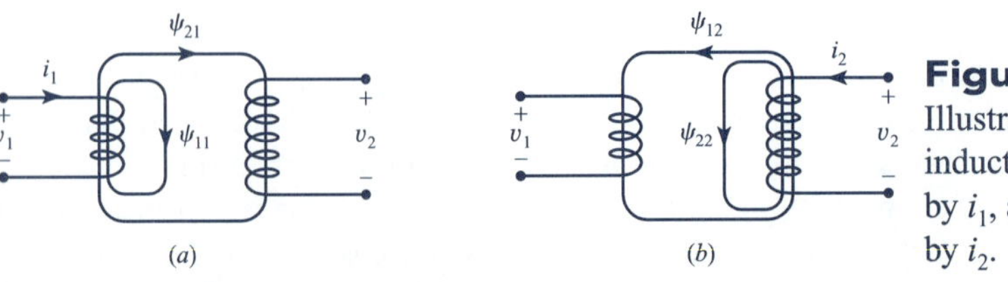

Figure 5.13 Illustration of mutual inductance: (*a*) flux created by i_1, and (*b*) flux created by i_2.

Similar results are obtained if a current $i_2(t)$ is applied to coil 2 as shown in Fig. 5.13*b*. The total flux produced by $i_2(t)$ and passing through all N_2 turns of coil 2 is related to $i_2(t)$ by the self-inductance of coil 2:

$$\Phi_2 = L_2 i_2 \tag{5.44}$$

where $\Phi_2 = N_2(\psi_{22} + \psi_{12})$. The mutual inductance between coil 2 and coil 1 is the ratio of the flux passing through all N_1 turns of coil 1, $\Phi_{12} = N_1\psi_{12}$, and the current producing it, $i_2(t)$:

$$\Phi_{12} = M_{12} i_2 \tag{5.45}$$

According to Faraday's law, a voltage will be induced in coil 2 that is proportional to the time rate of change of the magnetic flux that passes that coil:

$$\begin{aligned} v_2(t) &= \frac{d\Phi_2}{dt} \\ &= L_2 \frac{di_2(t)}{dt} \end{aligned} \tag{5.46}$$

Similarly, a voltage will be induced in coil 1 that is proportional to the time rate of change of the magnetic flux that passes through that coil:

$$\begin{aligned} v_1(t) &= \frac{d\Phi_{12}}{dt} \\ &= M_{12} \frac{di_2(t)}{dt} \end{aligned} \tag{5.47}$$

If a current $i_1(t)$ is applied to coil 1 and a current $i_2(t)$ is also applied to coil 2, the total voltages appearing at the terminals will be the superposition of the above contributions:

$$\boxed{\begin{aligned} v_1(t) &= L_1 \frac{di_1(t)}{dt} + M_{12} \frac{di_2(t)}{dt} \\ v_2(t) &= M_{21} \frac{di_1(t)}{dt} + L_2 \frac{di_2(t)}{dt} \end{aligned}} \tag{5.48}$$

It certainly makes sense that these mutual inductances are equal,

$$M_{12} = M_{21} = M \tag{5.49}$$

and both are denoted by M. This intuitive fact can be directly shown using energy considerations*. Hence the relations shown in (5.48) simplify to

$$\boxed{\begin{aligned} v_1(t) &= L_1 \frac{di_1(t)}{dt} + M \frac{di_2(t)}{dt} \\ v_2(t) &= M \frac{di_1(t)}{dt} + L_2 \frac{di_2(t)}{dt} \end{aligned}} \tag{5.50}$$

---

*C.R. Paul, *Analysis of Linear Circuits,* McGraw-Hill, New York, 1989.

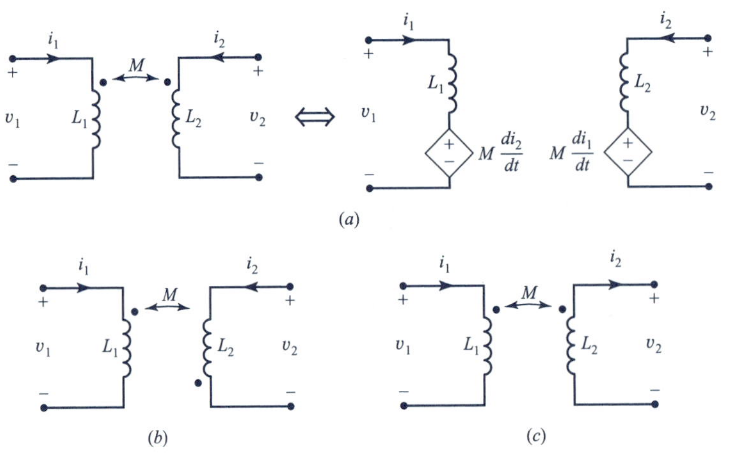

Figure 5.14 Illustration of writing the terminal relations for coupled inductors.

The circuit element symbol for these two coupled inductors is shown in Fig. 5.14*a*. Figure 5.14*a* also shows an alternative representation using controlled sources. The dots on the two inductances symbolize the relative orientation of the two coils. They provide the polarity of the mutually induced voltages according to the following rule:

> *A current in one coil, $i(t)$, that enters the dotted (undotted) end of that coil will induce a voltage in the other coil, $M\,di(t)/dt$, whose polarity is positive at the dotted (undotted) end of that coil.*

For example, consider the coupled coils shown in Fig. 5.14*b*. The terminal relations become

$$\begin{aligned} v_1(t) &= L_1 \frac{di_1(t)}{dt} - M \frac{di_2(t)}{dt} \\ v_2(t) &= -M \frac{di_1(t)}{dt} + L_2 \frac{di_2(t)}{dt} \end{aligned} \tag{5.51}$$

The polarities of the self terms, $L_1\,di_1(t)/dt$ and $L_2\,di_2(t)/dt$, are determined solely by whether the v and i are labeled with the passive sign convention in the usual fashion for an isolated inductor. The dots give the relative orientation of the two coils and hence only the sign of the mutual-inductance contribution. Figure 5.14*c* illustrates this point. The voltage of coil 1 is

$$v_1(t) = L_1 \frac{di_1(t)}{dt} - M \frac{di_2(t)}{dt} \tag{5.52a}$$

The self term is positive because v_1 and i_1 are labeled with the passive sign convention. The mutual term is negative because current i_2 enters the undotted end of coil 2 and hence will produce a voltage in coil 1 that is positive at the undotted end. But that voltage is opposite in polarity to the voltage of that coil, v_1, and is hence entered as a negative quantity in this equation for v_1. Similarly, the equation for v_2 becomes

$$v_2(t) = M \frac{di_1(t)}{dt} - L_2 \frac{di_2(t)}{dt} \tag{5.52b}$$

The self term is negative because v_2 and i_2 are not labeled with the passive sign convention. The mutual term is positive because current i_1 enters the dotted end of coil 1 and hence will produce a voltage in coil 2 that is positive at the dotted end. That voltage has the same polarity as the voltage of that coil, v_2, and is hence entered as a positive quantity in this equation for v_2.

The mutual inductance must satisfy the following relation:

$$k = \frac{M}{\sqrt{L_1 L_2}} < 1 \tag{5.53}$$

where k is said to be the *coupling coefficient* between the two coils. The coupling coefficient must be less than unity because not all of the magnetic flux produced in one coil will couple with the other coil, the difference being the leakage flux.

Example 5.5

Write the terminal relations for the coupled inductors shown in Fig. 5.15.

Solution First consider the two coupled inductors shown in Fig. 5.15*a*. The voltage of coil *a* becomes

$$v_a = -L_a \frac{di_a(t)}{dt} - M \frac{di_b(t)}{dt}$$

The self term is negative because v_a and i_a are not labeled with the passive sign convention. The mutual term is negative because i_b enters the undotted end of coil b and produces a voltage at the terminals of coil a that is therefore positive at the undotted end of that coil. But that induced voltage is opposite in polarity to v_a. The voltage of coil b becomes

$$v_b = L_b \frac{di_b(t)}{dt} + M \frac{di_a(t)}{dt}$$

The self term is positive because v_b and i_b are labeled with the passive sign convention. The mutual term is also positive because i_a enters the undotted end of coil a and produces a voltage at the terminals of coil b that is therefore positive at the undotted end of that coil. But that induced voltage has the same polarity as v_b.

Next consider the three coupled coils of Fig. 5.15*b*. The voltage expression for coil *a* is

$$v_a = -L_a \frac{di_a(t)}{dt} - M_{ab} \frac{di_b(t)}{dt} + M_{ac} \frac{di_c(t)}{dt}$$

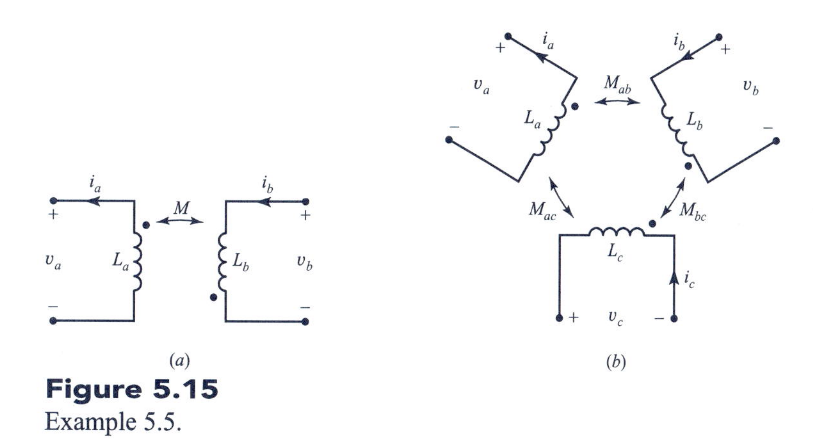

Figure 5.15
Example 5.5.

The self term is negative because v_a and i_a are not labeled with the passive sign convention. The first mutual term, M_{ab}, is negative because i_b enters the undotted end of coil b and produces a voltage at the terminals of coil a that is therefore positive at the undotted end of that coil. But that induced voltage is opposite in polarity to v_a, for which the equation is being written. The second mutual term, M_{ac}, is positive because i_c enters the dotted end of coil c and produces a voltage at the terminals of coil a that is therefore positive at the dotted end of that coil. But that induced voltage is of the same polarity as v_a, for which the equation is being written. The voltage expression for coil b is

$$v_b = M_{ab}\frac{di_a(t)}{dt} + L_b\frac{di_b(t)}{dt} - M_{bc}\frac{di_c(t)}{dt}$$

The self term is positive because v_b and i_b are labeled with the passive sign convention. The first mutual term, M_{ab}, is positive because i_a enters the undotted end of coil a and produces a voltage at the terminals of coil b that is therefore positive at the undotted end of that coil. But that induced voltage has the same polarity as v_b, for which the equation is being written. The second mutual term, M_{bc}, is negative because i_c enters the dotted end of coil c and produces a voltage at the terminals of coil b that is therefore positive at the dotted end of that coil. But that induced voltage opposite to the polarity as v_b, for which the equation is being written. The voltage expression for coil c is

$$v_c = M_{ac}\frac{di_a(t)}{dt} + M_{bc}\frac{di_b(t)}{dt} - L_c\frac{di_c(t)}{dt}$$

The self term is negative because v_c and i_c are not labeled with the passive sign convention. The first mutual term, M_{ac}, is positive because i_a enters the undotted end of coil a and produces a voltage at the terminals of coil c that is therefore positive at the undotted end of that coil. But that induced voltage has the same polarity as v_c, for which the equation is being written. The second mutual term, M_{bc}, is positive because i_b enters the undotted end of coil b and produces a voltage at the terminals of coil c that is therefore positive at the undotted end of that coil. But that induced voltage is of the same polarity as v_c, for which the equation is being written.

Exercise Problem 5.7

Write the terminal relations for the three coupled inductors shown in Fig. E5.7.

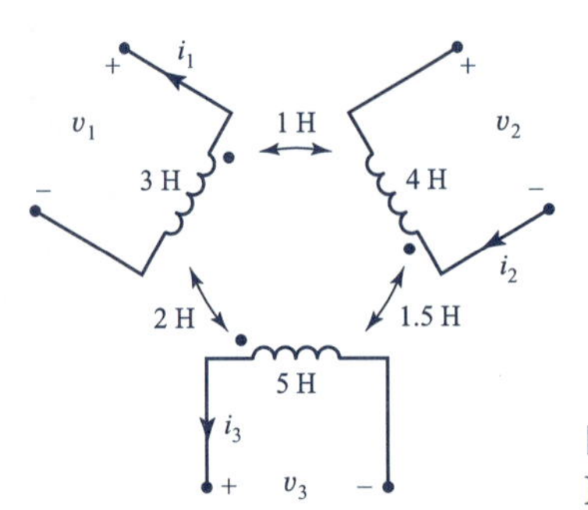

Figure E5.7
Exercise Problem 5.7.

Answer: $$v_1 = -3\frac{di_1}{dt} + \frac{di_2}{dt} - 2\frac{di_3}{dt}$$

$$v_2 = \frac{di_1}{dt} - 4\frac{di_2}{dt} + 1.5\frac{di_3}{dt}$$

$$v_3 = -2\frac{di_1}{dt} + 1.5\frac{di_2}{dt} - 5\frac{di_3}{dt}$$

5.3.1 The Ideal Transformer Coupled coils are used to construct *transformers*. For example, consider winding two coils on a common core as shown in Fig. 5.16*a*. In actual transformers, the material used to construct the core is usually a ferrous material such as iron or steel that has a large permeability. Iron and steel have relative permeabilities (relative to that of air) on the order of 2000. This large permeability of the core tends to provide a lower-reluctance path than air, thereby confining the majority of the magnetic flux to the core. Hence only a small portion of the flux produced by each coil leaks out into the air, and virtually all the flux produced in one coil passes through the other coil. An *ideal transformer* is a pair of such coupled coils satisifying the following three properties:

1. *The transformer has* unity coupling coefficient

$$k = 1 \tag{5.54a}$$

which means that all of the magnetic flux produced by each coil passes through the other coil and there is no leakage flux.

2. *The transformer is* lossless:

$$p_1 = v_1 i_1 = -p_2 = -v_2 i_2 \tag{5.54b}$$

3. *The self-inductances are infinite, but their ratio is finite. The inductance of a coil is given by (5.24) where the factor $4\pi \times 10^{-7}$ is the permeability of free space and μ_r is the relative permeability of the core material. If the permeability of the core is infinite, these self-inductances will also be infinite. Taking their ratios gives, according to (5.24),*

$$\frac{L_1}{L_2} = \left(\frac{N_1}{N_2}\right)^2 \tag{5.54c}$$

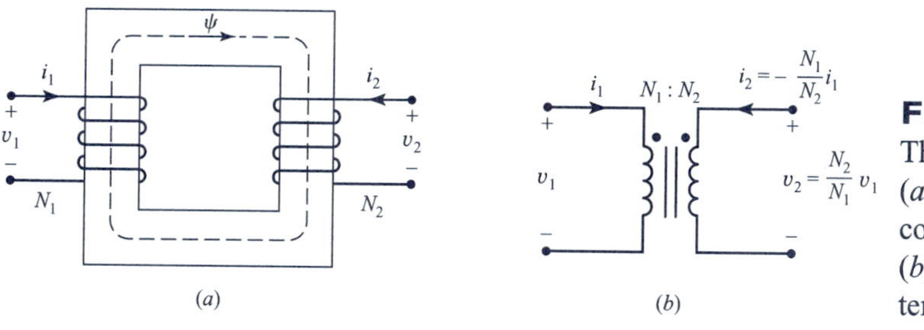

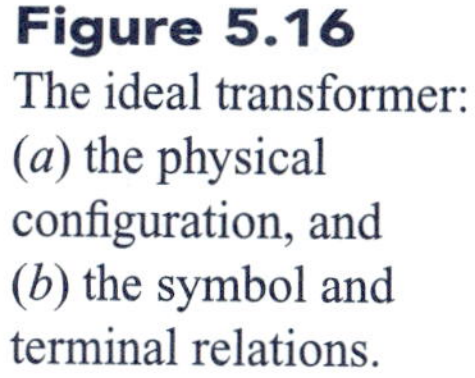

Figure 5.16 The ideal transformer: (*a*) the physical configuration, and (*b*) the symbol and terminal relations.

If each turn of coil 1 produces magnetic flux ψ and there are N_1 turns in coil 1, the total flux passing through coil 1 is $N_1\psi$. Hence the induced voltage at the terminals of coil 1 is

$$
\begin{aligned}
v_1 &= \frac{d}{dt}(N_1\psi) \\
&= N_1\frac{d\psi}{dt}
\end{aligned}
\tag{5.55}
$$

and the voltage induced at the terminals of coil 2 is

$$
\begin{aligned}
v_2 &= \frac{d}{dt}(N_2\psi) \\
&= N_2\frac{d\psi}{dt}
\end{aligned}
\tag{5.56}
$$

Taking the ratio of these two voltages gives

$$\frac{v_1}{v_2} = \frac{N_1}{N_2} \tag{5.57}$$

If $N_2 > N_1$, then $v_2 = (N_2/N_1)v_1$ is larger than v_1 and voltage v_1 is said to be *stepped up*. A common use of transformers is to step up or step down voltages. These voltages must be time-varying, since the coupling results from Faraday's law, which requires that the magnetic flux be time-varying. Applying the lossless attribute given in (5.54b) gives the relations between the currents:

$$\frac{i_1}{i_2} = -\frac{N_2}{N_1} \tag{5.58}$$

Observe that this is the reciprocal of the voltage relation: if the voltage is stepped up, the current is stepped down by the same turns ratio. The minus sign provides that the current into coil 1 flows out of coil 2. The circuit symbol of an ideal transformer is given in Fig. 5.16*b* along with the above voltage and current relations. The pair of vertical lines symbolizes that the core is a high-permeability material such as iron.

In addition to providing ideal step-up or step-down of voltages and currents, the ideal transformer provides another useful function: resistance reflection. Consider Fig. 5.17, wherein a resistance R is placed across the terminals of the second coil of an ideal transformer. The equivalent resistance seen at the terminals of the first coil is

$$R_{eq} = \frac{v_1}{i_1}$$

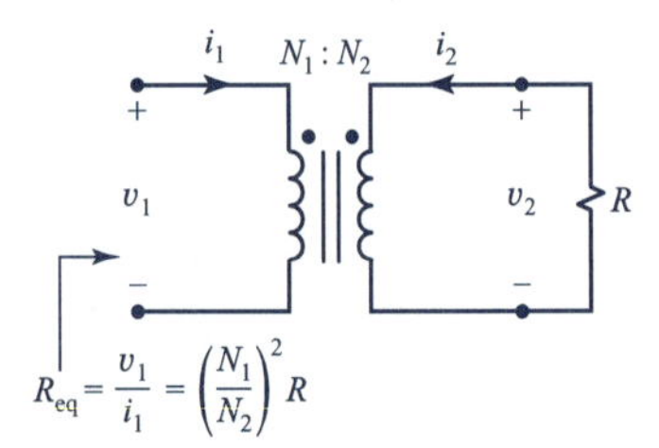

Figure 5.17
Illustration of the reflection of resistance through an ideal transformer.

Substituting

$$v_1 = \frac{N_1}{N_2} v_2$$

$$i_1 = -\frac{N_2}{N_1} i_2$$

along with the terminal relation for R

$$v_2 = -Ri_2$$

gives the equivalent resistance as

$$\begin{aligned} R_{\text{eq}} &= \frac{v_1}{i_1} \\ &= \left(\frac{N_1}{N_2}\right)^2 R \end{aligned} \tag{5.59}$$

Hence the resistance across the second coil appears larger or smaller looking into the first coil by a factor of the square of the turns ratio. This can be used for matching for maximum power transfer, as the following example shows.

Example 5.6

Consider a sinusoidal source having a 100-Ω source resistance that is attached to a load of 4 Ω as shown in Fig. 5.18*a*. Clearly maximum power will not be supplied to the load, since the source and load impedances are not equal. Design a matching circuit composed of an ideal transformer such that maximum power will be transferred from the source to the load.

Solution Inserting an ideal transformer between the source and the load having a turns ratio of 5:1 will reflect the load resistance to give 100 Ω seen by the source looking into the input of coil 1, as shown in Fig. 5.18*b*. Hence maximum power will be transferred from the source to everything to the right (the transformer and the load). But the transformer is ideal (lossless), and hence maximum power is transferred to the load. The voltage at the terminals of the source can be found using voltage division in Fig. 5.18*b* as

$$\begin{aligned} v_S &= \frac{100\ \Omega}{100\ \Omega + 100\ \Omega} 10 \sin \omega t \\ &= 5 \sin \omega t \end{aligned}$$

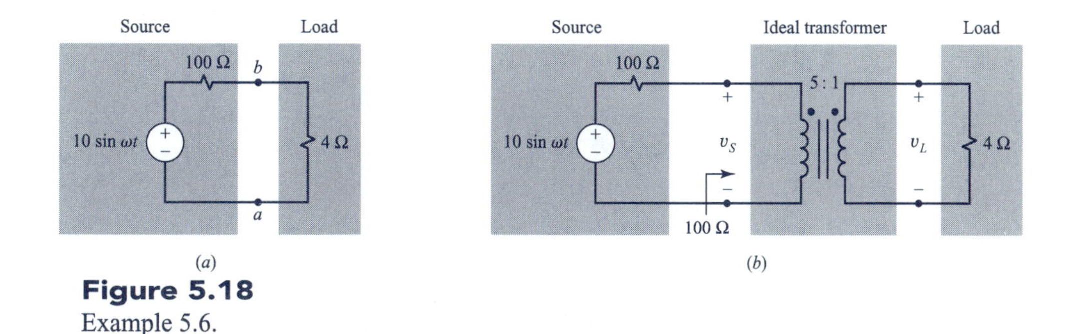

Figure 5.18
Example 5.6.

The voltage across the load can be found from this result using the transformer turns ratio as

$$v_L = \tfrac{1}{5} v_S$$
$$= \sin \omega t$$

Exercise Problem 5.8

Determine the equivalent resistance seen looking into the two terminals of the circuit of Fig. E5.8.

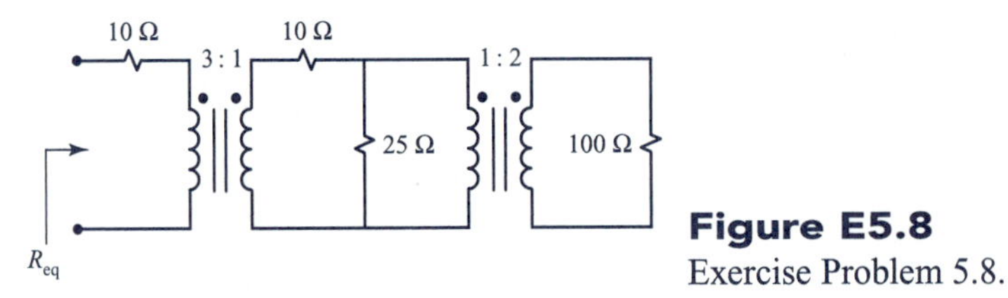

Figure E5.8
Exercise Problem 5.8.

Answer: 212.5 Ω.

5.4 Response of the Energy Storage Elements to DC Sources

The principle of superposition applies to linear circuits, which may contain linear resistors, controlled sources, capacitors, and inductors. The energy storage elements considered in this chapter are linear elements: the derivative and integral operations in their terminal relationships are linear operations. Hence we may apply superposition to determine the *portions* of the responses of any current or voltage due to each of the independent sources. In this section we will examine the portions of the responses to *dc sources,* i.e., sources whose output (voltage or current) are constants independent of time. In the next chapter we will examine the portions of the responses to sources that have sinusoidal time variation, and in Chapter 7 we will examine the response of circuits to sources that have arbitrary time variation.

First let us examine the response of a capacitor to a dc source. We would expect the *form* of the current and voltage response to be dc (constant) also. Thus we assume the form of those responses to be $i_C(t) = I_C$ and $v_C(t) = V_C$ as shown in Fig. 5.19*a*. Substituting these into the capacitor terminal relations yields

$$I_C = C\frac{d}{dt}V_C$$
$$= 0 \tag{5.60}$$

since the derivative of a constant is zero. Therefore *the response of a capacitor to a dc source is obtained by replacing the capacitor with an open circuit* as shown in Fig. 5.19*a*. Similarly, the

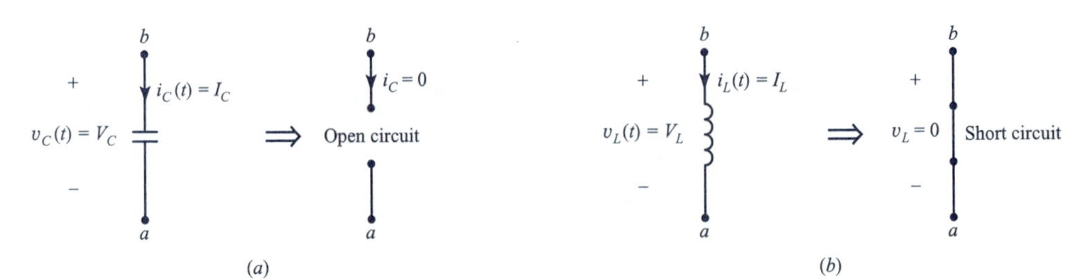

Figure 5.19
Response of (*a*) a capacitor, and (*b*) an inductor to a dc source.

response of an inductor to a dc source is assumed to be dc (constant) also: $i_L(t) = I_L$ and $v_L(t) = V_L$. Substituting these into the terminal relation for an inductor yields

$$\begin{aligned} V_L &= L\frac{d}{dt}I_L \\ &= 0 \end{aligned} \tag{5.61}$$

since the derivative of a constant is zero. Therefore *the response of an inductor to a dc source is obtained by replacing the inductor with a short circuit* as shown in Fig. 5.19*b*.

Example 5.7

Determine the voltage v_x and current i_x in the circuit of Fig. 5.20*a*.

Solution Since both sources are dc, we may leave them in one circuit. Replacing the capacitors with open circuits and the inductors with short circuits, we obtain the circuit of Fig. 5.20*b*. From this circuit we obtain $v_x = 0$ and $i_x = -2$ A.

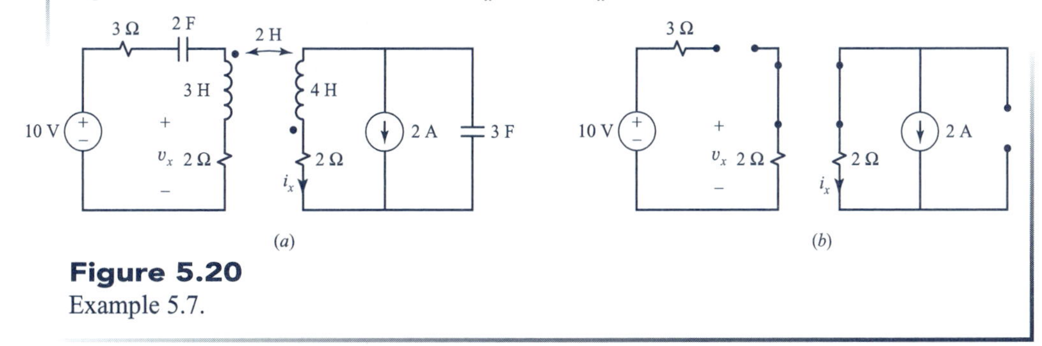

Figure 5.20
Example 5.7.

Exercise Problem 5.9

Determine current i_x and voltage v_x in the circuit of Fig. E5.9.

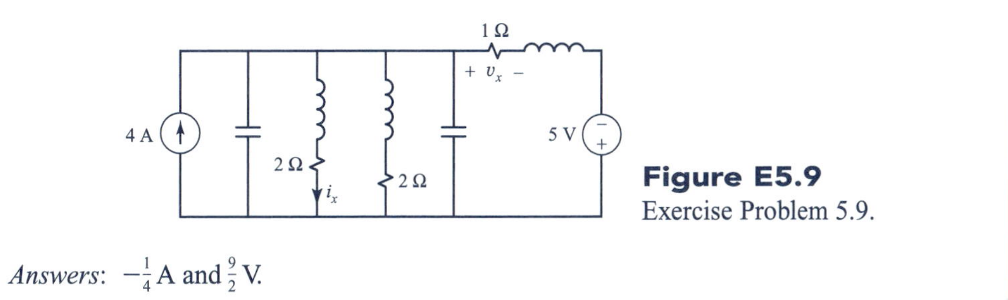

Figure E5.9
Exercise Problem 5.9.

Answers: $-\frac{1}{4}$ A and $\frac{9}{2}$ V.

5.5 The Differential Equations of a Circuit

If a circuit contains only resistors, controlled sources, and independent sources, the governing equations relating any element current or voltage to any of the independent sources will be purely *algebraic,* i.e., will not contain derivatives of those currents or voltages. If, however, the circuit contains, in addition to these elements, at least one energy storage element, an inductor or a capacitor, then the governing equations relating any element current or voltage to any of the independent sources will be *differential equations,* i.e., will contain derivatives of those currents or voltages with respect to time. The important point to be observed here is that the introduction of inductors or capacitors into a circuit will cause the governing equations to become differential equations, which are more difficult to solve than the algebraic equations that result from resistive circuits. We will devote the remaining chapters of this text to examining the solution to those differential equations.

The first fundamental task in the solution of a circuit is the determination of the governing equation relating a desired voltage or current to an independent source. In this section we will present a simple method that can be used to rapidly and easily determine the differential equation relating any current or voltage to a source. The method builds on the resistive circuit analysis techniques which we now have great facility with. It utilizes the concept of a *differential operator D*. Multiplying a time function (on the left) by the differential operator means that we differentiate the function:

$$Df(t) \Rightarrow \frac{df(t)}{dt} \tag{5.62}$$

Similarly, dividing a function by the differential operator means that we integrate the function:

$$\frac{1}{D} f(t) \Rightarrow \int f(t)\, dt \tag{5.63}$$

Multiplying (on the left) $1/D$ by D nullifies the integral operation:

$$D\left[\frac{1}{D} f(t)\right] \Rightarrow f(t) \tag{5.64}$$

This differential operator obeys most of the rules of algebra and can be manipulated as though it were an algebraic variable:

$$\begin{aligned}
D^n f(t) &\Rightarrow \frac{d^n f(t)}{dt^n} \\
\frac{1}{D^m} f(t) &\Rightarrow \underbrace{\int \cdots \int f(\tau)\, d\tau}_{m\text{-fold integral}} \\
D[f_1(t) + f_2(t)] &\Rightarrow \frac{df_1(t)}{dt} + \frac{df_2(t)}{dt} \\
(D + a)(D + b) f(t) &= [D^2 + (a + b)D + ab] f(t) \\
&\Rightarrow \frac{d^2 f(t)}{dt^2} + (a + b)\frac{df(t)}{dt} + abf(t) \\
D^n[D^m f(t)] &= D^{n+m} f(t) \\
&\Rightarrow \frac{d^{n+m} f(t)}{dt^{n+m}} \\
D^n\left(\frac{1}{D^m}\right) f(t) &= D^{n-m} f(t) \qquad n \geq m \\
&\Rightarrow \frac{d^{n-m} f(t)}{dt^{n-m}}
\end{aligned} \tag{5.65}$$

The terminal relations for the resistor, inductor, and capacitor can be written in a compact manner using the differential operator as

$$v_R(t) = Ri_R(t) \tag{5.66a}$$

$$\begin{aligned}
v_L(t) &= L\frac{di_L(t)}{dt} \\
&\Rightarrow LDi_L(t)
\end{aligned} \tag{5.66b}$$

$$v_C(t) = \frac{1}{C}\int_{-\infty}^{t} i_C(\tau)\, d\tau$$
$$\Rightarrow \frac{1}{CD} i_C(t) \tag{5.66c}$$

The important observation here is that we can treat these elements as being resistors having resistance values of

$$R_R = R \tag{5.67a}$$

$$R_L = LD \tag{5.67b}$$

$$R_C = \frac{1}{CD} \tag{5.67c}$$

Hence we can replace the elements with these "resistors" and use all our previous resistive circuit analysis techniques to obtain the *transfer function* relating the voltage or current of interest to the source(s). This transfer function relating the current or voltage of interest, $y(t)$, to the source $x(t)$ will be of the form of the ratio of two polynomials in D:

$$y(t) = K\frac{D^m + b_1 D^{m-1} + \cdots + b_m}{D^n + a_1 D^{n-1} + \cdots + a_n} x(t) \tag{5.68}$$

The differential equation is obtained by multiplying both sides by the denominator to yield

$$(D^n + a_1 D^{n-1} + \cdots + a_n)\, y(t) = K(D^m + b_1 D^{m-1} + \cdots + b_m)\, x(t) \tag{5.69}$$

Operating on the variables gives the differential equation

$$\frac{d^n y(t)}{dt^n} + a_1 \frac{d^{n-1} y(t)}{dt^{n-1}} + \cdots + a_n y(t) = K\left(\frac{d^m x(t)}{dt^m} + b_1 \frac{d^{m-1} x(t)}{dt^{m-1}} + \cdots + b_m x(t)\right) \tag{5.70}$$

Since we presumably know the functional form of $x(t)$ (it is the value of an independent source), we can perform the required differentiations on the right-hand side of this equation.

Example 5.8

Determine the differential equation relating the current $i(t)$ to the source in the circuit of Fig. 5.21*a*.

Solution Without the use of the differential operator, determining the circuit differential equations is generally not a methodical process, i.e., the correct and optimum steps are not so obvious. In order to demonstrate this so that the utility of the differential operator technique will be appreciated, we will first determine the differential equation by "brute force" using KVL, KCL, and the element relations. First label the elements with their voltages and currents as shown in Fig. 5.21*b*. Applying KVL gives

$$10 \cos 3t = 2i + v_C \tag{a}$$

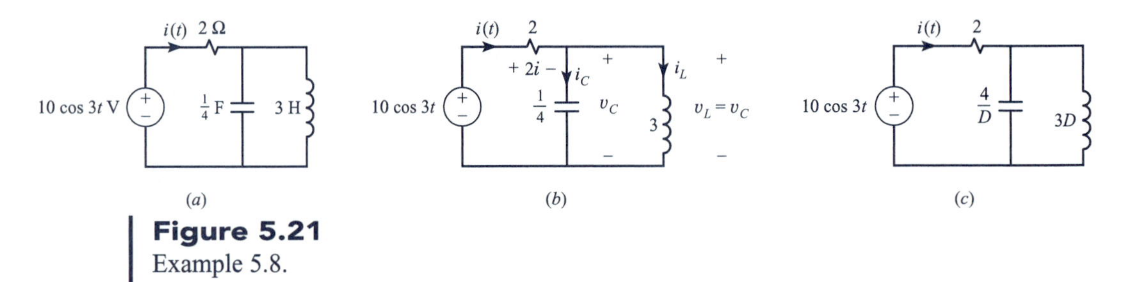

Figure 5.21
Example 5.8.

and

$$v_C = v_L = 3\frac{di_L}{dt} \tag{b}$$

Applying KCL gives

$$i = \frac{1}{4}\frac{dv_C}{dt} + i_L \tag{c}$$

Differentiating (c) once and substituting (b) gives

$$\begin{aligned}\frac{di}{dt} &= \frac{1}{4}\frac{d^2v_C}{dt^2} + \frac{di_L}{dt} \\ &= \frac{1}{4}\frac{d^2v_C}{dt^2} + \frac{1}{3}v_C\end{aligned} \tag{d}$$

Substituting (a) into (d) gives one equation in i:

$$\begin{aligned}\frac{di}{dt} &= \frac{1}{4}\frac{d^2}{dt^2}(10\cos 3t - 2i) + \frac{1}{3}(10\cos 3t - 2i) \\ &= -\frac{1}{2}\frac{d^2i}{dt^2} - \frac{2}{3}i + \frac{1}{4}\frac{d^2}{dt^2}(10\cos 3t) + \frac{1}{3}(10\cos 3t)\end{aligned} \tag{e}$$

Rearranging this gives the differential equation in standard form:

$$\begin{aligned}\frac{d^2i}{dt^2} + 2\frac{di}{dt} + \frac{4}{3}i &= \frac{1}{2}\frac{d^2}{dt^2}(10\cos 3t) + \frac{2}{3}(10\cos 3t) \\ &= -45\cos 3t + \tfrac{20}{3}\cos 3t \\ &= -\tfrac{115}{3}\cos 3t\end{aligned}$$

Clearly, obtaining the differential equation for even a simple circuit can be quite tedious.

Replacing the elements with their differential-operator "resistances" as shown in Fig. 5.21*c*, we obtain, using resistive circuit analysis techniques,

$$\begin{aligned}i(t) &= \frac{10\cos 3t}{2 + (4/D)\,\|\,(3D)} \\ &= \frac{10\cos 3t}{2 + \dfrac{(4/D)(3D)}{4/D + 3D}} \\ &= \frac{3D^2 + 4}{6D^2 + 12D + 8} \times 10\cos 3t\end{aligned}$$

Multiplying through by the denominator gives

$$(6D^2 + 12D + 8)i(t) = (3D^2 + 4)\,10\cos 3t$$

Operating on the variables gives the differential equation

$$\begin{aligned}6\frac{d^2i(t)}{dt^2} + 12\frac{di(t)}{dt} + 8i(t) &= 3\frac{d^2}{dt^2}(10\cos 3t) + 4(10\cos 3t)\\ &= -270\cos 3t + 40\cos 3t\\ &= -230\cos 3t\end{aligned}$$

or, dividing both sides by the coefficient of the highest derivative to place it in standard form,

$$\frac{d^2i(t)}{dt^2} + 2\frac{di(t)}{dt} + \frac{4}{3}i(t) = -\frac{115}{3}\cos 3t$$

giving the same result as obtained by brute force previously. The differential-operator technique clearly provides a very efficient and straightforward method for determining the circuit differential equations.

Exercise Problem 5.10

Determine the differential equation relating the voltage $v(t)$ to the source in the circuit of Fig. E5.10.

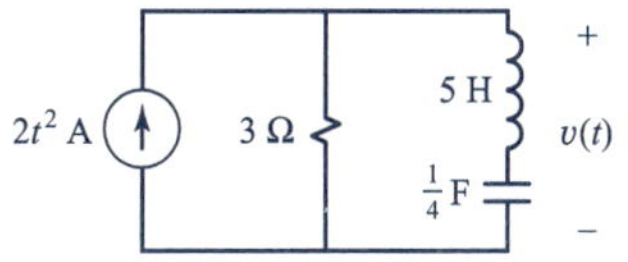

Figure E5.10
Exercise Problem 5.10.

Answer: $d^2v(t)/dt^2 + \frac{3}{5}\,dv(t)/dt + \frac{4}{5}v(t) = 12 + \frac{24}{5}t^2$.

The remainder of this text will be devoted to solving the circuit differential equations for general circuits. We will devote Chapter 6 exclusively to the solution of the differential equations for independent sources that have sinusoidal time variation, whereas Chapter 7 will be devoted to the solution of the differential equations for circuits that contain independent sources that have arbitrary time variation. In all these cases we will develop methods of solution that utilize our resistive-circuit analysis techniques. The solution for dc sources likewise evolved into a resistive-circuit analysis solution. Hence the resistive-circuit analysis techniques developed earlier are critical to the solution of all other circuit analysis problems.

5.6 The Op-Amp Differentiator and Integrator

The operational amplifier can be used to construct circuits that differentiate or integrate a signal. For example, consider the circuit of Fig. 5.22*a*. Recognizing this as being in the form of an inverting amplifier and substituting the operator resistance for the capacitor, we obtain

$$\begin{aligned}\frac{v_{\text{out}}(t)}{v_{\text{in}}(t)} &= -\frac{R_f = \dfrac{1}{CD}}{R_i = R}\\ &= -\frac{1}{RCD}\end{aligned} \tag{5.71}$$

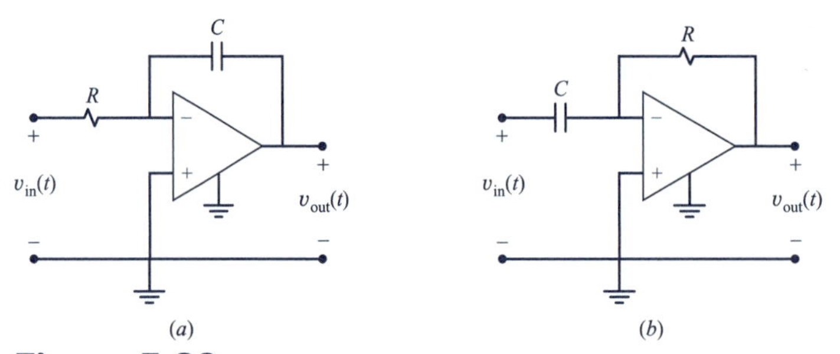

Figure 5.22

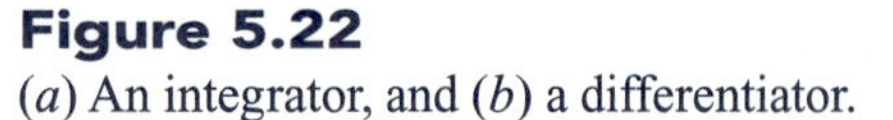

(*a*) An integrator, and (*b*) a differentiator.

Writing this out gives

$$
\begin{aligned}
v_{out}(t) &= -\frac{1}{RCD} v_{in}(t) \\
&= -\frac{1}{RC} \int_{-\infty}^{t} v_{in}(\tau)\, d\tau
\end{aligned}
\tag{5.72}
$$

Figure 5.22*b* shows a differentiator. Again recognizing this as an inverting amplifier and replacing the capacitor with its operator resistance gives

$$
\begin{aligned}
\frac{v_{out}(t)}{v_{in}(t)} &= -\frac{R_f = R}{R_i = \dfrac{1}{CD}} \\
&= -RCD
\end{aligned}
\tag{5.73}
$$

Writing this out gives

$$
\begin{aligned}
v_{out}(t) &= -RCDv_{in}(t) \\
&= -RC\frac{dv_{in}(t)}{dt}
\end{aligned}
\tag{5.74}
$$

Exercise Problem 5.11

The voltage waveform shown in Fig. E5.11 is first applied to an integrator and then to a differentiator. In both cases, $R = 1$ kΩ and $C = 5$ μF. Determine the output voltages at $t = 2$ ms.

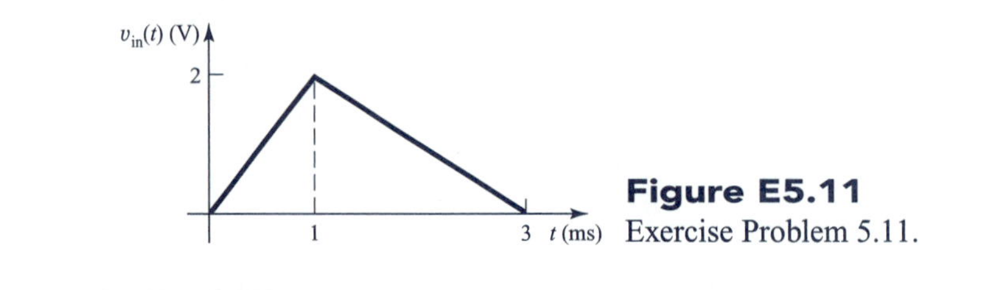

Figure E5.11
Exercise Problem 5.11.

Answers: −0.5 V and 5 V.

5.7 PSPICE Applications

The previous applications of PSPICE involved resistive circuits excited by dc sources. This used the .DC module of PSPICE. We now extend this to circuits that (a) have capacitors or inductors in them and (b) can have time-varying sources applied. This extension uses the .TRAN module of PSPICE.

First we address how to include capacitors and inductors in PSPICE. The terminal relation for the capacitor voltage is given in (5.9) as

$$v_C(t) = \frac{1}{C}\int_0^t i_C(\tau)\,d\tau + v_C(0) \tag{5.75}$$

All PSPICE solutions start at $t = 0$. This can be represented as the series combination of a capacitor with zero voltage at $t = 0$ and a dc voltage source whose value is the *initial condition* at time $t = 0$, $v_C(0)$, as shown in Fig. 5.23*a*. Similarly, the terminal relation for the inductor current is given in (5.27) as

$$i_L(t) = \frac{1}{L}\int_0^t v_L(\tau)\,d\tau + i_L(0) \tag{5.76}$$

This can be represented as the parallel combination of an inductor with zero current at $t = 0$ and a dc current source whose value is the *initial condition* at time $t = 0$, $i_L(0)$, as shown in Fig. 5.23*b*.

The PSPICE coding for these elements is of the form

CXXX N1 N2 c IC=$v_C(0)$
LXXX N1 N2 l IC=$i_L(0)$

as shown in Fig. 5.24. The item IC= gives the initial (at $t = 0$) capacitor voltage and inductor current. The polarity of these initial conditions conforms to the usual rules; the capacitor voltage is assumed positive at the first-named node, N1, and the inductor current is assumed to flow from the first-named node, N1, to the last-named node, N2.

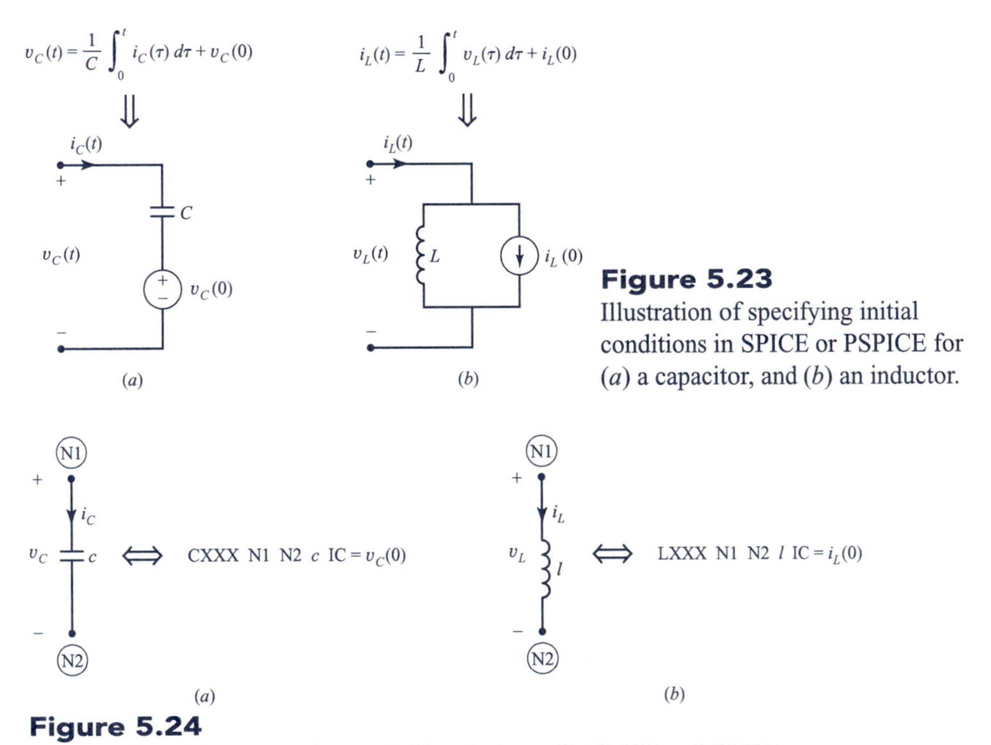

Figure 5.23
Illustration of specifying initial conditions in SPICE or PSPICE for (*a*) a capacitor, and (*b*) an inductor.

Figure 5.24
Specification of (*a*) a capacitor, and (*b*) an inductor in SPICE or PSPICE.

Mutual inductances are similarly defined:

KXXX L1 L2 k

The code for mutual inductances is K, and L1 and L2 are the names of the two inductors that are coupled. These two inductors must be defined elsewhere in the circuit specification statements. Mutual inductances are specified not by the mutual inductance M but by the coefficient of coupling:

$$k = \frac{M}{\sqrt{L1 \times L2}}$$

The dot convention is implemented by the following rule: *define the inductors L1 and L2 with the first-named nodes in their specification statements on the dotted end of the respective inductor.* In this case, the coefficient of coupling k is entered as a positive number. If the dot is not on the first-named node of one of the inductors but is on the other, then k is entered as a negative number. If for both inductors the dot is not on the first-named node, then k is entered as a positive number.

The next modification is the control statement. Since the result will be a solution that varies with time, we cannot use the .DC control statement as we did for dc resistive circuits. The control statement for time-varying waveforms is the .TRAN statement:

.TRAN *print step* *final time* [*results delay* [*max step size*]] UIC

PSPICE essentially solves the differential equations of the circuit by discretizing them and using a numerical approximation. For example (the PSPICE discretization is actually more sophisticated than this), a simple discretization of the differential equation

$$\frac{dy(t)}{dt} + ay(t) = b$$

is

$$\frac{y(t + \Delta t) - y(t)}{\Delta t} + ay(t) = b$$

where the time interval is discretized into segments 0, Δt, $2\Delta t$, $3\Delta t$,.... Rewriting this gives

$$y(t + \Delta t) = (1 - a\,\Delta t)\,y(t) + b\,\Delta t$$

This is solved in a "bootstrapping" fashion in that this expression is solved for the succeeding values using the solution for the preceding time:

$$\begin{aligned} y(\Delta t) &= (1 - a\,\Delta t)\,y(0) + b\,\Delta t \\ y(2\Delta t) &= (1 - a\,\Delta t)\,y(\Delta t) + b\,\Delta t \\ y(3\Delta t) &= (1 - a\,\Delta t)\,y(2\,\Delta t) + b\,\Delta t \\ &\vdots \end{aligned}$$

The accuracy of the solution increases as the discretization time step Δt is made smaller.

The *print step* is the time interval between desired solutions. For example, a print step of 5 ms means that we want an output every 5 ms. The *final time* is the last time at which we want PSPICE to obtain a solution. We would choose a value for this well past the point where any interesting changes are taking place in the solution. Generally, this requires trial and error: arbitrarily choose a final time, run the program, and rerun it with a better choice.

The other two entries are optional. The *results delay* is the time past $t = 0$ for which we do not want an output. The solution *always* starts at $t = 0$. The final item, *max step size*, is also optional (but if we include it, we must precede it with a *results delay*, e.g. 0). Contrary to the above rudimentary discretization scheme where a fixed discretization Δt is used, PSPICE uses a variable step size to increase accuracy. For some problems, PSPICE may choose a step size that is too large to give accurate results. In this case, in order to produce better accuracy, we can specify a maximum step size that PSPICE is allowed to use. Ordinarily PSPICE is capable of choosing an adequate step size, so that we rarely use the *results delay* and *max step size* parameters and they are omitted.

The final entry UIC tells PSPICE to use the initial conditions specified on the element specification statements with the IC= entries. SPICE and PSPICE are set up to calculate the initial capacitor voltages and inductor currents *unless* the UIC item is included in the .TRAN statement. But it is a good idea to get in the habit of explicitly including the initial conditions for the inductors and capacitors and using the UIC statement. In Chapter 7, we will find that we must include them.

The final item we need to examine is the specification of arbitrary waveforms that vary with time. We will consider the three most common forms: the piecewise linear waveform, the pulse, and the sinusoid. The piecewise linear waveform is shown in Fig. 5.25, and the source specification statement (for either an independent voltage source or an independent current source) is

$$\left.\begin{matrix}\text{VXXX}\\ \text{IXXX}\end{matrix}\right\} \quad \text{N1} \quad \text{N2} \quad \text{PWL(0}\ V_0\ t_1\ V_1\ t_2\ V_2\ t_3\ V_3 \cdots)$$

This draws straight lines between the points that are defined by the time and the value of the waveform at that time point. Observe that the function holds the last-named value and does not return it to zero. The next waveform is the periodic pulse given by

$$\left.\begin{matrix}\text{VXXX}\\ \text{IXXX}\end{matrix}\right\} \quad \text{N1} \quad \text{N2} \quad \text{PULSE(}V_1\ V_2\ \text{TD TR TF PW PERIOD)}$$

and is shown in Fig. 5.26. This waveform is periodic in that it repeats itself over intervals of length PERIOD. The delay is TD, the rise time is TR, the fall time is TF, the pulse width is PW. The last useful waveform is the sinusoid defined by

$$V_0 + V_A e^{-(t-\text{TD})\text{THETA}} \sin\left[2\pi\,\text{FREQ}(t - \text{TD})\right]$$

This represents a sinusoid that may have an exponentially decaying amplitude. The specification statement is

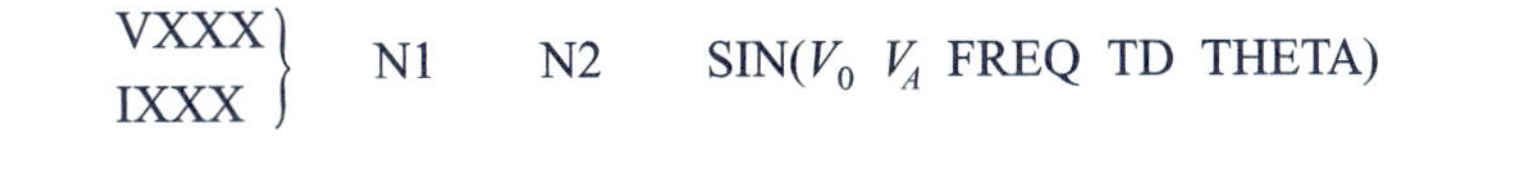

$$\left.\begin{matrix}\text{VXXX}\\ \text{IXXX}\end{matrix}\right\} \quad \text{N1} \quad \text{N2} \quad \text{SIN(}V_0\ V_A\ \text{FREQ TD THETA)}$$

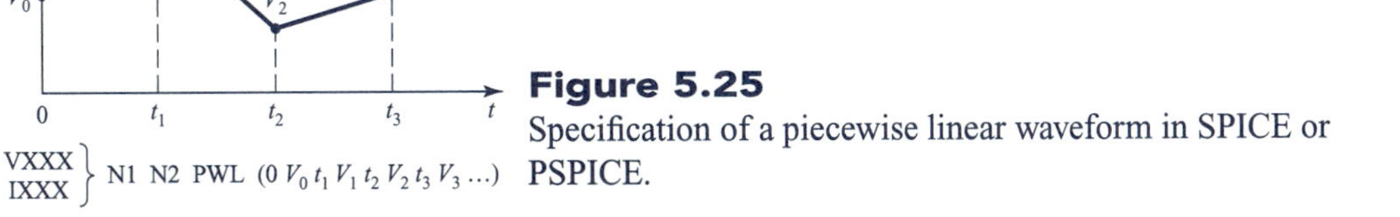

Figure 5.25
Specification of a piecewise linear waveform in SPICE or PSPICE.

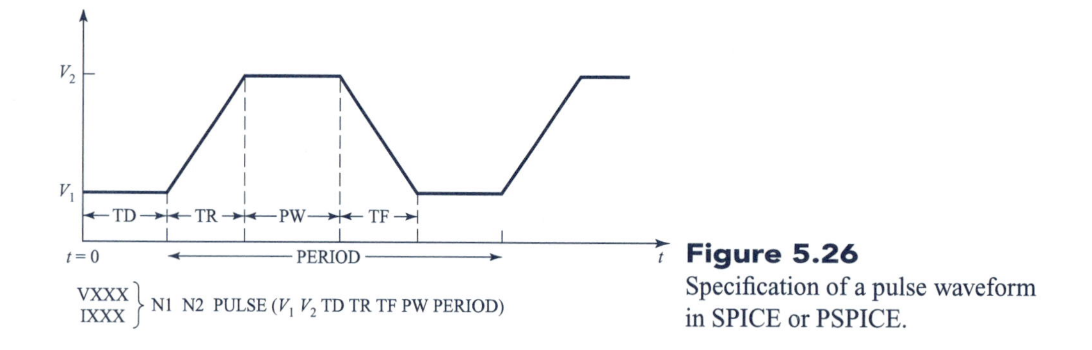

Figure 5.26
Specification of a pulse waveform in SPICE or PSPICE.

For example, in order to specify a sinusoid as 10 sin $(2\pi \times 10^3 t)$ we would write

```
VXXX }
IXXX }   N1     N2     SIN(0  10  1E3  0  0)
```

And last we describe one of the most important features in PSPICE that was not originally available in SPICE: PROBE. The probe command is simply

```
.PROBE
```

It acts like the probe of an oscilloscope or voltmeter in that any voltage or current in the circuit can be displayed versus time in a screen plot. Once the solution is completed, a time-axis screen is produced, and the user can then select which solution voltages or currents are to be displayed. This is an extraordinarily powerful feature of PSPICE in that very detailed plots of the time history of a solution can be obtained. The following examples show these features.

Example 5.9

Use PSPICE to plot the capacitor current for Exercise Problem E5.1 shown in Fig. E5.1.

Solution The circuit with nodes labeled is shown in Fig. 5.27, and the PSPICE program is

```
EXAMPLE 5.9
VS        1     0     PWL(0 0 1M 2 2M 0 3M -2 4M -2 5M 1 6M 0)
C         1     0     10U    IC=0
.TRAN     0.1M  6M    UIC
.PROBE
*THE CURRENT IS I(C)
.END
```

Figure 5.28*a* shows the voltage-source waveform, and Fig. 5.28*b* shows the capacitor current. Since

$$i_C(t) = (10\ \mu\text{F})\frac{dv_S(t)}{dt}$$

we can confirm the results by noting, for example, that from $t = 1$ ms to $t = 3$ ms, the slope of $v_S(t)$ is -2 V/ms, so that over this time interval the capacitor current should be $10\ \mu\text{F} \times (-2\ \text{V/ms}) = -20$ mA.

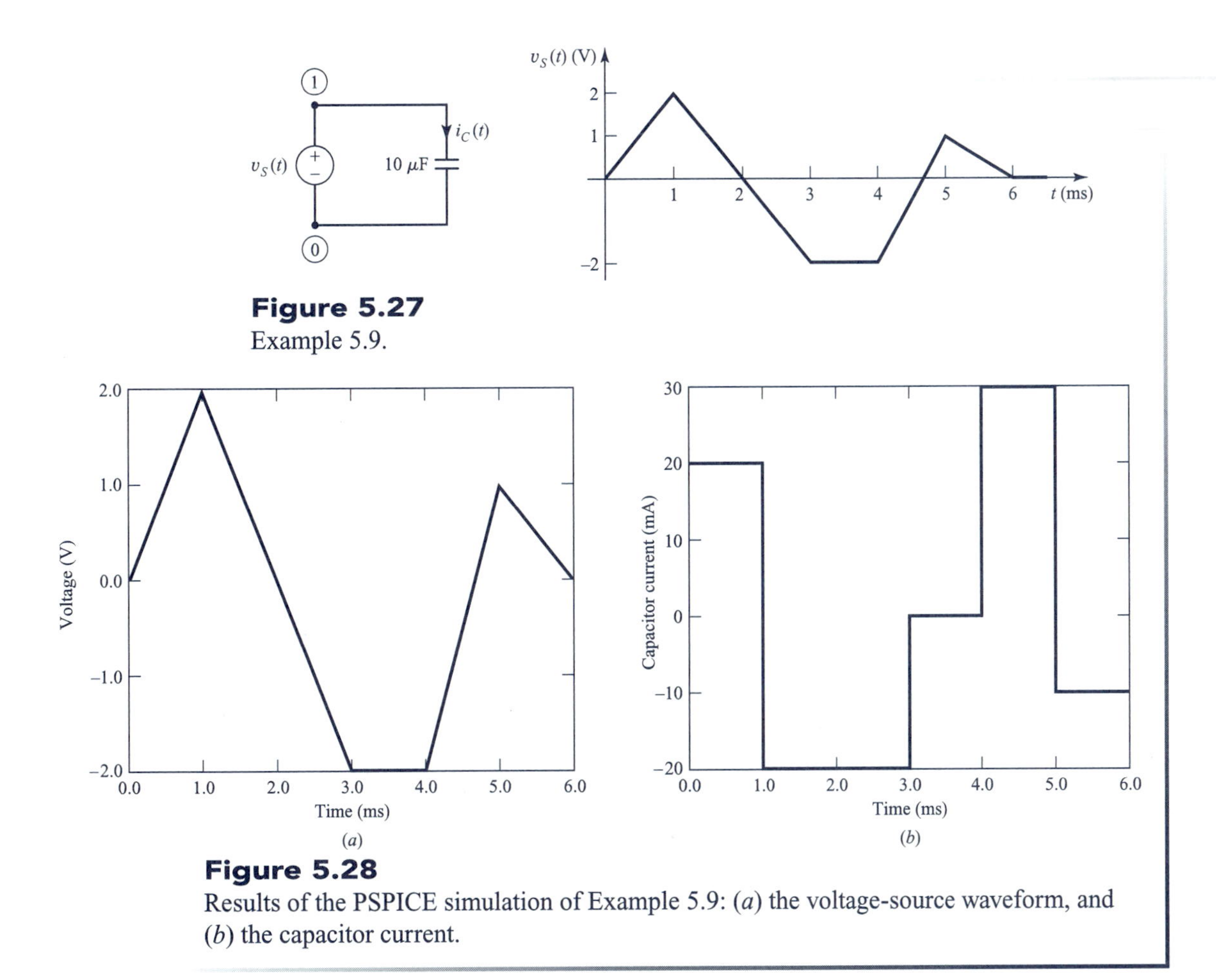

Figure 5.27
Example 5.9.

Figure 5.28
Results of the PSPICE simulation of Example 5.9: (*a*) the voltage-source waveform, and (*b*) the capacitor current.

There are two important restrictions on the use of SPICE and PSPICE. The first is that *every node in the circuit must have a dc path to the reference node.* Capacitors are treated as open circuits and inductors are treated as short circuits in order to determine whether this condition exists. The second is that *the circuit cannot contain a loop consisting solely of voltage sources and/or inductors, nor may it contain a supernode cut solely by current sources and/or capacitors.* The next example illustrates this restriction and a method for overcoming it.

Example 5.10

Replace the voltage source in the previous example with a current source having the same waveform, and use PSPICE to plot the capacitor voltage.

Solution The circuit with nodes labeled is shown in Fig. 5.29. Observe that this violates one of the above rules: there is a supernode at the top that is penetrated by only current sources and capacitors. If we coded this and submitted it to PSPICE, we would receive an error message. In order to get around this restriction, we add a resistor in parallel, and hence the supernode is not cut *solely* by current sources and capacitors. But this is not the original problem, so how do we choose the value of this resistor to give a solution that closely approximates the desired original solution? In Chapter 7 we will find that the answer is to choose R such that the *time constant RC is much larger than the time interval of interest.* Hence if we choose $R = 100\ \text{M}\Omega$, the time constant will be 1000 s. This is a bit of overkill, but as long as PSPICE encounters no numerical errors, it is acceptable. We can check this by

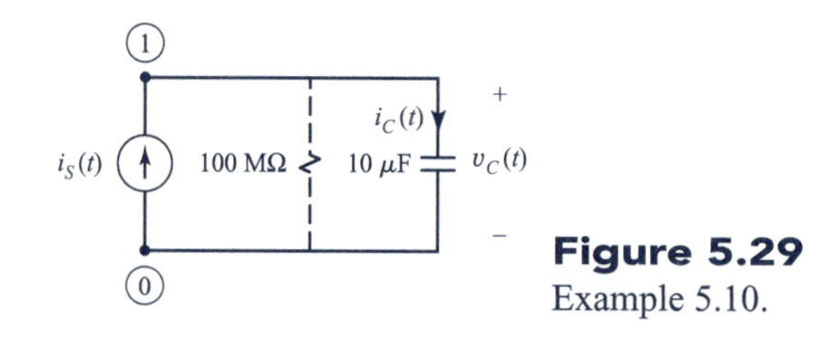

Figure 5.29
Example 5.10.

plotting the current through the capacitor to insure that it is approximately that of the current source, i.e., virtually no current goes through the resistor. The PSPICE program is

```
EXAMPLE 5.10
IS        0      1     PWL(0 0 1M 2 2M 0 3M -2  4M -2 5M 1 6M 0)
C         1      0     10U    IC=0
R         1      0     100MEG
.TRAN     0.1M   6M    UIC
.PROBE
*THE VOLTAGE IS V(1)
.END
```

Observe that we have simply changed the voltage source to a current source by replacing VS with IS. Observe also that this necessitates reordering the node numbers, since the current flow is from node 0 to node 1. The remaining PWL statement uses the same waveform as for the voltage source of the previous example. Figure 5.30*a* shows the capacitor voltage, which one can check using

$$v_C(t) = \frac{1}{C}\int_0^t i_S(\tau)\,d\tau$$

For example, at $t = 2$ ms, the area under the $i_S(t)$ curve is 2×10^{-3} A-s. Hence the voltage at $t = 2$ ms should be 2×10^{-3} A-s/10 μF $= 200$ V, as it is. To demonstrate that the added resistor is not affecting the solution, we plot the current through the capacitor in Fig. 5.30*b*, which is virtually identical to the current-source current.

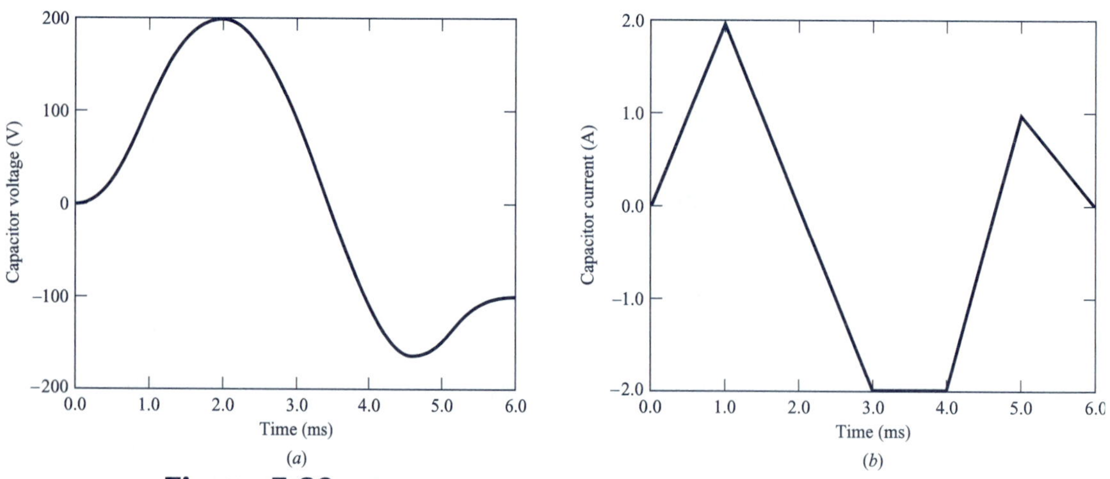

Figure 5.30
Results of the PSPICE simulation of Example 5.10: (*a*) the capacitor voltage, and (*b*) the capacitor current.

The next two examples show the duals to the above two problems using a 10-mH inductor and the same source waveforms.

Example 5.11

Use PSPICE to plot the inductor voltage in the circuit of Fig. 5.31.

Solution The circuit with the nodes labeled is shown, and the PSPICE program is

```
EXAMPLE 5.11
IS        0      1      PWL(0 0 1M 2 2M 0 3M -2 4M -2 5M 1 6M 0)
L         1      0      10M    IC=0
.TRAN     0.1M   6M     UIC
.PROBE
*THE VOLTAGE IS V(1)
.END
```

The voltage across the inductor is shown in Fig. 5.32*a* and can be confirmed from

$$v_L(t) = L\frac{di_S(t)}{dt}$$

For example, at $t = 2$ ms, the voltage is 10 mH $\times$ (-2 V/ms) = -20 V. The current through the inductor is shown in Fig. 5.32*b*; it is the same as the current-source current.

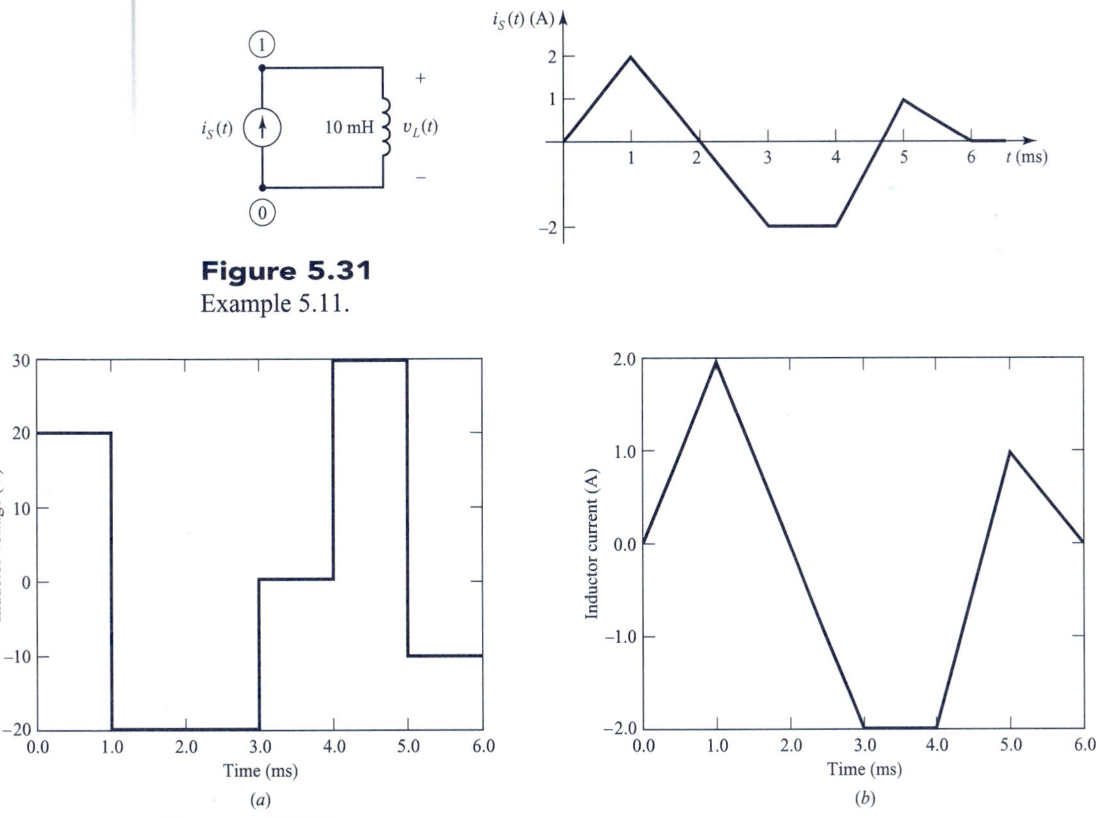

Figure 5.31
Example 5.11.

Figure 5.32
Results of the PSPICE simulation of Example 5.11: (*a*) the inductor voltage, and (*b*) the inductor current.

Example 5.12

Replace the current source in the previous example with a voltage source having the same waveform, and use PSPICE to plot the inductor current.

Solution The circuit with nodes labeled is shown in Fig. 5.33. Observe that this again violates one of the above rules: there is a loop consisting solely of voltage sources and inductors. In order to get around this restriction, we add a resistor in series to break the loop. But this is again not the original problem, so how do we choose the value of this resistor to give a solution that closely approximates the desired original solution? In Chapter 7 we will find that the answer is to choose R such that the *time constant L/R is much larger than the time interval of interest.* Hence if we choose $R = 1\ \mu\Omega$, the time constant will be 10,000 s. Again, this is a bit of overkill, but as long as PSPICE encounters no numerical errors, it is acceptable. We can check this by plotting the voltage across the inductor to insure that it is approximately that of the voltage source, i.e., virtually no voltage is dropped across the resistor. The PSPICE program is

```
EXAMPLE 5.12
VS        1      0      PWL(0 0 1M 2 2M 0 3M -2 4M -2 5M 1 6M 0)
L         2      0      10M    IC=0
R         1      2      1U
.TRAN     0.1M   6M     UIC
.PROBE
*THE CURRENT IS I(L)
.END
```

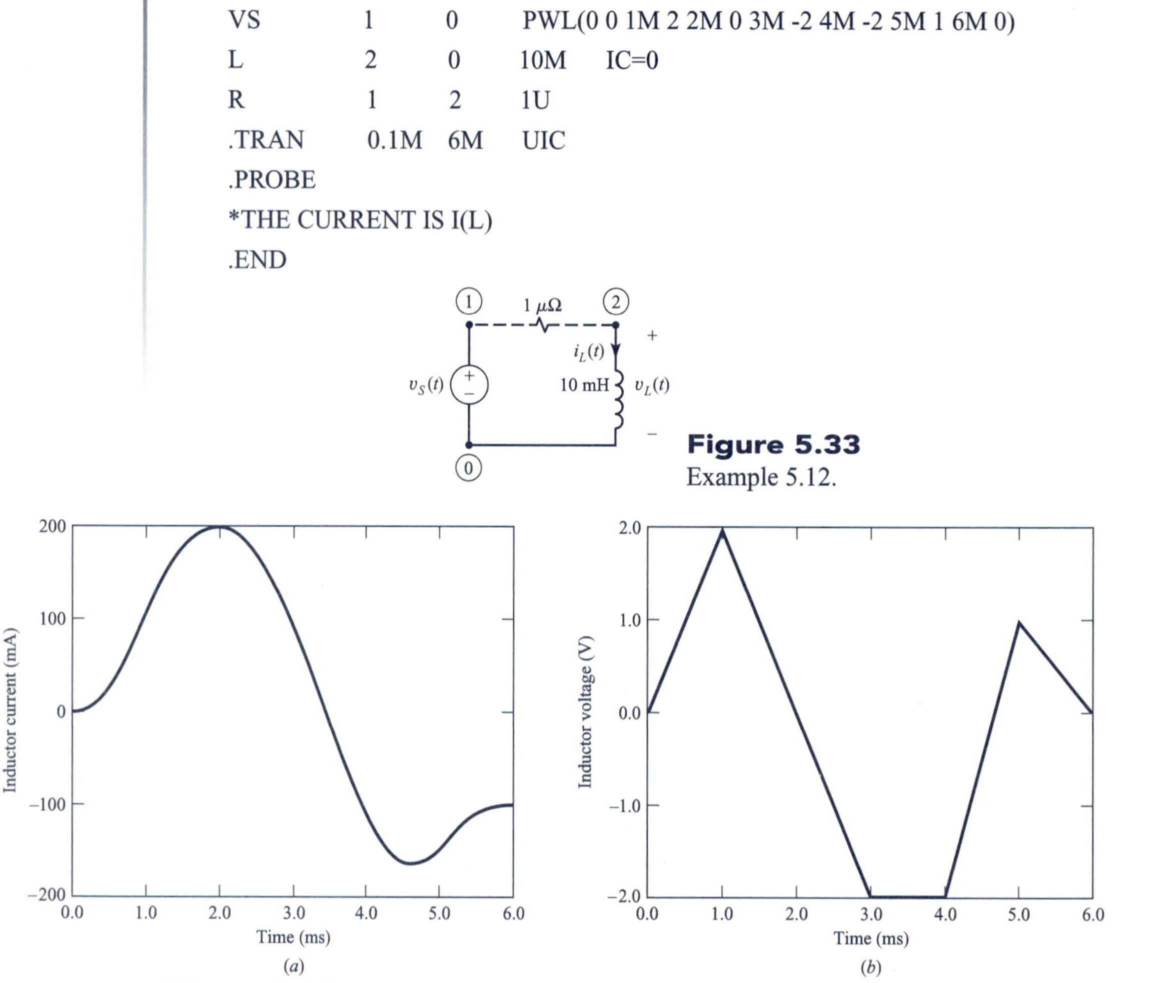

Figure 5.33
Example 5.12.

Figure 5.34
Results of the PSPICE simulation of Example 5.12: (*a*) the inductor current, and (*b*) the inductor voltage.

Figure 5.34a shows the inductor current, which one can check using

$$i_L(t) = \frac{1}{L}\int_0^t v_S(\tau)\, d\tau$$

For example, at $t = 2$ms, the area under the $v_S(t)$ curve is 2×10^{-3} V-s. Hence the voltage at $t = 2$ ms should be 2×10^{-3} V-s/10 mH = 200 mA, as it is. To demonstrate that the added resistor is not affecting the solution, we plot the voltage across the inductor in Fig. 5.34b; it is virtually identical to the voltage-source voltage.

As a final example we will combine these results.

Example 5.13

Plot the voltage across the current source in the circuit shown in Fig. 5.35.

Solution The nodes are numbered as shown. Observe that we have added a 100-kΩ resistor in parallel with the capacitor to break the current source-capacitor supernode. The PSPICE program is

```
EXAMPLE 5.13
IS        0     1     PWL(0 0 1U 5M 2U 0)
C         3     0     1U      IC=0
L         2     3     1U      IC=0
R         1     2     2
RFIX      3     0     100K
.TRAN     0.1U  3U    UIC
.PROBE
*THE VOLTAGE IS V(1)
.END
```

The result is shown in Fig. 5.36. The resistor voltage, the inductor voltage, and the capacitor voltage are plotted. The total voltage is, by KVL, the sum of these. The reader should verify that these are correct by computing, for example, the individual voltages at $t = 1$ µs. These are 10, 5, and 2.5 mV for a total (just before $t = 1$ µs) of 17.5 mV.

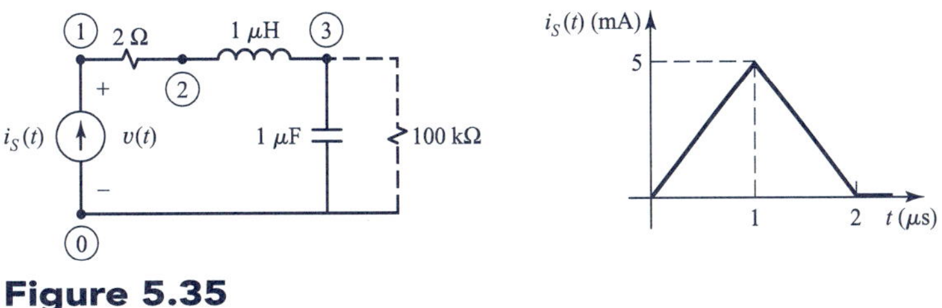

Figure 5.35
Example 5.13.

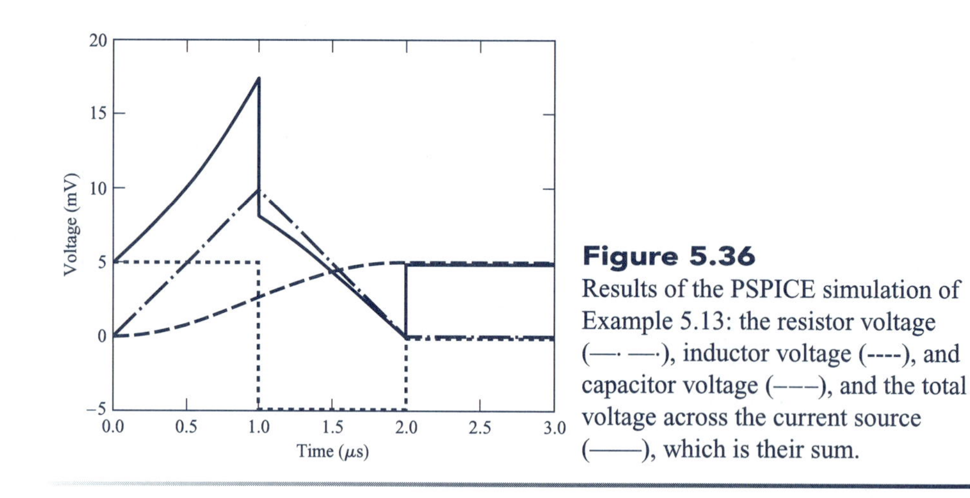

Figure 5.36
Results of the PSPICE simulation of Example 5.13: the resistor voltage (—· —·), inductor voltage (----), and capacitor voltage (–––), and the total voltage across the current source (——), which is their sum.

5.8 Application Examples

In this section we will give two practical applications of capacitors and inductors. The first is a circuit to measure time, and the second illustrates the nonideal effects of inductance of conductors in causing the power supply voltage to drop in digital logic circuits.

5.8.1 An Electronic Timer

Consider the integrator shown in Fig. 5.37. The output voltage was derived as

$$v_o(t) = -\frac{1}{RC}\int_{t_0}^{t} V_S\, dt + v_o(t_0)$$

If the input voltage V_S is constant (dc), then it can be removed from the integrand, leaving

$$v_o(t) = -\frac{V_S}{RC}(t - t_0) + v_o(t_0)$$

If we provide a switch across C that opens at t_0, then the initial output voltage is zero, since by the virtual short-circuit condition, the capacitor voltage equals the output voltage. Hence $v_o(t_0) = 0$. Therefore the output voltage is a measure of the time interval after the switch is opened. The output voltage can be read with a voltmeter and we have the ratio between that voltage and the elapsed time as

$$\frac{V_S}{RC} \qquad \text{(V/s)}$$

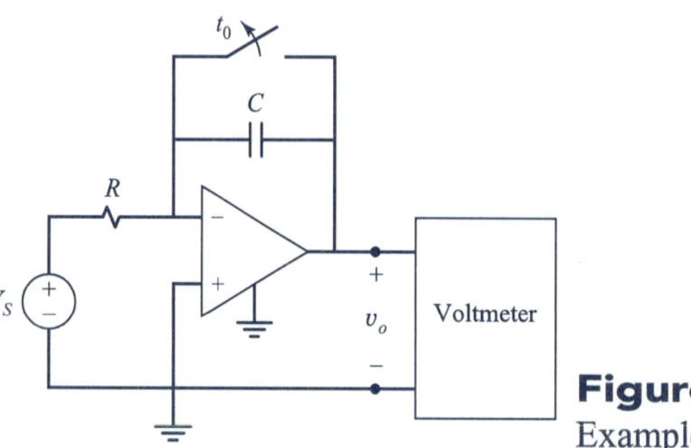

Figure 5.37
Example of a timer using an op-amp integrator.

Of course, because of the minus sign, this voltage is increasingly negative. Nevertheless suppose we wanted 1-V output to represent 1 ms. Hence we must choose

$$\frac{V_S}{RC} = \frac{1\text{ V}}{1\text{ ms}} = 10^3\text{ V/s}$$

A suitable design (one of many) would be to choose $V_S = 5$ V, $R = 5$ kΩ, $C = 1$ μF. A reading of −5 V on the voltmeter would indicate that 5 ms has elapsed.

5.8.2 DC Power Distribution Sag in Digital Logic Circuits

Digital logic circuits operate on two discrete voltage levels to form zeros and ones in encoding data in binary format. Typical transistor–transistor logic (TTL) uses levels of 0 V (zero) and 5 V (one). All logic gates are supplied with a 5-V dc power supply to power them. The voltage must be maintained or else the gate may not function properly. This dc voltage is distributed throughout the circuit on *lands,* which are rectangular-cross-section conductors. These conductors form loops on the printed circuit board (PCB) and hence possess inductance. Typical land inductances are on the order of 15 nH/in.

Consider the model of such a power distribution circuit shown in Fig. 5.38*a*. The distribution land, several inches in length, from the dc power supply to the +5 terminal of the logic gate may have an inductance of 40 nH. The two states of the logic gate are ON when it appears as an open circuit between the +5 terminal and ground and draws very little current, and OFF when it appears as a short circuit between the +5 terminal and ground. Typical TTL logic gates draw some 30 mA in the OFF state. Suppose this logic gate changes state by 30 mA in 1ns as shown. This current change flows through the land inductance and generates a voltage across the land inductance of

$$\begin{aligned} v_L &= L\frac{di_G}{dt} \\ &= (40\text{ nH})\frac{30\text{ mA}}{1\text{ ns}} \\ &= 1.2\text{ V} \end{aligned}$$

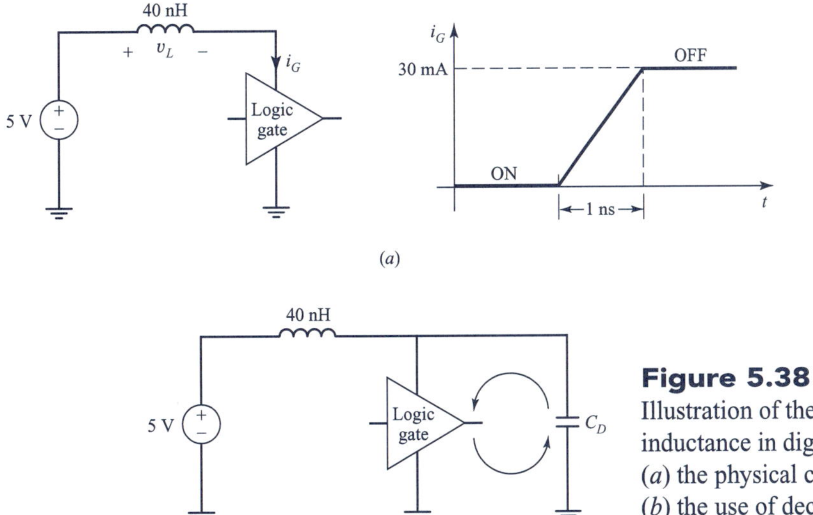

Figure 5.38
Illustration of the effect of land inductance in digital circuits: (*a*) the physical configuration, and (*b*) the use of decoupling capacitors to mitigate the effect of land inductance.

This subtracts from the 5 V of the dc power supply, so that the voltage supplied to the gate drops to $5 - 1.2 = 3.8$ V. Hence the logic gate may not perform correctly for this low a supplied voltage.

In order to mitigate the effect of the land inductance, *decoupling capacitors* C_D are placed across and very close to the power-supply terminals of the logic gate as shown in Fig. 5.38*b*. During quiescent conditions, charge is slowly transferred from the dc power supply to that capacitor. During switching, that charge is drawn from the decoupling capacitor through the logic gate. But since the lands connecting the decoupling capacitor to the logic-gate power-supply terminals are very short, there is very little inductive voltage drop across them. Hence the voltage supplied to the power-supply terminals remains nearly +5V during switching. We can approximately determine the required value of C_D for this problem. Suppose the power supply voltage at the logic-gate terminals is allowed to sag only 0.1 V. Hence using (5.9) we have

$$\begin{aligned} 0.1 &= \frac{1}{C}\int_0^{1\text{ns}} (30 \times 10^6 t)\, dt \\ &= \frac{1}{C}[15 \times 10^{-12}] \end{aligned}$$

giving a required value of $C_D = 150$ pF.

Problems

Section 5.1 The Capacitor

5.1-1 Sketch the current $i(t)$ in the circuit of Fig. P5.1-1. *Answers*: $-2 < t < 0, i = 0.3$ A; $0 < t < 1, i = -0.8$ A; $1 < t < 2, i = 0$ A; $2 < t < 3, i = 0.4$ A; $t > 3, i = 0$ A.

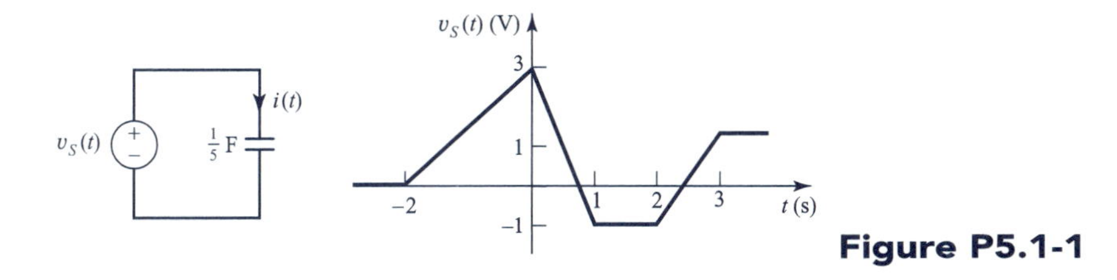

Figure P5.1-1

5.1-2 Sketch the current $i(t)$ in the circuit of Fig. P5.1-2. *Answers*: $0 < t < 2, i = 0.75$ A; $2 < t < 3, i = -1.5$ A; $3 < t, i = 0$ A.

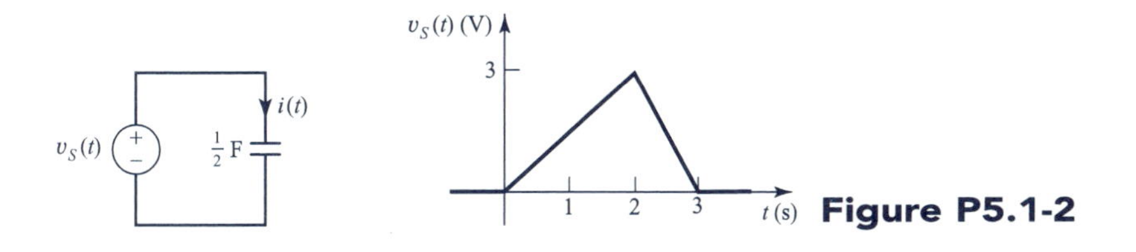

Figure P5.1-2

5.1-3 Sketch the current $i(t)$ in the circuit of Fig. P5.1-3. *Answers*: $-1 < t < 0, i = \frac{1}{3}t + \frac{5}{6}$A; $0 < t < 2, i = -\frac{1}{3}t - \frac{1}{6}$A; $2 < t < 4, i = -\frac{1}{3}$A; $4 < t < 5, i = \frac{1}{3}t - \frac{7}{6}$A; $t > 5, i = 0$ A.

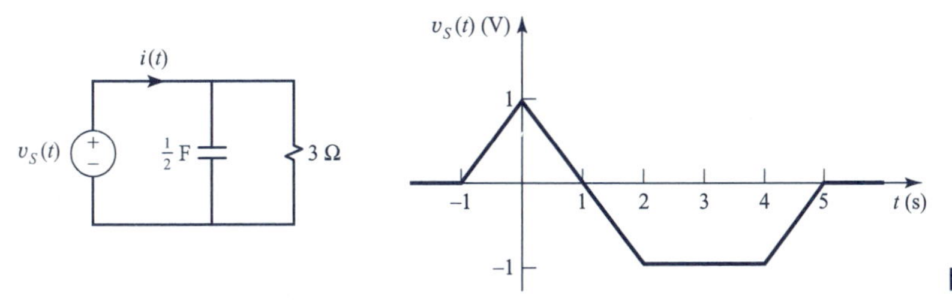

Figure P5.1-3

5.1-4 Sketch the voltage $v(t)$ in the circuit of Fig. P5.1-4. *Answers*: $0 < t < 1, v = 4t$ V; $1 < t < 2, v = 8t - 4$ V; $2 < t < 3, v = -8t + 28$ V; $t > 3, v = 4$ V.

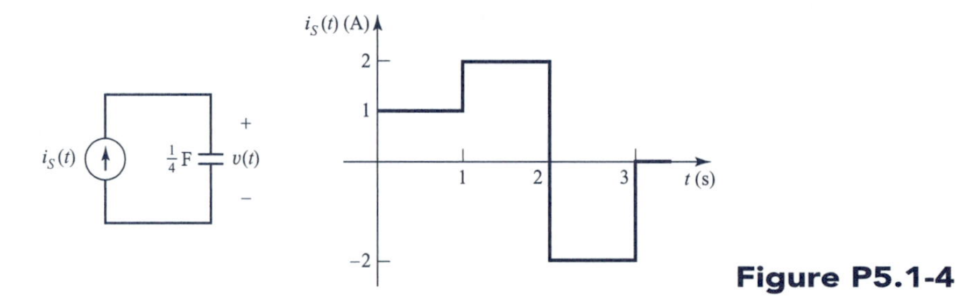

Figure P5.1-4

5.1-5 Sketch the voltage $v(t)$ in the circuit of Fig. P5.1-5. *Answers*: $-1 < t < 0, v = 3t + 3$ V; $0 < t < 1, v = -\frac{3}{2}t + 3$ V; $1 < t < 3, v = \frac{3}{2}$ V; $3 < t < 5, v = \frac{3}{2}t - 3$ V; $t > 5, v = \frac{9}{2}$ V.

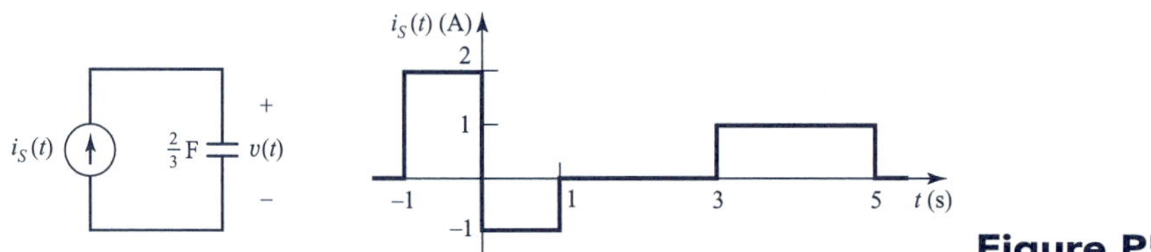

Figure P5.1-5

5.1-6 Sketch the voltage $v(t)$ in the circuit of Fig. P5.1-6. *Answers*: $0 < t < 1, v = 6t + 4$ V; $1 < t < 2, v = -3t + 7$ V; $t > 2, v = 3$ V.

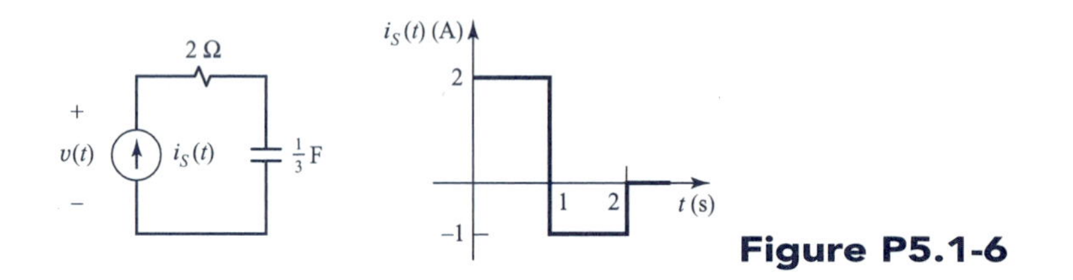

Figure P5.1-6

5.1-7 Determine the equivalent capacitance C_{eq} at terminals *ab* in Fig. P5.1-7. *Answer:* $\frac{28}{15}$ F.

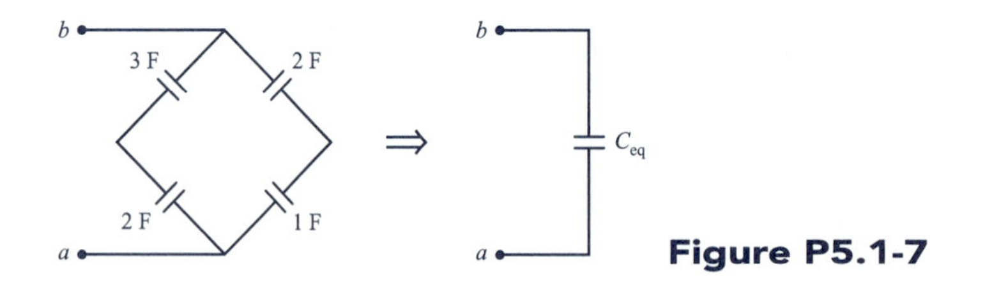

Figure P5.1-7

5.1-8 Determine the equivalent capacitance C_{eq} at terminals *ab* in Fig. P5.1-8. *Answer:* $\frac{29}{64}$ F.

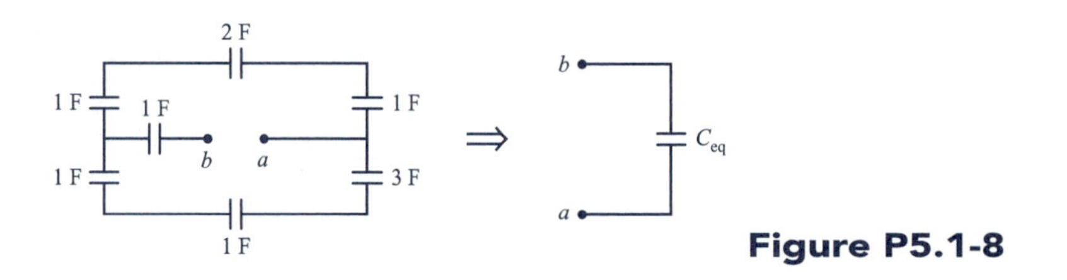

Figure P5.1-8

5.1-9 Determine the equivalent capacitance C_{eq} at terminals *ab* in Fig. P5.1-9. *Answer:* $\frac{34}{37}$ F.

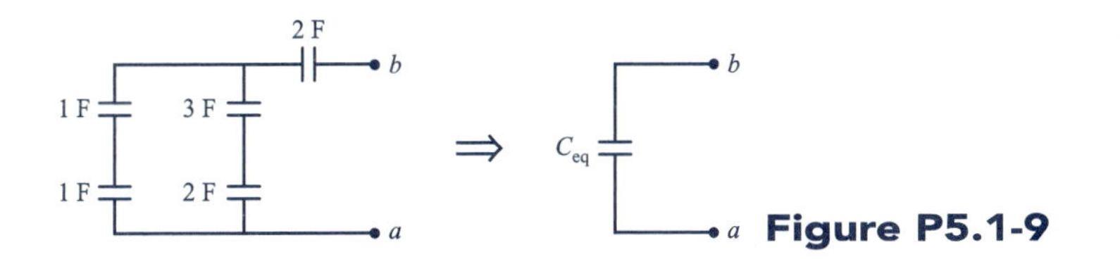

Figure P5.1-9

Section 5.2 The Inductor

5.2-1 Sketch the voltage $v(t)$ in the circuit of Fig. P5.2-1. *Answers:* $0 < t < 1$, $v = 2$ V; $1 < t < 2$, $v = 0$ V; $2 < t < 4$, $v = -2$ V; $4 < t < 5$, $v = 2$ V; $t > 5$, $v = 0$ V.

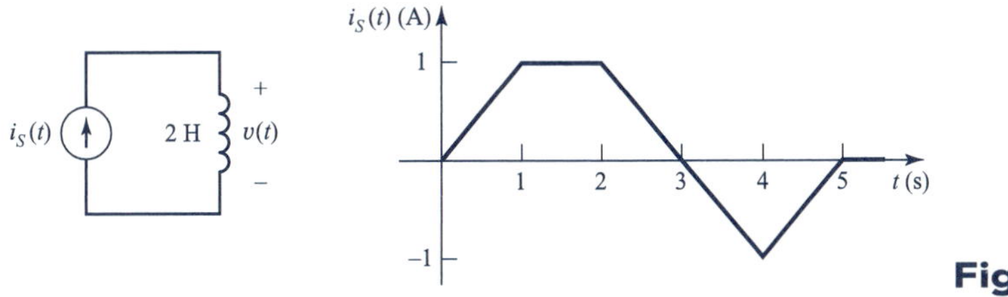

Figure P5.2-1

5.2-2 Sketch the voltage $v(t)$ in the circuit of Fig. P5.2-2. *Answers:* $-1 < t < 1$, $v = 2$ V; $1 < t < 2$, $v = -8$ V; $2 < t < 3$, $v = 0$ V; $3 < t < 4$, $v = 4$ V; $t > 4$, $v = 0$ V.

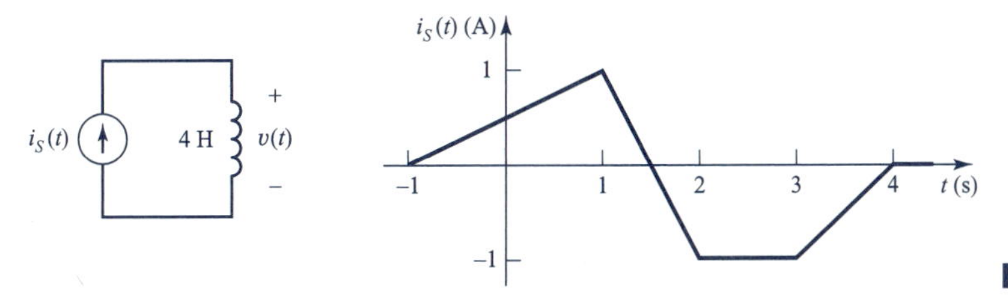

Figure P5.2-2

5.2-3 Sketch the voltage $v(t)$ in the circuit of Fig. P5.2-3. *Answers:* $-1 < t < 0$, $v = 4t + 8$ V; $0 < t < 3$, $v = -2t + 2$ V; $3 < t < 4$, $v = 2t - 6$ V; $t > 4$, $v = 0$ V.

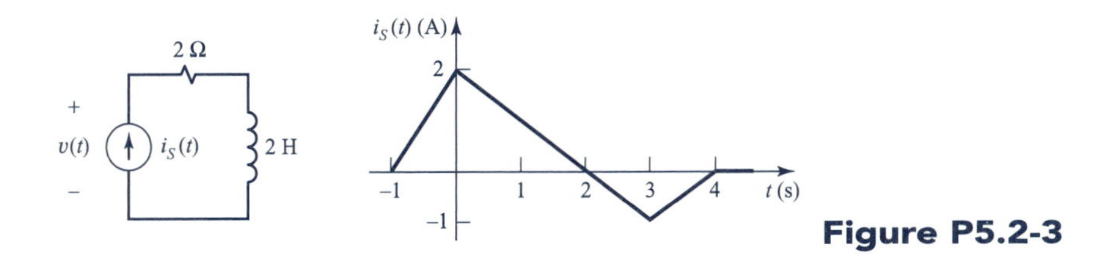

Figure P5.2-3

5.2-4 Sketch the current $i(t)$ in the circuit of Fig. P5.2-4. *Answers:* $-1 < t < 0$, $i = 4t + 4$ A; $0 < t < 1$, $i = 2t + 4$ A; $1 < t < 3$, $i = -2t + 8$ A; $t > 3$, $i = 2$ A.

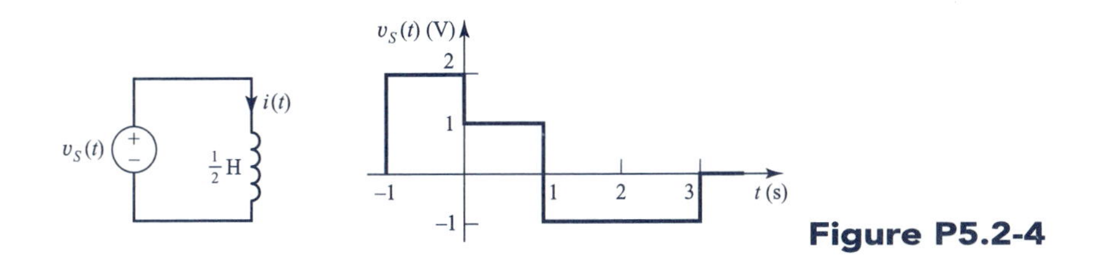

Figure P5.2-4

5.2-5 Sketch the current $i(t)$ in the circuit of Fig. P5.2-5. *Answers:* $-1 < t < 0, i = \frac{1}{2}t + \frac{1}{2}$ A; $0 < t < 1, i = -\frac{t}{4} + \frac{1}{2}$ A; $1 < t < 3, i = \frac{1}{4}$ A; $3 < t < 5, i = \frac{1}{4}t - \frac{1}{2}$ A; $t > 5, i = \frac{3}{4}$ A.

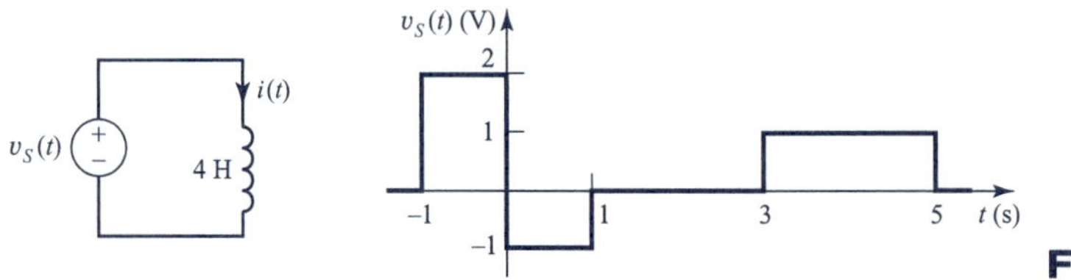

Figure P5.2-5

5.2-6 Sketch the current $i(t)$ in the circuit of Fig. P5.2-6. *Answers:* $0 < t < 1$, $i = 1 + t + \frac{1}{2}t^2$ A; $1 < t < 2, i = t - t^2/2$ A; $t > 2, i = 1$ A.

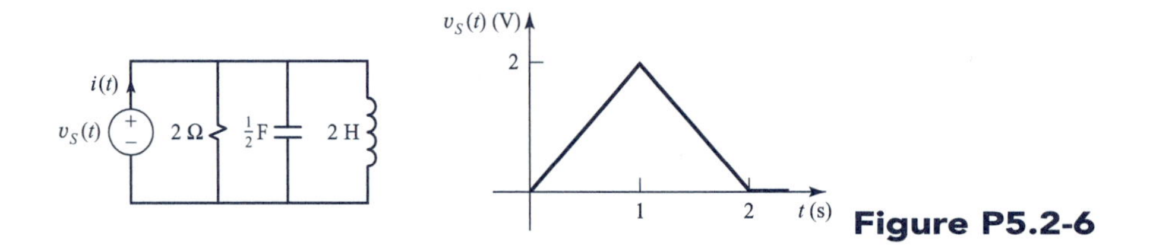

Figure P5.2-6

5.2-7 Determine the equivalent inductance L_{eq} at terminals ab in Fig. P5.2-7. *Answer:* $\frac{12}{7}$ H.

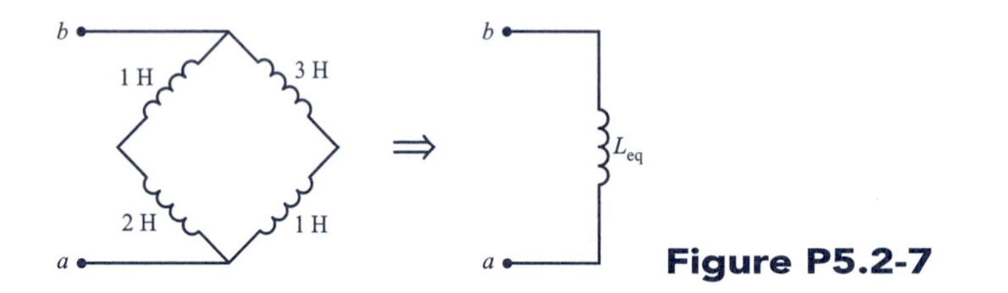

Figure P5.2-7

5.2-8 Determine the equivalent inductance L_{eq} at terminals ab in Fig. P5.2-8. *Answer:* $\frac{22}{5}$ H.

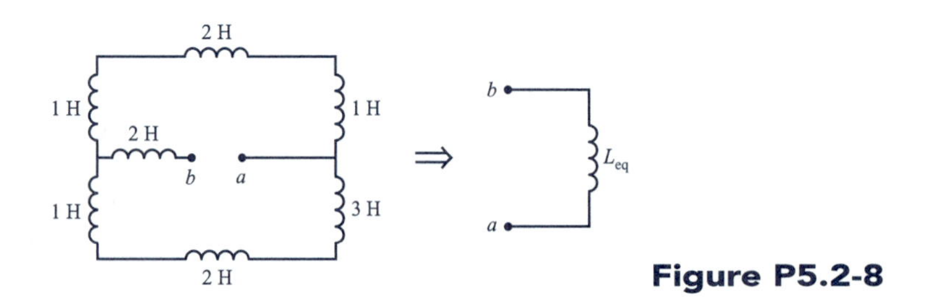

Figure P5.2-8

5.2-9 Determine the equivalent inductance L_{eq} at terminals ab in Fig. P5.2-9. *Answer:* $\frac{24}{7}$ H.

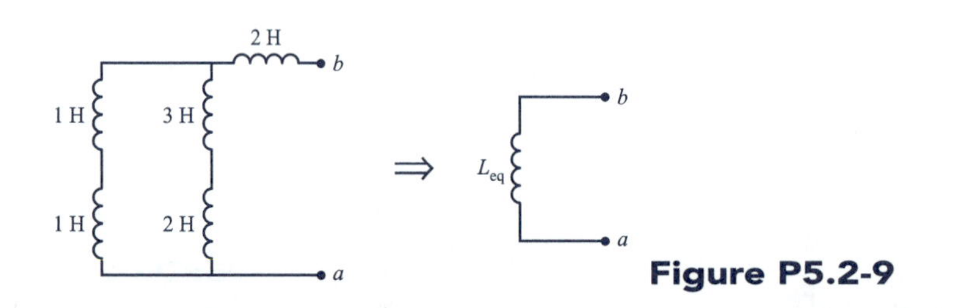

Figure P5.2-9

Section 5.3 Mutual Inductance

5.3-1 Write the terminal voltage–current relations for the coupled inductors of Fig. P5.3-1.
Answers: $v_1 = -L_1\, di_1/dt + M\, di_2/dt$, $v_2 = -M\, di_1/dt + L_2\, di_2/dt$.

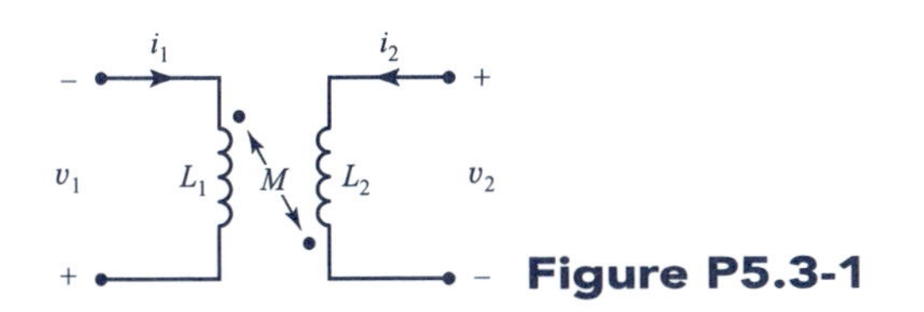

Figure P5.3-1

5.3-2 Write the terminal voltage–current relations for the coupled inductors of Fig. P5.3-2.
Answers: $v_1 = -L_1\, di_1/dt - M\, di_2/dt$, $v_2 = -M\, di_1/dt - L_2\, di_2/dt$.

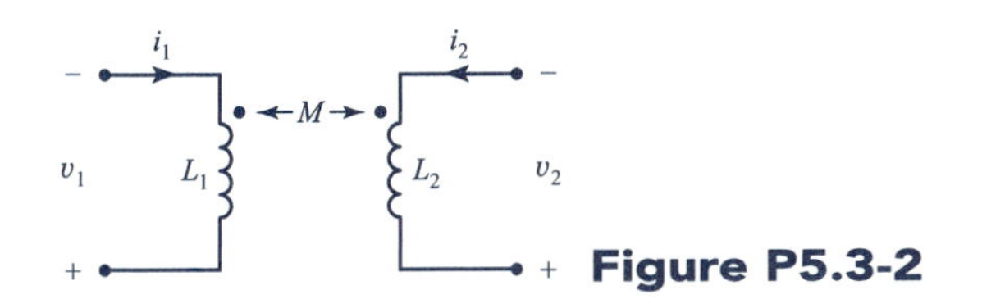

Figure P5.3-2

5.3-3 Write the terminal voltage–current relations for the coupled inductors of Fig. P5.3-3.
Answers: $v_1 = -L_1\, di_1/dt + M_{12}\, di_2/dt + M_{13}\, di_3/dt$, $v_2 = -M_{12}\, di_1/dt + L_2\, di_2/dt + M_{23}\, di_3/dt$, $v_3 = M_{13}\, di_1/dt - M_{23}\, di_2/dt - L_3\, di_3/dt$.

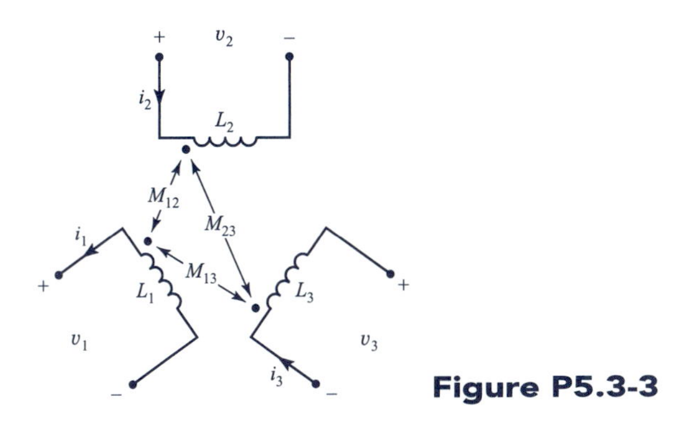

Figure P5.3-3

5.3-4 Write the terminal voltage–current relations for the coupled inductors of Fig. P5.3-4.
Answers: $v_1 = -L_1\, di_1/dt + M_{12}\, di_2/dt - M_{13}\, di_3/dt$, $v_2 = -M_{12}\, di_1/dt + L_2\, di_2/dt - M_{23}\, di_3/dt$, $v_3 = M_{13}\, di_1/dt - M_{23}\, di_2/dt + L_3\, di_3/dt$.

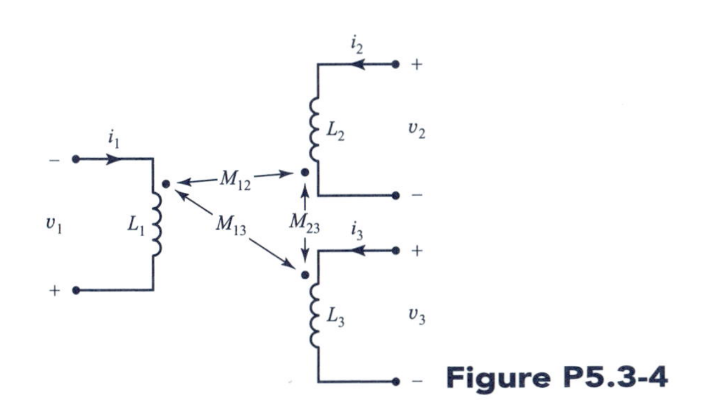

Figure P5.3-4

5.3-5 Determine voltage $v_{out}(t)$ in the circuit of Fig. P5.3-5. *Answer:* $-0.81 \sin 3t$ V.

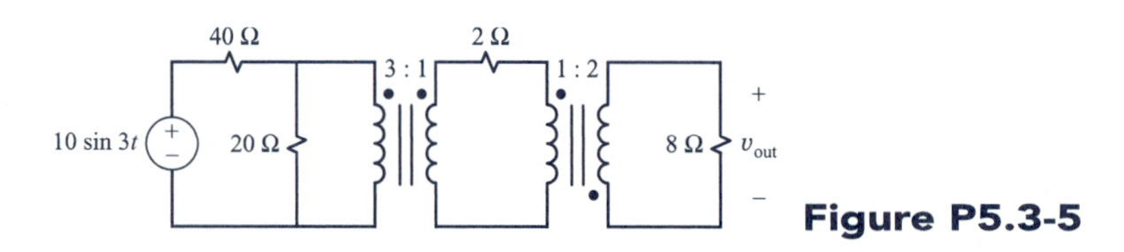

Figure P5.3-5

Section 5.4 Response of the Energy Storage Elements to DC Sources

5.4-1 Determine current I in the circuit of Fig. P5.4-1. *Answer:* 6 A.

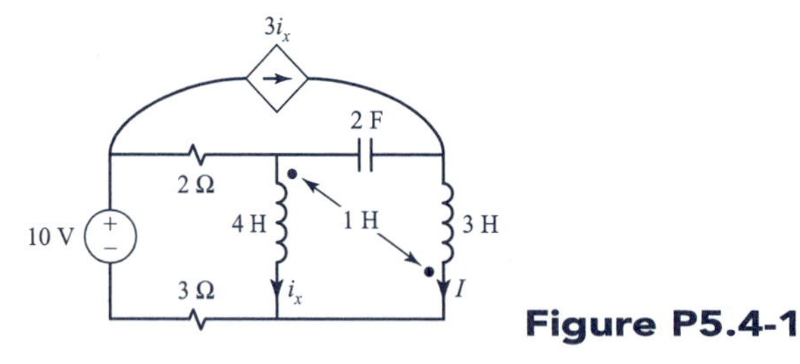

Figure P5.4-1

5.4-2 Determine current I in the circuit of Fig. P5.4-2. *Answer:* $\frac{4}{5}$ A.

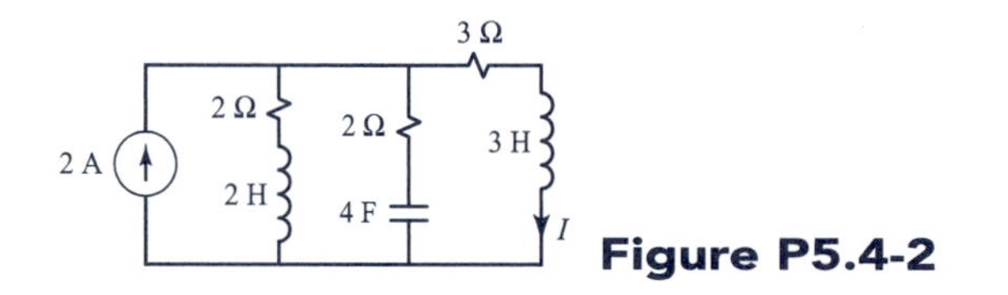

Figure P5.4-2

5.4-3 Determine current I in the circuit of Fig. P5.4-3. *Answer:* -2 A.

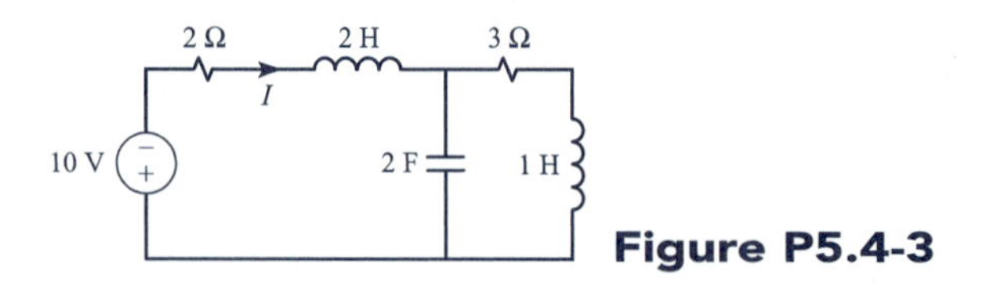

Figure P5.4-3

5.4-4 Determine current I in the circuit of Fig. P5.4-4. *Answer:* 2 A.

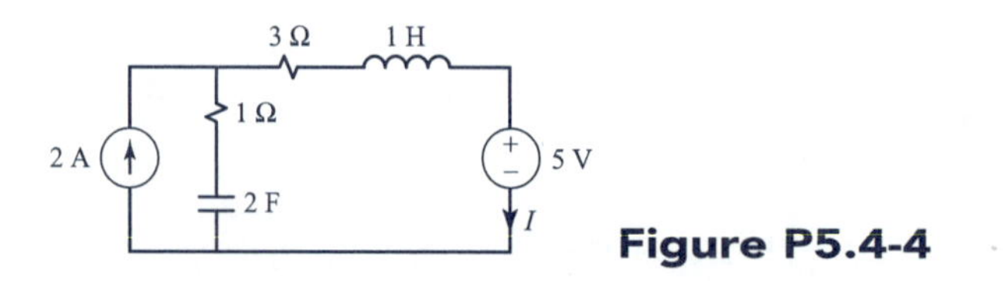

Figure P5.4-4

Section 5.5 The Differential Equations of a Circuit

5.5-1 Determine the differential equation governing current $i(t)$ in the circuit of Fig. P5.5-1.
Answer: $d^2i(t)/dt^2 + 2\, di(t)/dt + \frac{5}{2}i(t) = 5 \sin 2t$.

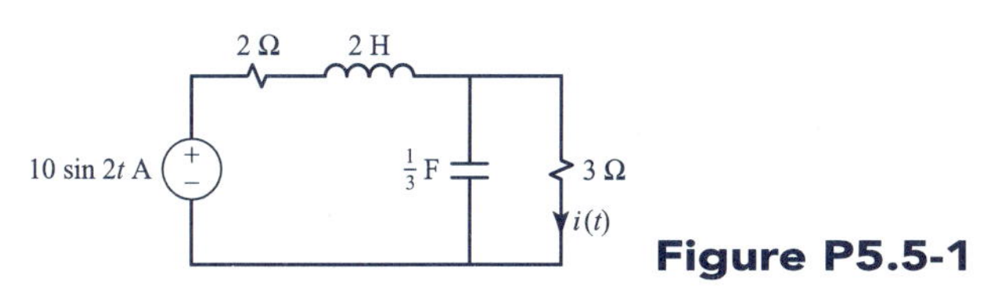

Figure P5.5-1

5.5-2 Determine the differential equation governing current $i(t)$ in the circuit of Fig. P5.5-2.
Answer: $d^2i(t)/dt^2 + 4\, di(t)/dt + 2i(t) = 0$.

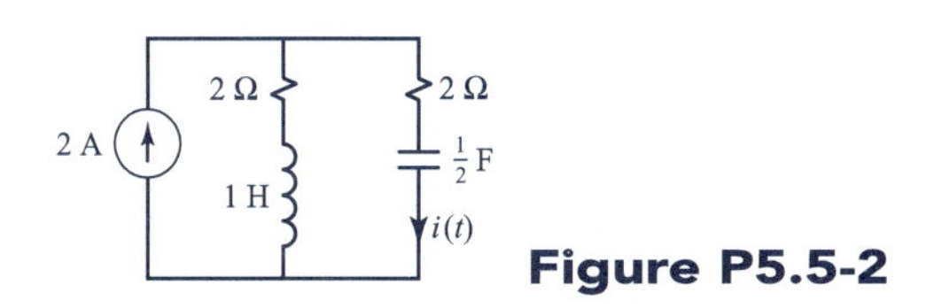

Figure P5.5-2

5.5-3 Determine the differential equation governing current $i(t)$ in the circuit of Fig. P5.5-3.
Answer: $d^2i(t)/dt^2 + \frac{3}{2} di(t)/dt + \frac{3}{2} i(t) = 8$.

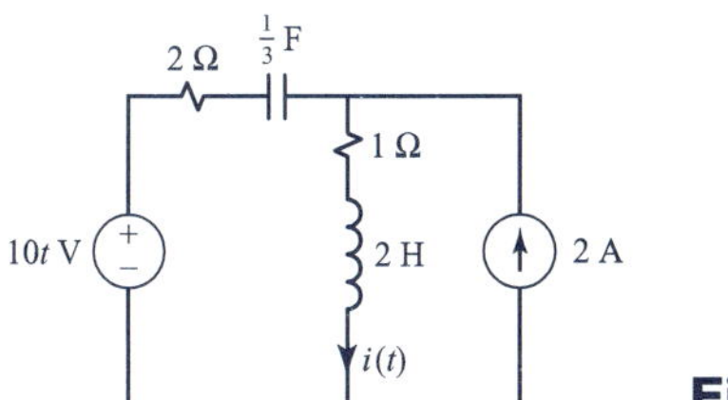

Figure P5.5-3

5.5-4 Determine the differential equation governing current $i(t)$ in the circuit of Fig. P5.5-4.
Answer: $d^2i(t)/dt^2 + di(t)/dt + 6i(t) = 6 \cos 2t$.

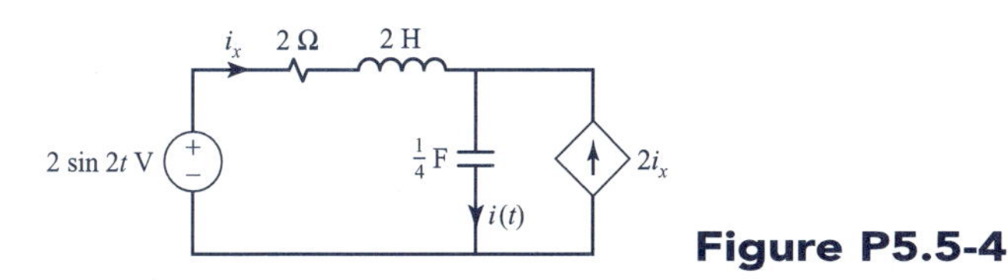

Figure P5.5-4

Section 5.6 The Op-Amp Differentiator and Integrator

5.6-1 Sketch the output voltage for the circuit of Fig. P5.6-1. *Answers:* $0 < t < 1$ ms, $v_{out} = -(10^6/2)t^2$ V; 1 ms $< t <$ 2 ms, $v_{out} = -1000t + 0.5$ V; 2 ms $< t <$ 3 ms, $v_{out} = 2.5 - 3000t + (10^6/2)t^2$; 3 ms $< t$, $v_{out} = -2$ V.

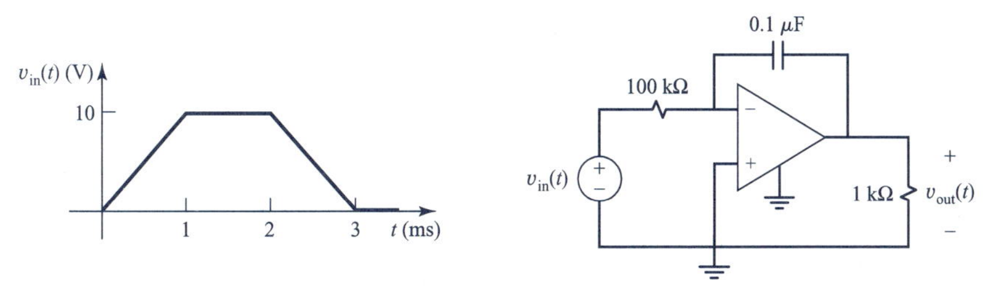

Figure P5.6-1

5.6-2 Sketch the output voltage for the circuit of Fig. P5.6-2. *Answers:* $0 < t < 1$ μs, $v_{out} = -(10^8/2)t^2$ V; 1 μs $< t <$ 3 μs, $v_{out} = 75 \times 10^{-6} - 150t + 25 \times 10^6 t^2$ V; 3 μs $< t$, $v_{out} = -150$ μV.

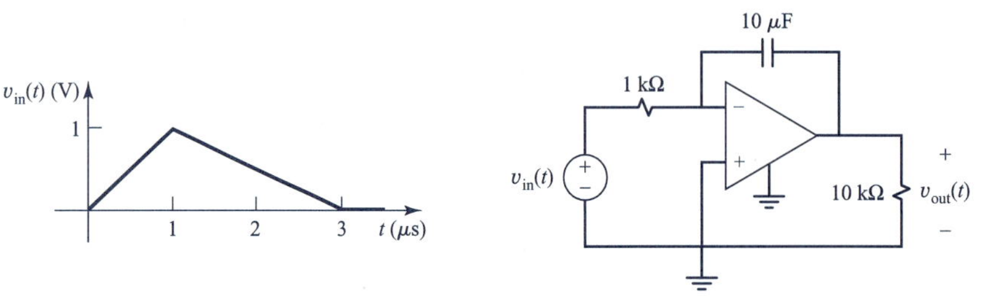

Figure P5.6-2

5.6-3 Interchange the resistor and the capacitor in the circuit of Fig. P5.6-1, and sketch the output voltage. *Answers:* $0 < t < 1$ ms, $v_{out} = -100$ V; 1 ms $< t <$ 2 ms, $v_{out} = 0$ V; 2 ms $< t <$ 3 ms, $v_{out} = 100$ V; 3 ms $< t$, $v_{out} = 0$ V.

5.6-4 Interchange the resistor and the capacitor in the circuit of Fig. P5.6-2, and sketch the output voltage. *Answers:* $0 < t < 1$ μs, $v_{out} = -10$ kV; 1 μs $< t <$ 3 μs, $v_{out} = 5$ kV; 3 μs $< t$, $v_{out} = 0$ V.

Section 5.7 PSPICE Applications

5.7-1 Use PSPICE to plot the current $i(t)$ in the circuit of Fig. P5.1-1.

5.7-2 Use PSPICE to plot the current $i(t)$ in the circuit of Fig. P5.1-2.

5.7-3 Use PSPICE to plot the current $i(t)$ in the circuit of Fig. P5.1-3.

5.7-4 Use PSPICE to plot the voltage $v(t)$ in the circuit of Fig. P5.1-4.

5.7-5 Use PSPICE to plot the voltage $v(t)$ in the circuit of Fig. P5.1-5.

5.7-6 Use PSPICE to plot the voltage $v(t)$ in the circuit of Fig. P5.1-6.

5.7-7 Use PSPICE to plot the voltage $v(t)$ in the circuit of Fig. P5.2-1.

5.7-8 Use PSPICE to plot the voltage $v(t)$ in the circuit of Fig. P5.2-2.

5.7-9 Use PSPICE to plot the voltage $v(t)$ in the circuit of Fig. P5.2-3.

5.7-10 Use PSPICE to plot the current $i(t)$ in the circuit of Fig. P5.2-4.

5.7-11 Use PSPICE to plot the current $i(t)$ in the circuit of Fig. P5.2-5.

5.7-12 Use PSPICE to plot the current $i(t)$ in the circuit of Fig. P5.2-6.

CHAPTER 6

Sinusoidal Excitation of Circuits

A linear circuit composed of linear elements (resistors, controlled sources, capacitors, and inductors) can be analyzed using the principle of superposition. Suppose a linear circuit has the following three types of independent sources: (a) dc sources, (b) sinusoidal sources, and (c) sources having arbitrary time variation. We may determine, using superposition, the solutions (responses) for each element current and voltage as the sums of the corresponding responses of each current and voltage from three circuits, one containing only the dc sources, another containing the sinusoidal sources, and a third containing the sources having arbitrary time variation. We determined in the previous chapter a method for determining the portion of the solutions due to dc sources: replace capacitors with open circuits and inductors with short circuits, and solve the resulting resistive circuit. In this chapter we will develop a method for determining the portions of the responses due to sources having sinusoidal time variation. This method, the phasor method, will reduce the desired circuit to one containing complex-valued resistors. All of our previous resistive-circuit analysis methods (resistors in series and in parallel, voltage and current division, the direct method, source transformation, superposition, Thevenin and Norton equivalent circuits, and node-voltage and mesh-current equations) may be applied to solve those circuits with only the added burden of manipulating complex numbers. In the next chapter we will examine a method, the Laplace transform, for similarly solving circuits but whose sources have arbitrary time variation. This method will also reduce the circuit to one having resistorlike elements, and our resistive-circuit analysis methods can again be brought to bear in obtaining a solution. Hence our earlier resistive-circuit analysis methods and skills are extremely important and will be used over and over in various guises to solve circuits.

6.1 The Sinusoidal Source

The sinusoidal function represents a very important time variation that is useful in representing many practical sources. Figure 6.1*a* shows a sinusoid that has a sine variation, $A \sin \omega t$. This sinusoidal function achieves a maximum of A at values of $\omega t = \ldots, \pi/2, 5\pi/2, \ldots$ and a value of $-A$ at values of $\omega t = \ldots, -\pi/2, 3\pi/2, \ldots$. The function is zero at $\omega t = \ldots, -\pi, 0, \pi, 2\pi, \ldots$. The quantity ω is called the *radian frequency* of the source, and its units are radians per second (rad/s). The *cyclic frequency*, f, is related to the radian frequency by $\omega = 2\pi f$, and its units are cycles per second (cycles/s) or hertz (Hz): 1 Hz = 1 cycle/s. A cosine waveform, $A \cos \omega t$, is shown in Fig. 6.1*b* and achieves a maximum of A at values of $\omega t = \ldots, 0, 2\pi, \ldots$ and a minimum of $-A$ at values of $\omega t = \ldots, -\pi, \pi, \ldots$. The function is zero at $\omega t = \ldots, -\pi/2, \pi/2, 3\pi/2, \ldots$. The argument of the sine and cosine functions, ωt, has the units of radians. We often refer to it in degrees, where 2π

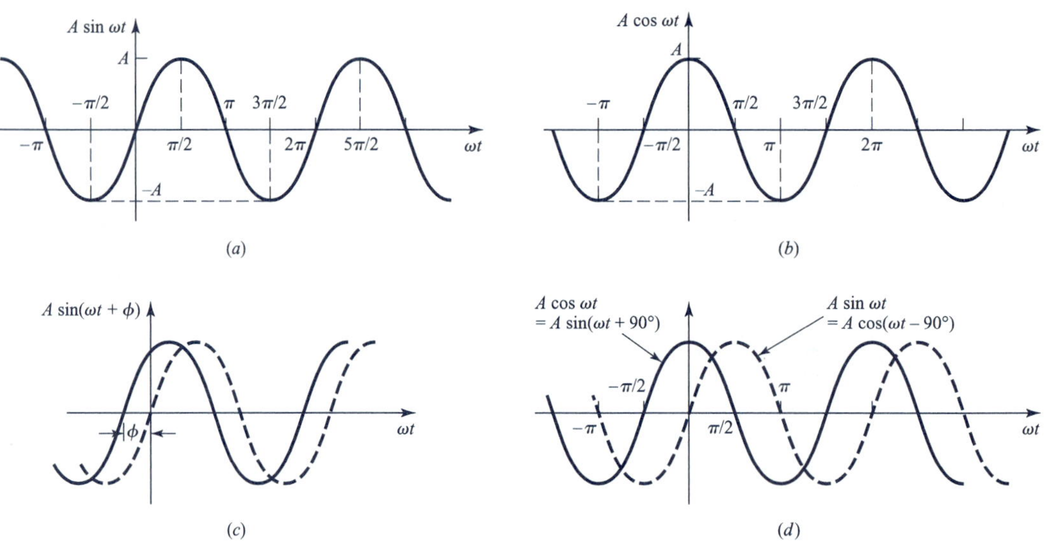

Figure 6.1
Illustration of sinusoidal waveforms: (*a*) $A \sin \omega t$, (*b*) $A \cos \omega t$, (*c*) $A \sin(\omega t + \phi)$, and (*d*) the relation between sin and cos: $A \sin \omega t = A \cos(\omega t - 90°)$, $A \cos \omega t = A \sin(\omega t + 90°)$.

radians equals 360°, so that 1 radian equals 57.3°, and π radians equals 180°. Note that the maxima and the minima of both functions are separated by 2π radians, as are all other corresponding points.

Adding a phase angle ϕ to the argument of the sine function shifts it *backward* by that phase angle, as indicated in Fig. 6.1*c*. This is easy to see and remember because of the following observation. Note that points on two sinusoidal waveforms are identical where the arguments of the sinusoidal functions are identical, i.e., $\sin \phi_1 = \sin \phi_2$ where $\phi_1 = \phi_2$. So adding ϕ to ωt to change $\sin \omega t$ to $\sin(\omega t + \phi)$ means that we must shift $\sin \omega t$ *backward* by ϕ in order to track corresponding points on the two waveforms. For example, $\sin \omega t = 0$ at $\omega t = 0$, so that $\sin(\omega t + \phi) = 0$ at $\omega t + \phi = 0$, that is, at $\omega t = -\phi$. The same observation applies to the cosine function. Consequently, the sine and cosine functions are related by either adding or subtracting $90° = \pi/2$ rad to their arguments as illustrated in Fig. 6.1*d*:

$$\cos \omega t = \sin(\omega t + 90°) \tag{6.1a}$$

$$\sin \omega t = \cos(\omega t - 90°) \tag{6.1b}$$

In addition, it is apparent that

$$-\sin \omega t = \sin(\omega t \pm 180°) \tag{6.2a}$$

$$-\cos \omega t = \cos(\omega t \pm 180°) \tag{6.2b}$$

Certain other trigonometric identities can also prove useful:

$$\sin(\omega t \pm \phi) = \sin \omega t \cos \phi \pm \cos \omega t \sin \phi \tag{6.3a}$$

$$\cos(\omega t \pm \phi) = \cos \omega t \cos \phi \mp \sin \omega t \sin \phi \tag{6.3b}$$

It is also important to remember that

$$\sin(-\phi) = -\sin\phi \tag{6.4a}$$

$$\cos(-\phi) = \cos\phi \tag{6.4b}$$

Example 6.1

Convert the cosine function $10\cos(\omega t - 60^\circ)$ to a sine function, and give its representation in terms of sine and cosine functions that have no phase angle.

Solution Converting to the sine function, we add 90° according to (6.1a):

$$\begin{aligned} 10\cos(\omega t - 60^\circ) &= 10\sin(\omega t - 60^\circ + 90^\circ) \\ &= 10\sin(\omega t + 30^\circ) \end{aligned}$$

Using (6.3b), we obtain

$$\begin{aligned} 10\cos(\omega t - 60^\circ) &= 10\cos\omega t\cos 60^\circ + 10\sin\omega t\sin 60^\circ \\ &= \underbrace{10\cos 60^\circ}_{5}\cos\omega t + \underbrace{10\sin 60^\circ}_{8.66}\sin\omega t \\ &= 5\cos\omega t + 8.66\sin\omega t \end{aligned}$$

Exercise Problem 6.1

Convert the following cosine or sine functions to an equivalent sine or cosine form: $\cos(\omega t + 60^\circ)$, $\sin(\omega t + 135^\circ)$, $\cos(\omega t + 230^\circ)$, and $\sin(\omega t - 30^\circ)$.

Answers: $\sin(\omega t + 150^\circ)$, $\cos(\omega t - 120^\circ)$, $\cos(\omega t + 45^\circ)$, and $\sin(\omega t - 40^\circ)$.

Exercise Problem 6.2

Convert the following forms to one that contains only a cosine or only a sine function: $3\cos\omega t - 4\sin\omega t$, $3\cos\omega t - 2\sin\omega t$, and $\cos\omega t - 2\sin\omega t$.

Answers: $3.61\sin(\omega t + 123.69^\circ)$, $2.236\cos(\omega t + 63.43^\circ)$, and $5\cos(\omega t + 53.13^\circ)$.

6.1.1 Representation of General Waveforms via the Fourier Series

The sine and cosine functions are *periodic functions,* which repeat themselves over intervals of length $\omega t = 2\pi$. A general periodic function satisfies the property that

$$x(t) = x(t \pm nT) \tag{6.5}$$

where T is the *period* of the waveform. A general periodic waveform is shown in Fig. 6.2. There are a number of important periodic waveforms associated with various independent voltage and current sources. It is possible to represent any periodic waveform (6.5) alternatively as an infinite

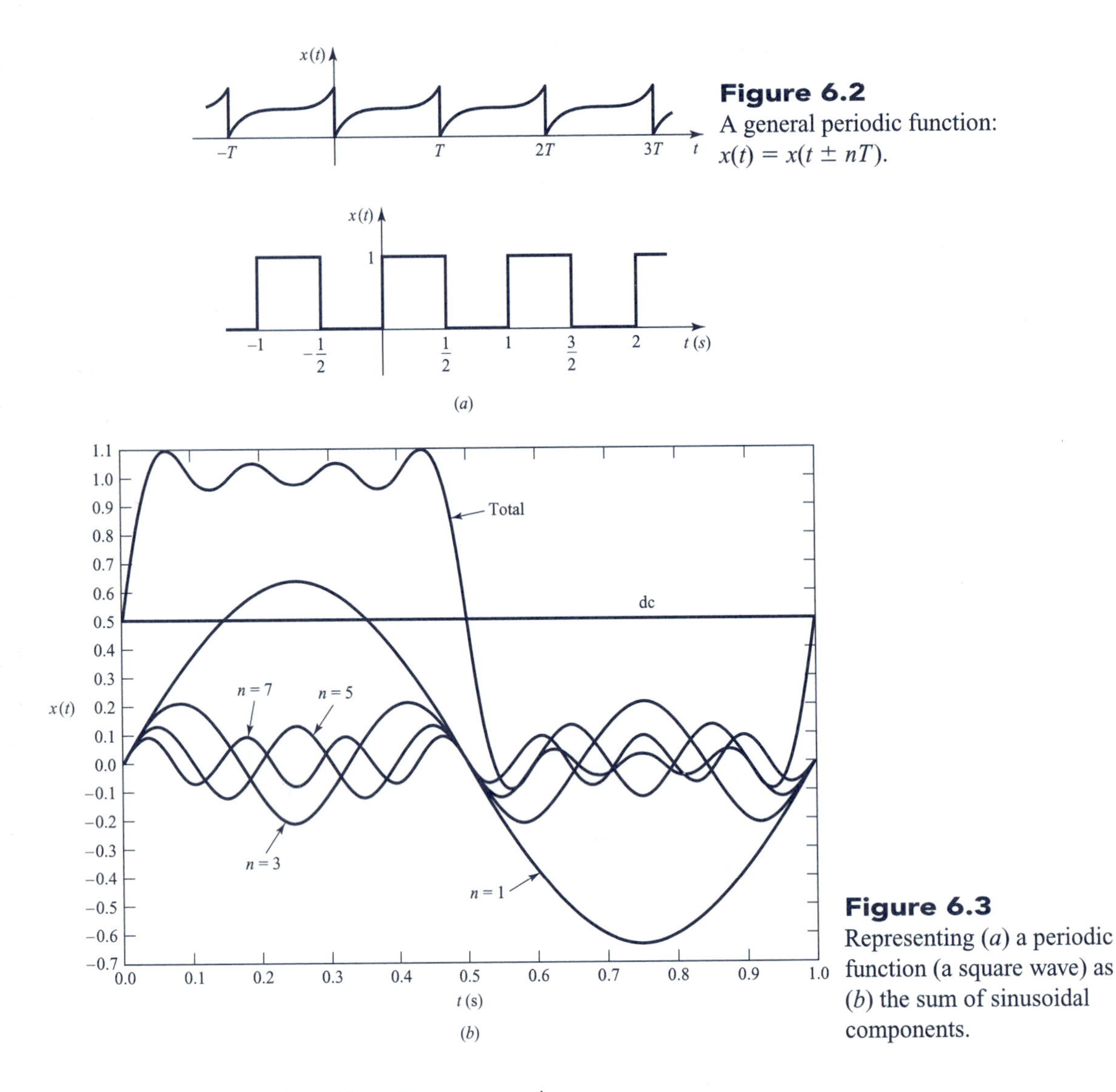

Figure 6.2 A general periodic function: $x(t) = x(t \pm nT)$.

Figure 6.3 Representing (*a*) a periodic function (a square wave) as (*b*) the sum of sinusoidal components.

sum of sinusoids: a *Fourier series*:*

$$\begin{aligned} x(t) = a_0 &+ a_1 \cos \omega_0 t + a_2 \cos 2\omega_0 t + a_3 \cos 3\omega_0 t + \cdots \\ &+ b_1 \sin \omega_0 t + b_2 \sin 2\omega_0 t + b_3 \sin 3\omega_0 t + \cdots \end{aligned} \tag{6.6}$$

where the coefficients $a_0, a_1, a_2, a_3, \ldots, b_1, b_2, b_3, \ldots$ are to be determined for the specific waveform to be represented. Note that the Fourier series represents the periodic waveform as the sum of cosine and sine functions whose radian frequencies are integer multiples of the repetition frequency of the waveform, $\omega_0 = 2\pi f_0$ and $f_0 = 1/T$.

For example, Fig. 6.3*a* shows a periodic *square wave* function that has unit amplitude and period $T = 1$s. The Fourier series of this waveform can be obtained as*

$$x(t) = \underbrace{\frac{1}{2}}_{a_0} + \underbrace{\frac{2}{\pi}}_{b_1} \sin 2\pi t + \underbrace{\frac{2}{3\pi}}_{b_3} \sin 6\pi t + \underbrace{\frac{2}{5\pi}}_{b_5} \sin 10\pi t + \underbrace{\frac{2}{7\pi}}_{b_7} \sin 14\pi t + \cdots$$

*C.R. Paul, *Analysis of Linear Circuits*, McGraw-Hill, 1989.

Figure 6.3*b* shows the result of combining the first seven terms, a_0, b_1, b_3, b_5, b_7. The coefficients of the even-numbered sine frequencies, b_2, b_4, b_6, ..., are zero, as are all the cosine coefficients, a_1, a_2, a_3, ... Increasing the number of sine terms used to represent the function provides a better representation, although theoretically an infinite number of terms must be used to provide an exact representation.

Virtually any *nonperiodic* function may similarly be represented as a sum of sinusoidal components with the Fourier transform.* Hence, with a few pathological exceptions that are not of practical interest, *any voltage or current source waveform may be represented as the sum of sinusoids even though the waveform is not sinusoidal.* This is illustrated in Fig. 6.4. Figure 6.4*b* shows that, according to (6.6), we may replace the source with the sum (series for a voltage source and parallel for a current source) of a dc component, a_0, and sine and cosine sources whose frequencies are multiples of the basic repetition frequency of the original periodic waveform of the source. Hence we arrive at the most important reason we investigate the response of circuits to sinusoidal waveforms:

> *Once we replace the waveform that has arbitrary time variation with the sum of sinusoidal sources representing the Fourier components of the waveform, we may apply superposition to determine the response to the original waveform as the sum of the responses to the individual sinusoidal components of that waveform.*

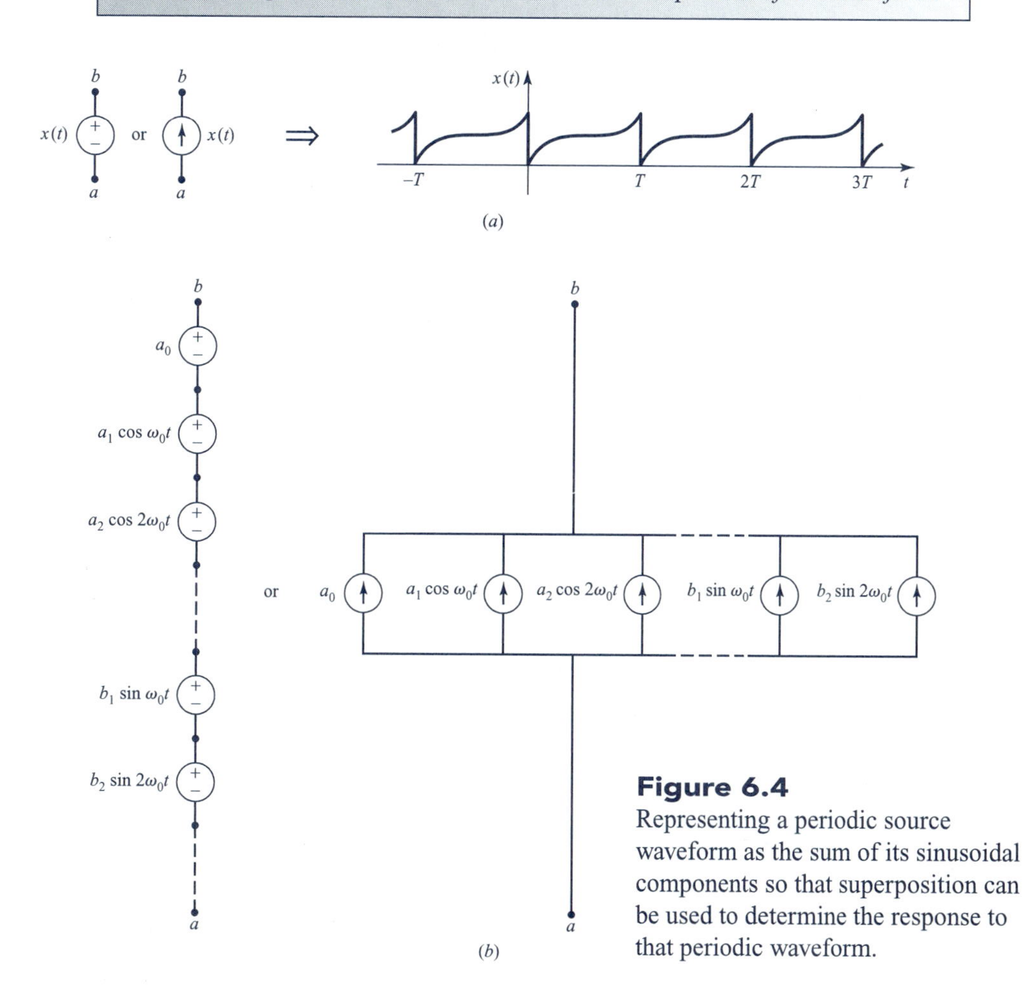

Figure 6.4
Representing a periodic source waveform as the sum of its sinusoidal components so that superposition can be used to determine the response to that periodic waveform.

*C.R. Paul, *Analysis of Linear Circuits,* McGraw-Hill, 1989.

Therefore, *if we can determine the response of a circuit to a sinusoidal source, we can, using Fourier methods, determine the response to any arbitrary source waveform using superposition.* It is for this primary reason that this chapter will be devoted to the analysis of circuits that contain sinusoidal sources.

6.1.2 Response of Circuits to Sinusoidal Sources

As an introduction to the *phasor method* which we will develop for analyzing the response of circuits to sinusoidal sources, consider the circuit shown in Fig. 6.5. We wish to determine the solution for the current $i(t)$. Writing KVL around the single loop circuit gives

$$2\,\frac{di(t)}{dt} + 3i(t) = 10 \sin 3t$$

Let us guess the *form* of the solution. First try

$$i(t) = I \sin 3t$$

where the coefficient I is to be determined. Substituting gives

$$6I \cos 3t + 3I \sin 3t = 10 \sin 3t$$

Matching coefficients of the sin and cos terms on both sides, we see that this cannot be satisfied for any choice of the coefficient I. Hence we make another choice for the form of the solution:

$$i(t) = I_c \cos 3t + I_s \sin 3t$$

Substituting into the differential equation gives

$$-6I_c \sin 3t + 6I_s \cos 3t + 3I_c \cos 3t + 3I_s \sin 3t = 10 \sin 3t$$

Matching the coefficients of the sin and cos terms on both sides yields the following two equations for the coefficients I_c and I_s:

$$\begin{aligned} 3I_s - 6I_c &= 10 \\ 6I_s + 3I_c &= 0 \end{aligned}$$

Solving these gives $I_s = 0.67$ and $I_c = -1.33$, so that the solution becomes

$$i(t) = 0.67 \sin 3t - 1.33 \cos 3t \text{ A}$$

This can alternatively be written, using the previous trigonometric identities, as

$$i(t) = 1.49 \sin(3t - 63.43°) \text{ A}$$

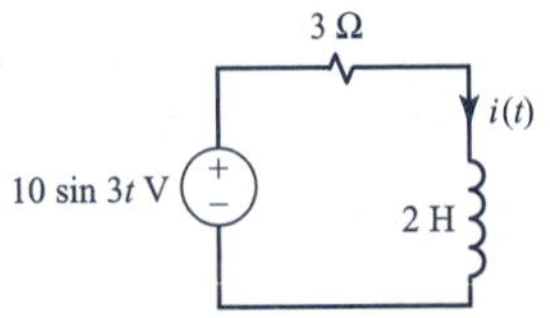

Figure 6.5
An example of a circuit that is driven by a sinusoidal source.

Although this procedure has obtained the solution, there are some important observations to be made:

1. The response $i(t)$ has the same frequency as the source, $\omega = 3$. This will be true for the response of any other linear circuit to a sinusoidal source.
2. The response $i(t)$ differs from the source as well as other currents and voltages only in magnitude (1.49) and phase ($-63.43°$). This will also be true for the response of any other linear circuit to a sinusoidal source. Hence *our task will be to determine the magnitude and the phase angle of the currents and voltages, since they will be the same trigonometric function as for the source (sin or cos) and their frequencies, ω, will be the same as for the source.*
3. Determining the required response consists in
   a. writing the differential equation relating the response to the source, and
   b. solving two equations in two unknowns in order to determine the solution coefficients, I_c and I_s.

In the phasor method to be developed in this chapter we will devise a simple way of determining the magnitude and phase of the solution. In addition, the method will avoid the two problems in observation (3) above: (a) writing the differential equation relating the response to the source and (b) solving two equations in two unknowns for the solution coefficients. The method will represent the circuit as one having complex-valued resistors so that we may use all of our previous resistive circuit analysis techniques with only the added burden of complex arithmetic.

6.2 Complex Numbers, Complex Algebra, and Euler's Identity

As indicated, the phasor method to be developed uses the algebra of complex numbers. Hence we need to become familiar with the manipulation of such numbers.

A complex number consists of two parts: a real part and an imaginary part. Complex numbers will be denoted with a caret (ˆ) throughout this text. For example, the complex number $\hat{C}$ may be represented by

$$\hat{C} = a + jb \tag{6.7}$$

where a and b are real numbers and j symbolizes the square root of -1:

$$j = \sqrt{-1} \tag{6.8}$$

The *real part* of $\hat{C}$ is denoted as

$$a = \text{Re}\,\hat{C} \tag{6.9a}$$

and the *imaginary part* is denoted as

$$b = \text{Im}\,\hat{C} \tag{6.9b}$$

The complex number $\hat{C}$ can be represented as a vector in the two-dimensional *complex plane* by plotting the real part a on the horizontal axis and the imaginary part b on the vertical axis, as shown in Fig. 6.6. The length of this vector is the magnitude of $\hat{C}$ and is denoted by

$$C = |\hat{C}| = \sqrt{a^2 + b^2} \tag{6.10a}$$

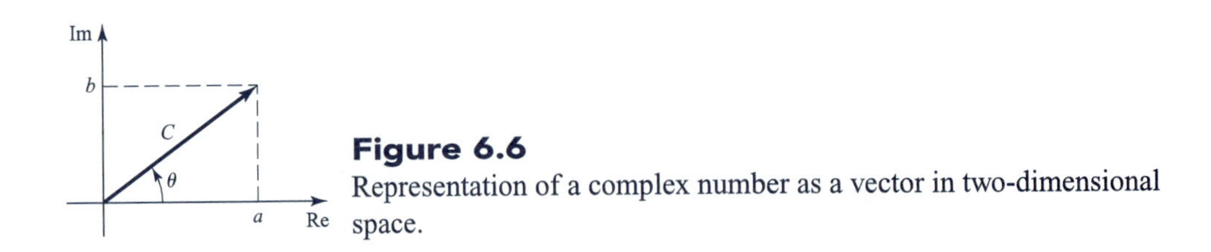

Figure 6.6
Representation of a complex number as a vector in two-dimensional space.

The angle θ is measured *counterclockwise from the positive real axis* and becomes

$$\theta = \tan^{-1}\left(\frac{b}{a}\right) \tag{6.10b}$$

The representation of $\hat{C}$ in terms of its real and imaginary parts as in (6.7) is referred to as the *rectangular form* of the complex number. An alternative representation is the *polar form* in terms of its magnitude and angle:

$$\hat{C} = C \angle \theta \tag{6.11}$$

From Fig. 6.6

$$a = C \cos \theta \tag{6.12a}$$

and

$$b = C \sin \theta \tag{6.12b}$$

Thus we may write the complex number as

$$\begin{aligned} \hat{C} &= \underbrace{C \cos \theta}_{a} + \underbrace{C \sin \theta}_{b} \\ &= C(\cos \theta + j \sin \theta) \end{aligned} \tag{6.13}$$

In this expression we may identify the *complex exponential* as

$$\begin{aligned} e^{j\theta} &= \cos \theta + j \sin \theta \\ &= 1 \angle \theta \end{aligned} \tag{6.14}$$

This very important expression is referred to as *Euler's identity* (pronounced "oiler's"). From it we may obtain the following common representations of the cosine and sine functions:

$$\sin \theta = \frac{e^{j\theta} - e^{-j\theta}}{2j} \tag{6.15a}$$

$$\cos \theta = \frac{e^{j\theta} + e^{-j\theta}}{2} \tag{6.15b}$$

as the reader should verify by using Euler's identity. These expressions for sine and cosine should be committed to memory, as they will be used in many instances.

Euler's identity can be proven in the following manner. Recall the infinite series expansion for the exponential:

$$e^x = 1 + \frac{x}{1!} + \frac{x^2}{2!} + \frac{x^3}{3!} + \cdots \tag{6.16}$$

and the infinite series expansions for sine and cosine:

$$\sin\theta = \theta - \frac{\theta^3}{3!} + \frac{\theta^5}{5!} - \frac{\theta^7}{7!} + \cdots \tag{6.17a}$$

$$\cos\theta = 1 - \frac{\theta^2}{2!} + \frac{\theta^4}{4!} - \frac{\theta^6}{6!} + \cdots \tag{6.17b}$$

Substituting $x = j\theta$ into (6.16), combining even and odd powers of θ, and identifying (6.17) gives Euler's identity in (6.14).

Now consider the addition, subtraction, multiplication, and division of two complex numbers. The *addition* or *subtraction* of complex numbers is best accomplished with the numbers in *rectangular form*. For two numbers

$$\hat{A} = a + jb \tag{6.18a}$$

$$\hat{B} = c + jd \tag{6.18b}$$

we define addition and subtraction by operation on the real and imaginary parts:

$$\boxed{\hat{A} + \hat{B} = (a + c) + j(b + d)} \tag{6.19}$$

$$\boxed{\hat{A} - \hat{B} = (a - c) + j(b - d)} \tag{6.20}$$

This represents the vector addition or subtraction in the complex plane as illustrated in Fig. 6.7. From this we observe two useful identities:

$$\boxed{\mathrm{Re}(\hat{A} + \hat{B}) = \mathrm{Re}\,\hat{A} + \mathrm{Re}\,\hat{B}} \tag{6.21a}$$

$$\boxed{\mathrm{Im}(\hat{A} + \hat{B}) = \mathrm{Im}\,\hat{A} + \mathrm{Im}\,\hat{B}} \tag{6.21b}$$

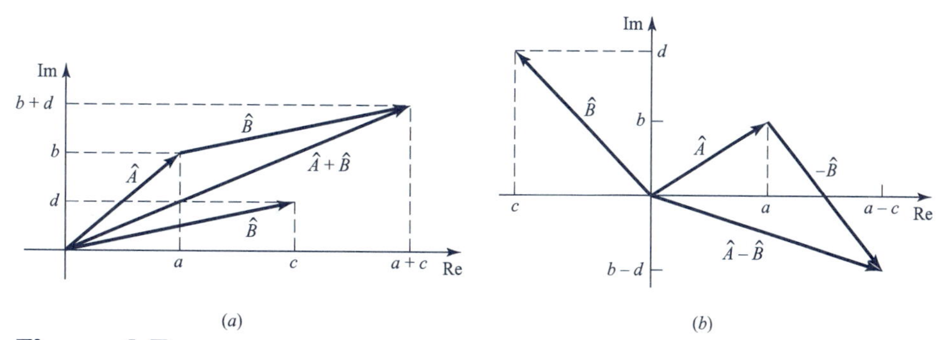

Figure 6.7
Illustration of (*a*) the sum and (*b*) the difference of two complex numbers as the addition and subtraction of vectors.

The *multiplication* or *division* of complex numbers is best accomplished with the numbers in polar form. For two numbers

$$\hat{A} = A \angle \theta_A \tag{6.22a}$$

$$\hat{B} = B \angle \theta_B \tag{6.22b}$$

multiplication and division are defined as

$$\boxed{\hat{A}\hat{B} = AB \angle (\theta_A + \theta_B)} \tag{6.23a}$$

$$\boxed{\frac{\hat{A}}{\hat{B}} = \frac{A}{B} \angle (\theta_A - \theta_B)} \tag{6.23b}$$

Multiplication and division of two complex numbers can also be performed with the numbers in rectangular form, although the labor is greater. Interpreting j as the square root of -1, we obtain the various powers of j as

$$\boxed{\begin{aligned} j &= \sqrt{-1} \\ j^2 &= -1 \\ j^3 &= -j \\ j^4 &= 1 \end{aligned}} \tag{6.24}$$

Hence

$$\begin{aligned} \hat{A}\hat{B} &= (a + jb)(c + jd) \\ &= (ac - bd) + j(ad + bc) \end{aligned} \tag{6.25}$$

To perform division it is helpful to define the *conjugate* of a complex number (denoted with an asterisk) as

$$\boxed{\begin{aligned} \hat{C} &= a + jb \\ \hat{C}^* &= a - jb \end{aligned}} \tag{6.26}$$

Hence to determine the conjugate we simply replace j with $-j$. With the number in polar form, the conjugate is simply obtained by negating the sign of the angle:

$$\boxed{\begin{aligned} \hat{C} &= C \angle \theta \\ \hat{C}^* &= C \angle -\theta \end{aligned}} \tag{6.27}$$

An important result is that *the product of a complex number and its conjugate yields the magnitude squared of that number*:

$$\boxed{\begin{aligned} \hat{C}\hat{C}^* &= a^2 + b^2 \\ &= C \angle \theta \, C \angle -\theta \\ &= C^2 \end{aligned}} \tag{6.28}$$

Division of two numbers that are in rectangular form can also be accomplished by multiplying numerator and denominator by the conjugate of the denominator:

$$\begin{aligned}\frac{\hat{A}}{\hat{B}} &= \frac{\hat{A}}{\hat{B}}\frac{\hat{B}*}{\hat{B}*} \\ &= \frac{\hat{A}\hat{B}*}{B^2} \\ &= \frac{(a+jb)(c-jd)}{c^2+d^2} \\ &= \left(\frac{ac+bd}{c^2+d^2}\right)+j\left(\frac{bc-ad}{c^2+d^2}\right)\end{aligned} \tag{6.29}$$

Several other useful identities can now be proven:

$$\begin{aligned}(\hat{A}*)* &= \hat{A} \\ (\hat{A}\hat{B})* &= \hat{A}*\hat{B}* \\ \left(\frac{\hat{A}}{\hat{B}}\right)^* &= \frac{\hat{A}*}{\hat{B}*} \\ (\hat{A}+\hat{B})* &= \hat{A}*+\hat{B}*\end{aligned} \tag{6.30}$$

Example 6.2

If

$$\begin{aligned}\hat{A} &= 2-j3 = 3.61\angle -56.31° \\ \hat{B} &= -3+j5 = 5.83\angle 120.96°\end{aligned}$$

determine

(a) $\hat{C} = \hat{A} + \hat{B}$

(b) $\hat{C} = \hat{A} - \hat{B}$

(c) $\hat{C} = \hat{A}\hat{B}$

(d) $\hat{C} = \dfrac{\hat{A}}{\hat{B}}$

(e) $\hat{C} = \hat{A} + \hat{B}* = (\hat{A}* + \hat{B})*$

(f) $\hat{C} = \hat{A}*\hat{B} = (\hat{A}\hat{B}*)*$

(g) $\hat{C} = \dfrac{\hat{A}}{\hat{B}*} = \left(\dfrac{\hat{A}*}{\hat{B}}\right)^*$

Solution

$$\begin{aligned}\text{(a) } \hat{C} &= \hat{A} + \hat{B} \\ &= (2-j3)+(-3+j5) \\ &= -1+j2 \\ &= 2.24\angle 116.57°\end{aligned}$$

$$
\begin{aligned}
\text{(b)}\ \hat{C} &= \hat{A} - \hat{B}\\
&= (2 - j3) - (-3 + j5)\\
&= 5 - j8\\
&= 9.43\angle - 57.99°
\end{aligned}
$$

$$
\begin{aligned}
\text{(c)}\ \hat{C} &= \hat{A}\hat{B}\\
&= (3.61\angle - 56.31°)(5.83\angle 120.96°)\\
&= 21.02\angle 64.65°
\end{aligned}
$$

or

$$
\begin{aligned}
\text{(c)}\ \hat{C} &= \hat{A}\hat{B}\\
&= (2 - j3)(-3 + j5)\\
&= 9 + j19
\end{aligned}
$$

Note that $9 + j19 = 21.02\angle 64.65°$, as the reader should verify.

$$
\begin{aligned}
\text{(d)}\ \hat{C} &= \frac{\hat{A}}{\hat{B}}\\
&= \frac{3.61\angle - 56.31°}{5.83\angle 120.96°}\\
&= 0.62\angle - 177.27°
\end{aligned}
$$

or

$$
\begin{aligned}
\text{(d)}\ \hat{C} &= \frac{\hat{A}}{\hat{B}}\\
&= \frac{2 - j3}{-3 + j5}\\
&= \left(\frac{2 - j3}{-3 + j5}\right)\left(\frac{-3 - j5}{-3 - j5}\right)\\
&= -0.62 - j0.03
\end{aligned}
$$

Note that $-0.62 - j0.03 = 0.62\angle - 177.27°$, as the reader should verify.

$$
\begin{aligned}
\text{(e)}\ \hat{C} &= \hat{A} + \hat{B}^*\\
&= (2 - j3) + (-3 - j5)\\
&= -1 - j8\\
&= 8.06\angle - 97.13°\\
&= (\hat{A}^* + \hat{B})^*\\
&= (2 + j3 - 3 + j5)^*\\
&= (-1 + j8)^*\\
&= -1 - j8
\end{aligned}
$$

$$
\begin{aligned}
\text{(f)}\ \hat{C} &= \hat{A}^*\hat{B}\\
&= (3.61\angle 56.31°)(5.83\angle 120.96°)\\
&= 21.02\angle 177.27°\\
&= (2 + j3)(-3 + j5)\\
&= -21 + j1\\
&= (\hat{A}\hat{B}^*)^*\\
&= [(3.61\angle - 56.31°)(5.83\angle - 120.96°)]^*\\
&= (21.02\angle - 177.27°)^*\\
&= 21.02\angle 177.27°
\end{aligned}
$$

(g) $$\hat{C} = \frac{\hat{A}}{\hat{B}^*}$$

$$= \frac{3.61\angle -56.31°}{5.83\angle -120.96°}$$

$$= 0.62 \angle 64.65°$$

$$= \frac{2 - j3}{-3 - j5}$$

$$= \left(\frac{2 - j3}{-3 - j5}\right)\left(\frac{-3 + j5}{-3 + j5}\right)$$

$$= \frac{9 + j19}{34}$$

$$= 0.26 + j0.56$$

$$= 0.62 \angle 64.65°$$

$$= \left(\frac{\hat{A}^*}{\hat{B}}\right)^*$$

$$= \left(\frac{3.61 \angle 56.31°}{5.83 \angle 120.96°}\right)^*$$

$$= (0.62\angle -64.65°)^*$$

$$= 0.62 \angle 64.65°$$

Complex operations can be used repeatedly to reduce an expression involving complex numbers. This will be the primary use of complex algebra here.

Example 6.3

Given

$$\hat{A} = 10 + j20$$
$$\hat{B} = 30 - j50$$
$$\hat{C} = 2 + j4$$
$$\hat{D} = 1 - j3$$
$$\hat{E} = 4 + j1$$

Evaluate the expression

$$\hat{F} = \frac{\hat{A} + \dfrac{\hat{B}}{\hat{C}}}{\hat{D}\hat{E}}$$

Solution First we evaluate $\hat{B}/\hat{C}$. Writing $\hat{B}$ and $\hat{C}$ in polar form gives

$$\frac{\hat{B}}{\hat{C}} = \frac{58.31\angle -59.04°}{4.47 \angle 63.43°}$$

$$= 13.04\angle -122.47°$$

Converting to rectangular form gives

$$\frac{\hat{B}}{\hat{C}} = -7.0 - j11.0$$

so that

$$\hat{A} + \frac{\hat{B}}{\hat{C}} = 3.0 + j9.0$$
$$= 9.49 \angle 71.57°$$

Finally

$$\hat{F} = \frac{\hat{A} + \dfrac{\hat{B}}{\hat{C}}}{\hat{D}\hat{E}}$$
$$= \frac{9.49 \angle 71.57°}{(1 - j3)(4 + j1)}$$
$$= \frac{9.49 \angle 71.57°}{(3.16\angle - 71.57°)(4.12 \angle 14.04°)}$$
$$= \frac{9.49 \angle 71.57°}{13.02\angle - 57.53°}$$
$$= 0.73 \angle 129.09°$$

Exercise Problem 6.3

Determine

$$\hat{E} = \frac{\hat{A} + \hat{B}}{\hat{C}} + \hat{D}$$

where $\hat{A} = 2 \angle 30°$, $\hat{B} = 3\angle - 60°$, $\hat{C} = 3 - j5$, and $\hat{D} = 1 \angle 100°$.

Answer: $\hat{E} = 0.35 + j1.32 = 1.36 \angle 75.3°$.

6.3 The Phasor (Frequency-Domain) Circuit

In this important section we will outline a very simple way of determining the response of currents and voltages in a circuit to a sinusoidal source. It obviates the two problems found in Section 6.1.2: (a) the need to derive the differential equation relating the current or voltage to the sinusoidal source, and (b) the need to solve two equations in two unknowns in order to determine the two coefficients in the assumed solution. We will essentially represent inductors and capacitors as complex-valued resistors, so that all of our previous resistive-circuit analysis techniques can be brought to bear in solving the problem, with only the additional burden of complex arithmetic. This technique, the *phasor method,* was invented by Charles Proteus Steinmetz around the turn of the century while working for the General Electric Company. Next to resistive-circuit analysis techniques, it is perhaps the most important skill that an electrical engineer possesses. The ubiquitous need to analyze the response of circuits to sinusoidal sources occurs throughout electrical engineering, and hence the reader needs to thoroughly develop this important skill.

6.3.1 Representation of Sinusoidal Sources with Euler's Identity The heart of the phasor method is the representation of a sinusoidal source with Euler's identity:

$$Ae^{j(\omega t + \phi)} = A\cos(\omega t + \phi) + jA\sin(\omega t + \phi) \tag{6.31}$$

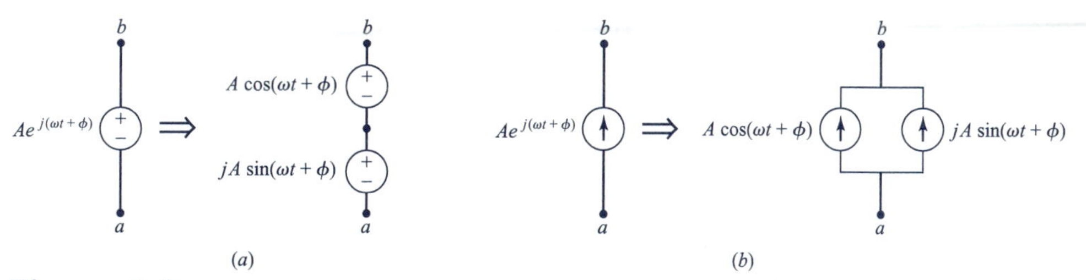

Figure 6.8
Representation of a complex exponential source as the connection of two sources (series for a voltage source and parallel for a current source) using Euler's identity.

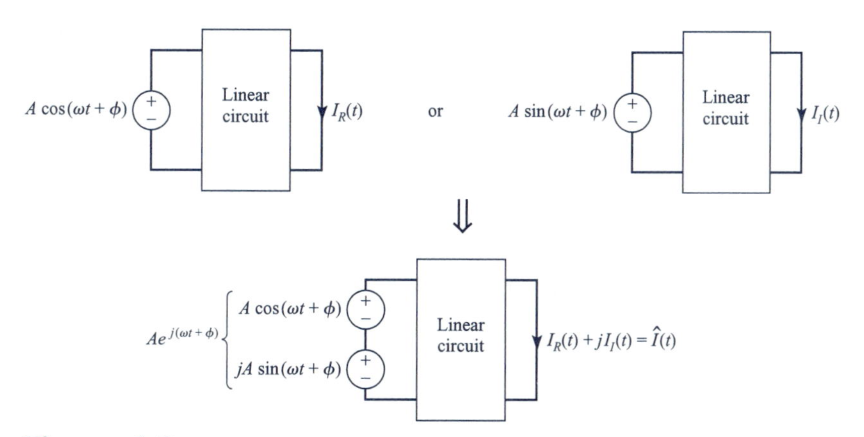

Figure 6.9
The key to the phasor method: in order to determine the response to either a sine or a cosine source, replace the source with the complex exponential source and use superposition.

Although we cannot physically generate such a source, we can mathematically represent a source using this identity as shown in Fig. 6.8. The heart of the idea is this. Suppose we wish to determine the response of a circuit to a voltage or current source whose value is *either* $A\cos(\omega t+\phi)$ *or* $A\sin(\omega t+\phi)$ as illustrated in Fig. 6.9. For illustration we are assuming that the source is a voltage source and the desired response is a current as, for example, in Fig. 6.5, although the method applies equally well to the determination of a voltage response and a sinusoidal current source. We solve instead for the response to (6.31) as illustrated in Fig. 6.9. Since Euler's identity in (6.31) shows that the $Ae^{j(\omega t+\phi)}$ source is the sum of $A\cos(\omega t+\phi)$ and $jA\sin(\omega t+\phi)$ sources, we simply use superposition, so that the response to the $A\cos(\omega t+\phi)$ source, $I_R(t)$, is the real part of the response to the $Ae^{j(\omega t+\phi)}$ source, $\hat{I}(t)=I_R(t)+jI_I(t)$. Similarly, the response to the $A\sin(\omega t+\phi)$ source, $I_I(t)$, is the imaginary part of the response to the $Ae^{j(\omega t+\phi)}$ source, $\hat{I}(t)=I_R(t)+jI_I(t)$. So superposition and Euler's identity are the heart of this very clever method. Hence, if the circuit is nonlinear, the method does not apply.

In order to demonstrate this method, consider the circuit of Fig. 6.5, which was solved directly in Section 6.1.2. Replace the actual source, $10\sin 3t$, with $10e^{j3t}$ as shown in Fig. 6.10. Writing the differential equation relating the response $\hat{I}(t)$ to the exponential source $10e^{j3t}$ yields

$$2\frac{d\hat{I}(t)}{dt}+3\hat{I}(t)=10e^{j3t}$$

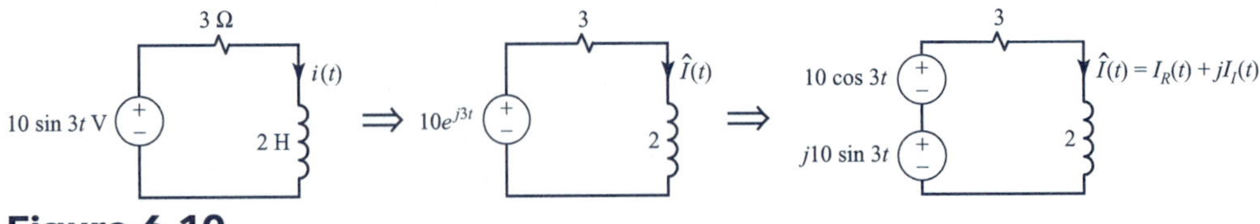

Figure 6.10
Solution of the circuit of Fig. 6.5 by replacing the original source with the complex exponential source and using Euler's identity and superposition.

We guess a form of the solution as

$$\hat{I}(t) = \hat{I}e^{j3t}$$

Substituting gives

$$2(j3)\hat{I}e^{j3t} + 3\hat{I}e^{j3t} = 10e^{j3t}$$

since

$$\frac{d}{dt}e^{j\omega t} = (j\omega)e^{j\omega t} \tag{6.32}$$

Canceling the common e^{j3t} on both sides gives

$$(j6 + 3)\hat{I} = 10$$

or

$$\begin{aligned}\hat{I} &= \frac{10}{3 + j6}\\ &= 1.49\angle - 63.43^\circ\end{aligned}$$

Since the original source was a sine source, we take the imaginary part of the result:

$$\begin{aligned}i(t) &= \text{Im}(\hat{I}e^{j3t})\\ &= \text{Im}(1.49\angle - 63.43^\circ e^{j3t})\\ &= \text{Im}(1.49e^{-j63.43^\circ}e^{j3t})\\ &= \text{Im}(1.49e^{j(3t - 63.43^\circ)})\\ &= \text{Im}(1.49\cos(3t - 63.43^\circ) + j1.49\sin(3t - 63.43^\circ))\\ &= 1.49\sin(3t - 63.43^\circ)\text{ A}\end{aligned}$$

as before.

Exercise Problem 6.4

Determine the voltage $v(t)$ in the circuit of Fig. E6.4.

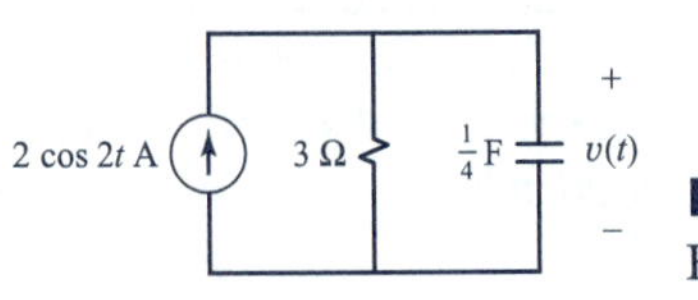

Figure E6.4
Exercise Problem 6.4.

Answer: $v(t) = 3.33\cos(2t - 56.31^\circ)$ V.

6.3.2 The Phasor Circuit The *phasor circuit* is formed by:

Step 1: Replacing the actual source, $A\cos(\omega t + \phi)$ or $A\sin(\omega t + \phi)$, with the exponential source, $Ae^{j(\omega t+\phi)} = Ae^{j\phi}e^{j\omega t} = (A \angle \phi)e^{j\omega t}$.

Step 2: Representing all response voltages and currents as having a similar form:

$$\hat{V}(t) = \underbrace{V \angle \theta_V}_{\hat{V}} e^{j\omega t} \quad \text{and} \quad \hat{I}(t) = \underbrace{I \angle \theta_I}_{\hat{I}} e^{j\omega t}.$$

Step 3: Canceling the factor $e^{j\omega t}$ in the source and in all response variables (the complex voltages and currents with the $e^{j\omega t}$ terms removed, $\hat{V}$ and $\hat{I}$, are referred to as the *phasor voltages and currents*).

Step 4: Replacing the resistors, inductors, and capacitors with their *impedances*.

The first step, replacing the actual sinusoidal source with the exponential source, was justified previously. The second step, assuming the response voltages and currents have the same form as the exponential source with an undetermined magnitude and phase, is valid because the circuit is linear and hence this form can be made to satisfy the differential equation relating the response to the complex exponential source, as was demonstrated in the previous subsection. The third step, canceling (removing) the common factor $e^{j\omega t}$, is done simply because it will be common to all variables and we need not carry this additional notational burden. The fourth and final step requires that we determine the relations between the current and voltage for the inductor and capacitor in the phasor circuit. The phasor circuit is also referred to as the *frequency-domain circuit*. The original circuit is called the *time-domain circuit*.

Once we replace the actual sinusoidal source with the complex exponential source, all currents and voltages in the circuit will have the same form as the complex exponential source. For example, consider the resistor shown in Fig. 6.11*a*. The ratio of the phasor voltage and phasor current gives the *impedance* of the resistor:

$$\begin{aligned} \hat{Z}_R &= \frac{\hat{V}_R}{\hat{I}_R} \\ &= R \end{aligned} \tag{6.33}$$

as shown in Fig. 6.11*a*. Similarly the phasor current and voltage of the inductor are related by

$$\begin{aligned} \hat{V}_L e^{j\omega t} &= L\frac{d}{dt}\hat{I}_L e^{j\omega t} \\ &= (j\omega L)\hat{I}_L e^{j\omega t} \end{aligned}$$

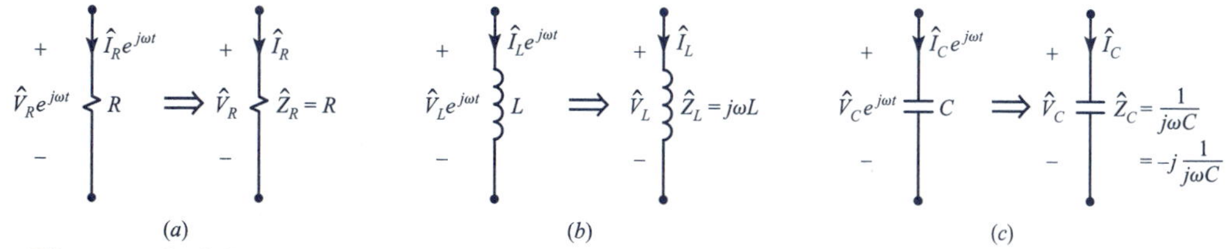

(*a*) (*b*) (*c*)

Figure 6.11

The phasor impedances for (*a*) the resistor, (*b*) the inductor, and (*c*) the capacitor.

because of the important fact that the derivative of $e^{j\omega t}$ is $(j\omega)e^{j\omega t}$ as shown in (6.32). Hence the *impedance* of the inductor is

$$\begin{aligned}\hat{Z}_L &= \frac{\hat{V}_L}{\hat{I}_L} \\ &= j\omega L\end{aligned} \tag{6.34}$$

as shown in Fig. 6.11*b*. Similarly the phasor current and voltage of the capacitor are related by

$$\begin{aligned}\hat{I}_C e^{j\omega t} &= C\frac{d}{dt}\hat{V}_C e^{j\omega t} \\ &= (j\omega C)\hat{V}_C e^{j\omega t}\end{aligned}$$

Hence the *impedance* of the capacitor is

$$\begin{aligned}\hat{Z}_C &= \frac{\hat{V}_C}{\hat{I}_C} \\ &= \frac{1}{j\omega C} \\ &= -j\frac{1}{\omega C}\end{aligned} \tag{6.35}$$

as shown in Fig. 6.11*c*. Note that in the impedance for a capacitor we have used the properties of $j = \sqrt{-1}$ to rewrite the result:

$$\frac{1}{j} = \frac{1}{j}\frac{j}{j} = -j$$

The imaginary part of the impedance is referred to as the *reactance* of the element and denoted with X. For example, we write the impedance of the inductor as

$$\hat{Z}_L = jX_L \tag{6.36a}$$

where the inductive reactance is

$$X_L = \omega L \tag{6.36b}$$

Similarly the impedance of a capacitor is written as

$$\hat{Z}_C = jX_C \tag{6.37a}$$

where the capacitive reactance is

$$X_C = -\frac{1}{\omega C} \tag{6.37b}$$

The term *impedance* simply indicates the resistance to current flow for a specific voltage applied across the element. Observe that the impedance of an inductor is zero at dc ($\omega = 0$), or, in other words, the inductor appears as a short circuit to dc. This was demonstrated in the previous chapter. The impedance of the inductor increases linearly with frequency, approaching an open circuit at infinite frequency ($\omega = \infty$). This also is a sensible result, because, as discussed in the previous chapter, the induced voltage that opposes current flow in an inductor depends, according to Faraday's law, on the time rate of change of the magnetic flux and hence is directly proportional to frequency. The capacitor, on the other hand, has an infinite impedance at dc ($\omega = 0$) and appears as an open circuit, which was proven in the previous chapter. The capacitor has zero impedance at infinite frequency ($\omega = \infty$) and appears as a short circuit. This also is a sensible result, since the voltage across a capacitor is inversely proportional to frequency.

The reciprocal of impedance is referred to as *admittance* and denoted as $\hat{Y}$. The admittances of the elements are

$$\begin{aligned}\hat{Y}_R &= \frac{1}{R}\\ &= G\end{aligned} \tag{6.38}$$

$$\begin{aligned}\hat{Y}_L &= \frac{1}{j\omega L}\\ &= -j\frac{1}{\omega L}\end{aligned} \tag{6.39}$$

$$\hat{Y}_C = j\omega C \tag{6.40}$$

The imaginary parts of the admittances are referred to as *susceptances* and denoted by B:

$$\begin{aligned}\hat{Y}_L &= jB_L\\ \hat{Y}_C &= jB_C\end{aligned} \tag{6.41a}$$

and the inductive and capacitive susceptances are

$$\begin{aligned}B_L &= -\frac{1}{\omega L}\\ B_C &= \omega C\end{aligned} \tag{6.41b}$$

Kirchhoff's laws also hold for the phasor currents and voltages. For example, KVL requires that in the time-domain circuit the algebraic sum of the voltages around all loops must sum to zero:

$$\sum v_i(t) = 0$$

Replacing the actual voltages with their phasor forms yields

$$\sum \hat{V}_i e^{j\omega t} = \left(\sum \hat{V}_i\right) e^{j\omega t} = 0$$

Hence

$$\sum \hat{V}_i = 0$$

KCL can be similarly shown to apply to the phasor currents in the phasor circuit.

For example, consider the circuit in Fig. 6.10, which was solved earlier. This is redrawn in the time domain in Fig. 6.12*a*. The phasor, or frequency-domain, circuit is shown in Fig. 6.12*b* and is formed by (a) replacing the time-domain source, $10 \sin 3t$, with its phasor equivalent, $10 \angle 0°$, and removing the factor e^{j3t}; (b) denoting the phasor response as $\hat{I}$ and removing the factor e^{j3t} from $\hat{I}(t) = \hat{I}e^{j3t}$; and (c) replacing the elements with their impedances, $\hat{Z}_R = 3$ and

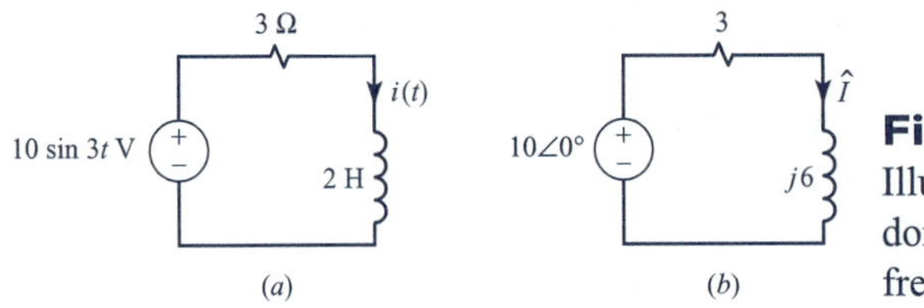

Figure 6.12 Illustration of converting (*a*) the time-domain circuit to (*b*) its phasor, or frequency-domain, equivalent.

$\hat{Z}_L = j\omega L = j(3)(2) = j6$. The resulting phasor circuit is a single-loop circuit, so that the phasor current can be found from

$$\begin{aligned}\hat{I} &= \frac{10 \angle 0^\circ}{3 + j6} \\ &= 1.49\angle - 63.43^\circ\end{aligned}$$

Hence, the time-domain current is found (since the source was a sine source) from

$$\begin{aligned}i(t) &= \mathrm{Im}(\hat{I}e^{j3t}) \\ &= \mathrm{Im}(1.49\angle - 63.43^\circ\, e^{j3t}) \\ &= \mathrm{Im}(1.49e^{-j63.43^\circ}\, e^{j3t}) \\ &= \mathrm{Im}(1.49e^{j(3t - 63.43^\circ)}) \\ &= \mathrm{Im}[1.49\cos(3t - 63.43^\circ) + j1.49\sin(3t - 63.43^\circ)] \\ &= 1.49\sin(3t - 63.43^\circ)\ \mathrm{A}\end{aligned}$$

In the following section we illustrate the straightforward use of the resistive circuit analysis techniques to solve the phasor circuit.

Exercise Problem 6.5

Determine the voltage $v(t)$ in the circuit of Fig. E6.4 using the phasor method.

Answer: $v(t) = 3.33\cos(2t - 56.31^\circ)$ V.

6.4 Applications of Resistive-Circuit Analysis Techniques in the Phasor Circuit

All of the resistive-circuit analysis methods may be applied to the phasor circuit. We simply treat the phasor circuit as a resistive circuit having complex-valued resistors (the impedances) and complex-valued currents and voltages, and solve using any of the resistive-circuit analysis techniques. Series and parallel reductions of impedances, voltage and current division, the direct method, source transformations, Thevenin and Norton equivalents, and node-voltage and mesh-current analysis are all available for use in solving the phasor circuit in exactly the same fashion as they were used to solve resistive circuits. The only difference is the need to use complex algebra.

Example 6.4

Determine the voltage $v(t)$ in the circuit of Fig. 6.13*a*.

Solution The phasor circuit is shown in Fig. 6.13*b*. It is formed by replacing the actual source, $2\cos(5t + 30^\circ)$, with the phasor source, $2 \angle 30^\circ e^{j5t}$. Similarly, the desired voltage, $v(t)$, is replaced with its assumed form, $\hat{V}e^{j5t}$. The factor e^{j5t} is common to all and is removed.

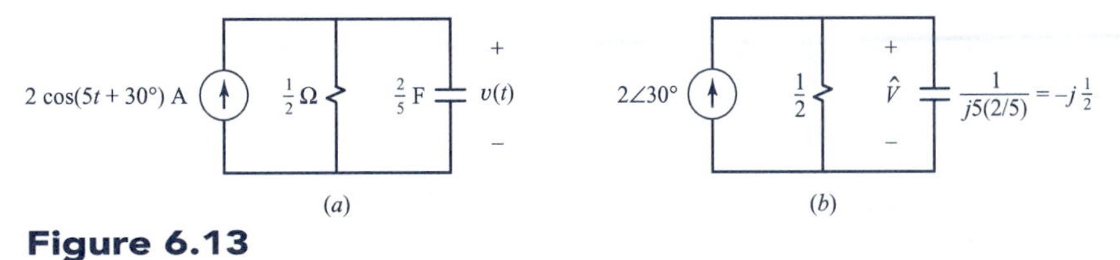

Figure 6.13
Example 6.4.

The elements are replaced with their impedances. The resistor's impedance is its resistance, but the capacitor's impedance is $1/j\omega C$. Substituting $\omega = 5$, $C = \frac{2}{5}$ F yields

$$\frac{1}{j\omega C} = \frac{1}{j5 \times \frac{2}{5}} = \frac{1}{j2} = -j\tfrac{1}{2}$$

The phasor circuit is a single-node-pair circuit, and the resistor and capacitor are in parallel, so that

$$\begin{aligned}\hat{V} &= \left[\left(\tfrac{1}{2}\right) \| \left(-j\tfrac{1}{2}\right)\right] \times 2 \angle 30^\circ \\ &= \frac{\frac{1}{2}\left(-j\frac{1}{2}\right)}{\frac{1}{2} - j\frac{1}{2}} \times 2 \angle 30^\circ \\ &= \frac{\frac{1}{4}\angle -90^\circ}{\frac{1}{2}\sqrt{2}\angle -45^\circ} \times 2 \angle 30^\circ \\ &= 0.707\angle -15^\circ\end{aligned}$$

Hence the time-domain voltage becomes

$$\begin{aligned}v(t) &= \mathrm{Re}(\hat{V}e^{j5t}) \\ &= \mathrm{Re}(0.707\angle -15^\circ e^{j5t}) \\ &= \mathrm{Re}(0.707e^{j(5t-15^\circ)}) \\ &= \mathrm{Re}[0.707\cos(5t-15^\circ) + j0.707\sin(5t-15^\circ)] \\ &= 0.707\cos(5t-15^\circ)\text{ V}\end{aligned}$$

and we have taken the real part of the result, since the source was a cosine.

Observe in this last example that we can bypass the laborious last step: taking the real or imaginary part. *If the source is a cosine, then the result is a cosine, and if the source is a sine, the result is a sine.*

Exercise Problem 6.6

Determine $v(t)$ in the circuit of Fig. E6.6.

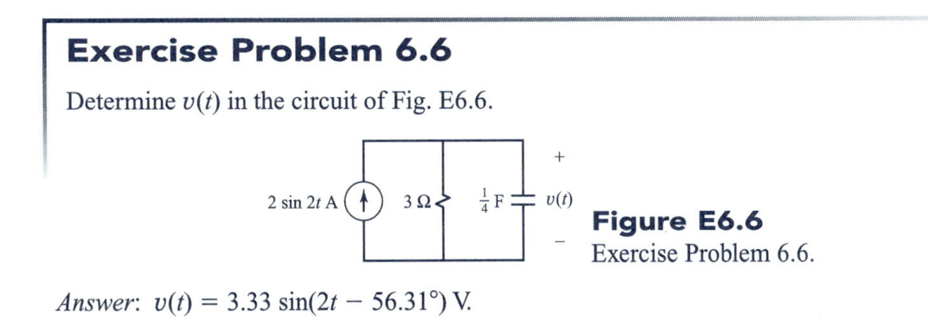

Figure E6.6
Exercise Problem 6.6.

Answer: $v(t) = 3.33\sin(2t - 56.31^\circ)$ V.

Example 6.5

Determine the current $i(t)$ and voltage $v(t)$ in the circuit of Fig. 6.14*a*.

Solution The phasor circuit is shown in Fig. 6.14*b*. This is a single-loop circuit, so that the current is

$$\hat{I} = \frac{2 \angle 30^\circ}{6 + j12 - j3}$$
$$= \frac{2 \angle 30^\circ}{6 + j9}$$
$$= 0.185\angle - 26.31^\circ$$

and the voltage can be found by voltage division as

$$\hat{V} = \frac{j12 - j3}{6 + j12 - j3} 2 \angle 30^\circ$$
$$= \frac{9 \angle 90^\circ}{6 + j9} 2 \angle 30^\circ$$
$$= 1.66 \angle 63.69^\circ$$

The time-domain current and voltage become (the source was a sine)

$$i(t) = 0.185 \sin(4t - 26.31^\circ) \text{ A}$$
$$v(t) = 1.66 \sin(4t + 63.69^\circ) \text{ V}$$

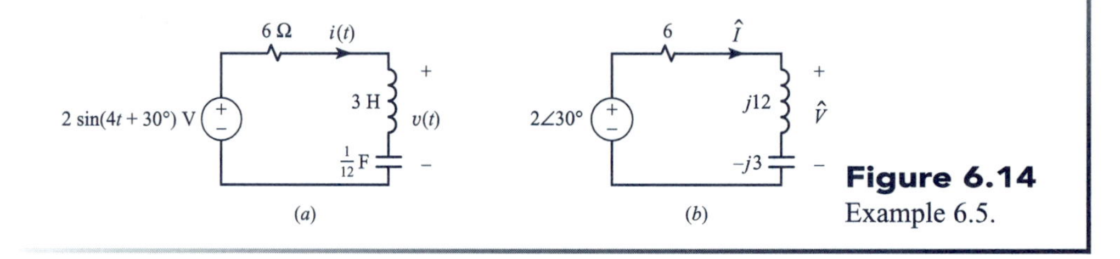

Figure 6.14
Example 6.5.

Exercise Problem 6.7

Determine the current $i(t)$ in the circuit of Fig. E6.7.

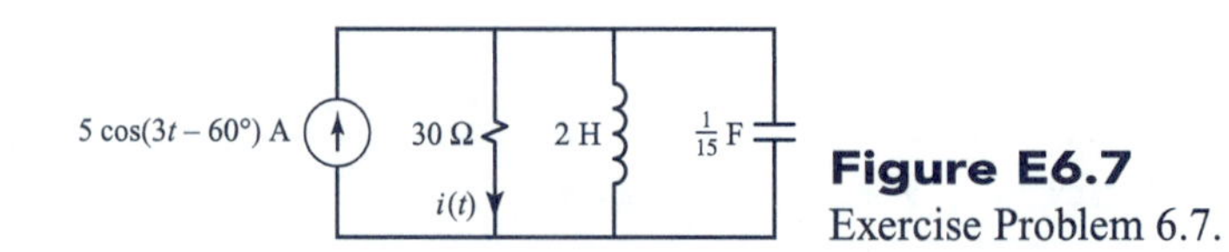

Figure E6.7
Exercise Problem 6.7.

Answer: $i(t) = 3.54 \cos(3t - 105^\circ)$ A.

Voltage and current division as well as impedances in series and parallel are all techniques that are available in solving phasor circuits in exactly the same fashion as for resistive circuits.

Example 6.6

Determine the current $i(t)$ in the circuit of Fig. 6.15*a*.

Solution The phasor circuit is shown in Fig. 6.15*b*. Observe that the resistor and inductor are in parallel and their equivalent impedance is

$$
\begin{aligned}
3 \| j3 &= \frac{(3)(j3)}{3 + j3} \\
&= \frac{9 \angle 90^\circ}{3\sqrt{2} \angle 45^\circ} \\
&= \frac{3}{\sqrt{2}} \angle 45^\circ \\
&= \tfrac{3}{2} + j\tfrac{3}{2}
\end{aligned}
$$

Once these are combined, we may use current division to obtain the current through the capacitor as

$$
\begin{aligned}
\hat{I} &= \frac{3\| j3}{3\| j3 - j1} 10\angle - 60^\circ \\
&= \frac{\frac{3}{\sqrt{2}} \angle 45^\circ}{\frac{3}{2} + j\frac{3}{2} - j1} 10\angle - 60^\circ \\
&= 13.42\angle - 33.43^\circ
\end{aligned}
$$

Hence the time-domain current becomes

$$i(t) = 13.42 \cos(3t - 33.43^\circ)\ \text{A}$$

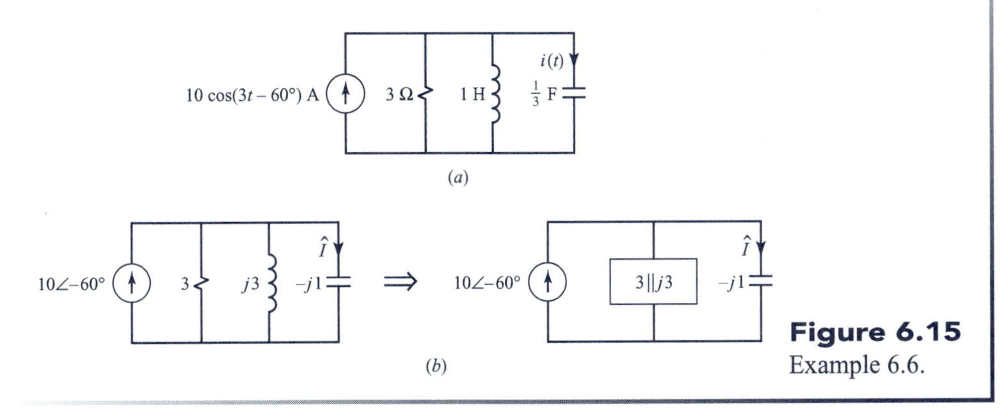

Figure 6.15
Example 6.6.

Exercise Problem 6.8

Determine $v(t)$ in the circuit of Fig. E6.8.

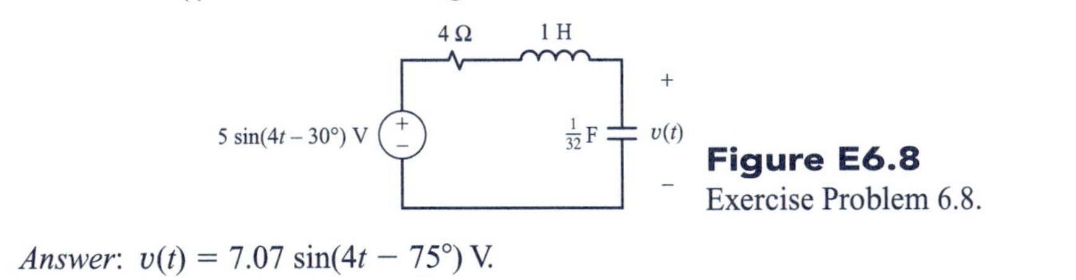

Figure E6.8
Exercise Problem 6.8.

Answer: $v(t) = 7.07 \sin(4t - 75^\circ)$ V.

Source transformations are equally useful in solving the phasor circuit.

Example 6.7

Determine $i(t)$ in the circuit of Fig. 6.16a using source transformations.

Solution The phasor circuit is shown in Fig. 6.16b and is transformed to a single-loop circuit as shown in Fig. 6.16c using a source transformation. The current source is transformed to a voltage source whose value is

$$\begin{aligned} j6 \times 5 \angle 75° &= 6 \angle 90° \times 5 \angle 75° \\ &= 30 \angle 165° \end{aligned}$$

Hence the current is

$$\begin{aligned} \hat{I} &= \frac{30 \angle 165°}{2 + j6 - j2} \\ &= 6.71 \angle 101.57° \end{aligned}$$

or in the time domain,

$$i(t) = 6.71 \sin(2t + 101.57°) \text{ A}$$

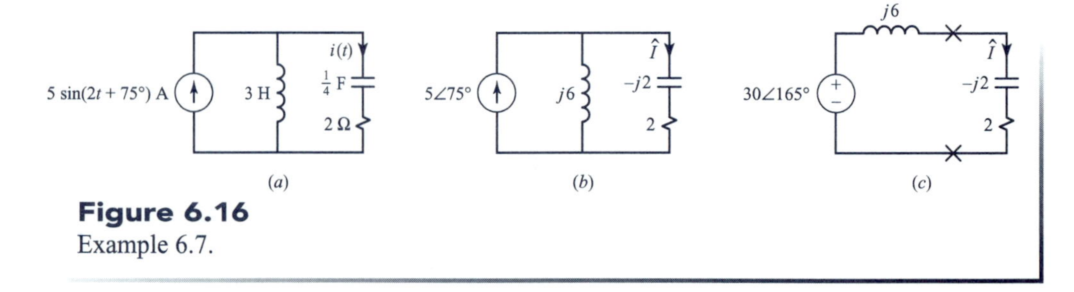

Figure 6.16
Example 6.7.

Exercise Problem 6.9

Determine current $i(t)$ in the circuit of Fig. E6.9.

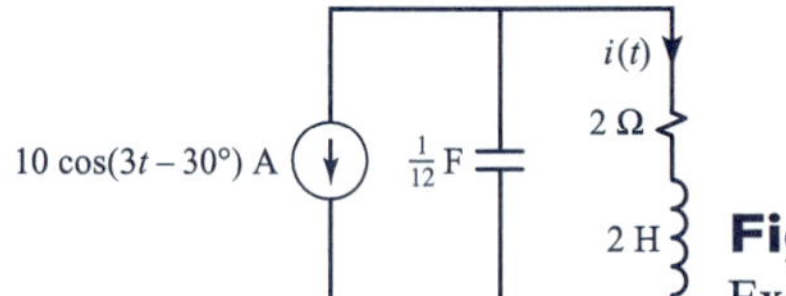

Figure E6.9
Exercise Problem 6.9.

Answer: $i(t) = -14.14 \cos(3t - 165°) \text{ A} = 14.14 \cos(3t + 15°) \text{ A}$.

The direct method is often a very effective means of solution.

Example 6.8

Determine voltage $v(t)$ in the circuit of Fig. 6.17a.

Solution The phasor circuit is shown in Fig. 6.17b. KCL is used to determine the current through the capacitor, and KVL, written around the outside loop, yields

$$10\angle -30° = (4 + j8)\hat{I}_x + (-j1)4\hat{I}_x$$

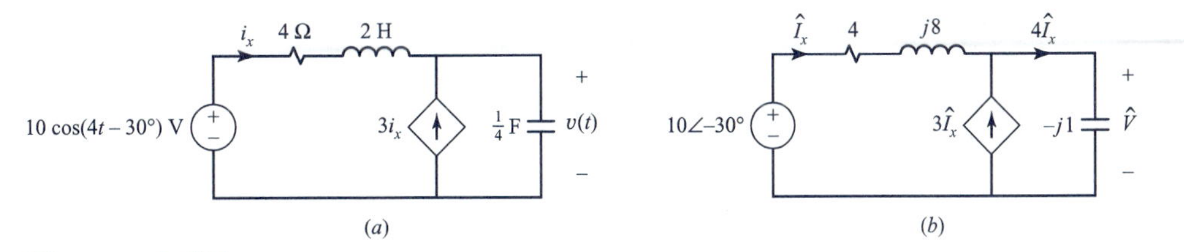

Figure 6.17
Example 6.8.

or

$$\hat{I}_x = \frac{10\angle - 30°}{4 + j4}$$

$$= \frac{10\angle - 30°}{4\sqrt{2}\angle 45°}$$

$$= 1.77\angle - 75°$$

Therefore

$$\hat{V} = -j1(4\hat{I}_x)$$

$$= -7.07\angle 15°$$

$$= 7.07\angle - 165°$$

Hence

$$v(t) = 7.07\cos(4t - 165°)\text{ V}$$

Exercise Problem 6.10

Determine current $i_x(t)$ in the circuit of Fig. E6.10.

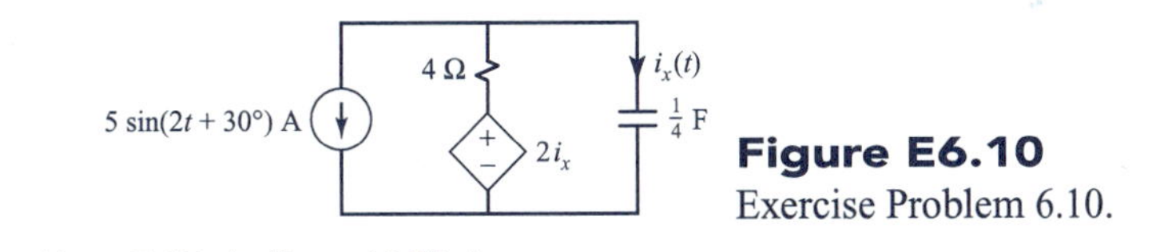

Figure E6.10
Exercise Problem 6.10.

Answer: $i_x(t) = 7.07\sin(2t - 105°)$ A.

Thevenin and Norton equivalents can also be useful, and there are no significant differences between their use in reducing the phasor circuit and their earlier use in purely resistive circuits.

Example 6.9

Solve for the voltage $v(t)$ in the circuit of Fig. 6.18*a* by reducing the phasor circuit at terminals *a* and *b* to a Thevenin equivalent.

Solution The phasor circuit is shown in Fig. 6.18*b*. The phasor open-circuit voltage is found from Fig. 6.18*c* as

$$\hat{V}_{\text{OC}} = \frac{2}{2 + j9}\,2\angle 10°$$

$$= 0.43\angle - 67.47°$$

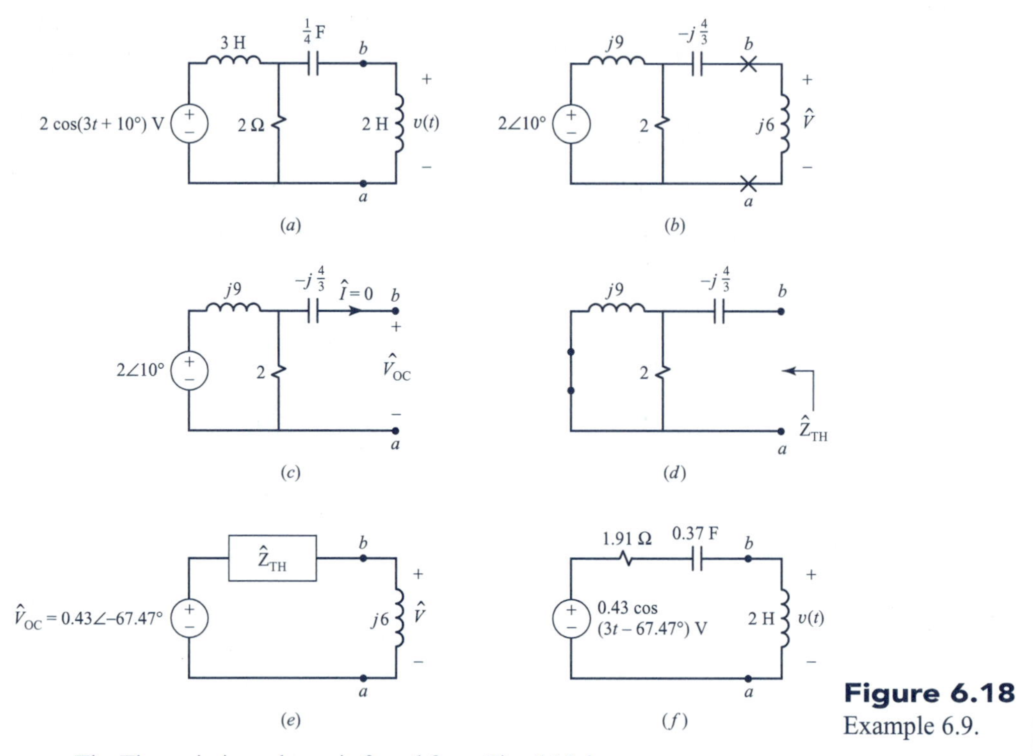

Figure 6.18
Example 6.9.

The Thevenin impedance is found from Fig. 6.18*d* as

$$\begin{aligned}\hat{Z}_{TH} &= -j\frac{4}{3} + 2 \| j9 \\ &= 2.11\angle - 25.52^\circ \\ &= 1.91 - j0.91\end{aligned}$$

Reattaching the load gives the circuit of Fig. 6.18*e*, from which we obtain

$$\begin{aligned}\hat{V} &= \frac{j6}{j6 + \hat{Z}_{TH}}\hat{V}_{OC} \\ &= \frac{j6}{\underbrace{j6 + 1.91 - j0.91}_{1.10\angle 20.53^\circ}} 0.43\angle - 67.47^\circ \\ &= 0.48\angle - 46.94^\circ\end{aligned}$$

Hence

$$v(t) = 0.48\cos(3t - 46.94^\circ)\text{ V}$$

Figure 6.18*f* shows the conversion of the Thevenin equivalent to the time domain. The open-circuit voltage becomes

$$v_{OC}(t) = 0.43\cos(3t - 67.47^\circ)\text{ V}$$

and the Thevenin impedance can be modeled as a 1.91-Ω resistor in series with a capacitor that has the value

$$\frac{1}{(\omega = 3)C} = 0.91$$

or

$$C = 0.37\text{ F}$$

Exercise Problem 6.11

Determine the current $i(t)$ in the inductor of Fig. E6.11 using a Thevenin equivalent reduction at terminals ab.

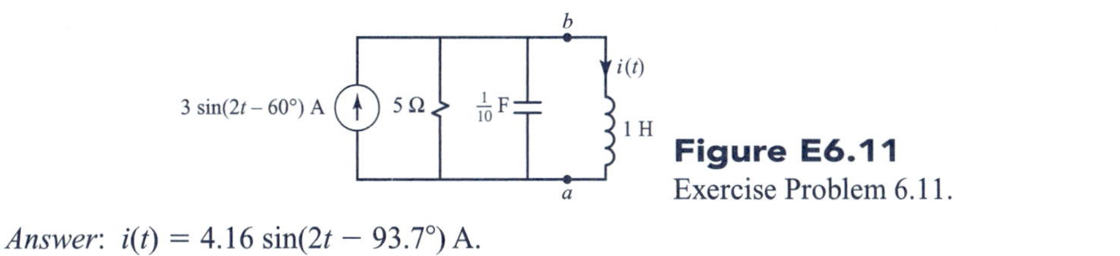

Figure E6.11
Exercise Problem 6.11.

Answer: $i(t) = 4.16 \sin(2t - 93.7°)$ A.

6.5 Circuits Containing More than One Sinusoidal Source

The previous examples have considered circuits having only one sinusoidal source. For a circuit having more than one sinusoidal source we may apply superposition *in the time domain* to yield time-domain circuits involving only one sinusoidal source in the usual fashion. Then we may solve each of these one-source circuits by reducing them to phasor circuits.

Example 6.10

Determine the current $i(t)$ in the circuit of Fig. 6.19 using superposition.

Solution Applying superposition, we obtain the two time-domain circuits shown in Fig. 6.19b and c. Each of these is solved by representing them as phasor circuits as shown. Hence we obtain

$$\begin{aligned}\hat{I}'\big|_{\omega=2} &= \frac{10\angle -30°}{4 + j4} \\ &= 1.77\angle -75°\end{aligned}$$

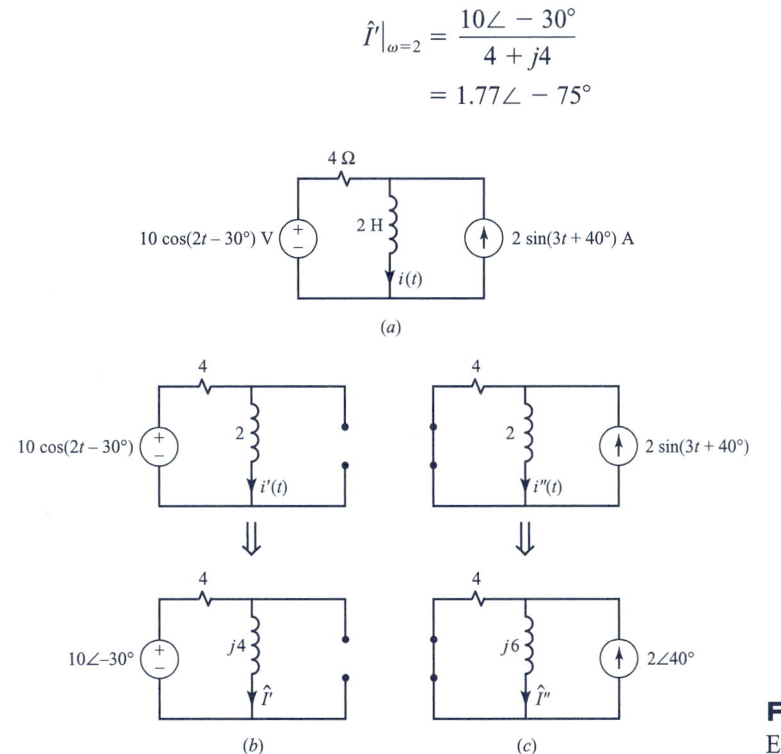

Figure 6.19
Example 6.10.

and

$$\hat{I}''\big|_{\omega=3} = \frac{4}{4 + j6} 2 \angle 40°$$
$$= 1.11\angle - 16.31°$$

Thus the time-domain solutions are

$$i'(t) = 1.77 \cos(2t - 75°) \text{ A}$$

and

$$i''(t) = 1.11 \sin(3t - 16.31°) \text{ A}$$

Thus the total solution is

$$i(t) = i'(t) + i''(t)$$
$$= 1.77 \cos(2t - 75°) + 1.11 \sin(3t - 16.31°) \text{ A}$$

Observe in this last example that the two components of the solution cannot be combined further, because their radian frequencies are different.

Exercise Problem 6.12

Determine the current $i(t)$ in the circuit of Fig. E6.12.

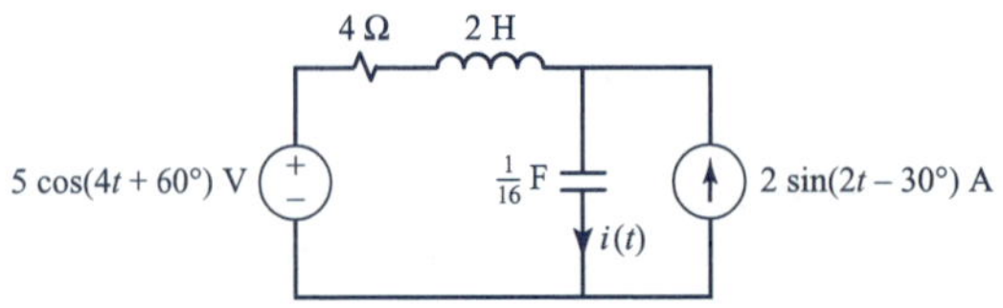

4 Ω 2 H

5 cos(4t + 60°) V

$\frac{1}{16}$ F

$i(t)$

2 sin(2t − 30°) A

Figure E6.12
Exercise problem 6.12.

Answer: $i(t) = 0.88 \cos(4t + 15°) + 2 \sin(2t + 60°)$ A.

6.5.1 Sources of the Same Frequency Sinusoidal sources having different radian frequencies cannot be combined into one phasor circuit, because of the need to compute the impedances, $j\omega L$ and $1/j\omega C$, in those phasor circuits. However, sinusoidal sources of the same radian frequency can be combined into the same phasor circuit; in that case there will be no such problem in computing the impedances, because the ω of both sources is the same. However, if we choose to place two or more sinusoidal sources that have the same radian frequency in the same phasor circuit, *they must all be of either cosine or sine form*. The reason for this restriction is that when we return the phasor solution to the time domain we must choose a cosine or a sine form.

Example 6.11

Determine the current $i(t)$ in the circuit of Fig. 6.20*a* by placing both sinusoidal sources in the same phasor circuit. Repeat that solution by using superposition in the time domain.

Solution Since both sources have the same radian frequency, they can be combined into one phasor circuit (if we wish). But both must be either sine or cosine form. We arbitrarily choose to convert the current source to a sine form to match that of the voltage source as shown in Fig. 6.20*a*. The phasor circuit containing both sources is shown in Fig. 6.20*b*. At this stage we have all the resistive-circuit analysis techniques available for use. We could

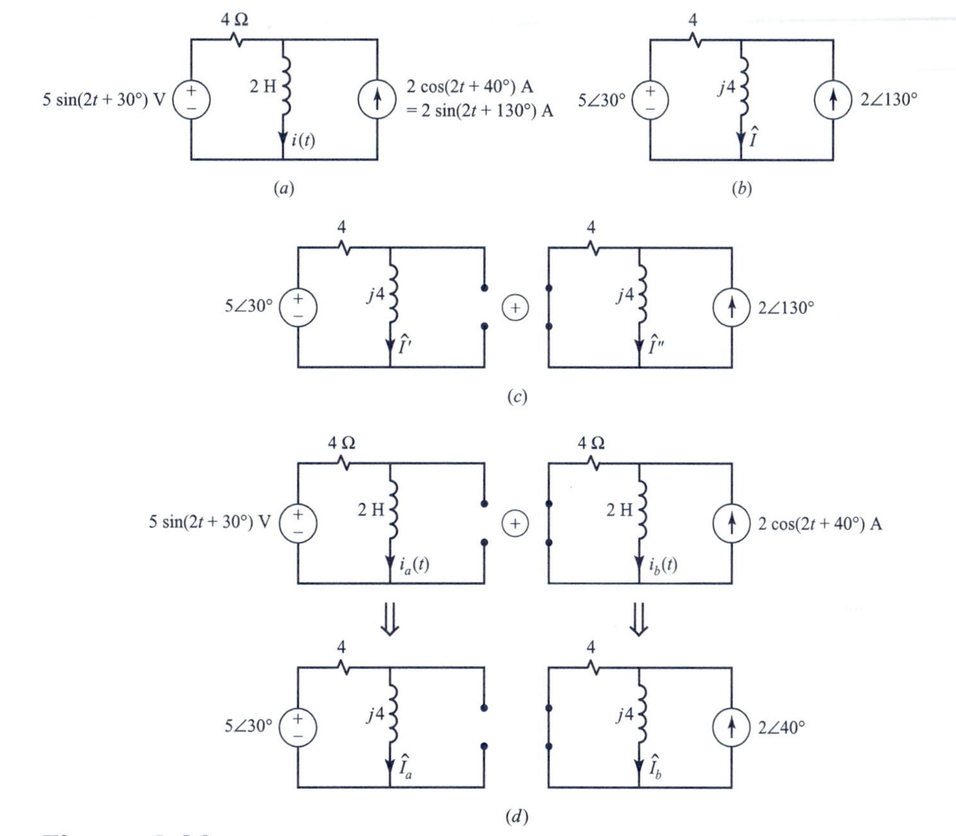

Figure 6.20
Example 6.11.

solve this phasor circuit using (a) a source transformation, (b) the direct method, (c) Thevenin and Norton equivalents, (d) node voltages or mesh currents, or (e) superposition *in the frequency domain*. Let us use superposition in the frequency domain. The two circuits are shown in Fig. 6.20*c*, from which we obtain

$$\begin{aligned} \hat{I}' &= \frac{5 \angle 30°}{4 + j4} \\ &= 0.88\angle - 15° \end{aligned}$$

and

$$\begin{aligned} \hat{I}'' &= \frac{4}{4 + j4} 2 \angle 130° \\ &= 1.41 \angle 85° \end{aligned}$$

Hence

$$\begin{aligned} \hat{I} &= \hat{I}' + \hat{I}'' \\ &= 0.88\angle - 15° + 1.41 \angle 85° \\ &= 1.53 \angle 50.38° \end{aligned}$$

Converting back to the time domain gives

$$i(t) = 1.53 \sin(2t + 50.38°) \text{ A}$$

Had we chosen to solve this by using superposition *in the time domain,* we would generate the two circuits shown in Fig. 6.20*d* (and would not need to convert the current source to a sine form), from which we would obtain the two corresponding phasor circuits giving

$$\hat{I}_a = \frac{5\angle 30^\circ}{4 + j4}$$
$$= 0.88\angle -15^\circ$$

and

$$\hat{I}_b = \frac{4}{4 + j4} 2\angle 40^\circ$$
$$= 1.41\angle -5^\circ$$

Transforming these each back to the time domain gives

$$i_a(t) = 0.88 \sin(2t - 15^\circ)\text{ A}$$

and

$$i_b(t) = 1.41 \cos(2t - 5^\circ)\text{ A}$$

so that

$$\begin{aligned} i(t) &= i_a(t) + i_b(t) \\ &= 0.88 \sin(2t - 15^\circ) + 1.41 \cos(2t - 5^\circ)\text{ A} \end{aligned}$$

Using trigonometric identities this can be converted to

$$\begin{aligned} i(t) &= i_a(t) + i_b(t) \\ &= 0.88 \sin(2t - 15^\circ) + 1.41 \cos(2t - 5^\circ) \\ &= 0.98 \sin 2t + 1.18 \cos 2t \\ &= 1.53 \sin(2t + 50.38^\circ)\text{ A} \end{aligned}$$

showing the equivalence to the previous result.

Exercise Problem 6.13

Determine current $i(t)$ in the circuit of Fig. E6.12 when the radian frequency of the current source has been changed to $\omega = 4$.

Answer: $i(t) = 2.88 \cos(4t - 85.62^\circ)$ A.

Node voltage equations and mesh current equations can also be effective in the solution of multiple-source *phasor* circuits.

Example 6.12

Determine current $i(t)$ in the circuit of Fig. 6.21*a* using mesh-current analysis in the phasor circuit.

Solution Since both sources have the same radian frequency, they may be placed in the same phasor circuit. However, they must both be sine or cosine. On arbitrarily choosing to convert the sine voltage source to a cosine source as shown in Fig. 6.21*a*, the resulting phasor circuit containing both sources is shown in Fig. 6.21*b*. Defining the phasor mesh currents as shown, we write the mesh current equations as

$$M_1: \quad j9\hat{I}_1 + j3\hat{I}_2 - j2(\hat{I}_1 - \hat{I}_2) = 10\angle -30^\circ$$
$$M_2: \quad -(-j2)(\hat{I}_1 - \hat{I}_2) + j6\hat{I}_2 + j3\hat{I}_1 + 3\hat{I}_2 = -5\angle -90^\circ = j5$$

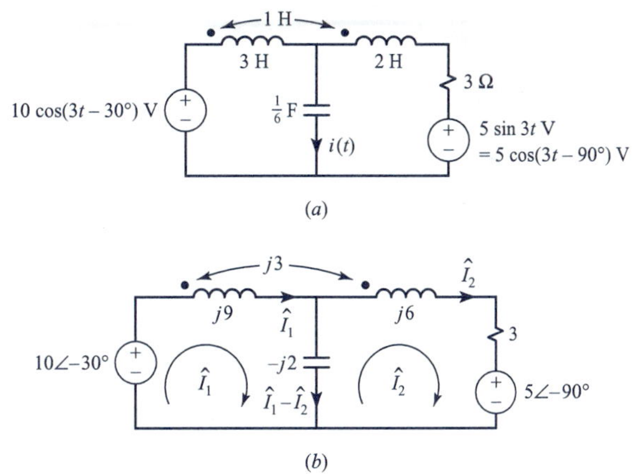

Figure 6.21
Example 6.12.

or

$$\text{M}_1: \quad j7\hat{I}_1 + j5\hat{I}_2 = 10\angle - 30^\circ$$

$$\text{M}_2: \quad j5\hat{I}_1 + (3 + j4)\,\hat{I}_2 = -5\angle - 90^\circ = j5$$

Solving these by Cramer's rule yields

$$\begin{aligned}\hat{I}_1 &= \frac{\begin{vmatrix} 10\angle - 30^\circ & j5 \\ -5\angle - 90^\circ & 3 + j4 \end{vmatrix}}{\begin{vmatrix} j7 & j5 \\ j5 & 3 + j4 \end{vmatrix}} \\ &= \frac{(10\angle - 30^\circ)(3 + j4) - (j5)(-5\angle - 90^\circ)}{-3 + j21} \\ &= \frac{73.65\,\angle\,15.47^\circ}{21.21\,\angle\,98.13^\circ} \\ &= 3.472\angle - 82.66^\circ\end{aligned}$$

and

$$\begin{aligned}\hat{I}_2 &= \frac{\begin{vmatrix} j7 & 10\angle - 30^\circ \\ j5 & -5\angle - 90^\circ \end{vmatrix}}{\begin{vmatrix} j7 & j5 \\ j5 & 3 + j4 \end{vmatrix}} \\ &= \frac{j7(-5\angle - 90^\circ) - (j5)(10\angle - 30^\circ)}{-3 + j21} \\ &= \frac{73.98\angle - 144.18^\circ}{21.21\,\angle\,98.13^\circ} \\ &= 3.488\,\angle\,117.69^\circ\end{aligned}$$

Hence the desired current is

$$\begin{aligned}\hat{I} &= \hat{I}_1 - \hat{I}_2 \\ &= 6.85\angle - 72.46^\circ\end{aligned}$$

Returning to the time domain, we have

$$i(t) = 6.85\cos(3t - 72.46^\circ)\text{ A}$$

Exercise Problem 6.14

Determine voltage $v(t)$ in the circuit of Fig. E6.14 using node-voltage equations in the phasor circuit.

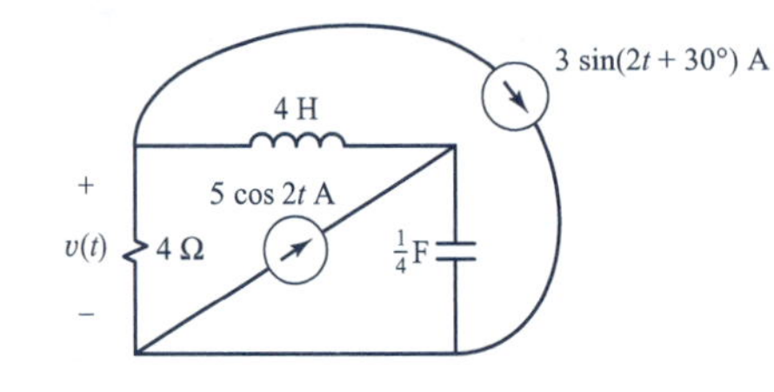

Figure E6.14
Exercise Problem 6.14.

Answer: $v(t) = 13.63\cos(2t + 174.32)$ V.

Exercise Problem 6.15

Determine the current $i(t)$ in the circuit of Fig. E6.15 by placing both sources in one phasor circuit and using source transformation.

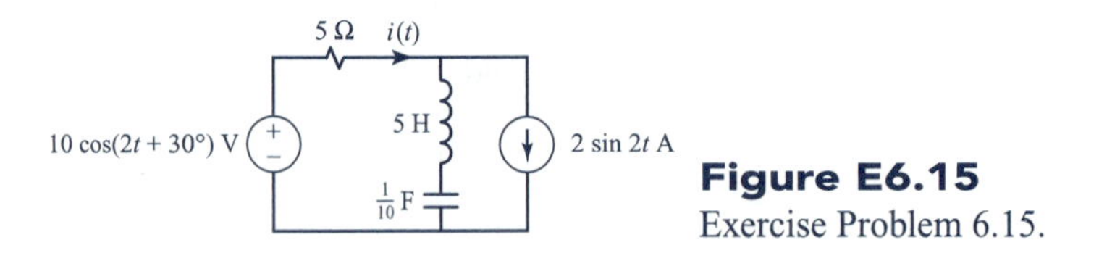

Figure E6.15
Exercise Problem 6.15.

Answer: $i(t) = 2.73\cos(2t - 30°)$ A.

6.6 Power

Now that we have obtained the solution for currents and voltages due to sinusoidal sources, let us consider the calculation of power. The instantaneous power delivered to an element shown in Fig. 6.22*a* is

$$\begin{aligned} p(t) &= v(t)i(t) \\ &= V\sin(\omega t + \theta_V)\, I\sin(\omega t + \theta_I) \\ &= \frac{VI}{2}\cos(\theta_V - \theta_I) - \frac{VI}{2}\cos(2\omega t + \theta_V + \theta_I) \end{aligned} \tag{6.42}$$

and we have used trigonometric identities to write the result as the sum of a constant term and a term that varies at twice the frequency of the individual voltage and current waveforms. This is plotted in Fig. 6.22*b*.

Of more interest than instantaneous power is the *average power* delivered to the element. The average value of a periodic waveform is defined as the area under the curve over a period, divided by the period. Hence the average power is

$$\begin{aligned} P_{AV} &= \frac{1}{T}\int_0^T p(t)\,dt \\ &= \frac{VI}{2}\cos(\theta_V - \theta_I) \end{aligned} \tag{6.43}$$

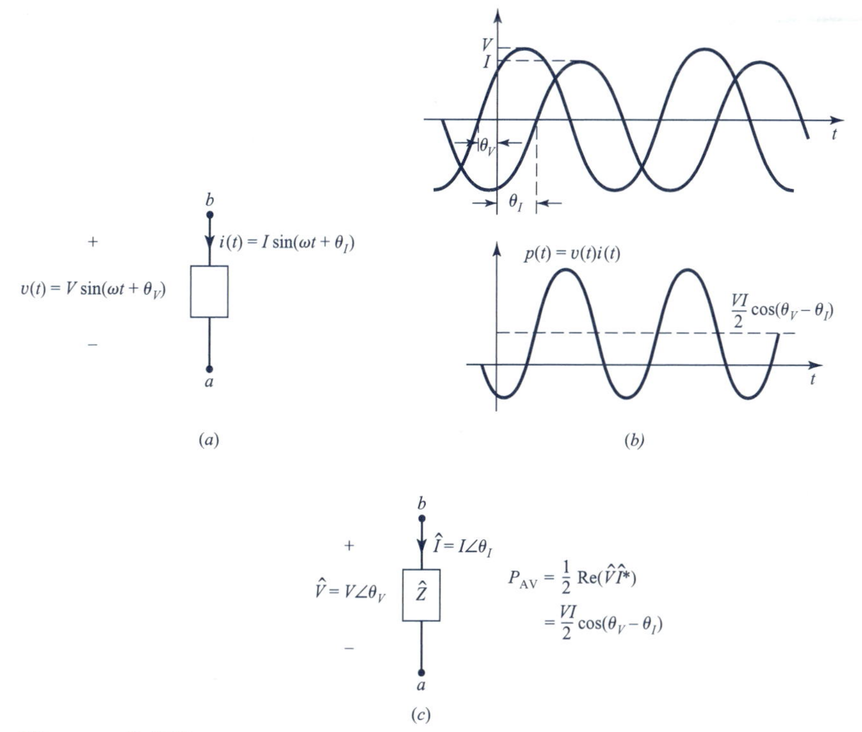

Figure 6.22
Illustration of average power delivered to an element: (*a*) the time-domain relation for the element, (*b*) plot of the instantaneous power, and (*c*) the average power delivered to an element in terms of its phasor voltage and phasor current.

and the units are watts (W). *Average power is real power dissipated by the element, usually in the form of heat, or converted to useful work as in an electric motor*. This can be computed in terms of a more general result. First let us define the *complex power* as

$$\begin{aligned}\hat{P} &= \tfrac{1}{2}\hat{V}\hat{I}^* \\ &= \tfrac{1}{2}(V\angle\theta_V)(I\angle-\theta_I) \\ &= \tfrac{1}{2}VI\angle(\theta_V-\theta_I) \\ &= \tfrac{1}{2}VI\cos(\theta_V-\theta_I)+j\tfrac{1}{2}VI\sin(\theta_V-\theta_I)\end{aligned} \tag{6.44}$$

The *magnitude* of the complex power, $\frac{1}{2}VI$, is referred to as *apparent* power, and its units are volt-amperes (VA). The real part of the complex power is the *average power*:

$$\begin{aligned}P_{\text{AV}} &= \text{Re}\,\hat{P} \\ &= \tfrac{1}{2}\text{Re}(\hat{V}\hat{I}^*) \\ &= \tfrac{1}{2}VI\cos(\theta_V-\theta_I)\end{aligned} \tag{6.45a}$$

whose units are watts (W), and the imaginary part of the complex power is referred to as *reactive power*:

$$\begin{aligned} Q &= \text{Im}\,\hat{P} \\ &= \tfrac{1}{2}\,\text{Im}(\hat{V}\hat{I}^*) \\ &= \tfrac{1}{2}\,VI\sin(\theta_V - \theta_I) \end{aligned} \tag{6.45b}$$

whose units are volt-amperes reactive (VAR). Reactive power represents energy storage in the reactive elements (inductors and capacitors).

The sum of the instantaneous powers delivered to all elements equals zero:

$$\sum_{\substack{\text{all circuit}\\ \text{elements}}} v_i(t)i_i(t) = 0 \tag{6.46}$$

The time integral of this is simply a statement of conservation of energy. This is referred to as Tellegen's theorem and is a consequence of KVL and KCL. Replacing the time functions with their phasor equivalents yields

$$\sum_{\substack{\text{all circuit}\\ \text{elements}}} \hat{V}_i\hat{I}_i = 0 \tag{6.47}$$

This relies on KVL and KCL, which are invariant under conjugation, so that we also have

$$\sum_{\substack{\text{all circuit}\\ \text{elements}}} \hat{V}_i\hat{I}_i^* = \sum_{\substack{\text{all circuit}\\ \text{elements}}} \hat{P}_i = 0 \tag{6.48}$$

So *in any circuit, conservation of complex power is achieved.* This also implies that *in any circuit, conservation of average power and conservation of reactive power are achieved*:

$$\sum_{\substack{\text{all circuit}\\ \text{elements}}} P_{i,\,\text{AV}} = 0 \tag{6.49}$$

$$\sum_{\substack{\text{all circuit}\\ \text{elements}}} Q_i = 0 \tag{6.50}$$

The reader is cautioned that it is *not true* that apparent power (the magnitude of the complex power) is conserved. Only the real and imaginary parts of the complex power are conserved.

Example 6.13

Determine the average power and reactive power delivered by the source in the circuit of Fig. 6.23*a*. Determine the average power and reactive power delivered to each element, and show that conservation of complex power, average power, and reactive power is achieved.

Solution The phasor current leaving the positive terminal of the source is determined from the phasor circuit of Fig. 6.23*b* as

$$\begin{aligned} \hat{I} &= \frac{10\angle -30^\circ}{2 + j8 - j3} \\ &= 1.86\angle -98.2^\circ \end{aligned}$$

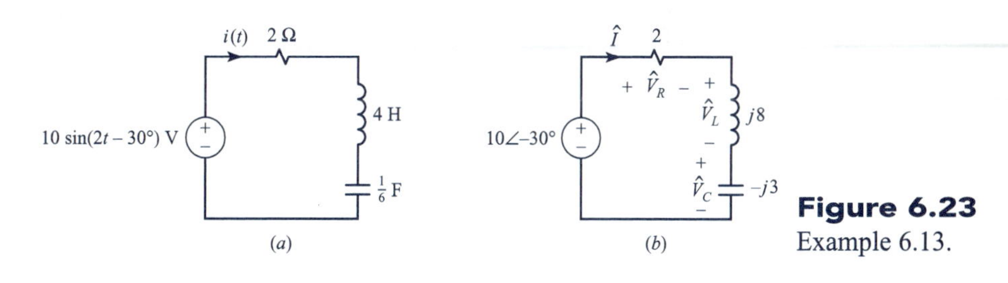

Figure 6.23
Example 6.13.

Thus the average power delivered by the source is

$$\begin{aligned} P_{\text{AV, source}} &= \tfrac{1}{2}\,\text{Re}[(10\angle-30^\circ)(1.86\angle-98.2^\circ)^*] \\ &= \tfrac{1}{2}\,\text{Re}[(10\angle-30^\circ)(1.86\angle 98.2^\circ)] \\ &= 9.28\cos(-30^\circ + 98.2^\circ) \\ &= 3.45\ \text{W} \end{aligned}$$

The reactive power delivered by the source is

$$\begin{aligned} Q_{\text{source}} &= \tfrac{1}{2}\,\text{Im}[(10\angle-30^\circ)(1.86\angle-98.2^\circ)^*] \\ &= \tfrac{1}{2}\,\text{Im}[(10\angle-30^\circ)(1.86\angle 98.2^\circ)] \\ &= 9.28\sin(-30^\circ + 98.2^\circ) \\ &= 8.62\ \text{VAR} \end{aligned}$$

and the complex power delivered by the source is

$$\begin{aligned} \hat{P}_{\text{source}} &= P_{\text{AV, source}} + jQ_{\text{source}} \\ &= 3.45 + j8.62\ \text{VA} \end{aligned}$$

The complex power delivered to the resistor, inductor, and capacitor are found by determining the voltages across them:

$$\begin{aligned} \hat{V}_R &= 2\hat{I} \\ &= 3.71\angle-98.2^\circ \\ \hat{V}_L &= j8\hat{I} \\ &= 14.86\angle-8.2^\circ \\ \hat{V}_C &= -j3\hat{I} \\ &= 5.57\angle-188.2^\circ \end{aligned}$$

Thus the complex power delivered to each element is

$$\begin{aligned} \hat{P}_R &= \tfrac{1}{2}\hat{V}_R\hat{I}^* \\ &= 3.45 + j0 \\ \hat{P}_L &= \tfrac{1}{2}\hat{V}_L\hat{I}^* \\ &= 0 + j13.79 \\ \hat{P}_C &= \tfrac{1}{2}\hat{V}_C\hat{I}^* \\ &= 0 - j5.17 \end{aligned}$$

which shows that conservation of complex power is achieved, as are conservation of average power and conservation of reactive power:

$$\underbrace{\hat{P}_{\text{source}}}_{3.45 + j8.62} = \underbrace{\hat{P}_R}_{3.45 + j0} + \underbrace{\hat{P}_L}_{0 + j13.79} + \underbrace{\hat{P}_C}_{0 - j5.17}$$

$$\underbrace{P_{\text{AV, source}}}_{3.45} = \underbrace{P_{\text{AV, }R}}_{3.45} + \underbrace{P_{\text{AV, }L}}_{0} + \underbrace{P_{\text{AV, }C}}_{0}$$

$$\underbrace{Q_{\text{source}}}_{8.62} = \underbrace{Q_R}_{0} + \underbrace{Q_L}_{13.79} + \underbrace{Q_C}_{-5.17}$$

Observe that apparent power (the magnitude of the complex power) is not conserved, since the apparent power delivered by the source here is

$$|\hat{P}_{\text{source}}| = \sqrt{(3.45)^2 + (8.62)^2} = 9.28 \text{ VA}$$
$$= \tfrac{1}{2}(10)(I = 1.86)$$

which does not equal the sum of the apparent powers delivered to the elements:

$$\underbrace{|\hat{P}_R|}_{3.45} + \underbrace{|\hat{P}_L|}_{13.79} + \underbrace{|\hat{P}_C|}_{5.17} = 22.41 \text{ VA}.$$

Exercise Problem 6.16

Determine the average and reactive powers delivered by the source in the circuit of Fig. E6.6. Check that complex power, average power, and reactive power are conserved.

Answer: 1.85 W and −2.77 VAR.

6.6.1 Power Relations for the Elements

Now let us examine the power relations for the elements. First consider the resistor shown in Fig. 6.24. For the resistor, the voltage and current are in phase:

$$\hat{V}_R = R\hat{I}_R \tag{6.51}$$

so that $\theta_V - \theta_I = 0°$. Hence the average power is

$$P_{\text{AV},R} = \frac{1}{2}\frac{V_R^2}{R} = \tfrac{1}{2} I_R^2 R \tag{6.52}$$

and the reactive power is zero for the resistor:

$$Q_R = 0 \tag{6.53}$$

This shows that the resistor dissipates power, in the form of heat, as expected.

For the inductor shown in Fig. 6.25, the voltage *leads* the current by 90°, so that $\theta_V - \theta_I = 90°$:

$$\hat{V}_L = j\omega L\hat{I}_L = (\omega L \angle 90°)\hat{I}_L \tag{6.54}$$

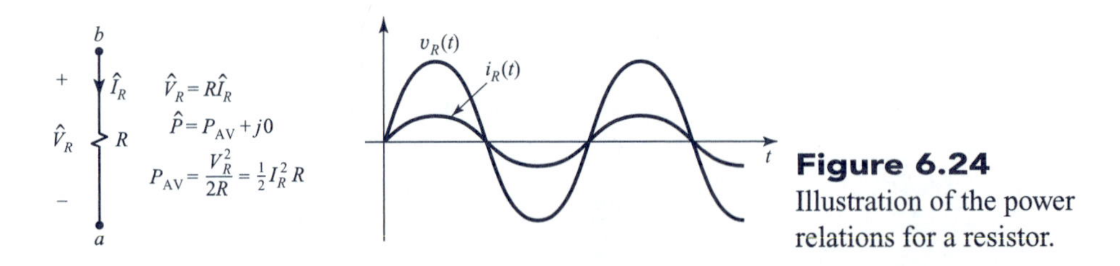

Figure 6.24
Illustration of the power relations for a resistor.

Thus the average power is zero:

$$\begin{aligned} P_{\text{AV},L} &= \tfrac{1}{2} V_L I_L \cos 90^\circ \\ &= 0 \end{aligned} \tag{6.55}$$

and the reactive power is

$$\begin{aligned} Q_L &= \tfrac{1}{2} V_L I_L \sin 90^\circ \\ &= \tfrac{1}{2} V_L I_L \end{aligned} \tag{6.56}$$

This shows that the inductor dissipates no real power but has a reactive power that is *positive* associated with it. Figure 6.25 illustrates why this is the case. When the current and voltage have the same sign, the inductor is storing $\frac{1}{2} Li^2$ J of energy, but when the voltage and current have opposite signs, this energy is being returned to the circuit that is attached to its terminals. This energy exchange takes place at twice the frequency of the sinusoidal excitation.

For the capacitor shown in Fig. 6.26, the current *leads* the voltage by 90°, so that $\theta_V - \theta_I = -90°$:

$$\begin{aligned} \hat{I}_C &= j\omega C \hat{V}_C \\ &= (\omega C \angle 90^\circ)\hat{V}_C \end{aligned} \tag{6.57a}$$

or

$$\begin{aligned} \hat{V}_C &= \frac{1}{j\omega C}\hat{I}_C \\ &= -j\frac{1}{\omega C}\hat{I}_C \\ &= \left(\frac{1}{\omega C}\angle - 90^\circ\right)\hat{I}_C \end{aligned} \tag{6.57b}$$

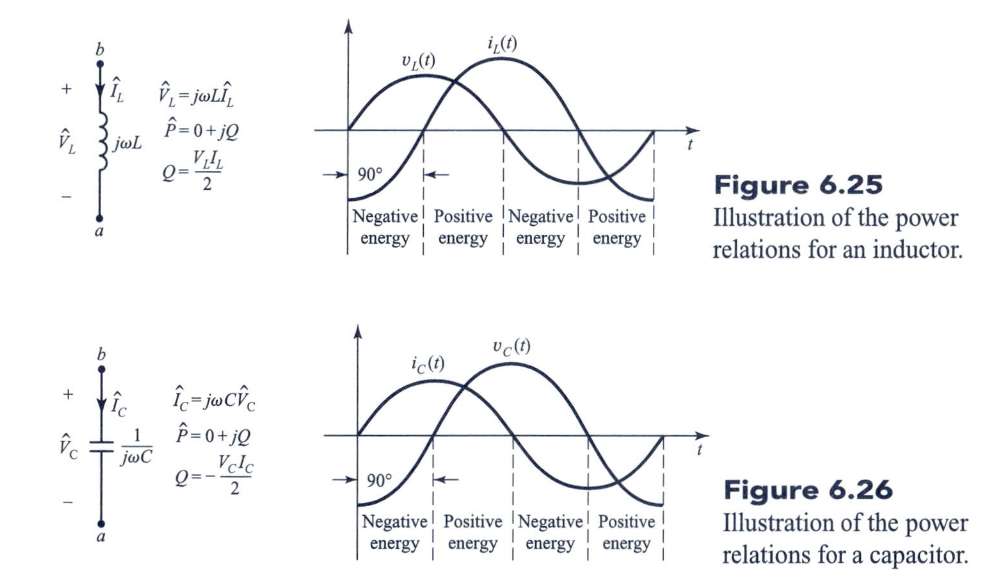

Figure 6.25 Illustration of the power relations for an inductor.

Figure 6.26 Illustration of the power relations for a capacitor.

Like the inductor, the average power delivered to the capacitor is zero:

$$\begin{aligned} P_{AV,C} &= \tfrac{1}{2} V_C I_C \cos(-90°) \\ &= 0 \end{aligned} \tag{6.58}$$

but the reactive power is

$$\begin{aligned} Q_C &= \tfrac{1}{2} V_C I_C \sin(-90°) \\ &= -\tfrac{1}{2} V_C I_C \end{aligned} \tag{6.59}$$

Observe that although the inductor and the capacitor dissipate no average power, the sign of the reactive power term is positive for the inductor but negative for the capacitor. Figure 6.26 illustrates why the capacitor dissipates no average power. When the current and voltage have the same sign, the capacitor is storing $\frac{1}{2} Cv^2$ J of energy, but when the voltage and current have opposite signs, this energy is being returned to the circuit that is attached to its terminals. This energy exchange again takes place at twice the frequency of the sinusoidal excitation.

Resistors dissipate power and have no reactive power associated with them. If one knows the *magnitude* of the current through a resistor, I_R, or voltage across it, V_R, then the average power delivered to the resistor can be easily calculated as shown in (6.52). Similarly, the average power delivered to an inductor or capacitor is zero, so we need no longer calculate this. (See Example 6.13.) Controlled sources may be delivering or absorbing average power.

Example 6.14

Check conservation of average and reactive powers for the circuit shown in Fig. 6.27*a*.

Solution The phasor circuit is shown in Fig. 6.27*b*. From that we determine, writing KVL around the loop,

$$10 \angle 30° = (j9 - j3)\hat{I}_x + 1(3\hat{I}_x)$$

or

$$\begin{aligned} \hat{I}_x &= \frac{10 \angle 30°}{3 + j6} \\ &= 1.49\angle - 33.43° \end{aligned}$$

The complex power delivered by the independent source is

$$\begin{aligned} \hat{P}_{\text{indep source}} &= \tfrac{1}{2}(10 \angle 30°)(1.49\angle - 33.43°)^* \\ &= 3.33 + j6.67 \text{ VA} \end{aligned}$$

The complex powers delivered to the resistor, inductor and capacitor are

$$\begin{aligned} \hat{P}_R &= \tfrac{1}{2}(3I_x)^2 1 \\ &= 10 \\ \hat{P}_L &= \tfrac{1}{2}(j9\hat{I}_x)(\hat{I}_x^*) \\ &= j10 \\ \hat{P}_C &= \tfrac{1}{2}(-j3\hat{I}_x)(\hat{I}_x^*) \\ &= -j3.33 \end{aligned}$$

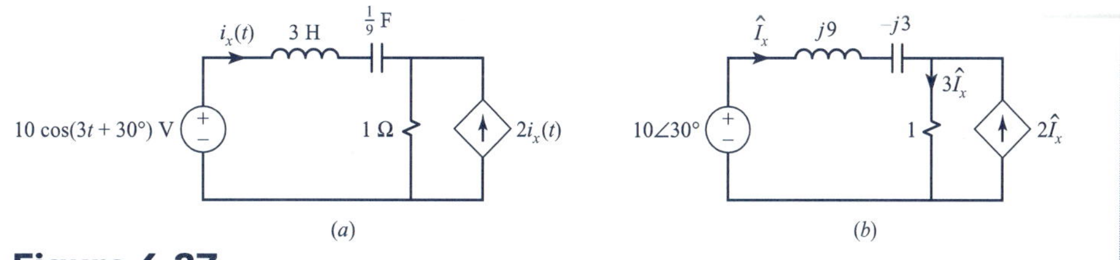

Figure 6.27
Example 6.14.

The complex power delivered *by* the controlled source is

$$\begin{aligned}\hat{P}_{\text{cont source}} &= \tfrac{1}{2}(3\hat{I}_x)(2\hat{I}_x)^* \\ &= 6.67 + j0 \text{ VA}\end{aligned}$$

since the voltage across it is the voltage across the resistor. Hence the independent source and the controlled source are delivering a total of 3.33 + 6.67 = 10 W average power, and the resistor is absorbing all of that average power. The independent source is delivering 6.67 VAR of reactive power, and all of that is being supplied to the inductor and the capacitor (10 VAR − 3.33 VAR = 6.67 VAR). The controlled source is delivering no reactive power.

Exercise Problem 6.17

Determine the average and reactive powers delivered by the source in the circuit of Fig. E6.17. Check conservation of complex, average, and reactive powers.

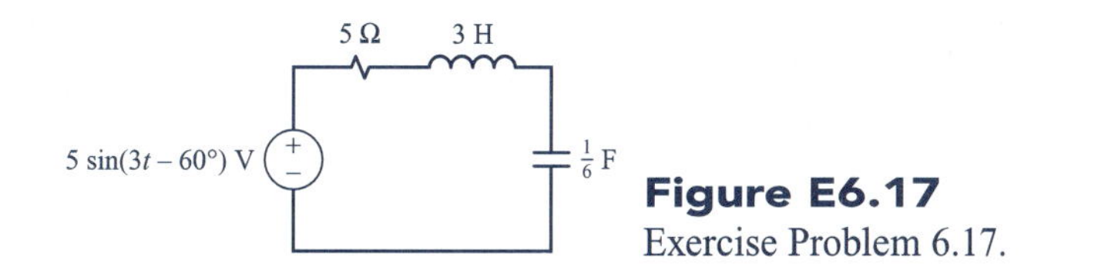

Figure E6.17
Exercise Problem 6.17.

Answer: 0.845 W and 1.182 VAR.

6.6.2 Power Factor

The ability of an impedance to absorb power is determined by its *power factor*. For example, consider the impedance $\hat{Z}$ shown in Fig. 6.28, having a phasor voltage $\hat{V} = V \angle \theta_V$ across it and a phasor current $\hat{I} = I \angle \theta_I$ through it. The complex, average, and reactive powers delivered to the impedance are

$$\begin{aligned}\hat{P} &= \tfrac{1}{2}\hat{V}\hat{I}^* \\ P_{\text{AV}} &= \frac{VI}{2}\cos(\theta_V - \theta_I) \\ Q &= \frac{VI}{2}\sin(\theta_V - \theta_I)\end{aligned} \tag{6.60a}$$

and

$$\hat{P} = P_{\text{AV}} + jQ \tag{6.60b}$$

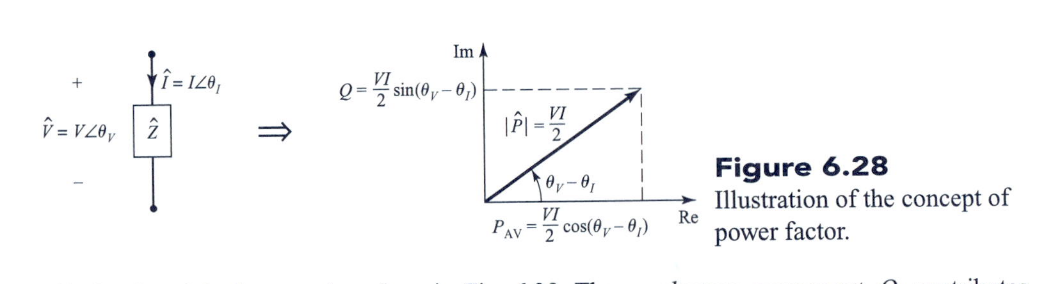

Figure 6.28 Illustration of the concept of power factor.

This is plotted in the complex plane in Fig. 6.28. The *quadrature component, Q*, contributes nothing to the average power consumption of the impedance. For a given voltage and current, the average power delivered to the impedance is a maximum when $\theta_V = \theta_I$, i.e., the impedance is purely resistive. The ability of an impedance to absorb average power depends on the *power factor*:

$$\text{pf} = \cos(\theta_V - \theta_I) \tag{6.61}$$

Hence the average power delivered to an impedance can be written in terms of the power factor of that impedance as

$$P_{AV} = \frac{1}{2} VI \times \text{pf} \tag{6.62}$$

An impedance that is purely reactive has a pf of zero, since $\theta_V - \theta_I = \pm 90°$. Hence the power factor varies between the extremes

$$0 \leq \text{pf} \leq 1 \tag{6.63}$$

If one is interested in maximizing the average power delivered to an impedance, the pf of that impedance must be unity, in which case we say that the impedance has *unity power factor*. The power factor is said to be *leading* or *lagging* depending on whether the current leads or lags the voltage. Hence *an inductive load has a lagging power factor and a capacitive load has a leading power factor*.

Example 6.15

Determine the average and reactive powers delivered to the load impedance of Fig. 6.29*a* and the power factor of that load.

Solution The phasor circuit is drawn in Fig. 6.29*b*. The phasor current and voltage of the load are

$$\begin{aligned}\hat{V}_L &= \frac{5 + j9 - j2}{4 + j6 + 5 + j9 - j2} 100 \angle 0° \\ &= 54.41\angle - 0.84°\end{aligned}$$

$$\begin{aligned}\hat{I}_L &= \frac{100 \angle 0°}{4 + j6 + 5 + j9 - j2} \\ &= 6.32\angle - 55.3°\end{aligned}$$

The average power delivered to the load can be computed by simply computing the average power delivered to the 5-Ω resistor:

$$\begin{aligned}P_{\text{AV, load}} &= \tfrac{1}{2}|\hat{I}_L|^2 \times 5 \\ &= \tfrac{1}{2}(6.32)^2 \times 5 \\ &= 100 \text{ W}\end{aligned}$$

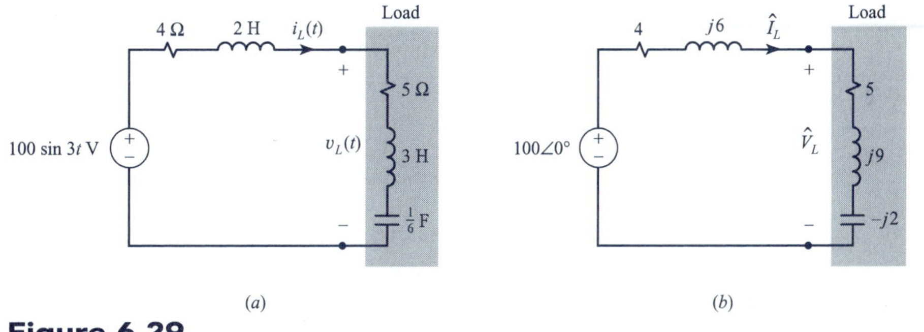

Figure 6.29
Example 6.15.

The reactive power delivered to the load is

$$\begin{aligned} Q_{\text{load}} &= \tfrac{1}{2} V_L I_L \sin(\theta_V - \theta_I) \\ &= \tfrac{1}{2}(54.41)(6.32)\sin(-0.84^\circ + 55.3^\circ) \\ &= 140 \text{ VAR} \end{aligned}$$

This could also be obtained by directly computing the complex power delivered to the load:

$$\begin{aligned} \hat{P}_{\text{load}} &= \tfrac{1}{2}\hat{V}_L \hat{I}_L^* \\ &= \tfrac{1}{2}(54.41\angle -0.84^\circ)(6.32\angle -55.3^\circ)^* \\ &= 100 + j140 \text{ VA} \end{aligned}$$

The power factor of the load is

$$\begin{aligned} \text{pf} &= \cos(\theta_V - \theta_I) \\ &= \cos(-0.84^\circ + 55.3^\circ) \\ &= 0.581 \end{aligned}$$

and is said to be *lagging,* since the current lags the voltage.

Let us now consider the distribution of power from a source to a load. Figure 6.30 shows a typical power distribution circuit. The source is represented as an ideal source having no impedance. This is typical of power distribution and is said to be an "ínfinite bus" in that the voltage supplied to consumers is virtually independent of the current drawn by each consumer. The parameter R_{line} represents the resistance of the conductors of the power distribution transmission line and is usually quite small. Some of the voltage generated by the source is dropped across the power-line resistance, but that is usually very small compared to the load voltage. Typical consumer loads consist of motors such as are contained in air conditioners, as well as lights and electric stoves. Hence it is typical to represent consumer loads as a resistance in series with an inductance. The inductance represents the windings of the motor, whereas the resistor represents the resistance of those windings as well as other devices such as household lighting and electric

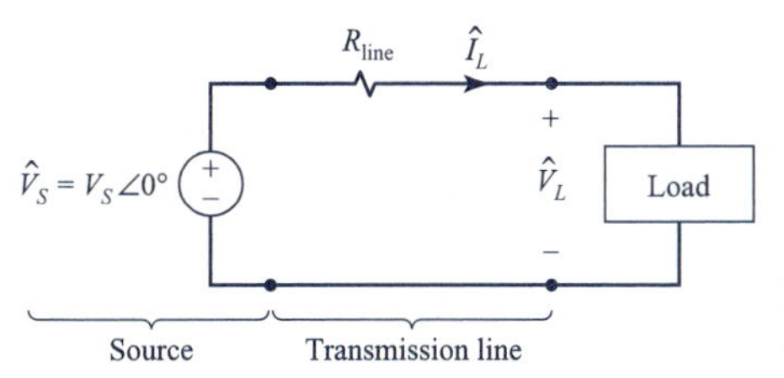

Figure 6.30
Illustration of a power transmission line connecting a source to a load.

stoves. The consumer is charged for the average power consumed by the load. The load requires a certain total apparent power, $VI/2$ VA. However, only a portion of that apparent power,

$$\frac{VI}{2}\underbrace{\cos(\theta_V - \theta_I)}_{\text{pf}}$$

is consumed in the load and thus charged to the consumer. Power factors that are less than unity require higher transmission-line currents, which generate losses in the transmission-line resistance, which the power company and not the consumer pays for. Hence, power companies desire the consumer loads to be as close to unity power factor as possible.

Example 6.16

Suppose that the load voltage in the power distribution problem of Fig. 6.30 is 170 V, the line resistance is 0.1 Ω, and the load requires 10 kW of average (useful) power. Examine the line losses for a load power factor of unity and for a power factor of 0.7 lagging.

Solution The load current is obtained from

$$P_{\text{AV}} = \tfrac{1}{2} V_L I_L\,(\text{pf})$$

For unity power factor this is

$$\begin{aligned} I_L &= \frac{2(10\text{ kW})}{(170\text{ V})(1)} \\ &= 117.65\text{ A} \end{aligned}$$

and for a power factor of 0.7 it is

$$\begin{aligned} I_L &= \frac{2(10\text{ kW})}{(170\text{ V})(0.7)} \\ &= 168.07\text{ A} \end{aligned}$$

The powers consumed in the line losses for each of these cases are

$$\begin{aligned} P_{\text{AV, line}} &= \tfrac{1}{2} I_L^2 R_{\text{line}} \\ &= \begin{cases} 692.04\text{ W} & (\text{unity pf}) \\ 1412.33\text{ W} & (\text{pf} = 0.7) \end{cases} \end{aligned}$$

Hence, the powers required to be generated by the power company are

$$P_{\text{AV, gen}} = \begin{cases} 10{,}692.04\text{ W} & (\text{unity pf}) \\ 11{,}412.33\text{ W} & (\text{pf} = 0.7) \end{cases}$$

The power company therefore is required to generate some 720 W more power to supply the customer load if the pf is 0.7 than if it is unity because of the higher line current and attendant losses in the line resistance. This illustrates why power companies may give discounts to consumers who have power factors closer to unity.

Power factors that are not unity can be brought closer to unity by adding a corrective network across its input terminals. For example, in the previous example, a capacitor can be placed across the load and correct the power factor of the load to unity. A capacitor is used because the load is inductive (lagging power factor). This is fairly common, since most industrial loads appear inductive for reasons stated above.

Example 6.17

In Example 6.16 place a capacitor across the load and determine the value of that capacitor to correct the power factor of the parallel combination of the capacitor and the original load to unity if the power frequency is 60 Hz.

Solution We assume that the load is required to consume the same average power of 10 kW and the load voltage is still 170 V. Hence the current into the load is

$$\hat{I}_L = 168.07\angle - 45.57°$$

The angle of the current is determined as $\theta_I = -\cos^{-1}(0.7) = -45.57°$ and is negative, since the power factor is lagging. The current through the added capacitor is

$$\hat{I}_C = j\omega C \times 170 \angle 0°$$

Hence the total current to the parallel combination (which is the line current) is

$$\begin{aligned}\hat{I}_{\text{line}} &= \hat{I}_L + \hat{I}_C \\ &= 168.07\angle - 45.57° + j\omega C \times 170 \angle 0° \\ &= 117.66 - j120.02 + j(2\pi \times 60)C \times 170\end{aligned}$$

Solving this to cancel out the reactive component gives

$$C = 1873\ \mu\text{F}$$

Exercise Problem 6.18

The voltage across a load is 170 V, and the load requires 5 kW of average power and has a power factor of 0.8 leading. Place an inductor across the load, and determine the value of that inductor to correct the power factor to unity if the frequency is 60 Hz.

Answer: $L = 10$ mH.

6.6.3 Maximum Power Transfer Consider the typical source–load configuration shown in Fig. 6.31. The impedance of the source, $\hat{Z}_S$, is fixed. We wish to determine the value of the load impedance, $\hat{Z}_L$, such that maximum average power will be delivered to that load. In order to determine that we represent the source and load impedances with real and imaginary parts as

$$\hat{Z}_S = R_S + jX_S \tag{6.64a}$$
$$\hat{Z}_L = R_L + jX_L \tag{6.64b}$$

The load current is

$$\begin{aligned}\hat{I}_L &= \frac{\hat{V}_S}{\hat{Z}_S + j\hat{Z}_L} \\ &= \frac{\hat{V}_S}{(R_S + R_L) + j(X_S + X_L)}\end{aligned} \tag{6.65}$$

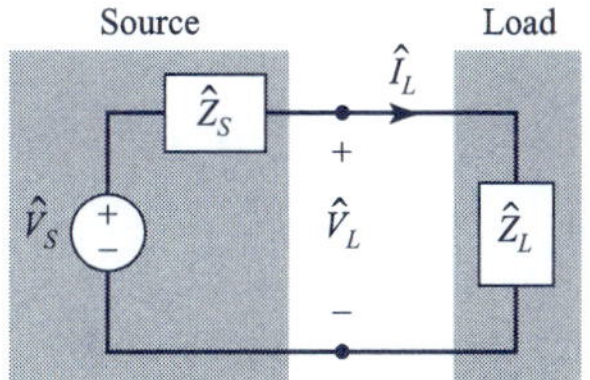

Figure 6.31
Illustration of the typical source-load configuration for determining maximum power transfer.

The average power delivered to the load is

$$P_{\text{AV, load}} = \tfrac{1}{2}|\hat{I}_L|^2 R_L$$
$$= \frac{1}{2}\frac{V_S^2 R_L}{(R_S + R_L)^2 + (X_S + X_L)^2} \tag{6.66}$$

Since the reactances can be negative, we choose

$$X_L = -X_S \tag{6.67}$$

leaving

$$P_{\text{AV, load}} = \frac{1}{2}\frac{V_S^2 R_L}{(R_S + R_L)^2} \tag{6.68}$$

However, the resistances cannot be negative, so we differentiate this expression with respect to R_L and set that result equal to zero to determine the required value of R_L to maximize this expression. Not surprisingly, the optimum value is $R_L = R_S$. Hence the desired value of $\hat{Z}_L$ that will maximize the average power delivered to it is the *conjugate of the source impedance*:

$$\boxed{\hat{Z}_L = \hat{Z}_S^*} \tag{6.69}$$

In this case we say that *the load is matched to the source*. This maximum power delivered to the load becomes

$$\boxed{P_{\text{AV, load, max}} = \frac{V_S^2}{8R_S}} \tag{6.70}$$

Exercise Problem 6.19

For the circuit of Fig. 6.29, replace the load with one that will absorb the maximum average power from the source. Determine that maximum power.

Answers: $4 - j6\ \Omega$ and 312.5 W.

6.6.4 Superposition of Average Power In all of our previous computations of average power, the circuit contained only one independent source. How do we compute average power when the circuit contains more than one (sinusoidal) source? Consider the circuit shown in Fig. 6.32*a* containing two sinusoidal sources attached to a circuit that contains no other independent sources. One element is extracted from the circuit, and we want to determine the instantaneous power delivered to it. By superposition in the time domain, the element's current and voltage will be composed of the sums of contributions due to all of the sources, as shown in Fig. 6.32*b* and *c*. Each of these contributions will also be sinusoidal at the frequency of the source producing it. As we now know, these responses will differ from the source producing it only in magnitude and phase. Denote these contributions as

$$V_S \sin(\omega_1 t + \phi_1) \Rightarrow \begin{cases} i'(t) = I_1 \sin(\omega_1 t + \theta_{I1}) \\ v'(t) = V_1 \sin(\omega_1 t + \theta_{V1}) \end{cases} \tag{6.71a}$$

$$I_S \sin(\omega_2 t + \phi_2) \Rightarrow \begin{cases} i''(t) = I_2 \sin(\omega_2 t + \theta_{I2}) \\ v''(t) = V_2 \sin(\omega_2 t + \theta_{V2}) \end{cases} \tag{6.71b}$$

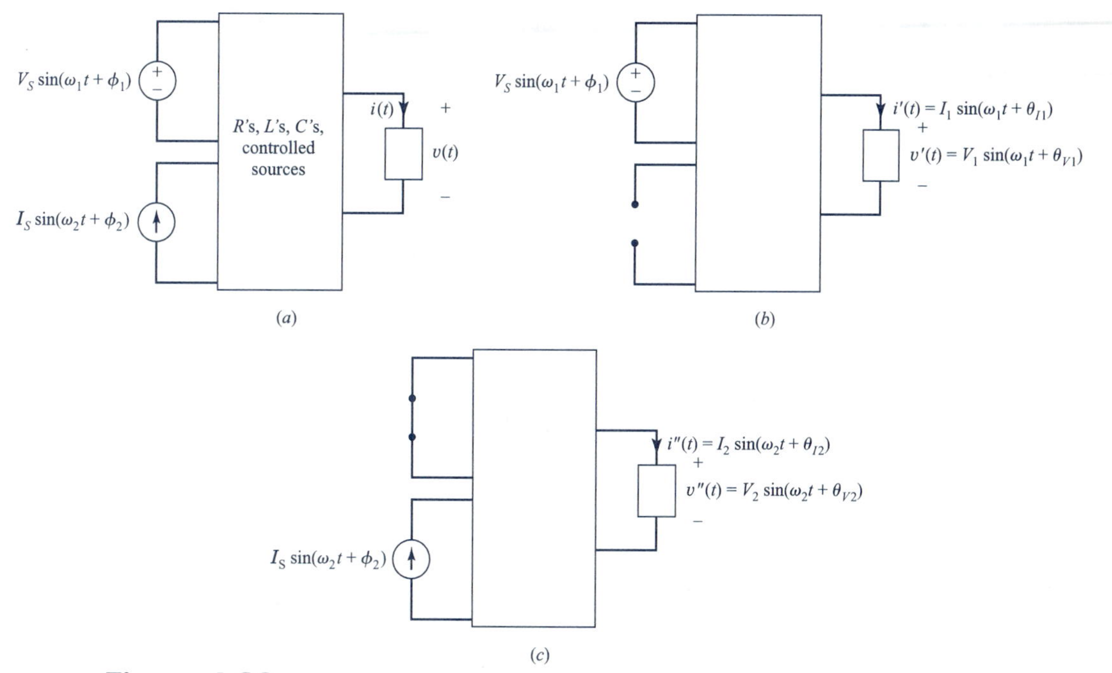

Figure 6.32
Illustration of the principle of superposition of average power for sinusoidal sources of different frequencies.

The instantaneous power delivered to the element is

$$\begin{aligned} p(t) &= v(t)i(t) \\ &= (v' + v'')(i' + i'') \\ &= (v'i' + v''i'') + (v'i'' + v''i') \end{aligned} \tag{6.72}$$

Substituting (6.71) gives

$$\begin{aligned} p(t) = &[V_1 \sin(\omega_1 t + \theta_{V1})\, I_1 \sin(\omega_1 t + \theta_{I1}) + V_2 \sin(\omega_2 t + \theta_{V2})\, I_2 \sin(\omega_2 t + \theta_{I2})] \\ &+ [V_1 \sin(\omega_1 t + \theta_{V1})\, I_2 \sin(\omega_2 t + \theta_{I2}) + V_2 \sin(\omega_2 t + \theta_{V2})\, I_1 \sin(\omega_1 t + \theta_{I1})] \end{aligned} \tag{6.73}$$

Using the trigonometric identity

$$\sin A \sin B = \tfrac{1}{2}\cos(A - B) - \tfrac{1}{2}\cos(A + B)$$

(6.73) reduces to

$$\begin{aligned} p(t) = &\underbrace{\frac{V_1 I_1}{2}\cos(\theta_{V1} - \theta_{I1})}_{P_{\text{AV1}}} + \underbrace{\frac{V_2 I_2}{2}\cos(\theta_{V2} - \theta_{I2})}_{P_{\text{AV2}}} \\ &- \frac{V_1 I_1}{2}\cos(2\omega_1 t + \theta_{V1} + \theta_{I1}) - \frac{V_2 I_2}{2}\cos(2\omega_2 t + \theta_{V2} + \theta_{I2}) \\ &+ \frac{V_1 I_2}{2}\cos[(\omega_1 - \omega_2)t + \theta_{V1} - \theta_{I2}] - \frac{V_1 I_2}{2}\cos[(\omega_1 + \omega_2)t + \theta_{V1} + \theta_{I2}] \\ &+ \frac{V_2 I_1}{2}\cos[(\omega_2 - \omega_1)t + \theta_{V2} - \theta_{I1}] - \frac{V_2 I_1}{2}\cos[(\omega_2 + \omega_1)t + \theta_{V2} + \theta_{I1}] \end{aligned} \tag{6.74}$$

The first two terms are the average powers delivered individually by the sources. Let us now suppose that the two frequencies are integer multiples of some frequency ω as $\omega_1 = n\omega$ and $\omega_2 = m\omega$. The above expression for instantaneous power becomes

$$
\begin{aligned}
p(t) = {} & \underbrace{\frac{V_1 I_1}{2}\cos(\theta_{V1} - \theta_{I1})}_{P_{AV1}} + \underbrace{\frac{V_2 I_2}{2}\cos(\theta_{V2} - \theta_{I2})}_{P_{AV2}} \\
& - \frac{V_1 I_1}{2}\cos(2n\omega t + \theta_{V1} + \theta_{I1}) - \frac{V_2 I_2}{2}\cos(2m\omega t + \theta_{V2} + \theta_{I2}) \\
& + \frac{V_1 I_2}{2}\cos[(n - m)\omega t + \theta_{V1} - \theta_{I2}] - \frac{V_1 I_2}{2}\cos[(n + m)\omega t + \theta_{V1} + \theta_{I2}] \\
& + \frac{V_2 I_1}{2}\cos[(m - n)\omega t + \theta_{V2} - \theta_{I1}] - \frac{V_2 I_1}{2}\cos[(m + n)\omega t + \theta_{V2} + \theta_{I1}]
\end{aligned} \tag{6.75}
$$

Averaging this over the common period $T = 2\pi/\omega$ gives

$$
\begin{aligned}
P_{AV} &= \frac{1}{T}\int_0^T p(t)dt \\
&= \begin{cases} P_{AV1} + P_{AV2} & \text{if} \quad n \neq m \\ P_{AV1} + P_{AV2} + \dfrac{V_1 I_2}{2}\cos(\theta_{V1} - \theta_{I2}) + \dfrac{V_2 I_1}{2}\cos(\theta_{V2} - \theta_{I1}) & \text{if} \quad n = m \end{cases}
\end{aligned} \tag{6.76}
$$

where

$$
\begin{aligned}
P_{AV1} &= \frac{V_1 I_1}{2}\cos(\theta_{V1} - \theta_{I1}) = \tfrac{1}{2}\mathrm{Re}(\hat{V}_1 \hat{I}_1^*) \\
P_{AV2} &= \frac{V_2 I_2}{2}\cos(\theta_{V2} - \theta_{I2}) = \tfrac{1}{2}\mathrm{Re}(\hat{V}_2 \hat{I}_2^*)
\end{aligned} \tag{6.77}
$$

Hence (6.76) shows that if $n \neq m$, then the average power delivered by both sources is the sum of the average powers delivered individually by each source. If $n = m$, there are additional terms. This shows that

> *We may superimpose the average powers delivered by sources of different frequencies, but we may not, in general, apply superposition to average power if the sources are of the same frequency.*

Although we may not, in general, superimpose the average powers delivered by sources of the same frequency, there are some special cases where we may do so. For example, suppose the circuit attached to the sources is purely resistive. In this case, there are no phase differences (other than a negative sign, which amounts to a phase angle of 180°) between the current or voltage and the source producing it: $\theta_{V1} = \theta_{I1} = \phi_1$ and $\theta_{V2} = \theta_{I2} = \phi_2$. Hence, $\theta_{V1} - \theta_{I2} = \phi_1 - \phi_2$ and $\theta_{V2} - \theta_{I1} = \phi_2 - \phi_1$. The cosine of these angles is zero if $\phi_1 - \phi_2 = \pm 90°$. This will be the case, for example, where both sources have the same phase angle and one is a sine and the other is a cosine. For this special case we may superimpose average powers even though the frequencies of the sources are the same. However, this is a rather rare occurrence, so that it is best not to superimpose average powers due to sources of the same frequency. For sources of the same frequency we may convert all to either cosine or sine form, place them in one phasor circuit, and compute the *total* phasor voltage and current associated with the element and hence the total average power delivered to that element. This will also take care of the above special case.

This result assumed that the frequencies of the two sources are integral multiples of some lower frequency, so that the resulting sinusoids in the power expression had a common period over which we could average the result. This result may be extended to sources whose frequencies are not so related by allowing the period over which we average to extend to infinity:

$$P_{AV} = \lim_{T \to \infty} \frac{1}{T} \int_0^T p(t)\, dt$$

These results are somewhat expected, for the following reason. Since we may not include sources of different frequencies in the same phasor circuit, it is reasonable to expect that we should be able to determine the average power supplied by each source by combining the average powers supplied by the individual sources in their individual phasor circuits, i.e., apply superposition to average powers. On the other hand, since we may include sources of the same frequency in the same phasor circuit so long as we convert both to sine or cosine form, it is sensible to expect that we might not be able to apply superposition to sources of the same frequency when both are sine or cosine form (and are placed in the same phasor circuit). In general, we may not apply superposition to instantaneous powers. The case of *average powers* due to sinusoidal sources of different frequencies represents an important special exception to this general rule. Observe that a dc source may be considered a source of zero frequency ($\omega = 0$). Hence we may not apply superposition of power to dc sources, as we already know. But we may apply superposition of power to a dc source and a sinusoidal source of nonzero frequency.

Example 6.18

Determine the average power delivered by the two sources in the circuit of Fig. 6.33*a*.

Solution The two phasor circuits are shown in Fig. 6.33*b* and *c*. In the phasor circuit for the voltage source ($\omega = 2$) in Fig. 6.33*b* we compute

$$\begin{aligned}\hat{I}' &= \frac{10 \angle 30^\circ}{2 + j4 + 1 - j1} \\ &= 2.357\angle -15^\circ\end{aligned}$$

Hence the average power delivered by the voltage source is

$$\begin{aligned}P'_{AV} &= \tfrac{1}{2}\,\mathrm{Re}(10 \angle 30^\circ \hat{I}'^*) \\ &= \tfrac{1}{2} \times 10 \times 2.357 \times \cos(30^\circ + 15^\circ) \\ &= 8.333 \text{ W}\end{aligned}$$

This may be confirmed by computing the average powers delivered to the two resistors:

$$\begin{aligned}P'_{AV} &= P'_{AV,\,2\,\Omega} + P'_{AV,\,1\,\Omega} \\ &= \tfrac{1}{2}|\hat{I}'|^2\, 2 + \tfrac{1}{2}|\hat{I}'|^2\, 1 \\ &= 8.333 \text{ W}\end{aligned}$$

In the phasor circuit for the voltage source ($\omega = 3$) in Fig. 6.33*c* we compute, by current division,

$$\begin{aligned}\hat{I}''_x &= \frac{1 - j\frac{2}{3}}{2 + j6 + 1 - j\frac{2}{3}}\, 3\angle -60^\circ \\ &= 0.589\angle -154.33^\circ\end{aligned}$$

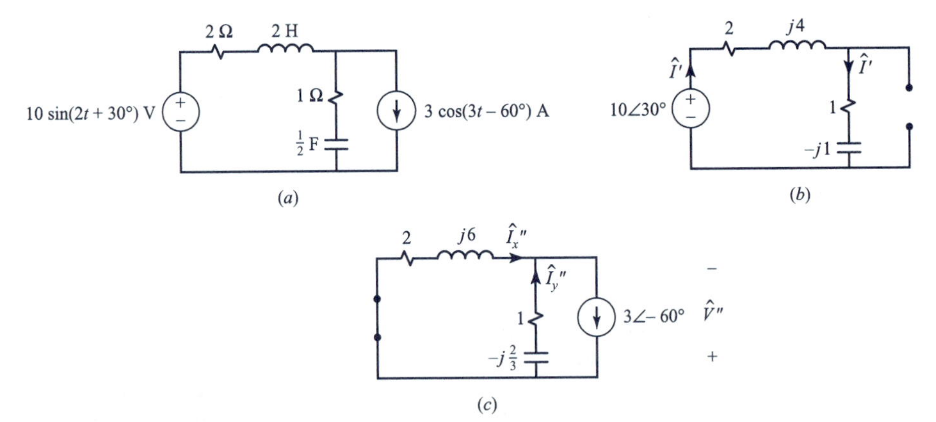

Figure 6.33
Example 6.18.

and

$$\hat{I}_y'' = \frac{2 + j6}{2 + j6 + 1 - j\frac{2}{3}} 3\angle - 60^\circ$$
$$= 3.101\angle - 49.08^\circ$$

Thus the voltage across the current source is

$$\hat{V}'' = (2 + j6)\hat{I}_x''$$
$$= 3.727\angle - 82.77^\circ$$

Hence the average power delivered by the current source is

$$P_{AV}'' = \tfrac{1}{2}\,\mathrm{Re}(\hat{V}''3 \angle\, 60^\circ)$$
$$= \tfrac{1}{2} \times 3.727 \times 3 \times \cos(-82.77^\circ + 60^\circ)$$
$$= 5.154 \text{ W}$$

This may be again confirmed by computing the average powers delivered to the two resistors:

$$P_{AV}'' = P_{AV,\,2\,\Omega}'' + P_{AV,\,1\,\Omega}''$$
$$= \tfrac{1}{2}|\hat{I}_x''|^2\, 2 + \tfrac{1}{2}|\hat{I}_y''|^2\, 1$$
$$= 5.154 \text{ W}$$

Since the frequencies of the two sources are not the same, the total average power delivered by the sources is the sum of the average powers delivered individually by each source: 8.333 + 5.154 = 13.487 W. This total average power is consumed by the resistors.

Exercise Problem 6.20

Determine the average power delivered by the two sources of Fig. E6.20.

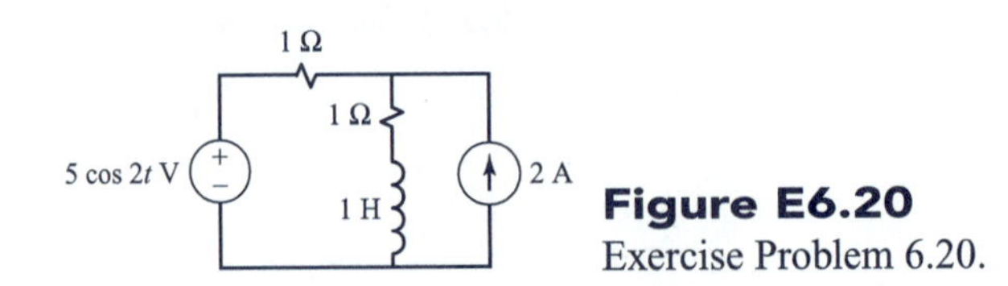

Figure E6.20
Exercise Problem 6.20.

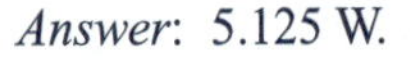
Answer: 5.125 W.

Example 6.19

Determine the average power delivered by the two sources in Fig. 6.34*a*.

Solution Since both sources have the same frequency, we cannot use superposition to determine the average powers delivered by them. However, we may include both sources in one phasor circuit. Converting the current source to cosine form gives the phasor circuit shown in Fig. 6.34*b*. The total average power delivered by the sources is equal to the average power delivered to the 2-Ω resistor. This may be determined by computing the current through that resistor. In order to compute that current we may use superposition in the frequency domain. The needed circuits are shown in Fig. 6.34*c* and *d*. In Fig. 6.34*c* we obtain

$$\begin{aligned}\hat{I}' &= \frac{10\angle 0°}{2 + j4 - j2} \\ &= 3.536\angle -45°\end{aligned}$$

From Fig. 6.34*d* we obtain

$$\begin{aligned}\hat{I}'' &= -\frac{-j2}{2 + j4 - j2}\, 5\angle -60° \\ &= 3.536\angle -15°\end{aligned}$$

The total phasor current is

$$\begin{aligned}\hat{I} &= \hat{I}' + \hat{I}'' \\ &= 3.536\angle -45° + 3.536\angle -15° \\ &= 6.831\angle -30°\end{aligned}$$

Hence, the average power delivered to the resistor and by the sources is

$$\begin{aligned}P_{AV} &= \tfrac{1}{2}|\hat{I}|^2 \times 2 \\ &= 46.66 \text{ W}\end{aligned}$$

Note that we may not superimpose the average powers delivered to the resistors by the individual sources:

$$\tfrac{1}{2}|\hat{I}'|^2 \times 2 + \tfrac{1}{2}|\hat{I}''|^2 \times 2 = 25 \neq 46.66$$

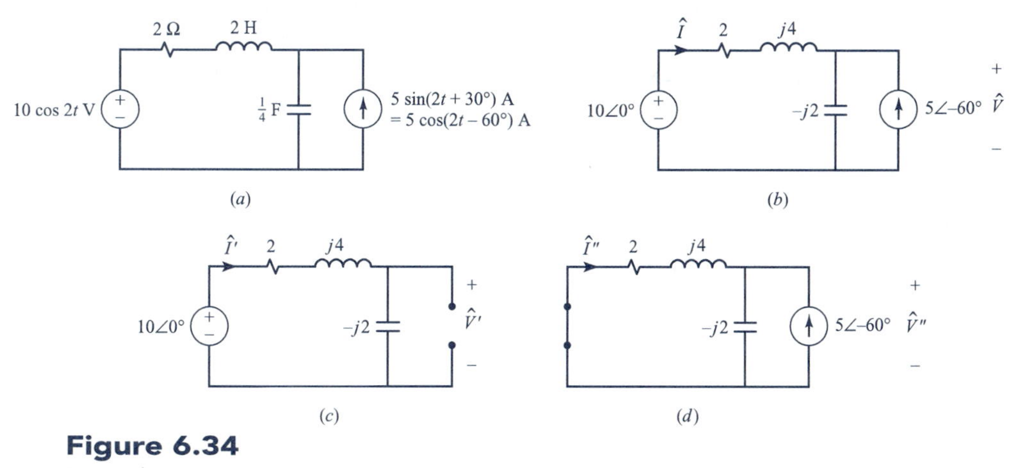

Figure 6.34
Example 6.19.

We can compute this total average power by directly computing the average powers delivered by the sources from Fig. 6.34*b*. The voltage across the current source is

$$\begin{aligned}\hat{V} &= 10 \angle 0° - (2 + j4)\hat{I} \\ &= 22.88\angle - 132.63°\end{aligned}$$

The average power delivered by the voltage source is

$$\begin{aligned}P_{\text{AV, voltage source}} &= \tfrac{1}{2}\,\text{Re}[(10 \angle 0°)\hat{I}^*] \\ &= 29.58 \text{ W}\end{aligned}$$

The average power delivered by the current source is

$$\begin{aligned}P_{\text{AV, current source}} &= \tfrac{1}{2}\,\text{Re}[\hat{V}(5 \angle 60°)] \\ &= 17.08 \text{ W}\end{aligned}$$

The total average power delivered by the sources is

$$\begin{aligned}P_{\text{AV, sources}} &= 29.58 + 17.08 \\ &= 46.66 \text{ W}\end{aligned}$$

which equals the average power delivered to the resistor.

Exercise Problem 6.21

Determine the average power delivered by the two sources in Fig. E6.21.

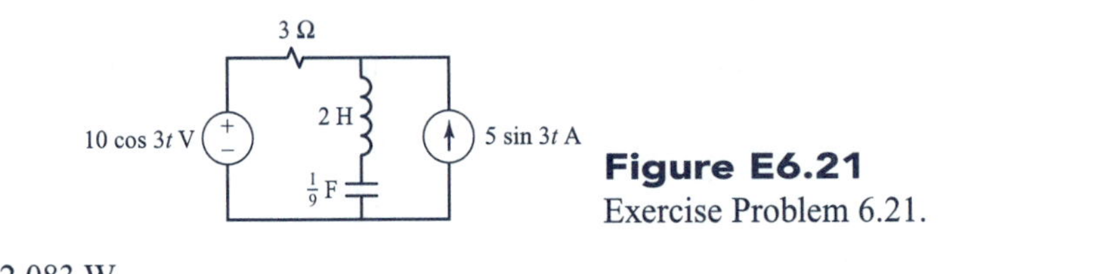

Figure E6.21
Exercise Problem 6.21.

Answer: 2.083 W.

6.6.5 Effective (RMS) Values of Periodic Waveforms

Let us now consider the average power delivered by more general periodic waveforms (of which the sinusoid is one). Consider applying a current source to a resistor as shown in Fig. 6.35. Suppose the waveform is periodic as shown in Fig. 6.2, so that it has the property

$$i(t) = i(t \pm nT) \qquad n = 0, 1, 2, 3, \ldots \tag{6.78}$$

and T is the period of the waveform. The instantaneous power delivered to the resistor is

$$p(t) = i^2(t)R \tag{6.79}$$

The average power delivered to the resistor is

$$\begin{aligned}P_{\text{AV}} &= \frac{1}{T}\int_0^T i^2(t)R\,dt \\ &= R\underbrace{\left[\frac{1}{T}\int_0^T i^2(t)\,dt\right]}_{I_{\text{eff}}^2}\end{aligned} \tag{6.80}$$

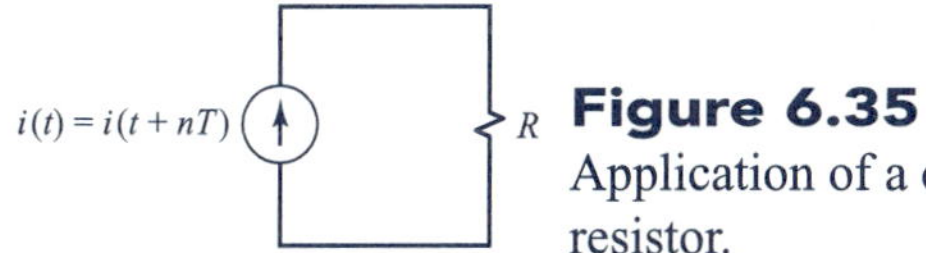

Figure 6.35
Application of a current source having a periodic waveform to a resistor.

Hence the average power delivered to the resistor by this periodic waveform can be viewed as equivalent to that produced by a dc waveform whose value is

$$I_{\text{eff}} = \sqrt{\frac{1}{T}\int_0^T i^2(t)\,dt} \qquad (6.81)$$

This is said to be the *effective value* of the waveform or the *root-mean-square* (rms) value of the waveform. A periodic voltage waveform has an rms value defined in the same way and for the same purpose.

Example 6.20

Determine the rms value of the current waveform shown in Fig. 6.36 and the average power this would deliver to a 3-Ω resistor.

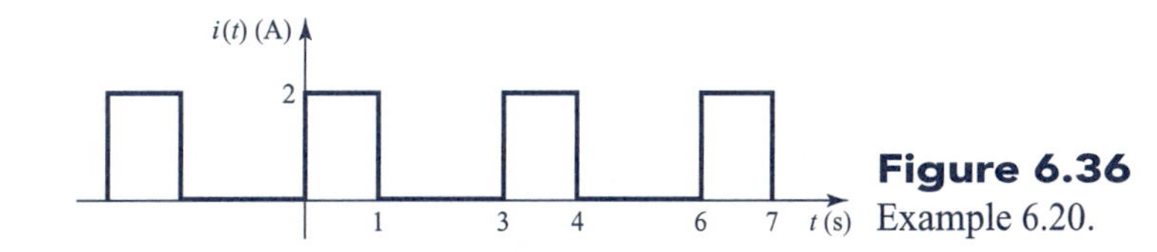

Figure 6.36
Example 6.20.

Solution The rms value of the waveform is

$$I_{\text{rms}} = \sqrt{\frac{1}{3}\int_0^1 4\,dt}$$
$$= 1.155 \text{ A}$$

Hence the average power delivered to the 3-Ω resistor is

$$P_{\text{AV}} = I_{\text{rms}}^2 \times 3\ \Omega$$
$$= 4 \text{ W}$$

Exercise Problem 6.22

For the circuit of Fig. E6.22 determine the rms voltage of the source and the average power delivered to the 5-Ω resistor.

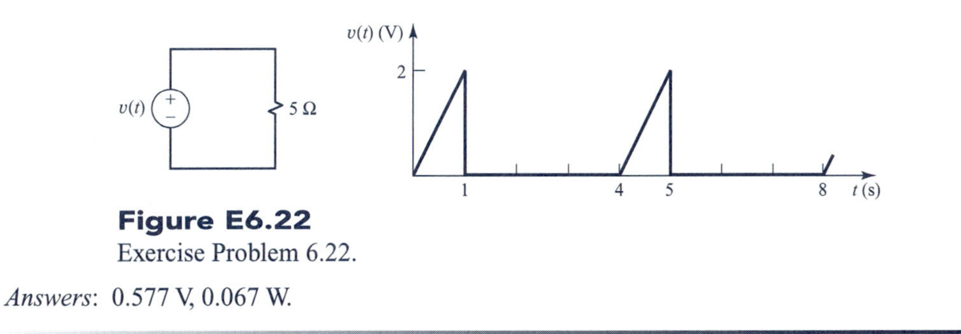

Figure E6.22
Exercise Problem 6.22.

Answers: 0.577 V, 0.067 W.

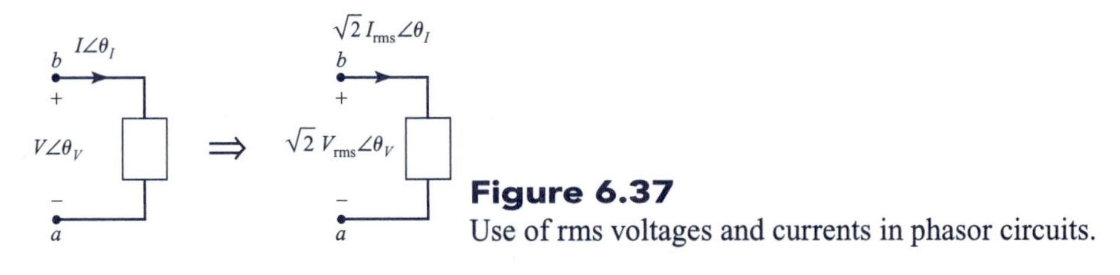

Figure 6.37
Use of rms voltages and currents in phasor circuits.

The sinusoid $x(t) = X\sin(\omega t + \phi)$ has a rms value of

$$\begin{aligned} X_{rms} &= \sqrt{\frac{1}{T}\int_0^T [X\sin(\omega t + \phi)]^2\,dt} \\ &= \frac{X}{\sqrt{2}} \\ &= 0.707\,X \end{aligned} \tag{6.82}$$

Hence the average power delivered to a resistor by a sinusoidal voltage or current waveform is

$$\begin{aligned} P_{AV,R} &= \frac{1}{2}\frac{V^2}{R} = \frac{V_{rms}^2}{R} \\ &= \tfrac{1}{2}I^2R = I_{rms}^2R \end{aligned} \tag{6.83}$$

In general, the average power delivered to an element shown in Fig. 6.37 is

$$\begin{aligned} P_{AV} &= \tfrac{1}{2}\,\mathrm{Re}(\hat{V}\hat{I}^*) \\ &= \mathrm{Re}\left(\frac{\hat{V}}{\sqrt{2}}\,\frac{\hat{I}^*}{\sqrt{2}}\right) \\ &= \mathrm{Re}(\hat{V}_{rms}\,\hat{I}^*_{rms}) \end{aligned} \tag{6.84}$$

Therefore, *if sinusoidal voltages and currents are specified in their rms values rather than their peak values, the factor* $\frac{1}{2}$ *is removed from all average-power expressions.* This is the only effect of specifying voltage and current values in rms rather than peak as we have been doing. However, the time-domain expressions require a magnitude that is the rms magnitude multiplied by $\sqrt{2}$:

$$\begin{aligned} X\sin(\omega t + \phi) &= \sqrt{2}\,X_{rms}\sin(\omega t + \phi) \\ &\neq X_{rms}\sin(\omega t + \phi) \end{aligned}$$

since X is the peak value of the waveform. Common household voltages are specified as 120 V. This is its rms value, and the peak value is 170 V.

Example 6.21

Determine the average power delivered by the source and the time-domain current $i(t)$ in the circuit of Fig. 6.38*a*.

Solution The phasor circuit is drawn in Fig. 6.38*b*, where we have specified the source sinusoid in terms of its rms rather than peak value. The phasor current is

$$\begin{aligned} \hat{I}_{rms} &= \frac{7.07\angle 30^\circ}{4 + j4} \\ &= 1.25\angle -15^\circ \end{aligned}$$

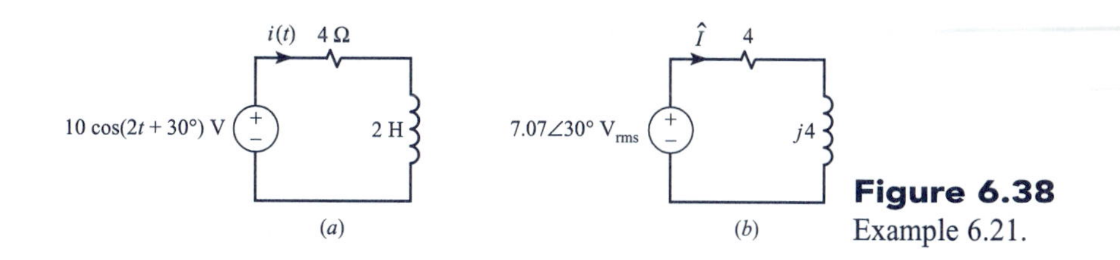

Figure 6.38
Example 6.21.

Hence the average power delivered by the source is

$$
\begin{aligned}
P_{\text{AV, source}} &= \text{Re}\, \hat{V}_{\text{rms}} \hat{I}^*_{\text{rms}} \\
&= \text{Re}[(7.07 \angle 30°)(1.25 \angle 15°)] \\
&= 6.25 \text{ W} \\
&= \tfrac{1}{2} \text{Re}[(10 \angle 30°)(1.25 \sqrt{2} \angle 15°)]
\end{aligned}
$$

The first power expression uses rms values of the sinusoids and therefore requires no factor $\frac{1}{2}$. The second power expression uses peak values and requires a factor $\frac{1}{2}$. Both give the same value of average power. The time-domain current is

$$
\begin{aligned}
i(t) &= \sqrt{2} I_{\text{rms}} \cos(2t - 15°) \\
&= 1.77 \cos(2t - 15°) \text{ A}
\end{aligned}
$$

One final point should be made. If a current is composed of the sum of sinusoidal components of *different* frequencies:

$$i(t) = I_1 \sin(\omega_1 t + \theta_1) + I_2 \sin(\omega_2 t + \theta_2) + I_3 \sin(\omega_3 t + \theta_3) + \cdots \tag{6.85}$$

then the rms, or effective, value is

$$I_{\text{rms}} = \sqrt{I^2_{1,\text{rms}} + I^2_{2,\text{rms}} + I^2_{3,\text{rms}} + \cdots} \tag{6.86}$$

This can be demonstrated by recalling that if the current is applied to a resistor R, the average power delivered to the resistor is

$$
\begin{aligned}
P_{\text{AV}} &= \frac{I_1^2}{2} R + \frac{I_2^2}{2} R + \frac{I_3^2}{2} R + \cdots \\
&= (I^2_{1,\text{rms}} + I^2_{2,\text{rms}} + I^2_{3,\text{rms}} + \cdots) R
\end{aligned} \tag{6.87}
$$

because we can superimpose average powers for sources of different frequencies. But in terms of the rms value of $i(t)$,

$$
\begin{aligned}
P_{\text{AV}} &= \frac{I^2}{2} R \\
&= I^2_{\text{rms}} R
\end{aligned} \tag{6.88}
$$

Comparing (6.87) and (6.88), we obtain (6.86). The result similarly applies to a voltage

$$v(t) = V_1 \sin(\omega_1 t + \theta_1) + V_2 \sin(\omega_2 t + \theta_2) + V_3 \sin(\omega_3 t + \theta_3) + \cdots \tag{6.89}$$

as

$$V_{\text{rms}} = \sqrt{V^2_{1,\text{rms}} + V^2_{2,\text{rms}} + V^2_{3,\text{rms}} + \cdots} \tag{6.90}$$

6.7 Phasor Diagrams

Recall that the phasor voltages and currents have the factor $e^{j\omega t}$ removed. To go back to the time domain, we multiply the phasor voltage or current by this factor and take the real or imaginary part of the result, according as the original source was a cosine or a sine:

$$v(t) = \text{Re}(\hat{V}e^{j\omega t}) \tag{6.91a}$$

or

$$v(t) = \text{Im}(\hat{V}e^{j\omega t}) \tag{6.91b}$$

This process of relating the phasor (frequency-domain) quantities to their time-domain equivalents can be easily visualized in the complex plane. Not only does this add insight, but it also can serve as a visual check on the correctness of our phasor computations. For example, consider Fig. 6.39, where we have shown a unit-magnitude phasor in the complex plane. Multiplying this by the $e^{j\omega t}$ terms gives $1e^{j\omega t} = 1\cos\omega t + j1\sin\omega t$. The time-domain results can then be viewed as the *projections of this rotating phasor on the real or imaginary axes of the complex plane* as shown in Fig. 6.39.

Consider the circuit shown in Fig. 6.40*a*. The phasor circuit is drawn in Fig. 6.40*b*. From that circuit we obtain

$$\hat{I}_R = \hat{I}_C = \frac{\hat{V}_S = 10\angle 30^\circ}{3 - j3} = \frac{10}{3\sqrt{2}}\angle 75^\circ$$

The individual voltages across the resistor and the capacitor are

$$\hat{V}_R = 3\hat{I}_R = \frac{10}{\sqrt{2}}\angle 75^\circ$$

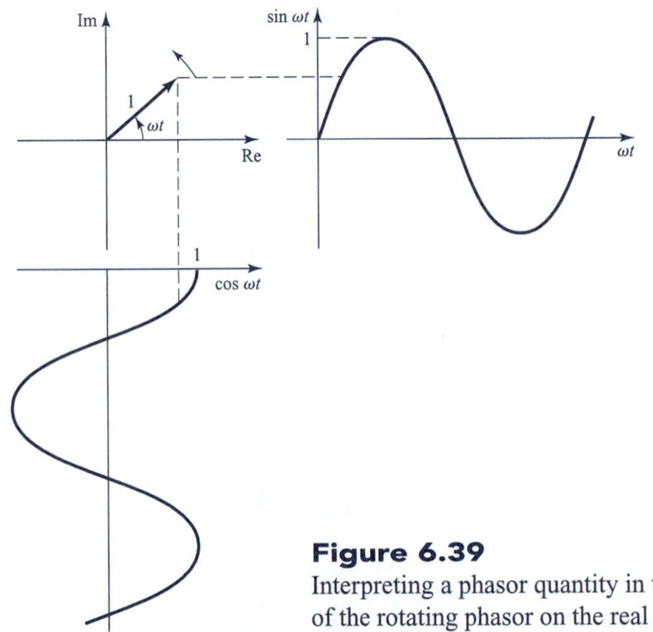

Figure 6.39
Interpreting a phasor quantity in the time domain as the projection of the rotating phasor on the real axis (for a cosine) or on the imaginary axis (for a sine).

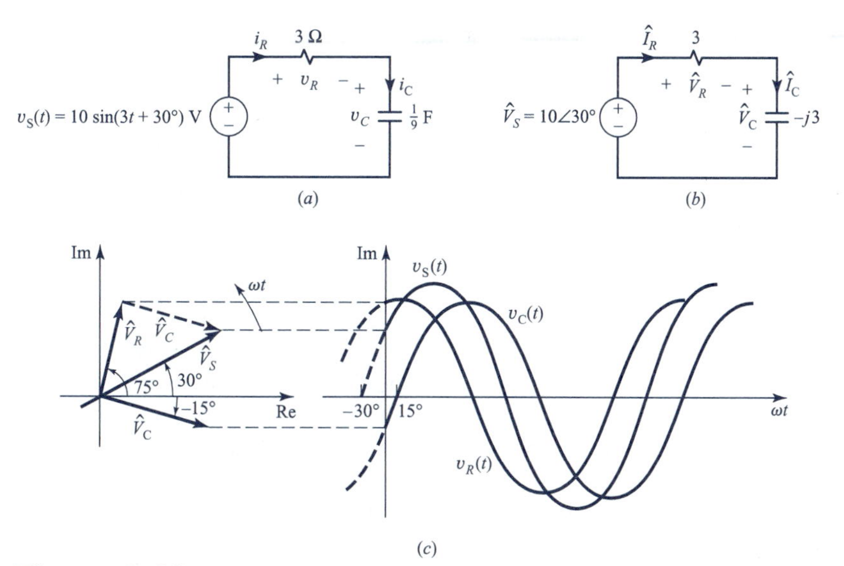

Figure 6.40
An example illustrating the visualization of phasor voltages and currents in a phasor diagram: (*a*) the time-domain circuit, (*b*) the phasor circuit, and (*c*) the phasor diagram.

and

$$\hat{V}_C = -j3\hat{I}_C$$
$$= \frac{10}{\sqrt{2}}\angle - 15°$$

Plotting these in the complex plane as shown in Fig. 6.40*c*, we see that the vector addition visually confirms that KVL is satisfied:

$$\hat{V}_S = \hat{V}_R + \hat{V}_C$$

Multiplication of these phasor quantities by $e^{j\omega t}$ amounts to adding the phase angle $\angle\, \omega t$ to all vectors so that they rotate counterclockwise at ω rad/s. Since the source was a sine form, we take the imaginary part of these to return to the time domain. This amounts to taking the projections of the vectors on the imaginary axis as they rotate, giving the resulting time-domain voltages shown in Fig. 6.40*c*. Hence the voltages are displaced in time according to their phase angles.

Exercise Problem 6.23

Determine the currents through the resistor and the inductor in the circuit of Fig. E6.23, draw the phasor diagram, and confirm visually that they satisfy KCL.

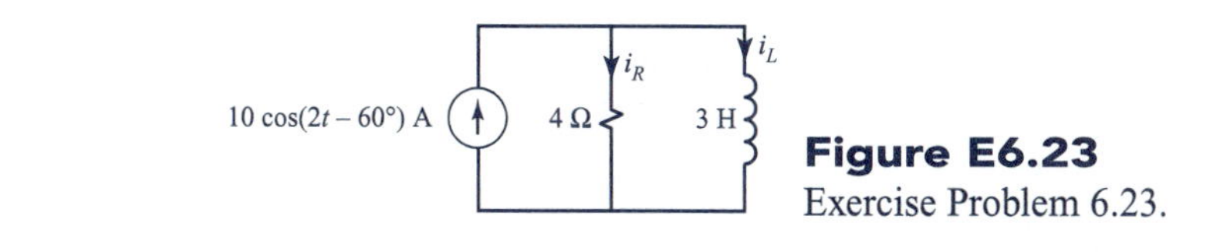

Figure E6.23
Exercise Problem 6.23.

Answer: $i_R(t) = 8.321 \cos(2t - 26.31°)$ A, $i_L(t) = 5.547 \cos(2t - 116.31°)$ A.

6.8 Frequency Response of Circuits

Previously we have been interested in determining the response of a circuit to a *single-frequency* sinusoid. Except for commercial power distribution problems, there is little interest in single-frequency signals, since they carry no information. It was pointed out at the beginning of this chapter that our true interest in the response of a circuit to a sinusoidal source stems from the important fact that practical waveforms (voice, music, radio transmissions, etc.) can be viewed with Fourier methods as being composed of sinusoidal components of various frequencies. Hence, in order to determine how a circuit *processes*, or modifies, these waveforms, we can alternatively investigate how the circuit *individually processes* these single-frequency, sinusoidal components. Again, this is the primary reason we investigate the response of *linear* circuits to single-frequency, sinusoidal sources. If the circuit is linear, we may determine the response to the general waveform alternatively by adding the responses to its individual, single-frequency components, using superposition. Although this seems to be a rather circuitous way of determining the response to some general waveform, it has the advantages that it is often simpler than a direct solution for the general waveform, and, more importantly, it clearly shows how the circuit has processed (altered) the individual sinusoidal components to yield the resulting response waveform. Investigating how the circuit modifies the individual sinusoidal components of a general waveform provides great insight into why the response waveform has occurred and, perhaps, how to redesign the circuit to give some more desirable response. An example of such a use is the design of *electrical filters*. Radio transmissions consist of the information (voice or music) superimposed on some carrier frequency. When we tune a radio we are simply moving a *bandpass filter* to that carrier frequency which passes the associated information and rejects other information associated with other carrier frequencies. Without the bandpass filter, modern radio communication would not be possible, since the radio output would be a noisy combination of all radio transmissions that are being simultaneously broadcast.

6.8.1 Transfer Functions

In order to investigate how a circuit processes signals having multiple frequency components, we require a mathematical relation between the phasor output and the phasor input (source) *as a function of frequency*. Such a relationship is called a *transfer function*. In order to compute a transfer function, we draw the phasor circuit and leave the radian frequency of the source, ω, undesignated. An economical method of doing so is to use the shorthand technique of letting the symbol p stand for $j\omega$. Hence we replace the elements with the impedances as shown in Fig. 6.41. Henceforth we will assume that there is only one independent source, the *input*, and we wish to investigate the response of a voltage or current, the *output*. Therefore we will symbolically denote the process as a *system* with a single input (the independent source), $x(t)$, and an output (the desired response voltage or current), $y(t)$, as shown in Fig. 6.42*a*. The phasor system (circuit) is symbolically denoted as shown in Fig. 6.42*b*. The *transfer function* is the ratio of the phasor output to the phasor input and is denoted as $\hat{H}(p)$:

$$\boxed{\hat{H}(p) = \frac{\hat{Y}}{\hat{X}}} \tag{6.92}$$

Once we determine this transfer function, we substitute

$$p \rightarrow j\omega$$

to obtain the desired transfer function $\hat{H}(j\omega)$.

$L \Rightarrow pL$ $\quad C \Rightarrow \frac{1}{pC}$ $\quad R \Rightarrow R$

Figure 6.41
Element impedances for computing transfer functions where $p = j\omega$.

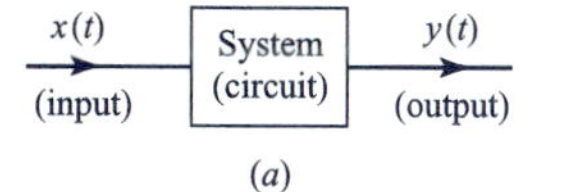

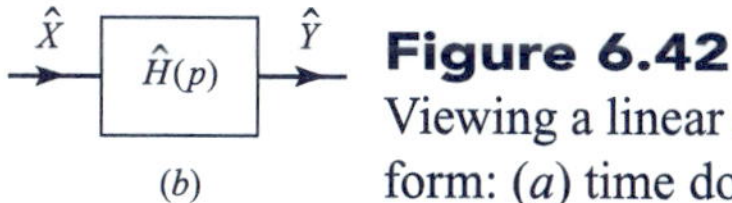

Figure 6.42
Viewing a linear system or electric circuit in block diagram form: (*a*) time domain, and (*b*) frequency domain.

Example 6.22

Determine the transfer function for the circuit of Fig. 6.43*a* where the input is the source and the output is the voltage across the capacitor.

Solution The phasor circuit is drawn in Fig. 6.43*b* with p denoting $j\omega$. From that circuit we determine, by voltage division,

$$\hat{V} = \frac{\frac{1}{pC}}{R + pL + \frac{1}{pC}}\hat{V}_S$$

$$= \frac{1}{p^2 LC + pRC + 1}\hat{V}_S$$

$$= \frac{1}{LC}\frac{1}{p^2 + \frac{R}{L}p + \frac{1}{LC}}\hat{V}_S$$

Hence the transfer function becomes

$$\hat{H}(p) = \frac{\hat{V}}{\hat{V}_S}$$

$$= \frac{1}{LC}\frac{1}{p^2 + \frac{R}{L}p + \frac{1}{LC}}$$

Although not strictly necessary, we have written the polynomial in the denominator so that the coefficient of the highest power of p, p^2, is unity. We will follow this practice in writing all future transfer functions. Finally we substitute $p \to j\omega$ to obtain

$$\hat{H}(j\omega) = \frac{\hat{V}}{\hat{V}_S}$$

$$= \frac{1}{LC}\frac{1}{(j\omega)^2 + \frac{R}{L}(j\omega) + \frac{1}{LC}}$$

$$= \frac{1}{LC}\frac{1}{\left(\frac{1}{LC} - \omega^2\right) + j\omega\frac{R}{L}}$$

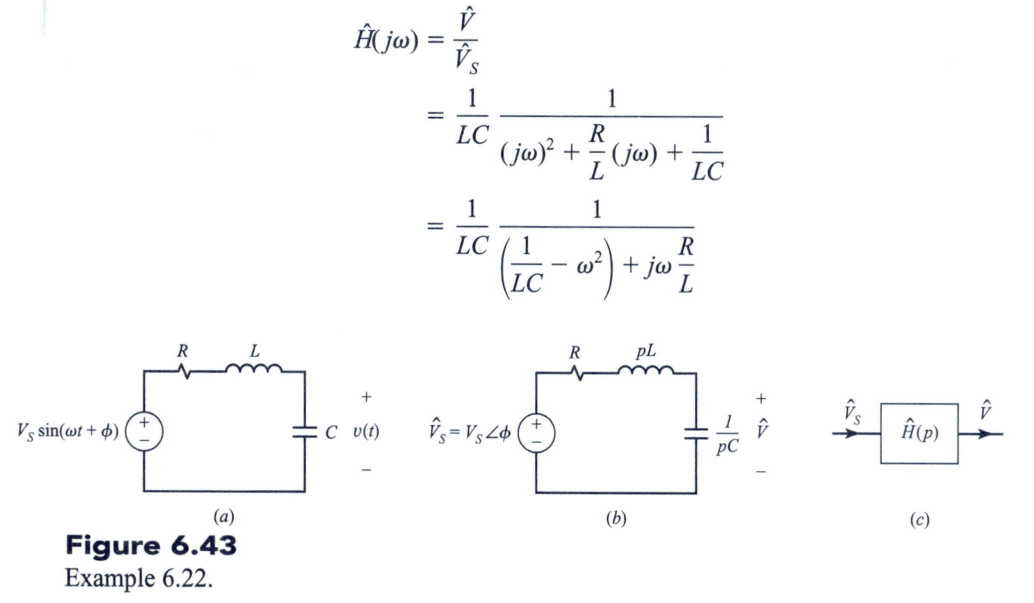

Figure 6.43
Example 6.22.

The transfer function has a magnitude and a phase angle

$$|\hat{H}(j\omega)| = H(j\omega)$$
$$= \frac{1}{LC}\frac{1}{\sqrt{\left(\frac{1}{LC}-\omega^2\right)^2+\left(\omega\frac{R}{L}\right)^2}}$$
$$\angle\,\hat{H}(j\omega) = -\tan^{-1}\left(\frac{\omega\frac{R}{L}}{\frac{1}{LC}-\omega^2}\right)$$

Figure 6.43*c* illustrates visualizing this as a single-input, single-output system.

Exercise Problem 6.24

Determine the transfer function for the circuit of Fig. E6.24 where the input is the source and the output is the current through the capacitor.

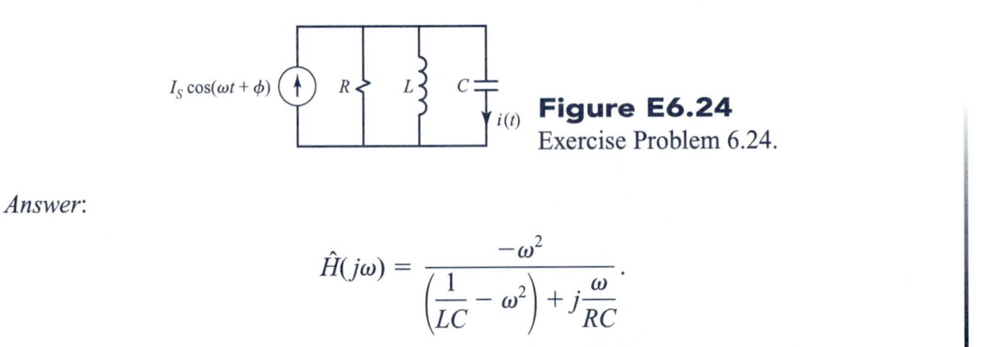

Figure E6.24
Exercise Problem 6.24.

Answer:

$$\hat{H}(j\omega) = \frac{-\omega^2}{\left(\frac{1}{LC}-\omega^2\right)+j\frac{\omega}{RC}}.$$

The utility of the transfer function is that it allows a rapid sketch of the magnitude and phase angle of the response *as the frequency of the input (the source) is varied*. Since the transfer function relates the output and input as

$$\hat{Y}(j\omega) = \hat{H}(j\omega)\hat{X}(j\omega) \tag{6.93}$$

then

$$|\hat{Y}| = |\hat{H}||\hat{X}| \tag{6.94a}$$

and

$$\angle\,\hat{Y} = \angle\,\hat{H} + \angle\,\hat{X} \tag{6.94b}$$

Example 6.23

Sketch the magnitude and phase angle of the resistor current in the circuit of Fig. 6.44*a* to a unity-amplitude and zero-phase source as the frequency is varied from dc to infinity.

Solution The phasor circuit is drawn in Fig. 6.44*b*. From that we obtain the transfer function as

$$\hat{H} = \frac{\hat{I}}{\hat{I}_S} = \frac{\dfrac{1}{pC}}{R + \dfrac{1}{pC}}$$

$$= \frac{1}{RC} \frac{1}{p + \dfrac{1}{RC}}$$

Substituting $p \to j\omega$ yields

$$\hat{H}(j\omega) = \frac{1}{RC} \frac{1}{\dfrac{1}{RC} - j\omega}$$

so that

$$|\hat{H}| = \frac{1}{RC} \frac{1}{\sqrt{\left(\dfrac{1}{RC}\right)^2 + \omega^2}}$$

$$\angle \hat{H} = -\tan^{-1}(-\omega RC)$$

At dc ($\omega = 0$), we have

$$|\hat{H}(0)| = 1$$

$$\angle \hat{H}(0) = 0°$$

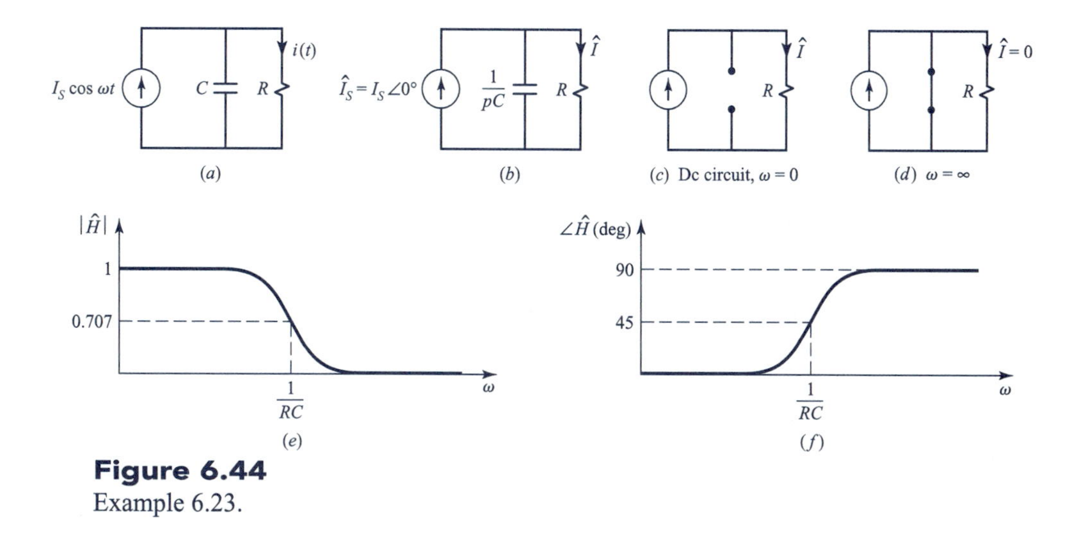

Figure 6.44
Example 6.23.

This may be confirmed from the dc circuit (capacitor an open circuit) shown in Fig. 6.44*c*. At $\omega = \infty$, the response is

$$|\hat{H}(\infty)| = 0$$
$$\angle \hat{H}(\infty) = 90°$$

The magnitude of this result is confirmed by replacing the capacitor with a short circuit (since its impedance at $\omega = \infty$ is zero) as shown in Fig. 6.44*d*. When the real and imaginary parts of the denominator are equal, $\omega = 1/RC$, the magnitude of the transfer function is $1/\sqrt{2}$, and the phase angle is 45°. This is plotted in Fig. 6.44*e* and *f*. From this result we see that high frequency components of the source (frequencies above $\omega = 1/RC$) are not passed to the output, but low frequency components are passed essentially unchanged in magnitude and angle to the output. Hence we might classify this as a *lowpass filter*.

Exercise Problem 6.25

Sketch the magnitude and phase angle of the resistor voltage in the circuit of Fig. E6.25*a* to a unity amplitude and zero phase source as the frequency is varied from dc to infinity.

Answer: The result is shown in Fig. E6.25*b*.

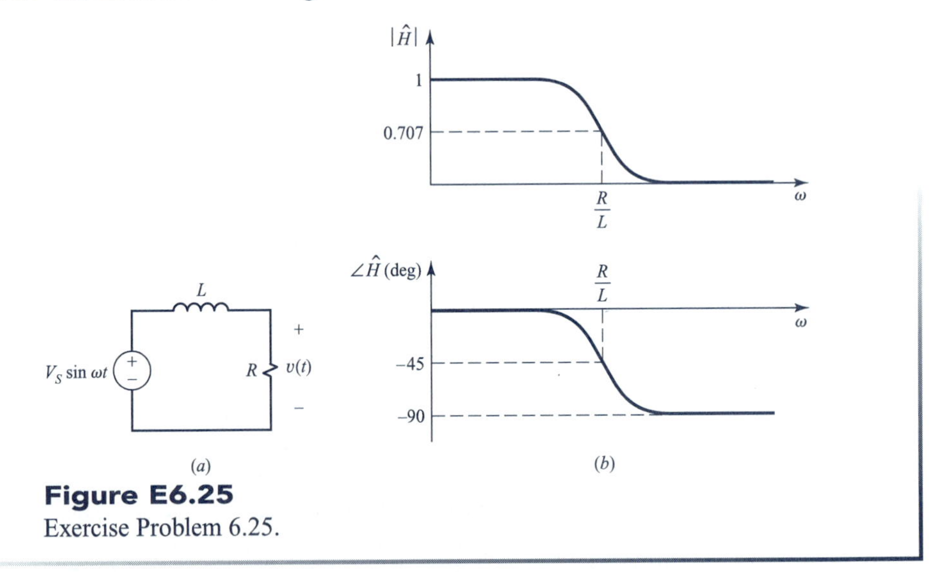

Figure E6.25
Exercise Problem 6.25.

6.8.2 Resonance The combination of an inductor and a capacitor in series or in parallel can provide an interesting and useful phenomenon known as *resonance*. For example, consider the series connection of an inductor and a capacitor shown in Fig. 6.45*a*. The phasor circuit is shown in Fig. 6.45*b*, and from that we obtain the equivalent impedance of the series connection as

$$\begin{aligned}\hat{Z}_{eq} &= j\omega L + \frac{1}{j\omega C}\\ &= j\omega L - j\frac{1}{\omega C}\\ &= j\left(\omega L - \frac{1}{\omega C}\right)\end{aligned} \tag{6.95}$$

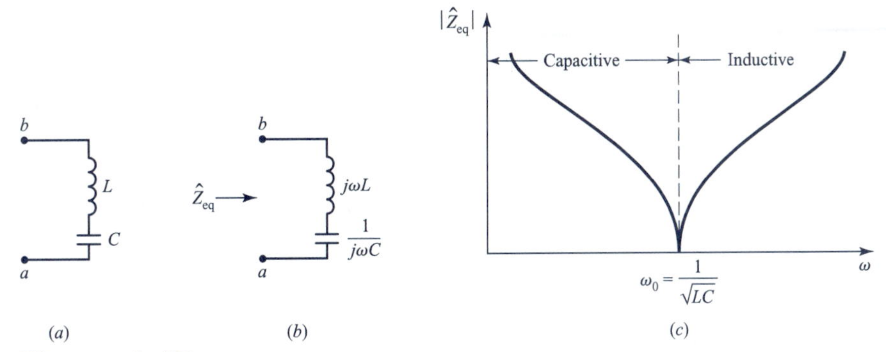

Figure 6.45
Illustration of series resonance: (*a*) the series *LC* circuit, (*b*) the phasor equivalent, and (*c*) a plot of the magnitude of the series impedance versus radian frequency showing that the impedance is zero at $\omega_0 = 1/\sqrt{LC}$, where the series combination acts like a short circuit.

Observe that at a frequency of

$$\omega_0 = \frac{1}{\sqrt{LC}} \tag{6.96}$$

the impedance of this series combination is zero. Hence if the series connection is excited at the *resonant frequency* ω_0 the series combination is equivalent to a *short circuit.* Below this resonant frequency, the phase angle of the impedance of the series combination approaches $-90°$ because the term $-j1/\omega C$ is much larger than the term $j\omega L$. Hence below the resonant frequency, the series combination appears capacitive. Above the resonant frequency the reverse is true and the phase angle of the impedance of the series combination is $+90°$. Hence, above the resonant frequency the series combination appears inductive. The magnitude of the frequency response of this series combination is sketched in Fig. 6.45*c*.

Now consider the parallel combination of a capacitor and an inductor shown in Fig. 6.46*a*. The equivalent admittance (reciprocal of impedance) is the sum of the admittances of the parallel combination:

$$\begin{aligned}\hat{Y}_{eq} &= \frac{1}{j\omega L} + j\omega C \\ &= -j\frac{1}{\omega L} + j\omega C\end{aligned} \tag{6.97}$$

At the resonant frequency given by (6.96), the admittance of this parallel combination is zero. Hence, if the parallel connection is excited at the *resonant frequency* of ω_0, the parallel combination is equivalent to an *open circuit.* Below this resonant frequency, the phase angle of the admittance of the parallel combination is $-90°$ because the term $-j1/\omega L$ is dominant and the parallel combination appears inductive, whereas above the resonant frequency the phase angle of the admittance of the combination is $+90°$ because the term $j\omega C$ is dominant and the parallel combination appears capacitive. The magnitude of the frequency response of the admittance of this parallel combination is sketched in Fig. 6.46*c*.

The phenomenon allows the selective passage or rejection of certain frequency components of an input to the output. This gives the opportunity to construct *electrical filters*.

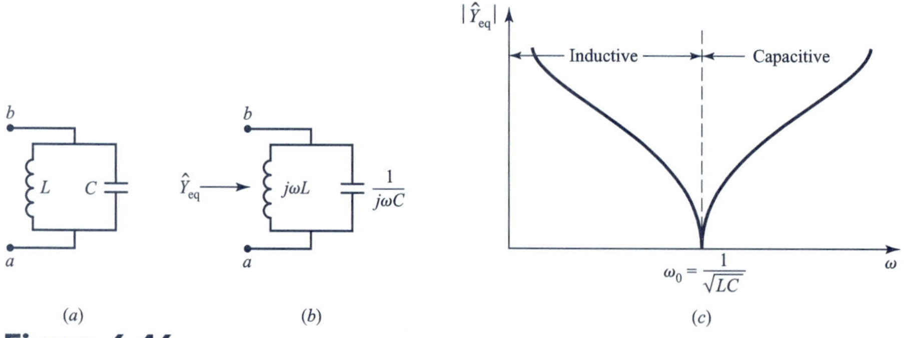

Figure 6.46
Illustration of parallel resonance: (*a*) the parallel *LC* circuit, (*b*) the phasor equivalent, and (*c*) a plot of the magnitude of the parallel admittance versus radian frequency showing that the admittance is zero at $\omega_0 = 1/\sqrt{LC}$ where the parallel combination acts like an open circuit.

Exercise Problem 6.26

Determine the resonant frequency in hertz for the series *LC* circuit if $L = 20$ mH and $C = 100$ pF.

Answer: 112.54 kHz.

Exercise Problem 6.27

Determine the resonant frequency in hertz for the parallel *LC* circuit if $L = 10$ μH and $C = 50$ pF.

Answer: 7.118 MHz.

6.8.3 Elementary Electrical Filters We have indicated previously the need to selectively pass or reject certain frequency components of an input signal, as in radio transmission and reception. Electrical circuits that perform this function are referred to as *filters*. Figure 6.47 shows four common classes of filters. Figure 6.47*a* shows the magnitude of the frequency response of the transfer function for a *lowpass filter*. The solid line denotes the *ideal* frequency response having a sharp *cutoff* at frequency ω_0. Unfortunately, it is not possible to construct a filter having this ideal shape, so that practical filters have a more gradual cutoff as indicated by the dashed line. Figure 6.47*b*, *c*, and *d* show the frequency responses of *highpass, bandpass*, and *bandreject filters*, respectively.

Figure 6.48 shows the implementation of these various filters. The transfer function of the *lowpass filter* shown in Fig. 6.48*a* is obtained as

$$\begin{aligned} \hat{H}_{\text{lowpass}} &= \frac{R}{R + j\omega L} \\ &= \frac{R}{L} \frac{1}{j\omega + \frac{R}{L}} \end{aligned} \tag{6.98}$$

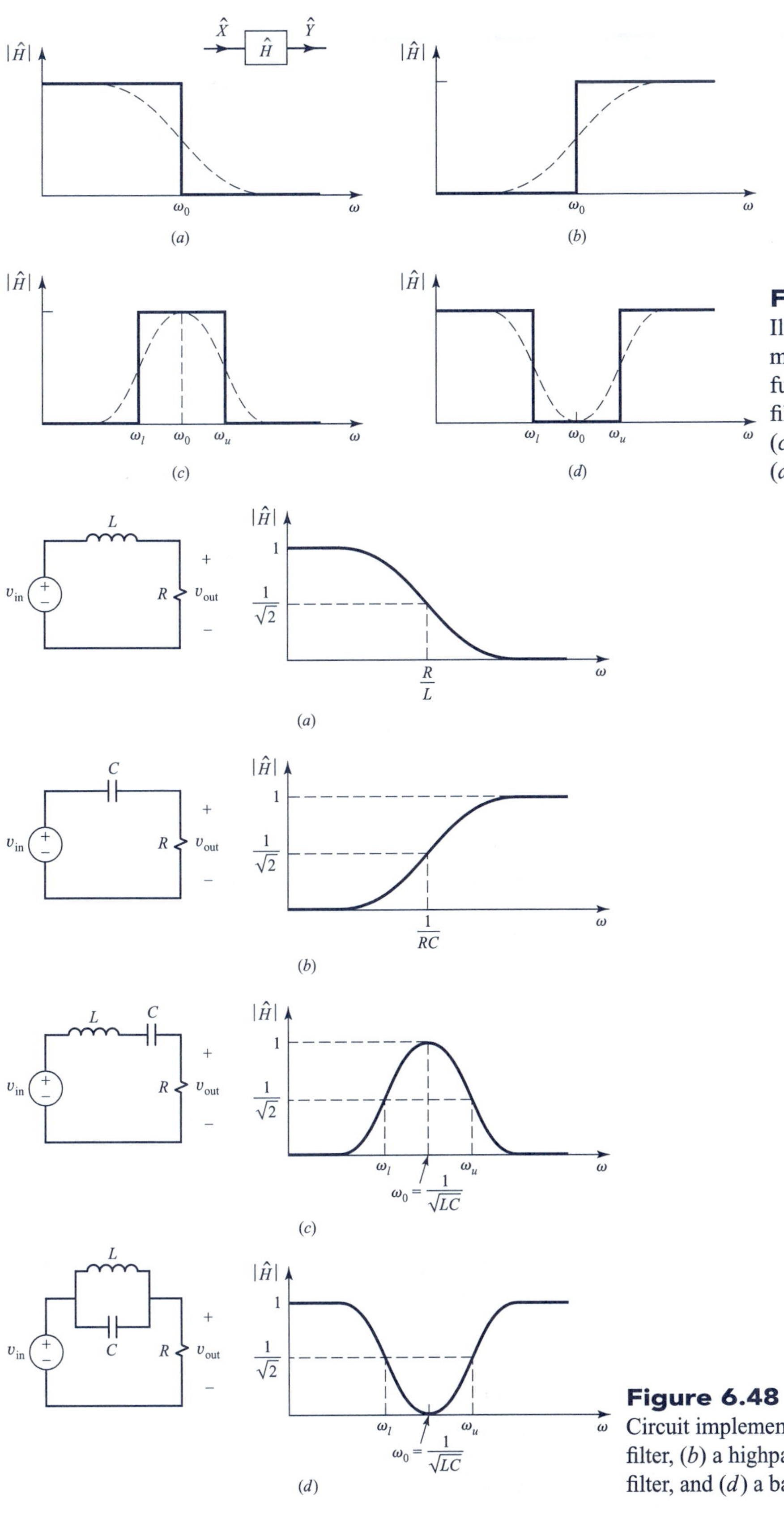

Figure 6.47 Illustration of the magnitude of the transfer function for (*a*) a lowpass filter, (*b*) a highpass filter, (*c*) a bandpass filter, and (*d*) a bandreject filter.

Figure 6.48 Circuit implementation of (*a*) a lowpass filter, (*b*) a highpass filter, (*c*) a bandpass filter, and (*d*) a bandreject filter.

This frequency response can be rapidly sketched directly from the circuit by observing that at dc ($\omega = 0$) the inductor is a short circuit and the input voltage source is directly connected to the output, and at $\omega = \infty$ the inductor is an open circuit and the input voltage source is isolated from the output. From the transfer function in (6.98), the real and imaginary parts of the denominator are equal at $\omega_0 = R/L$ and hence the magnitude of the transfer function is $1/\sqrt{2}$.

The transfer function of the *highpass filter* shown in Fig. 6.48*b* is obtained as

$$\begin{aligned}\hat{H}_{\text{highpass}} &= \frac{R}{R + \dfrac{1}{j\omega C}} \\ &= \frac{j\omega}{j\omega + \dfrac{1}{RC}}\end{aligned} \tag{6.99}$$

This frequency response can be rapidly sketched directly from the circuit by observing that at dc ($\omega = 0$) the capacitor is an open circuit and the input voltage source is isolated from the output, and at $\omega = \infty$ the capacitor is a short circuit and the input voltage source is directly connected to the output. From the transfer function in (6.99), the real and imaginary parts of the denominator are equal at $\omega_0 = 1/RC$ and hence the magnitude of the transfer function is $1/\sqrt{2}$.

The transfer function of the *bandpass filter* shown in Fig. 6.48*c* is obtained as

$$\begin{aligned}\hat{H}_{\text{bandpass}} &= \frac{R}{R + j\omega L + \dfrac{1}{j\omega C}} \\ &= \frac{j\omega RC}{(1-\omega^2 LC) + j\omega RC}\end{aligned} \tag{6.100}$$

This frequency response can be rapidly sketched directly from the circuit by observing that at dc ($\omega = 0$) the capacitor is an open circuit and at $\omega = \infty$ the inductor is an open circuit, so that at these extremes in frequency the input voltage source is isolated from the output. From the transfer function in (6.100), the magnitude is unity when $\omega_0 = 1/\sqrt{LC}$. This is the resonant frequency of the series LC combination where its impedance is zero. The lower and upper *cutoff frequencies*, ω_l and ω_u, are where the response is $1/\sqrt{2} = 0.707$ of the maximum at ω_0. Rewriting (6.100) as

$$\hat{H}_{\text{bandpass}} = \frac{1}{-j\dfrac{1 - \omega^2 LC}{\omega RC} + 1}$$

and solving for the frequencies that make the imaginary part of the denominator equal to plus or minus unity gives

$$\begin{aligned}\omega_l &= -\frac{R}{2L} + \sqrt{\left(\frac{R}{2L}\right)^2 + \frac{1}{LC}} \\ \omega_u &= \frac{R}{2L} + \sqrt{\left(\frac{R}{2L}\right)^2 + \frac{1}{LC}}\end{aligned} \tag{6.101}$$

The *bandwidth* is the difference between these two cutoff frequencies:

$$\begin{aligned}\text{BW} &= \omega_u - \omega_l \\ &= \frac{R}{L}\end{aligned} \tag{6.102}$$

An important measure of the ability of the filter to pass a band of frequencies is its Q, which is the ratio of the filter center frequency to its bandwidth:

$$\begin{aligned} Q &= \frac{\omega_0}{\text{BW}} \\ &= \frac{1}{R}\sqrt{\frac{L}{C}} \end{aligned} \tag{6.103}$$

Hence, increasing R lowers the Q and hence degrades the filter.

The transfer function of the *bandreject filter* shown in Fig. 6.48*d* is obtained as

$$\begin{aligned} \hat{H}_{\text{bandreject}} &= \frac{R}{R + \dfrac{(j\omega L)\left(\dfrac{1}{j\omega C}\right)}{j\omega L + \dfrac{1}{j\omega C}}} \\ &= \frac{1-\omega^2 LC}{(1-\omega^2 LC) + j\omega\dfrac{L}{R}} \end{aligned} \tag{6.104}$$

This frequency response can again be rapidly sketched directly from the circuit by observing that at dc ($\omega = 0$) the inductor is a short circuit and at $\omega = \infty$ the capacitor is a short circuit, so that at these extremes in frequency the input voltage source is directly connected to the output. From the transfer function in (6.104), the magnitude is zero when $\omega_0 = 1/\sqrt{LC}$. This is the resonant frequency of the parallel LC combination, where its admittance is zero (infinite impedance). The lower and upper cutoff frequencies, ω_l and ω_u, are again the frequencies where the response is $1/\sqrt{2} = 0.707$ of the maximum. Rewriting (6.104) as

$$\hat{H}_{\text{bandreject}} = \frac{1}{1 + j\dfrac{\omega L/R}{1-\omega^2 LC}}$$

and solving for the frequencies that make the imaginary part of the denominator equal to plus or minus unity gives

$$\begin{aligned} \omega_l &= -\frac{1}{2RC} + \sqrt{\left(\frac{1}{2RC}\right)^2 + \frac{1}{LC}} \\ \omega_u &= \frac{1}{2RC} + \sqrt{\left(\frac{1}{2RC}\right)^2 + \frac{1}{LC}} \end{aligned} \tag{6.105}$$

The bandwidth is the difference between these two cutoff frequencies:

$$\begin{aligned} \text{BW} &= \omega_u - \omega_l \\ &= \frac{1}{RC} \end{aligned} \tag{6.106}$$

and the Q is

$$\begin{aligned} Q &= \frac{\omega_0}{\text{BW}} \\ &= R\sqrt{\frac{C}{L}} \end{aligned} \tag{6.107}$$

These simple discussions of filters, while elementary, provide insight into the operation of more elaborate but practical filters. Although these implementations give the desired general shape, the cutoff is not particularly sharp. Practical filters are more elaborate than these, and their quality is judged by the sharpness of the cutoff, ideally approaching an abrupt transition.

Exercise Problem 6.28

Design a bandpass filter to have a bandwidth of 10 kHz and a center frequency of 1MHz. Determine the Q of your design.

Answers: $R = 1\ \text{k}\Omega$, $L = 15.9$ mH, $C = 1.59$ pF, $Q = 100$.

Exercise Problem 6.29

Design a bandreject filter to have a bandwidth of 20 kHz and a center frequency of 5 MHz. Determine the Q of your design.

Answers: $R = 1\ \text{k}\Omega$, $L = 0.1273\ \mu$H, $C = 7.958$ nF, $Q = 250$.

6.8.4 Active (Op-Amp) Filters The bandpass filter of Fig. 6.48*c* and the bandreject filter of Fig. 6.48*d* both require inductors in order to produce the filter characteristic. These rely on the phenomenon of resonance. Modern electronic technology allows the construction of an enormous number of circuit elements on extremely small *chips*, as in *very large-scale integrated* (VLSI) circuits. Resistors, capacitors, diodes, transistors, and op amps can be easily fabricated in large numbers on these chips. However, inductors having sufficiently large values cannot be so fabricated. To fabricate such filters in VLSI form requires an alternative to the inductor. Bandpass and bandreject filters can be fabricated on VLSI structures in large numbers using R's, C's, and op amps. These are called *active filters*, since the op amp internally employs active devices (transistors).

Figure 6.49 shows the construction of a bandpass active filter using an op amp. Using the elementary properties of an inverting op-amp circuit, the transfer function is

$$\begin{aligned}\hat{H}(p) &= \frac{\hat{V}_{\text{out}}}{\hat{V}_{\text{in}}} \\ &= -\frac{\hat{Z}_f}{\hat{Z}_i} \qquad (6.108)\end{aligned}$$

where

$$\hat{Z}_i = R_i + \frac{1}{pC_i} \qquad (6.109a)$$

$$\begin{aligned}\hat{Z}_f &= R_f \left\| \frac{1}{pC_f} \right. \\ &= \frac{R_f}{1 + pR_fC_f} \qquad (6.109b)\end{aligned}$$

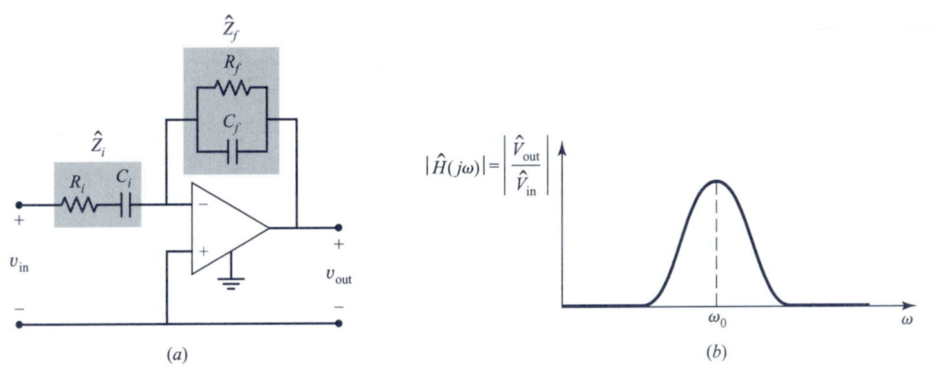

Figure 6.49
An op-amp (active) bandpass filter.

Substituting gives

$$\hat{H}(p) = -\frac{pR_fC_i}{(1 + pR_iC_i)(1 + pR_fC_f)} \tag{6.110}$$

Substituting $p = j\omega$ and rearranging gives

$$\hat{H}(j\omega) = -\frac{j\omega R_fC_i}{(1 - \omega^2 R_iC_iR_fC_f) + j\omega(R_iC_i + R_fC_f)} \tag{6.111}$$

Comparing this with the bandpass transfer function given in (6.100) shows that this is indeed a bandpass filter.

Exercise Problem 6.30

Determine the transfer function and type of active filter produced by the circuit of Fig. E6.30.

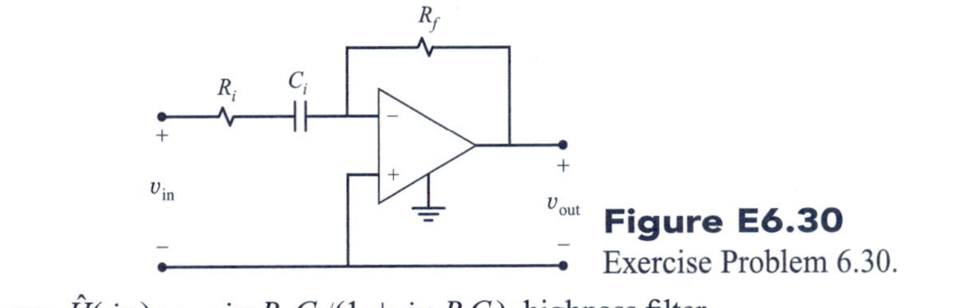

Figure E6.30
Exercise Problem 6.30.

Answers: $\hat{H}(j\omega) = -j\omega R_fC_i/(1 + j\omega R_iC_i)$, highpass filter.

6.9 Commercial Power Distribution

Electric power is generated by converting energy in one form (burning of coal or oil, nuclear, hydroelectric) to electric power using a *generator*. The frequency of the voltage and current is 60 Hz in the US (50 Hz in Europe). In this section we will investigate the transmission of that electric power from the power generation plant to the consumer.

Electric power is generated as a *three-phase* voltage, which can be represented as shown in Fig. 6.50*a*. The power plant can be represented as three ideal phasor voltage sources that are 120° out of phase:

$$\begin{aligned} \hat{V}_{an} &= V_P \angle 0^\circ \\ \hat{V}_{bn} &= V_P \angle -120^\circ \\ \hat{V}_{cn} &= V_P \angle 120^\circ \end{aligned} \tag{6.112}$$

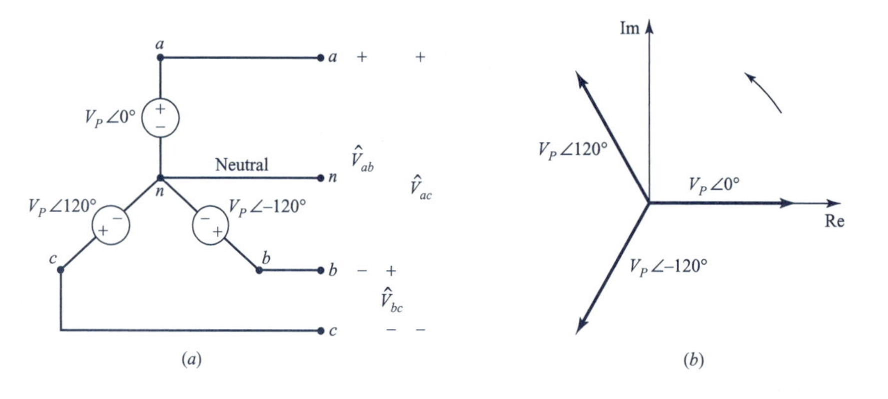

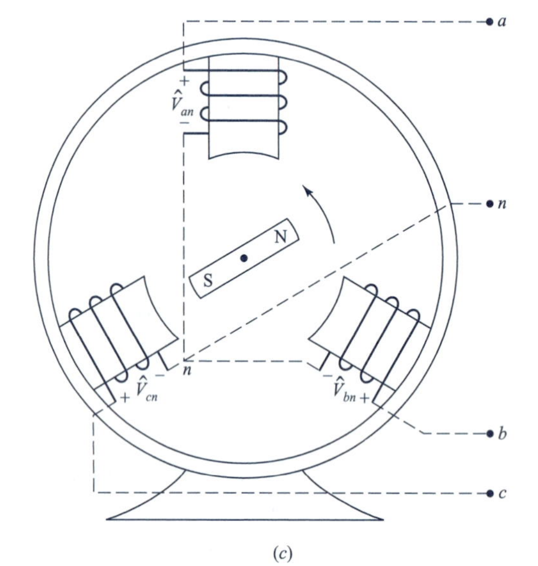

Figure 6.50
Representation of a three-phase power source: (*a*) the circuit model, (*b*) the phasor diagram, and (*c*) the physical construction of a three-phase generator.

They are represented in a phasor diagram as shown in Fig. 6.50*b*, where all phasor voltages rotate counterclockwise at the frequency of the source (60 Hz) and are called the *phase voltages*. Observe that the sum of the three phase voltages is zero:

$$V_P \angle 0° + V_P\angle - 120° + V_P \angle 120° = 0 \tag{6.113}$$

It is customary in the electric power industry to specify these voltages in their rms values, so that V_P is the rms value of the voltage. Hence, the peak values of these sinusoids are $\sqrt{2}V_P$ and their time-domain representations are

$$\begin{aligned} v_{an}(t) &= \sqrt{2}V_P \sin \omega t \text{ V} \\ v_{bn}(t) &= \sqrt{2}V_P \sin(\omega t - 120°) \text{ V} \\ v_{cn}(t) &= \sqrt{2}V_P \sin(\omega t + 120°) \text{ V} \end{aligned} \tag{6.114}$$

It is important to observe that these are *ideal sources*, i.e., they have zero source impedance, so that the output voltages are independent of the load. Hence the power distribution system from the generating plant to the consumer (town, area of the country, etc.) is referred to as an "infinite bus," meaning that variations in components of the load (one house in the town burns down) will not affect the transmission voltages to the other load.

The generators at the power plant are constructed as shown in Fig. 6.50*c*. Three windings are physically displaced by 120°. A rotating magnet, or *rotor* (a winding is placed on this rotating member and current is fed to it to produce the magnet) rotates at a certain speed. The magnetic field of the rotor penetrates the area enclosed by the stationary windings, producing, according to Faraday's law, voltages in those windings, which are displaced in time by 120°, and hence the three-phase voltages are produced. Rotation of the rotor is obtained by several means. Burning coal, for example, to produce steam to rotate a turbine that rotates the rotor is one common method. Nuclear power functions in the same manner. Hydroelectric means use the falling water at a dam to rotate the rotor. Hence mechanical energy is converted to electric energy. The connection of the three coils with a common point, or *neutral,* is referred to as a *wye connection,* since it resembles the letter Y. This is the most common way of interconnecting the coils.

The three-phase voltages are sent to the load via three conductors, which constitute a *transmission line* as shown in Fig. 6.50*a*. The neutral may or may not be carried with the other conductors. The *line-to-line* voltages between the conductors of the transmission line are the phasor sums of the phase voltages:

$$\begin{aligned}\hat{V}_{ab} &= \hat{V}_{an} - \hat{V}_{bn} = V_P\angle 0° - V_P\angle -120° = \sqrt{3}V_P\angle 30°\\ \hat{V}_{ac} &= \hat{V}_{an} - \hat{V}_{cn} = V_P\angle 0° - V_P\angle 120° = \sqrt{3}V_P\angle -30°\\ \hat{V}_{bc} &= \hat{V}_{bn} - \hat{V}_{cn} = V_P\angle -120° - V_P\angle 120° = \sqrt{3}V_P\angle -90°\end{aligned} \tag{6.115}$$

Hence the line-to-line voltages are $\sqrt{3}$ larger than the phase voltages: $V_L = \sqrt{3}\,V_P$.

6.9.1 Wye-Connected Loads Let us now investigate the transmission of power from the generator to the load. Figure 6.51*a* shows a wye-connected load. The three impedances making up that load, $\hat{Z}_L$, are identical. Hence the load is said to be *balanced*. This is a common

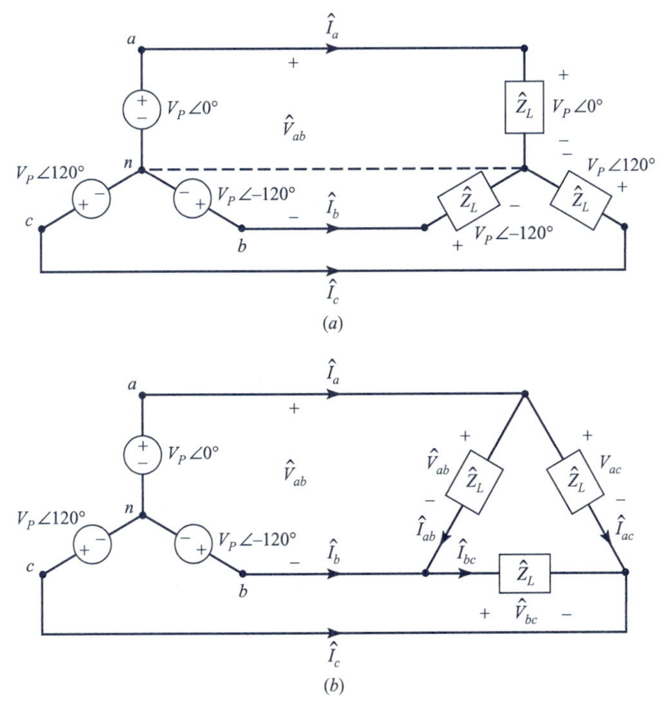

Figure 6.51 Two common load connections: (*a*) the wye-connected load, and (*b*) the delta-connected load.

situation, and the power companies attempt to balance the load. For this balanced situation, the voltages across the loads are the phase voltages, regardless of whether the neutral is attached. To show this let us compute the line currents with the neutral connecting the two central points of the generator and the load. In this case the voltages across the individual loads are the phase voltages, and hence the line currents are

$$\begin{aligned}\hat{I}_a &= \frac{V_P \angle 0^\circ}{\hat{Z}_L}\\ \hat{I}_b &= \frac{V_P \angle -120^\circ}{\hat{Z}_L}\\ \hat{I}_c &= \frac{V_P \angle 120^\circ}{\hat{Z}_L}\end{aligned} \tag{6.116}$$

The sum of these currents is again zero according to (6.113):

$$\begin{aligned}\hat{I}_a + \hat{I}_b + \hat{I}_c &= \frac{V_P \angle 0^\circ + V_P \angle -120^\circ + V_P \angle 120^\circ}{\hat{Z}_L}\\ &= 0\end{aligned} \tag{6.117}$$

Hence the current returning through the neutral wire is zero, and the neutral wire may be removed. Therefore, *for a balanced wye-connected load, the voltages across the individual loads are the respective phase voltages whether the neutral is connected or not.* Hence the power delivered to the individual loads can be easily calculated, and the total power delivered to the load is three times that delivered to an individual load, because the individual loads are identical:

$$\begin{aligned}P_{\text{AV, total}} &= 3\text{Re}(V_P \angle 0^\circ \hat{I}_a^*)\\ &= 3\text{Re}\left(V_P \angle 0^\circ \frac{V_P \angle 0^\circ}{\hat{Z}_L^*}\right)\\ &= 3\frac{V_P^2}{Z_L}\cos\theta_{Z_L}\end{aligned} \tag{6.118}$$

where the load impedance is denoted as $\hat{Z}_L = Z_L \angle \theta_{Z_L}$. Observe that, because the phase voltages V_P are specified in their rms values, there is no factor $1/2$ in the power expression.

The line-to-line voltages are more accessible than the phase voltages. Writing the power expression in terms of the line-to-line voltages $V_L = \sqrt{3}V_P$ and the line current I_L (which is also the current through the individual loads) gives

$$\begin{aligned}P_{\text{AV, total}} &= 3\frac{V_L}{\sqrt{3}} I_L \cos\theta_{Z_L}\\ &= \sqrt{3}V_L I_L \cos\theta_{Z_L}\end{aligned} \tag{6.119}$$

Example 6.24

Consider a balanced, wye-connected load where each load impedance is $\hat{Z}_L = 50 + j50\ \Omega$ and the phase voltages are 120 V. Determine the total average power delivered to the load.

Solution The line currents are

$$\hat{I}_a = \frac{120 \angle 0°}{50 + j50} = 1.7\angle - 45° \text{ A}$$

$$\hat{I}_b = \frac{120\angle - 120°}{50 + j50} = 1.7\angle - 165° \text{ A}$$

$$\hat{I}_c = \frac{120 \angle 120°}{50 + j50} = 1.7 \angle 75° \text{ A}$$

Hence, the average power delivered to each load is

$$\begin{aligned} P_{\text{AV}} &= \text{Re}[(120 \angle 0°)(1.7 \angle 45°)] \\ &= 120 \times 1.7 \times \cos(45°) \\ &= 144 \text{ W} \end{aligned}$$

The total average power delivered to the load is

$$P_{\text{AV, total}} = 3 \times 144 = 432 \text{ W}$$

Example 6.25

If the line voltage of a balanced, wye-connected load is 208 V and the total average power delivered to the load is 900 W, determine each load if their power factors are 0.8 leading.

Solution From (6.119),

$$900 \text{ W} = \sqrt{3}\, V_L I_L \cos \theta_{Z_L}$$

or

$$\begin{aligned} I_L &= \frac{900}{\sqrt{3}\,(208)(0.8)} \\ &= 3.12 \text{ A} \end{aligned}$$

The phase voltage is

$$\begin{aligned} V_P &= \frac{V_L}{\sqrt{3}} \\ &= 120 \text{ V} \end{aligned}$$

Thus the magnitude of the individual load impedances is

$$\begin{aligned} Z_L &= \frac{V_P}{I_L} \\ &= 38.43\ \Omega \end{aligned}$$

Since the power factor is 0.8 leading (current leads voltage; voltage lags current), $\theta_{Z_L} = -\cos^{-1} 0.8 = -36.87°$. Thus the individual loads are

$$\begin{aligned} \hat{Z}_L &= 38.43\angle - 36.87° \\ &= 30.74 - j23.06\ \Omega \end{aligned}$$

Exercise Problem 6.31

Three-phase power is supplied to a balanced, wye-connected load. The line-to-line voltage is 208V, and the load consumes a total power of 15 kW at a lagging power factor of 0.6. Determine the transmission-line currents and the individual loads.

Answers: 69.4 A and $1.04 + j1.38\ \Omega$.

6.9.2 Delta-Connected Loads Another common way of interconnecting the individual three-phase loads is the delta connection shown in Fig. 6.51*b*. In this case the voltages across the individual loads are the line-to-line voltages, which are given in (6.115):

$$\begin{aligned}\hat{V}_{ab} &= \sqrt{3}V_P \angle 30^\circ \\ \hat{V}_{ac} &= \sqrt{3}V_P \angle -30^\circ \\ \hat{V}_{bc} &= \sqrt{3}V_P \angle -90^\circ\end{aligned} \tag{6.120}$$

Therefore, the currents through the individual loads are

$$\begin{aligned}\hat{I}_{ab} &= \frac{\hat{V}_{ab}}{\hat{Z}_L} = \frac{\sqrt{3}V_P \angle 30^\circ}{\hat{Z}_L} \\ \hat{I}_{ac} &= \frac{\hat{V}_{ac}}{\hat{Z}_L} = \frac{\sqrt{3}V_P \angle -30^\circ}{\hat{Z}_L} \\ \hat{I}_{bc} &= \frac{\hat{V}_{bc}}{\hat{Z}_L} = \frac{\sqrt{3}V_P \angle -90^\circ}{\hat{Z}_L}\end{aligned} \tag{6.121}$$

The line currents are, by KCL,

$$\begin{aligned}\hat{I}_a &= \hat{I}_{ab} + \hat{I}_{ac} = \frac{3V_P \angle 0^\circ}{\hat{Z}_L} \\ \hat{I}_b &= \hat{I}_{bc} - \hat{I}_{ab} = \frac{3V_P \angle -120^\circ}{\hat{Z}_L} \\ \hat{I}_c &= -\hat{I}_{bc} - \hat{I}_{ac} = \frac{3V_P \angle 120^\circ}{\hat{Z}_L}\end{aligned} \tag{6.122}$$

Hence, in a delta-connected load, the line currents are $\sqrt{3}$ times the individual load currents. Since the individual loads are identical (balanced), the total power delivered to the load is three times the power delivered to each component of the load:

$$\begin{aligned}P_{\text{AV, total}} &= 3\,\mathrm{Re}(\hat{V}_{ab}\hat{I}^*_{ab}) \\ &= 3\,\mathrm{Re}\left((\sqrt{3}V_P \angle 30^\circ)\,\frac{\sqrt{3}V_P \angle -30^\circ}{\hat{Z}^*_L}\right) \\ &= 9\frac{V_P^2}{Z_L}\cos\theta_{Z_L}\end{aligned} \tag{6.123}$$

Again, the line-to-line voltages are more accessible than the phase voltages. Writing the above power expression in terms of the line-to-line voltages, $V_L = \sqrt{3}V_P$, gives

$$\begin{aligned} P_{\text{AV, total}} &= 9\frac{V_L}{\sqrt{3}}\frac{I_L}{3}\cos\theta_{Z_L} \\ &= \sqrt{3}V_L I_L \cos\theta_{Z_L} \end{aligned} \tag{6.124}$$

which is the same expression in terms of the line-to-line voltage and the line current as for the wye-connected load given in (6.119).

Example 6.26

A balanced, delta-connected load has a line-to-line voltage of 208 V, and the total average power delivered to the load is 600 W. Determine the individual loads if they have a lagging power factor of 0.7.

Solution The average power delivered to each individual load is 200 W, so that

$$200 = V_L \frac{I_L}{\sqrt{3}} \cos\theta_{Z_L}$$

or

$$\begin{aligned} I_L &= \frac{200(\sqrt{3})}{208(0.7)} \\ &= 2.38 \text{ A} \end{aligned}$$

Thus the magnitudes of the individual load impedances are

$$\begin{aligned} Z_L &= \frac{V_L}{I_L/\sqrt{3}} \\ &= 151.42 \end{aligned}$$

Since the power factor is 0.7 lagging,

$$\begin{aligned} \hat{Z}_L &= 151.42 \angle 45.57° \\ &= 106 + j108.14\ \Omega \end{aligned}$$

Exercise Problem 6.32

Repeat Exercise Problem 6.31 where the load is delta-connected.

Answers: 69.4 A and $3.12 + j4.15\ \Omega$.

6.10 PSPICE Applications

The .AC module of PSPICE performs the *phasor analysis* of a circuit that is driven by sinusoidal sources. The format of that control command is

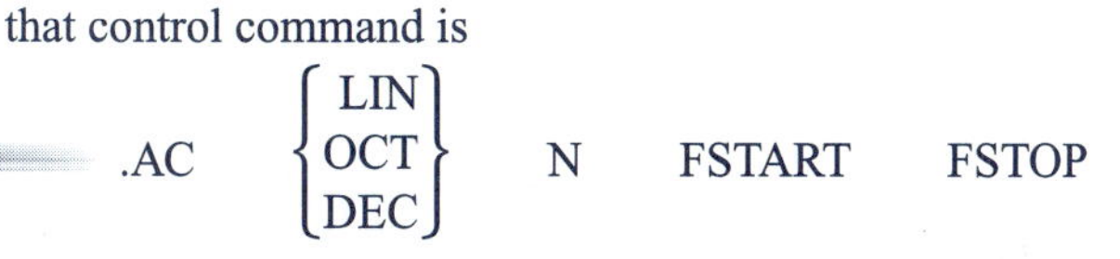

$$\text{.AC} \quad \begin{Bmatrix} \text{LIN} \\ \text{OCT} \\ \text{DEC} \end{Bmatrix} \quad \text{N} \quad \text{FSTART} \quad \text{FSTOP}$$

This is like the .DC command in that it is meant to sweep a frequency range from FSTART to FSTOP and recompute the solution at frequencies in that range. The item LIN means that N points in that range will be chosen for computation and they will be equally spaced on a linear axis from FSTART to FSTOP. For example, if we choose LIN, N=10, FSTART=100 Hz, and

FSTOP=1000 Hz, then 10 frequency points will be chosen for computation: 100, 200, 300, 400, 500, 600, 700, 800, 900, 1000 Hz. If the item OCT is chosen, the frequencies will be chosen so that there are N frequencies equally spaced on a *logarithmic* axis per *octave* (power of two). If the item DEC is chosen, the frequencies will be chosen so that there are N frequencies equally spaced on a logarithmic axis per *decade* (powers of ten). Ordinarily, in plotting the frequency response of a circuit such as the filters of Section 6.8.3, we choose a logarithmic frequency axis so that a wide frequency range can be accommodated on one graph. For example, suppose we wished to plot a response versus frequency from 1Hz to 1kHz. If we use a linear frequency scale, the region from 1 to 10 Hz (a decade) could be plotted in, say, one inch. Then the next inch would accommodate 11 to 20 Hz, the next inch would accommodate 21 to 30 Hz, and so forth. Clearly we would run out of paper before we got the last frequency of 1 kHz plotted. In order to circumvent this, we plot instead the logarithm (to the base 10) of the frequency. Hence the first frequency would be $\log_{10}(1) = 0$ and the last frequency would be plotted as $\log_{10}(1000) = 3$. Hence a linear scale from 0 to 3 would encompass this frequency range. Similarly we could plot a rather wide frequency range from 1 kHz to 100 MHz using a linear scale of 3 to 8. If OCT or DEC is chosen, the frequencies will be equally spaced on a log scale.

Since the analysis is the phasor analysis, PSPICE solves the phasor circuit and gives as the solution the phasor (magnitude and phase) voltages and currents. Hence if more than one sinusoidal source is to be used, they all must be of the same frequency and must all be sine or all be cosine functions. See Section 6.5.1 for a review of this. So we designate the sources with their magnitude and their phase as we would do in the phasor circuit. The specification statements for the independent sources are

$$\left.\begin{matrix}\text{VXXX}\\ \text{IXXX}\end{matrix}\right\}\quad \text{N1}\quad \text{N2}\quad \text{AC}\quad mag\quad phase$$

The phase is in degrees, and if it is omitted, the default is zero degrees. The inductors and capacitors are specified as before except that the IC= item is not used:

$$\left.\begin{matrix}\text{LXXX}\\ \text{CXXX}\end{matrix}\right\}\quad \text{N1}\quad \text{N2}\quad value$$

Example 6.27

Determine the current $i(t)$ in the circuit of Fig. 6.52*a* using PSPICE.

Solution The PSPICE circuit with nodes labeled is shown in Fig. 6.52*b*. The PSPICE program becomes

```
EXAMPLE 6.27
VS      1    0    AC   10   30
R       1    2    4
L       2    3    3M
C       3    4    250U
VTEST   4    0
.AC     LIN  1    318.31    318.31
.PRINT  AC   IM(VTEST)      IP(VTEST)
*THE MAGNITUDE OF THE CURRENT IS IM(VTEST) AND
*THE PHASE IS IP(VTEST)
.END
```

Note that the radian frequency of 2000 rad/s must be converted to cyclic frequency: $2000/2\pi = 318.31$Hz. The results are IM(VTEST)=1.768E+00 and IP(VTEST)=1.500E+01.

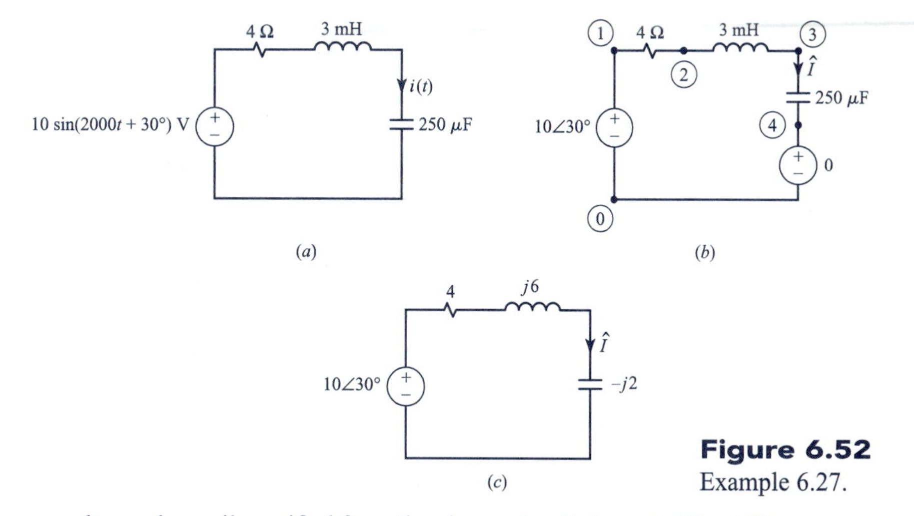

Figure 6.52
Example 6.27.

These results can be easily verified from the phasor circuit shown in Fig. 6.52*c* as

$$\begin{aligned}\hat{I} &= \frac{10 \angle 30^\circ}{4 + j6 - j2} \\ &= \frac{10 \angle 30^\circ}{4\sqrt{2} \angle 45^\circ} \\ &= 1.768\angle - 15^\circ\end{aligned}$$

Hence the time-domain current is

$$i(t) = 1.768 \sin(2000t - 15^\circ)\text{ A}$$

Next we show the solution for the frequency response of a circuit.

Example 6.28

Plot the frequency response of the bandpass filter shown in Fig. 6.48*c* as the source varies from 1 to 100 MHz, using PSPICE. Use element values of $R = 100\ \Omega$, $L = 159\ \mu\text{H}$, and $C = 1.6$ pF.

Solution Fig. 6.53 shows the PSPICE circuit. The PSPICE program is

```
EXAMPLE 6.28
VS      1    0    AC    1    0
R       3    0    100
L       1    2    159U
C       2    3    1.6P
.AC     DEC  50   1MEG   100MEG
.PROBE
*THE MAGNITUDE OF THE OUTPUT IS VM(3) AND THE PHASE IS VP(3)
.END
```

The magnitude of the frequency response is plotted in Fig. 6.54*a*. We have plotted the magnitude in decibels by using VDB(3). This computes

$$\text{VDB(3)} = 20 \log_{10}\text{VM(3)}$$

Figure 6.54*b* shows what we get if we request VM(3): the data are highly compressed outside the bandpass region and we get little resolution there. The phase is plotted in Fig. 6.54*c*. The

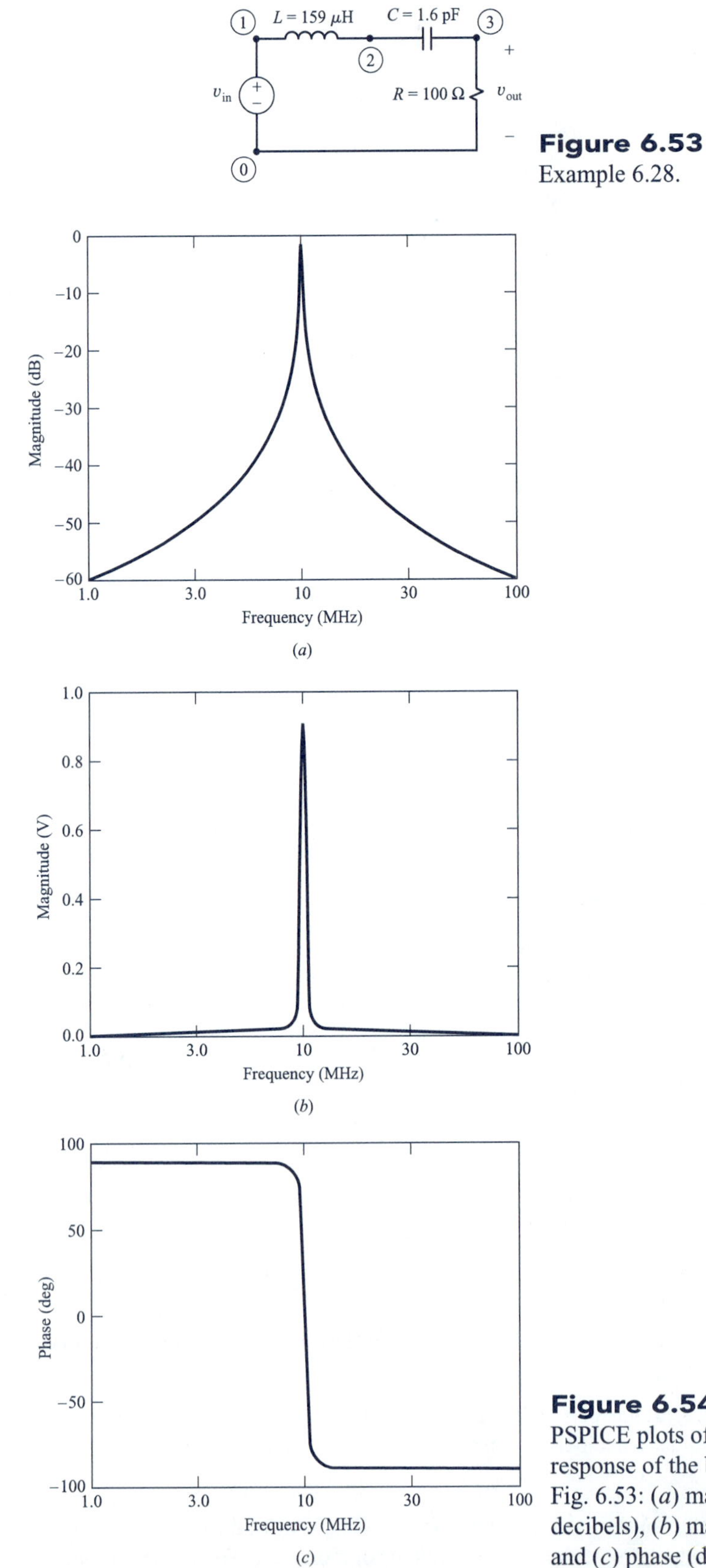

Figure 6.53
Example 6.28.

Figure 6.54
PSPICE plots of the frequency response of the bandpass filter of Fig. 6.53: (*a*) magnitude (in decibels), (*b*) magnitude (absolute), and (*c*) phase (degrees).

resonant frequency is 10 MHz. The phase is +90° below the resonant frequency due to the dominance of the capacitor in this range and is −90° above the resonant frequency due to the dominance of the inductor in this range. These observations bear out the important points about series resonance shown in Fig. 6.45.

6.11 MATLAB Applications

We will solve the above two PSPICE examples using MATLAB. First we have to discuss some additional features of MATLAB that weren't needed until now. The first important feature is that MATLAB handles complex numbers no differently than real numbers. Complex numbers are defined by appending the symbol i or the symbol j to the imaginary part. For example, the number $3+j4$ can be defined in MATLAB as

```
>>      c=3+4j
c=
        3.0000+4.0000i
```

or

```
>>      c=3+4i
c=
        3.0000+4.0000i
```

The magnitude of a complex number is obtained with the abs function:

```
>>      cmag=abs(c)
cmag=
        5.0000
```

The angle (in radians) is obtained with the angle command:

```
>>      cangle=angle(c)*180/pi
cangle=
        5.313e0001
```

Observe that *angles in MATLAB are always given in radians*. If we desire the angle in degrees we must multiply by the conversion factor $180/\pi$. Here we use the built in MATLAB command pi for π.

Example 6.29

Determine the current i in Example 6.27 shown in Fig. 6.52, using MATLAB.

Solution First we label the element "impedances" and the circuit diagram in terms of p as in Fig. 6.55. Hence the current is

$$\hat{I} = \frac{1}{R + pL + \dfrac{1}{pC}} \hat{V}_S$$

$$= \frac{Cp}{LCp^2 + RCp + 1} \hat{V}_S$$

Substituting the element values $R = 4\ \Omega$, $L = 3$ mH, $C = 250\ \mu$F gives

$$\hat{I} = \frac{250 \times 10^{-6} p}{7.5 \times 10^{-7} p^2 + 1 \times 10^{-3} p + 1} 10 \angle 30°$$

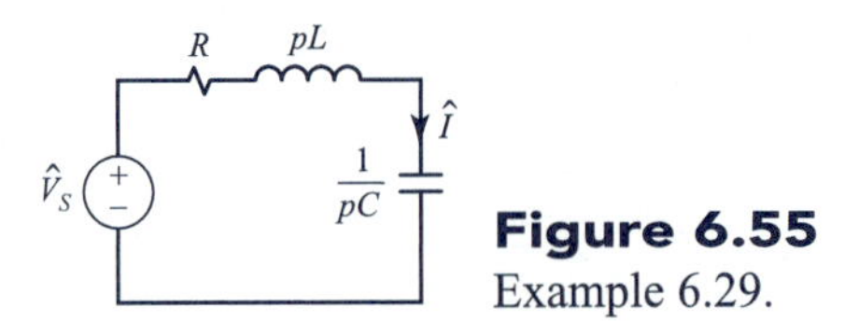

Figure 6.55
Example 6.29.

In order to evaluate this at $p = j\omega$ and $\omega = 2000$ we use the polyval(polynom,jw) function. This evaluates the polynomial polynom (defined previously) by substituting the variable (defined previously) $j\omega$ for p in that polynomial. We define the polynomial as an array of *coefficients* of the various powers of p in descending order. For example, the polynomial $6p^3 + 9p^2 + 5$ would be defined as polynom=[6,9,0,5]. Observe that the coefficient of p is zero and it must be included.

The next item to be discussed is defining a source in magnitude and phase form. Unfortunately, MATLAB does not provide directly for this, so we must define it in rectangular form. For example, the source in Example 6.27 is $\hat{V}_S = 10 \angle 30°$. This is defined in MATLAB with the commands

```
>>      vs=10*cos(30*pi/180)+10*sin(30*pi/180)*i
```

Note that the angles of the cosine and sine are presumed to be in radians, so one must convert the 30° to radians. Also note that we must multiply the sine by i using the asterisk, since MATLAB would not understand something like sin(3)i, though it would understand 5i.

With these commands we can show the MATLAB program to evaluate the above result:

```
>>        num=[250e-6 0]
num=
          2.5000e-004       0
>>        den=[7.5e-7,1e-3,1]
den=
          7.5000e-007       1.0000e-003       1.0000e+000
>>jw=2000j
jw=
          0+2.0000e+003i
>>h=polyval(num,jw)/polyval(den,jw)
h=
          1.2500e-001-1.2500e-001i
>>ang=30*pi/180
ang=
          5.2360e-001
>>vs=10*cos(ang)+10*sin(ang)*j
vs=
          8.6603e+000+5.0000e+000i
>>Iout=h*vs
Iout=
          1.7075e+000-4.5753e-001i
```

The final result, $\hat{I} = 1.7075 - j0.45753 = 1.768\angle - 15°$, is the same as was obtained with PSPICE in Example 6.27.

Next we use MATLAB to determine the frequency response of the bandpass filter in Fig. 6.53 that was obtained with PSPICE in Example 6.28. For this, in addition to the above new MATLAB commands, we need to discuss plotting. This is a powerful feature of MATLAB. There are two ways to plot a function that we will need. The first is useful in plotting a simple real function of one variable, e.g., $y(x) = 2e^{-3x} + 5$. We first need to define the values of the independent variable x. This is done in one of three ways:

$$x = x_1{:}\Delta x{:}x_2$$
$$x = \text{linspace}(x_1, x_2, n)$$
$$x = \text{logspace}(a, b, n)$$

The first form creates a vector or array x that consists of the entries

$$x = [x_1, x_1 + \Delta x, x_1 + 2\Delta x, \ldots, x_2]$$

The second form creates an array of n values of x that are equally spaced between x_1 and x_2:

$$x = \left[x_1, x_1 + \frac{x_2 - x_1}{n}, x_1 + 2\frac{x_2 - x_1}{n}, \ldots, x_2\right]$$

The third form creates an array of n values of x that are equally spaced on a logarithmic scale between $x_1 = 10^a$ and $x_2 = 10^b$. The simple way to plot the function $y(x) = 2e^{-3x} + 5$ is to define the array of values of y at each of the x values of that array:

```
>>y=2*exp(-3*x)+5;
```

Then we use the plot command and labeling of the axes to define the graph:

```
>>plot(x,y)
>>xlabel('The x axis')
>>ylabel('The y axis')
>>title('This is a test')
>>grid
```

Most of this is self-explanatory, and the grid command simply places an *xy* grid on the plot.

As discussed earlier, it is advantageous to plot frequency responses on a logarithmic frequency axis (the x axis) so that several decades can be accommodated. This necessitates plotting the y axis logarithmically also, or else the plot will have a strange shape. The simplest way to do this is to convert the y-axis variables to decibels (dB) by plotting $20 \log_{10} |\hat{Y}|$. To do this we use the freqs command freqs(num,den,w). This solves for the (complex-valued) result of the ratio of the two polynomials in p, num and den, that have been defined earlier at the values of *radian frequency* in the array w (which has been defined earlier). Note that the freqs command assumes that the first item is the numerator polynomial, the second item is the denominator polynomial, and the last item is the array of radian frequencies at which we desire a solution. Generally we are only interested in plots versus cyclic frequency f. First define the range using

```
f=logspace(a,b,n)
```

Then convert to radian frequency with

```
w=2*pi*f
```

which prepares another array w that has as it entries the entries in array f multiplied by 2π. The plot can be obtained with

```
>>semilogx (f,y)
```

Example 6.30

Sketch the frequency response for the bandpass filter of Fig. 6.53 that was solved using PSPICE in Example 6.28.

Solution Again, since MATLAB does not "solve a circuit," we must first determine the transfer function. Drawing the circuit with the elements replaced by their "impedances" in terms of the variable p as shown in Fig. 6.56, we obtain

$$
\begin{aligned}
H(p) &= \frac{\hat{V}_{\text{out}}}{\hat{V}_{\text{in}}} \\
&= \frac{R}{R + pL + \dfrac{1}{pC}} \\
&= \frac{RCp}{LCp^2 + RCp + 1}
\end{aligned}
$$

Substituting numerical values of $R = 100\ \Omega$, $L = 159\ \mu\text{H}$, $C = 1.6$ pF, we obtain

$$
H(p) = \frac{1.6 \times 10^{-10}p}{2.544 \times 10^{-16}p^2 + 1.6 \times 10^{-10}p + 1}
$$

The MATLAB program for plotting the magnitude and phase from 1 to 100 MHz is

```
>>num=[1.6e-10,0];
>>den=[2.544e-16,1.6e-10,1];
>>f=logspace(6,8,400);
>>w=2*pi*f;
>>H=freqs(num,den,w);
>>HdB=20*log10(abs(H));
>>semilogx(f,HdB)
>>xlabel('Frequency in Hz')
>>ylabel('Magnitude of Transfer Function in dB')
>>title('Frequency Response of Bandpass Filter, EXAMPLE 6.30')
>>grid
>>phase=angle(H)*180/pi;
>>semilogx(f,phase)
>>xlabel('Frequency in Hz')
>>ylabel('Angle of Transfer Function in degrees')
>>title('Frequency Response of Bandpass Filter, EXAMPLE 6.30')
>>grid
```

Placing a semicolon (;) after an entry will suppress printing out the results of that statement. This is useful when, as in this example, there are 400 entries in each vector. The resulting

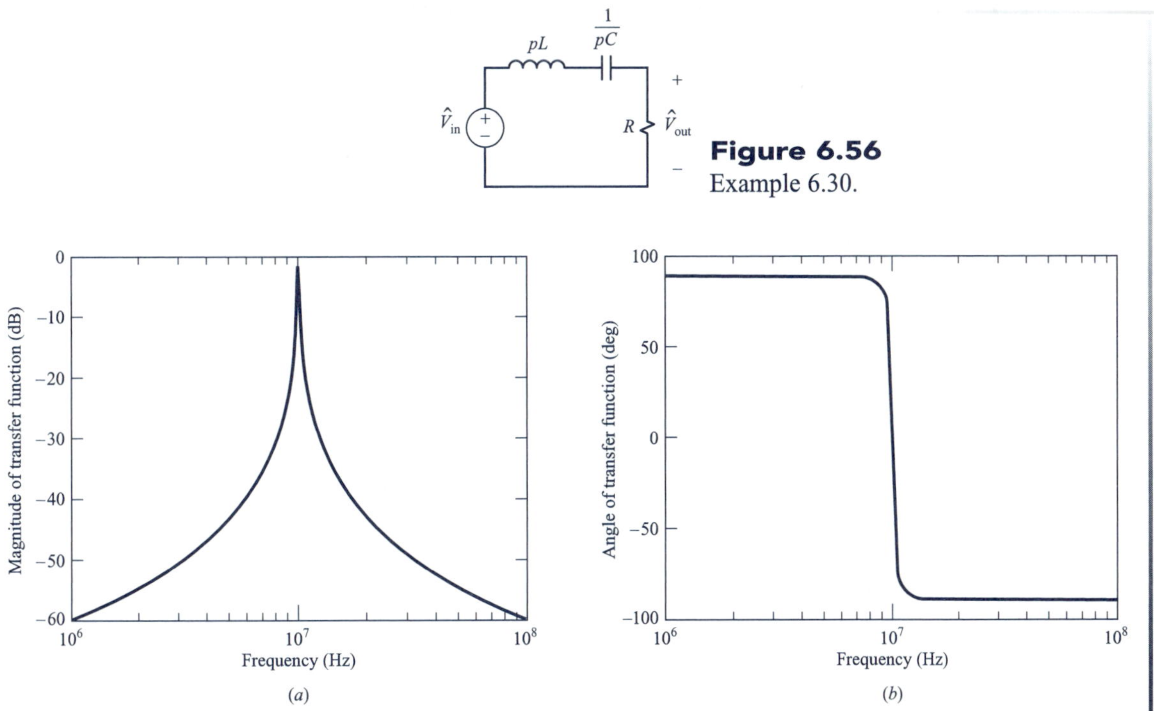

Figure 6.56
Example 6.30.

Figure 6.57
MATLAB plots of the frequency response of the bandpass filter of Fig. 6.53: (*a*) magnitude (in decibels), and (*b*) phase (degrees).

plots are shown in Fig. 6.57*a* and *b*. Comparing them with the corresponding ones generated with PSPICE and given in Fig. 6.54, we see that they are virtually identical.

6.12 Application Examples

In this final section we will give two examples of the application of the results of this chapter: an AM radio tuner and an illustration of crosstalk in communication circuits.

6.12.1 AM Radio Tuner Amplitude modulation (AM) is a method for translating voice, music, or other information from its low-frequency content to a higher *carrier frequency* to make the transmission from an antenna more efficient, as illustrated in Fig. 6.58. The information has a bandwidth of $2f_m$, where f_m is the frequency extent of the information (voice and music extend from about 100 Hz to 20 kHz). Numerous broadcast stations are transmitting simultaneously, and the need exists for filters to filter out the desired transmission and exclude the rest. Hence an AM radio tuner is a bandpass filter whose center frequency can be adjusted to tune in each station.

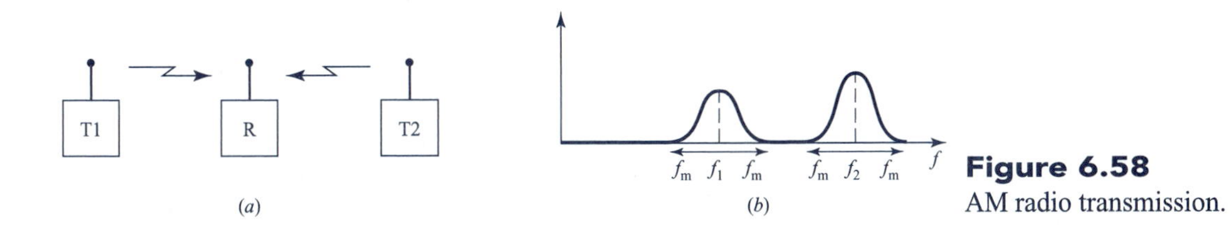

Figure 6.58
AM radio transmission.

Example 6.31

Two AM broadcast stations have carrier frequencies of 680 and 880 kHz. Both are transmitting voice and music whose frequency content extends from 0 to 20 kHz. Design a bandpass filter to pass the lower-frequency station and exclude the higher-frequency station.

Solution We need a bandpass filter, so we choose the series LC circuit shown in Fig. 6.59. We need a center frequency of

$$f_1 = 680\ \text{kHz} = \frac{1}{2\pi\sqrt{LC}}$$

and a bandwidth of

$$\text{BW} = 40\ \text{kHz} = \frac{R = 300\ \Omega}{2\pi L}$$

Solving these give values of $L = 1.194$ mH and $C = 45.9$ pF.

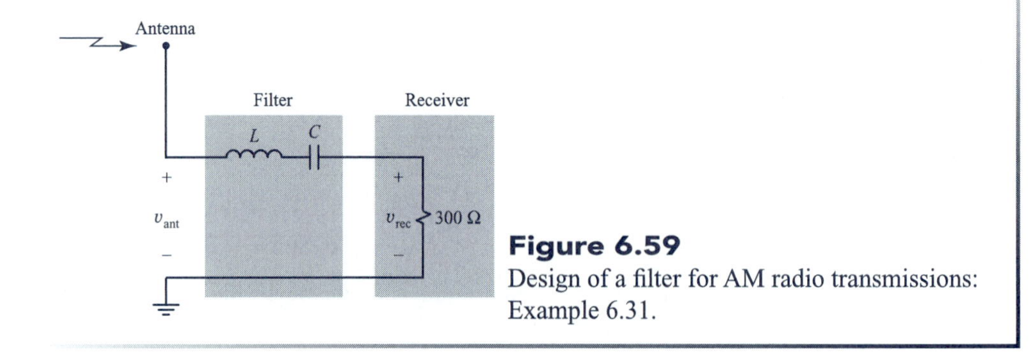

Figure 6.59
Design of a filter for AM radio transmissions: Example 6.31.

6.12.2 Crosstalk in Transmission Lines Many readers will have had the experience of talking on a telephone and hearing another conversation faintly in the background. This phenomenon is known as *crosstalk* and takes place because of unintended interaction, through electric and magnetic fields, between one telephone line and another. A typical scenario is shown in Fig. 6.60*a*. A *generator* circuit consisting of a pair of parallel wires (a transmission line) connects a source (modeled as a single-frequency sinusoid, although this is only representative of the components of a voice signal) to an intended receiver, modeled as a load resistor R_L. Running parallel to this transmission line is another transmission line, the *receptor* circuit, that connects two receivers, which are modeled as resistors R_{NE} and R_{FE}. The subscripts used here denote *near end* (NE) and *far end* (FE) with respect to the source and are common nomenclature in the telephone industry. The source causes a current, i_G, to pass down one wire and return on the other. This current generates a magnetic field in the vicinity of the line that passes through the loop of wire that constitutes the receptor line as shown in Fig. 6.60*b*. This unintended coupling represents a mutual inductance, M, between the two circuits and is modeled as shown in Fig. 6.60*d*. Similarly, the source voltage causes a voltage between the two wires of the generator circuit, v_G, which results in charge being induced on the receptor circuit wires as shown in Fig. 6.60*c* . This represents a mutual capacitance, C_M, between the two circuits and is modeled as shown in Fig. 6.60*d*. Each line has its own self-inductance and capacitance, L_G, L_R and C_G, C_R. The goal here is to compute the near-end and far-end induced (crosstalk) voltages v_{NE} and v_{FE}.

Figure 6.60*e* shows a circuit with typical values. The source is a 10-V, 10-kHz source. In order to compute these crosstalk voltages we could draw the phasor receptor circuit and write

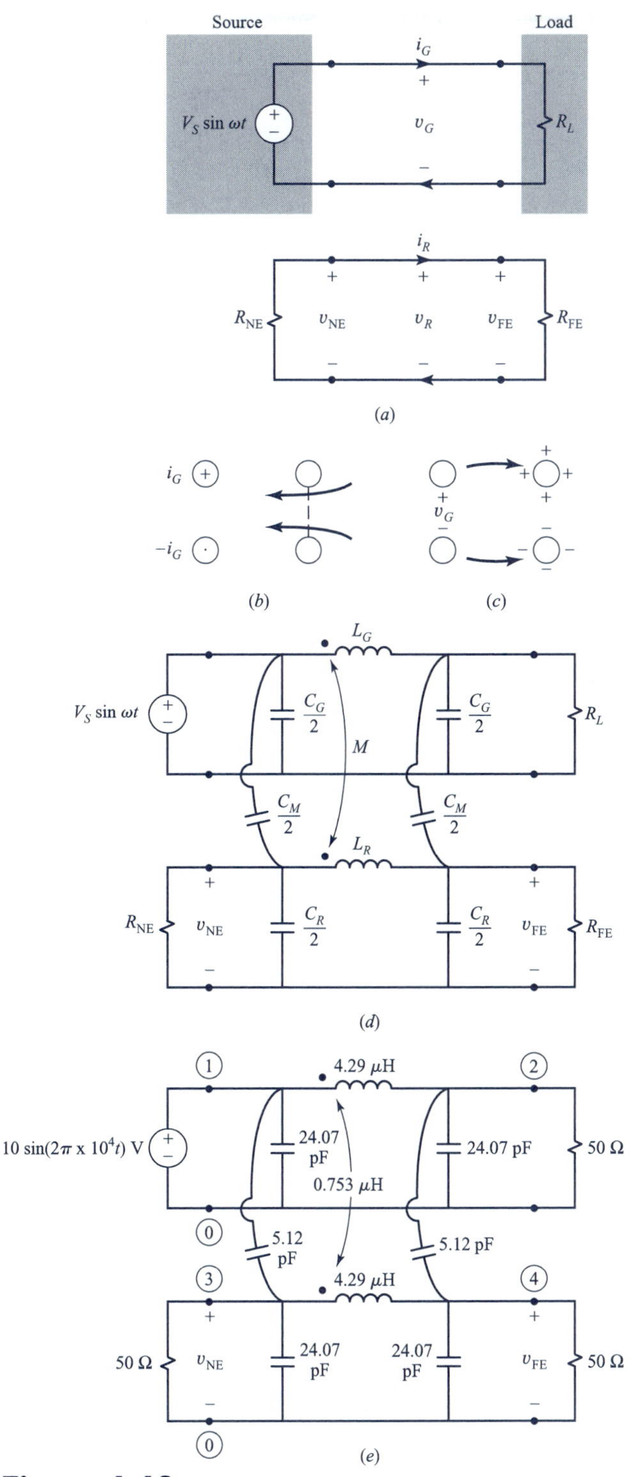

Figure 6.60
Illustration of the phenomenon of crosstalk between wire-connected circuits: (*a*) the physical configuration of two pairs of closely spaced wires, (*b*) magnetic field coupling, (*c*) electric field coupling, (*d*) a lumped-circuit model of the interaction between the two circuits, and (*e*) typical element values along with the PSPICE node labeling.

mesh-current or node-voltage equations. Clearly this would be a difficult task, so we choose to use PSPICE. The nodes are labeled as shown, and the PSPICE program becomes

```
CROSSTALK PROBLEM
VS 1 0 AC 10 0
CGL 1 0 24.07P
CRL 3 0 24.07P
CML 1 3 5.12P
CGR 2 0 24.07P
CRR 4 0 24.07P
CMR 2 4 5.12P
LG 1 2 4.29U
LR 3 4 4.29U
K LG LR 0.1754
RL 2 0 50
RNE 3 0 50
RFE 4 0 50
.AC DEC 1 10K 10K
.PRINT AC VM(3) VP(3) VM(4) VP(4)
.END
```

The results are

$$\hat{V}_{\text{NE}} = 4.889 \angle 89.55^\circ \text{ mV}$$
$$\hat{V}_{\text{FE}} = 4.567 \angle -90.47^\circ \text{ mV}$$

Hence a 10-V signal on one line can induce 5-mV signals at the ends of the other line. If the near- and far-end loads represent telephone receivers, this may be sufficient to be heard faintly in the background in those receivers.

Problems

Section 6.1 The Sinusoidal Source

6.1-1 Convert the following functions to equivalent functions by determining θ:

(a) $3\sin(2t - 30°) = 3\cos(2t + \theta)$

(b) $2\sin(3t + 45°) = 2\cos(3t + \theta)$

(c) $1\cos(t - 135°) = 1\sin(t + \theta)$

(d) $2\cos(2t + 215°) = 2\sin(2t + \theta)$

(e) $3\cos(3t - 65°) = 3\sin(3t + \theta)$

(f) $2\sin(t - 215°) = 2\cos(t + \theta)$

(g) $2\sin(3t - 30°) = 2\cos(3t + \theta)$

(h) $1\cos(2t - 250°) = 1\sin(2t + \theta)$

Sketch your results to verify that they are correct. *Answers*: (a) $\theta = -120°$, (b) $\theta = -45°$, (c) $\theta = -45°$, (d) $\theta = -55°$, (e) $\theta = 25°$, (f) $\theta = 55°$, (g) $\theta = -120°$, (h) $\theta = -160°$.

6.1-2 Convert the following trigonometric expressions to equivalent forms by giving M and θ or A and B as required:

(a) $2\cos 3t - 3\sin 3t = M\cos(3t + \theta)$

(b) $1\cos 2t + 2\sin 2t = M\sin(2t + \theta)$

(c) $-1\cos 2t + 4\sin 2t = M\cos(2t + \theta)$

(d) $2\cos 4t - 6\sin 4t = M\sin(4t + \theta)$

(e) $3\cos(t + 75°) = A\cos t + B\sin t$

(f) $-2\sin(t - 30°) = A\cos t + B\sin t$

(g) $-4\cos(t + 135°) = A\cos t + B\sin t$

Answers: (a) $M = 3.61$, $\theta = 56.31°$, (b) $M = 2.24$, $\theta = 26.57°$, (c) $M = 4.12$, $\theta = 255.96°$, (d) $M = 6.32$, $\theta = 161.57°$, (e) $A = 0.78$, $B = -2.9$, (f) $A = 1$, $B = -1.73$, (g) $A = 2.83$, $B = 2.83$.

Section 6.2 Complex Numbers, Complex Algebra, and Euler's Identity

6.2-1 For the following pairs of complex numbers $\hat{A}$ and $\hat{B}$, calculate $\hat{A} + \hat{B}$, $\hat{A}/\hat{B}$, and $\hat{A}\hat{B}$. Express your answer both in rectangular and polar form.

(a) $\hat{A} = 1 + j2, \hat{B} = -1 - j3$

(b) $\hat{A} = -1 + j1, \hat{B} = 2 - j1$

(c) $\hat{A} = 5 - j4, \hat{B} = 10 + j8$

(d) $\hat{A} = -3 + j7, \hat{B} = 8 - j14$

(e) $\hat{A} = 2 - j6, \hat{B} = -2 - j10$

(f) $\hat{A} = -10 - j1, \hat{B} = -4 + j2$

(g) $\hat{A} = 5 - j8, \hat{B} = -2 - j6$

(h) $\hat{A} = -3 + j10, \hat{B} = 8 - j2$

(i) $\hat{A} = 6 + j8, \hat{B} = -3 + j9$

(j) $\hat{A} = -2 - j10, \hat{B} = -12 + j20$

(k) $\hat{A} = 6 - j7, \hat{B} = 30 - j20$

(l) $\hat{A} = -8 - j2, \hat{B} = 10 + j15$

(m) $\hat{A} = -8 + j10, \hat{B} = -5 - j10$

Answers: (a) $1\angle -90°, 0.71\angle 171.87°, 7.07\angle -45°$, (b) $1\angle 0°, 0.63\angle 161.6°, 3.16\angle 108.43°$, (c) $15.52\angle 14.93°, 0.5\angle -77.32°, 82\angle 0°$, (d) $8.6\angle -54.46°, 0.47\angle 173.45°, 122.8\angle 52.94°$, (e) $16\angle -90°, 0.62\angle 29.74°, 64.5\angle -172.87°$, (f) $14.04\angle 175.91°, 2.25\angle -327.72°, 44.94\angle -20.85°$, (g) $14.32\angle -77.91°, 1.49\angle 50.44°, 59.67\angle -166.43°$, (h) $9.43\angle 57.99°, 1.27\angle 120.74°, 86.09\angle 92.66°$, (i) $17.26\angle 79.99°$,

$1.05\angle-55.3°, 94.87\angle 161.57°$, (j) $17.2\angle 144.46°, 0.44\angle-222.27°, 237.86\angle 19.65°$, (k) $45\angle-36.87°, 0.26\angle-15.71°, 332.42\angle-83.09°$, (l) $13.15\angle 81.25°, 0.46\angle-222.27°, 148.66\angle-109.65°$, (m) $13\angle 180°, 1.15\angle 245.22°, 143.18\angle 12.09°$.

6.2-2 For the complex numbers $\hat{A} = 1 - j3$, $\hat{B} = 2\angle-30°$, $\hat{C} = 2 + j1$, and $\hat{D} = 3\angle 150°$ evaluate the following expressions. Express your answer both in rectangular and in polar form.

(a) $\hat{A} + \hat{B}$ (c) $\hat{A}\hat{B}$ (e) $\hat{A}/\hat{B}$ (g) $1/\hat{A}$ (i) $\hat{B} + \hat{A}\hat{D}/\hat{C}$

(b) $\hat{B} + \hat{D}$ (d) $\hat{A}\hat{C}$ (f) $\hat{A}/\hat{C}$ (h) $\hat{A}\hat{C}^*$ (j) $\hat{B}\hat{C} - \hat{A}\hat{D}/\hat{C}$

Answers: (a) $2.732 - j4$, (b) $-0.866 + j0.5$, (c) $-1.268 - j6.196$, (d) $5 - j5$, (e) $1.183 - j1.049$, (f) $-0.2 - j1.4$, (g) $0.1 + j0.3$, (h) $-1 - j7$, (i) $4.352 + j2.337$, (j) $1.845 - j3.605$.

6.2-3 For the complex numbers $\hat{A} = 20\angle-135°$, $\hat{B} = 10 - j8$, $\hat{C} = 2 + j6$, and $\hat{D} = 15\angle 60°$ evaluate:

(a) $\hat{E} = \dfrac{\hat{A} + \hat{B}}{\hat{C}\hat{D}}$ (b) $\hat{F} = \dfrac{\hat{D}/\hat{B} + \hat{A}}{\hat{C}}$ (c) $\hat{G} = \dfrac{\hat{B}}{\hat{C} + \hat{D}} + \hat{A}$

Answers: (a) $0.24\angle-232.16°$, (b) $3.06\angle-209.36°$, (c) $20.51\angle-134.08°$.

6.2-4 Complex numbers having angles that are multiples of 45° are simple to write in either rectangular or polar form. Drawing a sketch in the complex plane facilitates this conversion. Express the following in rectangular or polar form:

(a) $3 - j3$ (c) $-1 + j$ (e) $3\angle 225°$ (g) $-2\angle-135°$

(b) $-2 - j2$ (d) $2\angle-135°$ (f) $1\angle 135°$

Answers: (a) $3\sqrt{2}\angle-45°$, (b) $2\sqrt{2}\angle-135°$, (c) $1\sqrt{2}\angle 135°$, (d) $-\sqrt{2} - j\sqrt{2}$, (e) $-3/\sqrt{2} - j3/\sqrt{2}$, (f) $-1/\sqrt{2} + j1/\sqrt{2}$, (g) $\sqrt{2} + j\sqrt{2}$.

6.2-5 Manipulating complex numbers whose angles are multiples of 90° or 45° can and should be done without a calculator. Evaluate the following expressions:

(a) $\hat{A} = j3 + \dfrac{j4}{-1 + j}$

(b) $\hat{A} = (1 - j1)(2\angle-135°) + \dfrac{-2 - j2}{2\angle-45°}$

(c) $\hat{A} = \dfrac{2\angle 135° - 3\angle 180°}{-3 - j3}$

(d) $\hat{A} = \dfrac{-3 + j3}{2\angle 45°} - \dfrac{(2 - j2)(1\angle-90°)}{-2j}$

Answers: (a) $2 + j1$, (b) $-2\sqrt{2} - j\sqrt{2}$, (c) $0.5\angle 176.73°$, (d) $-1 + j3.12$.

6.2-6 Use Euler's identity to show the following identities:

(a) $\sin 2A = 2\sin A\cos A$

(b) $\cos 2A = \cos^2 A - \sin^2 A$

(c) $\sin^2 A = \frac{1}{2}(1 - \cos 2A)$

(d) $\cos^2 A = \frac{1}{2}(1 + \cos 2A)$

(e) $\int \sin ax\, dx = -(1/a)\cos ax$

(f) $\sin A\cos B = \frac{1}{2}[\sin(A + B) + \sin(A - B)]$

Section 6.3 The Phasor (Frequency-Domain) Circuit

6.3-1 Draw the phasor circuit for the circuit of Fig. P6.3-1.

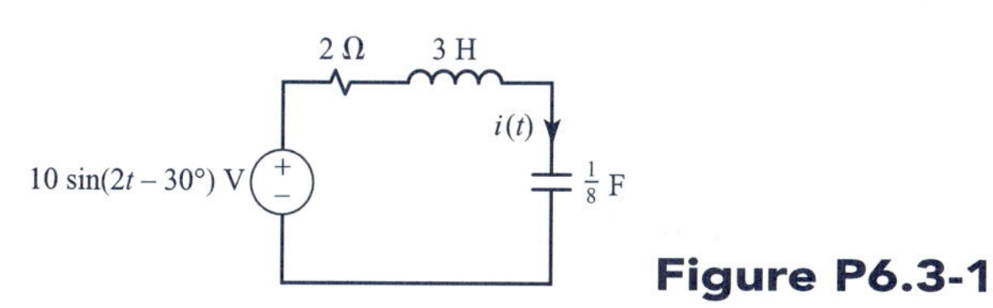

Figure P6.3-1

6.3-2 Draw the phasor circuit for the circuit of Fig. P6.3-2.

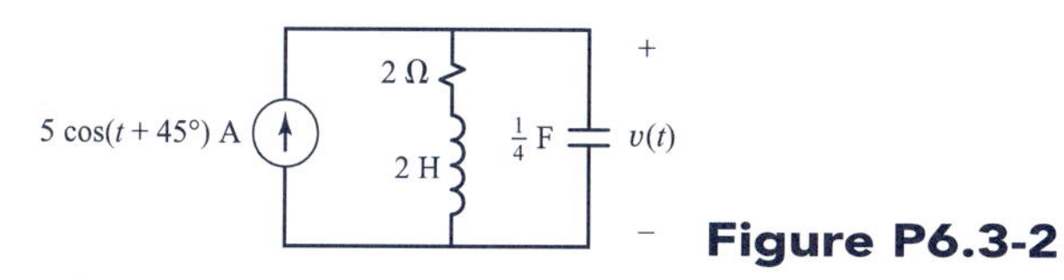

Figure P6.3-2

6.3-3 Draw the phasor circuit for the circuit of Fig. P6.3-3.

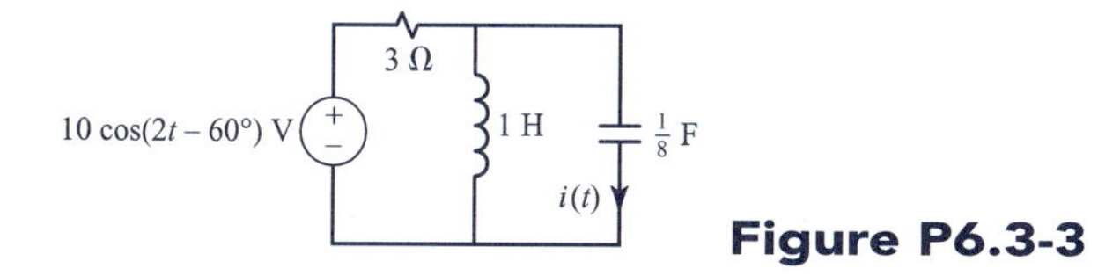

Figure P6.3-3

6.3-4 Draw the phasor circuit for the circuit of Fig. P6.3-4.

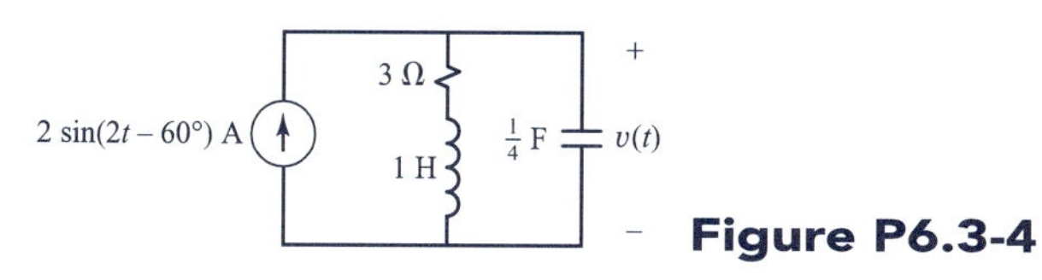

Figure P6.3-4

6.3-5 Draw the phasor circuit for the circuit of Fig. P6.3-5.

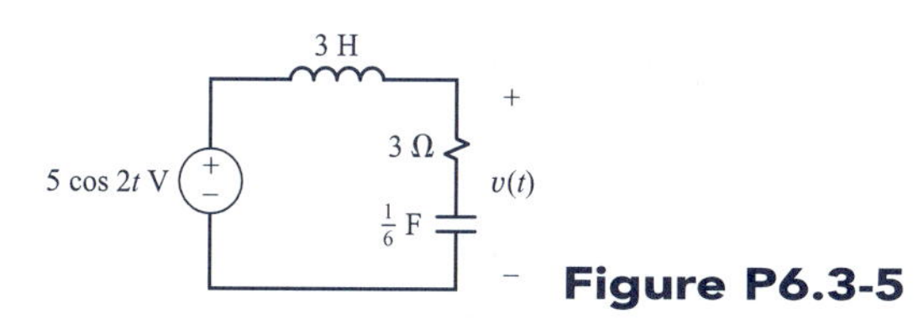

Figure P6.3-5

6.3-6 Draw the phasor circuit for the circuit of Fig. P6.3-6.

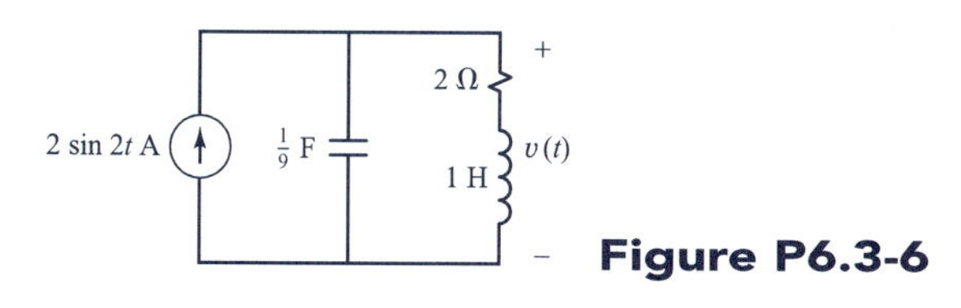

Figure P6.3-6

6.3-7 Draw the phasor circuit for the circuit of Fig. P6.3-7.

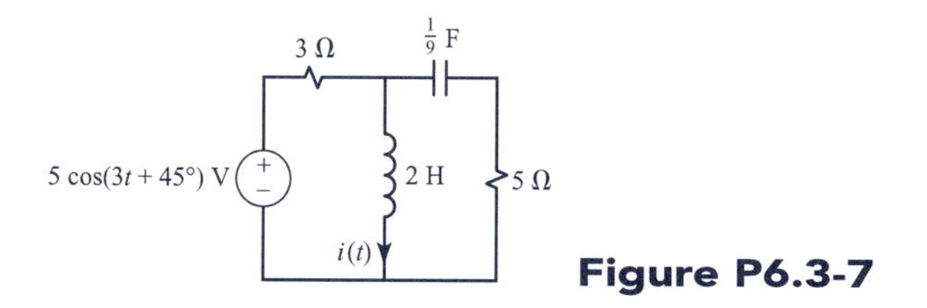

Figure P6.3-7

6.3-8 Draw the phasor circuit for the circuit of Fig. P6.3-8.

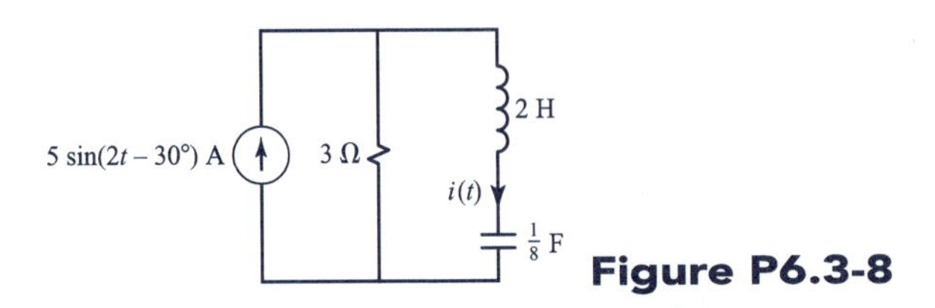

Figure P6.3-8

Section 6.4 Application of Resistive-Circuit Analysis Techniques in the Phasor Circuit

6.4-1 Determine the current $i(t)$ in the circuit of Fig. P6.3-1. *Answer*: $3.54 \sin(2t - 75°)$ A.

6.4-2 Determine the voltage $v(t)$ in the circuit of Fig. P6.3-2. *Answer*: $20 \cos(t + 45°)$ V.

6.4-3 Determine the current $i(t)$ in the circuit of Fig. P6.3-3. *Answer*: $2 \cos(2t + 66.87°)$ A.

6.4-4 Determine the voltage $v(t)$ in the circuit of Fig. P6.3-4. *Answer*: $4.81 \sin(2t - 116.31°)$ V.

6.4-5 Determine the voltage $v(t)$ in the circuit of Fig. P6.3-5. *Answer*: $5 \cos(2t - 90°)$ V.

6.4-6 Determine the voltage $v(t)$ in the circuit of Fig. P6.3-6. *Answer*: $7.95 \sin(2t + 6.34°)$ V.

6.4-7 Determine the current $i(t)$ in the circuit of Fig. P6.3-7. *Answer*: $0.57 \cos(3t - 35.73°)$ A.

6.4-8 Determine the current $i(t)$ in the circuit of Fig. P6.3-8. *Answer*: $5 \sin(2t - 30°)$ A.

6.4-9 Determine the current $i(t)$ in the circuit of Fig. P6.4-9 by using a Thevenin equivalent reduction in the phasor circuit. Repeat using a Norton reduction. *Answer*: $1.18 \sin(3t + 75°)$ A.

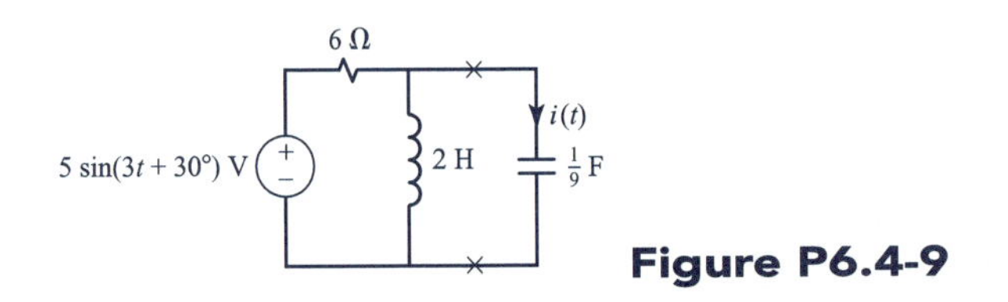

Figure P6.4-9

6.4-10 Determine the voltage $v(t)$ in the circuit of Fig. P6.4-10 by using a Thevenin equivalent reduction in the phasor circuit. Repeat using a Norton reduction.
Answer: $-2.83 \cos(2t + 15°)$ V.

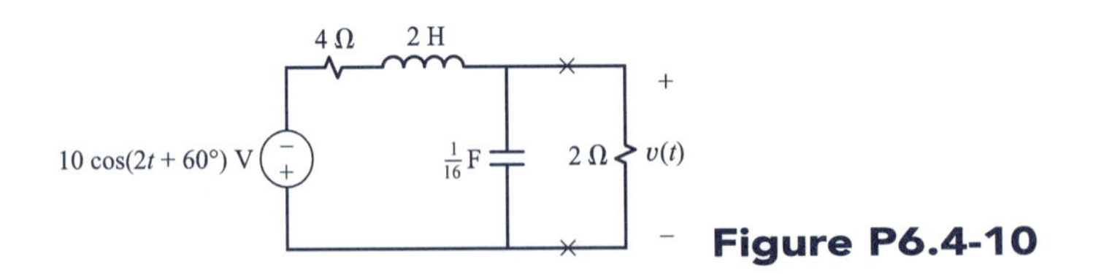

Figure P6.4-10

6.4-11 Determine the current $i(t)$ in the circuit of Fig. P6.4-11 by using a source transformation in the phasor circuit. *Answer*: $-2.24 \sin(3t - 146.57°)$ A.

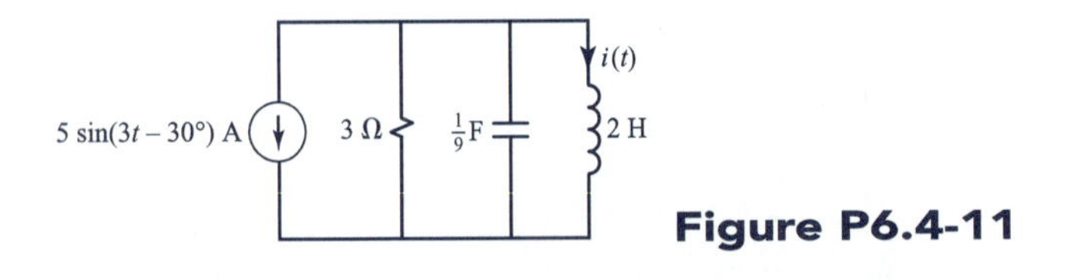

Figure P6.4-11

6.4-12 Determine the voltage $v(t)$ in the circuit of Fig. P6.4-12 by writing mesh-current equations in the phasor circuit. *Answer*: $3.16 \cos(2t - 71.57°)$ V.

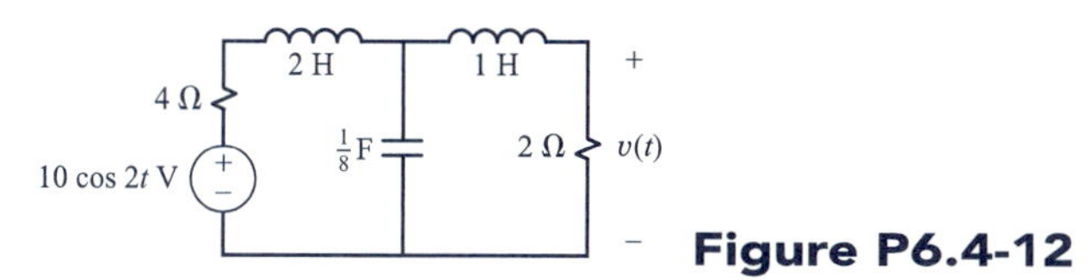

Figure P6.4-12

6.4-13 Determine the voltage $v(t)$ in the circuit of Fig. P6.4-13 by using voltage division in the phasor circuit. *Answer*: $35.36 \sin(4t - 75°)$ V.

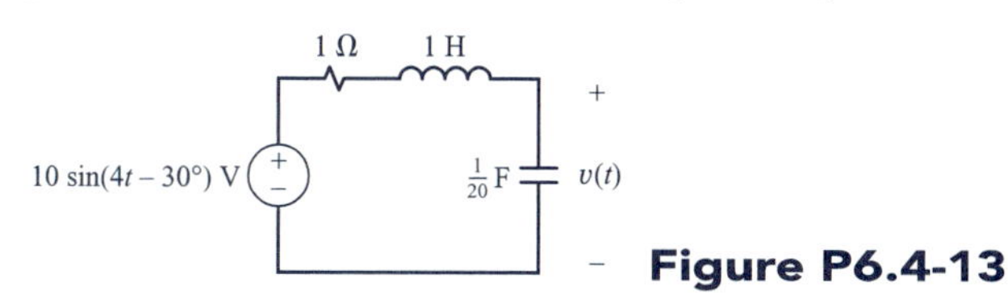

Figure P6.4-13

6.4-14 Determine the current $i(t)$ in the circuit of Fig. P6.4-14 by using current division in the phasor circuit. *Answer*: $-2.77 \cos(2t + 43.69°)$ A.

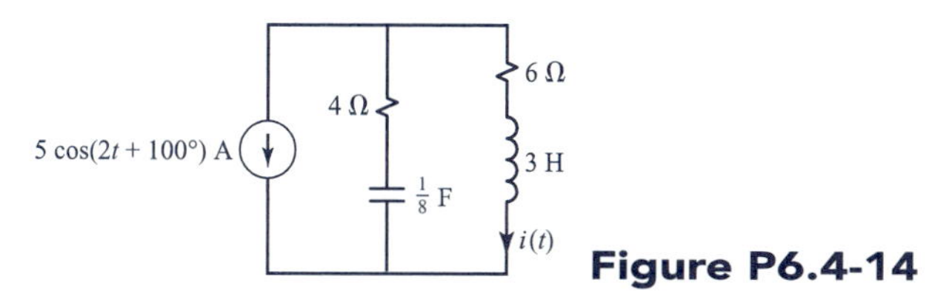

Figure P6.4-14

6.4-15 Determine the voltage $v(t)$ in the circuit of Fig. P6.4-15 by writing mesh-current equations in the phasor circuit. *Answer*: $2.5 \sin(2t - 90°)$ V.

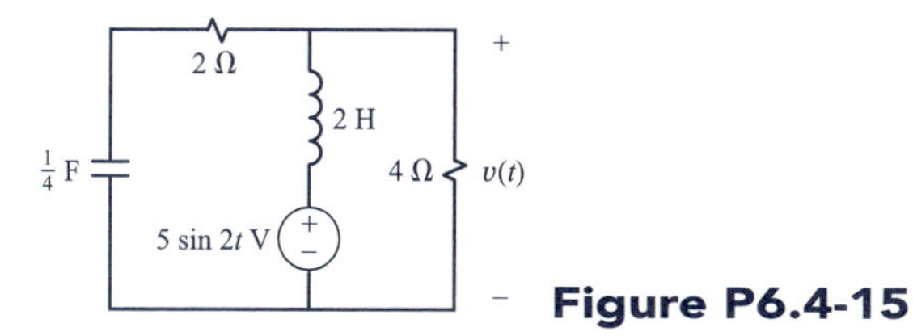

Figure P6.4-15

6.4-16 Determine the current $i(t)$ in the circuit of Fig. P6.4-16 by using current division in the phasor circuit. *Answer*: $-4.47 \cos(3t - 18.43°)$ A.

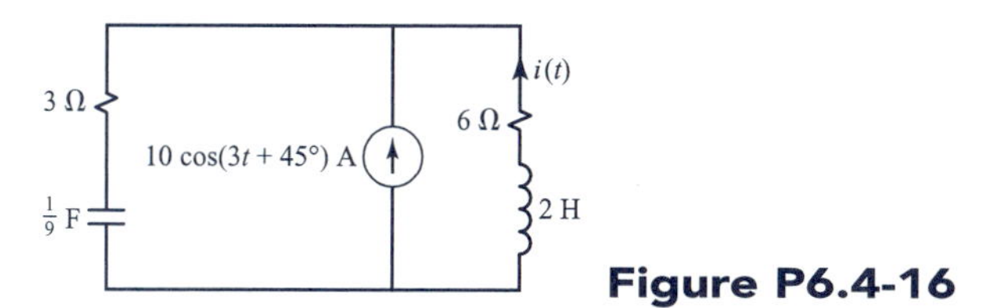

Figure P6.4-16

6.4-17 Determine the current $i(t)$ in the circuit of Fig. P6.4-17 by using a Thevenin reduction at the indicated terminals in the phasor circuit. Repeat using a Norton reduction.
Answer: $0.48 \sin(3t + 121.39°)$ A.

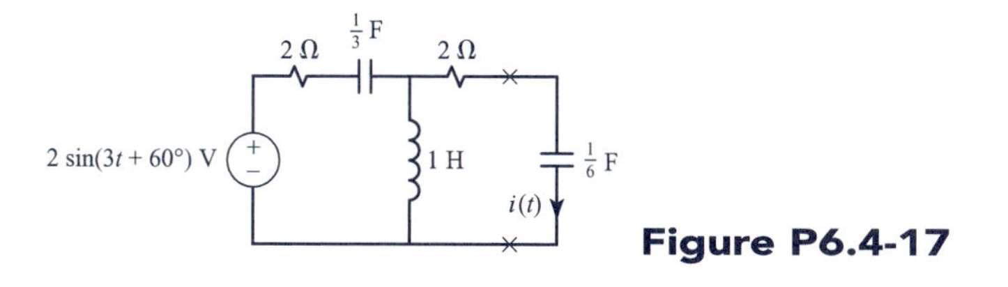

Figure P6.4-17

6.4-18 Determine the current $i(t)$ in the circuit of Fig. P6.4-18 by writing mesh-current equations in the phasor circuit. *Answer*: $-0.881\ \cos(3t - 27.39°)$ A.

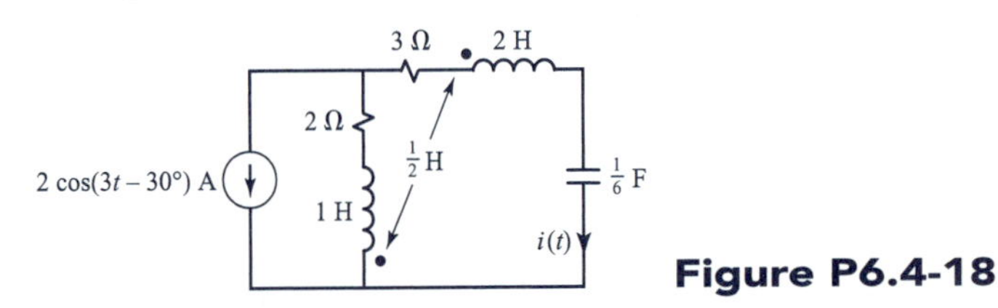

Figure P6.4-18

6.4-19 Determine the voltage $v(t)$ in the circuit of Fig. P6.4-19. *Answer*: $9.7\ \sin(3t + 15.96°)$ V.

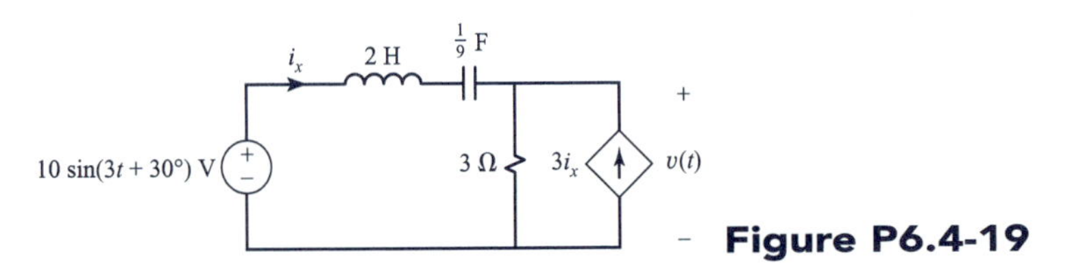

Figure P6.4-19

6.4-20 Determine the voltage $v(t)$ in the circuit of Fig. P6.4-20. *Answer*: $1.99\ \cos(2t - 49.76°)$ V.

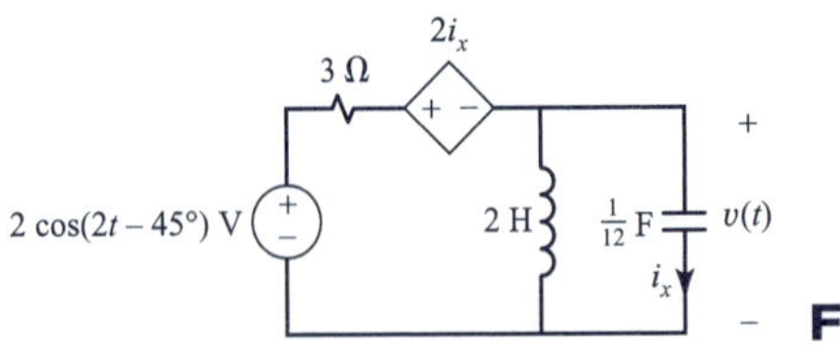

Figure P6.4-20

6.4-21 Determine the voltage $v(t)$ in the circuit of Fig. P6.4-21. *Answer*: $-4.47\ \sin(t + 86.57°)$ V.

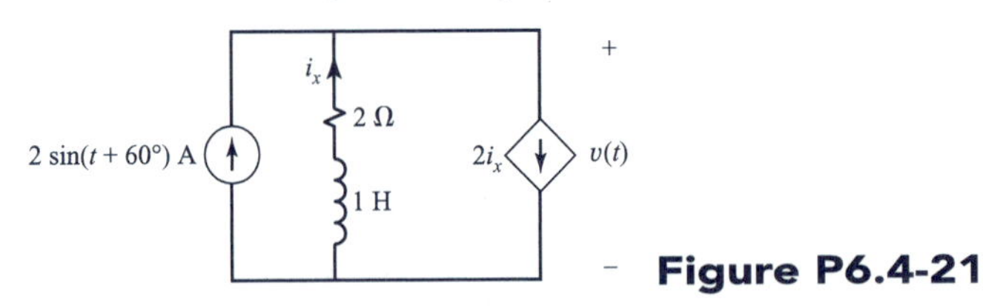

Figure P6.4-21

6.4-22 Determine the current $i(t)$ in the circuit of Fig. P6.4-22. *Answer*: $3.54\ \cos(2t - 135°)$ A.

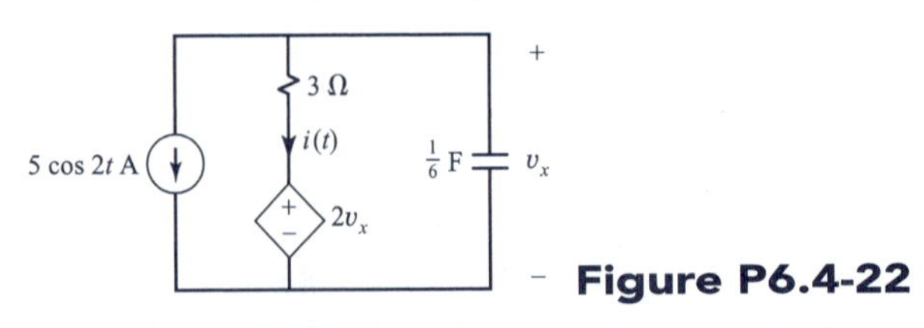

Figure P6.4-22

6.4-23 Determine the current $i(t)$ in the circuit of Fig. P6.4-23. *Answer*: $1.05\ \sin(3t - 108.43°)$ A.

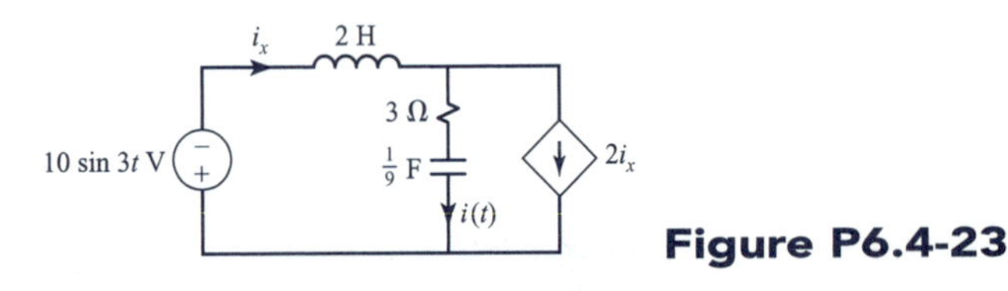

Figure P6.4-23

Section 6.5 Circuits Containing More than One Sinusoidal Source

6.5-1 Determine the current $i(t)$ in the circuit of Fig. P6.5-1 by placing both sources in the phasor circuit and writing node-voltage equations. *Answer*: $5.27 \sin(2t - 110.7°)$ A.

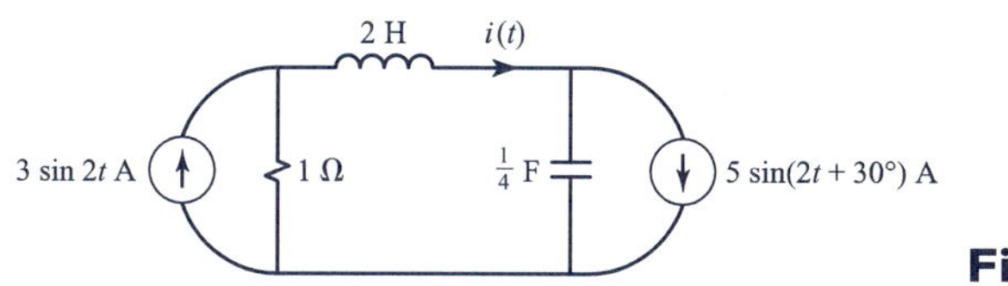

Figure P6.5-1

6.5-2 Determine the voltage $v(t)$ in the circuit of Fig. P6.5-2.
Answer: $5.55 \cos(2t + 56.31°) + 10 \sin(4t - 135°)$ V.

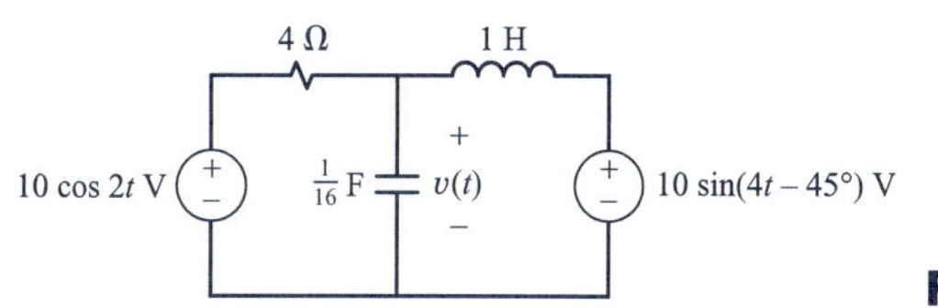

Figure P6.5-2

6.5-3 Determine the current $i(t)$ in the circuit of Fig. P6.5-3 by placing both sources in the phasor circuit and using superposition. *Answer*: $4.09 \cos(2t + 18.22°)$ A.

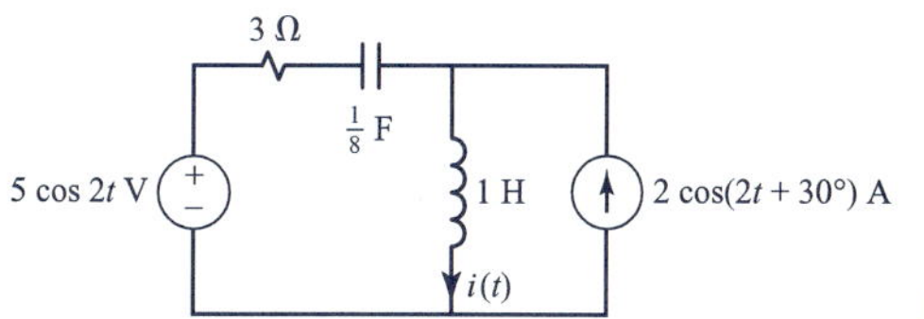

Figure P6.5-3

6.5-4 Determine the current $i(t)$ in the circuit of Fig. P6.5-4 by placing both sources in the phasor circuit and using superposition. *Answer*: $2.62 \cos(2t - 20.76°)$ A.

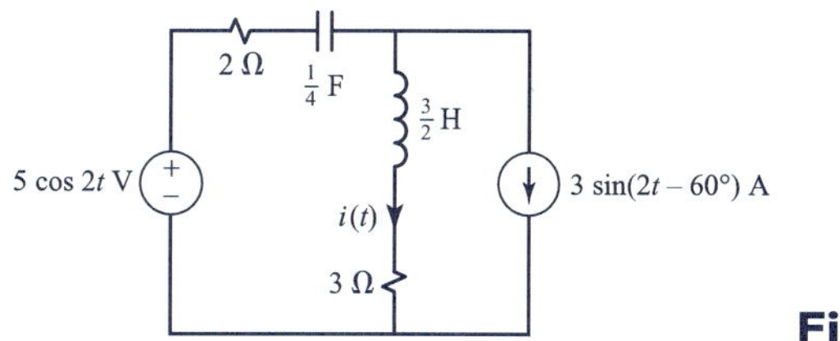

Figure P6.5-4

6.5-5 Determine the current $i(t)$ in the circuit of Fig. P6.5-5 by placing both sources in the phasor circuit and using superposition. *Answer*: $1.49 \sin(3t - 120.96°)$ A.

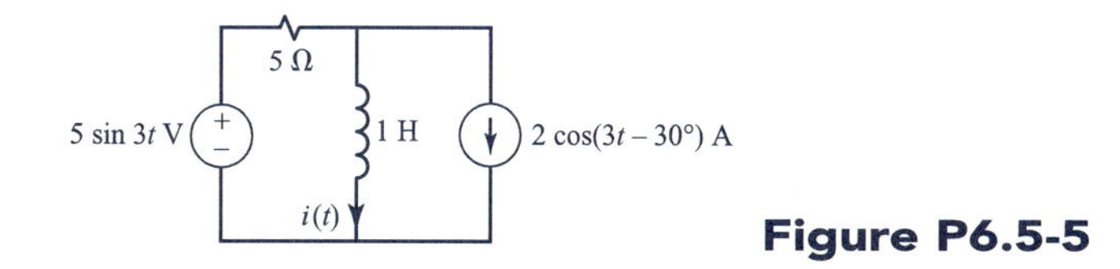

Figure P6.5-5

6.5-6 Determine the voltage $v(t)$ in the circuit of Fig. P6.5-6.
Answer: $-8.32 \sin(t - 33.69°) + 10$ V.

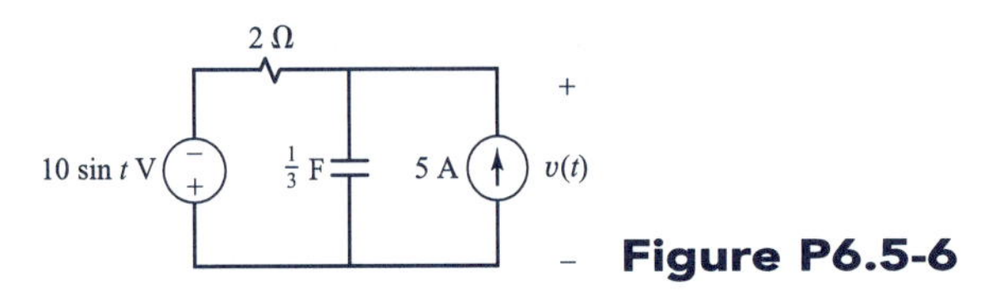

Figure P6.5-6

6.5-7 Determine the current $i(t)$ in the circuit of Fig. P6.5-7 by placing both sources in the phasor circuit and writing mesh-current equations. *Answer*: $0.64 \sin(3t + 163.56°)$ A.

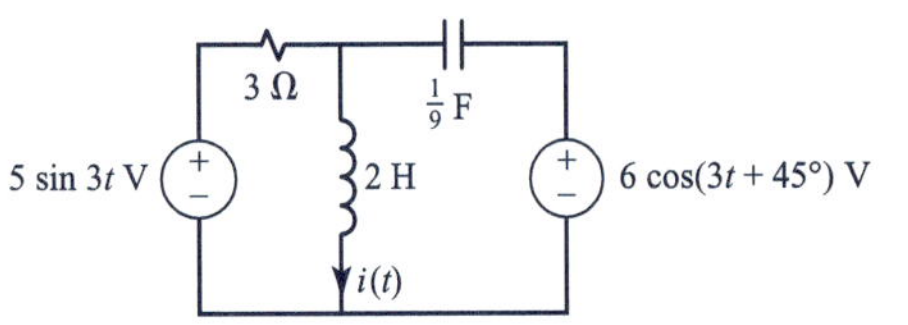

Figure P6.5-7

6.5-8 Determine the current $i(t)$ in the circuit of Fig. P6.5-8 by placing all sources in the phasor circuit and writing node-voltage equations. *Answer*: $5.98 \sin(t + 112.76°)$ A.

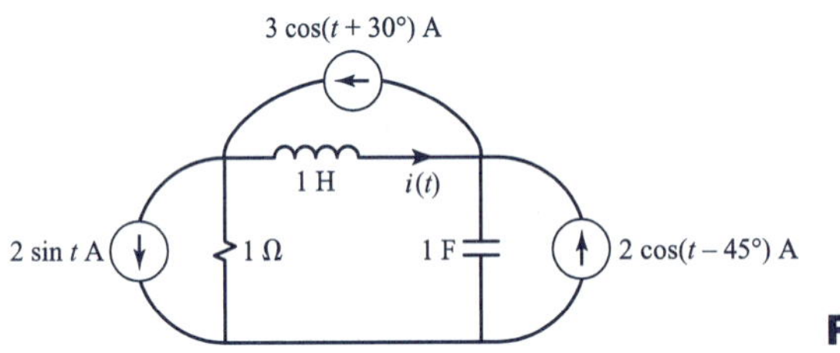

Figure P6.5-8

6.5-9 Determine the current $i_x(t)$ in the circuit of Fig. P6.5-9 by placing both sources in the phasor circuit and writing node-voltage equations. *Answer*: $0.27 \cos(2t + 10.3°)$ A.

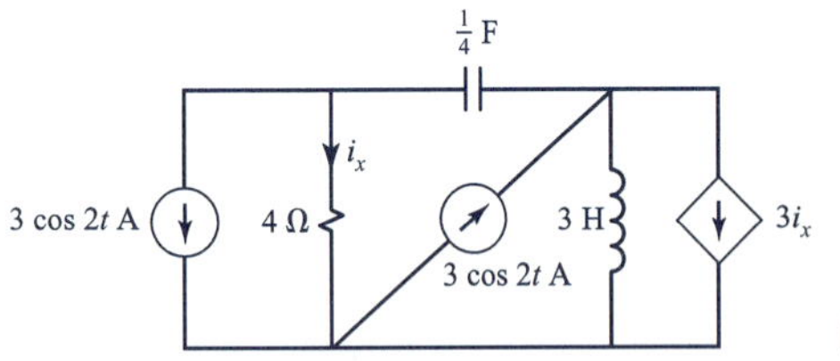

Figure P6.5-9

6.5-10 Determine the current $i_x(t)$ in the circuit of Fig. P6.5-10 by placing both sources in the phasor circuit and writing mesh-current equations. *Answer*: $1.04 \sin(2t - 34.24°)$ A.

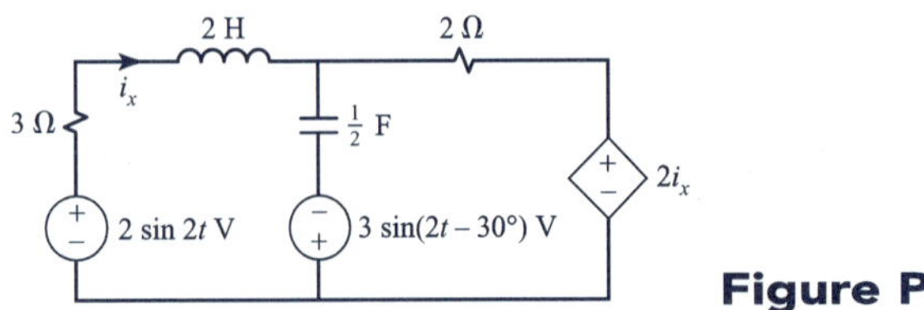

Figure P6.5-10

Section 6.6 Power

6.6-1 Determine the average power delivered to the elements in Fig. P6.6-1.
Answers: (a) 0 W, (b) 48.3 W, (c) 14.14 W, (d) 120.74 W.

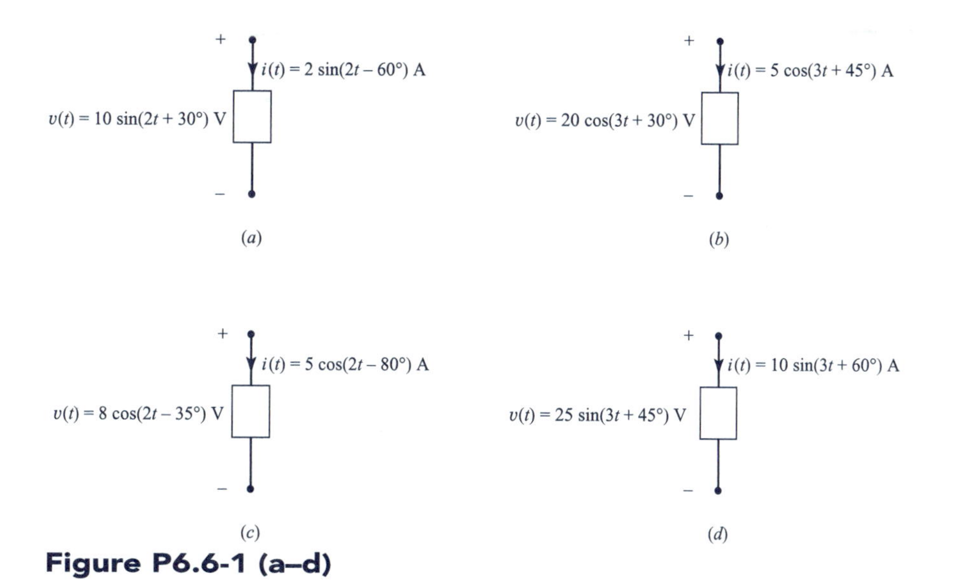

Figure P6.6-1 (a–d)

6.6-2 Determine the average power delivered by the independent source in the circuit of Fig. P6.6-2. Check that average power is conserved. *Answer*: 8.33 W.

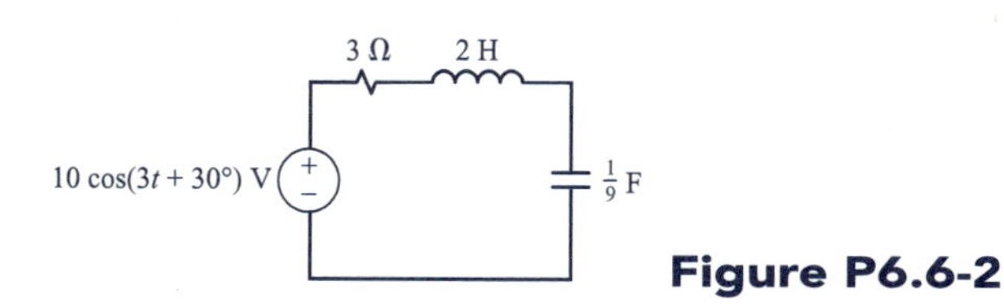

Figure P6.6-2

6.6-3 Determine the average power delivered by the independent source in the circuit of Fig. P6.6-3. Check that average power is conserved. *Answer*: 12.47 W.

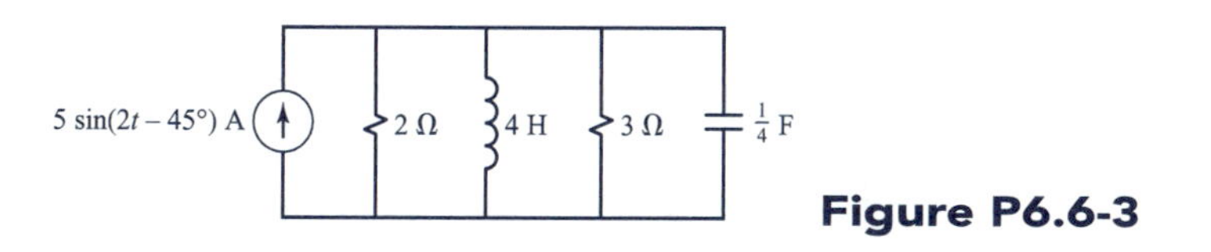

Figure P6.6-3

6.6-4 Determine the average power delivered by the independent source in the circuit of Fig. P6.6-4. Check that average power is conserved. *Answer*: 6.25 W.

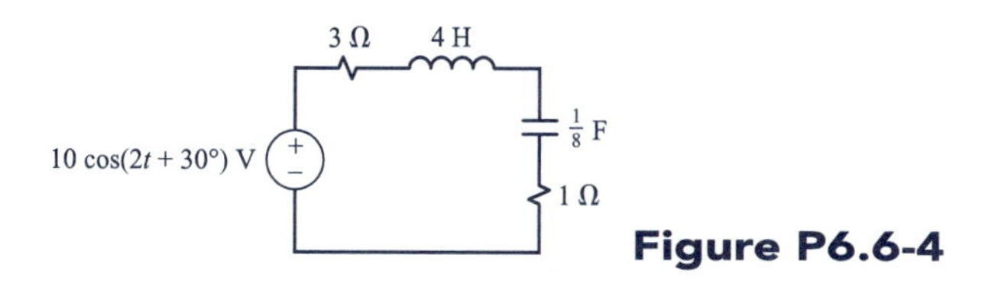

Figure P6.6-4

6.6-5 Determine the average power delivered by the independent source in the circuit of Fig. P6.6-5. Check that average power is conserved. *Answer*: 28.85 W.

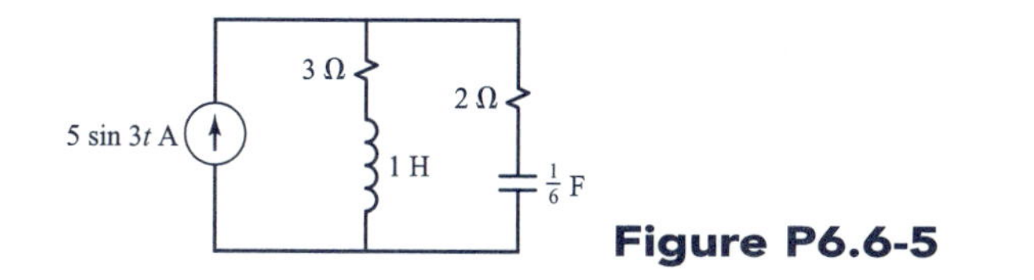

Figure P6.6-5

6.6-6 Determine the average power delivered by the independent source in the circuit of Fig. P6.6-6. Check that average power is conserved. *Answer*: 50 W.

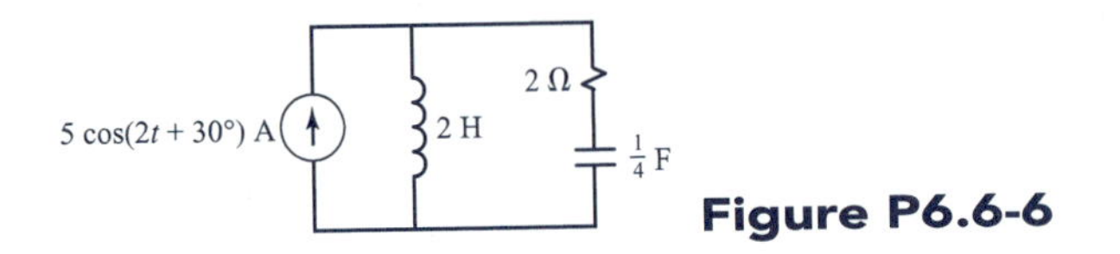

Figure P6.6-6

6.6-7 Determine the average power delivered by the independent source in the circuit of Fig. P6.6-7. Check that average power is conserved. *Answer*: 119.85 W.

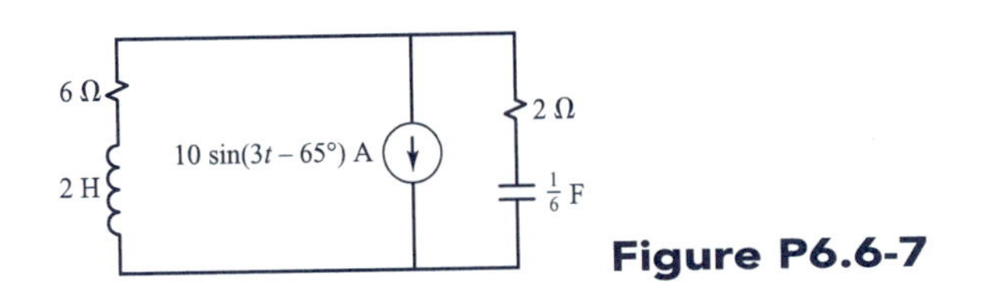

Figure P6.6-7

6.6-8 Determine the average power delivered by the independent source in the circuit of Fig. P6.6-8. Check that average power is conserved. *Answer*: 6.26 W.

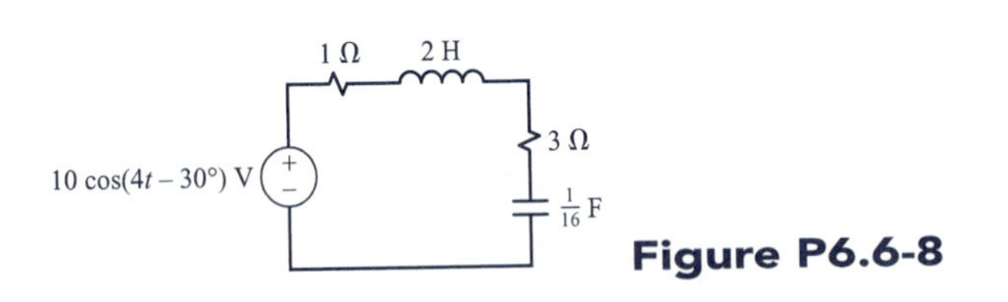

Figure P6.6-8

6.6-9 Determine the average power delivered by the independent source in the circuit of Fig. P6.6-9. Check that average power is conserved. *Answer*: 3.11 W.

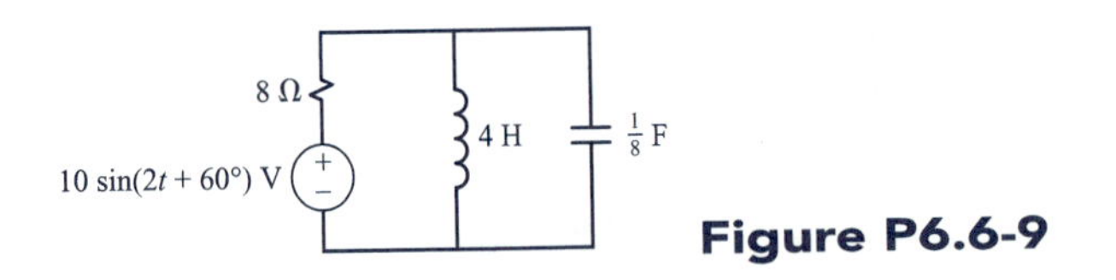

Figure P6.6-9

6.6-10 Determine the average power delivered by the independent source in the circuit of Fig. P6.6-10. Check that average power is conserved. *Answer*: 30 W.

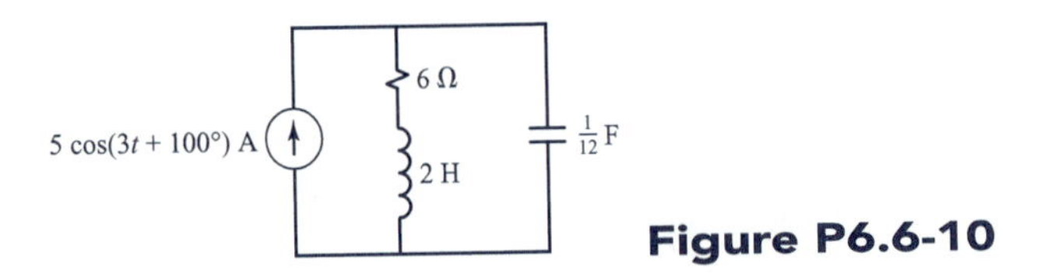

Figure P6.6-10

6.6-11 Determine the average power delivered by the independent source in the circuit of Fig. P6.6-11. Check that average power is conserved. *Answer*: 3.13 W.

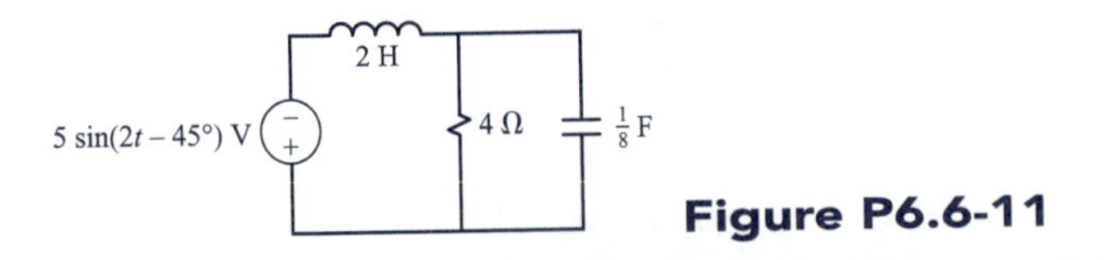

Figure P6.6-11

6.6-12 Determine the average power delivered by the independent source in the circuit of Fig. P6.6-12. Check that average power is conserved. *Answer*: 2.08 W.

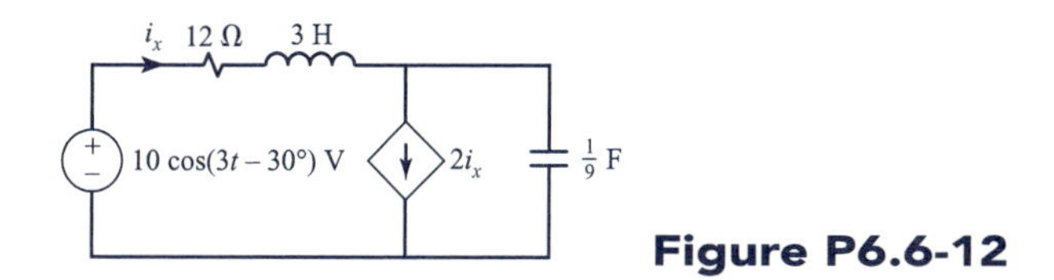

Figure P6.6-12

6.6-13 Determine the average power delivered by the independent source in the circuit of Fig. P6.6-13. Check that average power is conserved. *Answer*: 1.56 W.

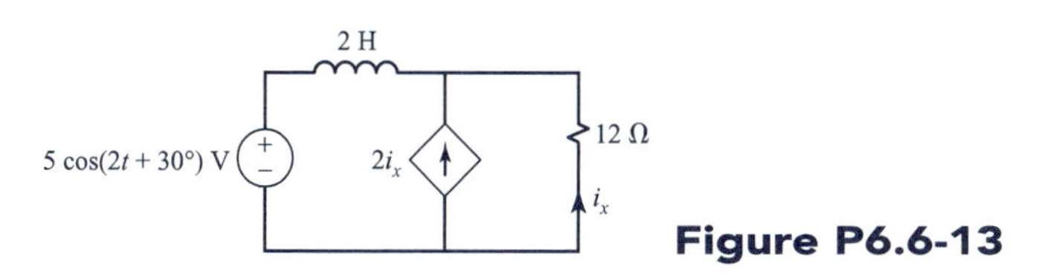

Figure P6.6-13

6.6-14 Determine the rms value of the voltage source waveform in Fig. P6.6-14 as well as the average power delivered to the resistor. *Answer*: 4.08 V, 8.33 W.

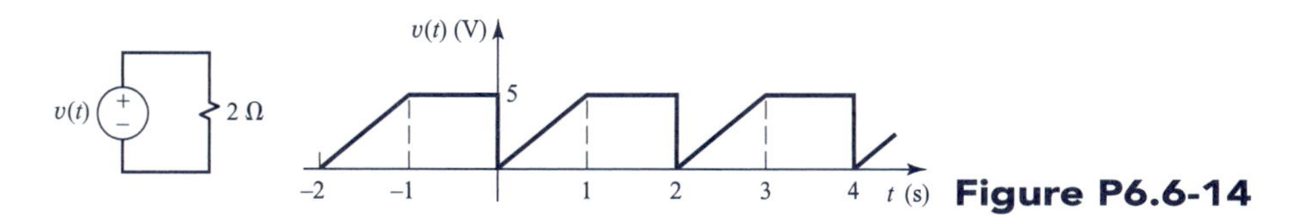

Figure P6.6-14

6.6-15 Determine the rms value of the voltage source waveform in Fig. P6.6-15 as well as the average power delivered to the resistor. *Answer*: 2.31 V, 2.67 W.

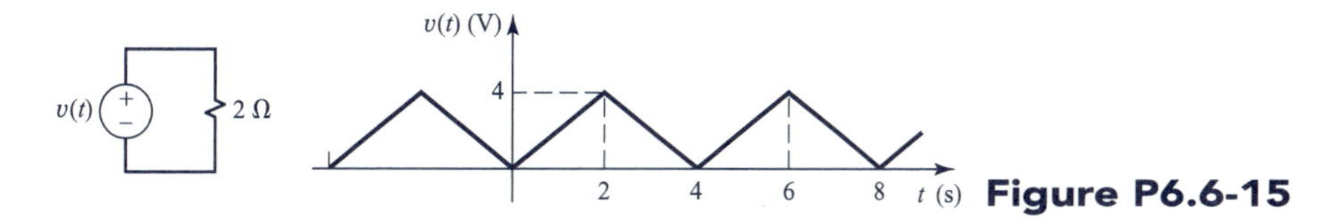

Figure P6.6-15

6.6-16 Determine the rms value of the current source waveform in Fig. P6.6-16 as well as the average power delivered to the resistor. *Answer*: 1.22 A, 4.5 W.

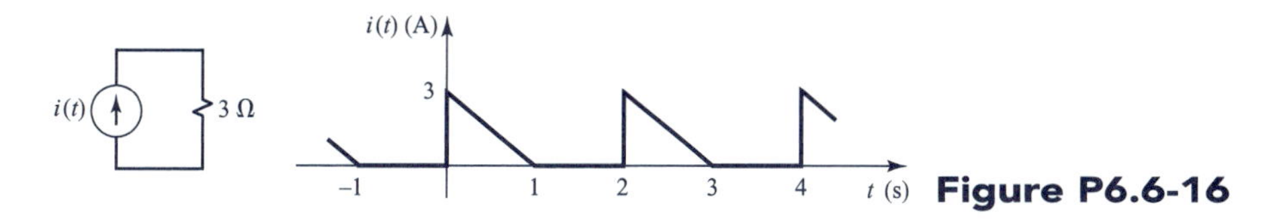

Figure P6.6-16

6.6-17 Determine the total average power delivered by the independent sources in the circuit of Fig. P6.6-17. *Answer*: 2.97 W, −0.83 W.

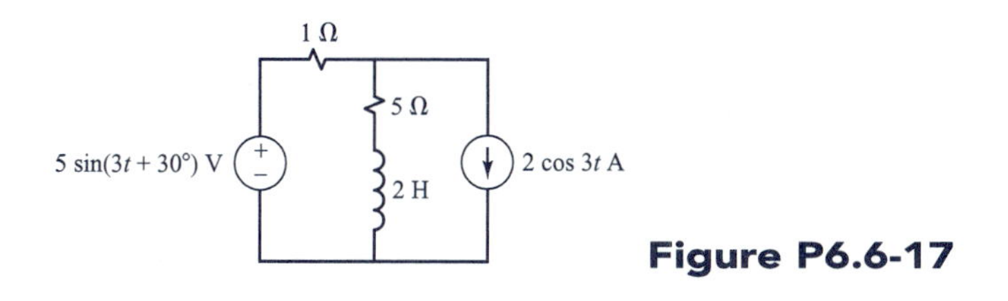

Figure P6.6-17

6.6-18 Determine the total average power delivered by the independent sources in the circuit of Fig. P6.6-18. *Answer*: 3.76 W, 12.5 W.

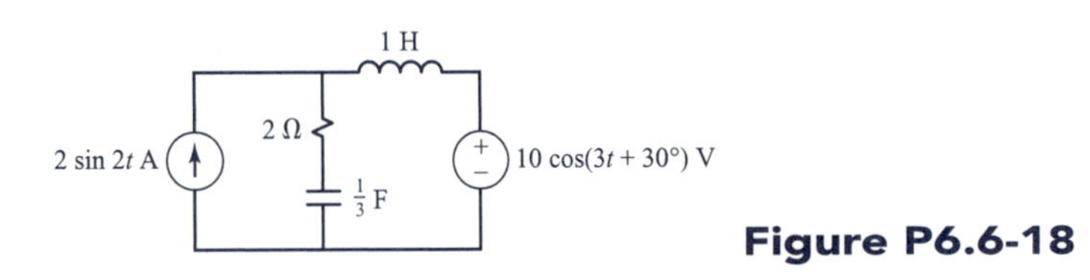

Figure P6.6-18

6.6-19 Determine the total average power delivered by the independent sources in the circuit of Fig. P6.6-19. *Answer*: 8.65 W, 0.5 W.

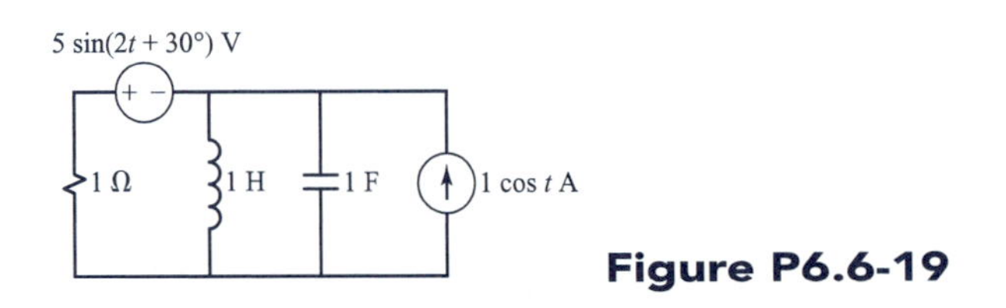

Figure P6.6-19

6.6-20 Determine the total average power delivered by the independent sources in the circuit of Fig. P6.6-20. *Answer*: 1.93 W, 2 W.

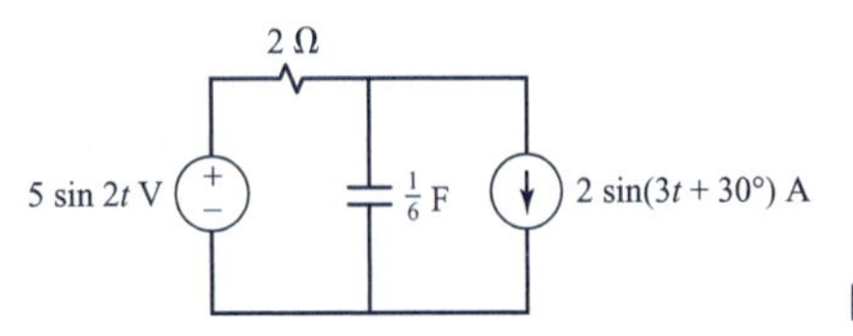

Figure P6.6-20

6.6-21 Determine the load impedance that will absorb the maximum average power in the circuit of Fig. P6.6-21, and determine that average power. *Answers*: $4 + j4\ \Omega$, 0.78 W.

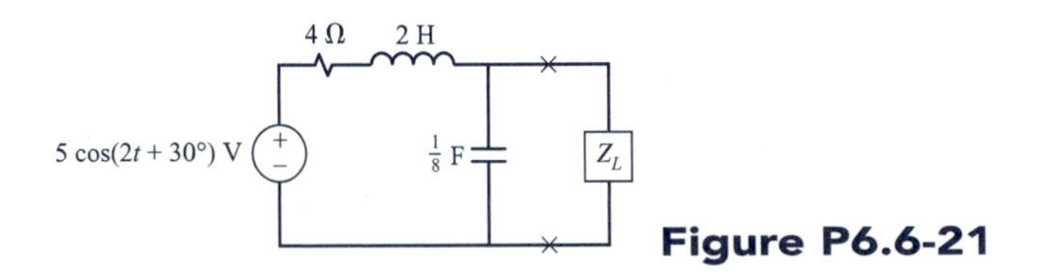

Figure P6.6-21

6.6-22 Determine the load impedance that will absorb the maximum average power in the circuit of Fig. P6.6-22, and determine that average power. *Answers*: $6 + j0\ \Omega$, 3 W.

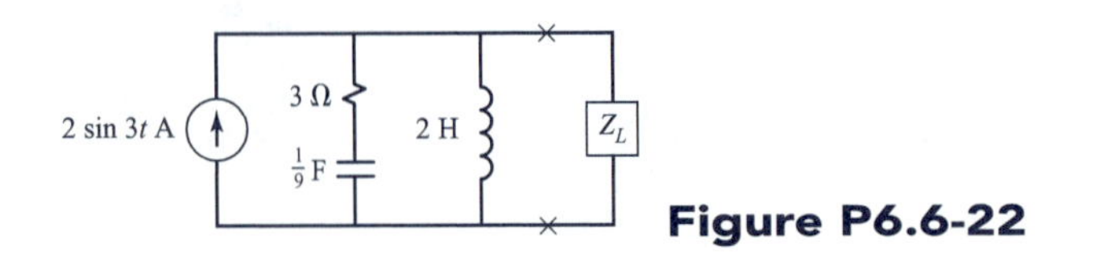

Figure P6.6-22

Section 6.7 Phasor Diagrams

6.7-1 Sketch the phasor diagram for the voltages in the circuit of Fig. 6.3-1.

6.7-2 Sketch the phasor diagram for the voltages in the circuit of Fig. 6.3-3.

Section 6.8 Frequency Response of Circuits

6.8-1 Replace the sinusoidal source in the circuit of Fig. 6.4-9 with a variable-frequency sinusoidal source, and determine the transfer function relating the current $i(t)$ to that source.
Answer: $\hat{H}(j\omega) = -2\omega^2/(54 - 12\omega^2 + j18\omega)$.

6.8-2 Replace the sinusoidal source in the circuit of Fig. 6.4-10 with a variable-frequency sinusoidal source, and determine the transfer function relating the voltage $v(t)$ to that source.
Answer: $\hat{H}(j\omega) = -8/(24 - \omega^2 + j10\omega)$.

6.8-3 Replace the sinusoidal source in the circuit of Fig. 6.4-11 with a variable-frequency sinusoidal source, and determine the transfer function relating the current $i(t)$ to that source.
Answer: $\hat{H}(j\omega) = -9/(9 - 2\omega^2 + j6\omega)$.

6.8-4 Replace the sinusoidal source in the circuit of Fig. 6.4-13 with a variable-frequency sinusoidal source, and determine the transfer function relating the voltage $v(t)$ to that source.
Answer: $\hat{H}(j\omega) = 20/(20 - \omega^2 + j\omega)$.

6.8-5 Determine the maximum current in the circuit of Fig. P6.8-5 and the frequency of the source at which this occurs. *Answers*: 5.033 kHz, 0.2 sin ωt A.

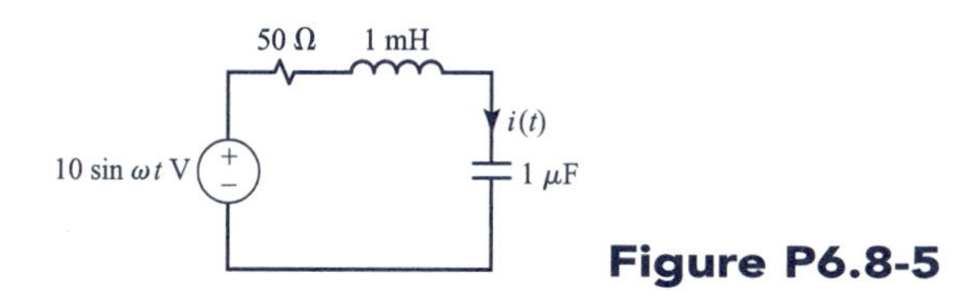

Figure P6.8-5

6.8-6 Determine the maximum voltage in the circuit of Fig. P6.8-6 and the frequency of the source at which this occurs. *Answers*: 15.915 MHz, 2 cos ωt kV.

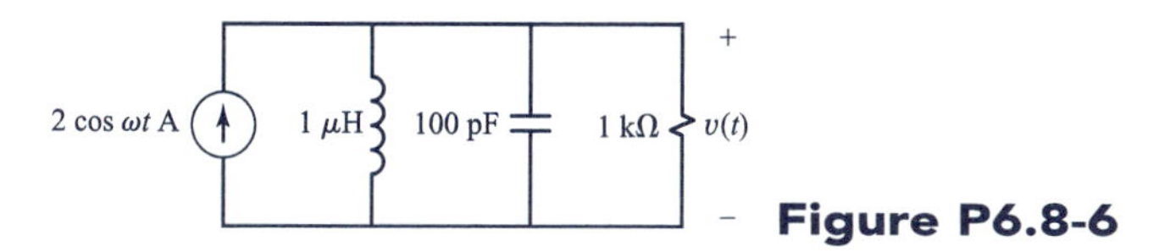

Figure P6.8-6

6.8-7 Determine the type of filter produced by the circuit of Fig. P6.8-7. Provide a sketch of the magnitude of the anticipated frequency response. *Answer*: Lowpass filter.

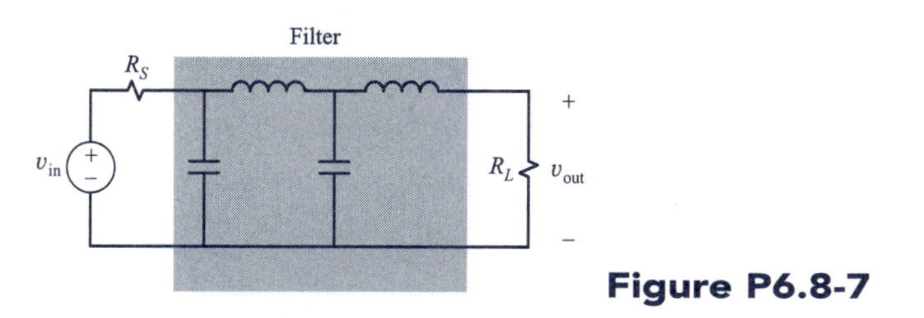

Figure P6.8-7

6.8-8 Determine the type of filter produced by the circuit of Fig. P6.8-8. Provide a sketch of the magnitude of the anticipated frequency response. *Answer*: Highpass filter.

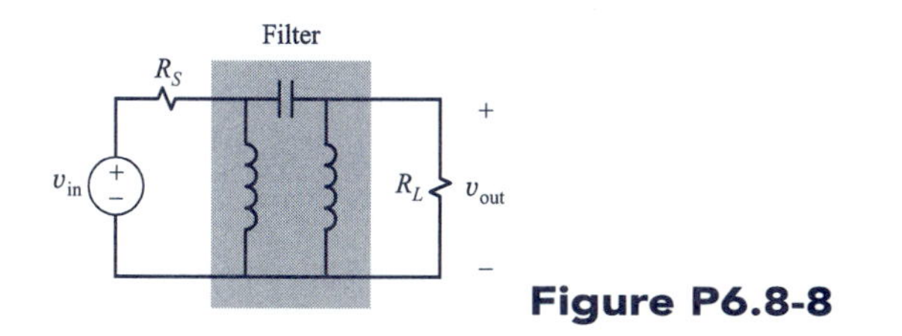

Figure P6.8-8

6.8-9 Determine the type of filter produced by the circuit of Fig. P6.8-9. Provide a sketch of the magnitude of the anticipated frequency response. *Answer*: Bandpass filter.

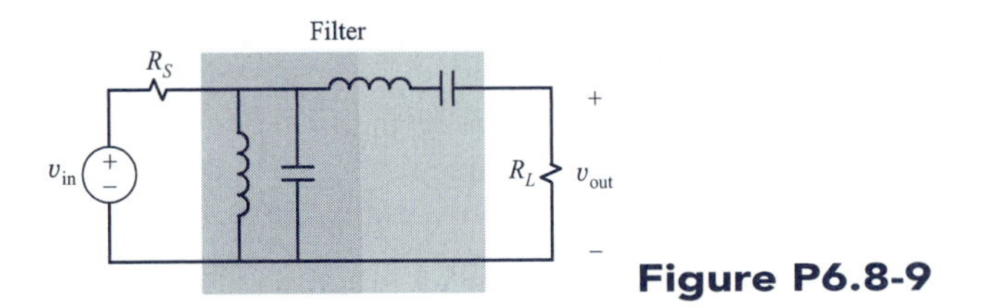

Figure P6.8-9

6.8-10 Design a lowpass filter having a cutoff frequency of 100 kHz and a 50-Ω load. *Answer*: $L = 79.6\ \mu H$.

6.8-11 Design a highpass filter having a cutoff frequency of 1MHz and a 100-Ω load. *Answer*: $C = 1.59$ nF.

6.8-12 Design a bandpass filter having a center frequency of 1 MHz, a bandwidth of 100 kHz, and a 100-Ω load. Determine the Q. *Answers*: $L = 159\ \mu H$, $C = 159$ pF, $Q = 10$.

6.8-13 Design a bandreject filter having a center frequency of 60 Hz, a bandwidth of 10 Hz, and a 50-Ω load. Determine the Q. *Answers*: $L = 22.1$ mH, $C = 318\ \mu F$, $Q = 6$.

Section 6.9 Commercial Power Distribution

6.9-1 For the three-phase circuit in Fig. P6.9-1 determine $i(t)$, and determine the total average power delivered to the load. *Answers*: 27.1 $\sin(120\pi t - 37°)$ A, 5527.6 W.

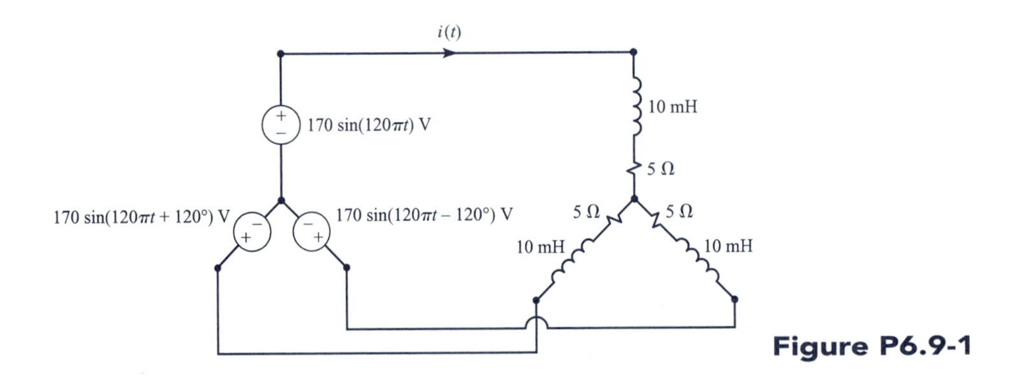

Figure P6.9-1

6.9-2 For the three-phase circuit in Fig. P6.9-2 determine $i(t)$, and determine the total average power delivered to the load. *Answers*: 8.03 $\cos(120\pi t + 35.92°)$ A, 1658.5 W.

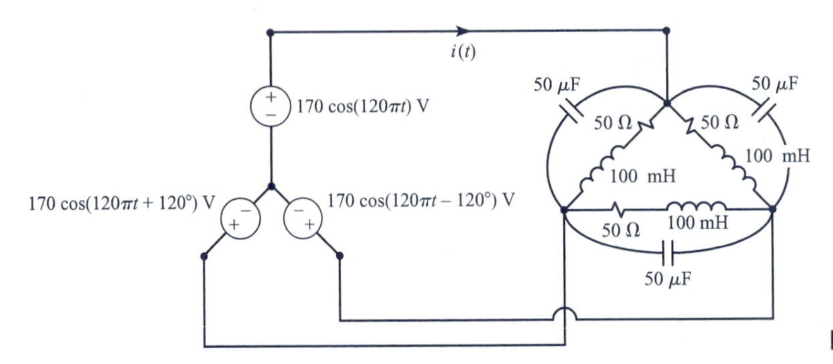

Figure P6.9-2

6.9-3 For the three-phase circuit in Fig. P6.9-3 determine $i_1(t)$, $i_2(t)$, $i_3(t)$, and determine the total average power delivered to the load. *Answers*: $i_1(t) = 55.86 \sin(120\pi t - 61.22°)$ A, $i_2(t) = 3.94 \sin(120\pi t + 151.03°)$ A, $i_3(t) = 52.57 \sin(120\pi t + 116.48°)$ A, 6752 W.

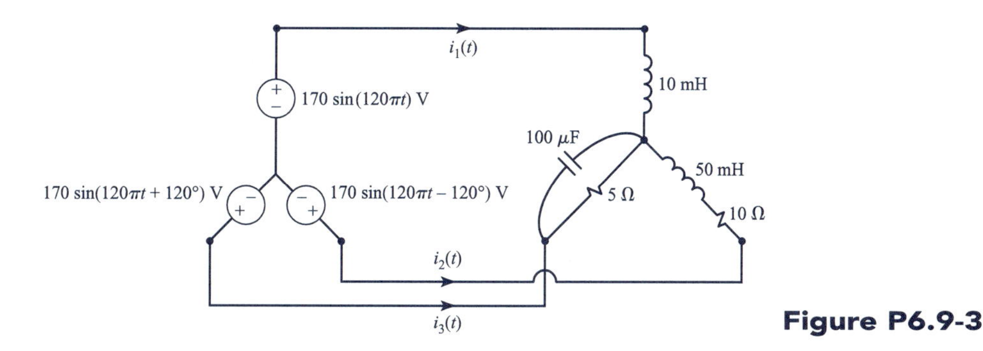

Figure P6.9-3

6.9-4 For the three-phase circuit in Fig. P6.9-4 determine $i(t)$, and determine the total average power delivered to the load. *Answers*: $0.67 \cos(120\pi t + 93.78°)$ A, 87.03 W.

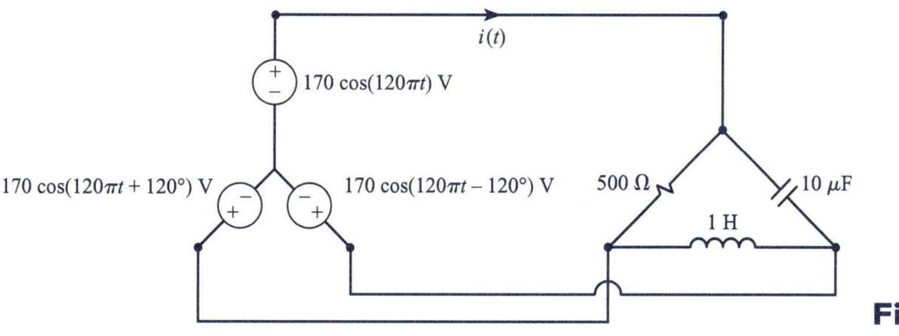

Figure P6.9-4

Section 6.10 PSPICE Applications

6.10-1 Use PSPICE to determine the voltage $v(t)$ in the circuit of Fig. P6.3-2.
Answer: $20 \cos(t + 45°)$ V.

6.10-2 Use PSPICE to determine the current $i(t)$ in the circuit of Fig. P6.3-3.
Answer: $2 \cos(2t + 66.87°)$ A.

6.10-3 Use PSPICE to determine the current $i(t)$ in the circuit of Fig. P6.3-7.
Answer: $0.57 \cos(3t - 35.73°)$ A.

6.10-4 Use PSPICE to determine the voltage $v(t)$ in the circuit of Fig. P6.4-10.
Answer: $2.83 \cos(2t - 165°)$ V.

6.10-5 Use PSPICE to determine the voltage $v(t)$ in the circuit of Fig. P6.4-12.
Answer: $3.16 \cos(2t - 71.57°)$ V.

6.10-6 Use PSPICE to determine the current $i(t)$ in the circuit of Fig. P6.4-18.
Answer: $0.881 \cos(3t + 152.6°)$ A.

6.10-7 Use PSPICE to determine the voltage $v(t)$ in the circuit of Fig. P6.4-19.
Answer: $9.7 \sin(3t + 15.96°)$ V.

6.10-8 Use PSPICE to determine the voltage $v(t)$ in the circuit of Fig. P6.4-20.
Answer: $1.993 \cos(2t - 49.74°)$ V.

6.10-9 Use PSPICE to plot the magnitude and phase of the lowpass filter response of Problem P6.8-10.

6.10-10 Use PSPICE to plot the magnitude and phase of the highpass filter response of Problem P6.8-11.

6.10-11 Use PSPICE to plot the magnitude and phase of the bandpass filter response of Problem P6.8-12.

6.10-12 Use PSPICE to plot the magnitude and phase of the bandreject filter response of Problem P6.8-13.

Section 6.11 MATLAB Applications

6.11-1 Use MATLAB to determine the voltage in the circuit of Fig. P6.3-2.
Answer: $20 \cos(t + 45°)$ V.

6.11-2 Use MATLAB to determine the current in the circuit of Fig. P6.3-3.
Answer: $2 \cos(2t + 66.87°)$ A.

6.11-3 Use MATLAB to sketch that magnitude and phase of the lowpass filter response of Problem P6.8-10.

6.11-4 Use MATLAB to sketch that magnitude and phase of the bandpass filter response of Problem P6.8-12.

CHAPTER 7

General Excitation of Circuits

In the previous chapters we have determined the *steady-state* portion of the response of a linear circuit to two types of independent source waveforms: the dc source and the sinusoidal source. The objective of this chapter is to extend that capability to determining the complete response of a linear circuit to these sources as well as independent sources whose waveforms have some general and arbitrary shape. We will investigate the response of a circuit when there has been some disturbance in it. The primary disturbances we will investigate are (a) the *switching operation,* where a switch opens or closes at $t = 0$ and hence changes the structure of the circuit, and (b) the *activation of an independent source,* where the source turns on (or turns off) at $t = 0$. We will find that there exists a transition time wherein the currents and voltages of the circuit are adjusting from their steady-state values prior to activation of the switch or the source, i.e., for $t < 0$, to their steady-state values after switch or source activation, i.e., for $t > 0$.

We first investigate the direct or *classical solution* of the circuit differential equations for some simple circuits: the first-order *RL* and *RC* circuits, and the second-order series *RLC* and parallel *RLC* circuits. This serves to illustrate some important principles and concepts that are common to all circuit solutions. The classical solution method becomes tedious for other than these simple circuits. A more direct and simpler solution method is the *Laplace transform,* which we investigate after the classical solution. The Laplace transform is a rather automatic method of solution that directly incorporates initial conditions and handles more complex independent sources. These advantages make it superior to the classical solution method for all but simple circuits.

7.1 First-Order Circuit Response

In this section we will outline the classical solution of circuits that contain only one energy storage element: an inductor or a capacitor. These circuits are called *first-order circuits* because the governing differential equations relating a current or voltage to the independent sources in that circuit will be first-order differential equations, i.e., the highest derivative will be the first derivative. These solutions bear many similarities to the solutions for higher-order circuits, i.e., those containing more than one energy storage element.

7.1.1 The *RL* Circuit Consider the circuit shown in Fig. 7.1*a*, where a switch closes at $t = 0$ connecting a dc voltage source to a resistor, which is in series with an inductor. Prior to the switch closure, the inductor current is zero, since there are no sources attached to it:

$$i_L(t) = 0 \qquad t < 0$$

After the switch closes ($t > 0$), the dc source is connected. Since the source is dc, we may determine a solution for the inductor current for $t > 0$ by replacing the inductor with a short circuit in the $t > 0$ circuit as shown in Fig. 7.1*b*, obtaining

$$i_L(t) = \frac{V_S}{R} \qquad t > 0$$

Figure 7.1*c* shows the result. Here we see a problem. In Chapter 5 we showed that

> *an inductor current cannot change instantaneously, i.e., inductor currents must be continuous in time.*

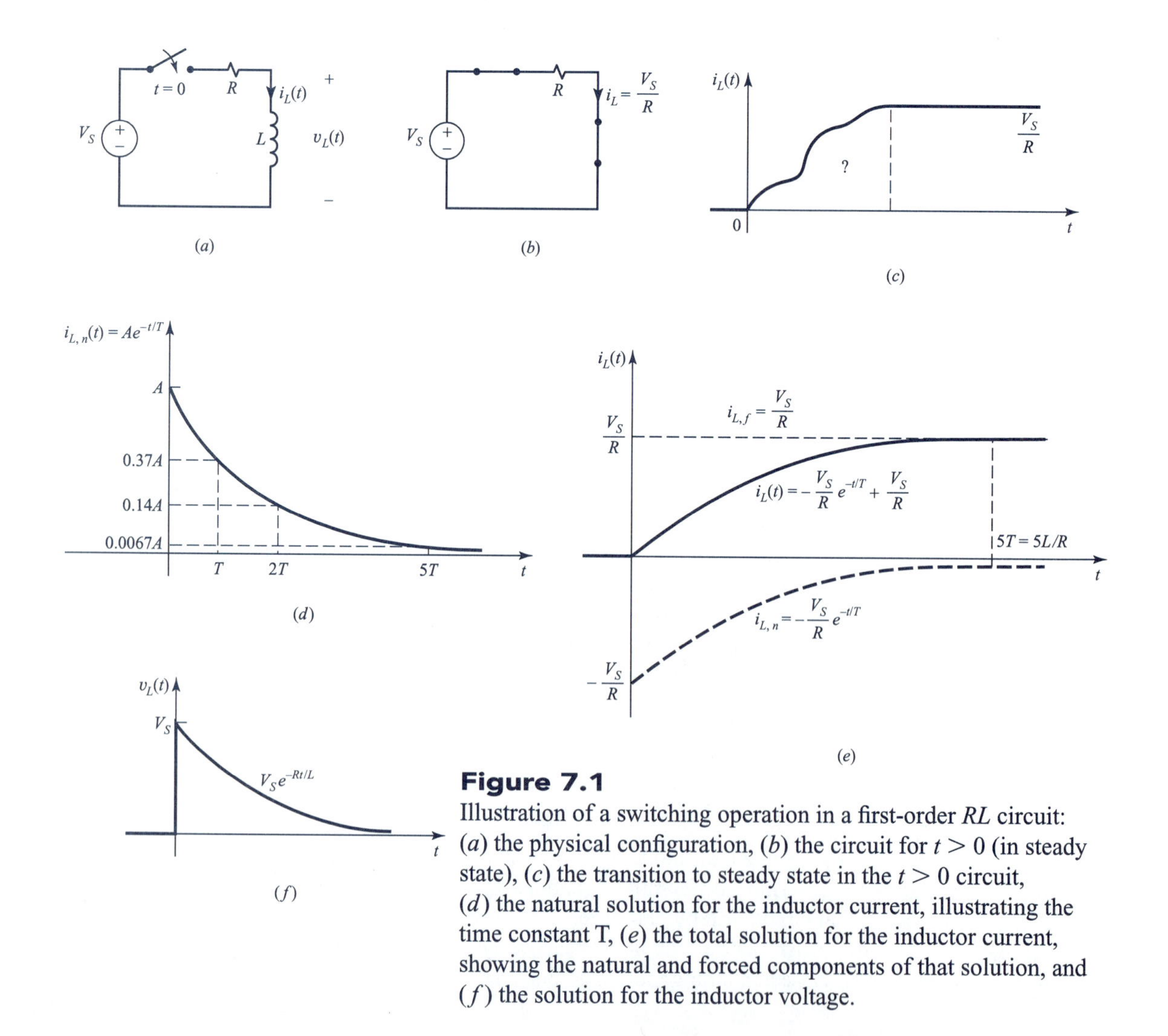

Figure 7.1
Illustration of a switching operation in a first-order *RL* circuit: (*a*) the physical configuration, (*b*) the circuit for $t > 0$ (in steady state), (*c*) the transition to steady state in the $t > 0$ circuit, (*d*) the natural solution for the inductor current, illustrating the time constant T, (*e*) the total solution for the inductor current, showing the natural and forced components of that solution, and (*f*) the solution for the inductor voltage.

This is due to the terminal relation of the inductor, $v_L(t) = L\, di_L(t)/dt$, which shows that if the inductor current changed value instantaneously, the inductor voltage would become infinite, which is an unrealistic situation. Hence we see that the inductor current cannot abruptly change from its value of 0 for $t < 0$ to its value of V_S/R for $t > 0$. There must be an interval of time where the inductor current is *smoothly* adjusting between these two values.

In order to determine the complete solution, we write the differential equation relating the inductor current to the source in the $t > 0$ circuit (after the switch has closed) by writing KVL around the loop, yielding

$$L\frac{di_L(t)}{dt} + Ri_L(t) = V_S \tag{7.1a}$$

or

$$\frac{di_L(t)}{dt} + \frac{R}{L}i_L(t) = 0 + \frac{V_S}{L} \tag{7.1b}$$

Observe that we have done something in this equation that may seem rather strange: we have added a zero to the right-hand side. Hence we are saying that there are two sources: one deactivated (set equal to zero) and one activated (set equal to its value of V_S). Because the circuit and differential equation are linear, the total solution consists, according to the principle of superposition, of the sum of the solutions due to these sources:

$$\frac{di_{L,n}(t)}{dt} + \frac{R}{L}i_{L,n}(t) = 0 \tag{7.2a}$$

$$\frac{di_{L,f}(t)}{dt} + \frac{R}{L}i_{L,f}(t) = \frac{V_S}{L} \tag{7.2b}$$

The solution $i_{L,n}(t)$ due to the zero-value source is referred to as the *natural solution,* since it is independent of the value of the source and only depends on the values of the resistor and the inductor. The solution $i_{L,f}(t)$ is referred to as the *forced solution,* since it is dependent on the value of the source, V_S: the *forcing function*. Because the differential equation is linear, the *total solution* is the sum of the natural and forced solutions:

$$i_L(t) = i_{L,n}(t) + i_{L,f}(t) \tag{7.3}$$

That this is true can be seen by adding the above solutions to obtain

$$\left[\frac{di_{L,n}(t)}{dt} + \frac{R}{L}i_{L,n}(t)\right] + \left[\frac{di_{L,f}(t)}{dt} + \frac{R}{L}i_{L,f}(t)\right] = [0] + \left[\frac{V_S}{L}\right]$$

which can be combined as

$$\frac{d[i_{L,n}(t) + i_{L,f}(t)]}{dt} + \frac{R}{L}[i_{L,n}(t) + i_{L,f}(t)] = \left[0 + \frac{V_S}{L}\right]$$

Now let us investigate the individual solutions. The natural solution is the solution to

$$\frac{di_{L,n}(t)}{dt} + \frac{R}{L}i_{L,n}(t) = 0 \tag{7.2a}$$

Any solution to this equation must be of a form such that its derivative gives back the same form. One such function is the exponential, so we *assume* a form of the natural solution as

$$i_{L,n}(t) = Ae^{pt} \tag{7.4}$$

where A and p are, as yet undetermined constants. Substituting gives

$$pAe^{pt} + \frac{R}{L}Ae^{pt} = 0$$

or, grouping terms,

$$\left(p + \frac{R}{L}\right)Ae^{pt} = 0$$

We cannot set $A = 0$, since that would be a trivial solution. The exponential is never zero. Hence this equation can only be satisfied if the factor in parentheses is zero, yielding

$$p = -\frac{R}{L} \tag{7.5}$$

Thus the natural solution is

$$\boxed{i_{L,n}(t) = Ae^{-Rt/L}} \tag{7.6}$$

This can be written in a more meaningful form by defining the *time constant*

$$\boxed{T = \frac{L}{R} \quad \text{s}} \tag{7.7}$$

which has the units of seconds. Hence the natural solution can be written as

$$\boxed{i_{L,n}(t) = Ae^{-t/T}} \tag{7.8}$$

For positive R and L the time constant is positive, and hence the natural solution eventually decays to zero and the circuit is said to be *stable*. It actually never reaches zero, but after a few time constants have elapsed, it is sufficiently close to zero to be ignored. The natural solution is sketched in Fig. 7.1*d*. After one time constant has elapsed, the natural solution is $Ae^{-1} = 0.37A$. After two and three time constants have elapsed, the natural solution is $Ae^{-2} = 0.14A$ and $Ae^{-3} = 0.05A$, respectively. After five time constants have elapsed, the natural solution is $Ae^{-5} = 0.0067A$.

The forced solution is the solution to

$$\frac{di_{L,f}(t)}{dt} + \frac{R}{L}i_{L,f}(t) = \frac{V_S}{L} \tag{7.2b}$$

According to the *method of undetermined constants* from courses in differential equations, we assume the forced solution to be a constant (since the right-hand side is a constant):

$$i_{L,f}(t) = M \tag{7.9}$$

Substituting gives

$$0 + \frac{R}{L}M = \frac{V_S}{L}$$

and hence

$$\boxed{i_{L,f}(t) = \frac{V_S}{R}} \tag{7.10}$$

Observe that this forced solution could have been more easily obtained directly from the circuit by replacing the inductor with a short circuit in the $t > 0$ circuit (since the source is dc) as shown in Fig. 7.1*b*. Adding the two solutions gives the total solution for $t > 0$ as

$$\begin{aligned} i_L(t) &= i_{L,n}(t) + i_{L,f}(t) \\ &= \underbrace{Ae^{-Rt/L}}_{\text{natural solution}} + \underbrace{\frac{V_S}{R}}_{\text{forced solution}} \end{aligned} \tag{7.11}$$

The natural solution is sometimes referred to as the *transient* solution, and the forced solution is sometimes referred to as the *steady-state* solution. We will use these terms as well.

Referring to the natural solution as the transient solution is sensible in that it eventually goes to zero, i.e., eventually goes away, so long as the time constant L/R is a positive number, as it is if R and L are positive. There are, however, circuits wherein the natural solution does not go away, but increases without bound, because the time constant is negative. Such circuits are referred to as *unstable* circuits. For those circuits, the term *transient* is inappropriate, since the solution does not go to zero.

Referring to the forced solution as the steady-state solution is also sensible in that it is what is left after the transient solution has decayed to zero (if the circuit is stable). In the previous chapters we have only been considering the steady-state solutions (for dc and sinusoidal sources). Hence we can obtain the steady-state (or forced) solution directly from the $t > 0$ circuit using the techniques of Chapters 5 and 6.

We have one remaining undetermined constant, A. In order to determine this and give a solution *valid for* $t > 0$, we need the inductor current *just after the switch closes,* i.e., at $t = 0^+$. But by continuity of inductor currents, that must be equal to the value of the inductor current *immediately before the switch closes,* i.e., at $t = 0^-$. Hence

$$i_L(0^+) = i_L(0^-) = 0$$

Applying this to the total solution yields

$$i_L(0^+) = 0 = A\underbrace{e^{-R\times 0^+/L}}_{1} + \frac{V_S}{R}$$

giving

$$A = -\frac{V_S}{R}$$

Thus the total solution, *valid for* $t > 0$, is

$$\boxed{i_L(t) = \underbrace{-\frac{V_S}{R}e^{-Rt/L}}_{\text{natural or transient solution}} + \underbrace{\frac{V_S}{R}}_{\text{forced or steady-state solution}} \qquad t > 0} \tag{7.12}$$

This solution is sketched in Fig. 7.1*e*. Observe that the natural, or transient, solution eventually decays to zero, leaving the forced, or steady-state, solution $i_{L,f} = V_S/R$. Hence the alternative terminology of transient and steady-state solutions is justifiable for this stable circuit.

Observe that we used the fact that the inductor current must be continuous in order to obtain the value of the inductor current at $t = 0^+$ from the value at $t = 0^-$, $i_L(0^-) = i_L(0^+)$. This then was used to evaluate the undetermined constant A in the solution. *None of the other circuit voltages or currents are necessarily continuous. So the continuity of inductor currents provides the only link between the* $t < 0$ *circuit and the* $t > 0$ *circuit.* This is why we chose to solve first for the inductor current rather than some other current or voltage in the circuit. Once we determine the inductor current, *all other voltages and currents in the circuit can then be obtained using the solution for the inductor current, KVL, KCL, and the element relations.* For example, the inductor voltage is obtained from its terminal relation as

$$\begin{aligned} v_L(t) &= L\frac{di_L(t)}{dt} \\ &= L\frac{d}{dt}\left(-\frac{V_S}{R}e^{-Rt/L} + \frac{V_S}{R}\right) \\ &= V_S e^{-Rt/L} \qquad t > 0 \end{aligned} \tag{7.13}$$

Observe that the inductor voltage is *discontinuous*: $v_L(0^-) = 0$ and $v_L(0^+) = V_S$. This is sketched in Fig. 7.1*f*. This result can also be obtained by writing KVL around the loop in the $t > 0$ circuit giving

$$v_L(t) = V_S - Ri_L(t)$$

This development for a series *RL* circuit can be easily extended to any other first-order circuit containing a single inductor. The key to doing so is to draw the $t > 0$ circuit in the form of the previous example. This is easily done by drawing the $t > 0$ circuit and reducing the circuit attached to the inductor to a Thevenin equivalent circuit as shown in Fig. 7.2*a*. Having done so, the general solution becomes, by matching with the previous development,

$$i_L(t) = Ae^{-R_{TH}t/L} + \frac{V_{OC}}{R_{TH}} \tag{7.14}$$

The *time constant* becomes

$$T = \frac{L}{R_{TH}} \tag{7.15}$$

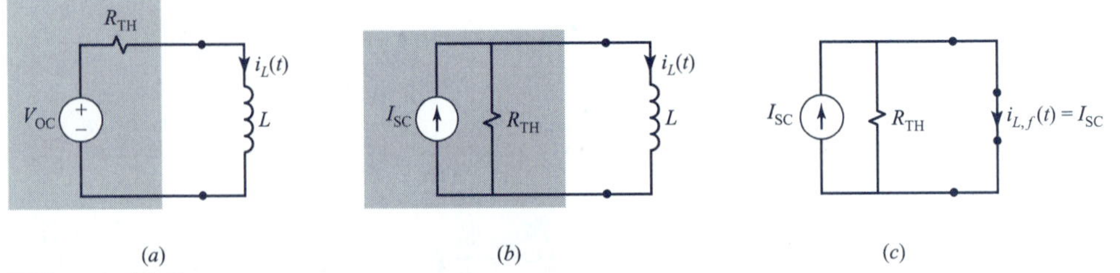

Figure 7.2
Reducing all first-order circuits containing an inductor to (*a*) a Thevenin equivalent and (*b*) a Norton equivalent, and (*c*) determining the forced solution for the inductor current from the Norton equivalent circuit.

The circuit is stable if the Thevenin resistance is positive, i.e., $R_{TH} > 0$. This will be the case for circuits that contain only positive resistors and inductors. If the circuit contains a controlled source, the Thevenin resistance may be negative, as we saw in Chapter 2. In this case the circuit would be classified as being unstable, and the terms *transient* and *steady state* lose their meaning, since the natural solution increases without bound. However, the terms *natural response* and *forced response* retain their meaning.

The only remaining task is to incorporate the initial condition, which may not be zero. Using continuity of inductor currents gives

$$\begin{aligned} i_L(0^-) &= i_L(0^+) \\ &= A + \frac{V_{OC}}{R_{TH}} \end{aligned}$$

or

$$A = \left[i_L(0^-) - \frac{V_{OC}}{R_{TH}} \right]$$

yielding the solution for $t > 0$:

$$\begin{aligned} i_L(t) &= \underbrace{\left[i_L(0^-) - \frac{V_{OC}}{R_{TH}} \right] e^{-R_{TH}t/L}}_{\substack{\text{natural or}\\ \text{transient}\\ \text{solution}}} + \underbrace{\frac{V_{OC}}{R_{TH}}}_{\substack{\text{forced or}\\ \text{steady-state}\\ \text{solution}}} \\ &= \underbrace{[i_L(0^-) - I_{SC}]e^{-R_{TH}t/L}}_{\substack{\text{natural or}\\ \text{transient}\\ \text{solution}}} + \underbrace{I_{SC}}_{\substack{\text{forced or}\\ \text{steady-state}\\ \text{solution}}} \end{aligned} \tag{7.16}$$

We have given an alternative form of the solution by replacing V_{OC}/R_{TH} with I_{SC}, i.e., converting to a Norton equivalent circuit as shown in Fig. 7.2*b*. Replacing the inductor with a short circuit as in Fig. 7.2*c* shows that the forced solution is indeed the short-circuit current.

Determining the inductor current prior to activation of the switch, $i_L(0^-)$, is a simple task if the $t < 0$ circuit contains only dc sources. We assume that the circuit was constructed at some time in the distant past so that all transients have decayed to zero before activation of the switch. Hence, $i_L(0^-)$ is the steady-state inductor current in the $t < 0$ circuit. We simply draw the $t < 0$ circuit and compute the steady-state value of the inductor current in that circuit. If the $t < 0$ circuit contains only dc sources, we determine that steady-state current by replacing the inductor with a short circuit. If the circuit contains a sinusoidal source, we compute the contribution to $i_L(0^-)$ due to this source using the phasor method of the preceeding chapter.

Observe the following important computational points. The Thevenin resistance, R_{TH}, needs to be calculated in order to determine the time constant given in (7.15). To complete the solution we only need to calculate the forced (steady-state) solution for the inductor current, $i_{L,f}(t)$, which can be easily calculated from the $t > 0$ circuit, either by replacing the inductor with a short circuit if the sources are dc, or by the phasor method if the sources are sinusoids.

Example 7.1

Determine the inductor current and resistor current for $t > 0$ in the circuit of Fig. 7.3*a*. The switch closes at $t = 0$.

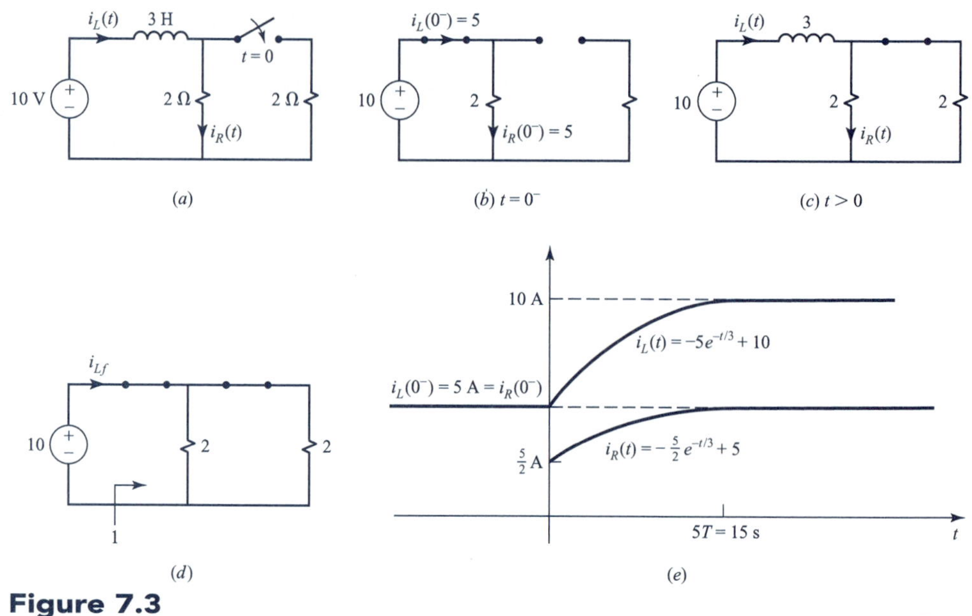

Figure 7.3
Example 7.1.

Solution The circuit at $t = 0^-$ is shown in Fig. 7.3*b*. It is constructed by opening the switch and replacing the inductor with a short circuit, since the source in that circuit is dc. This gives the value of the inductor current at $t = 0^-$ as

$$\begin{aligned} i_L(0^-) &= 5\ \text{A} \\ &= i_L(0^+) \end{aligned}$$

The $t > 0$ circuit is drawn in Fig. 7.3*c*, and the Thevenin equivalent resistance at the terminals of the inductor is $R_{TH} = 2\ \Omega \parallel 2\ \Omega = 1\ \Omega$. Hence the time constant is

$$\begin{aligned} T &= \frac{L}{R_{TH}} \\ &= \frac{3\ \text{H}}{1\ \Omega} \\ &= 3\ \text{s} \end{aligned}$$

Replacing the inductor with a short circuit in this $t > 0$ circuit yields the forced (steady-state) solution as shown in Fig. 7.3*d* as

$$\begin{aligned} i_{Lf}(t) &= \frac{10\ \text{V}}{2\ \Omega \parallel 2\ \Omega = 1\ \Omega} \\ &= 10\ \text{A} \end{aligned}$$

Hence the total solution becomes

$$i_L(t) = Ae^{-t/3} + 10$$

Applying the initial condition $i_L(0^+) = 5$ gives $A = 5 - 10 = -5$, so that the total solution becomes

$$\begin{aligned} i_L(t) &= (5 - 10)e^{-t/3} + 10 \\ &= -5e^{-t/3} + 10\ \text{A} \qquad t > 0 \end{aligned}$$

The resistor current is, by current division in the $t > 0$ circuit of Fig. 7.3*c*,

$$\begin{aligned} i_R(t) &= \tfrac{1}{2}\, i_L(t) \\ &= -\tfrac{5}{2} e^{-t/3} + 5\text{ A} \qquad t > 0 \end{aligned}$$

These solutions are plotted in Fig. 7.3*e*. Observe that the resistor current is discontinuous:

$$\begin{aligned} i_R(0^-) &= \frac{10\text{ V}}{2\ \Omega} = 5\text{ A} \\ i_R(0^+) &= \tfrac{1}{2}\, i_L(0^+) = \tfrac{5}{2}\text{ A} \end{aligned}$$

Exercise Problem 7.1

Determine the inductor current and resistor voltage for $t > 0$ in the circuit of Fig. E7.1.

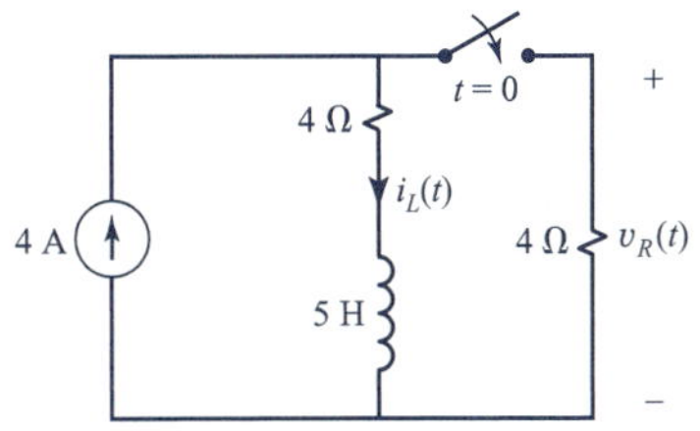

Figure E7.1
Exercise Problem 7.1.

Answers: $i_L(t) = 2e^{-8t/5} + 2\text{ A}$, $v_R(t) = 8 - 8e^{-8t/5}\text{ V}$.

7.1.2 The *RC* Circuit In a circuit containing a single capacitor, the procedure is essentially the same as for a circuit containing a single inductor. We first determine the differential equation relating the capacitor voltage to the independent source. We then write the solution as a sum of the natural or transient solution and the forced or steady-state solution. The final step is to evaluate the constant in the natural solution. This is accomplished using the important fact that

> *a capacitor voltage cannot change instantaneously, i.e., capacitor voltages must be continuous in time.*

This is due to the terminal relation of the capacitor, $i_C(t) = C\, dv_C(t)/dt$, which shows that if the capacitor voltage changed value instantaneously, the capacitor current would become infinite, which is an unrealistic situation. This is why we first solve for the capacitor voltage, regardless of the actual voltage or current of interest. Only the capacitor voltage is guaranteed to be continuous, and hence it is the only link between the $t < 0$ circuit and the $t > 0$ circuit. Again, all other voltages and currents in the circuit can then be obtained using the solution for the capacitor voltage, KVL, KCL, and the element relations.

Consider the parallel *RC* circuit shown in Fig. 7.4*a* consisting of the parallel combination of a dc current source I_S, a resistor R, and a capacitor C. The switch closes at $t = 0$. Since the capacitor is not connected to the source for $t < 0$, its voltage at $t = 0^-$ is zero, and by continuity of capacitor voltages we have

$$\begin{aligned} v_C(0^+) &= v_C(0^-) \\ &= 0 \end{aligned}$$

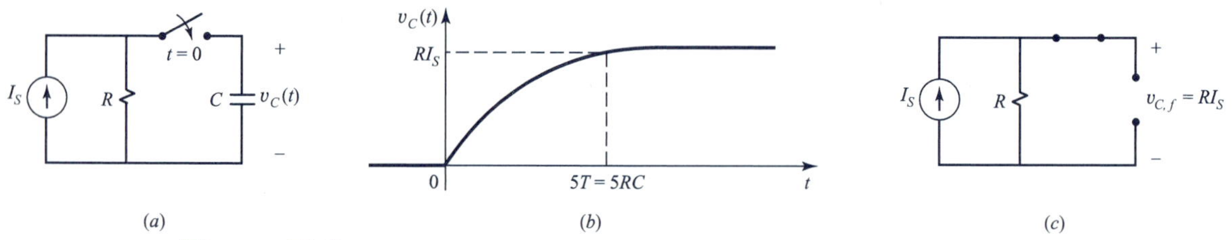

Figure 7.4
The first-order *RC* circuit: (*a*) the physical circuit, (*b*) the response of the capacitor voltage, and (*c*) determining the forced solution for the capacitor voltage.

The differential equation relating the capacitor voltage to the source for $t > 0$ is obtained by writing KCL as

$$C\frac{dv_C(t)}{dt} + \frac{v_C(t)}{R} = I_S \tag{7.17a}$$

or, writing in standard form,

$$\frac{dv_C(t)}{dt} + \frac{1}{RC}v_C(t) = \frac{I_S}{C} \tag{7.17b}$$

Comparing this differential equation with the corresponding one for the series *RL* circuit in (7.1), we see that the general solution is

$$\begin{aligned} v_C(t) &= Ae^{-t/RC} + RI_S \\ &= \underbrace{Ae^{-t/T}}_{\substack{\text{natural or} \\ \text{transient} \\ \text{solution}}} + \underbrace{RI_S}_{\substack{\text{forced or} \\ \text{steady-state} \\ \text{solution}}} \end{aligned} \tag{7.18}$$

where the time constant is

$$T = RC \quad \text{s} \tag{7.19}$$

Once again, since the source is dc, the forced or steady-state solution could be found directly by replacing the capacitor with an open circuit in the $t > 0$ circuit as shown in Fig. 7.4*c*, yielding $v_{C,f}(t) = RI_S$. Applying the initial condition $v_C(0^+) = v_C(0^-) = 0$ to (7.18) gives the solution for $t > 0$ as shown in Fig. 7.4*b*:

$$v_C(t) = -RI_S e^{-t/RC} + RI_S \tag{7.20}$$

This may be extended to more general circuits that contain only one capacitor and whose initial conditions are not zero by reducing the circuit attached to the capacitor to a Norton equivalent circuit as shown in Fig. 7.5*a*. This general solution becomes

$$\begin{aligned} v_C(t) &= \underbrace{[v_C(0^-) - R_{\text{TH}}I_{\text{SC}}]e^{-t/R_{\text{TH}}C}}_{\substack{\text{natural or} \\ \text{transient} \\ \text{solution}}} + \underbrace{R_{\text{TH}}I_{\text{SC}}}_{\substack{\text{forced or} \\ \text{steady-state} \\ \text{solution}}} \\ &= \underbrace{[v_C(0^-) - V_{\text{OC}}]e^{-t/R_{\text{TH}}C}}_{\substack{\text{natural or} \\ \text{transient} \\ \text{solution}}} + \underbrace{V_{\text{OC}}}_{\substack{\text{forced or} \\ \text{steady-state} \\ \text{solution}}} \end{aligned} \tag{7.21}$$

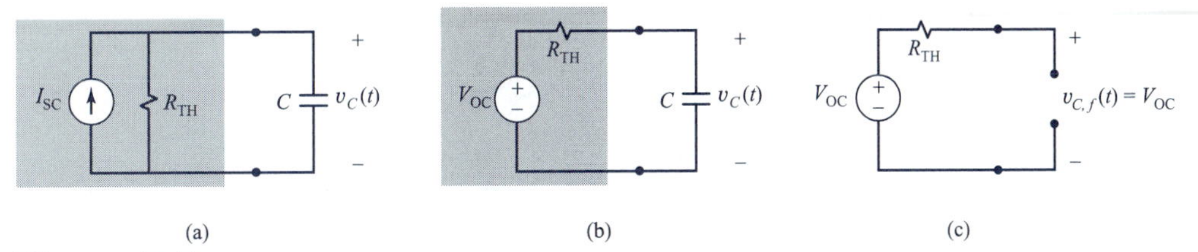

Figure 7.5
Reducing all first-order circuits containing a capacitor to (*a*) a Norton equivalent and (*b*) a Thevenin equivalent, and (*c*) determining the forced solution for the capacitor voltage from the Thevenin equivalent circuit.

We have given an alternative form of the solution by replacing $R_{TH}\,I_{SC}$ with V_{OC}, i.e., converting to a Thevenin equivalent circuit as shown in Fig. 7.5*b*. Replacing the capacitor with an open circuit as in Fig. 7.5*c* shows that the forced solution is indeed the open-circuit voltage.

Again observe the following important computational points. The Thevenin resistance R_{TH} needs to be calculated in order to determine the time constant $T = R_{TH}C$. To complete the solution we only need to calculate the forced solution for the capacitor voltage, which can be easily calculated from the $t > 0$ circuit, either by replacing the capacitor with an open circuit if the sources are dc or by the phasor method if the sources are sinusoids.

Example 7.2

Determine and sketch the solution for the capacitor voltage in the circuit of Fig. 7.6*a*. The switch closes at $t = 0$. Also determine the solutions for the capacitor current and the resistor current.

Solution The switch is open for $t < 0$, and we *assume* that the circuit was constructed at some distant time in the past, so that all transients (occurring when it was constructed) have died out, leaving the steady state. Since the source that is attached is dc, we obtain this steady-state value for $t < 0$ by replacing the capacitor with an open circuit as shown in Fig. 7.6*b*, giving the value of capacitor voltage immediately prior to switch closure as

$$v_C(0^-) = 10\text{ V}$$

The $t > 0$ circuit (after the switch closes) is shown in Fig. 7.6*c*. The Thevenin resistance seen by the capacitor is $R_{TH} = 4\,\Omega \,\|\, 4\,\Omega = 2\,\Omega$, and hence the time constant is

$$\begin{aligned} T &= R_{TH}C \\ &= 6\text{ s} \end{aligned}$$

The forced solution can be obtained by replacing the capacitor with an open circuit in the $t > 0$ circuit (since the source is dc) as shown in Fig. 7.6*d*, giving

$$v_{C,f}(t) = 5\text{ V}$$

Hence the total solution is

$$\begin{aligned} v_C(t) &= v_{C,n}(t) + v_{C,f}(t) \\ &= Ae^{-t/6} + 5 \end{aligned}$$

Applying the initial condition yields

$$v_C(0^-) = 10\text{ V} = v_C(0^+) = A + 5\text{ V}$$

Hence $A = 5$ V, and the total solution is

$$v_C(t) = 5e^{-t/6} + 5\text{ V} \qquad t > 0$$

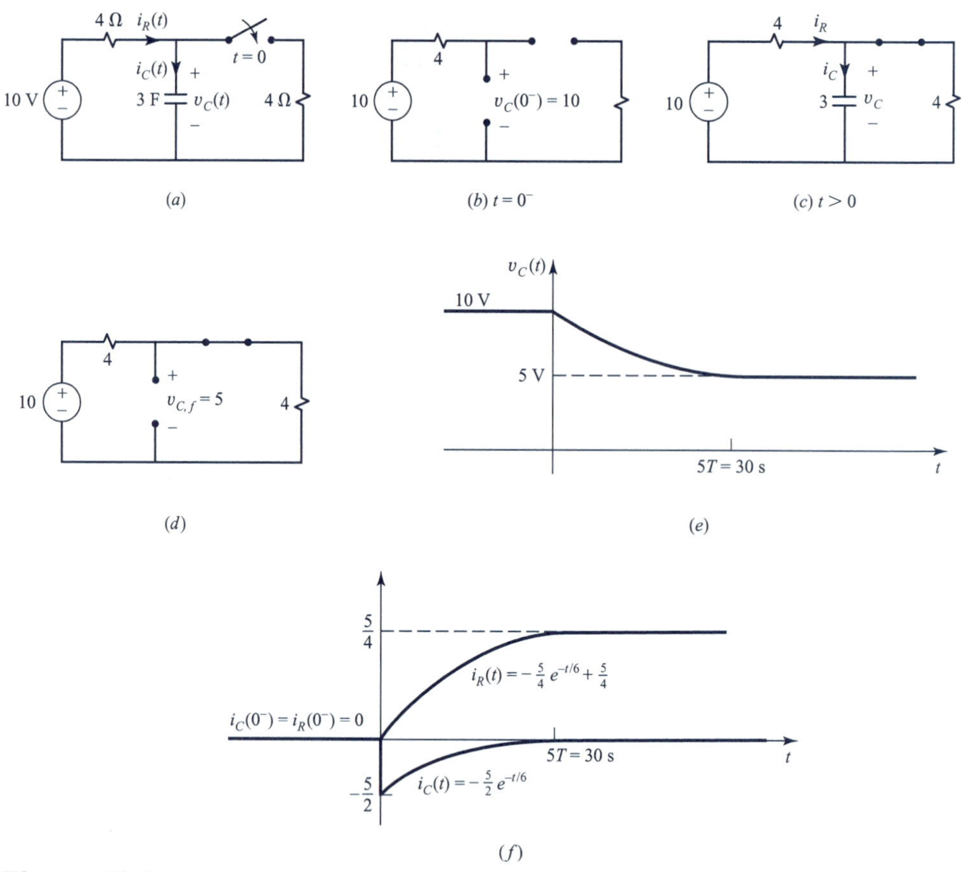

Figure 7.6
Example 7.2.

This is sketched in Fig. 7.6*e*.

Once we have determined the solution for the capacitor voltage in a circuit, we can easily determine other circuit variable solutions using this result and KVL, KCL, and the element relations. For example, the capacitor current is determined as

$$\begin{aligned} i_C(t) &= 3\frac{dv_C(t)}{dt} \\ &= 3\frac{d}{dt}(5e^{-t/6} + 5) \\ &= -\tfrac{5}{2}e^{-t/6}\,\text{A} \qquad t > 0 \end{aligned}$$

and the resistor current is obtained using KCL in the $t > 0$ circuit as

$$\begin{aligned} i_R(t) &= i_C(t) + \frac{v_C(t)}{4\,\Omega} \\ &= -\tfrac{5}{2}e^{-t/6} + \tfrac{5}{4}e^{-t/6} + \tfrac{5}{4} \\ &= -\tfrac{5}{4}e^{-t/6} + \tfrac{5}{4}\,\text{A} \qquad t > 0 \end{aligned}$$

The values of these variables at $t = 0^-$ can be determined from the $t = 0^-$ circuit in Fig. 7.6*b* as $i_C(0^-) = 0$ and $i_R(0^-) = 0$. These solutions are plotted in Fig. 7.6*f*. Observe that the capacitor current is discontinuous at $t = 0$.

Exercise Problem 7.2

Determine the capacitor voltage and current for $t > 0$ in the circuit of Fig. E7.2.

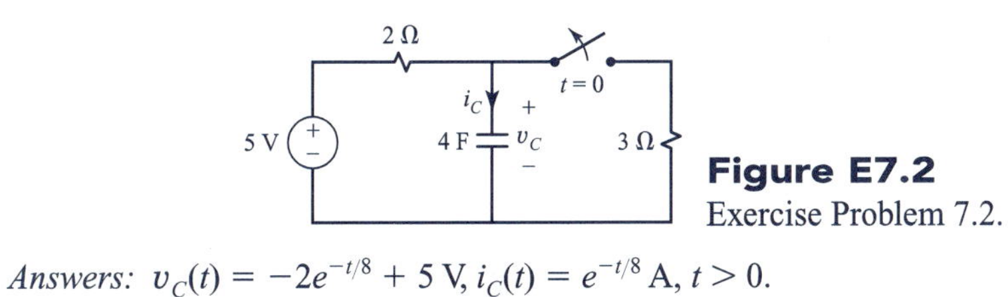

Figure E7.2
Exercise Problem 7.2.

Answers: $v_C(t) = -2e^{-t/8} + 5$ V, $i_C(t) = e^{-t/8}$ A, $t > 0$.

7.2 Second-Order Circuit Response

Next we investigate the solutions for circuits that contain two energy storage elements (two L's, two C's, or one L and one C). These are said to be *second-order* circuits because the differential equation relating any voltage or current to the sources is at most second order, i.e., the highest derivative in the differential equation is at most second order. There are a few cases of circuits containing two energy storage elements that are first order. For example, a circuit containing two inductors in series technically contains two energy storage elements. However, since the inductors are in series, they may be replaced by one equivalent inductance whose value is the sum of the values of the two inductors. Hence this would be a first-order circuit. Similarly, a circuit containing two capacitors in parallel technically contains two energy storage elements. However, since the capacitors are in parallel, they may be replaced by one equivalent capacitance whose value is the sum of the values of the two capacitors. Hence this would again be a first-order circuit. Most circuits containing two energy storage elements, though, are truly second-order circuits, as we will assume in the following.

Once again, we always solve for the inductor current and/or the capacitor voltage, regardless of the actual voltage or current of interest. The reason we do this is that only the inductor current and capacitor voltage are guaranteed to be continuous. Hence, the inductor current and capacitor voltage are the only links between the $t < 0$ and the $t > 0$ circuits. We will determine the inductor current and capacitor voltage at $t = 0^-$, $i_L(0^-)$ and $v_C(0^-)$, and transfer these to the $t = 0^+$ circuit, giving the initial conditions required in the solution to the differential equation, $i_L(0^-) = i_L(0^+)$ and $v_C(0^-) = v_C(0^+)$. Once we solve for the inductor current or capacitor voltage, we may determine the solution for any other voltage or current in the circuit using KVL, KCL, and the element relations as for first-order circuits. Hence determining the solutions for the inductor current and capacitor voltage in second-order circuits is again the crucial task.

7.2.1 The Series *RLC* Circuit Consider the series *RLC* circuit of Fig. 7.7*a* in which a dc source is connected to the series combination of a resistor, on inductor, and a capacitor. The switch closes at $t = 0$, connecting the dc voltage source to the series combination of a resistor, an inductor, and a capacitor. Since the source is not attached to the inductor or capacitor for $t < 0$, the inductor current and capacitor voltage are both zero for $t < 0$:

$$\begin{aligned} i_L(0^-) &= 0 \\ &= i_L(0^+) \\ v_C(0^-) &= 0 \\ &= v_C(0^+) \end{aligned}$$

The differential equations relating the capacitor voltage and the inductor current to the source are obtained by replacing the elements in the $t > 0$ circuit with their "operator resistances" using the

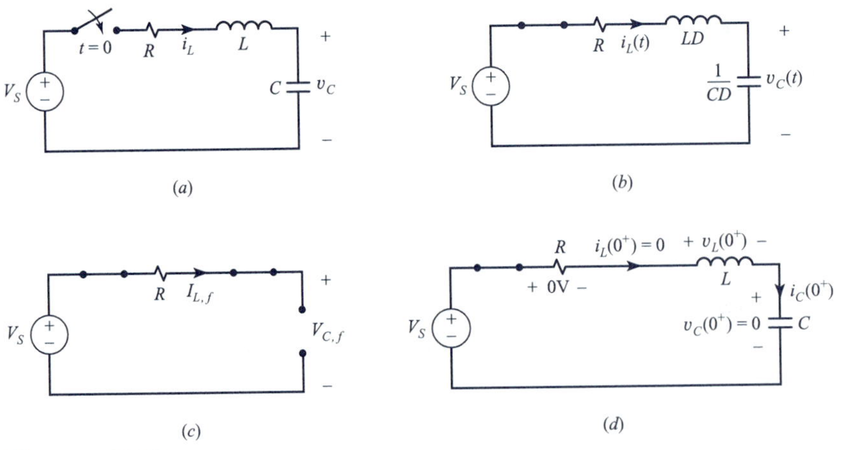

Figure 7.7
The second-order series *RLC* circuit: (*a*) the physical circuit, (*b*) determining the differential equations for the inductor current and capacitor voltage using the differential operator impedances, (*c*) determining the forced solutions for the inductor current and capacitor voltage, and (*d*) the $t = 0^+$ circuit (immediately after switch closure) for determining the derivatives of the inductor current and capacitor voltage at $t = 0^+$.

differential operator,

$$v_L(t) = L\frac{di_L(t)}{dt} \Rightarrow LDi_L(t)$$

$$i_C(t) = C\frac{dv_C(t)}{dt} \Rightarrow CDv_C(t)$$

as outlined in Section 5.5 of Chapter 5. This is shown in Fig. 7.7*b*, from which we obtain the results:

$$\begin{aligned} i_L(t) &= \frac{V_S}{R + LD + \dfrac{1}{CD}} \\ &= \frac{CD}{LCD^2 + RCD + 1} V_S \\ &= \frac{\dfrac{1}{L}D}{D^2 + \dfrac{R}{L}D + \dfrac{1}{LC}} V_S \end{aligned} \tag{7.22a}$$

$$\begin{aligned} v_C(t) &= \frac{\dfrac{1}{CD}}{R + LD + \dfrac{1}{CD}} V_S \\ &= \frac{1}{LCD^2 + RCD + 1} V_S \\ &= \frac{\dfrac{1}{LC}}{D^2 + \dfrac{R}{L}D + \dfrac{1}{LC}} V_S \end{aligned} \tag{7.22b}$$

From these we obtain the differential equations by multiplying through by the denominator and operating on the current and voltage:

$$\frac{d^2 i_L(t)}{dt^2} + \frac{R}{L}\frac{di_L(t)}{dt} + \frac{1}{LC} i_L(t) = \frac{1}{L}\frac{dV_S}{dt} = 0 \tag{7.23a}$$

$$\frac{d^2 v_C(t)}{dt^2} + \frac{R}{L}\frac{dv_C(t)}{dt} + \frac{1}{LC} v_C(t) = \frac{1}{LC} V_S \tag{7.23b}$$

These are second-order differential equations, because the highest derivative is second order. The task now becomes the solution of these differential equations.

As in the case of first-order circuits, because the differential equations are linear, the total solution is the sum of the natural or transient solution (with the right-hand side set equal to zero) and the forced or steady-state solution:

$$i_L(t) = i_{L,n}(t) + i_{L,f}(t) \tag{7.24a}$$

$$v_C(t) = v_{C,n}(t) + v_{C,f}(t) \tag{7.24b}$$

where the natural solutions satisfy

$$\frac{d^2 i_{L,n}(t)}{dt^2} + \frac{R}{L}\frac{di_{L,n}(t)}{dt} + \frac{1}{LC} i_{L,n}(t) = 0 \tag{7.25a}$$

$$\frac{d^2 v_{C,n}(t)}{dt^2} + \frac{R}{L}\frac{dv_{C,n}(t)}{dt} + \frac{1}{LC} v_{C,n}(t) = 0 \tag{7.25b}$$

and the forced solutions satisfy

$$\frac{d^2 i_{L,f}(t)}{dt^2} + \frac{R}{L}\frac{di_{L,f}(t)}{dt} + \frac{1}{LC} i_{L,f}(t) = 0 \tag{7.26a}$$

$$\frac{d^2 v_{C,f}(t)}{dt^2} + \frac{R}{L}\frac{dv_{C,f}(t)}{dt} + \frac{1}{LC} v_{C,f}(t) = \frac{1}{LC} V_S \tag{7.26b}$$

Once again, the forced solutions can more easily found directly from the $t > 0$ circuit by replacing the inductor with a short circuit and replacing the capacitor with an open circuit, since the sources here are dc. Hence, all that remains is to determine the natural solutions and evaluate any undetermined constants in those solutions.

The natural solution for either the inductor current or the capacitor voltage satisfies the following differential equation:

$$\frac{d^2 x_n(t)}{dt^2} + \frac{R}{L}\frac{dx_n(t)}{dt} + \frac{1}{LC} x_n(t) = 0 \tag{7.27}$$

Once again we require a function that when differentiated once and twice yields back the same form. Again, such a function is the exponential, so we assume a form

$$x_n(t) = Ae^{pt} \tag{7.28}$$

where A and p are, as yet, unknown constants. Substituting this into (7.27) yields

$$\left(p^2 + \frac{R}{L}p + \frac{1}{LC}\right)Ae^{pt} = 0$$

In order for this to be satisfied, the *characteristic equation* must equal zero:

$$p^2 + \frac{R}{L}p + \frac{1}{LC} = 0 \tag{7.29}$$

This second-order algebraic equation has two roots, p_1 and p_2, which satisfy

$$(p - p_1)(p - p_2) = 0 \tag{7.30}$$

Since the coefficients of the characteristic equation, R/L and $1/LC$, are real, the fundamental theorem of algebra provides that these two roots will be one of three forms:

1. *real and distinct,*
2. *real and repeated,*
3. *complex, with one the conjugate of the other.*

We now investigate the resulting form of the natural solution for each of these three cases.

1. Roots Real and Distinct: For this case, the form of the natural solution is the sum of two of the forms given in (7.28) with $p = p_1$ in one and $p = p_2$ in the other, and each term has its own undetermined constant:

$$x_n(t) = A_1e^{p_1t} + A_2e^{p_2t} \tag{7.31}$$

This case is referred to as the *overdamped case*. The reason for this designation will become apparent in the computed results to be obtained in Example 7.3.

2. Roots Real and Repeated: For this case, the two roots are equal, $p = p_1 = p_2$. It would not be appropriate to write the solution as for distinct roots, since the two parts can be combined, giving only one undetermined constant:

$$\begin{aligned} x_n(t) &= A_1e^{pt} + A_2e^{pt} \\ &= (A_1 + A_2)e^{pt} \end{aligned}$$

It can be shown that the following form cannot be reduced and satisfies (7.27):

$$\begin{aligned} x_n(t) &= A_1e^{pt} + A_2te^{pt} \\ &= (A_1 + A_2t)e^{pt} \end{aligned} \tag{7.32}$$

This case is referred to as the *critically damped* case. The reason for this designation will become apparent in the computed results to be obtained in Example 7.3.

3. Roots Complex and Conjugates of Each Other: In the final possibility, the roots are complex and one is the conjugate of the other:

$$p_1 = \alpha + j\beta$$
$$p_2 = \alpha - j\beta$$

Since these are distinct, we may write the form of the solution as

$$\begin{aligned} x_n(t) &= A_1 e^{(\alpha + j\beta)t} + A_2 e^{(\alpha - j\beta)t} \\ &= e^{\alpha t}(A_1 e^{j\beta t} + A_2 e^{-j\beta t}) \end{aligned} \tag{7.33}$$

Although this form is correct, it contains the complex j term, and we would like an explicit real form. This can be obtained by substituting Euler's identity:

$$e^{\pm j\beta t} = \cos \beta t \pm j \sin \beta t \tag{7.34}$$

and we have

$$\begin{aligned} x_n(t) &= e^{\alpha t}[A_1(\cos \beta t + j \sin \beta t) + A_2(\cos \beta t - j \sin \beta t)] \\ &= e^{\alpha t}[(A_1 + A_2) \cos \beta t + j(A_1 - A_2) \sin \beta t] \\ &= e^{\alpha t}(C_1 \cos \beta t + C_2 \sin \beta t) \end{aligned} \tag{7.35}$$

where we have defined two new undetermined constants:

$$C_1 = A_1 + A_2$$
$$C_2 = j(A_1 - A_2)$$

In order for the last form in (7.35) to yield a real time function, the constants C_1 and C_2 must be real. It is a simple matter to show that this implies *the two original undetermined constants must be conjugates of each other,* $A_2 = A_1^*$. This can be shown by forming $C_1 + jC_2 = 2A_2$ and $C_1 - jC_2 = 2A_1$. This case is referred to as the *underdamped* case. The reason for this designation will become apparent in the computed results to be obtained in Example 7.3.

The total solution is the sum of the natural solution and the forced solution (which can be found directly from the $t > 0$ circuit by replacing the inductor with a short circuit and the capacitor with an open circuit, since the source is dc) as one of three forms, depending on the roots of the characteristic equation:

$$\begin{aligned} x(t) &= A_1 e^{p_1 t} + A_2 e^{p_2 t} + X_f \\ x(t) &= (A_1 + A_2 t)e^{pt} + X_f \\ x(t) &= e^{\alpha t}(C_1 \cos \beta t + C_2 \sin \beta t) + X_f \end{aligned} \tag{7.36}$$

where X_f is the forced (steady-state) solution found from the $t > 0$ circuit.

To obtain the forced solution we draw the $t > 0$ circuit and replace the inductor with a short circuit and the capacitor with an open circuit as shown in Fig. 7.7*c*. This gives the forced solutions as

$$I_{L,f} = 0 \tag{7.37a}$$
$$V_{C,f} = V_S \tag{7.37b}$$

It now remains, as the final solution step, to determine the two undetermined constants in the natural solution in (7.36). The two required initial conditions are the value of $x(t)$ at $t = 0^+$, $x(0^+)$, and the derivative of $x(t)$ at $t = 0^+$, $dx(t)/dt|_{t=0^+} \equiv \dot{x}(0^+)$. Notice the shorthand notation for the derivative of $x(t)$ *evaluated at* $t = 0^+$: $\dot{x}(0^+)$. [Clearly this shorthand notation does not mean that we evaluate $x(t)$ at $t = 0^+$ and then take the derivative, since this would yield a result of zero every time, because $x(0^+)$ is a constant whose derivative is always zero.] Applying these to the various forms in (7.36) gives two equations, which can be solved for the two undetermined constants, giving the final solution. For distinct roots we obtain

$$\boxed{\begin{aligned} x(0^+) &= A_1 + A_2 + X_f \\ \dot{x}(0^+) &= p_1 A_1 + p_2 A_2 \end{aligned}} \tag{7.38a}$$

For repeated real roots we obtain

$$\boxed{\begin{aligned} x(0^+) &= A_1 + X_f \\ \dot{x}(0^+) &= pA_1 + A_2 \end{aligned}} \tag{7.38b}$$

For complex conjugate roots we obtain

$$\boxed{\begin{aligned} x(0^+) &= C_1 + X_f \\ \dot{x}(0^+) &= \alpha C_1 + \beta C_2 \end{aligned}} \tag{7.38c}$$

The two equations in (7.38) for each of the three cases can be solved for the two undetermined constants, yielding the complete solution for $t > 0$.

The values of the inductor current and capacitor voltage at $t = 0^+$ are repeated here:

$$\begin{aligned} i_L(0^-) &= 0 \\ &= i_L(0^+) \end{aligned} \tag{7.39a}$$

$$\begin{aligned} v_C(0^-) &= 0 \\ &= v_C(0^+) \end{aligned} \tag{7.39b}$$

The only remaining task is the determination of the derivatives of the inductor current and capacitor voltage at $t = 0^+$. To do this we draw the $t = 0^+$ as shown in Fig. 7.7*d* and label the initial conditions given in (7.39). The key here is to write the terminal relations for the inductor and the capacitor as

$$\frac{di_L(t)}{dt} = \frac{1}{L} v_L(t)$$

$$\frac{dv_C(t)}{dt} = \frac{1}{C} i_C(t)$$

Evaluating these at $t = 0^+$ gives

$$\boxed{\begin{aligned} \left.\frac{di_L(t)}{dt}\right|_{t=0^+} &= \frac{1}{L} v_L(0^+) \\ \left.\frac{dv_C(t)}{dt}\right|_{t=0^+} &= \frac{1}{C} i_C(0^+) \end{aligned}} \tag{7.40}$$

Hence, in order to determine the derivatives of the inductor current and capacitor voltage at $t = 0^+$, we must determine the *inductor voltage* and *capacitor current* at $t = 0^+$. These are obtained from the $t = 0^+$ circuit in Fig. 7.7*d* using KVL and KCL:

$$\begin{aligned} v_L(0^+) &= V_S - R\underbrace{i_L(0^+)}_{0} - \underbrace{v_C(0^+)}_{0} \\ &= V_S \\ i_C(0^+) &= \underbrace{i_L(0^+)}_{0} \\ &= 0 \end{aligned} \tag{7.41}$$

Hence, the derivatives of the inductor current and the capacitor voltage at $t = 0^+$ become

$$\begin{aligned} \dot{i}_L(0^+) &= \frac{1}{L} v_L(0^+) = \frac{V_S}{L} \\ \dot{v}_C(0^+) &= \frac{1}{C} i_C(0^+) = 0 \end{aligned} \tag{7.42}$$

where $\dot{i} \equiv di/dt$. Applying these to the distinct-real-root case in (7.38a), for example, gives (the forced solutions are $I_{L,f} = 0$ and $V_{C,f} = V_S$)

$$\begin{aligned} i_L(0^+) &= 0 = A_1 + A_2 \\ \dot{i}_L(0^+) &= \frac{V_S}{L} = p_1 A_1 + p_2 A_2 \end{aligned} \tag{7.43a}$$

and

$$\begin{aligned} v_C(0^+) &= 0 = B_1 + B_2 + V_S \\ \dot{v}_C(0^+) &= \frac{0}{C} = p_1 B_1 + p_2 B_2 \end{aligned} \tag{7.43b}$$

where we have used different symbols for the undetermined constants in the natural solutions for the inductor current and for the capacitor voltage to distinguish between the two. Solving (7.43) gives the undetermined constants. The undetermined constants for the repeated-real-root case in (7.38b) and the complex-conjugate-root case in (7.38c) can be determined in a similar fashion.

Example 7.3

Determine and sketch the solutions for the inductor current and capacitor voltage in the series *RLC* circuit of Fig. 7.7*a* for the following three cases:

1. $V_S = 10$ V, $R = 10\ \Omega$, $L = 2$ H, $C = \frac{1}{8}$ F,
2. $V_S = 10$ V, $R = 4\ \Omega$, $L = \frac{1}{2}$ H, $C = \frac{1}{8}$ F,
3. $V_S = 10$ V, $R = 2\ \Omega$, $L = 2$ H, $C = \frac{1}{25}$ F.

Solution *Case (a)*: For this case the characteristic equation given in (7.29) becomes

$$p^2 + 5p + 4 = (p + 1)(p + 4) = 0$$

Thus the roots are real and distinct:

$$\begin{aligned} p_1 &= -1 \\ p_2 &= -4 \end{aligned}$$

The forced solutions are

$$I_{L,f} = 0$$
$$V_{C,f} = V_S = 10$$

Hence the form of the total solutions is

$$i_L(t) = A_1 e^{-t} + A_2 e^{-4t}$$
$$v_C(t) = B_1 e^{-t} + B_2 e^{-4t} + 10$$

The initial conditions are

$$i_L(0^+) = 0$$
$$\dot{i}_L(0^+) = \frac{1}{L}[v_L(0^+) = V_S] = 5$$

and

$$v_C(0^+) = 0$$
$$\dot{v}_C(0^+) = \frac{1}{C}[i_C(0^+) = 0] = 0$$

Solving for the undetermined constants gives

$$0 = A_1 + A_2$$
$$5 = -A_1 - 4A_2$$

and

$$0 = B_1 + B_2 + 10$$
$$0 = -B_1 - 4B_2$$

So that $A_1 = \frac{5}{3}, A_2 = -\frac{5}{3}, B_1 = -\frac{40}{3}, B_2 = \frac{10}{3}$. Hence the total solutions valid for $t > 0$ are

$$i_L(t) = \tfrac{5}{3} e^{-t} - \tfrac{5}{3} e^{-4t}\ \text{A}$$
$$v_C(t) = -\tfrac{40}{3} e^{-t} + \tfrac{10}{3} e^{-4t} + 10\ \text{V}$$

Case (*b*): For this case the characteristic equation given in (7.29) becomes

$$p^2 + 8p + 16 = (p + 4)(p + 4) = 0$$

Thus the roots are real and repeated:

$$p_1 = -4$$
$$p_2 = -4$$

The forced solutions are unchanged:

$$I_{L,f} = 0$$
$$V_{C,f} = V_S = 10$$

Hence the form of the total solutions is

$$i_L(t) = A_1 e^{-4t} + A_2 t e^{-4t}$$
$$v_C(t) = B_1 e^{-4t} + B_2 t e^{-4t} + 10$$

The initial conditions are

$$i_L(0^+) = 0$$
$$\dot{i}_L(0^+) = \frac{1}{L}[v_L(0^+) = V_S] = 20$$

and

$$v_C(0^+) = 0$$

$$\dot{v}_C(0^+) = \frac{1}{C}[i_C(0^+) = 0] = 0$$

Solving for the undetermined constants gives

$$0 = A_1$$

$$20 = -4A_1 + A_2$$

and

$$0 = B_1 + 10$$

$$0 = -4B_1 + B_2$$

so that $A_1 = 0, A_2 = 20, B_1 = -10, B_2 = -40$. Hence the total solutions valid for $t > 0$ are

$$i_L(t) = 20te^{-4t} \text{ A}$$

$$v_C(t) = -10e^{-4t} - 40te^{-4t} + 10 \text{ V}$$

Case (*c*): For this case the characteristic equation given in (7.29) becomes

$$p^2 + p + 12.5 = (p + 0.5 + j3.5)(p + 0.5 - j3.5) = 0$$

Thus the roots are complex conjugate:

$$p_1 = -0.5 - j3.5$$

$$p_2 = -0.5 + j3.5$$

The forced solutions are unchanged:

$$I_{L,f} = 0$$

$$V_{C,f} = V_S = 10$$

Hence the form of the total solutions is

$$\begin{aligned} i_L(t) &= A_1 e^{(-0.5-j3.5)t} + A_2 e^{(-0.5+j3.5)t} \\ &= e^{-0.5t}(C_1 \cos 3.5t + C_2 \sin 3.5t) \\ v_C(t) &= B_1 e^{(-0.5-j3.5)t} + B_2 e^{(-0.5+j3.5)t} + 10 \\ &= e^{-0.5t}(D_1 \cos 3.5t + D_2 \sin 3.5t) + 10 \end{aligned}$$

The initial conditions are

$$i_L(0^+) = 0$$

$$\dot{i}_L(0^+) = \frac{1}{L}[v_L(0^+) = V_S] = 5$$

and

$$v_C(0^+) = 0$$

$$\dot{v}_C(0^+) = \frac{1}{C}[i_C(0^+) = 0] = 0$$

Solving for the undetermined constants gives

$$0 = C_1$$

$$5 = -0.5C_1 + 3.5C_2$$

and

$$0 = D_1 + 10$$

$$0 = -0.5D_1 + 3.5D_2$$

so that $C_1 = 0$, $C_2 = \frac{10}{7}$, $D_1 = -10$, $D_2 = -\frac{10}{7}$. Hence the total solutions valid for $t > 0$ are

$$i_L(t) = e^{-0.5t}(\tfrac{10}{7}\sin 3.5t)\ \text{A}$$
$$v_C(t) = e^{-0.5t}(-10\cos 3.5t - \tfrac{10}{7}\sin 3.5t) + 10\ \text{V}$$

The inductor current solutions are plotted in Fig. 7.8*a*, and the capacitor voltage solutions are plotted in Fig. 7.8*b*. Observe in these plots that for the case of real distinct roots the solution slowly converges to the steady-state value; hence this is referred to as the *overdamped* case. For the case of real, repeated roots, the solution converges rather rapidly to the steady-state solution and hence is referred to as the *critically damped* case. For the case of complex conjugate roots, the solution oscillates about the steady-state value, eventually converging to it. That is why this case is referred to as the *underdamped* case: there is insufficient damping to prevent oscillations, as in the ringing of a bell.

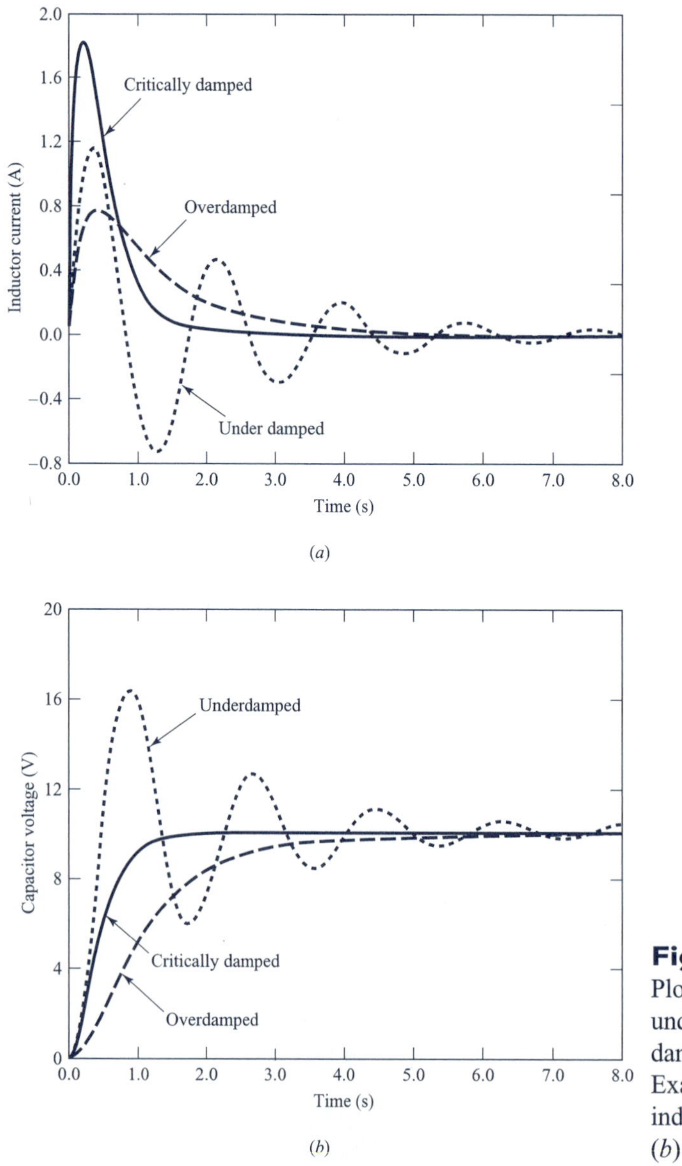

Figure 7.8 Plots of the overdamped, underdamped, and critically damped solutions of Example 7.3 for (*a*) the inductor current, and (*b*) the capacitor voltage.

Exercise Problem 7.3

For the series *RLC* circuit with $R = 3\ \Omega$, $L = \frac{1}{3}$ H, $C = \frac{1}{6}$ F, and $V_S = 10$ V determine the inductor current for $t > 0$.

Answer: $i_L(t) = 10e^{-3t} - 10e^{-6t}$ A.

Exercise Problem 7.4

For the series *RLC* circuit with $R = 2\ \Omega$, $L = \frac{1}{4}$ H, $C = \frac{1}{4}$ F, and $V_S = 5$ V determine the capacitor voltage for $t > 0$.

Answer: $v_C(t) = -5e^{-4t} - 20te^{-4t} + 5$ V.

Exercise Problem 7.5

For the series *RLC* circuit with $R = 6\ \Omega$, $L = 1$H, $C = \frac{1}{13}$ F, and $V_S = 10$ V determine the inductor current for $t > 0$.

Answer: $i_L(t) = 5e^{-3t} \sin 2t$ A.

7.2.2 The Parallel *RLC* Circuit Consider the parallel *RLC* circuit of Fig. 7.9*a*, where a dc current source is connected to the parallel combination of a resistor, an inductor, and a capacitor. The differential equations for the inductor current and capacitor voltage can again be obtained by replacing the elements with their "operator resistances" in the $t > 0$ circuit as described in Section 5.5 of Chapter 5. The resulting circuit is shown in Fig. 7.9*b*. From that circuit we obtain the inductor current, using current division, as

$$i_L(t) = \frac{R \left\| \dfrac{1}{CD} \right.}{R \left\| \dfrac{1}{CD} \right. + LD} I_S$$

$$= \frac{\dfrac{R}{1 + RCD}}{\dfrac{R}{1 + RCD} + LD} I_S$$

$$= \frac{R}{RLCD^2 + LD + R} I_S$$

Multiplying through by the denominator gives

$$(RLCD^2 + LD + R) i_L(t) = R\, I_S$$

Operating on the variables gives the differential equation:

$$\boxed{\frac{d^2 i_L(t)}{dt^2} + \frac{1}{RC}\frac{di_L(t)}{dt} + \frac{1}{LC} i_L(t) = \frac{1}{LC} I_S} \tag{7.44}$$

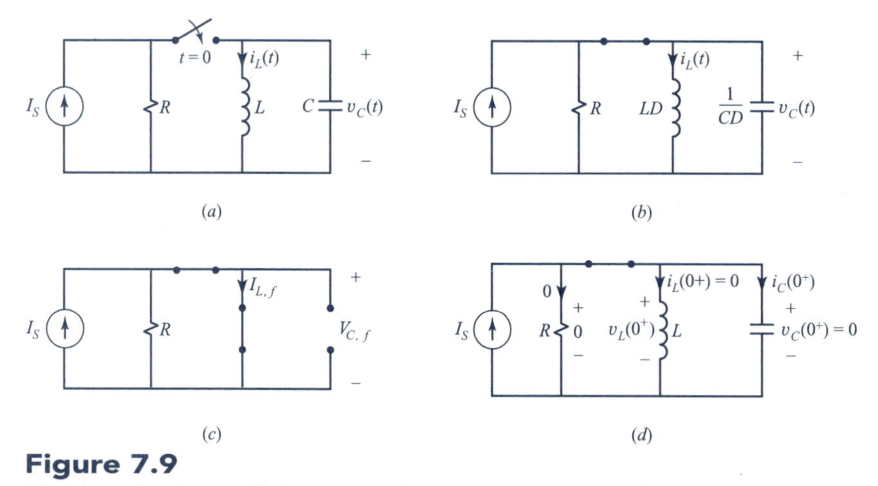

Figure 7.9
The second-order parallel *RLC* circuit: (*a*) the physical circuit, (*b*) determining the differential equations for the inductor current and capacitor voltage using the differential operator impedances, (*c*) determining the forced solutions for the inductor current and capacitor voltage, and (*d*) the $t = 0^+$ circuit (immediately after switch closure) for determining the derivatives of the inductor current and capacitor voltage at $t = 0^+$.

The differential equation for the capacitor voltage is similarly obtained from Fig. 7.9*b* as

$$\begin{aligned} v_C(t) &= \left(R \middle\| (LD) \middle\| \frac{1}{CD}\right) I_S \\ &= \frac{RLD}{RLCD^2 + LD + R} I_S \end{aligned}$$

Hence the differential equation becomes

$$\begin{aligned} \frac{d^2 v_C(t)}{dt^2} + \frac{1}{RC}\frac{dv_C(t)}{dt} + \frac{1}{LC} v_C(t) &= \frac{1}{C} D I_S \\ &= \frac{1}{C}\frac{d}{dt} I_S = 0 \end{aligned} \tag{7.45}$$

The characteristic equation for both cases is

$$p^2 + \frac{1}{RC} p + \frac{1}{LC} = 0 \tag{7.46}$$

Comparing this with the characteristic equation for the series *RLC* circuit given in (7.29), we see that the only difference is the coefficient of the p term. For the series *RLC* case it is R/L, and for the parallel *RLC* case it is $1/RC$. As a memory aid, these are recognized as being the reciprocals of the time constants for a series *RL* circuit and a parallel *RC* circuit, respectively.

The solution procedure for the parallel *RLC* case is virtually identical to that of the series *RLC* case:

1. Factor the characteristic equation given in (7.46) to determine the roots.
2. Based on whether those roots are real and distinct, real and repeated, or complex conjugate, construct the total solution consisting of the sum of the natural solution having one of the forms given in (7.36) and the forced solution.

3. Determine the forced solution directly from the circuit by replacing the inductor with a short circuit and the capacitor with an open circuit as shown in Fig. 7.9*c* (assuming the source is dc), obtaining

$$I_{L,f} = I_S \qquad (7.47)$$
$$V_{C,f} = 0$$

4. Determine the initial conditions from the $t < 0$ circuit and the $t = 0^+$ circuit as shown in Fig. 7.9*d*:

$$i_L(0^+) = i_L(0^-) = 0$$
$$\dot{i}_L(0^+) = \frac{1}{L}\left[v_L(0^+) = v_C(0^+) = 0\right] \qquad (7.48a)$$
$$= 0$$

and

$$v_C(0^+) = v_C(0^-) = 0$$
$$\dot{v}_C(0^+) = \frac{1}{C}\left(i_C(0^+) = I_S - \frac{v_C(0^+) = 0}{R} - \underbrace{i_L(0^+)}_{0}\right) \qquad (7.48b)$$
$$= \frac{I_S}{C}$$

5. Apply these initial conditions to the total solution to determine the two undetermined constants in that solution.

Example 7.4

Determine the inductor current for the parallel *RLC* circuit where $R = 1\ \Omega$, $L = \frac{3}{2}$ H, $C = \frac{1}{3}$ F and the source is sinusoidal, $I_S = 10\sin(3t + 30^\circ)$ A.

Solution The roots of the characteristic equation given in (7.46) are

$$p^2 + 3p + 2 = (p + 1)(p + 2) = 0$$

or

$$p_1 = -1, \qquad p_2 = -2$$

Therefore, the natural solution is

$$i_{L,n}(t) = A_1 e^{-t} + A_2 e^{-2t}$$

Since the source is sinusoidal, we obtain the forced response by using the phasor method of Chapter 6:

$$\hat{I}_L = \frac{R \left\| \frac{1}{j\omega C}\right.}{j\omega L + R\left\|\frac{1}{j\omega C}\right.} \times 10 \angle 30^\circ$$

$$= \frac{1\left\|\frac{1}{j3 \times \frac{1}{3}}\right.}{j3 \times \frac{3}{2} + 1\left\|\frac{1}{j3\times\frac{1}{3}}\right.} \times 10 \angle 30^\circ$$

$$= 1.75\angle -97.87^\circ$$

as the reader should verify. Hence the forced solution is

$$i_{L,f}(t) = 1.75 \sin(3t - 97.87^\circ)$$

Adding these gives the total solution for $t > 0$ as

$$i_L(t) = A_1 e^{-t} + A_2 e^{-2t} + 1.75 \sin(3t - 97.87^\circ) \qquad t > 0$$

We now need the initial conditions, $i_L(0^+)$ and $di_L/dt|_{t=0^+}$. These are obtained as usual from

$$i_L(0^+) = 0$$

and

$$\left.\frac{di_L}{dt}\right|_{t=0^+} = \frac{v_L(0^+)}{L} = \frac{0}{\frac{3}{2}} = 0$$

Applying these to the general solution gives

$$i_L(0^+) = 0 = A_1 + A_2 + 1.75 \sin(-97.87^\circ)$$

$$\left.\frac{di_L}{dt}\right|_{t=0^+} = 0 = -A_1 - 2A_2 + 5.26 \cos(-97.87^\circ)$$

Solving these gives

$$A_1 = 4.2, \qquad A_2 = -2.46$$

so that

$$i_L(t) = 4.2e^{-t} - 2.46e^{-2t} + 1.75 \sin(3t - 97.87^\circ)\text{ A} \qquad t > 0$$

Exercise Problem 7.6

For the parallel *RLC* circuit with $R = 2\ \Omega$, $L = \frac{8}{3}$ H, $C = \frac{1}{8}$ F, and $I_S = 2$ A determine the inductor current for $t > 0$.

Answer: $i_L(t) = -3e^{-t} + e^{-3t} + 2$ A.

Exercise Problem 7.7

For the parallel *RLC* circuit with $R = 5\ \Omega$, $L = \frac{6}{5}$ H, $C = \frac{1}{30}$ F, and $I_S = 5$ A determine the capacitor voltage for $t > 0$.

Answer: $v_C(t) = 37.5e^{-3t} \sin 4t$ V.

Although this procedure is the same for either the series or the parallel *RLC* circuit and is straightforward, it is nevertheless somewhat tedious. For structurally more complex second-order circuits (that are neither series *RLC* nor parallel *RLC* in form), it becomes even more so. One of the problems encountered in more general second-order circuits is the requirement to obtain the derivative of the inductor current and/or the derivative of the capacitor voltage at $t = 0^+$, followed by the evaluation of the two undetermined constants in the natural solution. For higher-order circuits containing more than two energy storage elements, these problems are compounded because there are correspondingly more undetermined constants in the natural solution, requiring higher-order derivatives at $t = 0^+$. For example, a third-order circuit containing three energy storage elements will contain three undetermined constants in the natural solution, requiring that we determine $x(0^+), \dot{x}(0^+), \ddot{x}(0^+)$. In the following sections, we will develop a simpler method for determining the total solution for any circuit, using the Laplace transform. This method automatically incorporates the required initial conditions into the solutions, and no simultaneous equations are required to be solved for the undetermined constants in the natural solution as was the case for the above methods. In addition, the Laplace transform allows us to consider more complicated independent source waveforms with no significant additional work.

7.3 The Laplace Transform

The previous sections have demonstrated the direct, or *classical*, solution of the circuit differential equations in determining the total solution in the presence of a switching operation. There are several distinct steps in that solution:

1. Draw the $t = 0^-$ circuit (prior to switch activation), and determine the value of inductor current or capacitor voltage at $t = 0^-$, $i_L(0^-)$ or $v_C(0^-)$, by replacing the inductor with a short circuit and the capacitor with an open circuit (if all sources in the $t < 0$ circuit are dc).
2. Draw the $t > 0$ circuit (after switch activation), and determine the differential equation relating the inductor current or capacitor voltage to the independent source(s).
3. From that differential equation determine the form of the natural solution.
4. Determine the forced solution by replacing the inductor with a short circuit and the capacitor with an open circuit (if the sources are all dc) in the $t > 0$ circuit. If a source is sinusoidal, determine the portion of the forced solution due to this source using the phasor methods of Chapter 6.
5. Form the total solution as the sum of the natural and forced solutions.
6. If the circuit is first-order, evaluate the undetermined constant in the natural solution by applying the initial condition: $i_L(0^-) = i_L(0^+)$ or $v_C(0^-) = v_C(0^+)$. If the circuit is second-order, determine the additional required initial conditions, $di_L/dt|_{t=0^+}$ and $dv_C/dt|_{t=0^+}$, and evaluate the two undetermined constants in the natural solution.
7. Determine any other current or voltage solutions from the solution for the inductor current or capacitor voltage along with KVL, KCL, and the element relations in the $t > 0$ circuit.

In the remaining sections of this chapter we will develop another solution technique, the *Laplace transform,* which avoids all of the above steps except step (1). Recall that the inductor current and/or the capacitor voltage are the only link between the $t < 0$ and $t > 0$ circuits, and their values at $t = 0^-$ must be determined regardless of the solution method used for the $t > 0$ circuit. The Laplace transform method is a rather automatic method for obtaining the solution for any circuit variable and at the same time automatically incorporating the required initial conditions. Furthermore, it allows a very straightforward way of handling a myriad of other circuit responses in addition to switching operations.

The Laplace transform is a process of transforming from the time domain to the Laplace domain, wherein derivative and integral operations are replaced with simpler algebraic operations. Once these are completed, we transform back to the time domain and have the desired solution. This is closely akin to the operations involving the logarithm. For example, suppose we wish to multiply two numbers as

$$c = a \times b$$

Taking the logarithm of both sides transforms the multiplication into a simpler addition operation as

$$\log_m c = \log_m a + \log_m b$$

where $\log_m$ is the logarithm to the base m. This is then converted back to the original domain by taking the antilog of this result:

$$\begin{aligned} c &= \text{antilog}(\log c) \\ &= m^{\log_m c} \end{aligned}$$

The solution process can be thought of as transforming from one *domain* to another in which the desired operations are simpler to perform than in the original domain. Once the operations are performed, we transform back to the original domain. Another example is the use of phasors to solve sinusoidal, steady-state circuits that was considered in the previous chapter. We first transform the time-domain circuit to the phasor (frequency-domain) circuit. Differentiation in the time domain, d^k/dt^k, is replaced by multiplication by $(j\omega)^k$ in the frequency domain. Once the phasor result is obtained, we transform back to the time domain by multiplying the phasor quantity by $e^{j\omega t}$ and taking the real or imaginary part of the result.

The Laplace transform of a function of t, $f(t)$, is defined as

$$\begin{aligned} \mathcal{L}\{f(t)\} &= F(s) \\ &= \int_{0^-}^{\infty} f(t)e^{-st}dt \end{aligned} \tag{7.49}$$

where the Laplace transform variable s is a complex variable:

$$s = \sigma + j\omega$$

We denote the transformation in (7.49) as

$$f(t) \Leftrightarrow F(s) \tag{7.50}$$

Observe that the lower limit of the integral in (7.49) is $t = 0^-$. This will become important in the evaluation of an important forcing function, the impulse function. The Laplace transform in (7.49) is said to be the *one-sided* transform and assumes that the function $f(t)$ is zero for $t < 0$:

$$f(t) = 0 \qquad t < 0$$

Virtually all functions that are of engineering interest have a Laplace transform. It will not be necessary or important to discuss convergence properties of the Laplace transform here. For such a discussion see C.R. Paul, *Analysis of Linear Circuits*, McGraw-Hill, 1989.

7.3.1 Important Properties of the Laplace Transform

There are some important properties of the Laplace transform that we may use in determining the transform of a function in terms of some previously determined transforms. The first property is that of *linearity*:

$$Af_1(t) + Bf_2(t) \Leftrightarrow AF_1(s) + BF_2(s) \tag{7.51}$$

This can be proven from the linearity of the Laplace transform operation:

$$\int_{0^-}^{\infty} [Af_1(t) + Bf_2(t)]e^{-st}\,dt = A\underbrace{\int_{0^-}^{\infty} f_1(t)e^{-st}\,dt}_{AF_1(s)} + B\underbrace{\int_{0^-}^{\infty} f_2(t)e^{-st}\,dt}_{BF_2(s)}$$

The second property is that of *time shift*:

$$\boxed{f(t - t_0) \Leftrightarrow e^{-st_0}F(s)} \tag{7.52}$$

that is, *the Laplace transform of a function that is shifted forward by t_0 (delayed in time by t_0) is the Laplace transform of the original unshifted function multiplied by e^{-st_0}*. This can be proven directly from

$$\begin{aligned}\int_{t_0}^{\infty} f(t - t_0)\,e^{-st}\,dt &= \int_{0^-}^{\infty} f(\lambda)\,e^{-s(\lambda+t_0)}d\lambda \\ &= e^{-st_0}\underbrace{\int_{0^-}^{\infty} f(\lambda)\,e^{-s\lambda}d\lambda}_{F(s)}\end{aligned}$$

where we have used a change of variables: $\lambda = t - t_0$. The third important property is the *s-shift property*:

$$\boxed{e^{-at}f(t) \Leftrightarrow F(s + a)} \tag{7.53}$$

that is, *the product of a time function with the exponential e^{-at} has as its transform the transform of the time function with s replaced by $s + a$*. This can be similarly proven quite easily from

$$\begin{aligned}\int_{0^-}^{\infty} (e^{-at}f(t))e^{-st}\,dt &= \int_{0^-}^{\infty} f(t)e^{-(s+a)t}\,dt \\ &= F(s + a)\end{aligned}$$

The fourth and final important property is the Laplace transform of the *derivative of a time function*:

$$\boxed{\mathcal{L}\left\{\frac{df(t)}{dt}\right\} \Leftrightarrow sF(s) - f(0^-)} \tag{7.54}$$

This can be proven using integration by parts from

$$\begin{aligned}\mathcal{L}\left\{\frac{df(t)}{dt}\right\} &= \int_{0^-}^{\infty} \frac{df(t)}{dt}e^{-st}dt \\ &= f(t)e^{-st}\Big|_{0^-}^{\infty} + s\int_{0^-}^{\infty} f(t)e^{-st}dt \\ &= -f(0^-) + sF(s)\end{aligned}$$

There are many more properties of the Laplace transform but these four are the most useful and will be adequate for all our future needs.

7.3.2 Transforms of Important Time Functions

The first important time function is the *unit step function* shown in Fig. 7.10*a*. The formal definition of the unit step function is

$$\boxed{u(t) = \begin{cases} 0 & t < 0 \\ 1 & t > 0 \end{cases}} \tag{7.55}$$

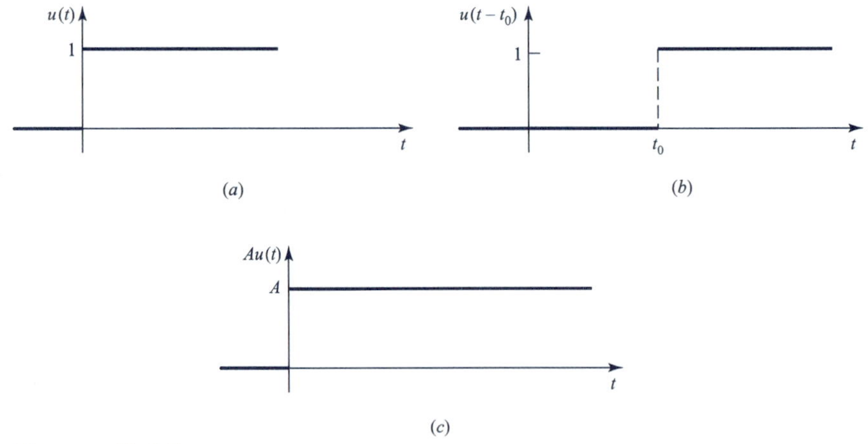

Figure 7.10
The unit impulse function: (*a*) $u(t)$, (*b*) $u(t - t_0)$, and (*c*) $Au(t)$.

Hence the unit step function is zero for $t < 0$ and is unity for $t > 0$. At $t = 0$ it is undefined. A unit step function that turns on at $t = t_0$, $u(t - t_0)$, is defined by

$$u(t - t_0) = \begin{cases} 0 & t < t_0 \\ 1 & t > t_0 \end{cases} \tag{7.56}$$

as is illustrated in Fig. 7.10*b*. Multiplication of the unit step function by some constant A changes the level by A as shown in Fig. 7.10*c*:

$$Au(t) = \begin{cases} 0 & t < 0 \\ A & t > 0 \end{cases} \tag{7.57}$$

The unit step function can also be used to "turn on" a function in that $f(t)u(t)$ is zero for $t < 0$ regardless of whether $f(t)$ is zero for $t < 0$. The Laplace transform of the unit step function is rather simple:

$$u(t) \Leftrightarrow \frac{1}{s} \tag{7.58}$$

This can be directly proven:

$$\begin{aligned}\int_{0^-}^{\infty} u(t)e^{-st}\,dt &= \int_{0^+}^{\infty} 1e^{-st}\,dt \\ &= -\frac{1}{s}\,e^{-st}\Big|_{0^+}^{\infty} = \frac{1}{s}\end{aligned}$$

There are some subtle convergence questions that we are omitting here. For example, if we substitute the definition of s into this last result, the proof requires that

$$\begin{aligned}\int_{0^-}^{\infty} u(t)e^{-st}dt &= \int_{0^+}^{\infty} 1e^{-(\sigma+j\omega)t}\,dt \\ &= -\frac{1}{s}\,e^{-\sigma t}e^{-j\omega t}\Big|_{0^+}^{\infty} \\ &= \frac{1}{s} + \lim_{t\to\infty}\left(-\frac{e^{-\sigma t}e^{-j\omega t}}{s}\right)\end{aligned}$$

In this last limit, the magnitude of $e^{-j\omega t}$ is unity, so that the limit goes to zero only for $\sigma > 0$. This is referred to as the *region of convergence* of the transform. This is of no practical consequence, since a property of complex variables known as *analytic continuation* can be used to extend the region of convergence everywhere except at $s = 0$. In all future transforms of time functions we will omit discussion of the region of convergence, since it similarly imposes no practical restrictions. The *s-shift property* in (7.53) gives the transform of an exponential:

$$e^{-at}u(t) \Leftrightarrow \frac{1}{s+a} \tag{7.59}$$

Note that the inclusion of the unit step function to "turn off" the exponential for $t < 0$ is an essential part of the result, since e^{-at} is not zero for $t < 0$ as is required for the Laplace transform of all time functions. An additional and important time function is

$$\frac{1}{n!}t^n e^{-at}\,u(t) \Leftrightarrow \frac{1}{(s+a)^{n+1}} \tag{7.60}$$

This can be proven using the following identity:

$$\int_0^\infty x^n e^{-ax}\,dx = \frac{n!}{a^{n+1}}$$

Hence

$$\begin{aligned}\mathscr{L}\left\{\frac{1}{n!}\,t^n e^{-at}u(t)\right\} &= \frac{1}{n!}\int_{0^-}^{\infty} t^n e^{-at}\,e^{-st}\,dt \\ &= \frac{1}{(s+a)^{n+1}}\end{aligned}$$

Several other important transforms can be developed using the above properties. For example,

$$\sin\omega t\,u(t) \Leftrightarrow \frac{\omega}{s^2+\omega^2} \tag{7.61}$$

$$\cos\omega t\,u(t) \Leftrightarrow \frac{s}{s^2+\omega^2} \tag{7.62}$$

Both of these can be determined using Euler's identity:

$$\begin{aligned}\mathscr{L}\{\sin\omega t\,u(t)\} &= \mathscr{L}\left\{\frac{e^{j\omega t}}{2j}\,u(t) - \frac{e^{-j\omega t}}{2j}\,u(t)\right\} \\ &= \frac{1}{2j}\,\mathscr{L}\{e^{j\omega t}\,u(t)\} - \frac{1}{2j}\,\mathscr{L}\{e^{-j\omega t}\,u(t)\} \\ &= \frac{1}{2j(s-j\omega)} - \frac{1}{2j(s+j\omega)} \\ &= \frac{(s+j\omega)-(s-j\omega)}{2j(s^2+\omega^2)} \\ &= \frac{\omega}{s^2+\omega^2}\end{aligned}$$

$$\begin{aligned}
\mathcal{L}\{\cos \omega t\, u(t)\} &= \mathcal{L}\left\{\frac{e^{j\omega t}}{2}u(t) + \frac{e^{-j\omega t}}{2}u(t)\right\} \\
&= \tfrac{1}{2}\mathcal{L}\{e^{j\omega t}\, u(t)\} + \tfrac{1}{2}\mathcal{L}\{e^{-j\omega t}\, u(t)\} \\
&= \frac{1}{2(s - j\omega)} + \frac{1}{2(s + j\omega)} \\
&= \frac{(s + j\omega) + (s - j\omega)}{2(s^2 + \omega^2)} \\
&= \frac{s}{s^2 + \omega^2}
\end{aligned}$$

Using the s-shift property in (7.53), we additionally obtain

$$e^{-at} \sin \omega t\, u(t) \Leftrightarrow \frac{\omega}{(s + a)^2 + \omega^2} \tag{7.63}$$

$$e^{-at} \cos \omega t\, u(t) \Leftrightarrow \frac{s + a}{(s + a)^2 + \omega^2} \tag{7.64}$$

The last time function for which we desire a Laplace transform is the *unit impulse function* shown in Fig. 7.11a. The formal definition of the unit impulse function is

$$\delta(t) = \begin{cases} 0 & t < 0 \\ 0 & t > 0 \\ \int_{0^-}^{0^+} \delta(t)\, dt = 1 \end{cases} \tag{7.65}$$

This is perhaps the most important time function we will encounter, and an intuitive discussion of its properties will be useful. The unit impulse function is nonzero only at $t = 0$. It can be viewed as having infinite height and zero width but containing unity *area*. It can be visualized as in Fig. 7.11b as the limit of a sequence of rectangular pulses as their width shrinks to zero and their height increases proportionally to contain unit area under the rectangle. An impulse of any area, or *strength*, can be obtained by multiplying the unit impulse by a constant A, as shown in Fig. 7.11c. As in the case of a unit step, a unit impulse that is activated at t_0 as illustrated in Fig. 7.11d is defined as

$$\delta(t - t_0) = \begin{cases} 0 & t < t_0 \\ 0 & t > t_0 \\ \int_{t_0^-}^{t_0^+} \delta(t - t_0)\, dt = 1 \end{cases} \tag{7.66}$$

An important property that we will frequently need is the *sifting property* of the impulse. *If we multiply any function by a unit impulse and integrate the product over any time that encompasses the impulse, the result will be the value of the function at the time of occurrence of the impulse*:

$$\int_{t_0^-}^{t_0^+} f(t)\delta(t - t_0)\, dt = f(t_0) \tag{7.67}$$

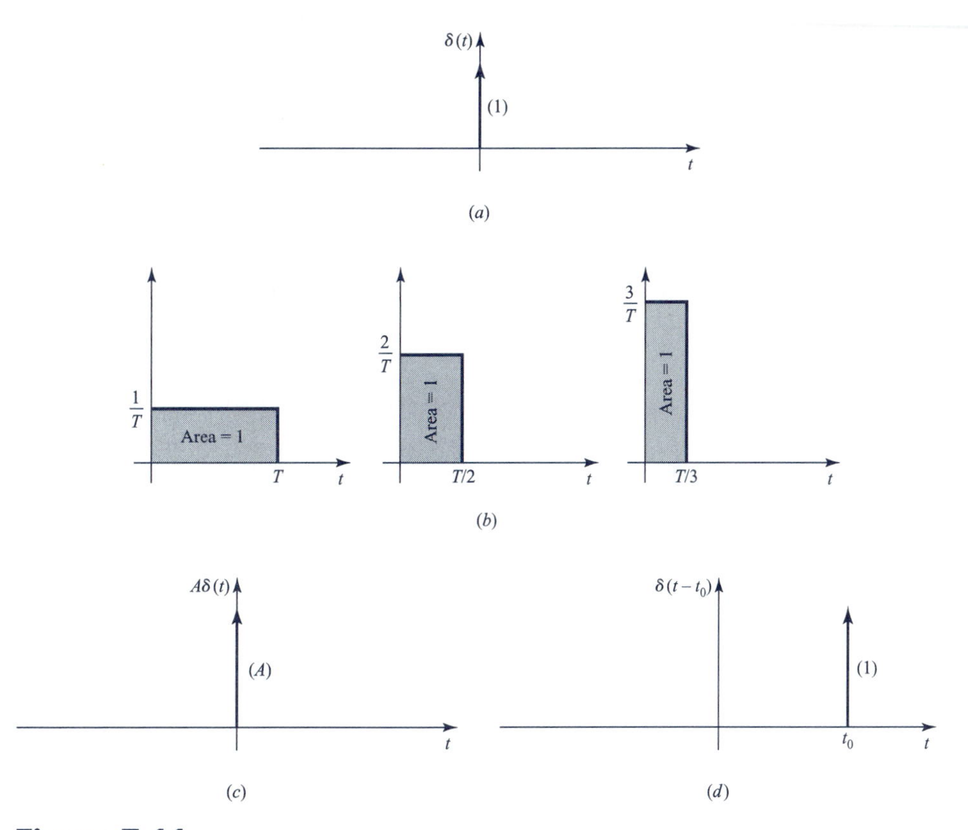

Figure 7.11
The unit impulse function: (*a*) $\delta(t)$, (*b*) approximating the unit impulse as a sequence of pulses whose width shrinks to zero but whose area remains unity, (*c*) $A\delta(t)$, and (*d*) $\delta(t - t_0)$.

This is evident because the integrand is nonzero only at t_0 because of its multiplication by the impulse:

$$\int_{t_0^-}^{t_0^+} f(t)\delta(t - t_0)\, dt = f(t_0)\int_{t_0^-}^{t_0^+} \delta(t - t_0)\, dt$$

$$= f(t_0) \tag{7.68}$$

The Laplace transform of the unit impulse function is unity:

$$\boxed{\delta(t) \Leftrightarrow 1} \tag{7.69}$$

This can be shown directly:

$$\int_{0^-}^{\infty} \delta(t)e^{-st}dt = e^{-st}\Big|_{t=0} \int_{0^-}^{0^+} \delta(t)\, dt$$

$$= 1$$

Because we initially defined the Laplace transform's lower limit as $t = 0^-$ rather than $t = 0^+$ or the ambiguous $t = 0$, the impulse is included in the integral. The impulse is one of the key reasons we did so.

Exercise Problem 7.8

Determine the Laplace transform of $f(t) = (t + e^{-at})u(t)$.

Answer: $F(s) = (s^2 + s + a)/s^2\,(s + a)$.

7.3.3 The Laplace Transforms of the *R*, *L*, and *C* Elements We are now ready to transform our circuits to the Laplace domain thereby simplifying their solution. First consider the linear resistor. The time-domain relation is

$$v_R(t) = R\,i_R(t)$$

Hence the Laplace transform is

$$\boxed{V_R(s) = R\,I_R(s)} \tag{7.70}$$

The resistor is said to have an *impedance* of

$$\boxed{Z_R(s) = R} \tag{7.71}$$

in the same way as in phasor circuits and is modeled as in Fig. 7.12*a*.

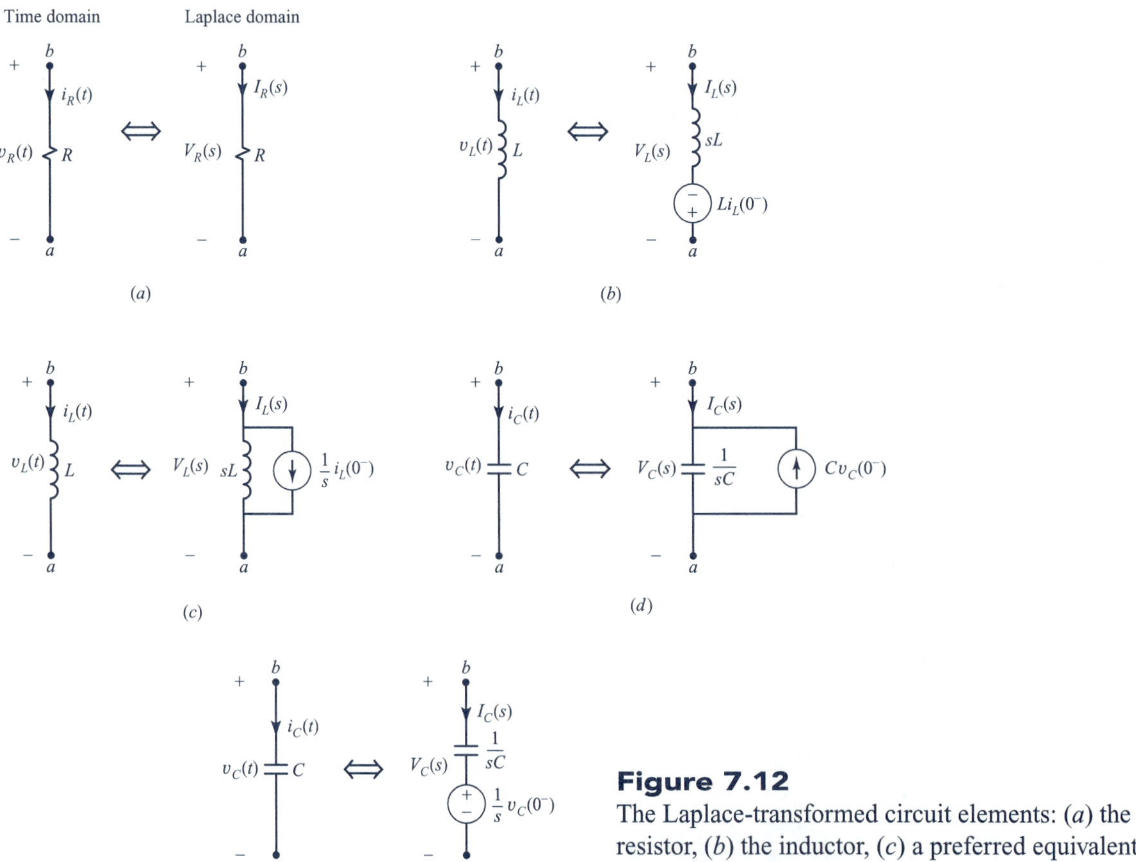

Figure 7.12
The Laplace-transformed circuit elements: (*a*) the resistor, (*b*) the inductor, (*c*) a preferred equivalent circuit of the inductor, (*d*) the capacitor, (*e*) a preferred equivalent circuit of the capacitor.

Next consider the inductor, whose voltage–current relation is

$$v_L(t) = L\frac{di_L(t)}{dt}$$

Substituting the derivative relation given in (7.54) yields

$$\boxed{V_L(s) = sLI_L(s) - Li_L(0^-)} \tag{7.72}$$

whose equivalent circuit, shown in Fig. 7.12*b*, consists of the *impedance* of the inductor,

$$\boxed{Z_L(s) = sL} \tag{7.73}$$

in series with a voltage source of value equal to the initial inductor current multiplied by the inductance value, $Li_L(0^-)$. This equivalent circuit simply comes from the result in (7.72). Hence the initial inductor current appears as an impulse voltage source. It is critical to observe (a) the polarity of the initial-condition voltage source [opposite to the transformed inductor voltage, $V_L(s)$] and (b) the fact that the transformed inductor voltage $V_L(s)$ is across the series combination and not just the inductor impedance. Failure to observe the proper polarity of the initial-condition source and failure to observe that *the transformed voltage is across the entire equivalent circuit and not just the inductor impedance* are very common (and serious) mistakes that are made in the use of the Laplace transform. Hence the reader should be alert to these two subtleties; otherwise there is no possibility for correct analysis of a circuit.

An alternative equivalent circuit is shown in Fig. 7.12*c*. It is obtained by inverting (7.72) as

$$\boxed{I_L(s) = \frac{1}{sL}V_L(s) + \frac{1}{s}i_L(0^-)} \tag{7.74}$$

The equivalent circuit consists of the parallel combination of the inductor impedance and a step current source whose value is the initial inductor current. Once again it is critical to observe (a) the polarity of the current source [in the direction of the transformed inductor current, $I_L(s)$] and (b) that the transformed inductor current $I_L(s)$ is through the entire combination and not just the inductor impedance. This alternative equivalent circuit could also have been obtained from that of Fig. 7.12*b* by using a source transformation.

The equivalent circuit of the inductor given in Fig. 7.12*c* is preferred by the author over that of Fig. 7.12*b*, since it is easily remembered with the following memory aid. The inductor current is the important variable of interest. The equivalent circuit of Fig. 7.12*c* shows (a) the initial condition as a current source that "turns on" as a step-function source and (b) that the direction of this initial condition current source is the same as that of the inductor current. The equivalent circuit of Fig. 7.12*b* does not share these logical memory aids, since (a) the inductor current initial condition is represented as a voltage source (an impulse) and (b) the polarity of this source is opposite to the inductor voltage. Each one can be converted to the other with a source transformation, but the circuit of Fig. 7.12*c* is more easily remembered.

Next we consider the equivalent circuit of the capacitor. The capacitor terminal relation is

$$i_C(t) = C\frac{dv_C(t)}{dt}$$

Transforming this using (7.54) gives

$$\boxed{I_C(s) = sC\,V_C(s) - C\,v_C(0^-)} \tag{7.75}$$

The equivalent circuit is shown in Fig. 7.12*d* and consists of a capacitor with an impedance of

$$Z_C(s) = \frac{1}{sC} \tag{7.76}$$

in parallel with an impulse current source of value $Cv_C(0^-)$. Again it is important to observe (a) the polarity of this initial condition source [opposite to the direction of the transformed capacitor current $I_C(s)$] and (b) the fact that the transformed capacitor current $I_C(s)$ is through the entire parallel combination and not just the capacitor impedance. An alternative equivalent circuit can be obtained, by using a source transformation on this circuit or by inverting (7.75), as

$$V_C(s) = \frac{1}{sC} I_C(s) + \frac{1}{s} v_C(0^-) \tag{7.77}$$

Hence this equivalent circuit consists of the series combination of the capacitor impedance and a step voltage source of value equal to the initial capacitor voltage, as shown in Fig. 7.12*e*. Once again it is crucial to observe (a) the polarity of the voltage source [in the direction of the transformed capacitor voltage $V_C(s)$] and (b) that the transformed capacitor voltage $V_C(s)$ is across the entire combination and not just across the capacitor impedance.

The equivalent circuit of the capacitor given in Fig. 7.12*e* is preferred by the author over that of Fig. 7.12*d*, since it is easily remembered with the following memory aid. The capacitor voltage is the important variable of interest. The equivalent circuit of Fig. 7.12*e* shows (a) the initial condition as a voltage source which "turns on" as a step-function source and (b) that the polarity of this initial-condition voltage source is the same as that of the capacitor voltage. The equivalent circuit of Fig. 7.12*d* does not share these logical memory aids, since (a) the capacitor voltage initial condition is represented as a current source (an impulse) and (b) the polarity of this source is opposite to the capacitor current. Each one can be converted to the other with a source transformation, but the circuit of Fig. 7.12*e* is more easily remembered and is therefore preferred by the author.

Finally, we consider the equivalent circuit of mutually coupled inductors. Consider two coupled inductors shown in Fig. 7.13*a*. The terminal relations are

$$\begin{aligned} v_1(t) &= L_1 \frac{di_1(t)}{dt} + M \frac{di_2(t)}{dt} \\ v_2(t) &= M \frac{di_1(t)}{dt} + L_2 \frac{di_2(t)}{dt} \end{aligned} \tag{7.78}$$

Transforming this using (7.54) yields

$$\begin{aligned} V_1(s) &= sL_1 I_1(s) - [L_1 i_1(0^-) + M i_2(0^-)] + sMI_2(s) \\ V_2(s) &= sL_2 I_2(s) - [L_2 i_2(0^-) + M i_1(0^-)] + sMI_1(s) \end{aligned} \tag{7.79}$$

The equivalent circuit is shown in Fig. 7.13*b* and consists of the self-impedance of the inductors in series with impulsive voltage sources. Again observe the polarity of these initial-condition sources as well as the fact that the total voltage is across the inductor impedance and the initial-condition source and not just across the inductor impedance. The sign of the mutual-impedance contributions and that of the initial-condition contribution through the mutual inductance are determined by the dot convention. The alternative equivalent circuit is shown in Fig. 7.13*c* and

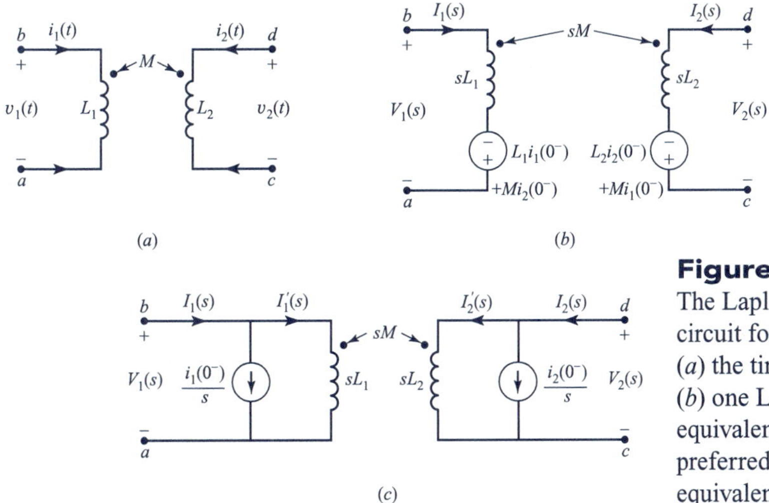

Figure 7.13 The Laplace-transformed circuit for mutual inductance: (*a*) the time-domain circuit, (*b*) one Laplace-transformed equivalent circuit, and (*c*) the preferred Laplace-transformed equivalent circuit.

can been obtained by rearranging (7.79) as

$$\begin{aligned} V_1(s) &= sL_1\left(I_1(s) - \frac{i_1(0^-)}{s}\right) + sM\left(I_2(s) - \frac{i_2(0^-)}{s}\right) \\ V_2(s) &= sM\left(I_1(s) - \frac{i_1(0^-)}{s}\right) + sL_2\left(I_2(s) - \frac{i_2(0^-)}{s}\right) \end{aligned} \tag{7.80}$$

Observing in Fig. 7.13*c* that the total currents through the inductors are

$$\begin{aligned} I_1'(s) &= I_1(s) - \frac{i_1(0^-)}{s} \\ I_2'(s) &= I_2(s) - \frac{i_2(0^-)}{s} \end{aligned} \tag{7.81}$$

we see that (7.80) yields the equivalent circuit of Fig. 7.13*c*. This latter equivalent circuit is preferred by the author, since the initial-condition current source is in the direction of the inductor current and consists solely of the initial current of that inductor as a step source.

With these basic transforms of the circuit elements along with the transforms of some typical independent source functions, we can now transform the circuit from the time domain to the Laplace domain by substituting the above equivalent circuits for the elements and transforming the independent source functions. Because of the linearity of the Laplace transform, KVL and KCL hold for the transformed voltages and currents:

$$\sum v_i(t) = 0 \Leftrightarrow \sum V_i(s) = 0 \tag{7.82}$$

$$\sum i_i(t) = 0 \Leftrightarrow \sum I_i(s) = 0 \tag{7.83}$$

Hence *all of our resistive-circuit analysis techniques (series and parallel combinations of elements, voltage and current division, source transformation, the direct method, superposition, Thevenin and Norton equivalents, and node-voltage and mesh-current equations) are equally valid for the Laplace-transformed circuit.* Thus, as in the case of phasor circuits, obtaining the transform of the desired voltage or current in the transformed circuit becomes a familiar resistive-circuit analysis problem that we have become proficient in solving.

Example 7.5

Determine the Laplace transform of the resistor current in the circuit of Fig. 7.14*a*.

Solution The transformed circuit is shown if Fig. 7.14*b*. Observe that the Laplace transform of a dc source is essentially a step function whose value is that of the dc source: $I_S \Leftrightarrow I_S/s$. The circuit is converted to a single-loop circuit in Fig. 7.14*c* using a source transformation. From that we obtain

$$I_R(s) = \frac{sL\left(\frac{I_S}{s} - \frac{i_L(0^-)}{s}\right) - \frac{v_C(0^-)}{s}}{sL + R + \frac{1}{sC}}$$

This can be written as a ratio of polynomials in s as

$$I_R(s) = [I_S - i_L(0^-)]\,\frac{s - \frac{v_C(0^-)}{LI_S - Li_L(0^-)}}{s^2 + \frac{R}{L}s + \frac{1}{LC}}$$

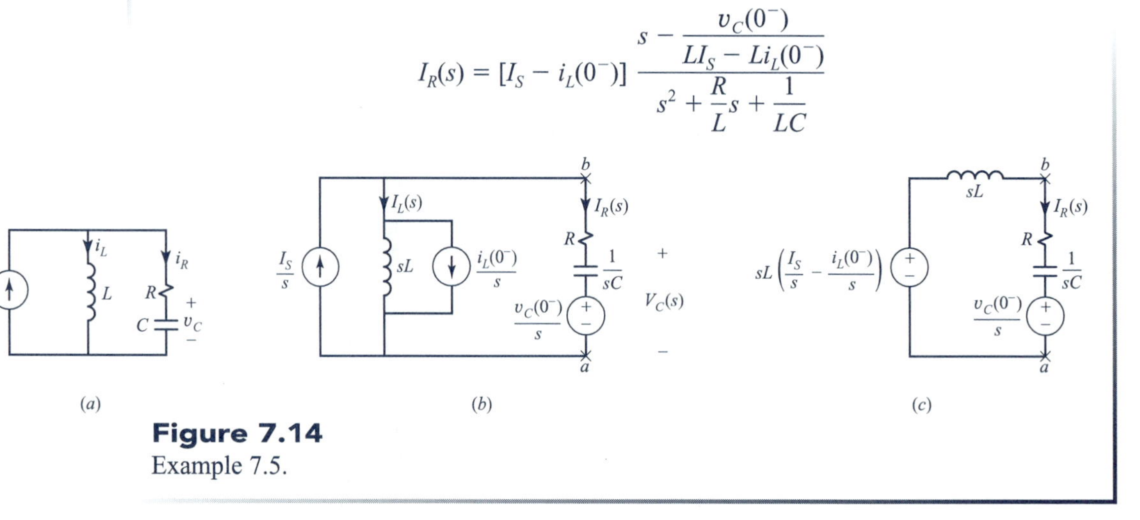

Figure 7.14
Example 7.5.

Example 7.6

Determine the Laplace transform of the inductor current and capacitor voltage in the circuit of Fig. 7.15*a* where the initial conditions are $i_L(0^-) = 2$ A and $v_C(0^-) = 3$ V.

Solution The transformed circuit is shown if Fig. 7.15*b*. Observe that the Laplace transform of an impulse source is the value of the source: $V_S\delta(t) \Leftrightarrow V_S$. The circuit is converted to a single-node-pair circuit in Fig. 7.15*b* using source transformations. From that we obtain

$$\begin{aligned} V_C(s) &= \frac{\frac{5}{3} - \frac{2}{s} + \frac{3}{4}}{\frac{1}{3} + \frac{1}{2s} + \frac{s}{4}} \\ &= \frac{29s - 24}{3s^2 + 4s + 6} \\ &= \frac{29}{3}\,\frac{s - \frac{24}{29}}{s^2 + \frac{4}{3}s + 2} \end{aligned}$$

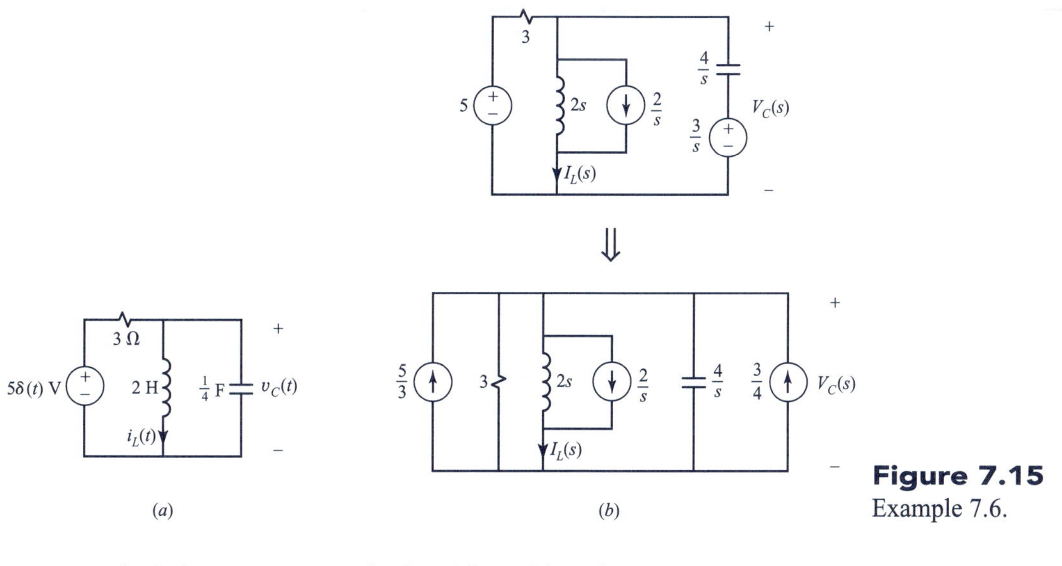

Figure 7.15 Example 7.6.

The inductor current can be found from this and KCL as

$$
\begin{aligned}
I_L(s) &= \frac{V_C(s)}{2s} + \frac{2}{s} \\
&= \frac{12s^2 + 45s}{2s(3s^2 + 4s + 6)} \\
&= \frac{6s + 22.5}{3s^2 + 4s + 6} \\
&= 2\,\frac{s + 3.75}{s^2 + \frac{4}{3}s + 2}
\end{aligned}
$$

Example 7.7

Determine the Laplace transform of the resistor current in the circuit of Fig. 7.16*a* where the initial condition is $v_C(0^-) = 2$ V.

Solution The transformed circuit is shown if Fig. 7.16*b*. The circuit is converted to a single-node-pair circuit in Fig. 7.16*b* using a source transformation. From that we obtain, using current division,

$$
\begin{aligned}
I_R(s) &= \frac{3 + \dfrac{1}{3s}}{3 + \dfrac{1}{3s} + 4}\left(\frac{2}{s} + \frac{6}{1 + 9s}\right) \\
&= \frac{24s + 2}{s(21s + 1)} \\
&= \frac{8}{7}\,\frac{s + \frac{1}{12}}{s^2 + \frac{1}{21}s}
\end{aligned}
$$

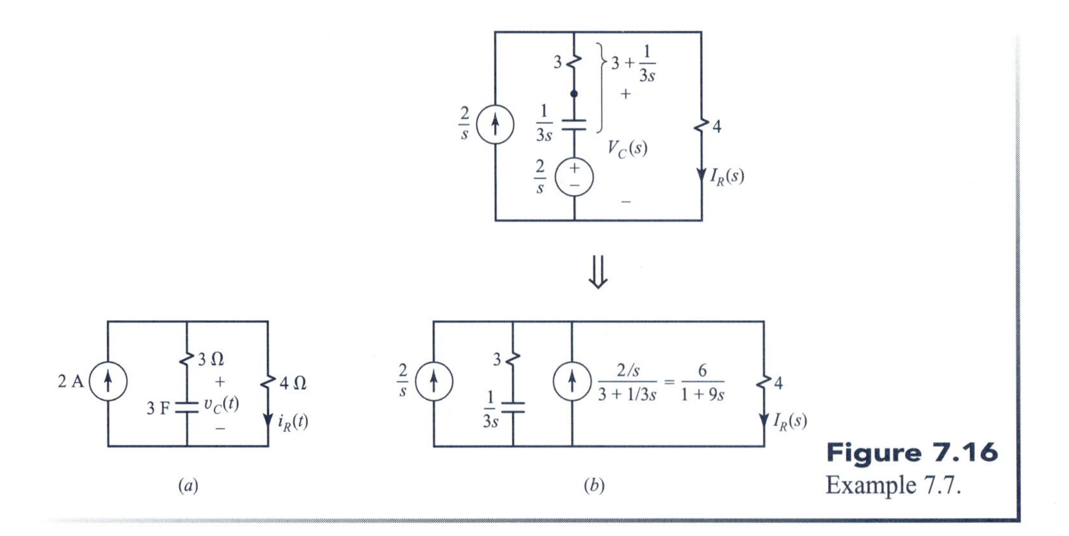

Figure 7.16
Example 7.7.

Exercise Problem 7.9

Determine the Laplace transform of the inductor current and the capacitor voltage of Fig. E7.9 where the initial conditions are $i_L(0^-) = 2$ A and $v_C(0^-) = 1$ V.

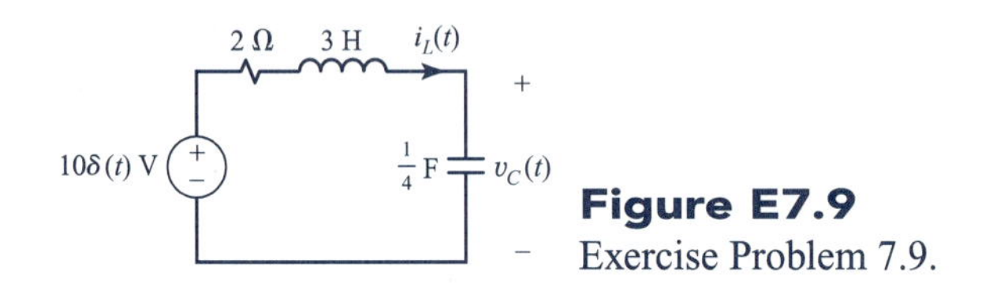

Figure E7.9
Exercise Problem 7.9.

Answer: $I_L(s) = (16s - 1)/(3s^2 + 2s + 4)$, $V_C(s) = (3s + 66)/(3s^2 + 2s + 4)$.

Exercise Problem 7.10

Determine the Laplace transform of the capacitor voltage of Fig. E7.10 where the initial condition is $v_C(0^-) = 2$ V.

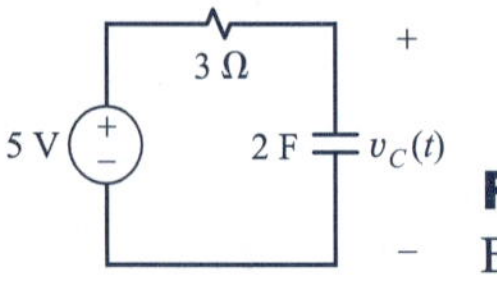

Figure E7.10
Exercise Problem 7.10.

Answer: $V_C(s) = (12s + 5)/s(6s + 1)$.

Exercise Problem 7.11

Determine the Laplace transform of the inductor current of Fig. E7.11 where the initial condition is $i_L(0^-) = 2$ A.

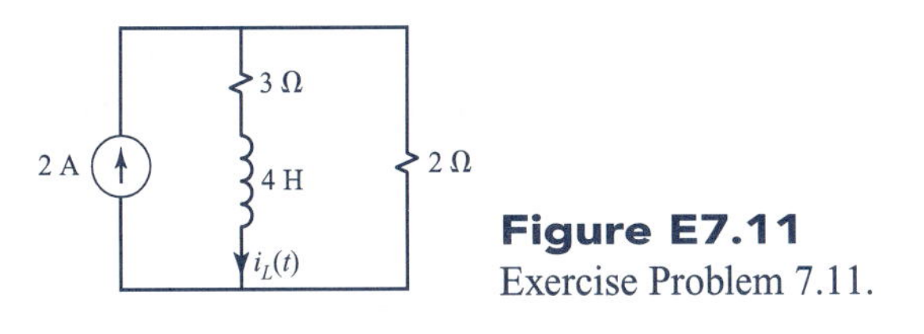

Figure E7.11
Exercise Problem 7.11.

Answer: $I_L(s) = (8s + 4)/s(4s + 5)$.

7.3.4 Determining Initial Conditions

The Laplace-transformed models of the inductors and capacitors in Fig. 7.12 require that we determine the inductor current and capacitor voltage at $t = 0^-$, $i_L(0^-)$ and $v_C(0^-)$, i.e., the *initial conditions*. The circuits we will be considering have some disturbance occurring at $t = 0$. Either a switch opens or closes or a source turns on or turns off at $t = 0$. Hence these initial conditions are the inductor current and capacitor voltage *immediately prior to the disturbance*, i.e., at $t = 0^-$. Again, we denote the time immediately after the disturbance has occurred as $t = 0^+$.

In the remainder of the text we will assume that the circuit for $t < 0$, prior to the disturbance, contains only dc sources. Hence, in order to obtain the initial conditions we replace inductors with short circuits and capacitors with open circuits in the $t < 0$ circuit and solve the resulting resistive circuit.

Example 7.8

Determine the inductor current and capacitor voltage in the circuit of Fig. 7.17*a* at $t = 0^-$. The switch opens at $t = 0$.

Solution The $t < 0$ circuit (prior to switch opening) is shown in Fig. 7.17*b*. Since the source in that circuit is dc, we replace the inductor with a short circuit and the capacitor with an open circuit and obtain

$$\begin{aligned} i_L(0^-) &= \frac{2\,\Omega}{4\,\Omega + 2\,\Omega} \times 5\text{ A} \\ &= \tfrac{5}{3}\text{ A} \\ v_C(0^-) &= (2\,\Omega \parallel 4\,\Omega) \times 5\text{ A} \\ &= \tfrac{20}{3}\text{ V} \end{aligned}$$

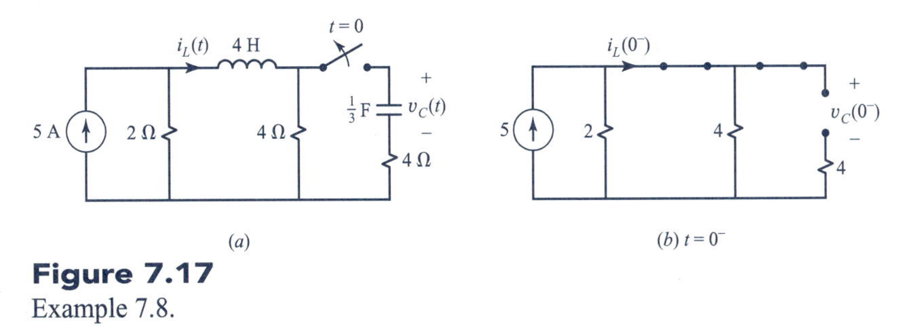

Figure 7.17
Example 7.8.

Example 7.9

Determine the inductor current and capacitor voltage in the circuit of Fig. 7.18a at $t = 0^-$. The switch opens at $t = 0$.

Solution The $t < 0$ circuit (prior to switch opening) is shown in Fig. 7.18b. Since the source in that circuit is dc, we replace the inductor with a short circuit and the capacitor with an open circuit and obtain

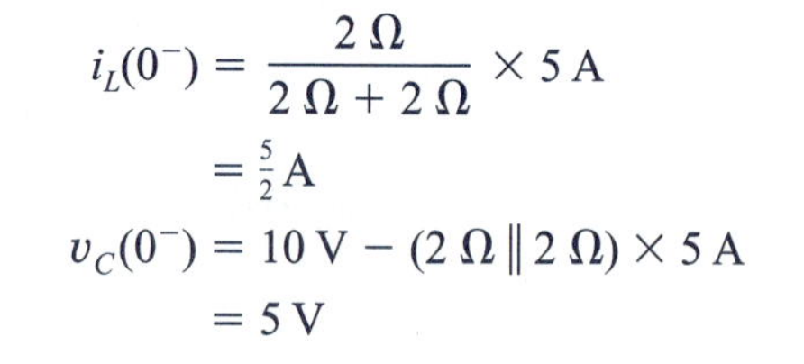

$$i_L(0^-) = \frac{2\,\Omega}{2\,\Omega + 2\,\Omega} \times 5\text{ A}$$
$$= \tfrac{5}{2}\text{ A}$$
$$v_C(0^-) = 10\text{ V} - (2\,\Omega \,\|\, 2\,\Omega) \times 5\text{ A}$$
$$= 5\text{ V}$$

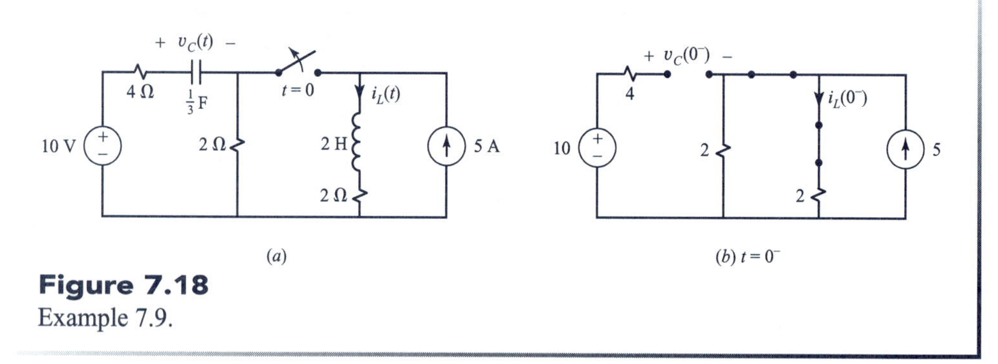

Figure 7.18
Example 7.9.

Example 7.10

Determine the inductor current in the circuit of Fig. 7.19a at $t = 0^-$. The sinusoidal source turns on at $t = 0$.

Solution Observe that the sinusoidal source is multiplied by the unit step function. Hence its value for $t < 0$ is zero, and so it is replaced by a short circuit in the $t < 0$ circuit as shown in Fig. 7.19b. The only source in that circuit is a dc source and hence the inductor is replaced with a short circuit, giving

$$i_L(0^-) = 1\text{ A}$$

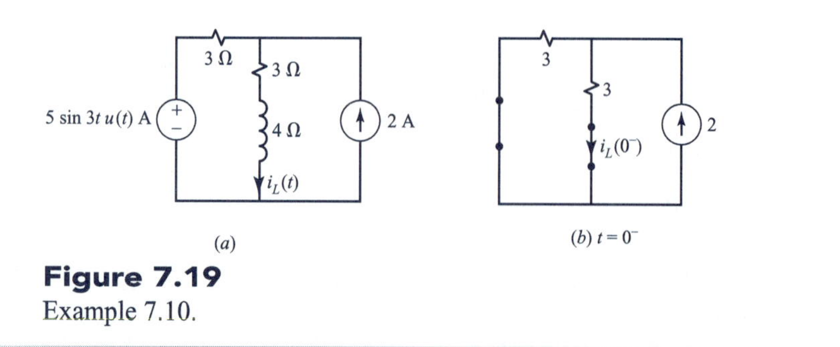

Figure 7.19
Example 7.10.

Exercise Problem 7.12

Determine the initial inductor current and capacitor voltage in the circuit of Fig. E7.12. The switch opens at $t = 0$.

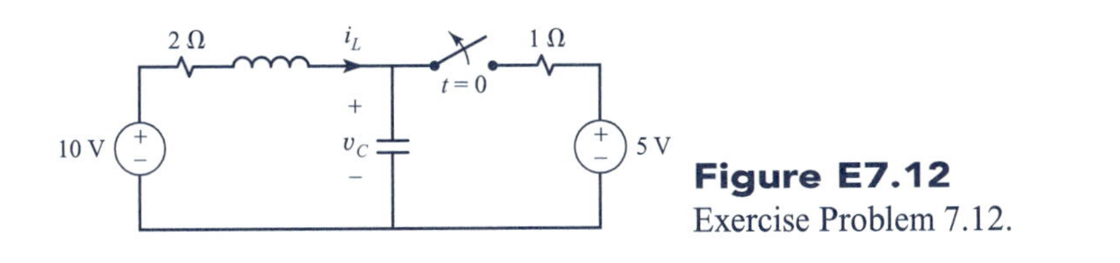

Figure E7.12
Exercise Problem 7.12.

Answers: $i_L(0^-) = \frac{5}{3}$ A, $v_C(0^-) = \frac{20}{3}$ V.

Exercise Problem 7.13

Determine the initial inductor current and capacitor voltage in the circuit of Fig. E7.13. The switch closes at $t = 0$.

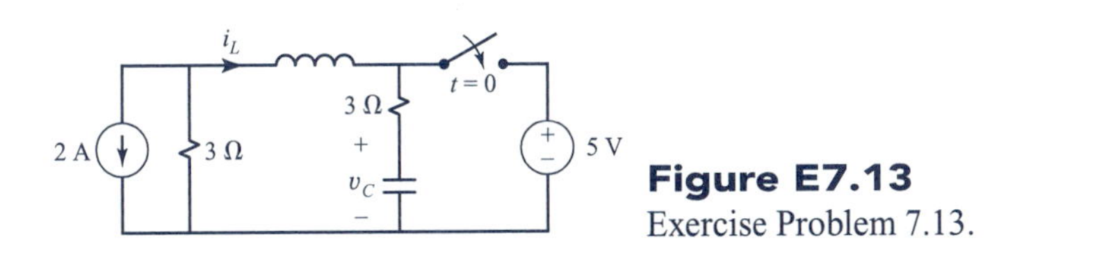

Figure E7.13
Exercise Problem 7.13.

Answers: $i_L(0^-) = 0$, $v_C(0^-) = -6$ V.

Exercise Problem 7.14

Determine the initial inductor current and capacitor voltage in the circuit of Fig. E7.14. The switch opens at $t = 0$.

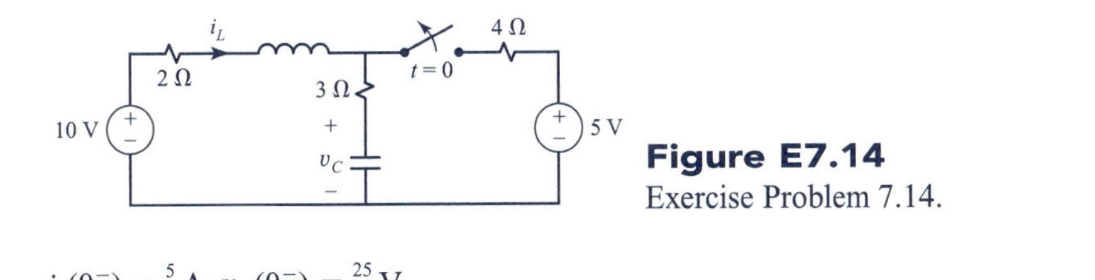

Figure E7.14
Exercise Problem 7.14.

Answers: $i_L(0^-) = \frac{5}{6}$ A, $v_C(0^-) = \frac{25}{3}$ V.

7.4 The Inverse Laplace Transform by Partial-Fraction Expansion

In order to obtain the Laplace transform of any voltage or current, we substitute the Laplace-transform equivalent circuits of the elements shown in Fig. 7.12 into the $t > 0$ circuit and transform all independent sources. This gives a circuit in which we can use all of our resistive-circuit analysis methods in order to determine the Laplace transform of the desired variable. The essential question once we obtain the Laplace transform of the desired variable is how to obtain the inverse Laplace transform of that function and hence return to the time domain. That final and important task is addressed in this section. We will use a simple method known as *the method of partial-fraction expansion* to expand that function as a sum of terms for which we know the inverse transforms.

Before we begin we need to state some important and fundamental facts about the Laplace transform of any circuit variable. These are proven in C.R. Paul, *Analysis of Linear Circuits*, McGraw-Hill, 1989.

> *The Laplace transform of any current or voltage of a circuit will be a ratio of two polynomials in s of the form*
>
> $$F(s) = K\frac{s^m + b_1 s^{m-1} + \cdots + b_m}{s^n + a_1 s^{n-1} + \cdots + a_n} \tag{7.84}$$
>
> *where the constant K and the coefficients a_i and b_i are real numbers.*

See Examples 7.5, 7.6, and 7.7. Observe that the coefficients of the highest powers of the numerator and denominator polynomials, s^m and s^n, are unity. It is good idea to achieve this form, for reasons that will soon become apparent. With the exception of a few cases of limited engineering interest, *the order of the numerator polynomial, m, is at most equal to the order of the denominator polynomial, n, i.e., $m \leq n$.* Another important fact is that

> *the denominator polynomial of (7.84) can be factored into n roots which are either (a) real and distinct (not appearing elsewhere), or (b) real and repeated, or (c) complex and in conjugate pairs. Hence the denominator polynomial of (7.84) can be factored into its n roots or poles, $p_1, p_2, \ldots, p_n$, as* (7.85)
>
> $$F(s) = K\frac{s^m + b_1 s^{m-1} + \cdots + b_m}{(s-p_1)(s-p_2)\cdots(s-p_n)}$$

This last property is a result of the fact that the coefficients in the denominator polynomial, a_i, are real numbers. The method of partial-fraction expansion allows (7.85) to be expressed as a sum of terms each of which is associated with a root of the denominator polynomial. The n roots of the denominator polynomial are commonly referred to as *poles*.

In order to partial-fraction-expand a ratio of polynomials in s such as in (7.84) or (7.85), the order of the numerator polynomial, m, must be less than the order of the denominator polynomial, n. If, in fact, the orders of the numerator and denominator polynomials are equal, we may use long division to write the polynomial as the sum

$$F(s) = K + \frac{N(s)}{D(s)} \qquad m = n \tag{7.86}$$

where the order of $N(s)$ is less than that of $D(s)$. The inverse transform of (7.86) is

$$f(t) = K\delta(t) + \mathcal{L}^{-1}\left\{\frac{N(s)}{D(s)}\right\} \qquad m = n \tag{7.87}$$

Hence, *if the numerator and denominator orders are equal, the time-domain response will contain an impulse whose strength is the constant K.* It is very important not to omit the impulse term, and it is a simple matter to determine if it exists in the inverse transform: if the orders of the numerator and denominator polynomials are equal it will be present; otherwise, it will not be present. Our remaining task will be to determine the partial-fraction expansion of the term $N(s)/D(s)$ and determine its inverse Laplace transform.

Because the coefficients of the denominator polynomial in (7.84) are real, the fundamental theorem of algebra again provides that the n roots (poles) of the denominator polynomial in (7.85), $p_1, p_2, \ldots, p_n$, will occur only as:

1. *real and distinct,*
2. *real and repeated, or*
3. *complex conjugate pairs.*

We now examine each of these cases.

7.4.1 Real and Distinct Poles Suppose that one of the poles is real and does not recur among the other poles: $p_i = a$. For this case the function can be expanded as

$$F(s) = \frac{A}{s-a} + F'(s) \tag{7.88}$$

where the remainder term, $F'(s)$, does not contain this pole. The task is to determine the coefficient A in this part of the expansion. To do so, multiply (7.88) by $s - a$ to obtain

$$(s-a)F(s) = A + (s-a)F'(s) \tag{7.89}$$

Multiplying $F(s)$ by $s - a$ removes that pole factor from the denominator. This is sometimes known as the *cover-up rule* in that to obtain $(s - a)\,F(s)$ we simply cover up $(s - a)$ in the denominator of $F(s)$. If we now let $s = a$, we obtain

$$A = [(s-a)F(s)]\big|_{s=a} \tag{7.90}$$

This result is due to the fact that $F'(s)$ does not contain the pole $s = a$, since the root was stipulated to be distinct, i.e., nonrepeated, and hence

$$[(s-a)F'(s)]\big|_{s=a} = 0$$

The contribution to the inverse transform of this is

$$f(t) = Ae^{at}\,u(t) + \mathscr{L}^{-1}\{F'(s)\} \tag{7.91}$$

Example 7.11

Determine the partial fraction expansion and inverse transform of the transform

$$F(s) = 5\,\frac{s+3}{s(s+1)(s+2)}$$

Solution First write the anticipated form of the expansion:

$$F(s) = \frac{A}{s} + \frac{B}{s+1} + \frac{C}{s+2}$$

Observe that since the order of the numerator polynomial ($m = 1$) is less than the order of the denominator polynomial ($n = 3$), there is no constant term in the expansion. Hence $f(t)$

will not contain an impulse. The coefficients are computed according to (7.90) as

$$A = \left(5\frac{s+3}{(s+1)(s+2)}\right)\bigg|_{s=0} = \frac{15}{2}$$

$$B = \left(5\frac{s+3}{s(s+2)}\right)\bigg|_{s=-1} = -10$$

$$C = \left(5\frac{s+3}{s(s+1)}\right)\bigg|_{s=-2} = \frac{5}{2}$$

Therefore, the partial-fraction expansion is

$$F(s) = \frac{\frac{15}{2}}{s} + \frac{-10}{s+1} + \frac{\frac{5}{2}}{s+2}$$

It is always a good idea to check such a result by multiplying out the expansion in order to see if it in fact equals the original $F(s)$. The inverse transform of this is

$$f(t) = \left(\tfrac{15}{2} - 10e^{-t} + \tfrac{5}{2}e^{-2t}\right)u(t)$$

Example 7.12

Determine the partial fraction expansion and inverse transform of the transform

$$F(s) = \frac{4s^2+3}{3s^2+12s} = \frac{4}{3}\frac{s^2+\frac{3}{4}}{s(s+4)}$$

Solution First write the anticipated form of the expansion:

$$F(s) = K + \frac{A}{s} + \frac{B}{s+4}$$

Note that since the orders of the numerator and denominator polynomials are equal, a constant is required in the expansion, which is

$$K = \tfrac{4}{3}$$

The remaining expansion coefficients are

$$A = \left(\frac{4s^2+3}{3(s+4)}\right)\bigg|_{s=0} = \frac{1}{4}$$

$$B = \left(\frac{4s^2+3}{3s}\right)\bigg|_{s=-4} = -\frac{67}{12}$$

Therefore, the partial-fraction expansion is

$$F(s) = \frac{4}{3} + \frac{\frac{1}{4}}{s} + \frac{-\frac{67}{12}}{s+4}$$

Again, it is a good idea to check this result by multiplying out the expansion in order to see if it in fact equals the original $F(s)$. The inverse transform of this is

$$f(t) = \tfrac{4}{3}\delta(t) + \left(\tfrac{1}{4} - \tfrac{67}{12}e^{-4t}\right)u(t)$$

Note that since the orders of the numerator and denominator polynomials are equal, the function contains an impulse, and the value of that impulse is the constant K in (7.84).

Exercise Problem 7.15

Determine the inverse transform of the transform

$$F(s) = \frac{6s^3 + 4s + 2}{2s^3 + 14s^2 + 20s}$$

Answer: $f(t) = 3\delta(t) + (\frac{1}{10} + \frac{9}{2}e^{-2t} - \frac{128}{5}e^{-5t})u(t)$.

Exercise Problem 7.16

Determine the inverse transform of the transform

$$F(s) = \frac{3s + 2}{4s^2 + 24s + 32}$$

Answer: $f(t) = (-\frac{1}{2}e^{-2t} + \frac{5}{4}e^{-4t})\,u(t)$.

7.4.2 Real and Repeated Poles Suppose that a pole $p_i = a$ is repeated q times, i.e., appears as $(s - a)^q$ in the denominator of $F(s)$. In this case, $F(s)$ may be expanded as

$$F(s) = \frac{A_1}{s - a} + \frac{A_2}{(s - a)^2} + \cdots + \frac{A_q}{(s - a)^q} + F'(s) \tag{7.92}$$

where $F'(s)$ does not contain this pole. If we multiply this by $(s - a)^q$, we obtain

$$(s - a)^q F(s) = (s - a)^{q-1} A_1 + (s - a)^{q-2} A_2 + \cdots + A_q + (s - a)^q F'(s) \tag{7.93}$$

This removes the repeated pole from the denominator of $F(s)$. Evaluating this at $s = a$ gives the coefficient A_q:

$$A_q = [(s - a)^q F(s)]\big|_{s=a} \tag{7.94}$$

Hence this coefficient can be simply evaluated with the cover-up rule, as for a distinct root. The remaining coefficients are obtained by differentiation. For example, suppose we differentiate (7.93) with respect to s:

$$\begin{aligned} \frac{d}{ds}[(s - a)^q F(s)] &= (q - 1)(s - a)^{q-2} A_1 + (q - 2)(s - a)^{q-3} A_2 + \cdots + A_{q-1} \\ &\quad + \frac{d}{ds}[(s - a)^q F'(s)] \end{aligned} \tag{7.95}$$

so that

$$A_{q-1} = \left(\frac{d}{ds}\left[(s - a)^q F(s)\right]\right)\Bigg|_{s=a} \tag{7.96}$$

In general, the coefficients are obtained from

$$A_{q-k} = \left(\frac{1}{k!}\frac{d^k}{ds^k}[(s-a)^q F(s)]\right)\Bigg|_{s=a} \tag{7.97}$$

The inverse transform is obtained according to (7.60) as

$$f(t) = \left(A_1 + A_2 t + A_3 \frac{1}{2!}t^2 + \cdots + A_q \frac{1}{(q-1)!}t^{q-1}\right) e^{at}\, u(t) + \mathcal{L}^{-1}(F'(s)) \tag{7.98}$$

Example 7.13

Determine the inverse transform of the following transform:

$$F(s) = 5\frac{s^2 + 4}{(s+1)\,(s+2)^2}$$

Solution First write the form of the expansion:

$$F(s) = \frac{B}{s+1} + \frac{A_1}{s+2} + \frac{A_2}{(s+2)^2}$$

Since the numerator polynomial is of order one less than the denominator polynomial, there is no constant term in the expansion. The expansion coefficient for the root $s = -1$ is obtained as before, since it is distinct:

$$\begin{aligned} B &= [(s+1)\,F(s)]\big|_{s=-1} \\ &= 5\left|\frac{s^2+4}{(s+2)^2}\right|_{s=-1} \\ &= 25 \end{aligned}$$

The coefficient of the highest power of the repeated root, A_2, can be found similarly:

$$\begin{aligned} A_2 &= [(s+2)^2\,F(s)]\big|_{s=-2} \\ &= 5\frac{s^2+4}{s+1}\bigg|_{s=-2} \\ &= -40 \end{aligned}$$

The coefficient A_1 can be found using (7.96):

$$\begin{aligned} A_1 &= \left(\frac{d}{ds}\left[(s+2)^2\,F(s)\right]\right)\bigg|_{s=-2} \\ &= \left[\frac{d}{ds}\left(5\frac{s^2+4}{s+1}\right)\right]\bigg|_{s=-2} \\ &= \left[5\frac{(s+1)\,(2s) - (s^2+4)\,(1)}{(s+1)^2}\right]\bigg|_{s=-2} \\ &= -20 \end{aligned}$$

This differentiation is actually unnecessary. We already know B and A_2. So substitute them into the expansion:

$$F(s) = \frac{25}{s+1} + \frac{A_1}{s+2} - \frac{40}{(s+2)^2}$$

To determine A_1 we can simply substitute any convenient value of s (except $s = -1$ or $s = -2$). For example, let $s = 0$, giving

$$F(0) = \frac{25}{1} + \frac{A_1}{2} - \frac{40}{2^2}$$

But

$$F(0) = 5\frac{(4)}{(1)(2)^2}$$
$$= 5$$

Hence

$$5 = 25 + \frac{A_1}{2} - 10$$

or

$$A_1 = -20$$

as was obtained with differentiation. As before, it is a good idea to multiply out the expansion to verify the correctness of the expansion coefficients. Hence, the inverse transform is

$$f(t) = (25e^{-t} - 20e^{-2t} - 40te^{-2t})\,u(t)$$

Exercise Problem 7.17

Determine the inverse transform of the following transform:

$$F(s) = \frac{3s^2 + 4}{s^2 + 6s + 9}$$

Answer: $f(t) = 3\delta(t) + (-18 + 31t)e^{-3t}\,u(t)$.

Exercise Problem 7.18

Determine the inverse transform of the following transform:

$$F(s) = \frac{3s + 4}{2s^2 + 16s + 32}$$

Answer: $f(t) = (\frac{3}{2}e^{-4t} - 4te^{-4t})u(t)$.

7.4.3 Complex Conjugate Poles

The final possibility is complex-valued poles. Because the coefficients of the denominator polynomial are real numbers, any complex pole must be accompanied by its conjugate. Hence we assume the two poles are

$$p_i, p_{i+1} = \alpha + j\beta, \alpha - j\beta$$

For this case the function can be expanded as

$$F(s) = \frac{\hat{A}}{s - \alpha - j\beta} + \frac{\hat{A}^*}{s - \alpha + j\beta} + F'(s) \tag{7.99}$$

where the remainder term, $F'(s)$, does not contain this pole. Observe that the two unknown coefficients are also the conjugates of each other. This is due to the fact that the resulting time-domain function must be real. The task is to determine the complex expansion coefficient $\hat{A}$. Since the two poles are distinct, the earlier method of obtaining the coefficients is applicable here:

$$\hat{A} = (s - \alpha - j\beta)F(s)\big|_{s=\alpha+j\beta} \tag{7.100}$$

$$\hat{A}^* = (s - \alpha + j\beta)F(s)\big|_{s=\alpha-j\beta} \tag{7.101}$$

A very common mistake is to use the wrong pole to evaluate each of these expressions. For example, in (7.100) the factor $s - \alpha - j\beta$ is removed by the cover-up rule and hence we must substitute $s = \alpha + j\beta$ into the result. Hence the inverse transform of (7.99) is

$$\begin{aligned} f(t) &= \hat{A}e^{\alpha t}\, e^{j\beta t}\, u(t) + \hat{A}^* e^{\alpha t} e^{-j\beta t} u(t) + \mathcal{L}^{-1}\{F'(s)\} \\ &= e^{\alpha t}\,(\hat{A}e^{j\beta t} + \hat{A}^* e^{-j\beta t})u(t) + \mathcal{L}^{-1}\{F'(s)\} \end{aligned} \tag{7.102}$$

So determining the expansion for a complex conjugate pole pair is no different than for distinct real roots (so long as this pole pair is not repeated). Although the inverse transform in (7.102) is correct, it is not in a form that indicates that it is real (as it surely must be). Hence our final task is to place the result in a real form. To do so we have two choices. The first choice (preferred by the author) is to use Euler's identity. Denote the complex expansion coefficients as

$$\hat{A} = A_R + jA_I \tag{7.103}$$

where A_R is the real part of $\hat{A}$ and A_I is the imaginary part of $\hat{A}$. Write the desired term as

$$\begin{aligned} e^{\alpha t}(\hat{A}e^{j\beta t} + \hat{A}^* e^{-j\beta t})\,u(t) &= e^{\alpha t}\Big(A_R\underbrace{(e^{j\beta t} + e^{-j\beta t})}_{2\cos\beta t} + jA_I\underbrace{(e^{j\beta t} - e^{-j\beta t})}_{2j\sin\beta t}\Big)u(t) \\ &= 2e^{\alpha t}(A_R\cos\beta t - A_I\sin\beta t)u(t) \end{aligned} \tag{7.104}$$

Alternatively, we could write the expansion coefficient as

$$\hat{A} = Ae^{j\theta} \tag{7.105}$$

so that

$$\begin{aligned} e^{\alpha t}(\hat{A}e^{j\beta t} + \hat{A}^* e^{-j\beta t})u(t) &= e^{\alpha t}\Big(\underbrace{Ae^{j(\beta t+\theta)} + Ae^{-j(\beta t+\theta)}}_{2A\cos(\beta t+\theta)}\Big)u(t) \\ &= 2Ae^{\alpha t}\cos(\beta t + \theta)u(t) \end{aligned} \tag{7.106}$$

Example 7.14

Determine the inverse Laplace transform of the following transform

$$F(s) = 5\frac{s^2 + 2s + 3}{s^2 + 4s + 13}.$$

Solution Before we determine the expansion we should point out a quick way of determining the complex conjugate roots. It is referred to as the method of *completing the square*.

For example, the denominator term can be written as

$$s^2 + 4s + 13 = (s+2)^2 + 9$$
$$= (s+2)^2 + (3)^2$$

Hence, the complex poles are

$$p_1, p_2 = -2 - j3, -2 + j3$$

Observe that the numerator and denominator polynomials are of the same order. Hence the expansion will contain a constant which is 5. Thus the expansion is

$$F(s) = 5 + \frac{\hat{A}}{s+2+j3} + \frac{\hat{A}^*}{s+2-j3}$$

The expansion coefficients are

$$\begin{aligned}
\hat{A} &= (s+2+j3)\,F(s)\big|_{s=-2-j3} \\
&= 5\,\frac{s^2+2s+3}{s+2-j3}\bigg|_{s=-2-j3} \\
&= 5\,\frac{-6+j6}{-j6} \\
&= -5 - j5 \\
&= 5\sqrt{2}\angle -135^\circ \\
\hat{A}^* &= (s+2-j3)\,F(s)\big|_{s=-2+j3} \\
&= 5\,\frac{s^2+2s+3}{s+2+j3}\bigg|_{s=-2+j3} \\
&= 5\,\frac{-6-j6}{j6} + j5 \\
&= -5 + j5 \\
&= 5\sqrt{2}\angle 135^\circ
\end{aligned}$$

Hence the inverse transform is

$$\begin{aligned}
f(t) &= 5\delta(t) + e^{-2t}\left[(-5-j5)e^{-j3t} + (-5+j5)e^{j3t}\right]u(t) \\
&= 5\delta(t) + e^{-2t}\Big[-5\underbrace{\left(e^{-j3t}+e^{j3t}\right)}_{2\cos 3t} - j5\underbrace{\left(e^{-j3t}-e^{j3t}\right)}_{-2j\sin 3t}\Big]u(t) \\
&= 5\delta(t) + e^{-2t}(-10\cos 3t - 10\sin 3t)\,u(t)
\end{aligned}$$

Alternatively, we may write

$$\begin{aligned}
f(t) &= 5\delta(t) + e^{-2t}\left(5\sqrt{2}e^{-j135^\circ}e^{-j3t} + 5\sqrt{2}e^{j135^\circ}e^{j3t}\right)u(t) \\
&= 5\delta(t) + e^{-2t}\Big[5\sqrt{2}\underbrace{\left(e^{-j(3t+135^\circ)} + e^{j(3t+135^\circ)}\right)}_{2\cos(3t+135^\circ)}\Big]u(t) \\
&= 5\delta(t) + 10\sqrt{2}e^{-2t}\cos(3t+135^\circ)\,u(t)
\end{aligned}$$

One can confirm, using trigonometric identities, that these two forms are identical.

Exercise Problem 7.19

Determine the inverse transform of the following transform:

$$F(s) = \frac{2s}{s^2 + 2s + 17}$$

Answer: $f(t) = e^{-t}(2\cos 4t - 0.5\sin 4t)\,u(t)$, or $f(t) = 2.06e^{-t}\cos(4t + 14.04^\circ)\,u(t)$.

Exercise Problem 7.20

Determine the inverse transform of the following transform:

$$F(s) = \frac{s + 2}{2s^2 + 8s + 40}$$

Answer: $f(t) = \frac{1}{2}e^{-2t}\cos 4t\, u(t)$.

An alternative way of obtaining the inverse transform for a complex conjugate pole pair is as follows. Equation (7.99) shows that the contribution due to the complex conjugate pole pair can be written as

$$\begin{aligned} F(s) &= \frac{B(s - \alpha) + C\beta}{(s - \alpha - j\beta)(s - \alpha + j\beta)} + F'(s) \\ &= \frac{B(s - \alpha) + C\beta}{(s - \alpha)^2 + \beta^2} + F'(s) \end{aligned} \tag{7.107}$$

where $B = 2A_R$ and $C = -2A_I$. Once we obtain the expansion coefficients B and C, the inverse transform becomes, according to (7.63) and (7.64),

$$f(t) = (Be^{\alpha t}\cos\beta t + Ce^{\alpha t}\sin\beta t)u(t) + \mathcal{L}^{-1}\{F'(s)\} \tag{7.108}$$

Example 7.15

Determine the inverse transform of the transform of Example 7.14 using the above alternative method.

Solution We attempt to expand the transform as

$$\begin{aligned} F(s) &= 5\frac{s^2 + 2s + 3}{s^2 + 4s + 13} \\ &= 5 + \frac{B(s + 2) + C(3)}{(s + 2)^2 + (3)^2} \end{aligned}$$

Cross-multiplying gives

$$5s^2 + 10s + 15 = 5s^2 + 20s + 65 + Bs + 2B + 3C$$

Matching powers of s yields

$$\begin{aligned} 10 &= 20 + B \\ 15 &= 65 + 2B + 3C \end{aligned}$$

Solving these yields $B = -10$ and $C = -10$, so that

$$f(t) = 5\delta(t) + (-10e^{-2t}\cos 3t - 10e^{-2t}\sin 3t)u(t)$$

as before.

Exercise Problem 7.21

Determine the inverse transform of the transform in Exercise Problem 7.19 using the representation in (7.107).

Answer: $f(t) = e^{-t}(2\cos 4t - 0.5\sin 4t)u(t)$.

Exercise Problem 7.22

Determine the inverse transform of the transform

$$F(s) = \frac{s+1}{s^2 + 2s + 10}$$

Answer: $f(t) = e^{-t}\cos 3t\, u(t)$.

7.5 First-Order Circuit Response

In this section we will consider circuits that contain only one energy storage element: an inductor or a capacitor. These circuits are called *first-order circuits* because the governing differential equations relating a current or voltage to the independent sources in that circuit will be first-order differential equations, i.e., the highest derivative will be the first derivative. These were solved previously using classical solution methods for the differential equation. We will obtain those solutions using the Laplace transform method.

7.5.1 The *RL* Circuit Consider a circuit containing an inductor and a resistor which are connected in series to a dc voltage source as shown in Fig. 7.20*a*. The inductor has an initial current of $i_L(0^-)$. The Laplace-transformed circuit is shown in Fig. 7.20*b*. [Observe that the Laplace-transformed inductor current, $I_L(s)$, is through the parallel combination of the inductor impedance and the source representing its initial condition. A frequent mistake made is to assume that the current $I_L(s)$ is through the impedance sL only.] This is transformed to a single-loop circuit using a source transformation. Using superposition as shown in Fig. 7.20*c*, we obtain the contribution due to the initial inductor current $i_L(0^-)$ as

$$\begin{aligned} I'_L(s) &= \frac{Li_L(0^-)}{R + sL} \\ &= \frac{i_L(0^-)}{\left(s + \frac{R}{L}\right)} \end{aligned} \tag{7.109a}$$

and the contribution due to the dc voltage source, V_S, as

$$\begin{aligned} I''_L(s) &= \frac{V_S/s}{R + sL} \\ &= \frac{V_S/L}{s\left(s + \frac{R}{L}\right)} \\ &= \frac{V_S/R}{s} - \frac{V_S/R}{s + \frac{R}{L}} \end{aligned} \tag{7.109b}$$

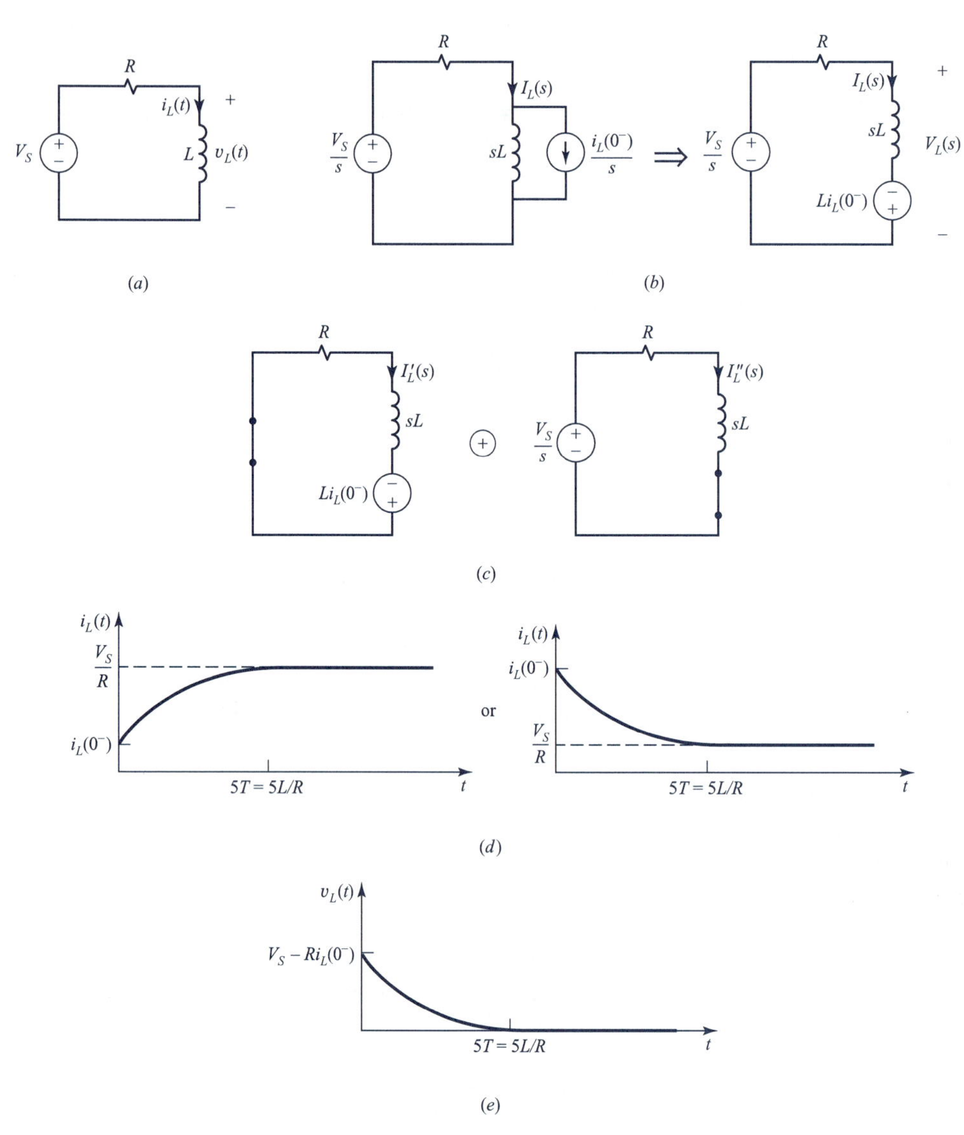

Figure 7.20 Solution of the first-order *RL* circuit with the Laplace transform method: (*a*) the time-domain circuit, (*b*) the Laplace transformed circuit, (*c*) using superposition to determine the portion of the response due to the inductor initial current and the portion of the response due to the open-circuit voltage source (which represents the effect of the actual sources in the circuit), (*d*) plot of the inductor current response, and (*e*) plot of the inductor voltage response.

The sum gives the transform of the inductor current as $I_L(s) = I_L'(s) + I_L''(s)$. The inverse transform is

$$i_L(t) = \underbrace{i_L(0^-)e^{-(R/L)t}}_{\text{zero-input solution, } i'(t)} + \underbrace{\frac{V_S}{R}(1 - e^{-(R/L)t})}_{\text{zero-state solution, } i''(t)} \qquad t > 0 \tag{7.110}$$

The *zero-input* portion of the solution represents the effect of the initial condition on the response when the actual source is deactivated. The *zero-state* portion of the solution represents the effect of the source on the response when the initial conditions are zero. The initial inductor current represents the initial stored energy in the circuit, $w = \frac{1}{2}Li_L^2(0^-)$, and is often referred to as giving the initial *state* of the circuit. This result has shown an important principle:

Initial conditions have the same effect as an independent source and should be treated no differently.

The Laplace transform method makes this immediately transparent, whereas the classical solution method does not. This result can also be written in the form of the sum of the natural and forced responses obtained with the classical solution of the differential equations:

$$
\begin{aligned}
i_L(t) &= \underbrace{\left(i_L(0^-) - \frac{V_S}{R}\right)e^{-Rt/L}}_{\substack{\text{natural or}\\ \text{transient}\\ \text{solution}}} + \underbrace{\frac{V_S}{R}}_{\substack{\text{forced or}\\ \text{steady-state}\\ \text{solution}}} \qquad t>0 \\
&= \underbrace{\left(i_L(0^-) - \frac{V_S}{R}\right)e^{-t/T}}_{\substack{\text{natural or}\\ \text{transient}\\ \text{solution}}} + \underbrace{\frac{V_S}{R}}_{\substack{\text{forced or}\\ \text{steady-state}\\ \text{solution}}} \qquad t>0
\end{aligned}
\tag{7.111}
$$

Again, observe in both cases that the inductor current approaches V_S/R as $t \to \infty$. This behavior can be confirmed by replacing the inductor with a short circuit. Also $i_L(0^-) = i_L(0^+)$ which confirms continuity of the inductor current. The second form of the solution in (7.111) has the natural solution written in terms of the *time constant*

$$T = \frac{L}{R} \qquad \text{s} \tag{7.112}$$

After about 5 time constants, the exponential becomes $e^{-5} = 0.0067$ and hence that portion of the solution goes to zero and the total solution approaches the forced or steady-state solution. Hence the use of the term *transient solution* is justified for this portion of the response. The solution in (7.111) is plotted in Fig. 7.20*d*.

Other voltage and current solutions can be obtained either directly from the Laplace-transformed circuit or directly from the inductor-current solution and KCL, KVL, and the element relations. For example, suppose we wish to obtain the inductor voltage. The transform of the inductor voltage can be obtained directly from Fig. 7.20*b* by superposition and voltage division as

$$
\begin{aligned}
V_L(s) &= \frac{sL}{R+sL}\left(\frac{V_S}{s}\right) - \frac{R}{R+sL}Li_L(0^-) \\
&= \frac{V_S - Ri_L(0^-)}{s + \dfrac{R}{L}}
\end{aligned}
\tag{7.113}
$$

The inverse transform of this is

$$v_L(t) = [V_S - Ri_L(0^-)]e^{-Rt/L} \qquad t>0 \tag{7.114}$$

This result is plotted in Fig. 7.20*e*. Alternatively we could differentiate the current expression in (7.110) or (7.111) using the element relation $v_L(t) = L\, di_L(t)/dt$ and obtain the same result. Observe in this result that the inductor voltage decays to zero, i.e., the steady-state value of the inductor voltage is zero. This is logical to expect, since the source in the $t > 0$ circuit, V_S, is dc and hence we replace the inductor in the $t > 0$ circuit with a short circuit to obtain this steady-state value.

Example 7.16

Determine the inductor current and resistor current for $t > 0$ in the circuit of Fig. 7.21*a*, which was solved in Example 7.1 by classical solution methods. The switch closes at $t = 0$.

Solution The circuit at $t = 0^-$ is shown in Fig. 7.21*b*. It is constructed by opening the switch and replacing the inductor with a short circuit, since the source in that circuit is dc. This gives the value of the inductor current at $t = 0^-$ as

$$i_L(0^-) = 5 \text{ A} \\ = i_L(0^+)$$

The Laplace-transformed $t > 0$ circuit is drawn in Fig. 7.21*c* and converted with a source transformation to a single-loop circuit. From that we obtain

$$I_L(s) = \frac{\frac{10}{s} + 15}{3s + 2 \parallel 2} \\ = \frac{5s + \frac{10}{3}}{s\left(s + \frac{1}{3}\right)}$$

and

$$I_R(s) = \tfrac{1}{2} I_L(s)$$

The inductor current is expanded in partial fractions to yield

$$I_L(s) = -\frac{5}{s + \frac{1}{3}} + \frac{10}{s}$$

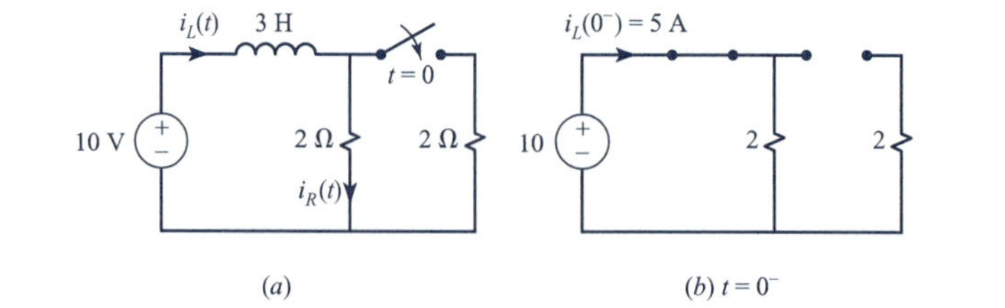

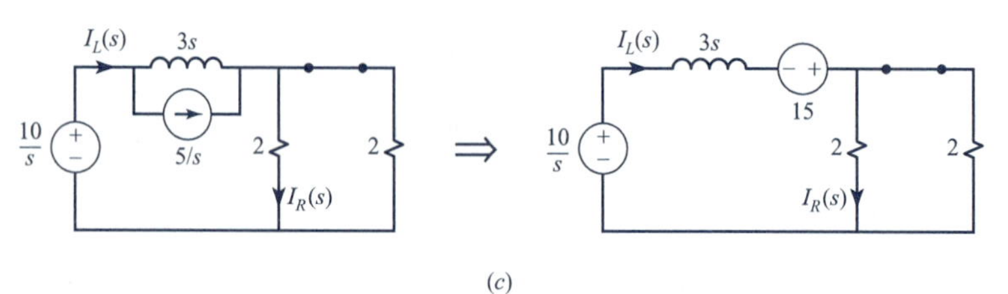

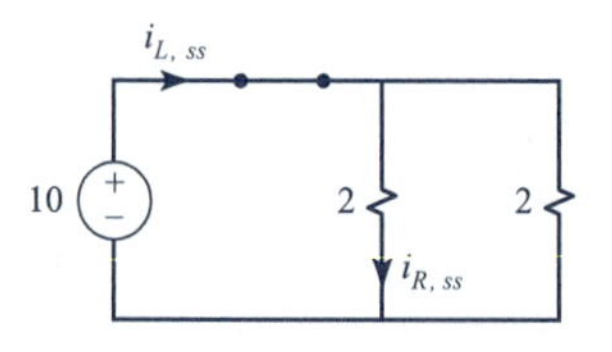

Figure 7.21
Example 7.16.

The inverse transform is

$$i_L(t) = -5e^{-t/3} + 10 \text{ A} \qquad t > 0$$

as was obtained in Example 7.1. Evidently the time constant is

$$T = \frac{L}{R}$$

$$= \frac{3 \text{ H}}{2 \,\|\, 2\Omega}$$

$$= 3 \text{ s}$$

The resistor current is

$$i_R(t) = \tfrac{1}{2}\, i_L(t)$$

$$= -\tfrac{5}{2}\, e^{-t/3} + 5 \text{ A} \qquad t > 0$$

which was also obtained in Example 7.1 As a check on this result we can directly compute the steady-state portion of the response from the $t > 0$ circuit by replacing the inductor with a short circuit as shown in Fig. 7.21*d*, obtaining $i_{L,ss} = 10$ A and $i_{R,ss} = 5$ A, as is obtained from the above solutions as $t \to \infty$. Also $i_L(0^-) = 5 \text{ A} = i_L(0^+)$.

Exercise Problem 7.23

Determine the inductor current and resistor voltage for $t > 0$ in the circuit of Fig. E7.1 using the Laplace transform.

Answer: $i_L(t) = 2e^{-8t/5} + 2$ A, $v_R(t) = 8 - 8e^{-8t/5}$ V, $t > 0$.

7.5.2 The *RC* Circuit In a circuit containing a single capacitor, the procedure is essentially the same as for a circuit containing a single inductor. We determine the capacitor voltage at $t = 0^-$, $v_C(0^-)$, by replacing the capacitor in the $t < 0$ circuit with an open circuit [assuming the source(s) in the $t < 0$ circuit are all dc]. Then we draw the transformed $t > 0$ circuit. We obtain the transform of the desired voltage or current in this circuit *using only resistive-circuit analysis techniques,* expand the result in partial fractions, and obtain the inverse transform valid for $t > 0$.

Consider a circuit containing one capacitor that is in parallel with a resistor and a dc current source as shown in Fig. 7.22*a*, where the inital capacitor voltage is $v_C(0^-)$. The transformed circuit is shown in Fig. 7.22*b* and is reduced, using a source transformation, to a single-node-pair circuit. Recognizing again that there are two sources, the initial capacitor voltage and the independent source, we obtain, by superposition, the contribution due to the initial capacitor voltage,

$$V_C'(s) = \left(R \,\|\, \frac{1}{sC}\right) \times Cv_C(0^-)$$

$$= \frac{v_C(0^-)}{s + \dfrac{1}{RC}} \tag{7.115a}$$

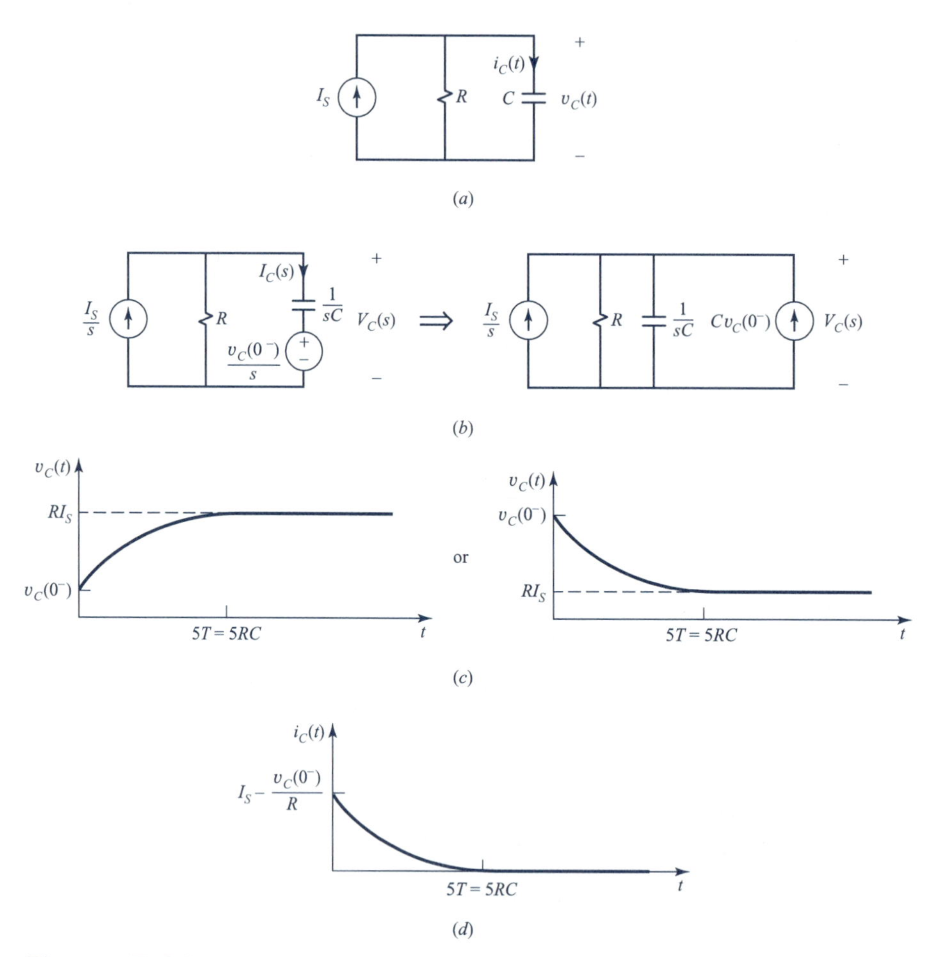

Figure 7.22
Solution of the first-order *RC* circuit with the Laplace transform method: (*a*) the time-domain circuit, (*b*) the Laplace transformed circuit, (*c*) plot of the capacitor voltage response, and (*d*) plot of the capacitor current response.

and the contribution due to the dc current source, I_S,

$$
\begin{aligned}
V_C''(s) &= \left(R \parallel \frac{1}{sC}\right) \times \frac{I_S}{s} \\
&= \frac{\frac{I_S}{C}}{s\left(s + \frac{1}{RC}\right)} \\
&= \frac{RI_S}{s} - \frac{RI_S}{s + \frac{1}{RC}}
\end{aligned}
\tag{7.115b}
$$

The total capacitor voltage is the sum of these contributions, $V_C(s) = V_C'(s) + V_C''(s)$, whose inverse transform is

$$v_C(t) = \underbrace{[v_C(0^-)]e^{-t/RC}}_{\text{zero-input solution, } v_C'(t)} + \underbrace{RI_S(1 - e^{-t/RC})}_{\text{zero-state solution, } v_C''(t)} \qquad t > 0 \tag{7.116}$$

which is again in the form of the zero-input solution and the zero-state solution. The Laplace transform method again makes it immediately clear that the initial voltage of the capacitor should be treated as though it were an actual source. Again this may be written in the form of the sum of the natural and forced solutions obtained with the classical solution of the differential equation:

$$\begin{aligned} v_C(t) &= \underbrace{[v_C(0^-) - RI_S]e^{-t/RC}}_{\text{natural or transient solution}} + \underbrace{RI_S}_{\text{forced or steady-state solution}} \qquad t > 0 \\ &= \underbrace{[v_C(0^-) - RI_S]e^{-t/T}}_{\text{natural or transient solution}} + \underbrace{RI_S}_{\text{forced or steady-state solution}} \qquad t > 0 \end{aligned} \tag{7.117}$$

Again, it is wise to check any result to insure that it (a) provides for continuity of capacitor voltage, and (b) yields the steady-state result that could be obtained from the $t > 0$ circuit by replacing the capacitor with an open circuit [assuming that the source(s) in the $t > 0$ circuit are dc]. The steady-state solution is $v_{C,ss} = RI_S$, which can be obtained by replacing the capacitor with an open circuit. The second form of the solution in (7.117) has the natural solution written in terms of the *time constant*

$$T = RC \qquad \text{s} \tag{7.118}$$

After about 5 time constants, the exponential becomes $e^{-5} = 0.0067$ and hence that portion of the solution goes to zero and the total solution approaches the forced or steady-state solution. Hence the use of the term *transient solution* is justified for that portion of the total response. The solution in (7.117) is plotted in Fig. 7.22*c*.

The capacitor current can be found from the transformed circuit or by differentiating (7.116) or (7.117):

$$\begin{aligned} i_C(t) &= C\frac{dv_C(t)}{dt} \\ &= \left(I_S - \frac{v_C(0^-)}{R}\right)e^{-t/RC} \qquad t > 0 \end{aligned} \tag{7.119}$$

From the transformed circuit in Fig. 7.22*b* we obtain, by superposition,

$$\begin{aligned} I_C(s) &= \frac{R}{R + \dfrac{1}{sC}} \times \frac{I_S}{s} - \frac{\dfrac{v_C(0^-)}{s}}{R + \dfrac{1}{sC}} \\ &= \frac{I_S - \dfrac{v_C(0^-)}{R}}{s + \dfrac{1}{RC}} \end{aligned} \tag{7.120}$$

whose inverse transform again gives (7.119). The solution for the capacitor current in (7.119) is plotted in Fig. 7.22*d*.

Example 7.17

Determine the capacitor voltage and resistor current for $t > 0$ in the circuit of Fig. 7.23*a*, which was solved using classical techniques in Example 7.2. The switch closes at $t = 0$.

Solution The initial capacitor voltage is found by replacing the capacitor with an open circuit in the $t < 0$ circuit as shown in Fig. 7.23*b*:

$$v_C(0^-) = 10 \text{ V}$$

The transformed circuit is shown in Fig. 7.23*c* and transformed to a single-loop circuit in Fig. 7.23*d* using a Thevenin reduction. From that circuit we obtain, by superposition,

$$\begin{aligned} V_C(s) &= \frac{\frac{1}{3s}}{2 + \frac{1}{3s}} \times \frac{5}{s} + \frac{2}{2 + \frac{1}{3s}} \times \frac{10}{s} \\ &= \frac{10s + \frac{5}{6}}{s\left(s + \frac{1}{6}\right)} \\ &= \frac{5}{s + \frac{1}{6}} + \frac{5}{s} \end{aligned}$$

The inverse transform of this is

$$v_C(t) = 5e^{-t/6} + 5 \text{ V} \qquad t > 0$$

Evidently the time constant is

$$\begin{aligned} T &= RC \\ &= 4 \| 4\ \Omega \times 3 \text{ F} \\ &= 6 \text{ s} \end{aligned}$$

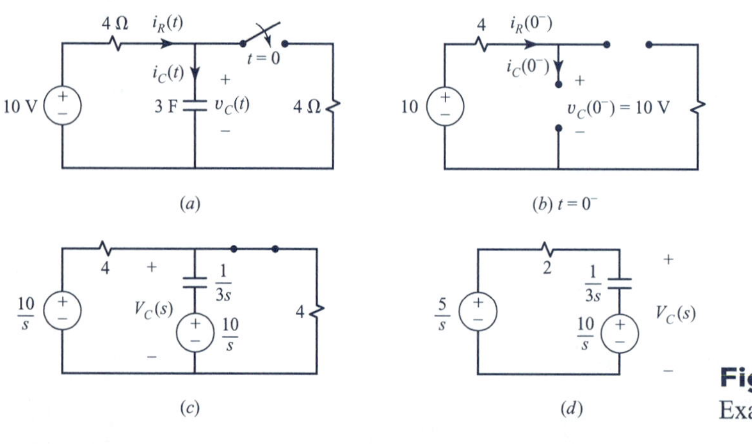

Figure 7.23
Example 7.17.

This solution should be checked for continuity of capacitor voltage, which it satisfies $[v_C(0^+) = 5 + 5 = 10 = v_C(0^-)]$. In addition, the steady-state value of the capacitor voltage is apparently 5V, which can be checked by replacing the capacitor with an open circuit in the $t > 0$ circuit. In order to determine the resistor current we observe that

$$i_R(t) = \frac{10\text{ V} - v_C(t)}{4\,\Omega}$$
$$= -\tfrac{5}{4}e^{-t/6} + \tfrac{5}{4}\text{A} \qquad t > 0$$

and the capacitor current is determined as

$$i_C(t) = 3\frac{dv_C(t)}{dt}$$
$$= 3\frac{d}{dt}\left(5e^{-t/6} + 5\right)$$
$$= -\tfrac{5}{2}e^{-t/6}\text{ A} \qquad t > 0$$

The values of these variables at $t = 0^-$ can be determined from the $t = 0^-$ circuit in Fig. 7.23*b* as $i_C(0^-) = 0$ and $i_R(0^-) = 0$. Observe that the capacitor current is discontinuous at $t = 0$.

Exercise Problem 7.24

Determine the capacitor voltage and current for $t > 0$ in the circuit of Fig. E7.2 using the Laplace transform.

Answers: $v_C(t) = \left(-2e^{-t/8} + 5\right)$ V, $i_C(t) = e^{-t/8}$ A, $t > 0$.

7.5.3 Additional Examples of First-Order Circuit Response

Example 7.18

Determine the inductor current and the resistor current in the circuit of Fig. 7.24*a*.

Solution The circuit for $t < 0$ is shown in Fig. 7.24*b*, where, because both sources are dc, we have replaced the inductor with a short circuit. From that circuit we obtain $i_L(0^-) = \frac{9}{5}$ A. The Laplace-transformed circuit for $t > 0$ is shown in Fig. 7.24*c*. Applying source transformations yields the single-loop circuit of Fig. 7.24*d*, from which we obtain

$$I_L(s) = \frac{\dfrac{10}{s} + \dfrac{27}{5}}{4 + 2 + 3s}$$
$$= \frac{9}{5}\,\frac{s + \frac{50}{27}}{s(s+2)}$$

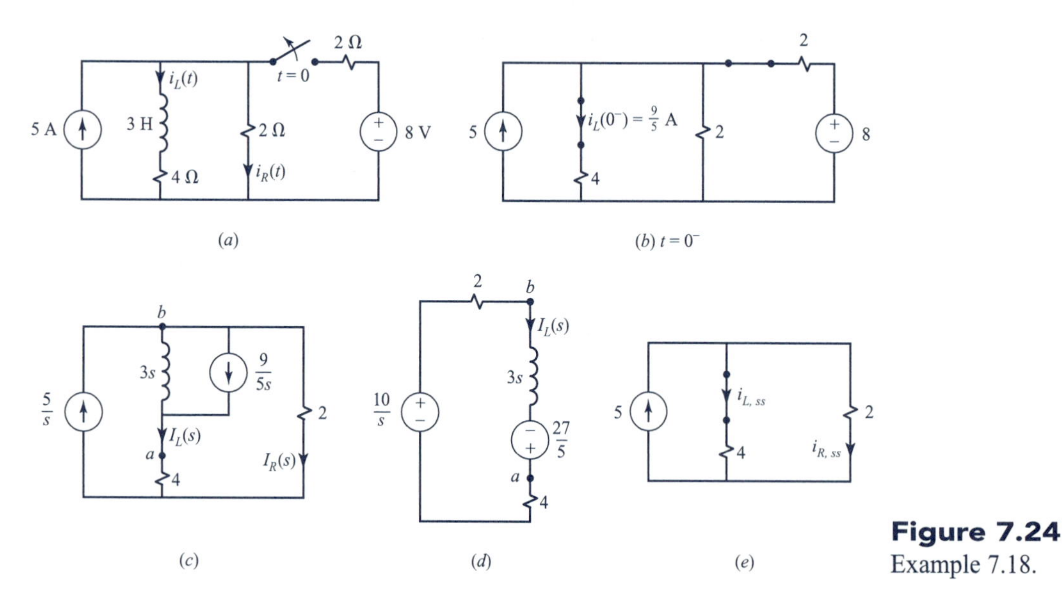

Figure 7.24
Example 7.18.

This is partial-fraction expanded as

$$I_L(s) = \frac{\frac{2}{15}}{s+2} + \frac{\frac{5}{3}}{s}$$

Hence the inverse transform is

$$i_L(t) = \tfrac{2}{15} e^{-2t} + \tfrac{5}{3}\,\text{A} \qquad t \geq 0$$

The resistor current is

$$\begin{aligned} i_R(t) &= 5 - i_L(t) \\ &= -\tfrac{2}{15} e^{-2t} + \tfrac{10}{3}\,\text{A} \qquad t > 0 \end{aligned}$$

The steady-state solutions here, $i_{L,ss} = \frac{5}{3}$ A and $i_{R,ss} = \frac{10}{3}$ A, can be confirmed by replacing the inductor with a short circuit in the $t > 0$ circuit as shown in Fig. 7.24*e*. The initial condition can be confirmed by evaluating the result at $t = 0^+$: $i_L(0^+) = \frac{2}{15} + \frac{5}{3} = \frac{9}{5}$ A.

Sinusoidal sources in the $t > 0$ circuit are also readily handled.

Example 7.19

Determine the capacitor voltage for $t > 0$ in the circuit of Fig. 7.25*a*.

Solution The $t < 0$ circuit is shown in Fig. 7.25*b*, where we have replaced the inductor with a short circuit, since in that circuit the source is dc. This yields $v_C(0^-) = 10$ V. The transformed $t > 0$ circuit is shown in Fig. 7.25*c*, from which we obtain, by superposition,

$$\begin{aligned} V_C(s) &= \frac{\frac{1}{3s}}{\frac{1}{3s} + 2 + 2} \frac{5}{s^2+1} + \frac{2+2}{\frac{1}{3s} + 2 + 2} \frac{10}{s} \\ &= 10 \frac{s^2 + \frac{25}{24}}{(s + \frac{1}{12})(s^2+1)} \end{aligned}$$

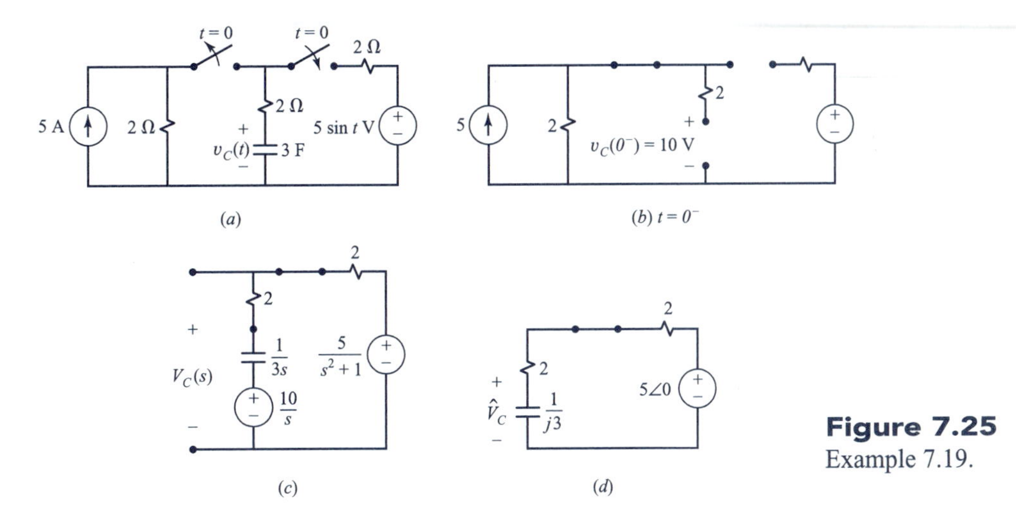

Figure 7.25
Example 7.19.

This is partial-fraction expanded as

$$V_C(s) = \frac{\frac{302}{29}}{s + \frac{1}{12}} + \frac{0.2076\,\angle 175.24°}{} + \frac{0.2076\angle -175.24°}{s - j}$$

Thus the inverse transform is

$$v_C(t) = \tfrac{302}{29}\,e^{-t/12} + 0.4152\cos(t - 175.24°)\text{ V} \qquad t \geq 0$$

The steady-state value $0.4152\cos(t - 175.24°)$ can be confirmed from the phasor circuit shown in Fig. 7.25*d*:

$$\begin{aligned}\hat{V}_C &= \frac{\frac{1}{j3}}{2 + 2 + \frac{1}{j3}}\,5\angle 0° \\ &= \frac{5\angle 0°}{1 + j12} \\ &= 0.4152\angle - 85.24°\end{aligned}$$

But this was for a sine function, and subtracting 90° to convert to the cosine function in the answer gives the same result. The initial condition can be confirmed by evaluating the result at $t = 0^+$: $v_C(0^+) = \frac{302}{29} + 0.4152\cos(-175.24°) = 10$ V.

Exercise Problem 7.25

Determine the resistor current for $t > 0$ in the circuit of Fig. E7.25.

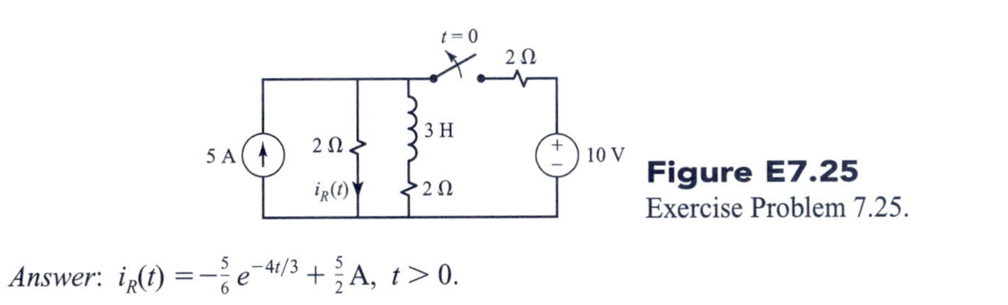

Figure E7.25
Exercise Problem 7.25.

Answer: $i_R(t) = -\frac{5}{6}e^{-4t/3} + \frac{5}{2}$A, $t > 0$.

Exercise Problem 7.26

Determine the resistor current for $t > 0$ in the circuit of Fig. E7.26.

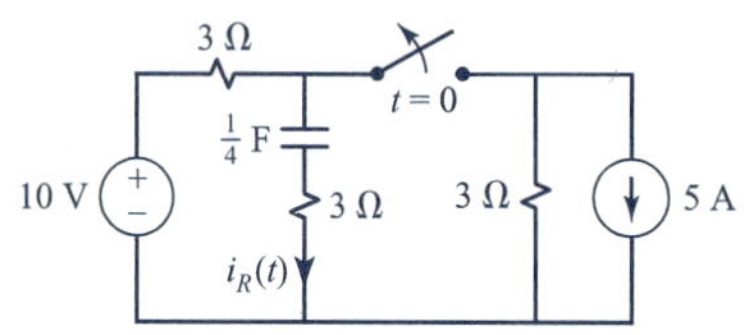

Figure E7.26
Exercise Problem 7.26.

Answer: $i_R(t) = \frac{25}{12} e^{-2t/3}$ A, $t > 0$.

7.6 Second-Order Circuit Response

Next we investigate the solutions for circuits that contain two energy storage elements (two L's, two C's or one L and one C). These are said to be *second-order* circuits because the differential equation relating any voltage or current to the sources is at most second order, i.e., the highest derivative in the differential equation is at most second order. In this section we will investigate the solution for second-order circuits by using the Laplace transform method. There are essentially no differences in the procedure from that for first-order circuits. We determine the initial inductor current and capacitor voltage, $i_L(0^-)$ and $v_C(0^-)$, transform the $t > 0$ circuit, solve for the transform of the desired variable using resistive-circuit analysis techniques, and finally inverse-transform that result using partial-fraction expansion.

Example 7.20

Determine and sketch the solutions for the inductor current and capacitor voltage in the series *RLC* circuit of Fig. 7.26*a* for the following three cases:

1. $V_S = 10$ V, $R = 10\ \Omega$, $L = 2$ H, $C = \frac{1}{8}$ F,
2. $V_S = 10$ V, $R = 4\ \Omega$, $L = \frac{1}{2}$ H, $C = \frac{1}{8}$ F,
3. $V_S = 10$ V, $R = 2\ \Omega$, $L = 2$ H, $C = \frac{1}{25}$ F.

This was solved using classical methods in Example 7.3. The switch closes at $t = 0$.

Solution Because the source is not attached for $t < 0$, the initial conditions are zero, i.e., $i_L(0^-) = 0$ and $v_C(0^-) = 0$. The transformed circuit is shown in Fig. 7.26*b*, from which we obtain

$$I_L(s) = \frac{\frac{V_S}{s}}{R + sL + \frac{1}{sC}} = \frac{1}{s^2 + \frac{R}{L}s + \frac{1}{LC}} \times \frac{V_S}{L}$$

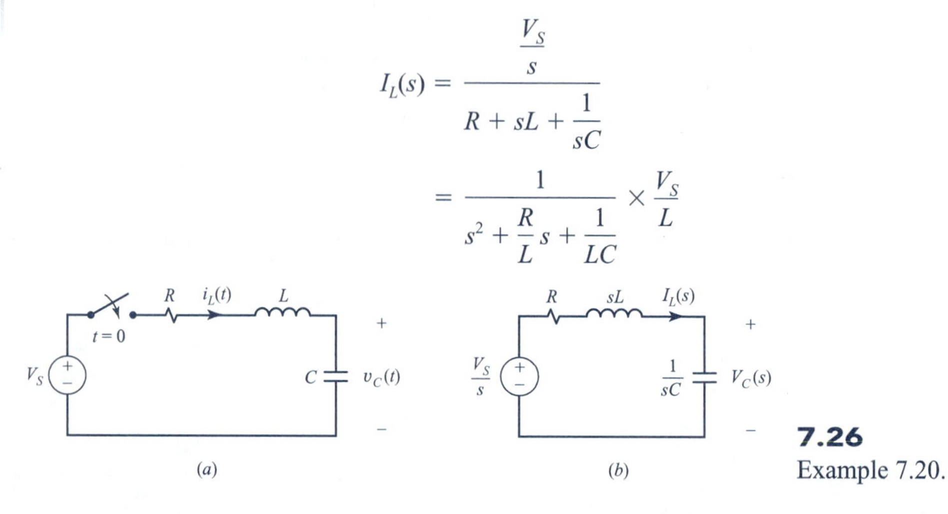

7.26
Example 7.20.

and

$$V_C(s) = \frac{\frac{1}{sC}}{R + sL + \frac{1}{sC}} \times \frac{V_S}{s}$$

$$= \frac{1}{s\left(s^2 + \frac{R}{L}s + \frac{1}{LC}\right)} \times \frac{V_S}{LC}$$

The different forms of the solution will depend on the roots of the characteristic equation

$$s^2 + \frac{R}{L}s + \frac{1}{LC}$$

Case (a): For this case the characteristic equation becomes

$$s^2 + 5s + 4 = (s + 1)(s + 4) = 0$$

Thus the roots are real and distinct:

$$p_1 = -1$$

$$p_2 = -4$$

The above transformed variables become

$$I_L(s) = \frac{1}{(s + 1)(s + 4)} \times \frac{10}{2}$$

$$= \frac{\frac{5}{3}}{s + 1} - \frac{\frac{5}{3}}{s + 4}$$

$$V_C(s) = \frac{1}{s(s + 1)(s + 4)} \times \frac{10}{\frac{1}{4}}$$

$$= -\frac{\frac{40}{3}}{s + 1} + \frac{\frac{10}{3}}{s + 4} + \frac{10}{s}$$

Hence, the solutions valid for $t \geq 0$ are

$$i_L(t) = \tfrac{5}{3}e^{-t} - \tfrac{5}{3}e^{-4t}\text{ A}$$

$$v_C(t) = -\tfrac{40}{3}e^{-t} + \tfrac{10}{3}e^{-4t} + 10\text{ V}$$

as obtained in Example 7.3.

Case (b): For this case the characteristic equation becomes

$$s^2 + 8s + 16 = (s + 4)(s + 4) = 0$$

Thus the roots are repeated:

$$p_1 = -4$$
$$p_2 = -4$$

The above transformed variables become

$$I_L(s) = \frac{1}{(s+4)^2} \times \frac{10}{\frac{1}{2}}$$
$$= \frac{20}{(s+4)^2}$$
$$V_C(s) = \frac{1}{s(s+4)^2} \times \frac{10}{\frac{1}{16}}$$
$$= -\frac{10}{s+4} - \frac{40}{(s+4)^2} + \frac{10}{s}$$

Hence, the solutions valid for $t \geq 0$ are

$$i_L(t) = 20te^{-4t}\ \text{A}$$
$$v_C(t) = -10e^{-4t} - 40te^{-4t} + 10\ \text{V}$$

as obtained in Example 7.3.

Case (*c*): For this case the characteristic equation becomes

$$s^2 + s + 12.5 = (s + 0.5 + j3.5)\,(s + 0.5 - j3.5) = 0$$

Thus the roots are complex conjugate:

$$p_1 = -0.5 - j3.5$$
$$p_2 = -0.5 + j3.5$$

The above transformed variables become

$$I_L(s) = \frac{1}{(s + 0.5 + j3.5)\,(s + 0.5 - j3.5)} \times \frac{10}{2}$$
$$= \frac{j\frac{5}{7}}{s + 0.5 + j3.5} + \frac{-j\frac{5}{7}}{s + 0.5 - j3.5}$$
$$V_C(s) = \frac{1}{s(s + 0.5 + j3.5)\,(s + 0.5 - j3.5)} \times \frac{10}{\frac{2}{25}}$$
$$= \frac{-5 - j\frac{5}{7}}{s + 0.5 + j3.5} + \frac{-5 + j\frac{5}{7}}{s + 0.5 - j3.5} + \frac{10}{s}$$

Hence, the solutions valid for $t \geq 0$ are

$$i_L(t) = e^{-0.5t}\left(\tfrac{10}{7} \sin 3.5t\right) \text{A}$$
$$v_C(t) = e^{-0.5t}\,(-10\ \cos 3.5t - \tfrac{10}{7} \sin 3.5t) + 10\ \text{V}$$

as were obtained in Example 7.3.

In all three examples, the reader should verify continuity of inductor currents and capacitor voltages, $i_L(0^-) = i_L(0^+)$ and $v_C(0^-) = v_C(0^+)$, as well as their steady-state values as $t \rightarrow \infty$, $v_{C,ss} = 10$ V and $i_{L,ss} = 0$ A.

The inductor current solutions are plotted in Fig. 7.8*a*, and the capacitor voltage solutions are plotted in Fig. 7.8*b*. Observe in these plots that for the case of real distinct roots the solution slowly converges to the steady-state value and hence this is referred to as the *overdamped* case. For the case of real, repeated roots, the solution converges rather rapidly to the steady-state solution and hence is referred to as the *critically damped* case. For the case of complex conjugate roots, the solution oscillates about the steady-state value eventually converging to it. This is why this case is referred to as the *underdamped* case: there is insufficient damping to prevent oscillations in the same fashion as a bell rings.

Observe that the Laplace transform method is much simpler than the classical solution of the circuit differential equations as in Example 7.3. Here there is no need to derive the differential equations (although we are essentially doing so when we obtain the Laplace transform of the inductor current and capacitor voltage). The Laplace transform method automatically incorporates the initial conditions into the solution. In the classical solution of the circuit differential equations we had to obtain, in addition to the initial conditions $i_L(0^-)$ and $v_C(0^-)$, the derivatives of these at $t = 0^+$, $\frac{di_L}{dt}\big|_{t=0^+}$ and $\frac{dv_C}{dt}\big|_{t=0^+}$. The Laplace transform method bypasses the need to determine the derivatives at $t = 0^+$ and incorporates these automatically into the solution.

Exercise Problem 7.27

For the series *RLC* circuit with $R = 3\ \Omega$, $L = \frac{1}{3}$ H, $C = \frac{1}{6}$ F, and $V_S = 10$ V determine the inductor current for $t \geq 0$.

Answer: $i_L(t) = 10e^{-3t} - 10e^{-6t}$ A.

Exercise Problem 7.28

For the series *RLC* circuit with $R = 2\ \Omega$, $L = \frac{1}{4}$ H, $C = \frac{1}{4}$ F, and $V_S = 5$ V determine the capacitor voltage for $t \geq 0$.

Answer: $v_C(t) = -5e^{-4t} - 20te^{-4t} + 5$ V.

Exercise Problem 7.29

For the series *RLC* circuit with $R = 6\ \Omega$, $L = 1$ H, $C = \frac{1}{13}$ F, and $V_S = 10$ V determine the inductor current for $t \geq 0$.

Answer: $i_L(t) = 5e^{-3t} \sin 2t$ A.

Example 7.21

Determine the inductor current and capacitor voltage for the parallel *RLC* circuit shown in Fig. 7.27*a*, where $R = 1\ \Omega$, $L = \frac{3}{2}$ H, $C = \frac{1}{3}$ F, and the source is $I_S = 10$ A. The switch closes at $t = 0$.

Solution The initial conditions are zero because the source is not connected for $t < 0$. The transformed circuit is shown in Fig. 7.27*b*, from which we obtain

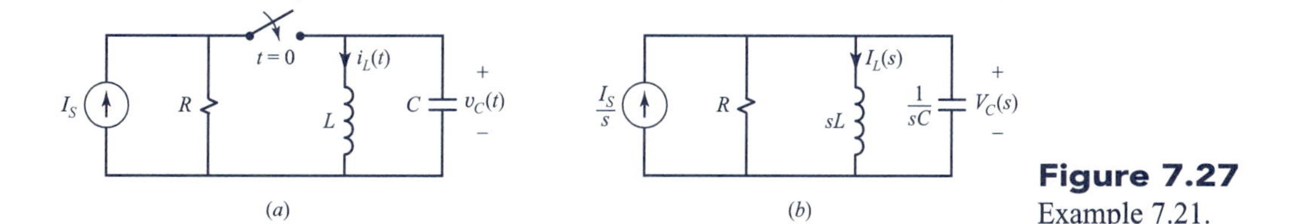

Figure 7.27 Example 7.21.

$$V_C(s) = \frac{I_S/s}{\frac{1}{R} + \frac{1}{sL} + sC}$$

$$= \frac{1}{s^2 + \frac{1}{RC}s + \frac{1}{LC}} \times \frac{I_S}{C}$$

$$= \frac{30}{s^2 + 3s + 2}$$

$$= \frac{30}{s+1} - \frac{30}{s+2}$$

and

$$I_L(s) = \frac{V_C(s)}{sL}$$

$$= \frac{1}{s\left(s^2 + \frac{1}{RC}s + \frac{1}{LC}\right)} \times \frac{I_S}{LC}$$

$$= \frac{20}{s(s^2 + 3s + 2)}$$

$$= -\frac{20}{s+1} + \frac{10}{s+2} + \frac{10}{s}$$

Therefore, the solutions valid for $t \geq 0$ are

$$v_C(t) = 30e^{-t} - 30e^{-2t} \text{ V}$$

$$i_L(t) = -20e^{-t} + 10e^{-2t} + 10 \text{ A}$$

Observe that these solutions satisfy continuity of inductor current and capacitor voltage, $v_C(0^-) = v_C(0^+) = 0$ and $i_L(0^-) = i_L(0^+) = 0$. In addition, they correctly give the steady-state responses, $v_{C,ss} = 0$ and $i_{L,ss} = 10$ A, which can be verified directly from the $t > 0$ circuit by replacing the inductor with a short circuit and the capacitor with an open circuit.

Exercise Problem 7.30

For the parallel *RLC* circuit with $R = 2\ \Omega$, $L = \frac{8}{3}$ H, $C = \frac{1}{8}$ F, and $I_S = 2$ A determine the inductor current for $t \geq 0$.

Answer: $i_L(t) = -3e^{-t} + e^{-3t} + 2$ A.

Exercise Problem 7.31

For the parallel *RLC* circuit with $R = 5\ \Omega$, $L = \frac{6}{5}$ H, $C = \frac{1}{30}$ F, and $I_S = 5$ A determine the capacitor voltage for $t \geq 0$.

Answer: $v_C(t) = 37.5e^{-3t} \sin 4t$ V.

Example 7.22

Determine the inductor current and capacitor voltage for $t > 0$ in the circuit of Fig. 7.28*a*.

Solution The $t < 0$ circuit is drawn in Fig. 7.28*b*, where we have replaced the inductor with a short circuit and the capacitor with an open circuit, since the source is dc. From this we determine the initial conditions: $i_L(0^-) = 2$ A and $v_C(0^-) = 4$ V. The transformed $t > 0$ circuit is shown in Fig. 7.28*c*. A source transformation converts this to a single-loop circuit as shown in Fig. 7.28*d*, from which we obtain

$$I_L(s) = \frac{\frac{10}{s} + 2 - \frac{4}{s}}{3 + s + \frac{2}{s}}$$

$$= 2\frac{s + 3}{s^2 + 3s + 2}$$

and

$$V_C(s) = \frac{2}{s}I_L(s) + \frac{4}{s}$$

$$= 4\frac{s + 3}{s(s^2 + 3s + 2)} + \frac{4}{s}$$

$$= 4\frac{s^2 + 4s + 5}{s(s^2 + 3s + 2)}$$

Partial-fraction expanding these gives

$$I_L(s) = \frac{4}{s + 1} - \frac{2}{s + 2}$$

$$V_C(s) = -\frac{8}{s + 1} + \frac{2}{s + 2} + \frac{10}{s}$$

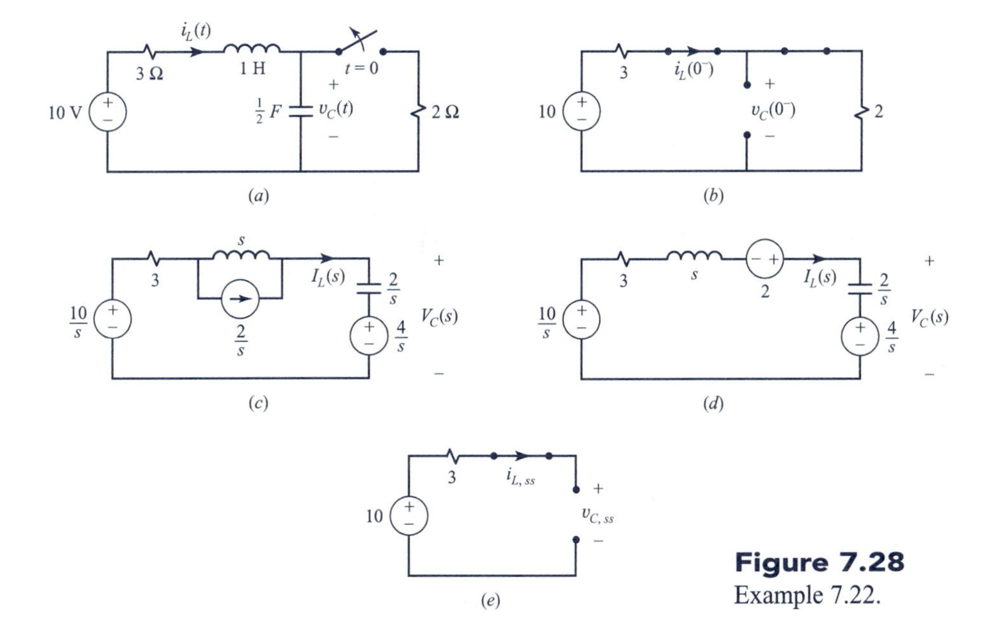

Figure 7.28
Example 7.22.

The inverse transforms are

$$i_L(t) = 4e^{-t} - 2e^{-2t}\text{ A} \qquad t \geq 0$$
$$v_C(t) = -8e^{-t} + 2e^{-2t} + 10\text{ V} \qquad t \geq 0$$

As a check on these results we obtain from them $i_L(0^+) = 2$ A, $v_C(0^+) = 4$ V, which satisfy continuity of inductor current and capacitor voltage. Similarly we obtain the steady-state solutions directly from these results as $i_{L,ss} = 0$ and $v_{C,ss} = 10$ V. These are confirmed by replacing the inductor with a short circuit and the capacitor with an open circuit in the $t > 0$ circuit as shown in Fig. 7.28*e*.

Example 7.23

Determine the inductor current and capacitor voltage for $t > 0$ in the circuit of Fig. 7.29*a*.

Solution The $t < 0$ circuit is drawn in Fig. 7.29*b*, where we have replaced the inductor with a short circuit and the capacitor with an open circuit, since the source is dc. From this we determine the initial conditions: $i_L(0^-) = 2.5$ A and $v_C(0^-) = 5$ V. The transformed $t > 0$

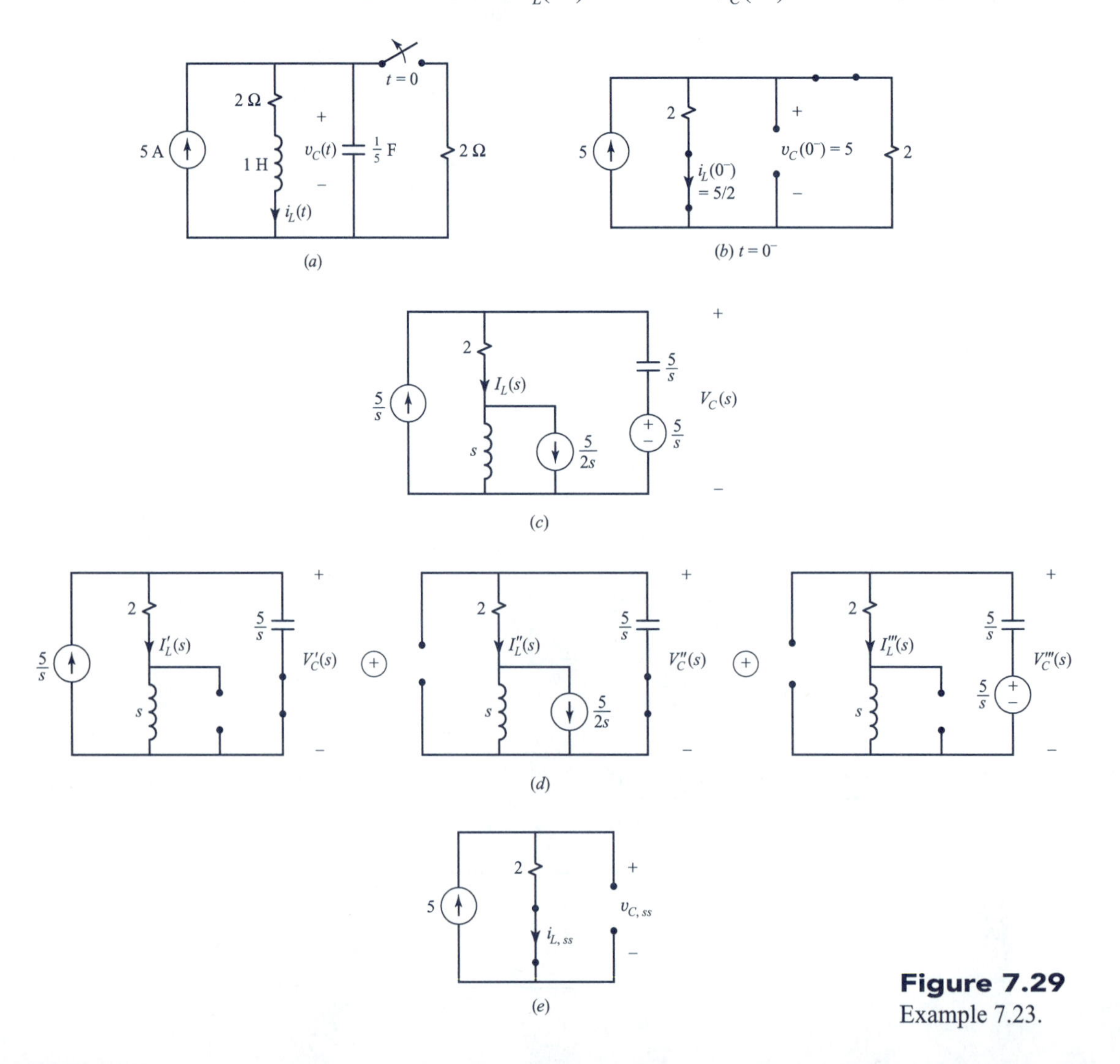

Figure 7.29
Example 7.23.

circuit is shown in Fig. 7.29*c*. The transformed inductor current and capacitor voltage are obtained using superposition as shown in Fig. 7.29*d*:

$$I_L'(s) = \frac{\frac{5}{s}}{2 + s + \frac{5}{s}} \frac{5}{s}$$

$$= 25 \frac{1}{s(s^2 + 2s + 5)}$$

$$V_C'(s) = \frac{(2 + s)\frac{5}{s}}{2 + s + \frac{5}{s}} \frac{5}{s}$$

$$= 25 \frac{s + 2}{s(s^2 + 2s + 5)}$$

$$I_L''(s) = \frac{s}{2 + s + \frac{5}{s}} \frac{5}{2s}$$

$$= 2.5 \frac{s}{s^2 + 2s + 5}$$

$$V_C''(s) = -\frac{s}{2 + s + \frac{5}{s}} \frac{5}{2s} \times \frac{5}{s}$$

$$= -\frac{25}{2} \frac{1}{s^2 + 2s + 5}$$

$$I_L'''(s) = \frac{\frac{5}{s}}{2 + s + \frac{5}{s}}$$

$$= 5 \frac{1}{s^2 + 2s + 5}$$

$$V_C'''(s) = \frac{2 + s}{2 + s + \frac{5}{s}} \frac{5}{s}$$

$$= 5 \frac{s + 2}{s^2 + 2s + 5}$$

Combining these gives

$$I_L(s) = \frac{2.5s^2 + 5s + 25}{s(s^2 + 2s + 5)}$$

$$V_C(s) = \frac{5s^2 + 22.5s + 50}{s(s^2 + 2s + 5)}$$

The poles are complex conjugates: $p_1, p_2 = -1 \pm j2$. Partial-fraction expanding these gives

$$I_L(s) = \frac{1.4 \angle -153.4^\circ}{s+1+j2} + \frac{1.4 \angle 153.4^\circ}{s+1-j2} + \frac{5}{s}$$

$$V_C(s) = \frac{3.125\angle -216.9^\circ}{s+1+j2} + \frac{3.125\angle 216.9^\circ}{s+1-j2} + \frac{10}{s}$$

The inverse transforms are

$$i_L(t) = 2.8e^{-t}\cos(2t + 153.4^\circ) + 5 \text{ A} \qquad t \geq 0$$

$$v_C(t) = 6.25e^{-t}\cos(2t + 216.9^\circ) + 10 \text{ V} \qquad t \geq 0$$

As a check on these results, we obtain from them $i_L(0^+) = 2.5$ A, $v_C(0^+) = 5$ V, which satisfy continuity of inductor current and capacitor voltage. Similarly we obtain the steady-state solutions directly from these results as $i_{L,ss} = 5$ A and $v_{C,ss} = 10$ V. These are confirmed by replacing the inductor with a short circuit and the capacitor with an open circuit in the $t > 0$ circuit as shown in Fig. 7.29*e*.

Mutual inductance is readily accommodated in the Laplace transform method.

Example 7.24

Determine the resistor current for $t > 0$ in the circuit of Fig. 7.30*a*.

Solution The $t < 0$ circuit is shown in Fig. 7.30*b*, where we replace the inductors with short circuits, since the source is dc. From this we obtain the initial current of the left inductor as 2A and the initial current of the right inductor as 0A. The $t > 0$ circuit (after switch closure) is transformed using the model in Fig. 7.13*c* in Fig. 7.30*c*. Because of the mutual inductance, the only feasible solution method is to write mesh-current equations. Defining mesh equations as shown, we obtain

$$\frac{10}{s} = 2I_1 + 2sI_3 - sI_2 + 3I_1 - 3I_2$$

$$3I_2 - 3I_1 + 3sI_2 - sI_3 + 3I_2 = 0$$

Noting that

$$I_3(s) = I_1(s) - \frac{2}{s}$$

and substituting gives

$$(5 + 2s)\,I_1(s) - (s+3)\,I_2(s) = \frac{10}{s} + 4$$

$$-(s+3)\,I_1(s) + (6 + 3s)\,I_2(s) = -2$$

Solving for $I_2(s) = I_R(s)$ gives

$$I_R(s) = I_2(s) = \frac{12s + 30}{s(5s^2 + 21s + 21)}$$

This is partial-fraction expanded, giving

$$I_R(s) = -\frac{1.369}{s + 1.64} - \frac{0.0596}{s + 2.56} + \frac{\frac{10}{7}}{s}$$

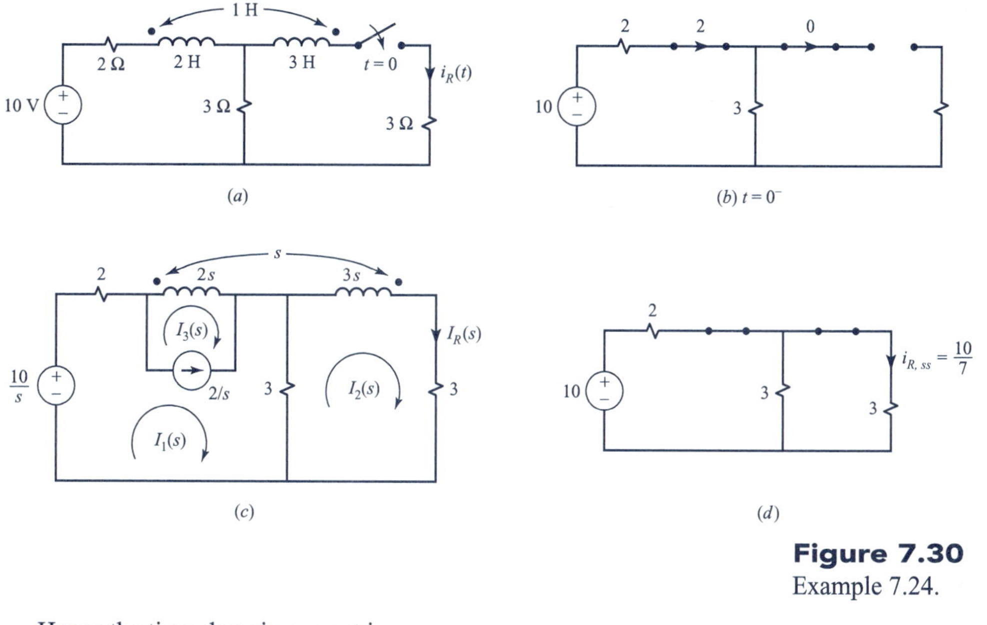

Figure 7.30
Example 7.24.

Hence the time-domain current is

$$i_R(t) = -1.369e^{-1.64t} - 0.0596e^{-2.56t} + \tfrac{10}{7}\text{ A} \qquad t > 0$$

As a check on these results, we obtain $i_R(0^+) = 0$ and $i_{R,ss} = \frac{10}{7}$ A directly from the equation. The first result is confirmed by continuity of inductor current through the right inductor, and the forced solution is confirmed from the circuit of Fig. 7.30*d*.

Example 7.25

Determine the resistor current for $t > 0$ in the circuit of Fig. 7.31*a*.

Solution The $t < 0$ circuit is shown in Fig. 7.31*b*. Observe that the sinusoidal source is multiplied by the unit step function. Hence its value is zero for $t < 0$, which is represented by a short circuit. Thus the only source present in the $t < 0$ circuit is the dc 8-V source, and hence the inductor is replaced with a short circuit and the capacitor with an open circuit, giving $i_L(0^-) = -1$ A and $v_C(0^-) = 3$ V. The $t > 0$ circuit transformed is shown in Fig. 7.31*c*. A source transformation on the inductor reduces this to the two-loop circuit shown in Fig. 7.31*d*. Writing mesh-current equations for this circuit yields

$$-\frac{30}{s^2+9} + \left(3 + s + \frac{5}{s}\right)I_1 - \frac{5}{s}I_2 + 1 + \frac{3}{s} = 0$$

$$-\frac{3}{s} + \left(\frac{5}{s} + 5\right)I_2 - \frac{5}{s}I_1 + \frac{8}{s} = 0$$

Collecting terms gives

$$\left(3 + s + \frac{5}{s}\right)I_1 - \frac{5}{s}I_2 = -1 - \frac{3}{s} + \frac{30}{s^2+9} = -\frac{s^3 + 3s^2 - 21s + 27}{s(s^2+9)}$$

$$-\frac{5}{s}I_1 + \left(\frac{5}{s} + 5\right)I_2 = -\frac{5}{s}$$

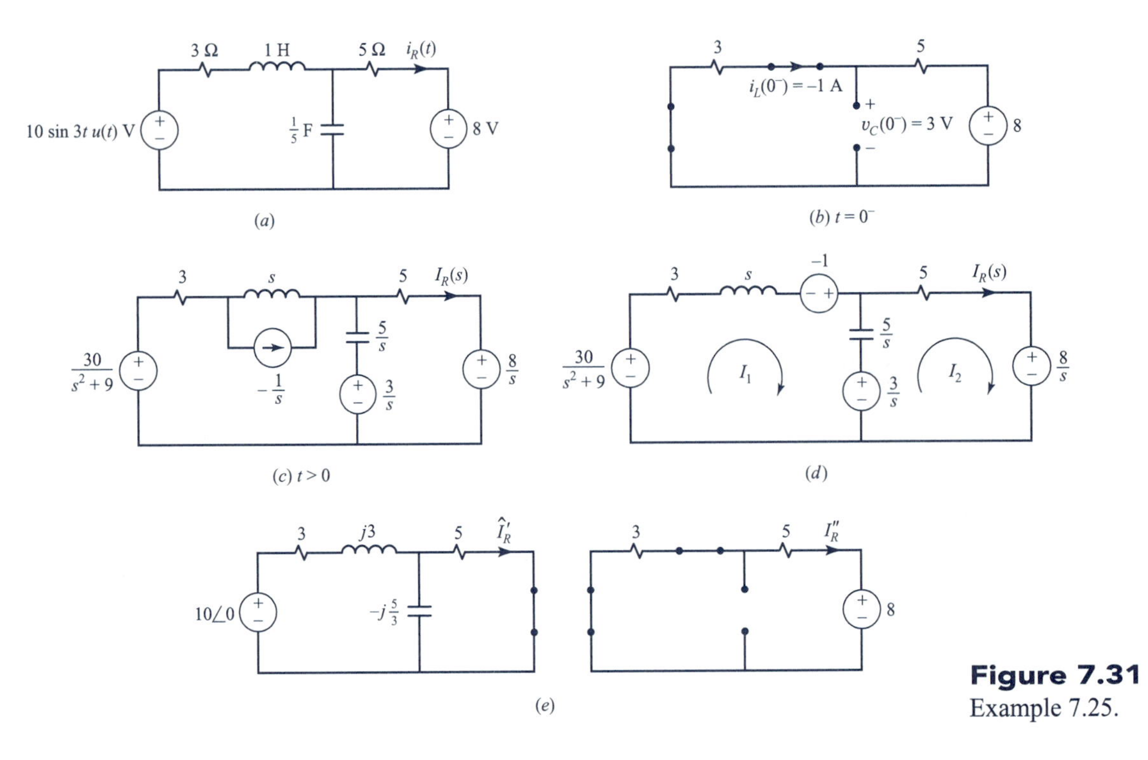

Figure 7.31 Example 7.25.

or

$$(s^2 + 3s + 5)I_1 - 5I_2 = -\frac{s^3 + 3s^2 - 21s + 27}{s^2 + 9}$$

$$-I_1 + (s + 1)I_2 = -1$$

Solving gives

$$I_R(s) = I_2(s) = -\frac{s^4 + 4s^3 + 17s^2 + 6s + 72}{s(s + 2 + j2)(s + 2 - j2)(s + j3)(s - j3)}$$

This is expanded as

$$I_R(s) = \frac{0.623\angle 48.36^\circ}{} + \frac{0.623\angle -48.36^\circ}{s + 2 - j2} - \frac{0.415\angle 4.77^\circ}{s + j3} - \frac{0.415\angle -4.77^\circ}{s - j3} - \frac{1}{s}$$

Hence, the time-domain solution is

$$i_R(t) = 1.246e^{-2t}\cos(2t - 48.36^\circ) - 0.83\cos(3t - 4.77^\circ) - 1 \text{ A} \qquad t > 0$$

This can again be confirmed by calculating directly from it $i_R(0^+) = -1$ A and $i_{R,ss} = -1 - 0.83\cos(3t - 4.77^\circ)$ A. The first result can be confirmed by observing that $i_R(0^+) = [v_C(0^+) - 8]/5\ \Omega = (3 - 8)/5 = -1$ A. The steady-state solution can be confirmed from Fig. 7.31*e*. The steady-state response due to the sinusoidal source is

$$\hat{I}'_R = \frac{-j\frac{5}{3}}{5 - j\frac{5}{3}} \frac{10\angle 0^\circ}{3 + j3 + 5\| - j\frac{5}{3}}$$

$$= 0.8305\angle -94.76^\circ$$

Realizing that this is for a sine function, we obtain the portion of the forced response computed above. The contribution to the forced response due to the 8-V dc source is computed from Fig. 7.31*e* as -1 A, which confirms this portion of the forced response computed above.

Exercise Problem 7.32

Determine the resistor current for $t > 0$ in the circuit of Fig. E7.32.

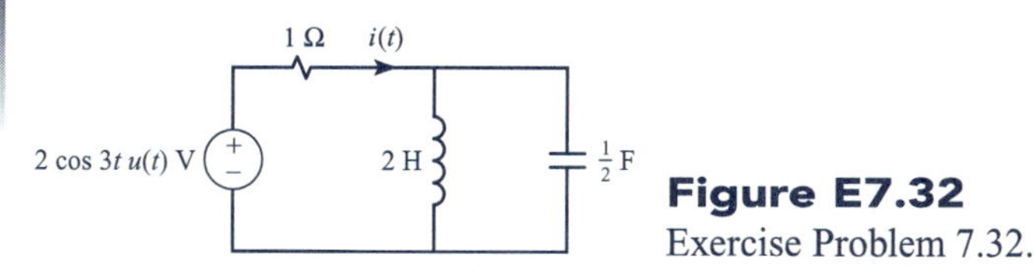

Figure E7.32
Exercise Problem 7.32.

Answer: $i_R(t) = 0.72e^{-t} - 0.4te^{-t} + 1.6\cos(3t + 36.87°)$ A, $t > 0$.

Exercise Problem 7.33

Determine the voltage $v(t)$ for $t > 0$ in the circuit of Fig. E7.33.

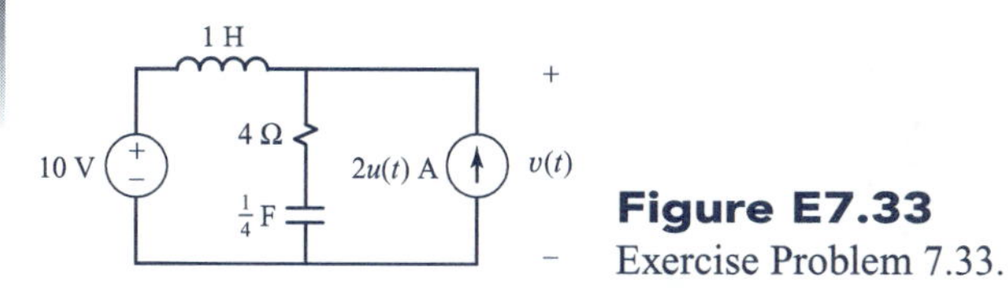

Figure E7.33
Exercise Problem 7.33.

Answer: $v(t) = 8e^{-2t} - 8te^{-2t} + 10$ V, $t > 0$.

7.7 The Step and Impulse Responses

In this section we will concentrate on calculating the response of a circuit that contains *only one independent source* that is either a *unit step function* or a *unit impulse function*. It turns out that

> *the unit-impulse response of a circuit is the most important response of that circuit in that the response to any other waveform can be found from the unit-impulse response alone.*

Hence, the ability to compute the impulse response of a circuit is a very important skill, which the reader should master. It is important to point out that when we speak of the step or impulse response of a circuit, it is assumed that there is only one independent source and that source has either a unit-step or a unit-impulse waveform. Furthermore, *it is assumed that all initial conditions are zero:* $i_L(0^-) = v_C(0^-) = 0$. The reason for this latter restriction to zero initial conditions is that

> *nonzero initial conditions—inductor currents or capacitor voltages—are essentially no different than independent sources.*

This is rather evident when we look at the Laplace-transformed equivalent circuits for inductors and capacitors shown in Fig. 7.12. Each of these equivalent circuits contains the initial condition as either a step or an impulse source, depending on the particular representation. Therefore, in order to have only one independent source in the circuit (the actual independent source) we must require that the inductor current and capacitor voltage at $t = 0^-$ be zero, so that the equivalent circuits for the inductor and the capacitor do not contain any independent sources and hence *the step and impulse responses of a circuit are determined for zero initial conditions.*

This latter fact means that the solution will be somewhat simpler than in circuits having switching operations, wherein the initial conditions may not be zero.

The unit-step and unit-impulse responses bear a close relationship that is often useful. The unit impulse is (in a loose but justifiable mathematical sense) the *derivative* of the unit step function:

$$\delta(t) = \frac{du(t)}{dt} \tag{7.121a}$$

Similarly, the unit step function is the *integral* of the unit impulse function:

$$u(t) = \int_{-\infty}^{t} \delta(\tau)\, d\tau \tag{7.121b}$$

Thus it appears reasonable to expect that the responses to the unit step and to the unit impulse are similarly related. For example, suppose we denote the response of a circuit variable (a current or a voltage) to the unit step as $y_u(t)$, and to the unit impulse as $y_\delta(t)$.
Then

$$y_\delta(t) = \frac{dy_u(t)}{dt} \tag{7.122a}$$

and

$$y_u(t) = \int_{-\infty}^{t} y_\delta(\tau)\, d\tau \tag{7.122b}$$

This can be proven in the following manner. Consider the circuit represented in *block diagram* form in Fig. 7.32*a*, with a single input (one independent source and zero initial conditions) $x(t)$ and a single output (desired current or voltage) $y(t)$. The impulse response in the time domain is denoted as $h(t)$, so that

$$\begin{aligned} x(t) &= \delta(t) \\ y(t) &= h(t) \end{aligned} \tag{7.123}$$

In the Laplace transform domain this corresponds to

$$\begin{aligned} X(s) &= 1 \\ Y(s) &= H(s) \end{aligned} \tag{7.124}$$

and $H(s)$ is the *transfer function,* which we also computed in Section 6.8.1 of Chapter 6:

$$Y(s) = H(s)\, X(s) \tag{7.125}$$

Hence the transforms of the step and impulse responses are

$$Y_\delta(s) = H(s) \times 1 \tag{7.126a}$$

$$Y_u(s) = H(s) \times \frac{1}{s} \tag{7.126b}$$

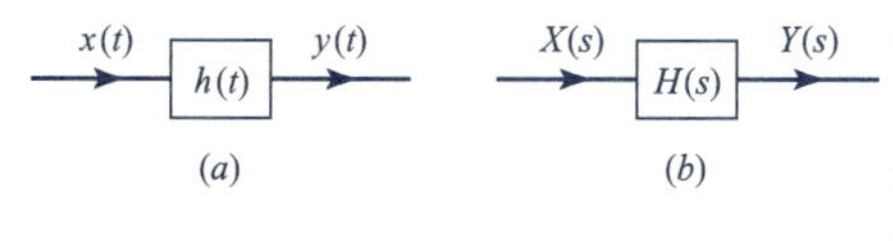

Figure 7.32
Viewing a circuit with one independent source and zero initial conditions as a single-input (the source), single-output (the desired voltage or current) system in (*a*) the time domain, and (*b*) the Laplace transform domain.

Therefore, the transformed step and impulse responses are related as

$$\boxed{Y_\delta(s) = sY_u(s)} \tag{7.127}$$

By recalling the fundamental transform of the derivative operation,

$$\frac{dy(t)}{dt} \Leftrightarrow sY(s) - \underbrace{y(0^-)}_{0} \tag{7.128}$$

we see that the result in (7.127) essentially proves the time-domain result in (7.122a) and also demonstrates why the initial conditions must be zero for the result to apply.

The steady-state response for the step-function excitation is the response with inductors replaced with short circuits and the capacitors replaced with open circuits. This is because the step source is essentially a dc source for $t > 0$. *The steady-state response for the impulse source is zero.* This is because the impulse source is zero for $t > 0$, and hence the $t > 0$ circuit does not contain a source. The Laplace transform automatically computes the total response, so that we do not need to directly compute the forced (steady-state) response using these observations. However, they can serve as a useful quick check of the Laplace transform result (at least the forced-response part of it).

There is one final observation about the impulse response of a circuit: *for an impulse source excitation, the inductor currents and capacitor voltages are no longer necessarily continuous.* For example, consider a first-order circuit having an impulse source. Because the source is an impulse and hence is zero for $t > 0$, the steady-state response is zero, and hence the total response, for example, for the inductor current is solely the zero-input response: $i_{L,\delta}(t) = i_L(0^+)\, e^{-Rt/L}$. Because the initial conditions for the impulse response are required to be zero $[i_{L,\delta}(0^-) = 0]$, the inductor current at $t = 0^+$ cannot be zero $[i_{L,\delta}(0^+) \neq 0]$: otherwise the solution for $t > 0$ would be zero. Similar remarks apply to the capacitor voltages. Hence we cannot check continuity of inductor currents and capacitor voltages for circuits that contain an impulse source.

Example 7.26

Determine the unit-step and unit-impulse responses of the inductor current and capacitor voltage in a series *RLC* circuit having $R = 3\ \Omega$, $L = 1$ H, $C = \frac{1}{2}$ F. Show that they are related as in (7.122).

Solution The Laplace-transformed circuit for the unit-step response is shown in Fig. 7.33*a*. From that we obtain

$$I_{L,u}(s) = \frac{\frac{1}{s}}{3 + s + \frac{2}{s}}$$

$$= \frac{1}{s^2 + 3s + 2}$$

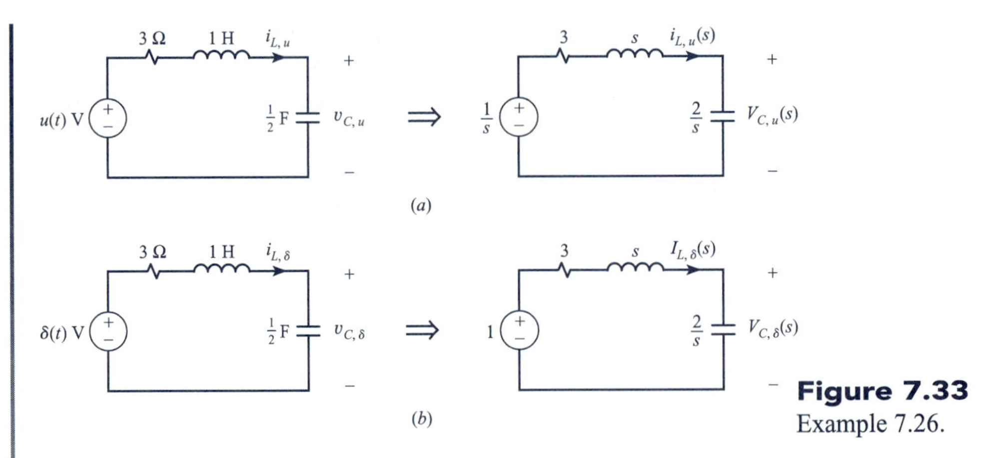

Figure 7.33
Example 7.26.

and

$$V_{C,u}(s) = \frac{\frac{2}{s}}{3 + s + \frac{2}{s}} \times \frac{1}{s}$$
$$= \frac{2}{s(s^2 + 3s + 2)}$$

These are expanded by partial fractions as

$$I_{L,u}(s) = \frac{1}{s+1} - \frac{1}{s+2}$$

and

$$V_{C,u}(s) = -\frac{2}{s+1} + \frac{1}{s+2} + \frac{1}{s}$$

The inverse transforms are

$$i_{L,u}(t) = (e^{-t} - e^{-2t})u(t)\ \text{A}$$

and

$$v_{C,u}(t) = (-2e^{-t} + e^{-2t} + 1)u(t)\ \text{V}$$

Note that these expressions are made valid for all time by appending $u(t)$ to them. The circuit for the impulse response is shown in Fig. 7.33*b*. Observe that the only difference between the transformed circuits is in the source function ($1/s$ for the unit step and 1 for the unit impulse). Nevertheless, the transformed responses are

$$I_{L,\delta}(s) = \frac{1}{3 + s + \frac{2}{s}}$$
$$= \frac{s}{s^2 + 3s + 2}$$

and

$$V_{C,\delta}(s) = \frac{\frac{2}{s}}{3 + s + \frac{2}{s}} \times 1$$
$$= \frac{2}{s^2 + 3s + 2}$$

These are expanded by partial fractions as

$$I_{L,\delta}(s) = -\frac{1}{s+1} + \frac{2}{s+2}$$

and

$$V_{C,\delta}(s) = \frac{2}{s+1} - \frac{2}{s+2}$$

The inverse transforms are

$$i_{L,\delta}(t) = (-e^{-t} + 2e^{-2t})u(t) \text{ A}$$

and

$$v_{C,\delta}(t) = (2e^{-t} - 2e^{-2t})u(t) \text{ V}$$

Observe that the inductor current impulse response at $t = 0^+$ is not zero $[i_{L,\delta}(0^+) = 1 \text{ A} \neq 0]$ and hence the inductor current impulse response is not continuous. The capacitor current impulse response happens to be continuous in this circuit.

We now demonstrate that the responses are related as in (7.122):

$$\begin{aligned} i_{L,\delta}(t) &= \frac{d}{dt}\underbrace{(e^{-t} - e^{-2t})u(t)}_{i_{L,u}(t)} \\ &= (-e^{-t} + 2e^{-2t})u(t) + \underbrace{(e^{-t} - e^{-2t})}_{=0 \text{ at } t=0}\underbrace{\frac{d}{dt}u(t)}_{\delta(t)} \\ &= (-e^{-t} + 2e^{-2t})u(t) \end{aligned}$$

and

$$\begin{aligned} v_{C,\delta}(t) &= \frac{d}{dt}\underbrace{(-2e^{-t} + e^{-2t} + 1)u(t)}_{v_{C,u}(t)} \\ &= (2e^{-t} - 2e^{-2t})u(t) + \underbrace{(-2e^{-t} + e^{-2t} + 1)}_{=0 \text{ at } t=0}\underbrace{\frac{d}{dt}u(t)}_{\delta(t)} \\ &= (2e^{-t} - 2e^{-2t})u(t) \end{aligned}$$

Observe in these last two results that we have used the chain rule to differentiate the functions. Further observe that the derivative of the step function in the result is the impulse function. Hence, its coefficient is evaluated at $t = 0$, and for these cases the coefficient is zero and therefore the result does not contain an impulse. There are certain cases where this coefficient is not zero and the response contains an impulse. Similarly we can demonstrate (7.122b):

$$\begin{aligned} i_{L,u}(t) &= \int_{-\infty}^{t}\underbrace{(-e^{-\tau} + 2e^{-2\tau})u(\tau)}_{i_{L,\delta}}\,d\tau \\ &= \int_{0}^{t}(-e^{-\tau} + 2e^{-2\tau})\,d\tau\, u(t) \\ &= (e^{-\tau} - e^{-2\tau})\Big|_0^t u(t) \\ &= (e^{-t} - e^{-2t})u(t) \end{aligned}$$

and

$$\begin{aligned}
v_{C,u}(t) &= \int_{-\infty}^{t} \underbrace{(2e^{-\tau} - 2e^{-2\tau})u(\tau)}_{v_{C,\delta}}\, d\tau \\
&= \int_{0}^{t} (2e^{-\tau} - 2e^{-2\tau})\, d\tau\, u(t) \\
&= (-2e^{-\tau} + e^{-2\tau})\Big|_0^t u(t) \\
&= (-2e^{-t} + e^{-2t})u(t) - (-1)u(t) \\
&= (-2e^{-t} + e^{-2t} + 1)u(t)
\end{aligned}$$

Example 7.27

Determine the unit-step and unit-impulse responses for the capacitor voltage in a series *RLC* circuit having $R = 4\ \Omega$, $L = 1$ H, $C = \frac{1}{4}$ F. Show that they are related as in (7.122).

Solution The unit-step circuit is shown transformed in Fig. 7.34*a*, from which we obtain

$$\begin{aligned}
V_{C,u}(s) &= \frac{\frac{4}{s}}{4 + s + \frac{4}{s}} \times \frac{1}{s} \\
&= \frac{4}{s\,(s^2 + 4s + 4)} \\
&= -\frac{1}{s+2} - \frac{2}{(s+2)^2} + \frac{1}{s}
\end{aligned}$$

Hence the unit-step response is

$$v_{C,u}(t) = (-e^{-2t} - 2te^{-2t} + 1)u(t)\ \text{V}$$

The unit-impulse circuit is shown in Fig. 7.34*b*, from which we obtain

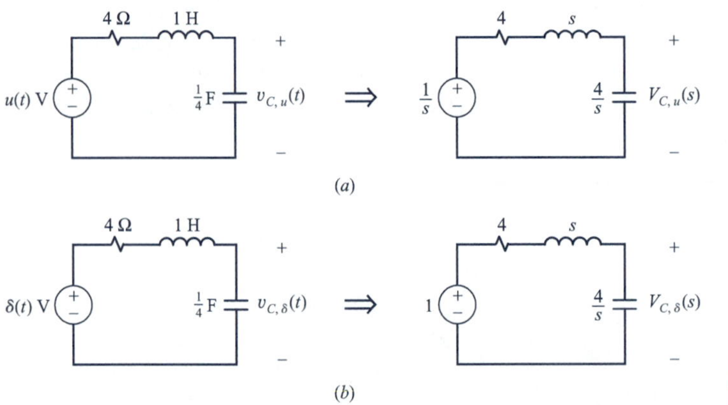

Figure 7.34
Example 7.27.

$$V_{C,\delta}(s) = \frac{\frac{4}{s}}{4 + s + \frac{4}{s}} \times 1$$

$$= \frac{4}{s^2 + 4s + 4}$$

$$= \frac{4}{(s+2)^2}$$

Hence the unit-impulse response is

$$v_{C,\delta}(t) = (4te^{-2t})u(t) \text{ V}$$

To show that these are related as in (7.122) we form

$$v_{C,\delta}(t) = \frac{d}{dt}\underbrace{(-e^{-2t} - 2te^{-2t} + 1)u(t)}_{v_{C,u}(t)}$$

$$= (2e^{-2t} - 2e^{-2t} + 4te^{-2t})u(t) + \underbrace{(-e^{-2t} - 2te^{-2t} + 1)}_{=0 \text{ at } t=0}\underbrace{\frac{d}{dt}u(t)}_{\delta(t)}$$

$$= (4te^{-2t})u(t)$$

and

$$v_{C,u}(t) = \int_{-\infty}^{t} \underbrace{(4\tau e^{-2\tau})u(\tau)}_{v_{C,\delta}(\tau)}\, d\tau$$

$$= \int_0^t (4\tau e^{-2\tau})\, d\tau\, u(t)$$

$$= \left(4\frac{e^{-2\tau}}{4}(-2\tau - 1)\right)\Bigg|_0^t u(t)$$

$$= \left[e^{-2t}(-2t - 1) + 1\right]u(t)$$

Example 7.28

Determine the unit-step and unit-impulse responses for the inductor current in a series RLC circuit having $R = 2\ \Omega$, $L = 1\text{H}$, $C = \frac{1}{5}$ F. Show that they are related as in (7.122).

Solution The unit-step circuit is shown transformed in Fig. 7.35*a*, from which we obtain

$$I_{L,u}(s) = \frac{\frac{1}{s}}{2 + s + \frac{5}{s}}$$

$$= \frac{1}{s^2 + 2s + 5}$$

$$= \frac{j\frac{1}{4}}{s + 1 + j2} - \frac{j\frac{1}{4}}{s + 1 - j2}$$

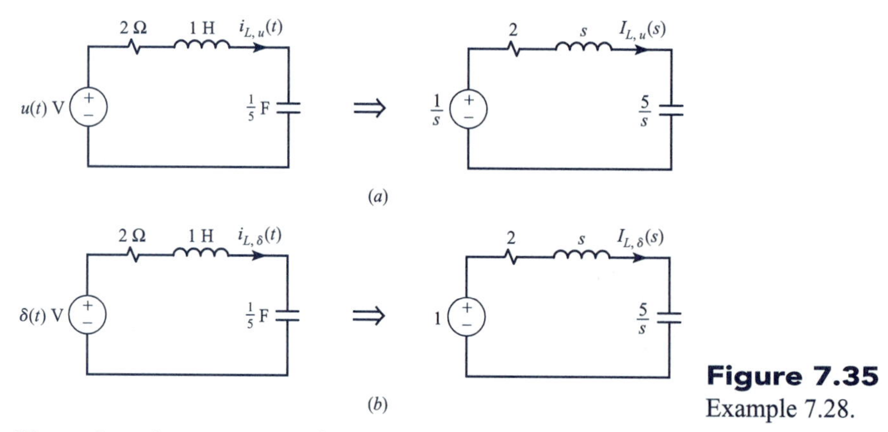

Figure 7.35
Example 7.28.

Hence the unit-step response is

$$i_{L,u}(t) = (\tfrac{1}{2} e^{-t} \sin 2t)u(t) \text{ A}$$

The unit-impulse circuit is shown in Fig. 7.35*b*, from which we obtain

$$\begin{aligned} I_{L,\delta}(s) &= \frac{1}{2 + s + \dfrac{5}{s}} \\ &= \frac{s}{s^2 + 2s + 5} \\ &= \frac{\frac{1}{2} - j\frac{1}{4}}{s + 1 + j2} + \frac{\frac{1}{2} + j\frac{1}{4}}{s + 1 - j2} \\ &= \frac{0.559\angle - 26.56°}{s + 1 + j2} + \frac{0.559\angle 26.56°}{s + 1 - j2} \end{aligned}$$

Hence the unit-impulse response is

$$\begin{aligned} i_{L,\delta}(t) &= e^{-t}(\cos 2t - \tfrac{1}{2} \sin 2t)u(t) \\ &= [1.118e^{-t} \cos (2t + 26.56°)]u(t) \text{ A} \end{aligned}$$

Observe that the inductor current impulse response at $t = 0^+$ is not zero $[i_{L,\delta}(0^+) = 1.118 \cos 26.56° \text{ A} \neq 0]$, and hence the inductor current impulse response is not continuous. To show that these are related as in (7.122) we form

$$\begin{aligned} i_{L,\delta}(t) &= \frac{d}{dt} \underbrace{(\tfrac{1}{2} e^{-t} \sin 2t)u(t)}_{i_{L,u}(t)} \\ &= (-\tfrac{1}{2} e^{-t} \sin 2t)u(t) + (e^{-t} \cos 2t)u(t) + \underbrace{(\tfrac{1}{2} e^{-t} \sin 2t)}_{= 0 \text{ at } t = 0} \underbrace{\frac{d}{dt} u(t)}_{\delta(t)} \\ &= e^{-t}(\cos 2t - \tfrac{1}{2} \sin 2t)u(t) \end{aligned}$$

and

$$\begin{aligned} i_{L,u}(t) &= \int_{-\infty}^{t} \underbrace{e^{-\tau}(\cos 2\tau - \tfrac{1}{2} \sin 2\tau)u(\tau)\, d\tau}_{i_{L,\delta}(\tau)} \\ &= \int_{0}^{t} e^{-\tau}(\cos 2\tau - \tfrac{1}{2} \sin 2\tau)\, d\tau\, u(t) \\ &= \left(\frac{e^{-\tau}}{5}(-\cos 2\tau + 2 \sin 2\tau) - \frac{1}{2}\frac{e^{-\tau}}{5}(-\sin 2\tau - 2\cos 2\tau)\right)\Bigg|_{0}^{t} u(t) \\ &= (\tfrac{1}{2} e^{-t} \sin 2t)u(t) \end{aligned}$$

Exercise Problem 7.34

Determine the step and impulse responses for the capacitor voltage in the circuit of Fig. E7.34. Show that they are related as in (7.122).

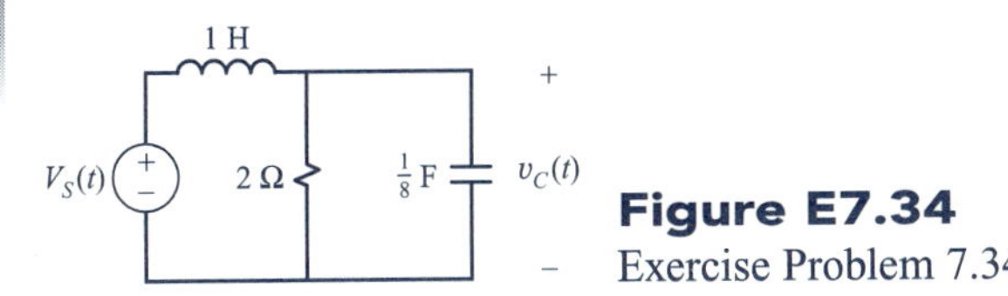

Figure E7.34
Exercise Problem 7.34.

Answers: $v_{C,u}(t) = (-e^{-2t}\cos 2t - e^{-2t}\sin 2t + 1)u(t)$ V and $v_{C,\delta}(t) = (4e^{-2t}\sin 2t)u(t)$ V.

Exercise Problem 7.35

Determine the step and impulse responses for current $i(t)$ in the circuit of Fig. E7.35. Show that these are related as in (7.122).

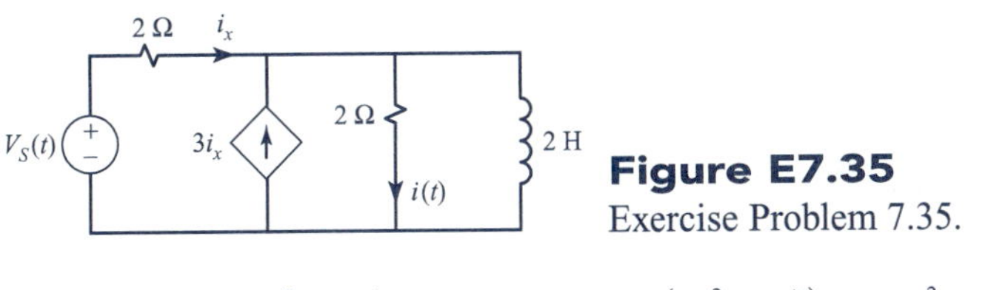

Figure E7.35
Exercise Problem 7.35.

Answers: $i_u(t) = \left(\frac{2}{5}e^{-t/5}\right)u(t)$ A and $i_\delta(\mathrm{t}) = \left(-\frac{2}{25}e^{-t/5}\right)u(t) + \frac{2}{5}\delta(t)$ A.

7.8 Convolution

In this section we will demonstrate an important property of the impulse response:

> *The response of a circuit to any input (source) waveform can be found from the impulse response of that circuit using the convolution integral.*

Represent the single-input (source), single-output (voltage or current response) circuit (system) as shown in Fig. 7.32*a*. If the input is a unit impulse, $x(t) = \delta(t)$, the output is referred to as the impulse response and is denoted as $y(t) = h(t)$. We will show that the response to any input can be found from

$$y(t) = \int_{0^-}^{t} h(t - \tau)x(\tau)\, d\tau \tag{7.129}$$

which is referred to as the *convolution integral*. The limits on the integral assume that $x(t) = 0$ for $t < 0$, as we have been assuming, and that $h(t) = 0$ for $t < 0$. This latter condition essentially states that the circuit (system) is *causal*, i.e., there is no response before the input (the impulse) is applied.

The convolution integral can be derived from the inverse transform of the product $Y(s) = H(s)X(s)$ as

$$
\begin{aligned}
y(t) &= \mathscr{L}^{-1}\{H(s)X(s)\} \\
&= \mathscr{L}^{-1}\left\{H(s)\underbrace{\int_{0^-}^{\infty} x(\tau)e^{-s\tau}\,d\tau}_{X(s)}\right\} \\
&= \mathscr{L}^{-1}\left\{\int_{0^-}^{\infty} \underbrace{e^{-s\tau}H(s)}_{\mathscr{L}\{h(t-\tau)\}}x(\tau)\,d\tau\right\} \\
&= \mathscr{L}^{-1}\left\{\int_{0^-}^{\infty} x(\tau)\left(\int_{0^-}^{\infty} h(t-\tau)e^{-st}\,dt\right)d\tau\right\} \\
&= \mathscr{L}^{-1}\left\{\int_{0^-}^{\infty} e^{-st}\left(\int_{0^-}^{\infty} x(\tau)h(t-\tau)\,d\tau\right)dt\right\} \\
&= \int_{0^-}^{\infty} x(\tau)h(t-\tau)\,d\tau
\end{aligned}
\tag{7.130}
$$

The upper limit of infinity in the result can be replaced with t, since, by causality, $h(t-\tau) = 0$ for $t < \tau$. Hence the convolution integral and $Y(s) = H(s)X(s)$ are transform pairs.

Example 7.29

Determine the response of the circuit in Fig. 7.36*a*.

Solution First we obtain the unit impulse response. The transformed circuit is shown in Fig. 7.36*b*. From that circuit we obtain

$$
\begin{aligned}
V_\delta(s) &= H(s) \\
&= \frac{6}{6+2s} = \frac{3}{s+3}
\end{aligned}
$$

Hence the impulse response is

$$h(t) = 3e^{-3t}u(t)\ \text{V}$$

Evaluating the convolution integral given in (7.129) requires that we construct two functions: $h(t-\tau)$ and $x(\tau)$. Forming $x(\tau)$ is simple: replace t with τ. Forming $h(t-\tau)$ is also simple. First form $h(-\tau)$ by flipping $h(t)$ about the $t = 0$ point and replacing t with $-\tau$. Then form $h(t-\tau)$ by simply adding t to the argument, i.e., relabeling the $h(0)$ point as t. This is illustrated in Fig. 7.36*c*. The convolution integration requires that we take the instantaneous product of the resulting two curves and determine the area under that product curve. This is illustrated in Fig. 7.36*d,* which gives

$$
\begin{aligned}
v(t) &= \int_{0^-}^{t} \underbrace{3e^{-3(t-\tau)}}_{h(t-\tau)}\,\underbrace{10\,d\tau}_{x(\tau)} \\
&= 30e^{-3t}\int_{0^-}^{t} e^{3\tau}\,d\tau \\
&= 10e^{-3t}[e^{3\tau}]_{0^-}^{t} \\
&= 10(1-e^{-3t})u(t)\ \text{V}
\end{aligned}
$$

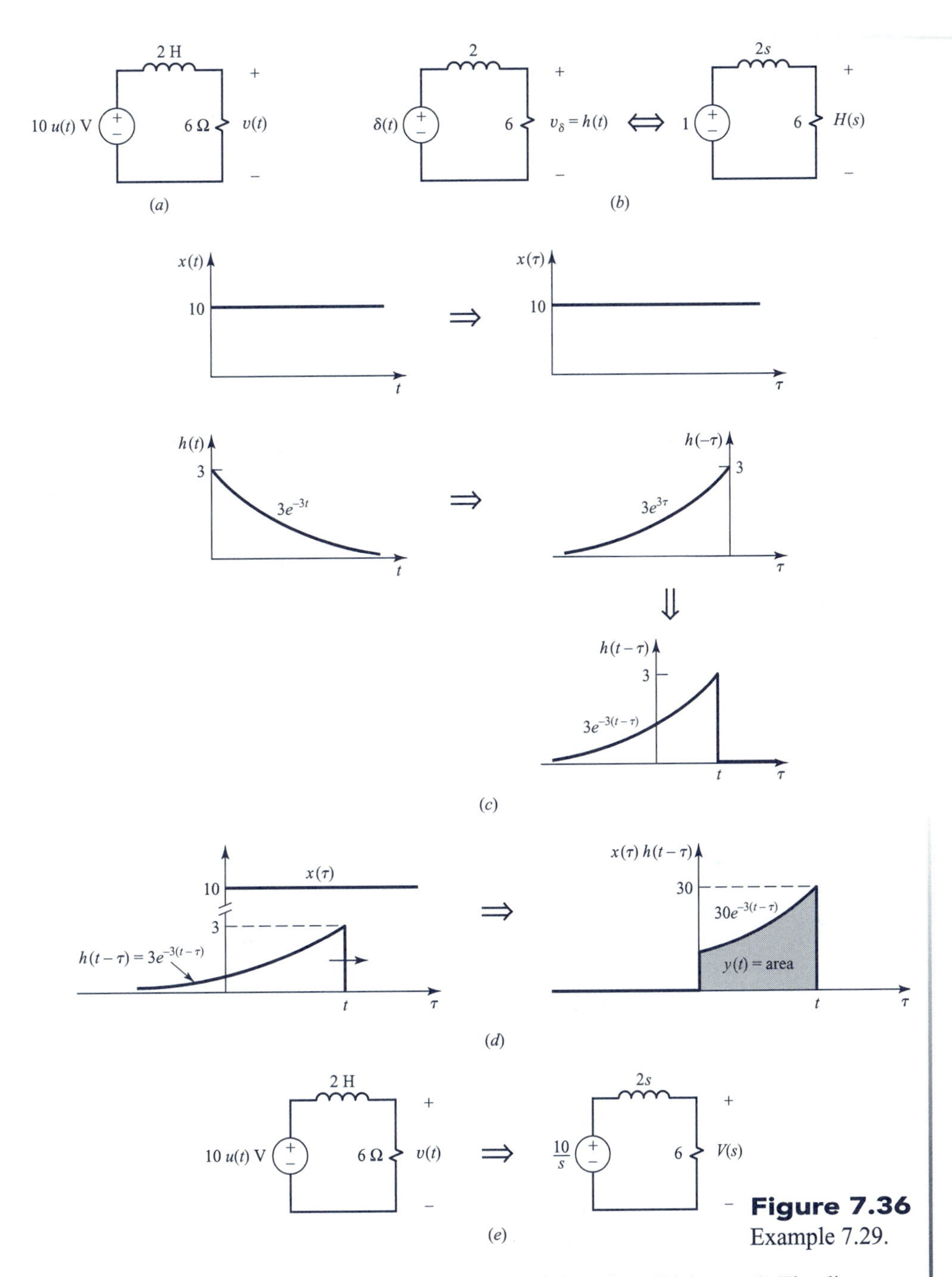

Figure 7.36
Example 7.29.

We have appended $u(t)$ to this result to point out that it is only valid for $t > 0$. The direct solution is indicated in Fig. 7.36e, where the transformed solution is

$$
\begin{aligned}
V(s) &= \frac{6}{6 + 2s}\frac{10}{s} \\
&= \frac{30}{s(s+3)} \\
&= -\frac{10}{s+3} + \frac{10}{s}
\end{aligned}
$$

The inverse transform of this is the same as the result obtained by convolution.

Exercise Problem 7.36

Determine the unit-step response for the current $i(t)$ in Fig. E7.35 using the convolution integral.

Answer: $i_u(t) = \left(\frac{2}{5}e^{-t/5}\right)u(t)$ A.

7.9 Application of Linearity and Time Invariance

Consider a circuit that has one input (an independent source) and a particular current or voltage response as the output. Again it is important to emphasize that *since initial conditions are represented as independent sources in the Laplace transform model of the elements, we must require that the initial conditions be zero if we are to have a system with one input*. A circuit fulfilling these stipulations is modeled as in Fig. 7.32, where we denote the input (the independent source) as $x(t)$ and denote the output (the desired current or voltage) as $y(t)$. Such a circuit (system) satisfies several principles, which can be used to reduce the solution effort for a general input signal.

The first important property, *linearity,* means that if the input (the independent source) is composed of the sum of two time functions, then the response is the sum of the responses to the individual time functions:

If
$$x_1(t) \Rightarrow y_1(t)$$
and
$$x_2(t) \Rightarrow y_2(t) \tag{7.131}$$
then
$$A_1 x_1(t) + A_2 x_2(t) \Rightarrow A_1 y_1(t) + A_2 y_2(t)$$

In other words, if $x_1(t)$ gives $y_1(t)$ and $x_2(t)$ gives $y_2(t)$, then a signal made up of a linear combination of these inputs has a response that is the same linear combination of the individual responses.

The circuits are also *time-invariant,* which basically means that the circuit element values (values of the R's, L's, C's and controlled sources) do not change with time, and hence the response to an input applied now is the same as the response to that input being applied at some time in the future. Hence if $y(t)$ is the response to $x(t)$, then the response to $x(t)$ being applied at some later time, $x(t - T)$, is the response to $x(t)$ shifted in time by T:

$$\begin{aligned} x(t) &\Rightarrow y(t) \\ x(t-T) &\Rightarrow y(t-T) \end{aligned} \tag{7.132}$$

There is one additional property, connected with linearity, that was discussed in the previous section with regard to step and impulse responses but applies to more general signals. Namely, the response to the derivative of an input is the derivative of the response to that input, and the response to the integral of an input is the integral of the response to that input:

If
$$x(t) \Rightarrow y(t)$$
then
$$\frac{dx(t)}{dt} \Rightarrow \frac{dy(t)}{dt} \tag{7.133}$$
and
$$\int_{-\infty}^{t} x(\tau)\,d\tau \Rightarrow \int_{-\infty}^{t} y(\tau)\,d\tau$$

These are very powerful properties that allow us to greatly simplify the solution for a wide variety of source waveforms.

Example 7.30

A pulse waveform source is applied to a first-order circuit as shown in Fig. 7.37*a*. Determine the response of the capacitor voltage to this waveform using (a) linearity and time invariance, (b) convolution, (c) differentiation, and (d) direct methods.

Solution First we decompose the pulse waveform into a sum of step functions as

$$v_S(t) = V_S u(t) - V_S u(t - T)$$

This decomposition is illustrated in Fig. 7.37*b*. If we know the response to a unit-step function, then the properties of linearity and time invariance allow us to obtain the response to the actual pulse source as the sum of these scaled and time-shifted responses to the unit-step input. In other words, if

$$v_S(t) = u(t) \Rightarrow v_{C,u}(t)$$

then

$$v_S(t) = V_S u(t) - V_S u(t - T) \Rightarrow V_S v_{C,u}(t) - V_S v_{C,u}(t - T)$$

So the first objective is to obtain the unit-step response of the circuit. This is obtained from the transformed circuit of Fig. 7.37*c* as

$$V_{C,u}(s) = \frac{\dfrac{1}{sC}}{R + \dfrac{1}{sC}} \times \frac{1}{s}$$

$$= \frac{\dfrac{1}{RC}}{\left(s + \dfrac{1}{RC}\right)s}$$

$$= \frac{-1}{s + \dfrac{1}{RC}} + \frac{1}{s}$$

The inverse transform of this is

$$v_{C,u}(t) = (1 - e^{-t/RC})u(t)$$

This can be checked by noting that the initial capacitor voltage is zero and the steady-state value is 1, which could be deduced by replacing the capacitor with an open circuit in the $t > 0$ circuit. Hence the properties of linearity and time invariance allow us to construct the response to the pulse function as

$$v_C(t) = V_S v_{C,u}(t) - V_S v_{C,u}(t - T)$$

$$= V_S(1 - e^{-t/RC})u(t) - V_S(1 - e^{-(t-T)/RC})u(t - T)$$

$$= \begin{cases} V_S(1 - e^{-t/RC}) & 0 \le t \le T \\ V_S e^{-t/RC}(e^{T/RC} - 1) & T \le t \end{cases}$$

This final form for the solution over the separate intervals of time is due to the fact that $u(t - T)$ is zero for $t < T$. The solution is sketched in Fig. 7.37*d* for two values of the pulse

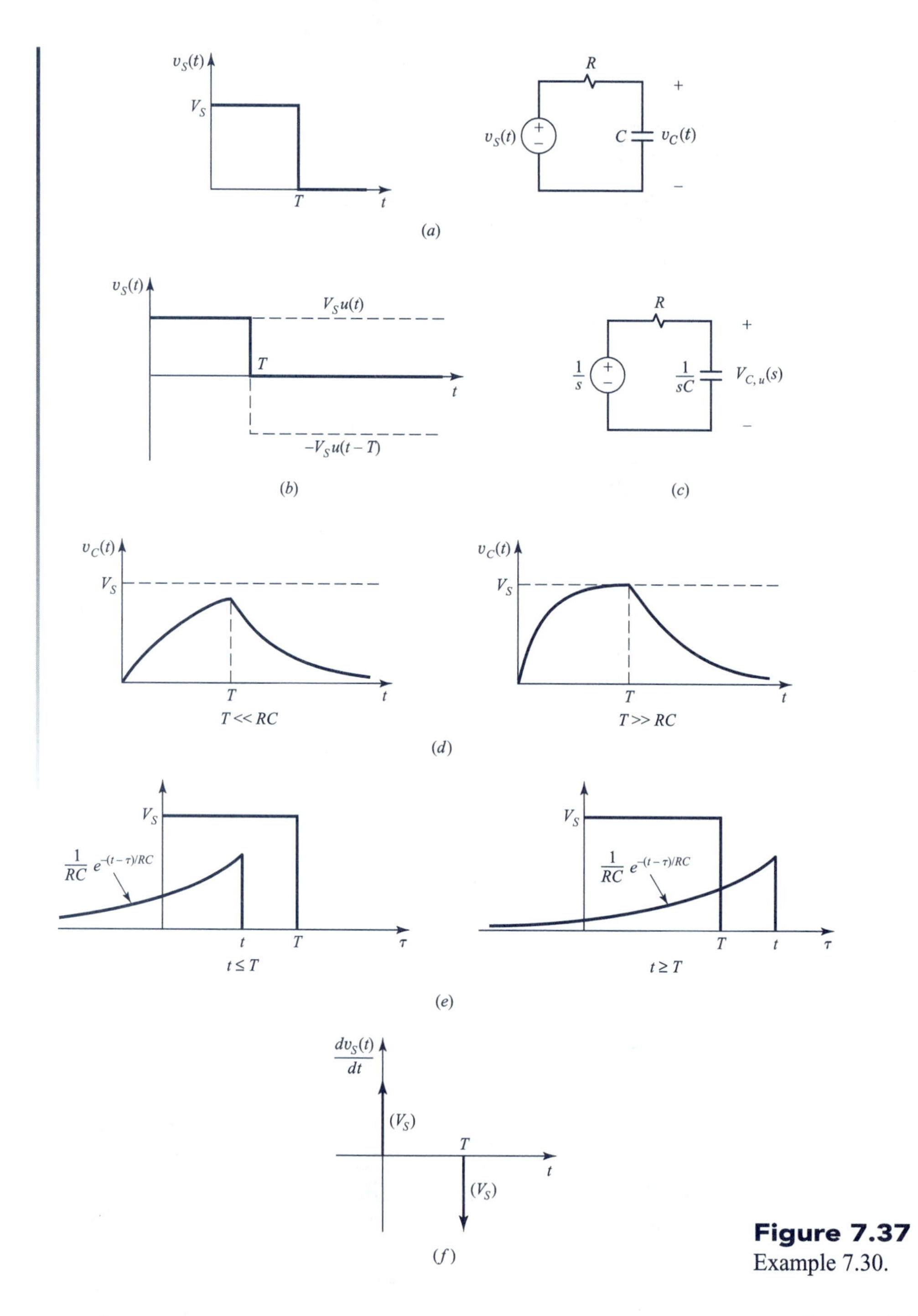

Figure 7.37
Example 7.30.

duration in relation to the circuit time constant. Observe that if the pulse duration is much less than a time constant ($T << RC$), the solution never reaches steady state before the pulse turns off. On the other hand, if the pulse duration is much longer than a time constant ($T >> RC$), the capacitor voltage almost reaches steady state before the pulse turns off. Hence we obtain two different waveforms for the capacitor voltage, depending on the relation of the pulse duration and the time constant of the circuit.

We next determine the solution using convolution. First we determine the impulse response. Replacing the source in Fig. 7.37*c* with the transform of a unit impulse, we obtain

$$V_{C,\delta}(s) = \frac{\frac{1}{sC}}{R + \frac{1}{sC}} \times 1$$

$$= \frac{\frac{1}{RC}}{s + \frac{1}{RC}}$$

and the impulse response is

$$h(t) = v_{C,\delta}(t)$$

$$= \frac{1}{RC} e^{-t/RC}\, u(t)$$

(This could also be obtained by differentiating the previously obtained unit-step response.) Hence the response to the pulse source is the convolution

$$v_C(t) = \int_0^t h(t - \tau)\, v_S(\tau)\, d\tau$$

$$= \int_0^t \left(\frac{1}{RC} e^{-(t-\tau)/RC}\right) u(t - \tau)\, [V_S u(\tau) - V_S u(\tau - T)]\, d\tau$$

This can be more easily evaluated by drawing the result as shown in Fig. 7.37*e*. There are two regions for t: $t \leq T$ and $t \geq T$. The solutions in these regions become

$$v_C(t) = \int_0^t \frac{V_S}{RC} e^{-(t-\tau)/RC}\, d\tau$$

$$= V_S (1 - e^{-t/RC}) \qquad 0 \leq t \leq T$$

and

$$v_C(t) = \int_0^T \frac{V_S}{RC} e^{-(t-\tau)/RC}\, d\tau$$

$$= V_S\, e^{-t/RC}(e^{T/RC} - 1) \qquad t \geq T$$

as before. Observe the limits of integration in the last convolution: the pulse function is zero for $t < 0$ and for $t > T$.

Figure 7.37*f* shows the derivative of $v_S(t)$. The derivative of each discontinuity is an impulse whose strength is the amount of *jump* in the discontinuity. The discontinuity at $t = T$ is a negative jump and hence is represented as a negative impulse. Hence we can write the derivative of this signal as

$$\frac{dv_S(t)}{dt} = V_S \delta(t) - V_S \delta(t - T)$$

Hence the derivative of the desired response is the response to this signal:

$$\frac{dv_C(t)}{dt} = V_S h(t) - V_S h(t - T)$$

where $h(t)$ is the unit impulse response determined previously as

$$h(t) = v_{C,\delta}(t)$$

$$= \frac{1}{RC} e^{-t/RC}\, u(t)$$

Therefore, the response to the derivative of the desired signal is

$$\frac{dv_C(t)}{dt} = \frac{V_S}{RC}e^{-t/RC}u(t) - \frac{V_S}{RC}e^{-(t-T)/RC}u(t-T)$$

In order to obtain the desired solution, we integrate this result as

$$\begin{aligned}
v_C(t) &= \int_{-\infty}^{t} \frac{dv_C(\tau)}{d\tau}\,d\tau \\
&= \int_{-\infty}^{t} \left(\frac{V_S}{RC}e^{-\tau/RC}u(\tau) - \frac{V_S}{RC}e^{-(\tau-T)/RC}u(\tau-T)\right)d\tau \\
&= \left(\int_{0}^{t} \frac{V_S}{RC}e^{-\tau/RC}\,d\tau\right)u(t) - \left(\int_{T}^{t} \frac{V_S}{RC}e^{-(\tau-T)/RC}\,d\tau\right)u(t-T) \\
&= -V_S(e^{-t/RC} - 1)u(t) + V_S(e^{-(t-T)/RC} - 1)u(t-T)
\end{aligned}$$

as before. Observe the effects of the unit step functions in the integrands in setting the limits of the integration. For example, $u(\tau - T)$ in the second integral is zero for $\tau < T$, and hence the lower limit of that integral becomes T instead of 0. Also, unit step functions are appended to the integrals to show their region of validity.

Finally we obtain the solution directly, using classical methods. We first obtain the result for $0 < t < T$ by observing that over this interval the source appears as a dc source. Hence the solution is

$$v_C(t) = \underbrace{Ae^{-t/RC}}_{\substack{\text{natural}\\ \text{solution}}} + \underbrace{V_S}_{\substack{\text{forced}\\ \text{solution}}} \qquad 0 < t < T$$

The initial voltage is $v_C(0^-) = v_C(0^+) = 0$. Applying this to evaluate the undetermined constant gives

$$v_C(t) = \underbrace{-V_S e^{-t/RC}}_{\substack{\text{natural}\\ \text{solution}}} + \underbrace{V_S}_{\substack{\text{forced}\\ \text{solution}}} \qquad 0 < t < T$$

Over the interval $t > T$ we observe that the pulse source is turned off and hence there is only a natural solution:

$$v_C(t) = Ae^{-t/RC} \qquad t > T$$

The initial condition is $v_C(T^-) = v_C(T^+)$, which is obtained from the solution over the previous time interval:

$$v_C(T^-) = -V_S e^{-T^-/RC} + V_S$$

Hence the undetermined constant is evaluated as

$$\begin{aligned}
v_C(T^+) &= Ae^{-T^+/RC} \\
&= v_C(T^-) \\
&= -V_S e^{-T^-/RC} + V_S
\end{aligned}$$

or

$$A = -V_S + V_S e^{T/RC}$$

giving

$$\begin{aligned}
v_C(t) &= (-V_S + V_S e^{T/RC})e^{-t/RC} \\
&= V_S e^{-t/RC}(e^{T/RC} - 1) \qquad t > T
\end{aligned}$$

as obtained before.

Exercise Problem 7.37

Determine the response for the capacitor current in the circuit of Fig. E7.37.

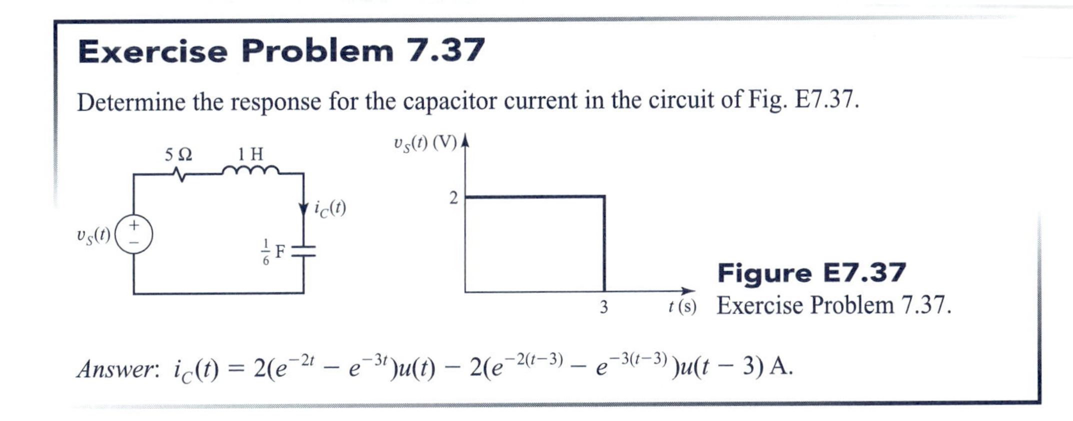

Figure E7.37 Exercise Problem 7.37.

Answer: $i_C(t) = 2(e^{-2t} - e^{-3t})u(t) - 2(e^{-2(t-3)} - e^{-3(t-3)})u(t-3)$ A.

7.10 PSPICE Applications

We will examine the direct time-domain solution using PSPICE. All of the needed functions were discussed in Section 5.7 of Chapter 5.

Example 7.31

Use PSPICE to plot the inductor current for $t > 0$ in the circuit of Fig. 7.38*a*. The switch opens at $t = 0$.

Solution The circuit at $t = 0^-$ for determining the initial conditions is shown in Fig. 7.38*b*. From that we determine the initial inductor current and capacitor voltage as

$$i_L(0^-) = 2 \text{ mA}$$
$$v_C(0^-) = 4 \text{ V}$$

The circuit with nodes numbered is shown in Fig. 7.38*c*, from which we obtain the PSPICE program as

```
EXAMPLE 7.31
IS       0     1     DC      10M
R        1     2     2K
VTEST    2     3
L        3     0     10M     IC=2M
C        1     0     100P    IC=4
.TRAN    .05U  50U   0   .05U    UIC
*INDUCTOR CURRENT IS I(VTEST) OR I(L)
.PROBE
.END
```

We have chosen to solve the circuit out to 50 μs in steps of 0.05 μs and have directed PSPICE to use a solution time step no larger that 0.05 μs as well as to use the initial conditions given for the inductor and capacitor. The result is shown in Fig. 7.39. The result starts at 2mA, the initial inductor current, and eventually converges to the steady-state value of 10mA, which can be confirmed by replacing the inductor with a short circuit and the

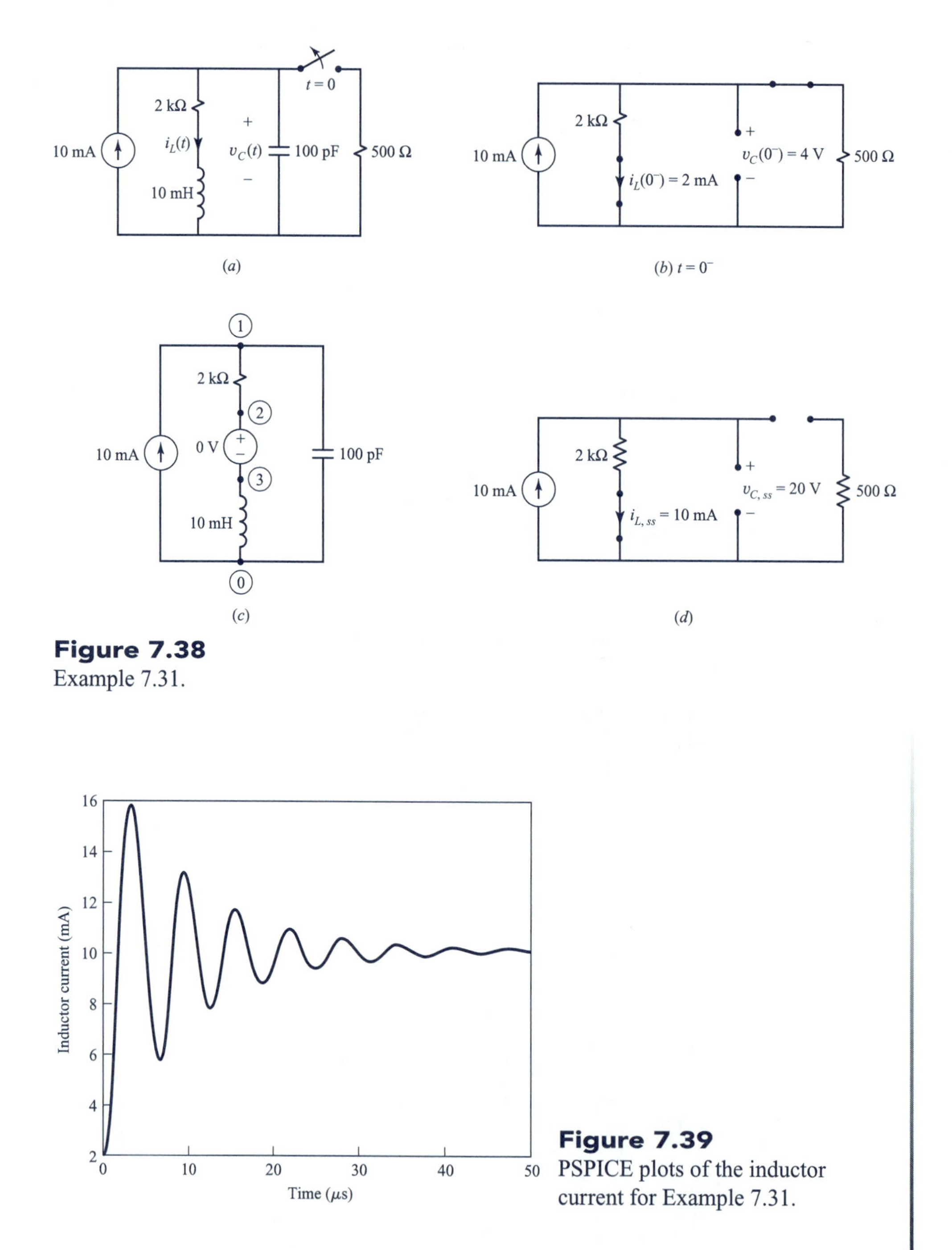

Figure 7.38
Example 7.31.

Figure 7.39
PSPICE plots of the inductor current for Example 7.31.

capacitor with an open circuit in the $t > 0$ circuit as shown in Fig. 7.38*d*. We see that the transient or natural solution is oscillatory, signaling that some poles of the Laplace transform are complex conjugates.

The last application of PSPICE that we will consider is the solution for the step and impulse responses. Recall that these are for zero initial conditions. The only new commands we must investigate in order to simulate the step and impulse responses are the specification of the step function and the impulse function. The step function is the easier of the two. We could specify it as a DC source:

```
VXXX }
IXXX }    N1    N2    DC    value
```

or as a piecewise linear waveform:

```
VXXX }
IXXX }    N1    N2    PWL(0 value)
```

Specifying the impulse function requires a bit more thought. Recall that the unit impulse can be thought of as a rectangular pulse with a very short width and a very large height such that the area contained under the rectangle is unity. So we could specify the unit impulse in the same fashion, but the important question remains: how short should the pulse width be? The answer is essentially that *the pulse duration must be much shorter than the time of any significant variations of current or voltage in the circuit*. For example, in a first-order circuit, variations of the currents and voltages during the transient time interval are on the order of the L/R or RC time constants. In addition, if the source varies with time (e.g., sinusoidally), then the significant variations of the steady-state solution are on the order of, say, $\frac{1}{10}$ of the period. These conditions must be observed when specifying the interval T as shown in Fig. 7.40.

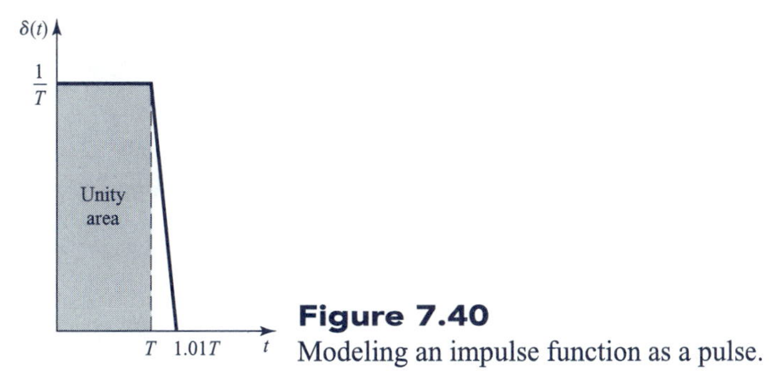

Figure 7.40
Modeling an impulse function as a pulse.

Example 7.32

Determine, using PSPICE, the unit step and unit impulse responses of the circuit shown in Fig. 7.41*a*.

Solution First we determine the unit-step response for the circuit. The PSPICE program with respect to the node numbering in Fig. 7.41*a* is

```
EXAMPLE 7.32
*UNIT STEP RESPONSE
VIN      1     0     DC     1
R        1     2     100
L        2     3     1M     IC=0
C        3     0     10U    IC=0
.TRAN    .1M   6M    0   .1M    UIC
.PROBE
.END
```

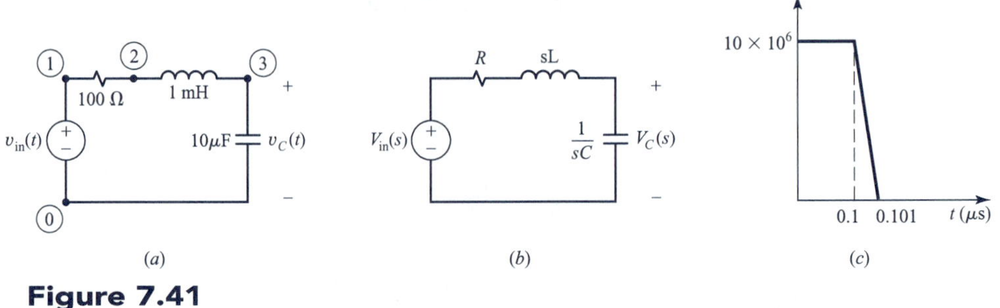

Figure 7.41
Example 7.32.

The result for the capacitor voltage is shown in Fig. 7.42*a*. The solution is zero at $t = 0$, as it should be, and approaches the steady-state value of 1 V, the value of the step-function input. In order to choose the proper impulse function width, it will be helpful to determine the poles of the transfer function. The Laplace transform circuit is shown in Fig. 7.41*b*. From that we obtain

$$V_C(s) = \frac{\frac{1}{sC}}{R + sL + \frac{1}{sC}} \times V_{in}(s)$$

$$= \frac{\frac{1}{LC}}{s^2 + \frac{R}{L}s + \frac{1}{LC}} V_{in}(s)$$

Substituting the numerical values of $R = 100\ \Omega$, $L = 1$ mH, $C = 10\ \mu$F gives

$$V_C(s) = \frac{1 \times 10^8}{s^2 + 1 \times 10^5 s + 1 \times 10^8} \times V_{in}(s)$$

The poles of this transfer function are

$$p_1 = -1.01 \times 10^3$$
$$p_2 = -9.899 \times 10^4$$

These have the units of frequency, so that their reciprocals have the units of time and hence can be interpreted as time constants. The smaller of the two, $1/p_2 = 10.1\ \mu$s, will govern the width of the simulated impulse, whereas the larger of the two, $1/p_1 \approx 1$ ms, tells us that we must choose a final time well past 1 ms. We have chosen 6 ms for this final time. The PSPICE program for the impulse response is

```
EXAMPLE 7.32
*UNIT IMPULSE RESPONSE
VIN     1     0     PWL(0 10E6 0.1E-6 10E6 0.101E-6 0)
R       1     2     100
L       2     3     1M       IC=0
C       3     0     10U      IC=0
.TRAN   .1M   6M    0   .1M    UIC
.PROBE
.END
```

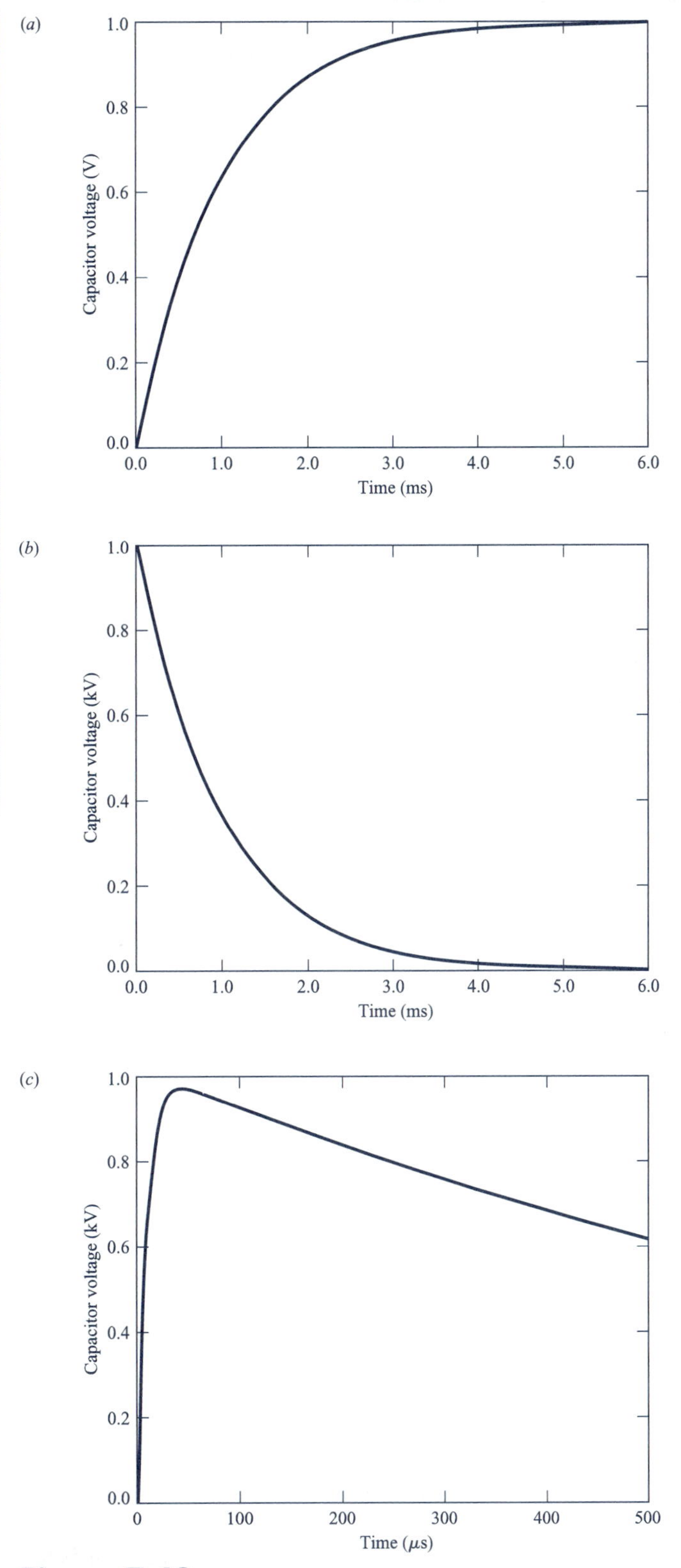

Figure 7.42

Results for the capacitor voltage in Example 7.32: (*a*) unit step response, (*b*) unit impulse response, and (*c*) unit impulse response showing the beginning time behavior.

The unit impulse is simulated with a width of 0.1 μs, which is $\frac{1}{100}$ of the smaller of the two time constants. The results of the simulation are shown in Fig. 7.42*b*. This can be rather easily checked for this problem by obtaining the partial-fraction expansion [with $V_{in}(s) = 1$] as

$$V_{C,\delta}(s) = \frac{1.021 \times 10^3}{s + 1.01 \times 10^3} - \frac{1.021 \times 10^3}{s + 9.899 \times 10^4}$$

The inverse Laplace transform of this is

$$v_{C,\delta}(t) = 1.021 \times 10^3(e^{-1.01\times10^3 t} - e^{-9.899\times10^4 t}) \qquad t > 0$$

This clearly goes to zero, as it should, and the initial value is

$$v_{C,\delta}(0^+) = 0$$

This can be seen more clearly by focusing the plot on the time interval from 0 to 0.5ms as shown in Fig. 7.42*c*.

7.11 MATLAB Applications

In this section we will use MATLAB to plot the solutions of the preceeding two PSPICE examples. We need to discuss some additional MATLAB commands for this. In Section 6.11 of Chapter 6 we discussed several commands that will be needed. Generation of the vector containing the solution time points can be done using

$$>>\text{t}=t_1{:}\Delta t{:}t_2$$

This generates a vector of time points of values $t_1, t_1 + \Delta t, t_1 + 2\,\Delta t, \ldots, t_2$. Another way of specifying these is with the linspace command:

$$>>\text{t=linspace}(t_1, t_2, n)$$

This generates n points equally spaced between t_1 and t_2. Recall the plotting commands:

```
>>plot(x,y)
>>xlabel('The x axis')
>>ylabel('The y axis')
>>title('This is a test')
>>grid
```

The result to be plotted can be represented as the equation of the graph (if it is known). For example, in order to plot $v(t) = 10 - 5e^{-3t}$ we would write

```
>>v=10-5*exp(-3*t)
```

This will generate a vector *v* that contains the values of *v* computed at the time points specified in the vector t that was created above.

The other, and more useful way for applications in circuit analysis is to derive the Laplace transform of the result and have MATLAB plot the solution directly from it. This will be our primary way of plotting the solution. Unfortunately, MATLAB, unlike PSPICE, does not "solve" the circuit. Instead we must obtain the Laplace transform of the result. For example, the result will be of the form of the ratio of two polynomials in s as

$$F(s) = \frac{b_m s^m + b_{m-1} s^{m-1} + \cdots + b_0}{a_n s^n + a_{n-1} s^{n-1} + \cdots + a_0} \tag{7.134}$$

or, in factored form in terms of poles and zeros, as

$$F(s) = K \frac{(s - z_1)(s - z_2) \cdots (s - z_m)}{(s - p_1)(s - p_2) \cdots (s - p_n)} \tag{7.135}$$

We generate vectors of the *coefficients* for the numerator polynomial and for the denominator polynomial in (7.134) (and arbitrarily name these vectors num and den) as

>>num=[b_m, b_{m-1}, ... , b_0]

>>den=[a_n, a_{n-1}, ... , a_0]

For example, suppose we wish to represent the function

$$F(s) = \frac{2s^2 + 5}{3s^3 + 2s}$$

This would be represented as

```
>>num=[2,0,5]
>>den=[3,0,2,0]
```

Note that the coefficients of the powers of s that are missing must be entered with a zero, and that the final entries in these vectors are *always* the coefficients of the zeroth powers of s, namely, b_0 and a_0. Otherwise MATLAB would have no way of knowing which powers of s these go with. The pole–zero form of the function given in (7.135) can be generated with

```
>>[z,p,k]=tf2zp(num,den)
```

This will generate and store the zeros in array z, the poles in array p, and the constant K in the (one-dimensional) array k, as in the form in (7.135). The command name means "transfer function to zero pole." Conversely, if we have arrays containing the zeros, the poles and the constant K, named, for example, z, p, and k, then MATLAB will construct the transfer function as in the form in (7.134) from

```
>>[num,den]=zp2tf(z,p,k)
```

This means "zero pole to transfer function." Note the important ordering of the entries in the argument of *tf2zp* (numerator and then denominator arrays) and in the argument of *zp2tf* (the array containing the zeros, the array containing the poles, and the array containing the constant K). MATLAB will also factor polynomials using the *roots* command:

```
>>roots(num)
>>roots(den)
```

And finally, MATLAB will construct the partial-fraction expansion coefficients with the *residue* command. The partial-fraction expansion is assumed to be

$$F(s) = K + \frac{r_1}{s - p_1} + \frac{r_2}{s - p_2} + \cdots + \frac{r_n}{s - p_n} \tag{7.136}$$

The expansion coefficients $r_1, r_2, \ldots, r_n$ are called *residues* in the mathematical literature. The command

```
>>[r,p,k]=residue(num,den)
```

returns these residues in array r, the poles in array p, and the constant K in array k. Repeated poles are handled in an obvious way. The command

```
>>[num,den]=residue(r,p,k)
```

accomplishes the reverse: it constructs the transfer function from the residues, poles, and constant K.

Finally, there are two important commands that determine, numerically, the inverse Laplace transform: the *impulse* and *step* commands. The command

```
>>f=impulse(num,den,t)
```

determines the values of the inverse transform at the values of t in that array and stores the results in the array f. This is useful in determining the inverse Laplace transform of any Laplace transform, regardless of whether it represents an impulse response. Since the Laplace transform of the unit impulse is 1, this command determines the inverse transform of any $F(s)$. The unit step response is obtained with the command

```
>>f=step(num,den,t)
```

The function represented by *num* and *den* is assumed to be the product of the impulse response and the transform of the unit step, $1/s$.

Example 7.33

Determine the solution for the inductor current in the circuit of Example 7.31 shown in Fig. 7.38*a*.

Solution The first task is to derive the Laplace transform of the inductor current. The initial values of the inductor current and capacitor voltage were determined in Example 7.31 to be

$$i_L(0^-) = 2 \text{ mA}$$
$$v_C(0^-) = 4 \text{ V}$$

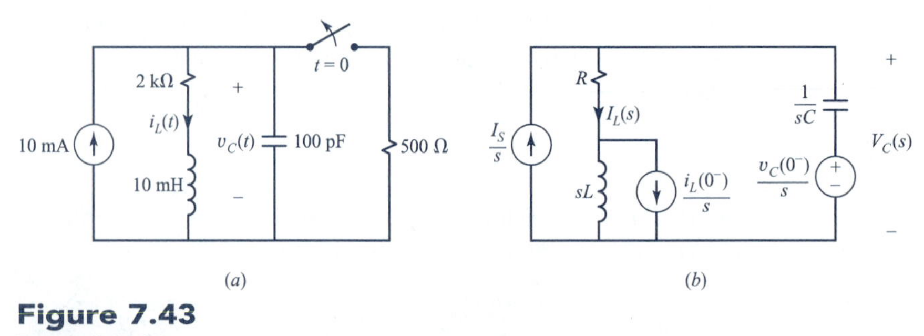

Figure 7.43
Example 7.33.

The Laplace-transformed circuit is shown, using symbols, in Fig. 7.43*b*. The Laplace transform of the inductor current is easily obtained with superposition:

$$I_L(s) = \frac{\frac{1}{sC}}{R + sL + \frac{1}{sC}} \times \frac{I_S}{s} + \frac{sL}{R + sL + \frac{1}{sC}} \times \frac{i_L(0^-)}{s} + \frac{1}{R + sL + \frac{1}{sC}} \times \frac{v_C(0^-)}{s}$$

This simplifies to

$$I_L(s) = \frac{LCi_L(0^-)\, s^2 + Cv_C(0^-)\, s + I_S}{LCs^3 + RCs^2 + s}$$

$$= \frac{i_L(0^-)\, s^2 + \frac{v_C(0^-)}{L} s + \frac{I_S}{LC}}{s\left(s^2 + \frac{R}{L}s + \frac{1}{LC}\right)}$$

Substituting the element and initial condition values gives

$$I_L(s) = \frac{2 \times 10^{-3}\, s^2 + 4 \times 10^2\, s + 1 \times 10^{10}}{s^3 + 2 \times 10^5\, s^2 + 1 \times 10^{12} s}$$

The poles of this are

$$p_1 = -1 \times 10^5 + j9.95 \times 10^5$$
$$p_2 = -1 \times 10^5 - j9.95 \times 10^5$$
$$p_3 = 0$$

The partial-fraction expansion of this expression is

$$I_L(s) = \frac{1 \times 10^{-2}}{s} + \frac{-4 \times 10^{-3} + j4.02 \times 10^{-4}}{s + 1 \times 10^5 - j9.95 \times 10^5} + \frac{-4 \times 10^{-3} - j4.02 \times 10^{-4}}{s + 1 \times 10^5 + j9.95 \times 10^5}$$

or

$$I_L(s) = \frac{1 \times 10^{-2}}{s} + \frac{4.02 \times 10^{-3}\angle 174.3°}{s + 1 \times 10^5 - j9.95 \times 10^5} + \frac{4.02 \times 10^{-3}\angle -174.3°}{s + 1 \times 10^5 + j9.95 \times 10^5}$$

From this we obtain the inverse Laplace transform as

$$i_L(t) = 1 \times 10^{-2} + 8.04 \times 10^{-3}\, e^{-10^5 t} \cos(9.95 \times 10^5 t + 174.3°) \text{ A}$$

The MATLAB program is

```
>> t=0:.1e-6:50e-6;
>> il=le-2+8.04e-3*exp(-1e5*t).*cos(9.95e5*t+174.3*pi/180);
>> plot(t,il)
>> xlabel('Time (seconds)')
>> ylabel('Inductor current (amps)')
>> title('Example 7.33')
>> num=[2e-3,4e2,1e10];
>> den=[1,2e5,1e12,0];
>> roots(num)
ans =
```

```
-1.0000e+005+ 2.2338e+006i
-1.0000e+005- 2.2338e+006i
>> roots(den)
ans =
          0
-1.0000e+005+ 9.9499e+005i
-1.0000e+005- 9.9499e+005i
>> [z,p,k]=tf2zp(num,den)
z =
-1.0000e+005+ 2.2338e+006i
-1.0000e+005- 2.2338e+006i
p =
          0
-1.0000e+005+ 9.9499e+005i
-1.0000e+005- 9.9499e+005i
k =
  2.0000e-003
>> [num,den]=zp2tf(z,p,k)
num =
          0 2.0000e-003  4.0000e+002  1.0000e+010
den =
  1.0000e+000  2.0000e+005  1.0000e+012        0
>> [r,p,k]=residue(num,den)
r =
-4.0000e-003+ 4.0202e-004i
-4.0000e-003- 4.0202e-004i
1.0000e-002
p =
```

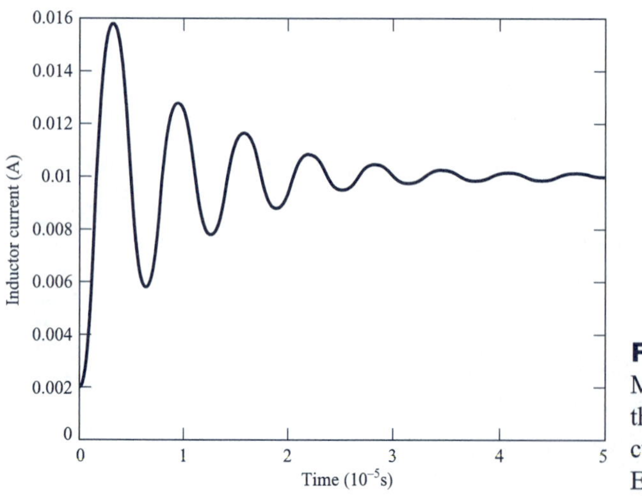

Figure 7.44 MATLAB plots of the inductor current of Example 7.33.

```
-1.0000e+005+ 9.9499e+005i
-1.0000e+005- 9.9499e+005i
        0
k =
   []
>> [num,den]=residue(r,p,k)
num =
     2.0000e-003 4.0000e+002 1.0000e+010
den =
   1.0000e+000 2.0000e+005 1.0000e+012      0
>> il=impulse(num,den,t);
>> plot(t,il)
>> xlabel('Time (seconds)')
>> ylabel('Inductor current (amps)')
>> title('Example 7.33')
```

The first line defines the time increments and places them in the array *t*. The next line computes the inductor current at each time point from the derived solution and puts the values in the array *il*. Note in forming the product of exp(-1e5*t) and cos(9.95e5*t+174.3*pi/180) that a period must be placed at the end of exp(-1e5*t), since exp(-1e5*t) and cos(9.95e5*t+174.3*pi/180) are both arrays (one-dimensional vectors) and the period denotes the dot product—the result of multiplying corresponding components together and then summing the products. The next few lines define the plot of this inductor current, which is shown in Fig. 7.44. This is virtually identical to the PSPICE-generated plot in Fig. 7.39. The next several lines define the coefficients of the numerator polynomial and denominator polynomial. Once these are defined, we compute the roots of the numerator and denominator with *roots(num)* and *roots(den)*. Then we compute the zeros and poles and constant K with the *tf2zp* command and reverse it with the *zp2tf* command. Next we compute the partial-fraction expansion with *[r,p,k]=residue(num,den)*. These should be compared with the partial-fraction expansion obtained by hand and given above. This is reversed with *[num,den]=residue(r,p,k)*, which gives back the Laplace transform. And finally we determine the numerical inversion of the Laplace transform with the *impulse(num,den,t)* command. This creates the array *il*, which is plotted, giving the same plot as shown in Fig. 7.44.

Example 7.34

Determine the impulse and step responses for the capacitor voltage in Fig. 7.41*a* that was obtained with PSPICE in Example 7.32.

Solution Again the first task is to determine the transforms of the impulse response (transfer function). This was determined in Example 7.32 and is repeated here:

$$V_C(s) = \frac{1 \times 10^8}{s^2 + 1 \times 10^5\, s + 1 \times 10^8} \times V_{\text{in}}(s)$$

The poles of this transfer function are

$$p_1 = -1.01 \times 10^3$$
$$p_2 = -9.899 \times 10^4$$

Hence the impulse response is

$$\begin{aligned} H(s) &= \frac{V_C(s)}{V_{in}(s) = 1} \\ &= \frac{1 \times 10^8}{s^2 + 1 \times 10^5\, s + 1 \times 10^8} \\ &= \frac{1.021 \times 10^3}{s + 1.01 \times 10^3} - \frac{1.021 \times 10^3}{s + 9.899 \times 10^4} \end{aligned}$$

and the inverse Laplace transform of this is

$$v_{C,\delta}(t) = 1.021 \times 10^3 \left(e^{-1.01\times10^3 t} - e^{-9.899\times10^4 t}\right) \qquad t > 0$$

The MATLAB program is

```
>> num=[1e8];
>> den=[1,1e5,1e8];
>> [r,p,k]=residue(num,den)
r =
 -1.0206e+003
  1.0206e+003
p =
 -9.8990e+004
 -1.0102e+003
k =
    []
>> t=0:.1e-3:6e-3;
>> h=impulse(num,den,t);
>> plot(t,h)
>> xlabel('Time (seconds)')
>> ylabel('Capacitor voltage (V)')
>> title('Example 7.34, Impulse Response')
```

This confirms the partial-fraction expansion above. The plot of the impulse response is shown in Fig. 7.45, which should be compared with the result in Fig. 7.42*b* that was obtained with PSPICE. The transform of the step response is obtained by multiplying the above impulse response by the transform of a unit step, $1/s$, to yield

$$\begin{aligned} V_{C,u}(s) &= H(s) \times \left(V_{in}(s) = \frac{1}{s}\right) \\ &= \frac{1 \times 10^8}{s^2 + 1 \times 10^5\, s + 1 \times 10^8} \times \frac{1}{s} \\ &= \frac{1 \times 10^8}{s^3 + 1 \times 10^5\, s^2 + 1 \times 10^8\, s} \\ &= \frac{1}{s} + \frac{1.031 \times 10^{-2}}{s + 9.899 \times 10^4} - \frac{1.011}{s + 1.01 \times 10^3} \end{aligned}$$

The inverse transform is

$$v_{C,u}(t) = 1 + 1.031 \times 10^{-2}\, e^{-9.899\times10^4 t} - 1.011 e^{-1.01\times10^3 t} \qquad t > 0$$

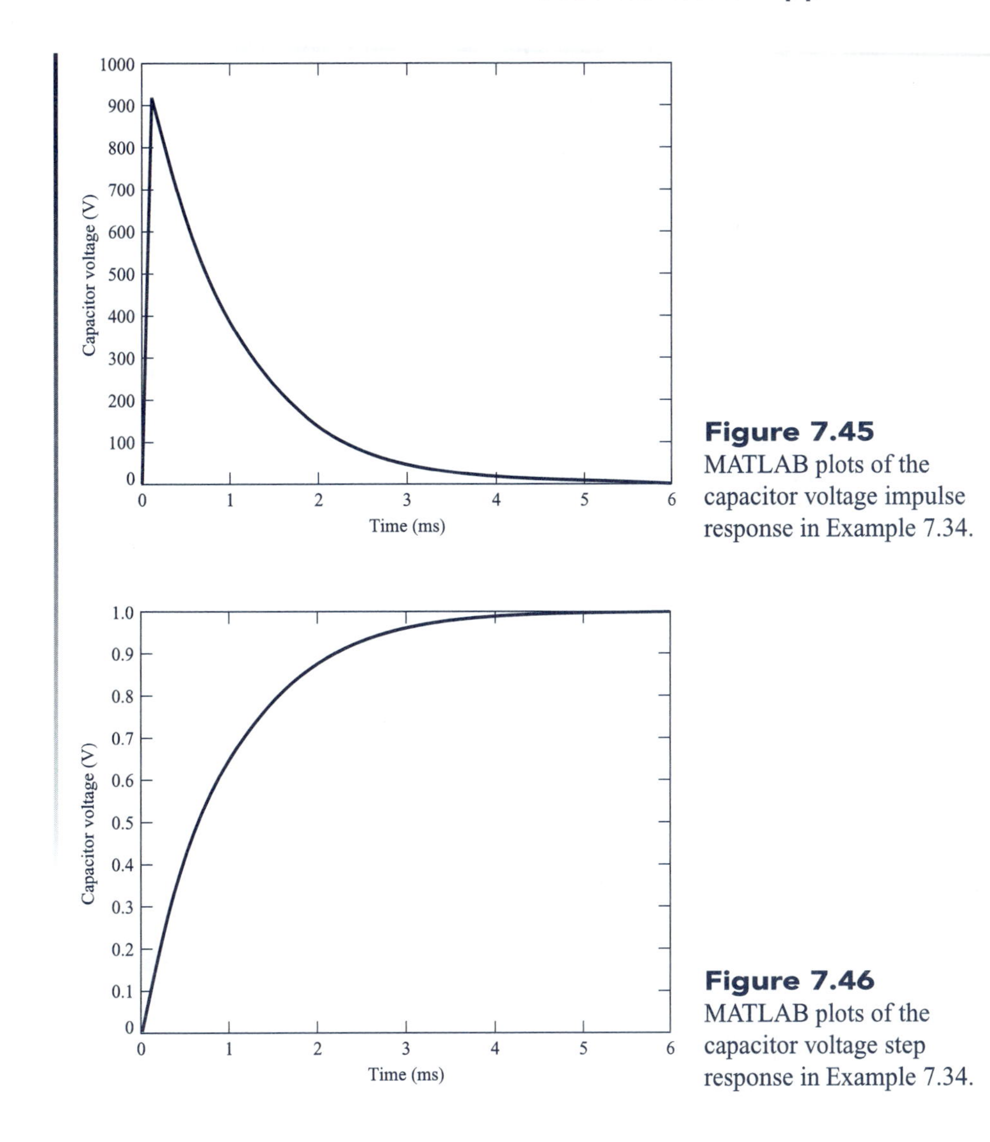

Figure 7.45
MATLAB plots of the capacitor voltage impulse response in Example 7.34.

Figure 7.46
MATLAB plots of the capacitor voltage step response in Example 7.34.

The MATLAB program is

```
>> num=[1e8];
>> den=[1,1e5,1e8,0];
>> [r,p,k]=residue(num,den)
r =
  1.0310e-002
-1.0103e+000
1.0000e+000
p =
-9.8990e+004
-1.0102e+003
        0
k =
    []
```

```
>> t=0:.1e-3:6e-3;
>> vcstep=impulse(num,den,t);
>> plot(t,vcstep)
>> xlabel('Time (seconds)')
>> ylabel('Capacitor voltage (V)')
>> title('Example 7.34, Step Response')
```

This confirms the above partial-fraction expansion. The step response is plotted in Fig. 7.46 and should be compared with the result obtained with PSPICE and shown in Fig. 7.42*a*.

7.12 Application Examples

This section includes some examples of the practical application of the methods and concepts of this chapter.

7.12.1 A Light Flasher The first example is the design of a circuit to cause a neon bulb to flash at a specified rate. A typical circuit for doing this is shown in Fig. 7.47*a*. The neon bulb appears as an open circuit until its voltage equals $V_{\text{breakdown}}$. This is a first-order RC circuit. Hence, the capacitor charges up exponentially at the rate of $e^{-t/RC}$, approaching the battery voltage V_{battery}. From our earlier results, once the battery is connected, the capacitor voltage increases as

$$v_C(t) = \left(1 - e^{-t/RC}\right)V_{\text{battery}}$$

This continues until $v_C(T) = V_{\text{breakdown}}$. Hence we can solve for the time T by combining these results as

$$\begin{aligned} V_{\text{breakdown}} &= v_C(T) \\ &= \left(1 - e^{-T/RC}\right)V_{\text{battery}} \end{aligned}$$

Solving this gives

$$T = RC \ln\left(\frac{1}{1 - \dfrac{V_{\text{breakdown}}}{V_{\text{battery}}}}\right)$$

Once the capacitor voltage equals $V_{\text{breakdown}}$, the neon bulb breaks down and becomes a short circuit discharging the capacitor, and the process continues periodically. For example, an

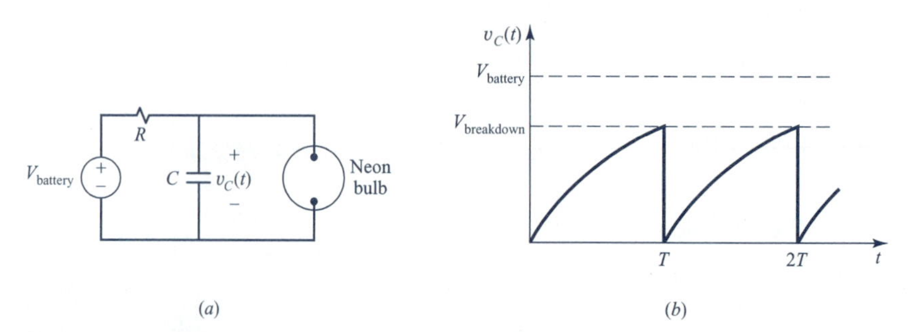

Figure 7.47
A light flasher: (a) the physical circuit, and (b) the resulting waveform of the capacitor voltage.

automobile turn signal may use an 8-V bulb and a 12-V battery. If a capacitor of 1 μF is used and a flash rate of 1 flash/s is desired, we solve the above relation to give a value for R of 910.2 kΩ.

7.12.2 Ringing in Digital Circuits As was explained in an application example in the previous chapter, electronic circuits are interconnected by conductors (wires, PCB lands). A parallel pair of these conductors constitutes a transmission line and has inductance and capacitance associated with it. This nonideal property of such a line can have a significant effect on signal transmission along it. As an example, let us consider a pair of closely spaced, insulated wires, as in a ribbon cable that is commonly used to interconnect digital logic circuits. The two wires are 28 gauge and are separated by 50 mils (1 mil = 0.0001 in). Figure 7.48*a* shows such an interconnection circuit. The inductance and capacitance of the transmission line (the ribbon cable) make this a second-order circuit, and the line is modeled as shown in Fig. 7.48*b*. Hence the input voltage to the second gate can be underdamped and have oscillatory behavior. This is a very serious situation for digital logic circuits, because the input voltage to the second gate can vary into regions that cause the logic gate to falsely interpret a 0 as a 1 or vice versa. It is common to insert a resistor R, as shown, to damp these oscillations.

Figure 7.48*c* shows the Laplace-transformed circuit. From that we obtain the transform of the input voltage to the second gate as (the reader should verify this)

$$V_{\text{in}}(s) = \frac{\dfrac{1}{LC}}{s^2 + s\left(\dfrac{R+50}{L} + \dfrac{1}{CR_{\text{in}}}\right) + \left(\dfrac{R+50+R_{\text{in}}}{LCR_{\text{in}}}\right)} V_S(s)$$

Substituting the values of $R = 0$, $R_{\text{in}} = 100\ \text{k}\Omega$, $L = 232\ \text{nH}$, $C = 4.45\text{pF}$ gives the roots of the denominator as

$$p_1, p_2 = -1.0888 \times 10^8 \pm j9.7839 \times 10^8$$

Hence we expect to see an underdamped (oscillatory) variation of the input voltage to the second gate. This oscillation would have a radian frequency of $\omega = 9.7839 \times 10^8$ rad/s, that is, a cyclic frequency of 155.7 MHz. The reciprocal of this is the period of the oscillation, 6.42ns.

We will solve this using PSPICE. The PSPICE circuit is shown in Fig. 7.48*d*. Also shown is a typical digital logic waveform. The voltage varies between 0 and 5V and has a rise time and a fall time of 5ns. The PSPICE coding is

```
RINGING CIRCUIT (R=0)
VS 1 0 PWL(0  0  5N  5  25N  5  30N  0)
RS 1 2 50
R 2 3 1.E-6
L 3 4 232N IC=0
C 4 0 4.45P IC=0
RIN 4 0 100K
.TRAN 0.1N 60N UIC
.PROBE
* THE VOLTAGE AT THE INPUT TO GATE #2 IS V(4)
.END
```

The input voltage to gate 2 with R=0 is shown in Fig. 7.49*a*. Observe the oscillatory response as the voltage changes logic state. This is called *ringing,* or *overshoot* and *undershoot*. For this case

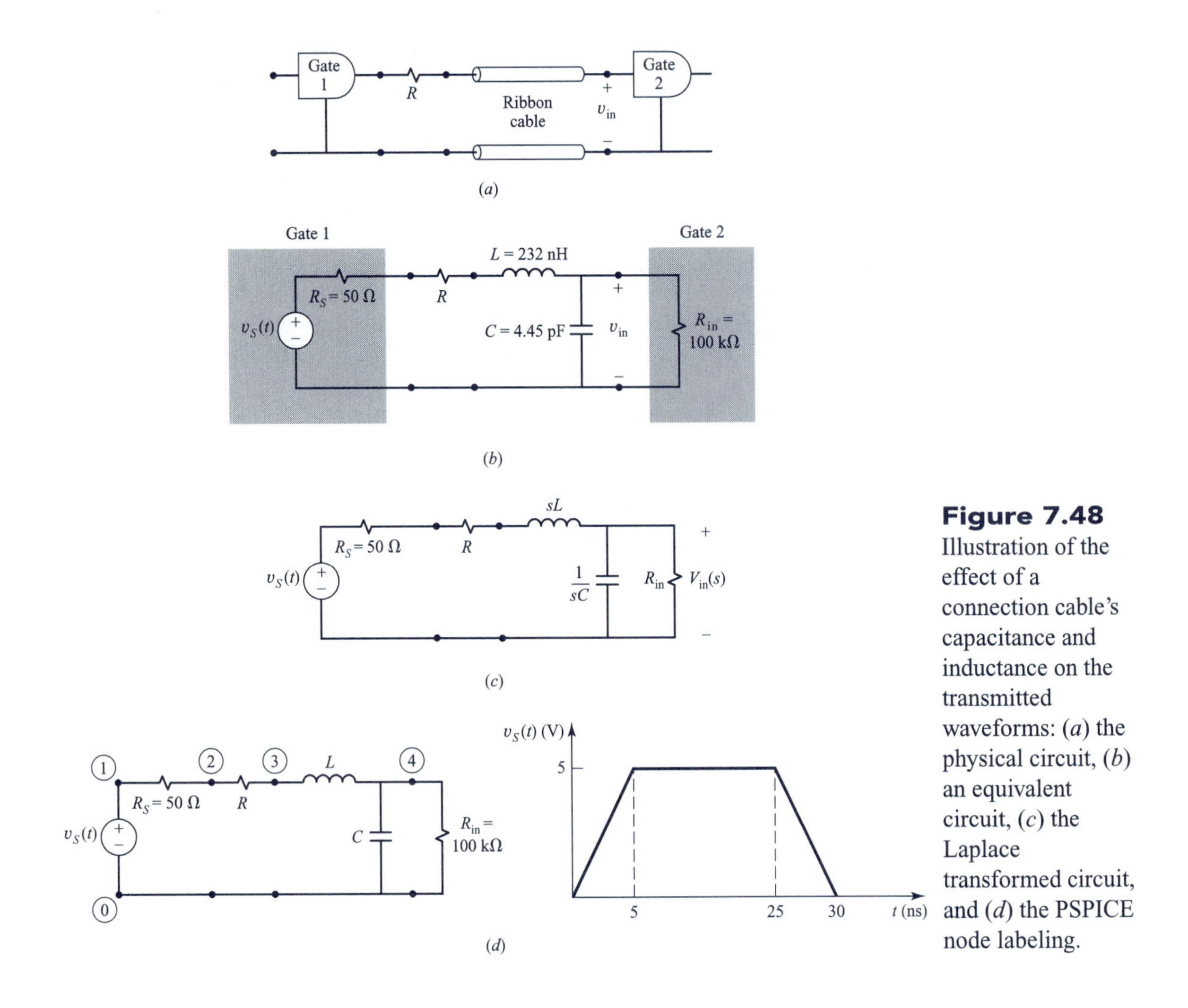

Figure 7.48 Illustration of the effect of a connection cable's capacitance and inductance on the transmitted waveforms: (*a*) the physical circuit, (*b*) an equivalent circuit, (*c*) the Laplace transformed circuit, and (*d*) the PSPICE node labeling.

the voltage in the 1 or high state should be 5 V, but it oscillates between 4 and 6 V. Similarly, the voltage in the 0 or low state should be 0 V but it oscillates between +1 and −1 V. This can cause severe logic problems. Also observe that the period of oscillation is about 6ns, as we predicted. Next we insert a value of R into the circuit to damp the oscillations. Figure 7.49*b* shows the result for $R = 100\ \Omega$, and Fig. 7.49*c* for $R = 250\ \Omega$. For $R = 250\ \Omega$, the oscillations are almost damped out and the logic gates should perform as expected.

7.12.3 Crosstalk in Digital Circuits In the application example in Section 6.12.2 of the previous chapter we discussed how the interaction between parallel conductors can unintentionally cause a signal from one pair to be coupled to another nearby pair and show up at the terminals of that pair. This is referred to as *crosstalk*. In this final example we will illustrate crosstalk for a digital logic circuit. Figure 7.50*a* shows a set of *lands* on a printed circuit board (PCB) that carry signals of two circuits. The lands are copper etchings on a fiberglass PCB. The dimensions are as shown. The equivalent circuit of that situation is shown in Fig. 7.50*b* along with the element values for the inductances and capacitances of the conductors. We will apply a trapezoidal pulse typical of a digital logic signal to the input of one line and compute, using

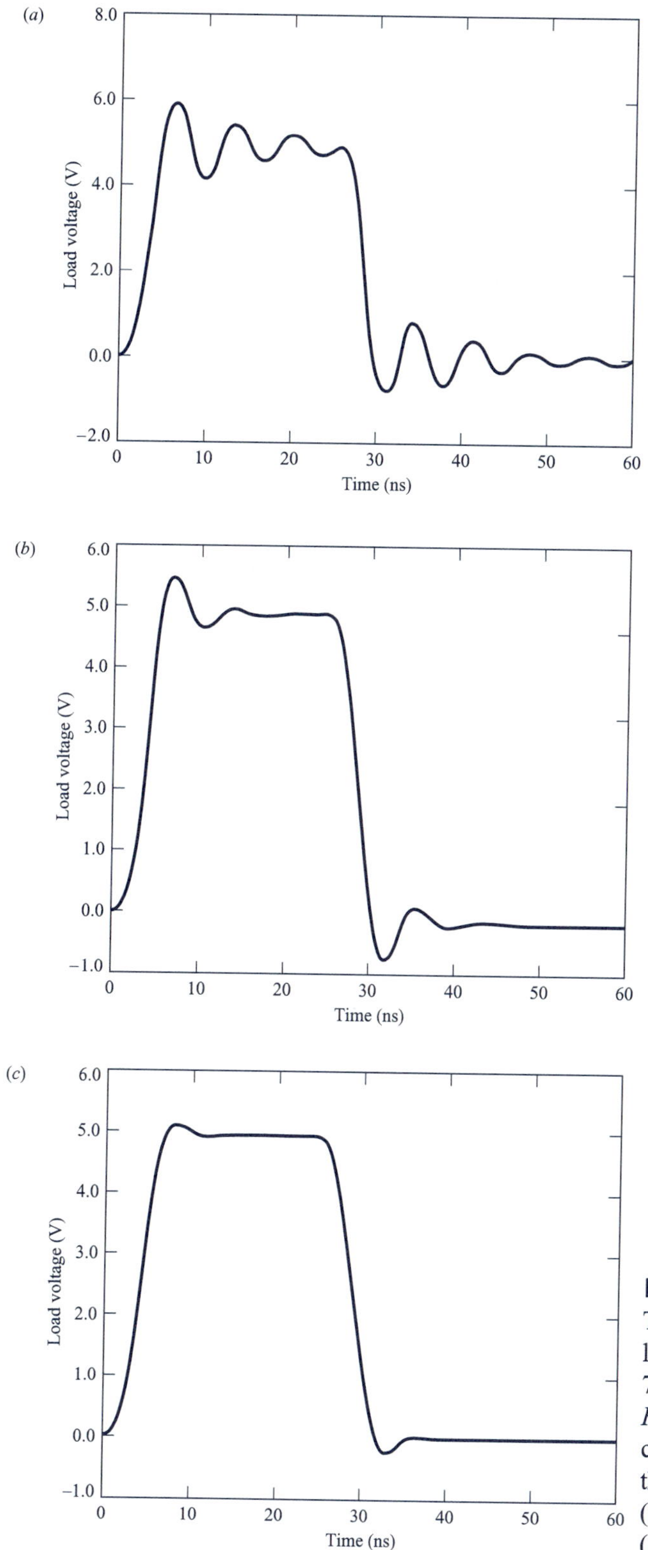

Figure 7.49
The PSPICE simulation of the load voltage of the circuit in Fig. 7.48 showing (*a*) the ringing for R=0Ω caused by the cable's capacitance and inductance, and the load voltage for (*b*) R=100Ω, and (*c*) R=250Ω.

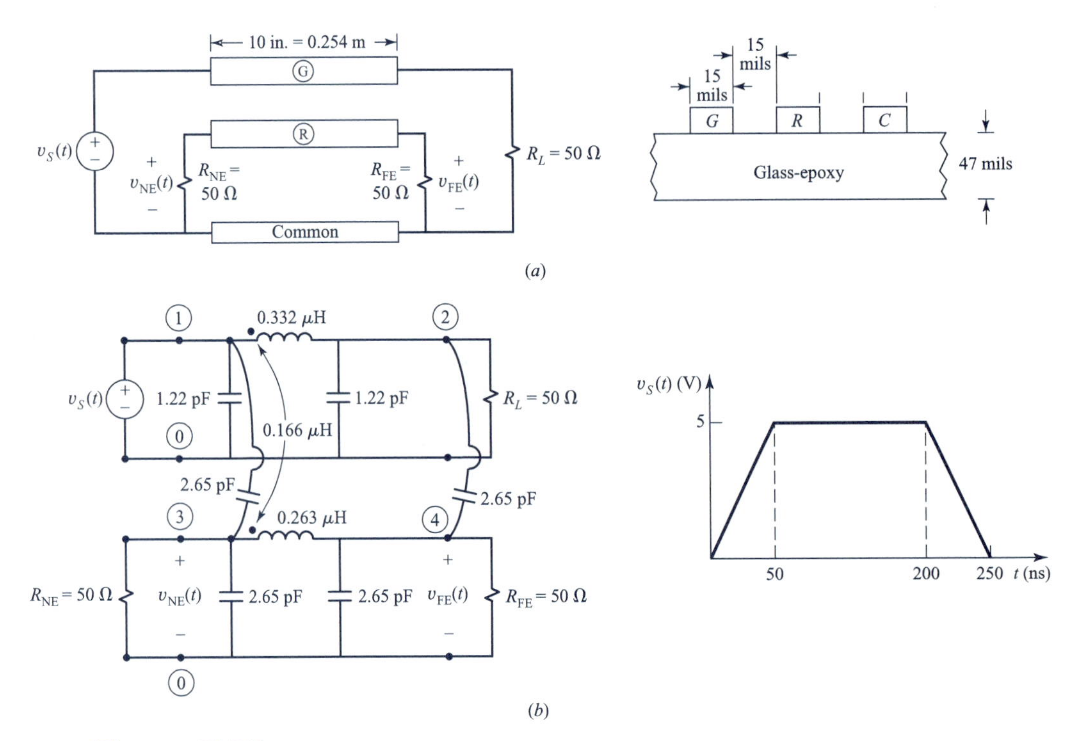

Figure 7.50
Illustration of crosstalk between transmission lines caused by their mutual capacitance and inductance: (*a*) the physical configuration, and (*b*) an equivalent circuit with typical values and nodes numbered for use in PSPICE.

PSPICE, the near-end and far-end crosstalk voltages, $v_{NE}(t)$ and $v_{FE}(t)$, that appear (unintentionally) at the terminations of the other line. Figure 7.50*b* shows that the pulse has rise and fall times of 50ns and varies between logic level 0 (0 V) and logic level 1 (5 V). The PSPICE code is

```
CROSSTALK PROBLEM (PCB)
VS 1 0 PWL(0 0 50N 5 200N 5 250N 0)
CGL 1 0 1.22P IC=0
CRL 3 0 2.65P IC=0
CML 1 3 2.65P IC=0
CGR 2 0 1.22P IC=0
CRR 4 0 2.65P IC=0
CMR 2 4 2.65P IC=0
LG 1 2 0.332U IC=0
LR 3 4 0.263U IC=0
K LG LR 0.5618
RL 2 0 50
RNE 3 0 50
RFE 4 0 50
.TRAN 0.5N 400N UIC
.PROBE
.END
```

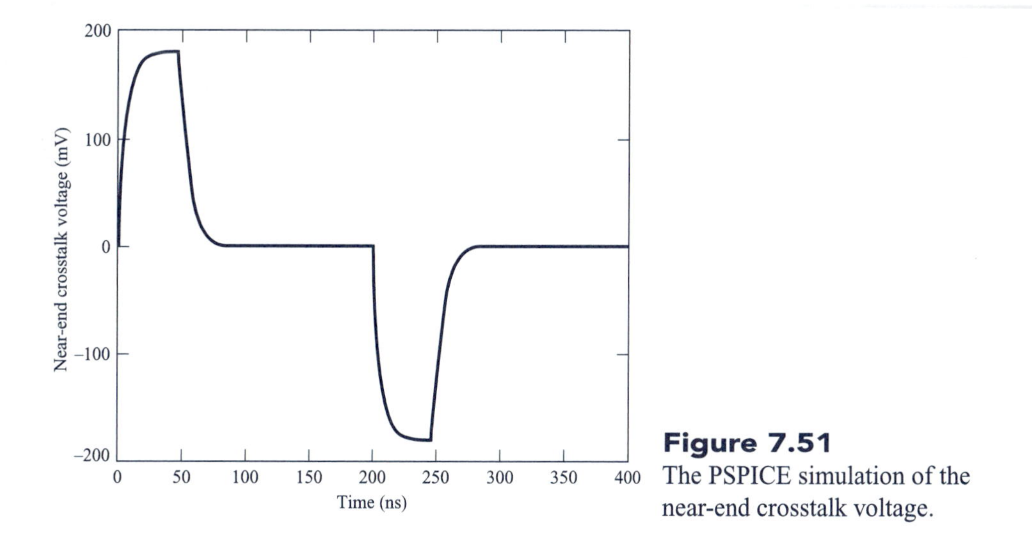

Figure 7.51
The PSPICE simulation of the near-end crosstalk voltage.

The near-end crosstalk voltage is plotted in Fig. 7.51. Observe that these pulses appear during the time the source voltage is changing. Their level is on the order of 180 mV. These and the other crosstalk results of Chapter 6 are verified experimentally in C.R. Paul, *Introduction to Electromagnetic Compatibility,* J Wiley-Interscience, New York, 1992.

Problems

Section 7.1 First-Order Circuit Response

7.1-1 Determine and sketch the solution for the current $i(t)$ for $t > 0$ in the circuit of Fig. P7.1-1.
Answer: $5e^{-2t/3} + 5$ A.

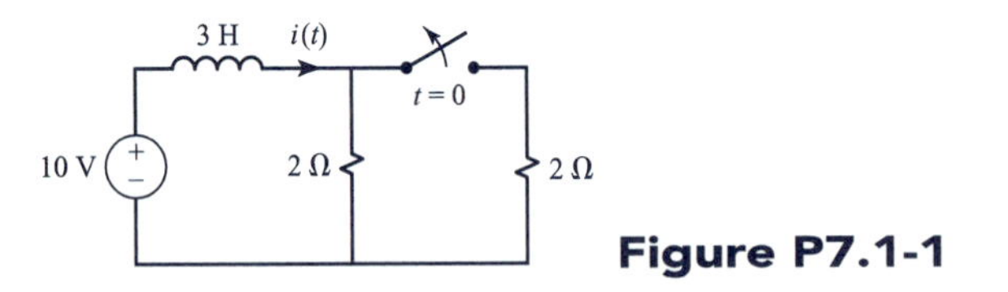

Figure P7.1-1

7.1-2 Determine and sketch the solution for the current $i(t)$ for $t > 0$ in the circuit of Fig. P7.1-2.
Answer: $-\frac{6}{21}e^{-t} + 1$ A.

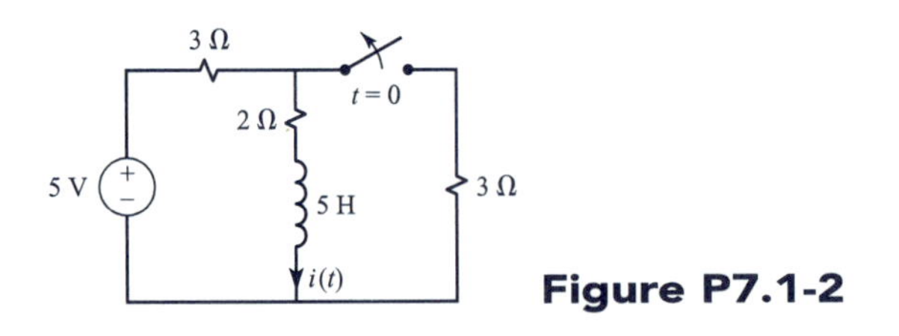

Figure P7.1-2

7.1-3 Determine and sketch the solution for the current $i(t)$ for $t > 0$ in the circuit of Fig. P7.1-3.
Answer: $-\frac{5}{8}e^{-2t/5} + \frac{5}{4}$ A.

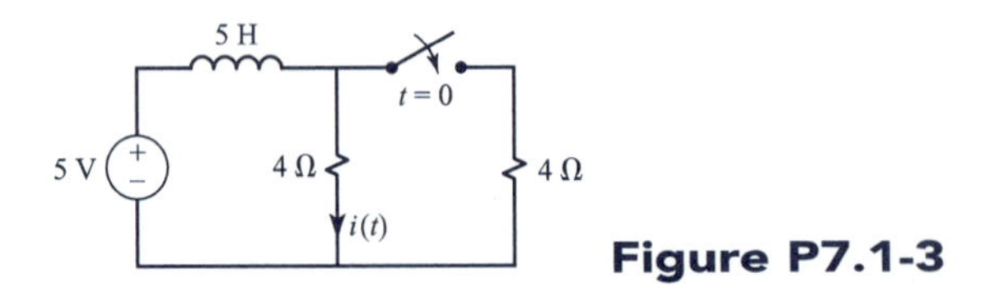

Figure P7.1-3

7.1-4 Determine and sketch the solution for the current $i(t)$ for $t > 0$ in the circuit of Fig. P7.1-4.
Answer: $2e^{-4t/3} + 2$ A.

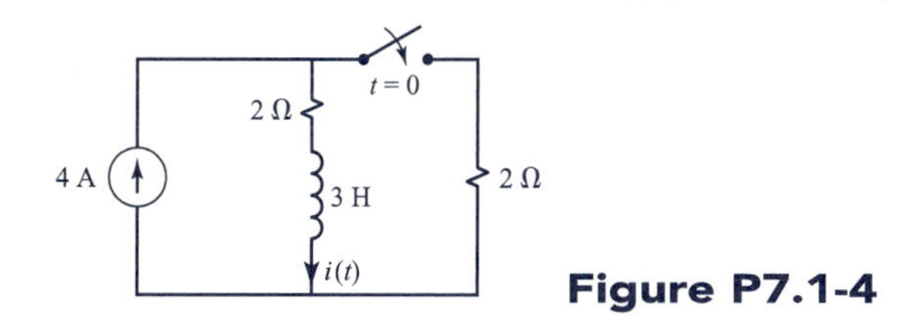

Figure P7.1-4

7.1-5 Determine and sketch the solution for the voltage $v(t)$ for $t > 0$ in the circuit of Fig. P7.1-5.
Answer: $-5e^{-t/6} + 10$ V.

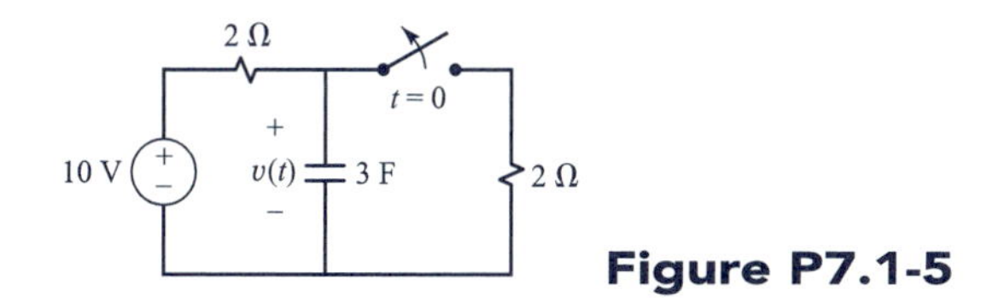

Figure P7.1-5

7.1-6 Determine and sketch the solution for the voltage $v(t)$ for $t > 0$ in the circuit of Fig. P7.1-6.
Answer: $-\frac{8}{3}e^{-9t/4} + \frac{8}{3}$ V.

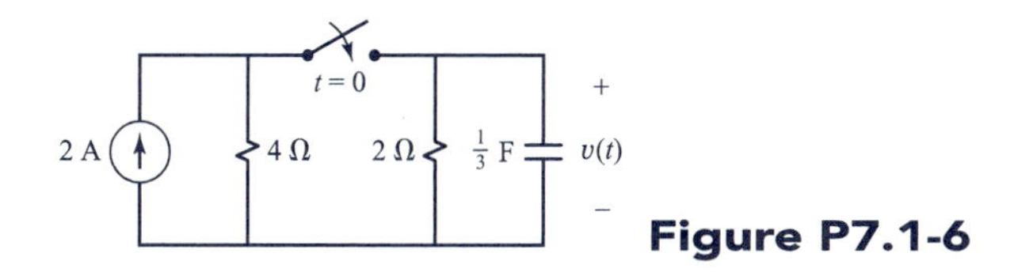

Figure P7.1-6

7.1-7 Determine and sketch the solution for the voltage $v(t)$ for $t > 0$ in the circuit of Fig. P7.1-7.
Answer: $\frac{5}{3}e^{-15t/4} + \frac{10}{3}$ V.

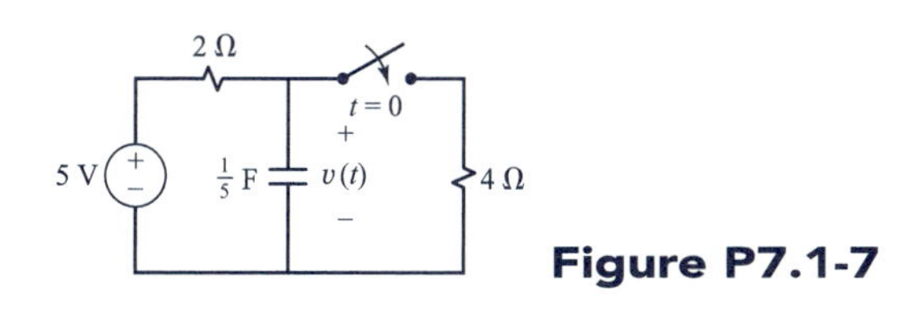

Figure P7.1-7

7.1-8 Determine and sketch the solution for the voltage $v(t)$ for $t > 0$ in the circuit of Fig. P7.1-8.
Answer: $\frac{8}{3}e^{-3t/20} + \frac{16}{3}$ V.

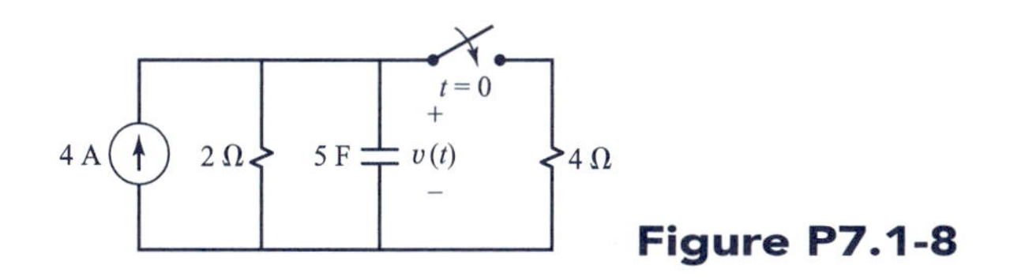

Figure P7.1-8

7.1-9 Determine and sketch the solution for the inductor current $i_L(t)$ for $t > 0$ in the circuit of Fig. P7.1-9. *Answer*: $5e^{-t/5}$ A.

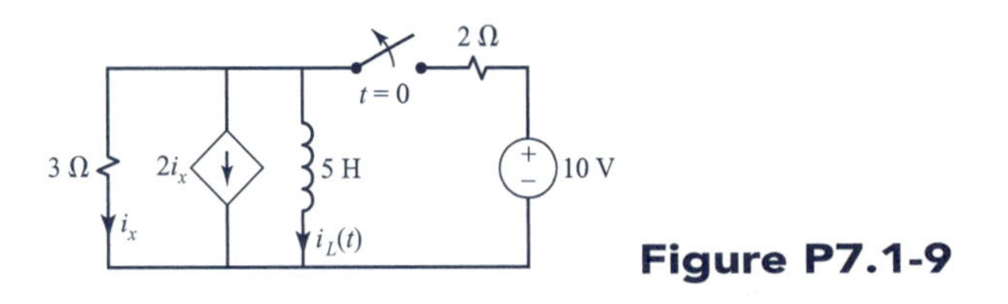

Figure P7.1-9

7.1-10 Determine and sketch the solution for the capacitor voltage $v_C(t)$ for $t > 0$ in the circuit of Fig. P7.1-10. *Answer*: $-\frac{25}{2}e^{-t} + 10$ V.

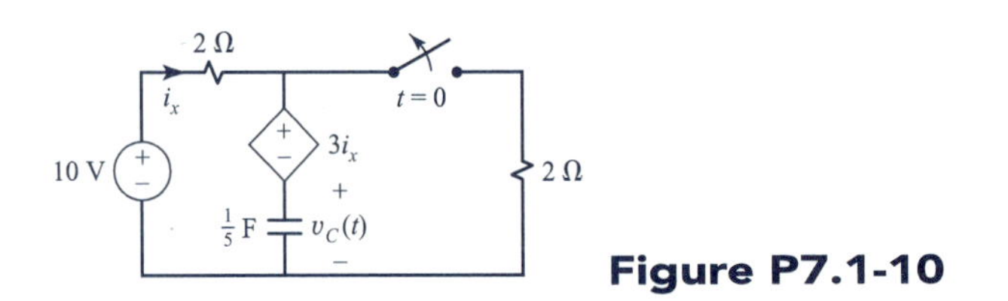

Figure P7.1-10

Section 7.2 Second-Order Circuit Response

7.2-1 For the circuit of Fig. P7.2-1, determine the solutions for the inductor current and capacitor voltage for $R = 4\ \Omega$, $L = 1$ H, $C = \frac{1}{3}$ F, $R_L = 4\ \Omega$.
Answers: $\frac{15}{8}e^{-t} - \frac{5}{8}e^{-3t}$ A, $-\frac{45}{8}e^{-t} + \frac{5}{8}e^{-3t} + 10$ V.

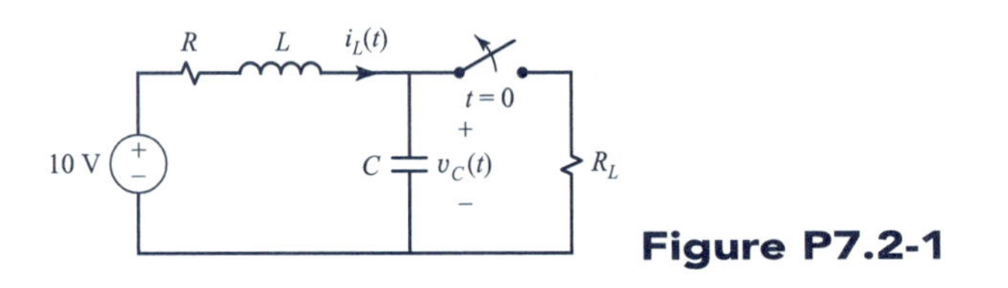

Figure P7.2-1

7.2-2 For the circuit of Fig. P7.2-1, determine the solutions for the inductor current and capacitor voltage for $R = 4\ \Omega$, $L = 1$ H, $C = \frac{1}{4}$ F, $R_L = 2\ \Omega$.
Answers: $\frac{5}{3}e^{-2t} + \frac{10}{3}te^{-2t}$ A, $-\frac{20}{3}e^{-2t} - \frac{20}{3}te^{-2t} + 10$ V.

7.2-3 For the circuit of Fig. P7.2-1, determine the solutions for the inductor current and capacitor voltage for $R = 8\ \Omega$, $L = 2$ H, $C = \frac{1}{26}$ F, $R_L = 4\ \Omega$.
Answers: $\frac{5}{6}e^{-2t}\cos 3t + \frac{5}{9}e^{-2t}\sin 3t$ A, $-\frac{20}{3}e^{-2t}\cos 3t + \frac{25}{9}e^{-2t}\sin 3t + 10$ V.

7.2-4 For the circuit of Fig. P7.2-4, determine the solutions for the inductor current and capacitor voltage for $R = 4\ \Omega$, $L = \frac{3}{2}$ H, $C = \frac{1}{12}$ F, $R_L = 4\ \Omega$.
Answers: $-4e^{-2t} + 2e^{-4t} + 2$ A, $12e^{-2t} - 12e^{-4t}$ V.

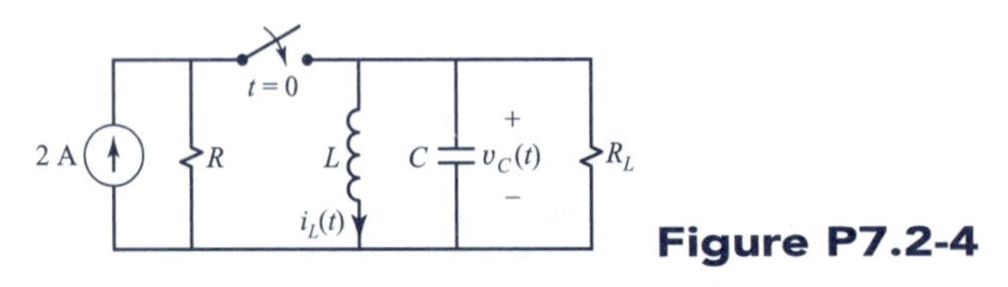

Figure P7.2-4

7.2-5 For the circuit of Fig. P7.2-4, determine the solutions for the inductor current and capacitor voltage for $R = 4\ \Omega$, $L = \frac{4}{5}$ H, $C = \frac{1}{20}$ F, $R_L = 4\ \Omega$.
Answers: $-2e^{-5t} - 10te^{-5t} + 2$ A, $40te^{-5t}$ V.

7.2-6 For the circuit of Fig. P7.2-4, determine the solutions for the inductor current and capacitor voltage for $R = 4\ \Omega$, $L = \frac{8}{29}$ H, $C = \frac{1}{8}$ F, $R_L = 4\ \Omega$.
Answers: $-2e^{-2t}\cos 5t - \frac{4}{5}e^{-2t}\sin 5t + 2$ A, $\frac{16}{5}e^{-2t}\sin 5t$ V.

Section 7.3 The Laplace Transform

7.3-1 Determine the Laplace transform of $f(t) = (a + bt)u(t)$. *Answer*: $a/s + b/s^2$.

7.3-2 Determine the Laplace transform of $f(t) = (e^{-at} - e^{-bt})u(t)$.
Answer: $(b - a)/[s^2 + (a + b)s + ab]$.

7.3-3 Determine the Laplace transform of $f(t) = e^{-at}\,u(t - T)$. *Answer*: $e^{-(s+a)T}/(s + a)$.

7.3-4 Determine the Laplace transform of the function shown in Fig. P7.3-4 by (a) direct integration and (b) writing it as a sum of step functions.
Answer: $(B/s)(1 - e^{-sT}) + (A/s)(e^{-sT} - e^{-2sT})$.

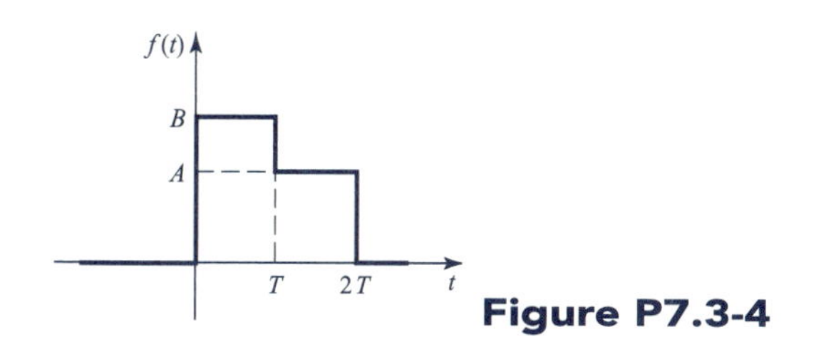

Figure P7.3-4

7.3-5 Determine the Laplace transform of the inductor current and the capacitor voltage in the circuit of Fig. P7.3-5 where $i_L(0^-) = 2$ A and $v_C(0^-) = 3$ V.
Answer: $I_L(s) = (2s + 1)/(s^2 + \frac{3}{2}s + 3)$, $V_C(s) = (3s^2 + \frac{33}{2}s + 15)/[s(s^2 + \frac{3}{2}s + 3)]$.

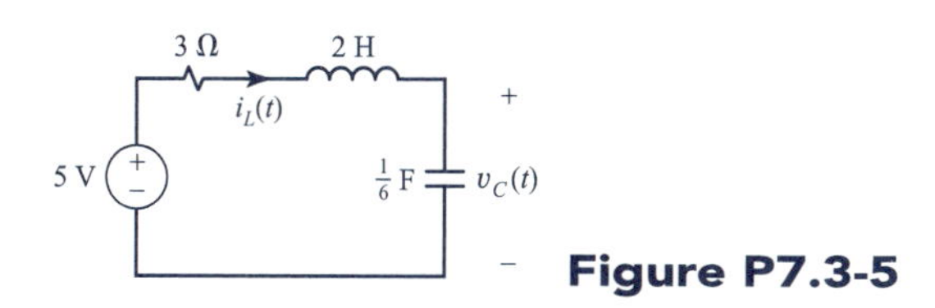

Figure P7.3-5

7.3-6 Determine the Laplace transform of the inductor current and the capacitor voltage in the circuit of Fig. P7.3-6 where $i_L(0^-) = 1$ A and $v_C(0^-) = 2$ V.
Answers: $I_L(s) = (s^3 + 2s^2 + 9s + 36)/[(s^2 + 9)(s^2 + 2s + 3)]$, $V_C(s) = (2s^3 + s^2 + 36s + 45)/[(s^2 + 9)(s^2 + 2s + 3)]$.

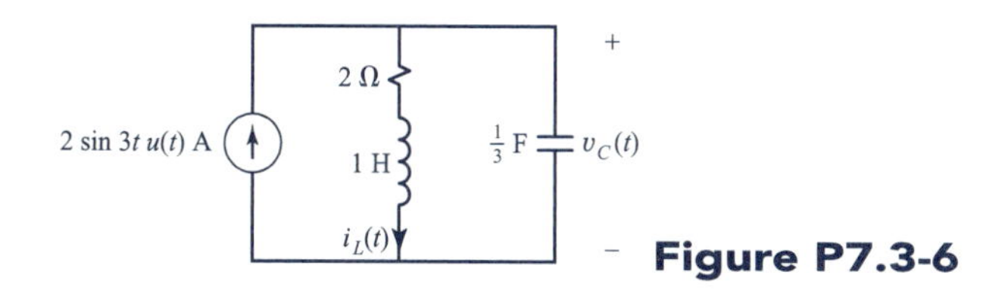

Figure P7.3-6

7.3-7 Determine the Laplace transform of the inductor current and the capacitor voltage in the circuit of Fig. P7.3-7 where $i_L(0^-) = 4$ A and $v_C(0^-) = 1$ V.
Answers: $I_L(s) = (8s^3 + 23s^2 + 2s + 6)/[s^2(2s^2 + 8s + 9)]$, $V_C(s) = (2s^4 + 32s^3 + 78s^2 + 6s + 18)/[s^2(s + 3)(2s^2 + 8s + 9)]$.

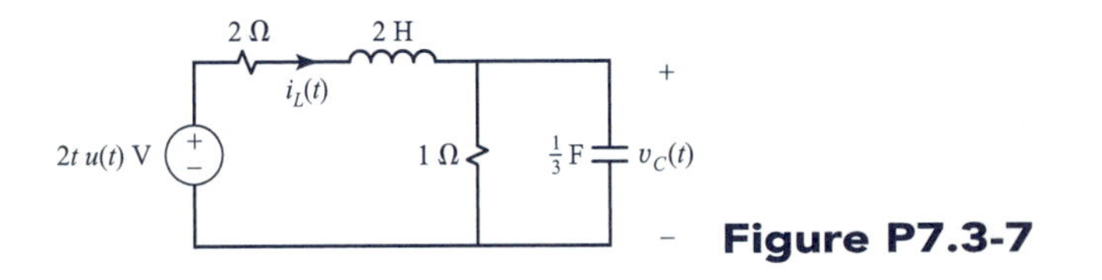

Figure P7.3-7

7.3-8 Determine the initial inductor current and capacitor voltage in the circuit of Fig. P7.3-8.
Answers: 1 A, 3 V.

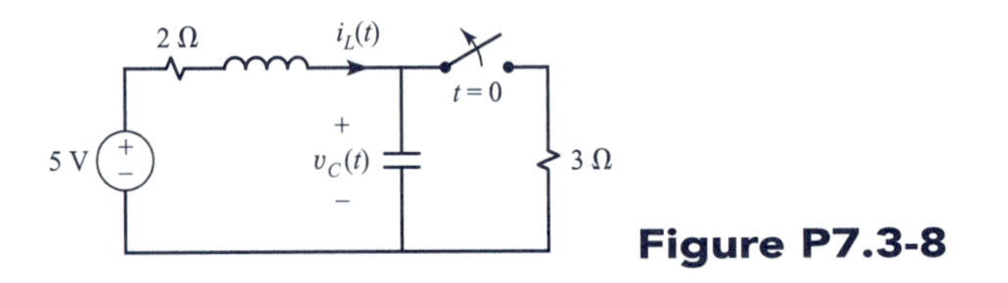

Figure P7.3-8

7.3-9 Determine the initial inductor current and capacitor voltage in the circuit of Fig. P7.3-9.
Answers: −1 A, 3 V.

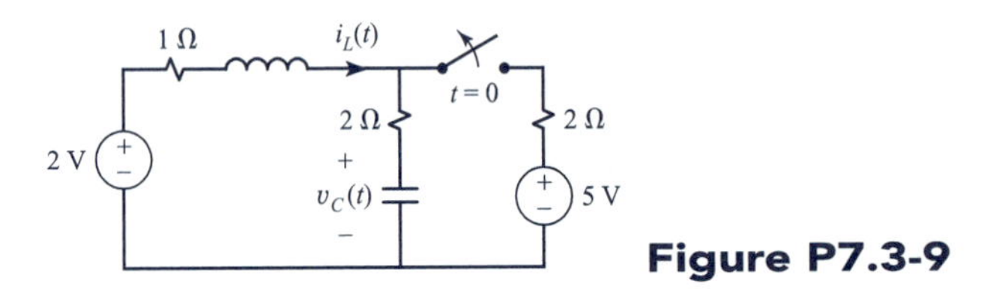

Figure P7.3-9

7.3-10 Determine the initial inductor current and capacitor voltage in the circuit of Fig. P7.3-10.
Answers: $\frac{5}{4}$ A, $\frac{5}{2}$ V.

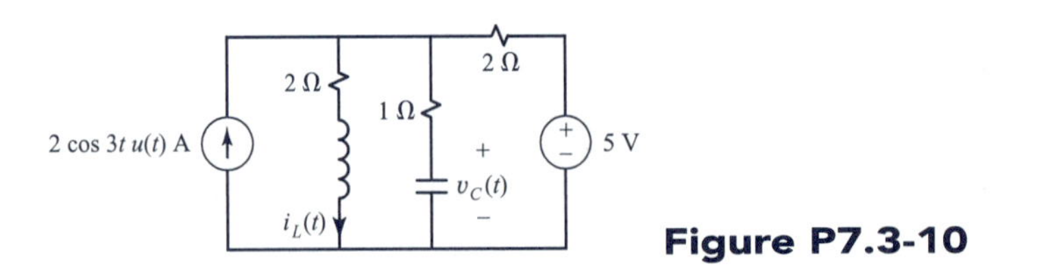

Figure P7.3-10

7.3-11 Determine the initial inductor current and capacitor voltage in the circuit of Fig. P7.3-11.
Answers: $-\frac{10}{3}$ A, $\frac{20}{3}$ V.

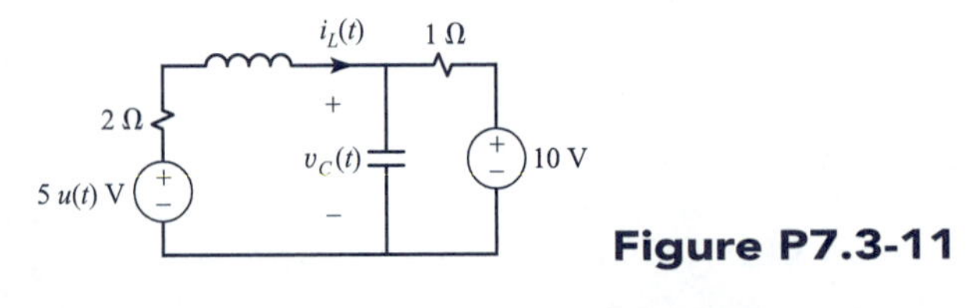

Figure P7.3-11

Section 7.4 The Inverse Laplace Transform by Partial-Fraction Expansion

7.4-1 Determine the inverse transform of $F(s) = (2s + 5)/(s^2 + 3s + 2)$.
Answer: $f(t) = (3e^{-t} - e^{-2t})u(t)$.

7.4-2 Determine the inverse transform of $F(s) = (3s + 2)/(s^2 + 2s + 1)$.
Answer: $f(t) = (3e^{-t} - te^{-t})u(t)$.

7.4-3 Determine the inverse transform of $F(s) = (3s + 4)/(s^2 + 2s + 5)$.
Answer: $f(t) = (3e^{-t} \cos 2t + \frac{1}{2} e^{-t} \sin 2t)u(t)$.

7.4-4 Determine the inverse transform of $F(s) = (4s^2 + 6s + 2)/(2s^2 + 16s + 30)$.
Answer: $f(t) = 2\delta(t) + (5e^{-3t} - 18e^{-5t})u(t)$.

7.4-5 Determine the inverse transform of $F(s) = (3s^2 + 4s + 3)/(2s^2 + 12s + 18)$.
Answer: $f(t) = \frac{3}{2}\delta(t) + (-7e^{-3t} + 9te^{-3t})u(t)$.

7.4-6 Determine the inverse transform of $F(s) = (3s^2 + 2s + 2)/(s^2 + 4s + 29)$.
Answer: $f(t) = 3\delta(t) - (10e^{-2t} \cos 5t + 13e^{-2t} \sin 5t)u(t)$.

7.4-7 Determine the inverse transform of $F(s) = (3s^2 + 2s + 1)/(2s^2 + 10s + 12)$.
Answer: $f(t) = \frac{3}{2}\delta(t) + (\frac{9}{2} e^{-2t} - 11e^{-3t})u(t)$.

7.4-8 Determine the inverse transform of $F(s) = 2s^2/(s^2 + 6s + 5)$.
Answer: $f(t) = 2\delta(t) + (\frac{1}{2} e^{-t} - \frac{25}{2} e^{-5t})u(t)$.

7.4-9 Determine the inverse transform of $F(s) = 4s^2/(s^2 + 6s + 9)$.
Answer: $f(t) = 4\delta(t) - (24e^{-3t} - 36te^{-3t})u(t)$.

7.4-10 Determine the inverse transform of $F(s) = (3s^2 + 2s + 1)/(s^2 + 4s + 4)$.
Answer: $f(t) = 3\delta(t) - (10e^{-2t} - 9te^{-2t})u(t)$.

7.4-11 Determine the inverse transform of $F(s) = (5s + 2)/(s^2 + 2s + 2)$.
Answer: $f(t) = (5e^{-t} \cos t - 3e^{-t} \sin t)u(t)$.

7.4-12 Determine the inverse transform of $F(s) = 9s^2/(s^2 + 4s + 13)$.
Answer: $f(t) = 9\delta(t) - (36e^{-2t} \cos 3t + 15e^{-2t} \sin 3t)u(t)$.

7.4-13 Determine the inverse transform of $F(s) = (3s^2 + 2s + 1)/(2s^2 + 8s + 6)$.
Answer: $f(t) = \frac{3}{2}\delta(t) + (\frac{1}{2} e^{-t} - \frac{11}{2} e^{-3t})u(t)$.

7.4-14 Determine the inverse transform of $F(s) = (2s^2 + 1)/(s^2 + 4s + 4)$.
Answer: $f(t) = 2\delta(t) - (8e^{-2t} - 9te^{-2t})u(t)$.

7.4-15 Determine the inverse transform of $F(s) = 2/(s^2 + 4s + 13)$.
Answer: $f(t) = \frac{2}{3} e^{-2t} \sin 3t\, u(t)$.

Section 7.5 First-Order Circuit Response

7.5-1 Determine and sketch the inductor current $i_L(t)$ for $t > 0$ in the circuit of Fig. P7.5-1.
Answer: $\frac{5}{2} - \frac{5}{6} e^{-t}$ A.

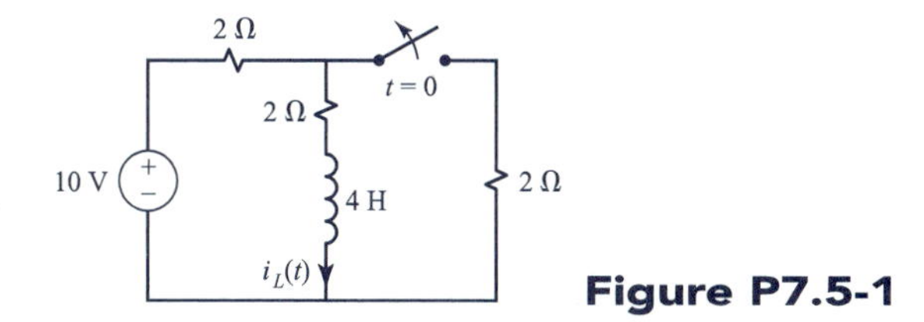

Figure P7.5-1

7.5-2 Determine and sketch the inductor current $i_L(t)$ for $t > 0$ in the circuit of Fig. P7.5-2.
Answer: $\frac{5}{4} + \frac{5}{12} e^{-16t/15}$ A.

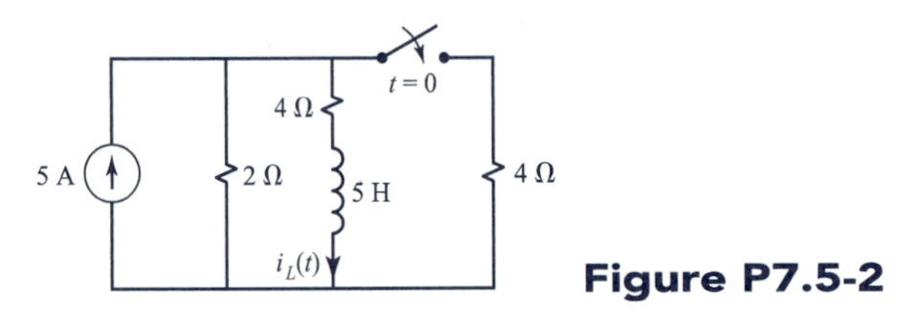

Figure P7.5-2

7.5-3 Determine and sketch the capacitor voltage $v_C(t)$ for $t > 0$ in the circuit of Fig. P7.5-3.
Answer: $10 - 5e^{-t/6}$ V.

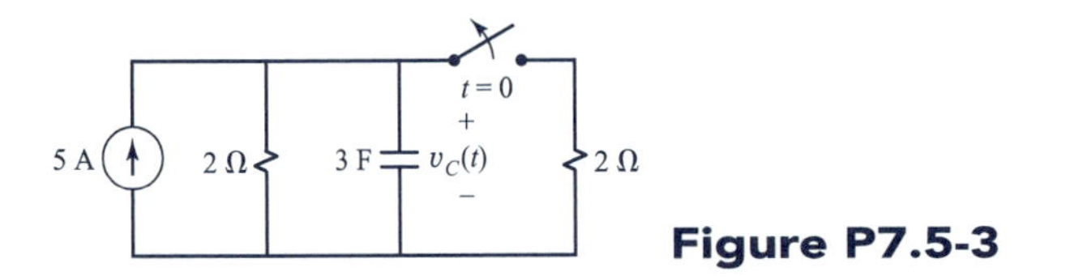

Figure P7.5-3

7.5-4 Determine and sketch the capacitor voltage $v_C(t)$ for $t > 0$ in the circuit of Fig. P7.5-4.
Answer: $5 + \frac{5}{3} e^{-t/5}$ V.

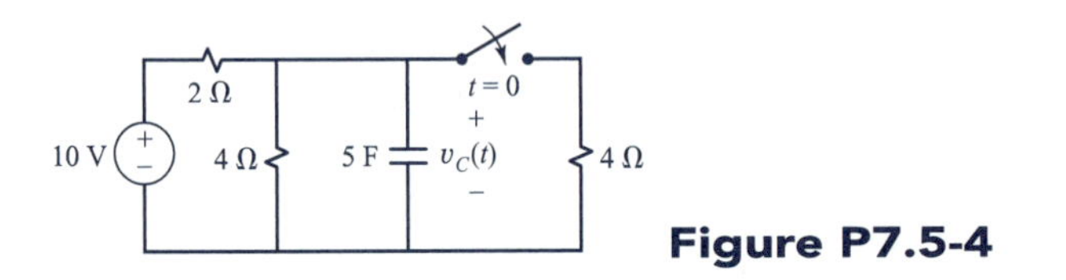

Figure P7.5-4

7.5-5 Determine and sketch the inductor current $i_L(t)$ for $t > 0$ in the circuit of Fig. P7.5-5.
Answer: $\frac{4}{3} e^{-3t} - \frac{1}{3} \cos 3t + \frac{1}{3} \sin 3t$ A.

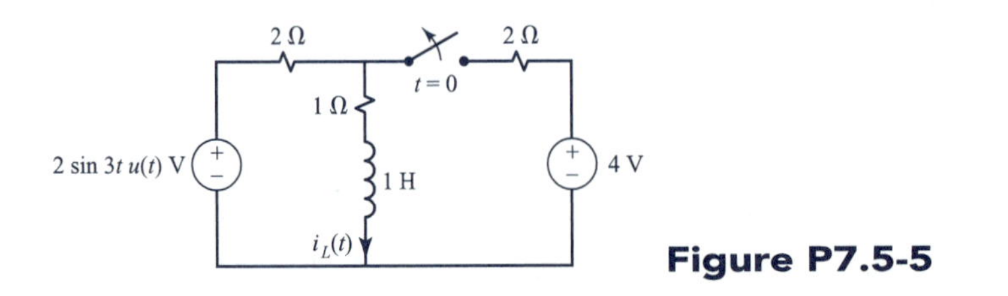

Figure P7.5-5

7.5-6 Determine and sketch the inductor current $i_L(t)$ for $t > 0$ in the circuit of Fig. P7.5-6.
Answer: $1.25 \times 10^{-3} (1 - e^{-2\times 10^6 t})$ A.

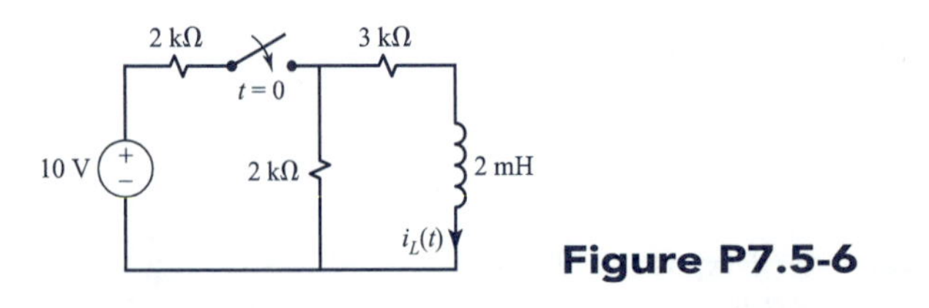

Figure P7.5-6

7.5-7 Determine and sketch the inductor current $i_L(t)$ for $t > 0$ in the circuit of Fig. P7.5-7.
Answer: $3 - 3e^{-5t/9}$ A.

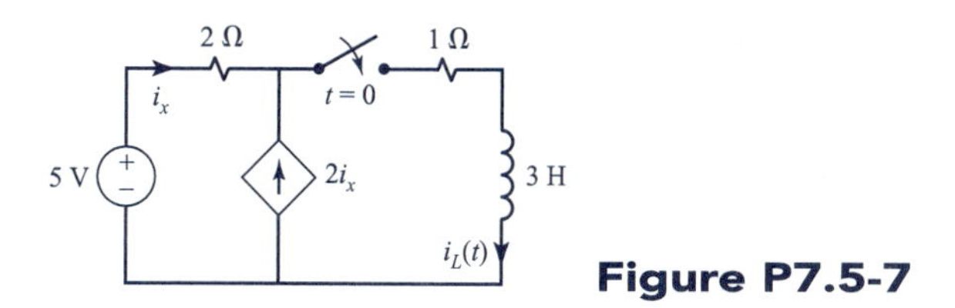

Figure P7.5-7

7.5-8 Determine and sketch the current $i(t)$ for $t > 0$ in the circuit of Fig. P7.5-8.
Answer: $\frac{5}{6} - \frac{5}{6}e^{-15t/4}$ A.

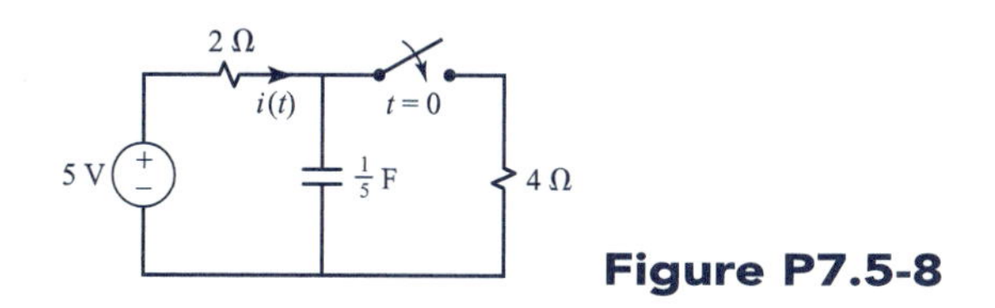

Figure P7.5-8

7.5-9 Determine and sketch the voltage $v(t)$ for $t > 0$ in the circuit of Fig. P7.5-9.
Answer: $4 - 4e^{-4t/3}$ V.

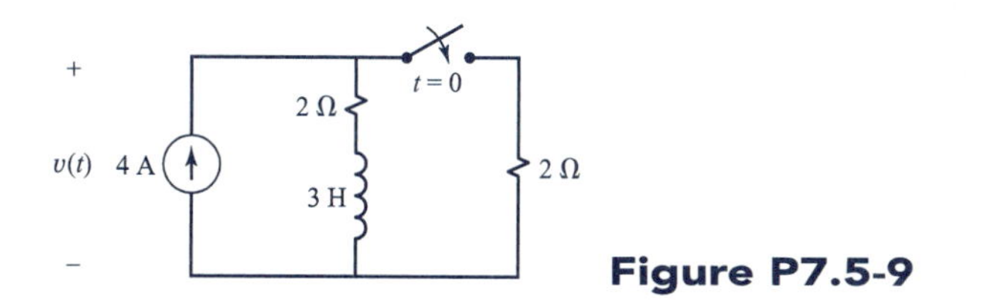

Figure P7.5-9

7.5-10 Determine and sketch the current $i(t)$ for $t > 0$ in the circuit of Fig. P7.5-10.
Answer: $\frac{5}{4} - \frac{5}{8}e^{-2t/5}$ A.

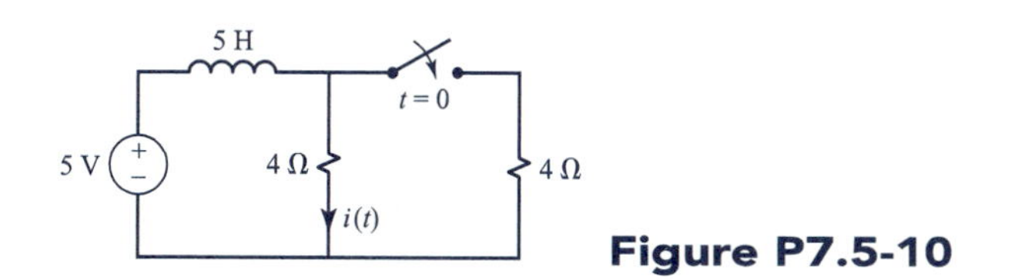

Figure P7.5-10

7.5-11 Determine and sketch the voltage $v(t)$ for $t > 0$ in the circuit of Fig. P7.5-11.
Answer: $5 - \frac{5}{3}e^{-2t}$ V.

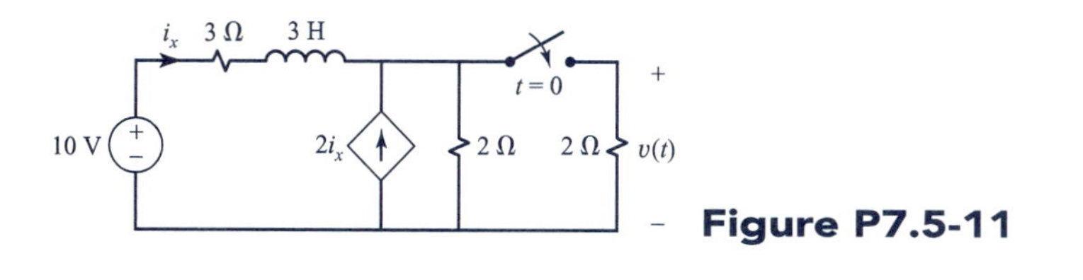

Figure P7.5-11

7.5-12 Determine and sketch the current $i(t)$ for $t > 0$ in the circuit of Fig. P7.5-12.
Answer: $-2 + 2e^{-t/4}$ A.

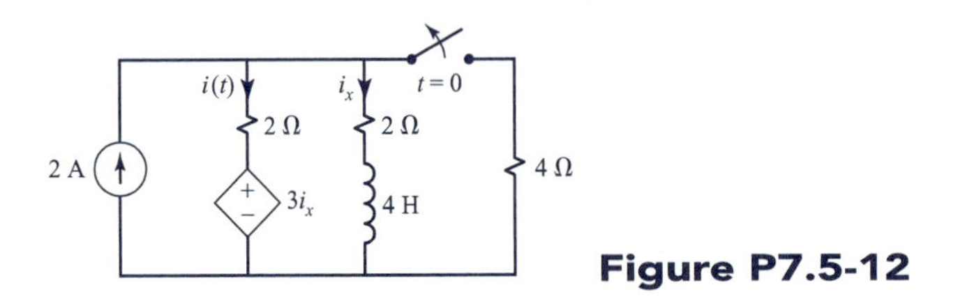

Figure P7.5-12

7.5-13 Determine and sketch the current $i_x(t)$ for $t > 0$ in the circuit of Fig. P7.5-13.
Answer: $-\frac{5}{7} + \frac{80}{63}e^{-7t/3}$ A.

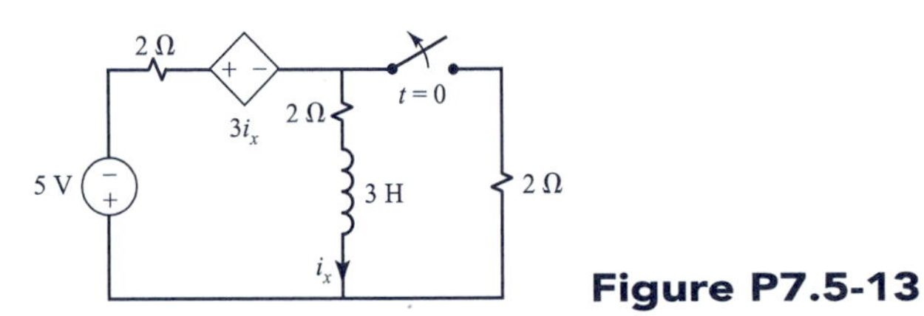

Figure P7.5-13

Section 7.6 Second-Order Circuit Response

7.6-1 Determine the inductor current and capacitor voltage for $t > 0$ in the circuit of Fig. P7.6-1 for $v_S(t) = 10$ V, $R = 4\ \Omega$, $L = 1$ H, $C = \frac{1}{3}$ F.
Answers: $5e^{-t} - 5e^{-3t}$ A, $10 - 15e^{-t} + 5e^{-3t}$ V.

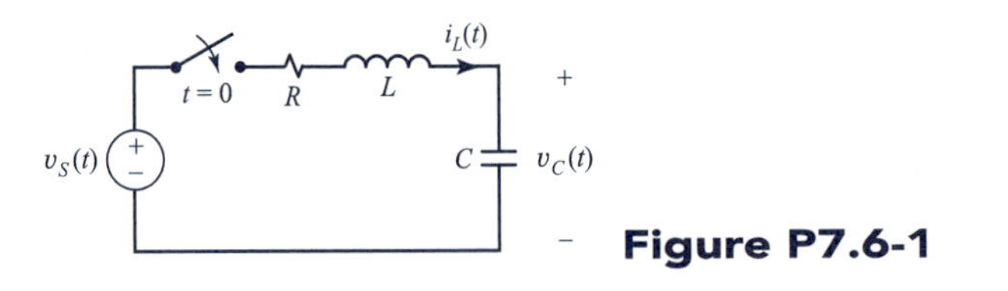

Figure P7.6-1

7.6-2 Determine the inductor current and capacitor voltage for $t > 0$ in the circuit of Fig. P7.6-1 for $v_S(t) = 5$ V, $R = 4\ \Omega$, $L = 1$ H, $C = \frac{1}{4}$ F. *Answers*: $5te^{-2t}$ A, $5 - 5e^{-2t} - 10te^{-2t}$ V.

7.6-3 Determine the inductor current and capacitor voltage for $t > 0$ in the circuit of Fig. P7.6-1 for $v_S(t) = 10$ V, $R = 8\ \Omega$, $L = 2$ H, $C = \frac{1}{26}$ F.
Answers: $\frac{5}{3}e^{-2t}\sin 3t$ A, $10 - 10e^{-2t}\cos 3t - \frac{20}{3}e^{-2t}\sin 3t$ V.

7.6-4 Determine the inductor current and capacitor voltage for $t > 0$ in the circuit of Fig. P7.6-4 for $i_S(t) = 2$ A, $R = 2\ \Omega$, $L = \frac{3}{2}$ H, $C = \frac{1}{12}$ F. *Answers*: $2 - 4e^{-2t} + 2e^{-4t}$ A, $12e^{-2t} - 12e^{-4t}$ V.

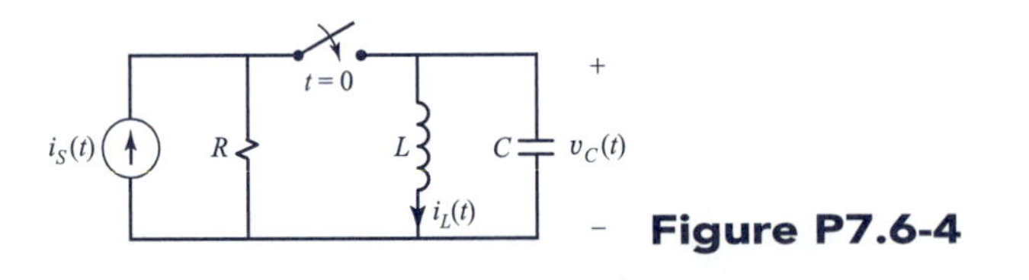

Figure P7.6-4

7.6-5 Determine the inductor current and capacitor voltage for $t > 0$ in the circuit of Fig. P7.6-4 for $i_S(t) = 5$ A, $R = 2\ \Omega$, $L = \frac{4}{5}$ H, $C = \frac{1}{20}$ F. *Answers*: $5 - 5e^{-5t} - 25te^{-5t}$ A, $100te^{-5t}$ V.

7.6-6 Determine the inductor current and capacitor voltage for $t > 0$ in the circuit of Fig. P7.6-4 for $i_S(t) = 1$ A, $R = 2\ \Omega$, $L = \frac{8}{29}$ H, $C = \frac{1}{8}$ F.
Answers: $1 - e^{-2t}\cos 5t - \frac{2}{5}e^{-2t}\sin 5t$ A, $\frac{8}{5}e^{-2t}\sin 5t$ V.

7.6-7 Determine the current $i(t)$ for $t > 0$ in the circuit of Fig. P7.6-7 for $L = \frac{1}{3}$ H, $C = \frac{3}{2}$ F.
Answer: $\frac{7}{2} - \frac{9}{2}e^{-t} + \frac{9}{2}e^{-2t}$ A.

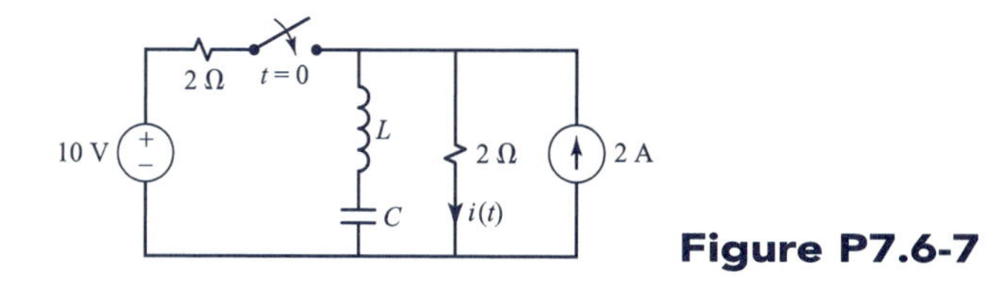

Figure P7.6-7

7.6-8 Determine the current $i(t)$ for $t > 0$ in the circuit of Fig. P7.6-7 for $L = \frac{1}{4}$ H, $C = 1$ F.
Answer: $\frac{7}{2} - 6te^{-2t}$ A.

7.6-9 Determine the current $i(t)$ for $t > 0$ in the circuit of Fig. P7.6-7 for $L = \frac{1}{2}$ H, $C = \frac{2}{5}$ F.
Answer: $\frac{7}{2} - \frac{3}{2}e^{-t}\sin 2t$ A.

7.6-10 Determine the inductor current for $t > 0$ in the circuit of Fig. P7.6-10 for $L = \frac{2}{5}$ H, $C = \frac{5}{8}$ F. *Answer*: $5 - \frac{10}{3}e^{-t} + \frac{5}{6}e^{-4t}$ A.

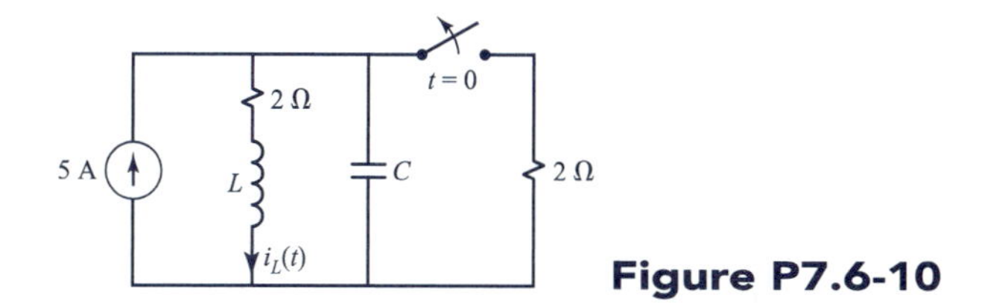

Figure P7.6-10

7.6-11 Determine the inductor current for $t > 0$ in the circuit of Fig. P7.6-10 for $L = \frac{1}{3}$ H, $C = \frac{1}{3}$ F. *Answer*: $5 - \frac{5}{2}e^{-3t} - \frac{15}{2}te^{-3t}$ A.

7.6-12 Determine the inductor current for $t > 0$ in the circuit of Fig. P7.6-10 for $L = \frac{1}{2}$ H, $C = \frac{1}{10}$ F. *Answer*: $5 - \frac{5}{2}e^{-2t}\cos 4t - \frac{5}{4}e^{-2t}\sin 4t$ A.

Section 7.7 The Step and Impulse Responses

7.7-1 Determine the unit-step and unit-impulse responses for current $i(t)$ in the circuit of Fig. P7.7-1. *Answers*: $i_\delta(t) = \left(-\frac{1}{2}e^{-t} + \frac{3}{2}e^{-3t}\right)u(t)$ A, $i_u(t) = \left(\frac{1}{2}e^{-t} - \frac{1}{2}e^{-3t}\right)u(t)$ A.

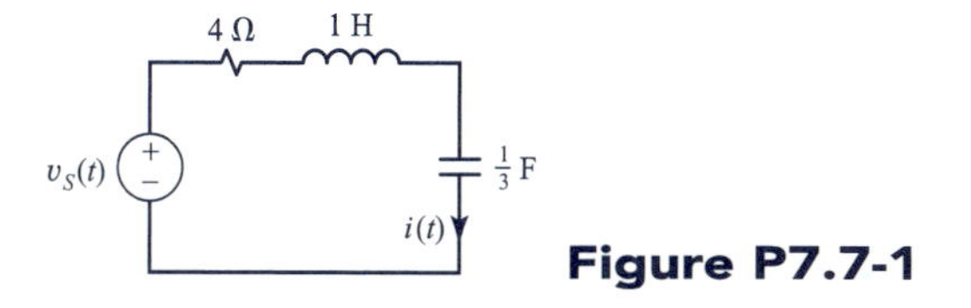

Figure P7.7-1

7.7-2 Determine the unit-step and unit-impulse responses for voltage $v(t)$ in the circuit of Fig. P7.7-2. *Answers:* $v_\delta(t) = (e^{-2t} - 2te^{-2t})u(t)$ V, $v_u(t) = te^{-2t}\,u(t)$ V.

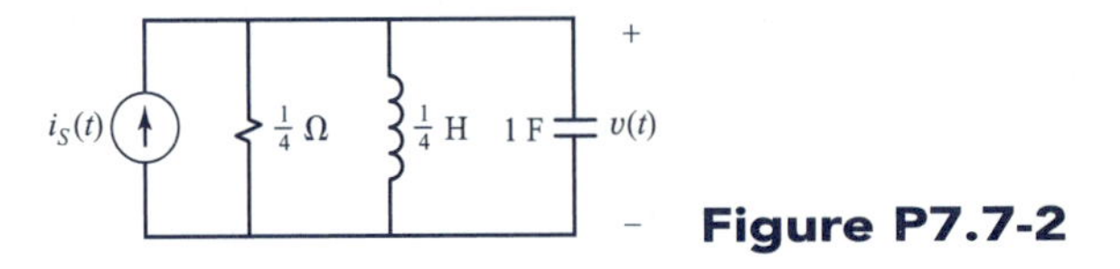

Figure P7.7-2

7.7-3 Determine the unit-step and unit-impulse responses for voltage $v(t)$ in the circuit of Fig. P7.7-3. *Answers:* $v_\delta(t) = \delta(t) - (4e^{-2t}\cos 3t + \frac{5}{3}e^{-2t}\sin 3t)u(t)$ V, $v_u(t) = \left(e^{-2t}\cos 3t - \frac{2}{3}e^{-2t}\sin 3t\right)u(t)$ V.

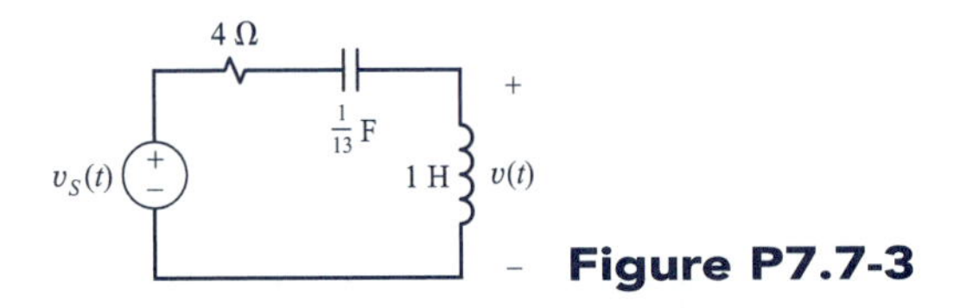

Figure P7.7-3

7.7-4 Determine the unit-step and unit-impulse responses for voltage $v(t)$ in the circuit of Fig. P7.7-4. *Answers:* $v_\delta(t) = 9te^{-3t}\,u(t)$ V, $v_u(t) = (1 - e^{-3t} - 3te^{-3t})u(t)$ V.

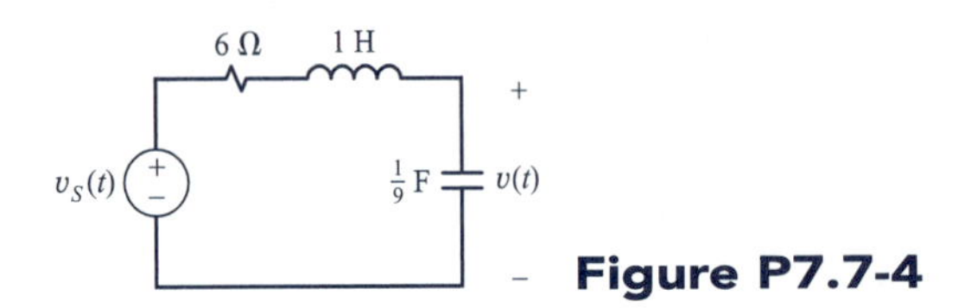

Figure P7.7-4

7.7-5 Determine the unit-step and unit-impulse responses for current $i(t)$ in the circuit of Fig. P7.7-5. *Answers:* $i_\delta(t) = \frac{5}{2}e^{-t}\sin 2t\,u(t)$ A, $i_u(t) = \left(1 - e^{-t}\cos 2t - \frac{1}{2}e^{-t}\sin 2t\right)u(t)$ A.

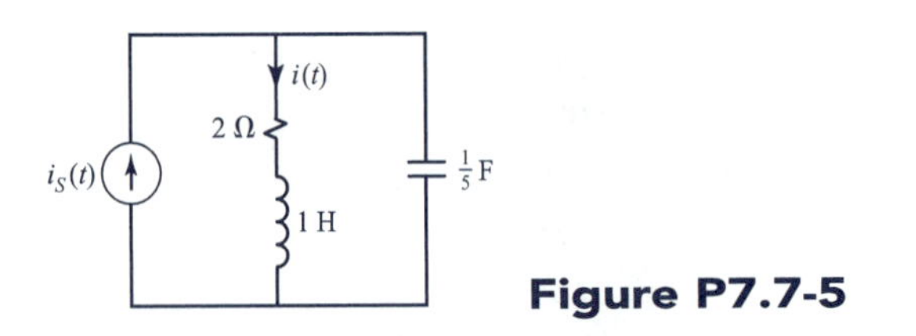

Figure P7.7-5

7.7-6 Determine the unit-step and unit-impulse responses for voltage $v(t)$ in the circuit of Fig. P7.7-6. *Answers:* $v_\delta(t) = \delta(t) - 10^6\,e^{-(5t/2)\times 10^6}\,u(t)$ V, $v_u(t) = \left(\frac{3}{5} + \frac{2}{5}e^{-(5t/2)\times 10^6}\right)u(t)$ V.

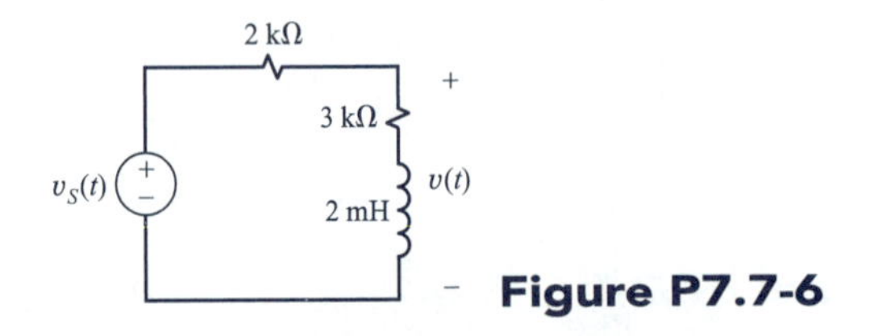

Figure P7.7-6

7.7-7 Determine the unit-step and unit-impulse responses for current $i(t)$ in the circuit of Fig. P7.7-7. *Answers*: $i_\delta(t) = \frac{1}{3}\,\delta(t) - \frac{1}{18} \times 10^3\, e^{-(t/6)\times 10^3}\, u(t)$ A, $i_u(t) = \frac{1}{3}\, e^{-(t/6)\times 10^3}\, u(t)$ A.

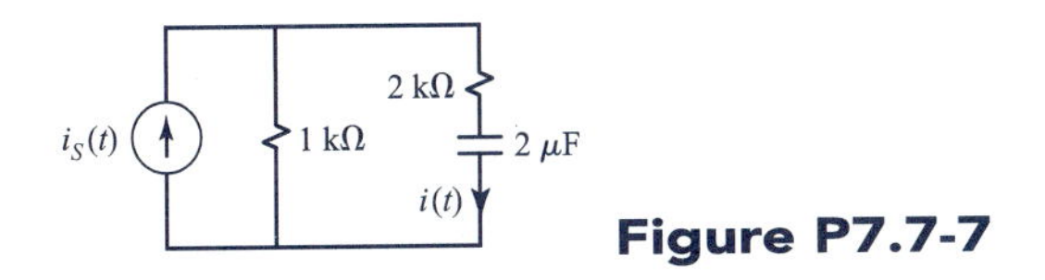

Figure P7.7-7

7.7-8 Determine the unit-step and unit-impulse responses for current $i(t)$ in the circuit of Fig. P7.7-8. *Answers*: $i_\delta(t) = \frac{1}{6}\, e^{-t/2}\, u(t)$ A, $i_u(t) = \left(\frac{1}{3} - \frac{1}{3}\, e^{-t/2}\right) u(t)$ A.

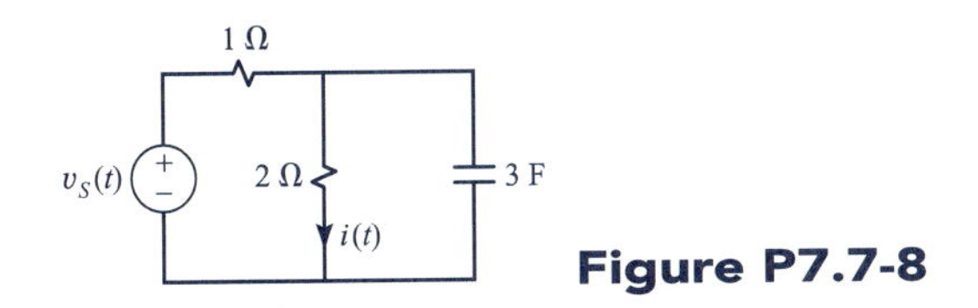

Figure P7.7-8

7.7-9 Determine the unit-step and unit-impulse responses for voltage $v(t)$ in the circuit of Fig. P7.7-9. *Answers*: $v_\delta(t) = \frac{3}{5} \times 10^6\, e^{-(t/5)\times 10^3}\, u(t)$ V, $v_u(t) = 3 \times 10^3 \left(1 - e^{-(t/5)\times 10^3}\right) u(t)$ V.

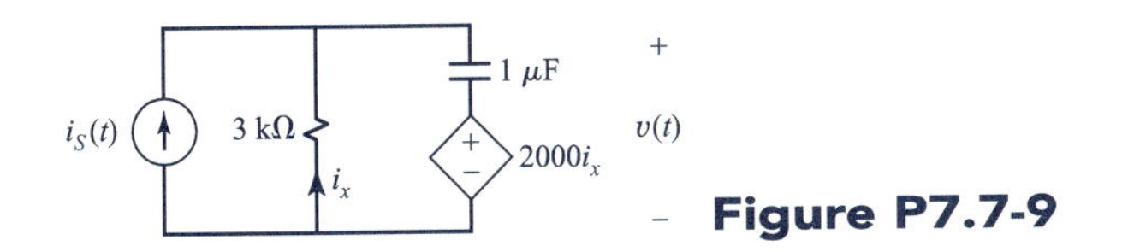

Figure P7.7-9

7.7-10 Determine the unit-step and unit-impulse responses for current $i(t)$ in the circuit of Fig. P7.7-10. *Answers*: $i_\delta(t) = \frac{1}{4}\,\delta(t) - \frac{1}{8}\, e^{-t}\, u(t)$ A, $i_u(t) = \left(\frac{1}{8} + \frac{1}{8}\, e^{-t}\right) u(t)$ A.

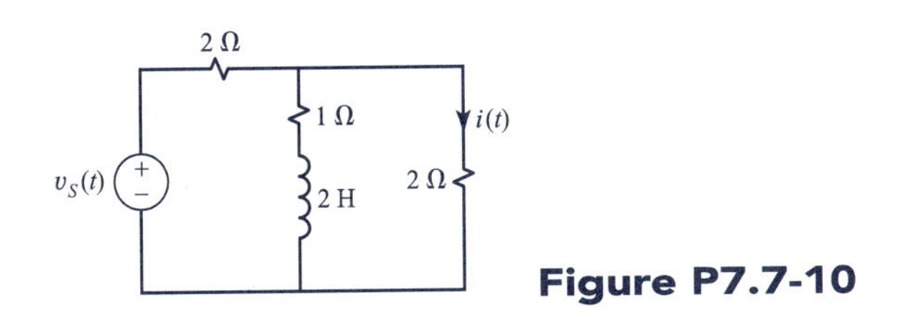

Figure P7.7-10

7.7-11 Determine the unit-step and unit-impulse responses for current $i(t)$ in the circuit of Fig. P7.7-11. *Answers*: $i_\delta(t) = \left(-2e^{-2t} + 3e^{-3t}\right) u(t)$ A, $i_u(t) = \left(e^{-2t} - e^{-3t}\right) u(t)$ A.

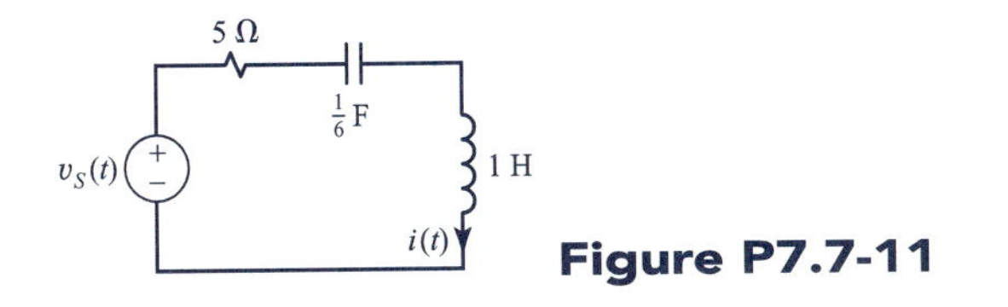

Figure P7.7-11

7.7-12 Determine the unit-step and unit-impulse responses for current $i(t)$ in the circuit of Fig. P7.7-12. *Answers*: $i_\delta(t) = \delta(t) - 9te^{-3t}\,u(t)$ A, $i_u(t) = \left(e^{-3t} + 3te^{-3t}\right)u(t)$ A.

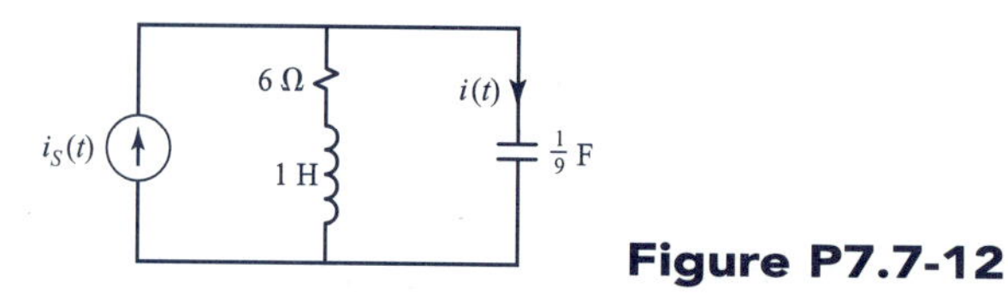

Figure P7.7-12

7.7-13 Determine the unit-step and unit-impulse responses for current $i(t)$ in the circuit of Fig. P7.7-13. *Answers*: $i_\delta(t) = \left(\frac{1}{2}e^{-2t}\cos 3t - \frac{1}{3}e^{-2t}\sin 3t\right)u(t)$ A, $i_u(t) = \frac{1}{6}e^{-2t}\sin 3t\,u(t)$ A.

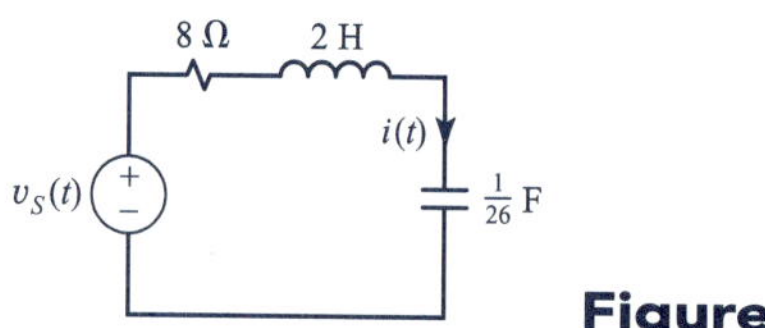

Figure P7.7-13

7.7-14 Determine the unit-step and unit-impulse responses for voltage $v(t)$ in the circuit of Fig. P7.7-14. *Answers*: $v_\delta(t) = 10\delta(t) + \left(\frac{2}{3}e^{-t} - \frac{128}{3}e^{-4t}\right)u(t)$ V, $v_u(t) = \left(-\frac{2}{3}e^{-t} + \frac{32}{3}e^{-4t}\right)u(t)$ V.

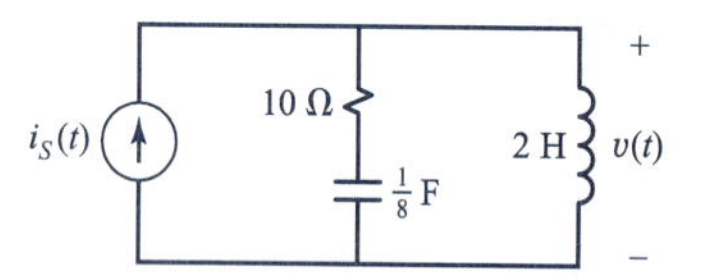

Figure P7.7-14

Section 7.8 Convolution

7.8-1 Determine the output of the linear system in Fig. P7.8-1 using convolution.
Answer: $0 \le t \le 1, y(t) = 2t$; $1 \le t \le 3, y(t) = 2$; $3 \le t \le 4, y(t) = 8 - 2t$; $4 \le t, y(t) = 0$.

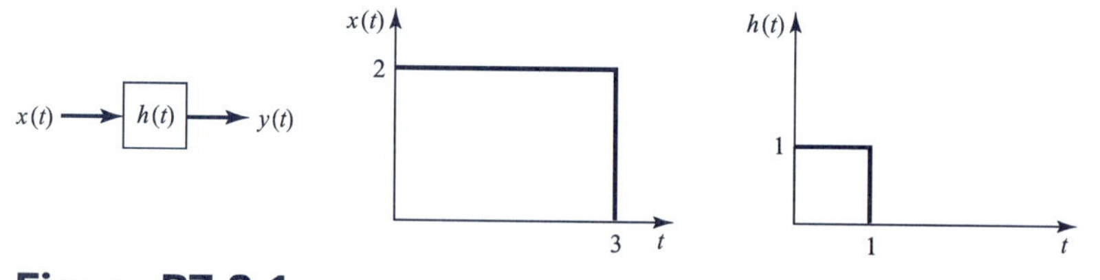

Figure P7.8-1

7.8-2 Determine the output of the linear system in Fig. P7.8-2 using convolution.
Answer: $0 \le t \le 1, y(t) = 3t$; $1 \le t \le 2, y(t) = 3t$; $2 \le t \le 3, y(t) = 18 - 6t$; $3 \le t, y(t) = 0$.

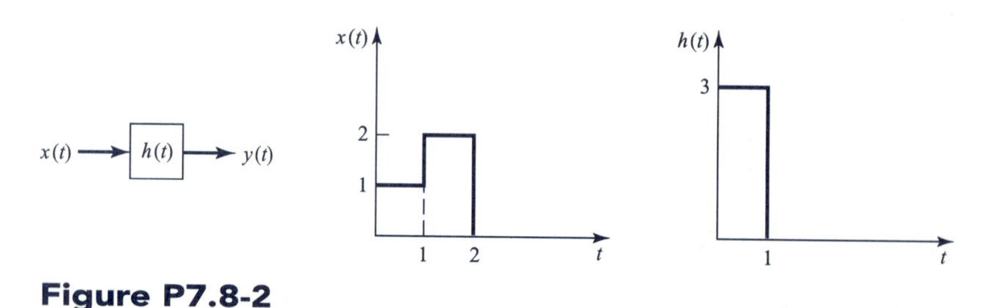

Figure P7.8-2

7.8-3 Determine the unit-step response of the inductor current in the circuit of Fig. P7.8-3 using convolution. *Answer*: $\left(\frac{1}{2} - \frac{1}{2}e^{-2t/3}\right)u(t)$ A.

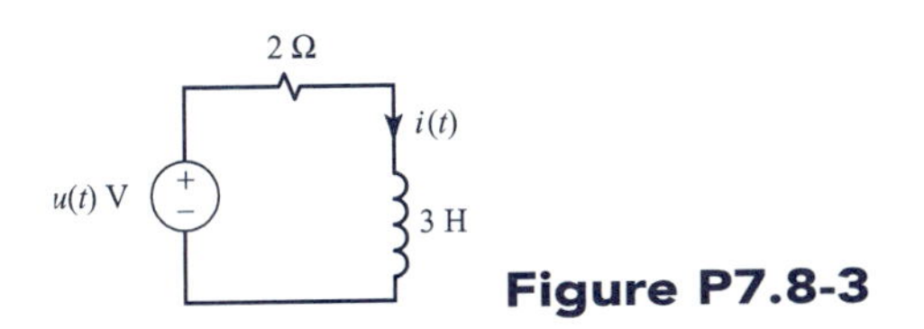

Figure P7.8-3

7.8-4 Determine the unit-step response of the capacitor voltage in the circuit of Fig. P7.8-4 using convolution. *Answer*: $(3 - 3e^{-4t/3})u(t)$ V.

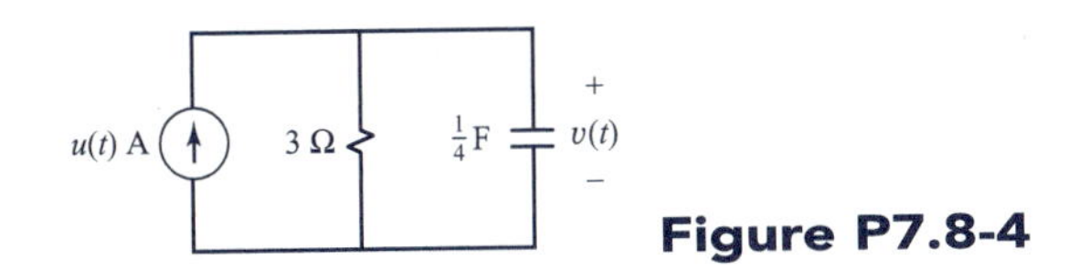

Figure P7.8-4

Section 7.9 Applications of Linearity and Time Invariance

7.9-1 Show that the unit-step and unit-impulse responses are related for the circuit of Problem 7.7-1. *Answer*: $y_\delta(t) = dy_u(t)/dt$.

7.9-2 Show that the unit-step and unit-impulse responses are related for the circuit of Problem 7.7-2. *Answer*: $y_\delta(t) = dy_u(t)/dt$.

7.9-3 Show that the unit-step and unit-impulse responses are related for the circuit of Problem 7.7-6. *Answer*: $y_\delta(t) = dy_u(t)/dt$.

7.9-4 Show that the unit-step and unit-impulse responses are related for the circuit of Problem 7.7-10. *Answer*: $y_\delta(t) = dy_u(t)/dt$.

7.9-5 Show that the unit-step and unit-impulse responses are related for the circuit of Problem 7.7-12. *Answer*: $y_\delta(t) = dy_u(t)/dt$.

7.9-6 Detemine the response of the voltage $v(t)$ in the circuit of Fig. P7.9-6 by (a) using the direct (classical) solution of the differential equations, and (b) using linearity and time invariance by writing $v_S(t)$ as a combination of step functions. *Answers*: (a) $0 \le t \le 1$, $v(t) = 2(1 - e^{-2t})$ V; $1 \le t \le 2$, $v(t) = e^{-2(t-1)} - 2e^{-2t} + 1$ V; $2 \le t$, $v(t) = e^{-2(t-1)} - 2e^{-2t} + e^{-2(t-2)}$ V;
(b) $v(t) = 2(1 - e^{-2t})\,u(t) - (1 - e^{-2(t-1)})\,u(t-1) - (1 - e^{-2(t-2)})\,u(t-2)$ V.

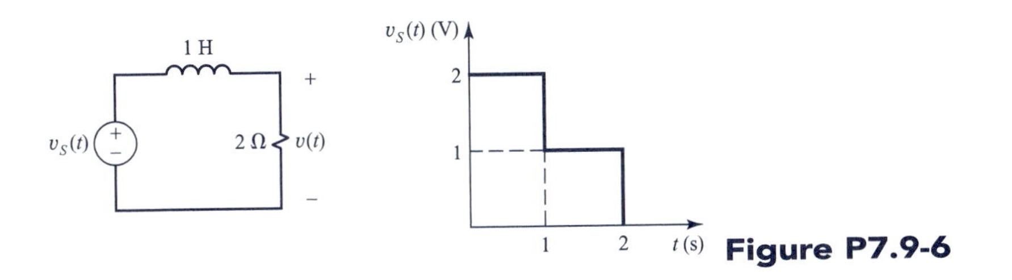

Figure P7.9-6

7.9-7 Detemine the response of the voltage, $v(t)$, in the circuit of Fig. P7.9-7 by (a) using the direct (classical) solution of the differential equations, and (b) using linearity and time invariance by writing $i_S(t)$ as a combination of step functions. *Answers*: (a) $0 \le t \le 1$, $v(t) = 4(1 - e^{-t})$ V; $1 \le t \le 2$, $v(t) = 4e^{-(t-1)} - 4e^{-t}$ V;

$2 \le t \le 3$, $v(t) = -2e^{-(t-2)} + 4e^{-(t-1)} - 4e^{-t} + 2$ V;
$3 \le t$, $v(t) = -2e^{-(t-2)} + 4e^{-(t-1)} - 4e^{-t} + 2e^{-(t-3)}$ V;
(b) $v(t) = 4(1 - e^{-t})\,u(t) - 4\,(1 - e^{-(t-1)})\,u(t-1) + 2\,(1 - e^{-(t-2)})\,u(t-2) - 2\,(1 - e^{-(t-3)})$ $u(t-3)$ V.

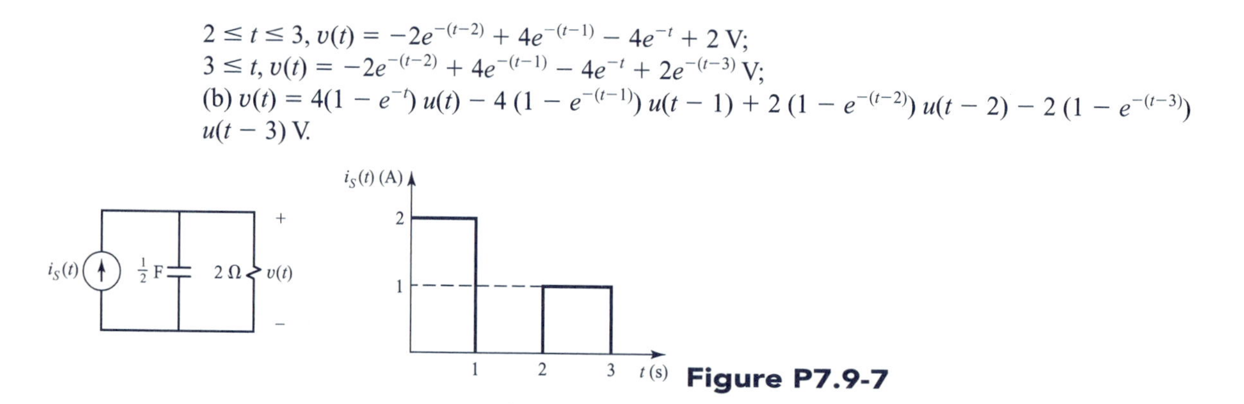

Figure P7.9-7

Section 7.10 PSPICE Applications

7.10-1 Use PSPICE to plot the response of the inductor current and capacitor voltage in the circuit of Problem 7.6-1.

7.10-2 Use PSPICE to plot the response of the inductor current and capacitor voltage in the circuit of Problem 7.6-2.

7.10-3 Use PSPICE to plot the response of the inductor current and capacitor voltage in the circuit of Problem 7.6-3.

7.10-4 Use PSPICE to plot the unit-step and unit-impulse responses for the circuit in Fig. P7.7-1.

7.10-5 Use PSPICE to plot the unit-step and unit-impulse responses for the circuit in Fig. P7.7-2.

7.10-6 Use PSPICE to plot the unit-step and unit-impulse responses for the circuit in Fig. P7.7-5.

Section 7.11 MATLAB Applications

7.11-1 Use MATLAB to determine the inverse Laplace transform of the inductor current and capacitor voltage in the circuit of Problem 7.6-1.
Answers: $5e^{-t} - 5e^{-3t}$ A, $10 - 15e^{-t} + 5e^{-3t}$ V.

7.11-2 Use MATLAB to determine the inverse Laplace transform of the inductor current and capacitor voltage in the circuit of Problem 7.6-2. *Answers*: $5te^{-2t}$ A, $5 - 5e^{-2t} - 10te^{-2t}$ V.

7.11-3 Use MATLAB to determine the inverse Laplace transform of the inductor current and capacitor voltage in the circuit of Problem 7.6-3.
Answers: $\frac{5}{3}e^{-2t}\sin 3t$ A, $10 - 10e^{-2t}\cos 3t - \frac{20}{3}e^{-2t}\sin 3t$ V.

APPENDIX

Solution of Linear Algebraic Equations

In this appendix we will provide techniques for solving the most common type of equations found in engineering applications: linear algebraic equations. Algebraic equations do not contain derivatives of the unknown variables, and linear equations have that unknown raised to a power no larger than one. An example of a linear algebraic equation is $ax(t) + b = c$ where $x(t)$ is the unknown to be solved for and a, b, and c are known. An example of a nonlinear algebraic equation is $ax^2(t) + b = c$, and an example of a nonalgebraic (differential) equation is $a\,dx(t)/dt + bx(t) = c$. This appendix is concerned with the simultaneous solution of systems of linear algebraic equations. For example, two equations in two unknowns, x_1 and x_2, can be written in the form

$$\begin{aligned} a_{11}x_1 + a_{12}x_2 &= b_1 \\ a_{21}x_1 + a_{22}x_2 &= b_2 \end{aligned} \tag{A.1}$$

There are numerous ways of solving these equations. For two equations in two unknowns as in (A.1) substitution is useful. For example, solve the first equation of the set for x_2 as

$$x_2 = \frac{b_1}{a_{12}} - \frac{a_{11}}{a_{12}}x_1 \tag{A.2}$$

Substitute this into the second equation of the set to yield the solution for x_1 as

$$\begin{aligned} x_1 &= \frac{b_2}{a_{21}} - \frac{a_{22}}{a_{21}}x_2 \\ &= \frac{b_2}{a_{21}} - \frac{a_{22}}{a_{21}}\left(\frac{b_1}{a_{12}} - \frac{a_{11}}{a_{12}}x_1\right) \end{aligned} \tag{A.3}$$

Solving this for x_1 and substituting that result into (A.2) gives the solutions for the two unknowns. Three equations in three unknowns can be similarly written as

$$\begin{aligned} a_{11}x_1 + a_{12}x_2 + a_{13}x_3 &= b_1 \\ a_{21}x_1 + a_{22}x_2 + a_{23}x_3 &= b_2 \\ a_{31}x_1 + a_{32}x_2 + a_{33}x_3 &= b_3 \end{aligned} \tag{A.4}$$

For more than two equations, the substitution method becomes unwieldy unless several of the coefficients a_{ij} are zero.

The set of two equations in two unknowns in (A.1) can be graphed as two straight lines in the two-dimensional plane consisting of axes x_1 and x_2. Where these two straight lines intersect is the solution. Hence we see that there can be three possibilities for the solution: (a) a unique solution (the two lines intersect at only one point), (b) no solution (the two lines are parallel and do not intersect), and (c) an infinite number of solutions (the two lines coincide). Similarly, the set of three equations in (A.4) can be graphed as three planes in three-dimensional space consisting of axes x_1, x_2, and x_3. As in the case of two equations, there may be (a) a unique solution, (b) no solution, or (c) an infinite number of solutions, depending on the intersection of the three planes.

A.1 Gauss Elimination

The basis for digital computer solution routines is the Gauss elimination technique. The basic idea is to reduce the equations to an equivalent set that has a triangular form. For the set of three equations in (A.4) we would try to reduce them to the form

$$\begin{aligned} c_{11}x_1 + c_{12}x_2 + c_{13}x_3 &= d_1 \\ c_{22}x_2 + c_{23}x_3 &= d_2 \\ c_{33}x_3 &= d_3 \end{aligned} \tag{A.5}$$

These are rather simple to solve by the method of *back substitution.* For example, solve the last equation for x_3 as

$$x_3 = \frac{d_3}{c_{33}} \tag{A.6}$$

Substitute this into the second equation and solve for x_2 as

$$\begin{aligned} x_2 &= \frac{d_2}{c_{22}} - \frac{c_{23}}{c_{22}}x_3 \\ &= \frac{d_2}{c_{22}} - \frac{c_{23}}{c_{22}}\left(\frac{d_3}{c_{33}}\right) \end{aligned} \tag{A.7}$$

Substituting these into the first equation gives the solution for x_1.

The essential question is: how do we reduce the original set in (A.4) to an *equivalent set* which has the triangular form in (A.5)? The answer is to use *elementary row operations,* which consist of (a) multiplication of any row by a nonzero number, (b) interchange any two equations, and (c) add or subtract any two equations and replace one equation with the result.

Example A.1

Consider the set of three linear algebraic equations

$$x + 3y - z = 4 \tag{1}$$
$$2x - 4y + 2z = 3 \tag{2}$$
$$3x + y + z = 2 \tag{3}$$

Solve these by Gauss elimination.

Solution Multiply the first equation by -2 and add the result to the second equation:

$$x + 3y - z = 4 \quad (1)$$
$$0 - 10y + 4z = -5 \quad (2')$$
$$3x + y + z = 2 \quad (3)$$

Note that this has removed x from the second equation. Now multiply the first equation by -3 and add the result to the third equation:

$$x + 3y - z = 4 \quad (1)$$
$$0 - 10y + 4z = -5 \quad (2')$$
$$0 - 8y + 4z = -10 \quad (3')$$

This has eliminated x from the third equation. Eliminating y from the third equation will achieve the triangular form. Multiply the second equation by $-\frac{8}{10}$ and add the result to the third equation:

$$x + 3y - z = 4 \quad (1)$$
$$0 - 10y + 4z = -5 \quad (2')$$
$$0 + 0 + \frac{4}{5}z = -6 \quad (3'')$$

Solving these by back substitution gives

$$z = \frac{-6}{4/5} = -\frac{15}{2}$$

Substitute this into the second equation and solve for y:

$$-10y + 4\left(-\tfrac{15}{2}\right) = -5$$

or

$$y = -\tfrac{5}{2}$$

Substitute these into the first equation and solve for x:

$$x + 3\left(-\tfrac{5}{2}\right) - \left(-\tfrac{15}{2}\right) = 4$$

from which we obtain

$$x = 4$$

A.2 Cramer's Rule

A second technique, which is very useful for hand calculations, is Cramer's rule. Cramer's rule requires the use of the concept of a determinant. To illustrate why this occurs, the solution for two equations in two unknowns from (A.2) and (A.3) is, as the reader should verify,

$$x_1 = \frac{a_{22}b_1 - a_{12}b_2}{a_{11}a_{22} - a_{12}a_{21}}$$
$$x_2 = \frac{a_{11}b_2 - a_{21}b_1}{a_{11}a_{22} - a_{12}a_{21}} \quad (A.8)$$

Note that the denominator of each solution is

$$a_{11}a_{22} - a_{12}a_{21} \quad (A.9)$$

This quantity is called the *determinant* of the coefficients of the unknowns and is denoted by

$$\begin{vmatrix} a_{11} & a_{12} \\ a_{21} & a_{22} \end{vmatrix} = a_{11}a_{22} - a_{12}a_{21} \tag{A.10}$$

Note that it is equal to *the product of the main-diagonal terms minus the product of the off-diagonal terms.* We now see a simple rule for computing the solutions in (A.8):

$$x_1 = \frac{\begin{vmatrix} b_1 & a_{12} \\ b_2 & a_{22} \end{vmatrix}}{\begin{vmatrix} a_{11} & a_{12} \\ a_{21} & a_{22} \end{vmatrix}} \qquad x_2 = \frac{\begin{vmatrix} a_{11} & b_1 \\ a_{21} & b_2 \end{vmatrix}}{\begin{vmatrix} a_{11} & a_{12} \\ a_{21} & a_{22} \end{vmatrix}} \tag{A.11}$$

Hence the solution is obtained as a ratio of determinants. The denominator is the determinant of the array of coefficients of the unknowns, while the numerator is the determinant of that array but with the right-hand-side knowns replacing the particular column of the array corresponding to the unknown being solved for, i.e., for x_1 we replace the first column of the array with the right-hand-side knowns.

The extension of Cramer's rule to more than two equations is identical but slightly more involved. For three equations, Cramer's rule gives the solutions as

$$x_1 = \frac{\begin{vmatrix} b_1 & a_{12} & a_{13} \\ b_2 & a_{22} & a_{23} \\ b_3 & a_{32} & a_{33} \end{vmatrix}}{\begin{vmatrix} a_{11} & a_{12} & a_{13} \\ a_{21} & a_{22} & a_{23} \\ a_{31} & a_{32} & a_{33} \end{vmatrix}} \qquad x_2 = \frac{\begin{vmatrix} a_{11} & b_1 & a_{13} \\ a_{21} & b_2 & a_{23} \\ a_{31} & b_3 & a_{33} \end{vmatrix}}{\begin{vmatrix} a_{11} & a_{12} & a_{13} \\ a_{21} & a_{22} & a_{23} \\ a_{31} & a_{32} & a_{33} \end{vmatrix}} \qquad x_3 = \frac{\begin{vmatrix} a_{11} & a_{12} & b_1 \\ a_{21} & a_{22} & b_2 \\ a_{31} & a_{32} & b_3 \end{vmatrix}}{\begin{vmatrix} a_{11} & a_{12} & a_{13} \\ a_{21} & a_{22} & a_{23} \\ a_{31} & a_{32} & a_{33} \end{vmatrix}} \tag{A.12}$$

Observe that the solution is still the ratio of two determinants. The denominator determinant is the determinant of the array of coefficients of the unknowns, and the numerator determinant is that array but with the column of right-hand-side knowns replacing the column of that array that corresponds to the unknown being solved for. Each of these 3×3 determinants can be reduced to the evaluation of 2×2 deteminants using the following rule:

$$\begin{vmatrix} a_{11} & a_{12} & a_{13} \\ a_{21} & a_{22} & a_{23} \\ a_{31} & a_{32} & a_{33} \end{vmatrix} = a_{11}\underbrace{(-1)^{1+1}}_{+}\begin{vmatrix} a_{22} & a_{23} \\ a_{32} & a_{33} \end{vmatrix} + a_{12}\underbrace{(-1)^{1+2}}_{-}\begin{vmatrix} a_{21} & a_{23} \\ a_{31} & a_{33} \end{vmatrix} + a_{13}\underbrace{(-1)^{1+3}}_{+}\begin{vmatrix} a_{21} & a_{22} \\ a_{31} & a_{32} \end{vmatrix} \tag{A.13}$$

The key to understanding this result is to observe that it is the expansion along the first row of the array. Each entry is the product of the coefficient of that row, a sign term that is -1 raised to a power equal to the sum of the subscripts of that term, and a 2×2 determinant that is what remains when we strike out the row and the column of that coefficient. The sign term can be more easily seen with the *checkerboard sign pattern*:

$$\begin{vmatrix} + & - & + \\ - & + & - \\ + & - & + \end{vmatrix} \tag{A.14}$$

We could alternatively expand along any other row or any column. For example, suppose we expand along the second column of the array. The result is

$$\begin{vmatrix} a_{11} & a_{12} & a_{13} \\ a_{21} & a_{22} & a_{23} \\ a_{31} & a_{32} & a_{33} \end{vmatrix} = a_{12}\underbrace{(-1)^{1+2}}_{-}\begin{vmatrix} a_{21} & a_{23} \\ a_{31} & a_{33} \end{vmatrix} + a_{22}\underbrace{(-1)^{2+2}}_{+}\begin{vmatrix} a_{11} & a_{13} \\ a_{31} & a_{33} \end{vmatrix} + a_{32}\underbrace{(-1)^{3+2}}_{-}\begin{vmatrix} a_{11} & a_{13} \\ a_{21} & a_{23} \end{vmatrix} \tag{A.15}$$

Example A.2

Solve the problem of Example A.1 using Cramer's rule.

Solution The equations are

$$x + 3y - z = 4 \tag{1}$$

$$2x - 4y + 2z = 3 \tag{2}$$

$$3x + y + z = 2 \tag{3}$$

The determinant of the array of coefficients is

$$\begin{aligned} \begin{vmatrix} 1 & 3 & -1 \\ 2 & -4 & 2 \\ 3 & 1 & 1 \end{vmatrix} &= 1\begin{vmatrix} -4 & 2 \\ 1 & 1 \end{vmatrix} - 3\begin{vmatrix} 2 & 2 \\ 3 & 1 \end{vmatrix} + (-1)\begin{vmatrix} 2 & -4 \\ 3 & 1 \end{vmatrix} \\ &= 1(-4-2) - 3(2-6) - 1(2+12) \\ &= -8 \end{aligned}$$

The solutions are

$$x = \frac{\begin{vmatrix} 4 & 3 & -1 \\ 3 & -4 & 2 \\ 2 & 1 & 1 \end{vmatrix}}{-8} = \frac{-32}{-8} = 4$$

$$y = \frac{\begin{vmatrix} 1 & 4 & -1 \\ 2 & 3 & 2 \\ 3 & 2 & 1 \end{vmatrix}}{-8} = \frac{20}{-8} = -\frac{5}{2}$$

$$z = \frac{\begin{vmatrix} 1 & 3 & 4 \\ 2 & -4 & 3 \\ 3 & 1 & 2 \end{vmatrix}}{-8} = \frac{60}{-8} = -\frac{15}{2}$$

Cramer's rule may be further generalized to more than three equations, but it is not worth doing, since the errors one is prone to make in evaluating larger determinants renders the method useless for the solution of more than three equations.

A.3 Matrix Algebra

Sets of linear algebraic equations can be written in a compact form as the product of matrices. A *matrix* is an ordered array of elements (numbers, functions, etc). For example, the set of two equations in (A.1) can be written compactly as

$$\mathbf{AX} = \mathbf{B} \tag{A.16}$$

where the matrices (denoted as boldface) are defined as

$$\begin{aligned} \mathbf{A} &= \begin{bmatrix} a_{11} & a_{12} \\ a_{21} & a_{22} \end{bmatrix} \\ \mathbf{X} &= \begin{bmatrix} x_1 \\ x_2 \end{bmatrix} \\ \mathbf{B} &= \begin{bmatrix} b_1 \\ b_2 \end{bmatrix} \end{aligned} \tag{A.17}$$

There are consistent rules of matrix algebra that define the addition, subtraction, multiplication, and division of matrices, as in the case of scalar algebra. A matrix having m rows and n columns is said to be $m \times n$. A *square matrix* is one in which the number of rows equals the number of columns. Two matrices may be added if the numbers of rows are the same and the numbers of columns are the same. In this case, the result is a matrix having the same dimensions as either matrix and whose elements are the sums of the corresponding elements of each matrix. For example,

$$\underbrace{\begin{bmatrix} a_{11} & a_{12} & a_{13} \\ a_{21} & a_{22} & a_{23} \end{bmatrix}}_{\mathbf{A}} + \underbrace{\begin{bmatrix} b_{11} & b_{12} & b_{13} \\ b_{21} & b_{22} & b_{23} \end{bmatrix}}_{\mathbf{B}} = \underbrace{\begin{bmatrix} a_{11}+b_{11} & a_{12}+b_{12} & a_{13}+b_{13} \\ a_{21}+b_{21} & a_{22}+b_{22} & a_{23}+a_{23} \end{bmatrix}}_{\mathbf{C}} \tag{A.18}$$

Subtraction of two matrices is similarly defined. Multiplication of two matrices as

$$\underset{m\times p}{\underbrace{\mathbf{A}}}\,\underset{p\times n}{\underbrace{\mathbf{B}}} = \underset{m\times n}{\underbrace{\mathbf{C}}} \tag{A.19}$$

is possible *only if* the number of columns of the first matrix equals the number of rows of the second matrix. In this case, the elements of the result are defined by

$$\begin{aligned} c_{ij} &= \sum_{k=1}^{p} a_{ik} b_{kj} \\ &= a_{i1}\, b_{1j} + a_{i2}\, b_{2j} + \cdots + a_{ip} b_{pj} \end{aligned} \tag{A.20}$$

For example,

$$\begin{bmatrix} a_{11} & a_{12} & a_{13} \\ a_{21} & a_{22} & a_{23} \end{bmatrix} \begin{bmatrix} b_{11} & b_{12} \\ b_{21} & b_{22} \\ b_{31} & b_{32} \end{bmatrix} = \begin{bmatrix} (a_{11}b_{11} + a_{12}b_{21} + a_{13}b_{31}) & (a_{11}b_{12} + a_{12}b_{22} + a_{13}b_{32}) \\ (a_{21}b_{11} + a_{22}b_{21} + a_{23}b_{31}) & (a_{21}b_{12} + a_{22}b_{22} + a_{23}b_{32}) \end{bmatrix} \tag{A.21}$$

Observe that the entry c_{ij} in **C** is the familiar dot product of the two vectors consisting of the ith row of **A** and the jth column of **B**:

$$
\begin{aligned}
c_{ij} &= [a_{i1} \quad a_{i2} \quad \cdots \quad a_{ip}]\begin{bmatrix} b_{1j} \\ b_{2j} \\ \vdots \\ b_{pj} \end{bmatrix} \\
&= a_{i1}b_{1j} + a_{i2}b_{2j} + \cdots + a_{ip}b_{pj}
\end{aligned} \tag{A.22}
$$

The reader should verify that the matrix equations in (A.16) and (A.17) indeed give the two equations of (A.1) using the above matrix multiplication result.

Example A.3

Evaluate the matrices **AB**, **CA**, **AD**, and **C**−2**D** given the following matrices:

$$\mathbf{A} = \begin{bmatrix} 3 & 1 \\ 2 & 1 \\ -1 & 0 \end{bmatrix}$$

$$\mathbf{B} = \begin{bmatrix} 1 & 1 \\ -1 & 2 \end{bmatrix}$$

$$\mathbf{C} = \begin{bmatrix} 1 & 2 & 1 \\ 0 & 1 & 1 \end{bmatrix}$$

$$\mathbf{D} = \begin{bmatrix} 2 & 1 & 0 \\ 3 & 1 & 1 \end{bmatrix}$$

Solution

$$\mathbf{AB} = \begin{bmatrix} 3 & 1 \\ 2 & 1 \\ -1 & 0 \end{bmatrix}\begin{bmatrix} 1 & 1 \\ -1 & 2 \end{bmatrix} = \begin{bmatrix} 2 & 5 \\ 1 & 4 \\ -1 & -1 \end{bmatrix}$$

$$\mathbf{CA} = \begin{bmatrix} 1 & 2 & 1 \\ 0 & 1 & 1 \end{bmatrix}\begin{bmatrix} 3 & 1 \\ 2 & 1 \\ -1 & 0 \end{bmatrix} = \begin{bmatrix} 6 & 3 \\ 1 & 1 \end{bmatrix}$$

$$\mathbf{AD} = \begin{bmatrix} 3 & 1 \\ 2 & 1 \\ -1 & 0 \end{bmatrix}\begin{bmatrix} 2 & 1 & 0 \\ 3 & 1 & 1 \end{bmatrix} = \begin{bmatrix} 9 & 4 & 1 \\ 7 & 3 & 1 \\ -2 & -1 & 0 \end{bmatrix}$$

$$\mathbf{C} - 2\mathbf{D} = \begin{bmatrix} 1 & 2 & 1 \\ 0 & 1 & 1 \end{bmatrix} - 2\begin{bmatrix} 2 & 1 & 0 \\ 3 & 1 & 1 \end{bmatrix} = \begin{bmatrix} -3 & 0 & 1 \\ -6 & -1 & -1 \end{bmatrix}$$

The final operation in matrix algebra, the matrix inverse, is the counterpart to division in scalar algebra. In scalar algebra, the multiplicative inverse of a scalar is defined to be the scalar that, when multiplied by the original scalar, yields unity:

$$
\begin{aligned}
(a^{-1})(a) &= 1 \\
(a)(a^{-1}) &= 1
\end{aligned} \tag{A.23}
$$

The inverse of a square ($n \times n$) matrix $\mathbf{A}$, $\mathbf{A}^{-1}$, is similarly defined by

$$\begin{aligned} \mathbf{A}\mathbf{A}^{-1} &= \mathbf{1}_n \\ \mathbf{A}^{-1}\mathbf{A} &= \mathbf{1}_n \end{aligned} \tag{A.24}$$

where $\mathbf{1}_n$ denotes the $n \times n$ identity matrix that has ones on the main diagonal and zeros elsewhere:

$$\mathbf{1}_n = \begin{bmatrix} 1 & 0 & \cdots & 0 \\ 0 & 1 & \ddots & \vdots \\ \vdots & \ddots & \ddots & 0 \\ 0 & \cdots & 0 & 1 \end{bmatrix} \tag{A.25}$$

For example,

$$\mathbf{1}_3 = \begin{bmatrix} 1 & 0 & 0 \\ 0 & 1 & 0 \\ 0 & 0 & 1 \end{bmatrix} \tag{A.26}$$

Note that the inverse of a matrix is only defined for square matrices, that is, those with the same number of rows as columns. Otherwise it would not be possible to multiply $\mathbf{A}$ and $\mathbf{A}^{-1}$ in either order as in (A.24).

The final task is to determine how to obtain $\mathbf{A}^{-1}$ from a given $\mathbf{A}$. To do this we denote the entries in $\mathbf{A}^{-1}$ as β_{ij}. Thus (A.24) becomes for a 3×3 matrix

$$\underbrace{\begin{bmatrix} a_{11} & a_{12} & a_{13} \\ a_{21} & a_{22} & a_{23} \\ a_{31} & a_{32} & a_{33} \end{bmatrix}}_{\mathbf{A}} \underbrace{\begin{bmatrix} \beta_{11} & \beta_{12} & \beta_{13} \\ \beta_{21} & \beta_{22} & \beta_{23} \\ \beta_{31} & \beta_{32} & \beta_{33} \end{bmatrix}}_{\mathbf{A}^{-1}} = \underbrace{\begin{bmatrix} 1 & 0 & 0 \\ 0 & 1 & 0 \\ 0 & 0 & 1 \end{bmatrix}}_{\mathbf{1}_3} \tag{A.27}$$

Multiplying this out gives three sets of equations:

$$\begin{aligned} \mathbf{A}\begin{bmatrix} \beta_{11} \\ \beta_{21} \\ \beta_{31} \end{bmatrix} &= \begin{bmatrix} 1 \\ 0 \\ 0 \end{bmatrix} \\ \mathbf{A}\begin{bmatrix} \beta_{12} \\ \beta_{22} \\ \beta_{32} \end{bmatrix} &= \begin{bmatrix} 0 \\ 1 \\ 0 \end{bmatrix} \\ \mathbf{A}\begin{bmatrix} \beta_{13} \\ \beta_{23} \\ \beta_{33} \end{bmatrix} &= \begin{bmatrix} 0 \\ 0 \\ 1 \end{bmatrix} \end{aligned} \tag{A.28}$$

These can be solved using Cramer's rule. The result can be summarized by the following rule for computing the inverse of $\mathbf{A}$.

Step 1: Form the *transpose* of $\mathbf{A}$, denoted by $\mathbf{A}^t$, by rotating the matrix about its main diagonal, i.e., $[\mathbf{A}^t]_{ij} = a_{ji}$. For a 3×3 matrix this becomes

$$\mathbf{A}^t = \begin{bmatrix} a_{11} & a_{21} & a_{31} \\ a_{12} & a_{22} & a_{32} \\ a_{13} & a_{23} & a_{33} \end{bmatrix} \tag{A.29}$$

For example,

$$\begin{bmatrix} 1 & 2 & 3 \\ 4 & 5 & 6 \\ 7 & 8 & 9 \end{bmatrix}^t = \begin{bmatrix} 1 & 4 & 7 \\ 2 & 5 & 8 \\ 3 & 6 & 9 \end{bmatrix} \tag{A.30}$$

Essentially then, *the rows of the original matrix become the columns of the transpose.* If **A** is symmetric, this step may be bypassed, since in this case $\mathbf{A}^t = \mathbf{A}$.

Step 2: Form the *adjoint* of **A**. This is obtained by replacing each element of $\mathbf{A}^t$ with its *cofactor*, which is the determinant of the matrix left by striking out the row and column of that element and multiplying it by the checkerboard sign. For example, the adjoint of **A** is formed from the transpose in (A.29) as

$$\text{adj}(\mathbf{A}) = \begin{bmatrix} +\begin{vmatrix} a_{22} & a_{32} \\ a_{23} & a_{33} \end{vmatrix} & -\begin{vmatrix} a_{12} & a_{32} \\ a_{13} & a_{33} \end{vmatrix} & +\begin{vmatrix} a_{12} & a_{22} \\ a_{13} & a_{23} \end{vmatrix} \\ -\begin{vmatrix} a_{21} & a_{31} \\ a_{23} & a_{33} \end{vmatrix} & +\begin{vmatrix} a_{11} & a_{31} \\ a_{13} & a_{33} \end{vmatrix} & -\begin{vmatrix} a_{11} & a_{21} \\ a_{13} & a_{23} \end{vmatrix} \\ +\begin{vmatrix} a_{21} & a_{31} \\ a_{22} & a_{32} \end{vmatrix} & -\begin{vmatrix} a_{11} & a_{31} \\ a_{12} & a_{32} \end{vmatrix} & +\begin{vmatrix} a_{11} & a_{21} \\ a_{12} & a_{22} \end{vmatrix} \end{bmatrix} \tag{A.31}$$

Step 3: Form the inverse of **A** by dividing each element of the adjoint by the determinant of **A**, $|\mathbf{A}|$:

$$\mathbf{A}^{-1} = \frac{1}{|\mathbf{A}|}\,\text{adj}(\mathbf{A}) \tag{A.32}$$

The inverse is also useful is solving sets of linear algebraic equations. For example, if those equations are written in matrix form as

$$\mathbf{AX} = \mathbf{B} \tag{A.33}$$

then multiplying both sides (on the left) by $\mathbf{A}^{-1}$ yields the solution as

$$\underbrace{\mathbf{A}^{-1}\mathbf{A}}_{\mathbf{1}_n}\mathbf{X} = \mathbf{A}^{-1}\mathbf{B} \tag{A.34}$$

or

$$\mathbf{X} = \mathbf{A}^{-1}\mathbf{B} \tag{A.35}$$

Example A.4

Solve the equations of Example A.1 using the matrix inverse.

Solution The equations are written in matrix form as

$$\underbrace{\begin{bmatrix} 1 & 3 & -1 \\ 2 & -4 & 2 \\ 3 & 1 & 1 \end{bmatrix}}_{\mathbf{A}} \underbrace{\begin{bmatrix} x \\ y \\ z \end{bmatrix}}_{\mathbf{X}} = \underbrace{\begin{bmatrix} 4 \\ 3 \\ 2 \end{bmatrix}}_{\mathbf{B}}$$

The inverse of **A** is formed as follows :
Step 1: Form the transpose:

$$\mathbf{A}^t = \begin{bmatrix} 1 & 2 & 3 \\ 3 & -4 & 1 \\ -1 & 2 & 1 \end{bmatrix}$$

Step 2: Form the adjoint of **A**:

$$\text{adj}(\mathbf{A}) = \begin{bmatrix} +\begin{vmatrix} -4 & 1 \\ 2 & 1 \end{vmatrix} & -\begin{vmatrix} 3 & 1 \\ -1 & 1 \end{vmatrix} & +\begin{vmatrix} 3 & -4 \\ -1 & 2 \end{vmatrix} \\ -\begin{vmatrix} 2 & 3 \\ 2 & 1 \end{vmatrix} & +\begin{vmatrix} 1 & 3 \\ -1 & 1 \end{vmatrix} & -\begin{vmatrix} 1 & 2 \\ -1 & 2 \end{vmatrix} \\ +\begin{vmatrix} 2 & 3 \\ -4 & 1 \end{vmatrix} & -\begin{vmatrix} 1 & 3 \\ 3 & 1 \end{vmatrix} & +\begin{vmatrix} 1 & 2 \\ 3 & -4 \end{vmatrix} \end{bmatrix}$$

$$= \begin{bmatrix} -6 & -4 & 2 \\ 4 & 4 & -4 \\ 14 & 8 & -10 \end{bmatrix}$$

Step 3: Form the inverse of **A**:

$$\mathbf{A}^{-1} = \frac{1}{|\mathbf{A}|}\,\text{adj}(\mathbf{A})$$

$$= \frac{1}{-8}\,\text{adj}(\mathbf{A})$$

$$= \begin{bmatrix} \frac{3}{4} & \frac{1}{2} & -\frac{1}{4} \\ -\frac{1}{2} & -\frac{1}{2} & \frac{1}{2} \\ -\frac{7}{4} & -1 & \frac{5}{4} \end{bmatrix}$$

Step 4: Form and evaluate (A.35), $\mathbf{X} = \mathbf{A}^{-1}\mathbf{B}$:

$$\underbrace{\begin{bmatrix} x \\ y \\ z \end{bmatrix}}_{\mathbf{X}} = \underbrace{\begin{bmatrix} \frac{3}{4} & \frac{1}{2} & -\frac{1}{4} \\ -\frac{1}{2} & -\frac{1}{2} & \frac{1}{2} \\ -\frac{7}{4} & -1 & \frac{5}{4} \end{bmatrix}}_{\mathbf{A}^{-1}} \underbrace{\begin{bmatrix} 4 \\ 3 \\ 2 \end{bmatrix}}_{\mathbf{B}}$$

$$= \begin{bmatrix} 4 \\ -\frac{5}{2} \\ -\frac{15}{2} \end{bmatrix}$$

as before

A particularly important and easy to remember inverse is for a 2×2 matrix:

$$\begin{bmatrix} a_{11} & a_{12} \\ a_{21} & a_{22} \end{bmatrix}^{-1} = \frac{1}{a_{11}a_{22} - a_{12}a_{21}} \begin{bmatrix} a_{22} & -a_{12} \\ -a_{21} & a_{11} \end{bmatrix} \tag{A.36}$$

In other words, the inverse is formed by swapping the main-diagonal terms, negating the off-diagonal terms, and dividing the resulting matrix by the determinant of **A.** Hence the solution of the set of two equations in (A.1) becomes

$$\begin{bmatrix} x_1 \\ x_2 \end{bmatrix} = \frac{1}{a_{11}a_{22} - a_{12}a_{21}} \begin{bmatrix} a_{22} & -a_{12} \\ -a_{21} & a_{11} \end{bmatrix} \begin{bmatrix} b_1 \\ b_2 \end{bmatrix}$$

$$= \frac{1}{a_{11}a_{22} - a_{12}a_{21}} \begin{bmatrix} a_{22}b_1 - a_{12}b_2 \\ a_{11}b_2 - a_{21}b_1 \end{bmatrix} \tag{A.37}$$

as before.

Index

A

Active filters, 326
Adjoint matrix, 493
Admittance, 279
Algebraic equations, 485
Alternating current (ac), 51
Ammeters, 116
Ampere, 7
Amplifier, 118
Amplitude modulation, 341
Analytic continuation, 391
Apparent power, 293
Automobile storage battery, 35
Average power, 292

B

Balanced load, 329
Balanced wye-wye connection, 329
Balanced delta-delta connection, 332
Bandpass filter, 322
Bandreject filter, 322
Bandwidth of bandpass filter, 324
Bandwidth of bandreject filter, 325
Bandwidth of filters, 324
Bipolar junction transistor, 96
Branch, 164
Bridge, Wheatstone, 150
Buffer amplifier, 203

C

Capacitance, 213
Capacitor, energy stored in, 215
Capacitor impedance in Laplace domain, 396
Capacitor impedance in phasor domain, 278
Capacitor response to dc source, 234
Capacitor, 213
Capacitor, continuity of voltage, 220
Capacitors in series and in parallel, 218
Causal, 443
Characteristic equation, 376, 384, 425
Charge, 2
Charge, conservation of, 14
Circuit analysis, 1, 10
Circuit loop, 18
Circuit model, 2
Circuit reduction, 73
Classical solution, 361
Compact disc, 1
Comparator, 204
Completing the square, 410
Complex conjugate, 270
Complex exponential, 268
Complex exponential source, 274
Complex numbers, polar form, 268
Complex numbers, rectangular form, 268
Complex numbers, 267
Complex plane, 267
Complex power, 293
Conductance, 56
Conjugate of a complex number, 270
Conservation of charge, 12, 14
Conservation of energy, 12
Conservation of power, 28, 294
Controlled sources, 94
Convolution, 443
Coulomb, 2
Coulomb's law, 2
Coupling coefficient, 229
Cover-up rule, 405, 407, 410
Cramer's rule, 487
Critically damped response, 376, 382, 426
Crosstalk in digital circuits, 465
Crosstalk, 342
Current, 7
Current, alternating (ac), 51
Current, direct (dc), 51
Current division, 80
Current source, controlled, 95
Current source, independent, 53

D

Damping, 382, 486
D'Arsonval meter, 116
Dc power supply, op amp, 191

Deactivated sources, 136
Decoupling capacitors in digital circuits, 252
Delta connection, 332
Dependent source, see controlled source,
Dependent sources, 94
Determinant, 487
Difference amplifier, 199
Differential equations, 235
Differential operator, 236, 373, 383
Differentiator, op amp, 239
Digital circuits, lands, 251
Digital circuits, logic gates, 251
Digital circuits, power supply sag, 251
Direct current (dc), 51
Direct method, 85
Dot convention, mutual inductance, 228

E

Effective values of current and voltage, 310
Electric circuit, 12
Electric field, 215
Electric filter, 316, 322
Electricity bill, 39
Electronic devices, 95
Electronic timer, 250
Element, 10
Elementary row operations, 486
Energy stored in a capacitor, 215
Energy stored in an inductor, 222
Energy, conservation of, 23, 294
Energy, consumption of, 39
Energy, generation of, 37, 327
Energy, 10
Equivalent circuits, 32
Euler's identity, 268, 377, 410

F

Farad, 213
Faraday's law, 220, 226, 232
Filters, active (op amp), 326
Filters, bandpass, 324
Filters, bandreject, 325
Filters, electric, 322
Filters, highpass, 324
Filters, lowpass, 320, 322
First-order circuit, 361, 413
Forced solution, 363, 375, 415, 419
Fourier series, 263
Frequency, cyclic, 261
Frequency, radian, 261
Frequency, resonant, 321
Frequency, cutoff, 322
Frequency, repetition, 264
Frequency response of circuits, 316
Frequency response using PSPICE, 335
Frequency-domain circuit, 277
Fundamental theorem of algebra, 376, 405

G

Gain of inverting amplifier, 195
Gain of noninverting amplifier, 198
Gauss elimination, 486
Ground, 38

H

Heater electric, 213
Henry, 220
Hertz, 262

I

Ideal op amp, 193
Ideal transformer, 231
Impedance in Laplace domain, 394
Impedance in phasor domain, 277
Impedance, 277
Impulse function, 392
Impulse response, 435
Independent current source, 53
Independent voltage source, 51
Induced voltage, 220
Inductance, 220
Inductor, energy stored in, 222
Inductor, response to dc source, 235
Inductor, 220
Inductors, continuity of current, 225
Inductors, leakage flux, 226
Inductors in series and in parallel, 224
Initial conditions, 401
Instantaneous power, 292
Instrumentation amplifier, 200
Integrated circuit, VLSI, 326
Integrator, op amp, 239
International system of units (SI), 2
Inverse Laplace transform, 403
Inverting amplifier, 194

J

Joule, 3
Junction field-effect transistor, 95

K

Kirchhoff's current law (KCL), 12, 13
Kirchhoff's voltage law (KVL), 12, 18

L

Lagging power factor, 300
Laplace transform, 387

Leading power factor, 300
Light flasher, 464
Lightning, 54
Line-to-line voltage, 329
Line current in three-phase systems, 330, 332
Linear circuit, 135
Linearity, 388, 446
Load, 153, 303, 329
Loading effect, 203
Logarithm, 387
Loop, 18
Lowpass filter, 322
Lumped circuit element, 10

M

Magnetic coupling, 226
Magnetic field, 8
Magnetic flux, 220
Magnitude of complex number, 267
MATLAB, abs command, 337
MATLAB, angle command, 337
MATLAB, complex numbers, 337
MATLAB, defining polynomials in, 338
MATLAB, defining solution points, 339
MATLAB, freqs command, 339
MATLAB, impulse command, 458
MATLAB, linspace command, 339
MATLAB, logspace command, 339
MATLAB, plotting a function, 339
MATLAB, polyval command, 338
MATLAB, residue command, 457
MATLAB, roots command, 457
MATLAB, semilog command, 340
MATLAB, specifying polynomials, 457
MATLAB, step command, 458
MATLAB, tf2zp command, 457
MATLAB, zp2tf command, 457
MATLAB, 175
Matrix, 490
Matrix algebra, 490
Matrix inverse, 491
Matrix transpose, 492
Maximum power transfer in dc circuits, 154
Maximum power transfer in ac circuits, 223
Mesh, 164
Mesh-current method, 164, 290, 432, 433
Method of undetermined coefficients, 364
Mutual inductance dot convention, 228
Mutual inductance, 226, 396, 290

N

Natural solution, 363, 375, 415, 419
Negative resistor, 56
Node, 10
Node-voltage method, 155
Noninverting amplifier, 198
Nonsimple loops, 85
Nonsimple nodes, 85
Norton equivalent circuit in first-order RC circuits, 370
Norton equivalent circuit in first-order RL circuits, 367
Norton equivalent circuit, 147

O

Ohm's law, 56
Ohmmeters, 116
Op amp buffer, 203
Op amp comparator, 204
Op amp difference amplifier, 199
Op amp differentiator, 239
Op amp ideal model, 193
Op amp integrator, 239
Op amp inverting amplifier, 194
Op amp modeling in PSPICE, 207
Op amp noninverting amplifier, 198
Op amp saturation current, 193
Op amp saturation voltage, 193
Op amp slew rate, 194
Op amp summer, 201
Op amp virtual short-circuit, 193
Op amp, 191
Open circuit, 136
Open-circuit voltage, 143
Operational amplifier see op amp,
Overdamped response, 376, 382, 426
Overshoot in digital circuits, 465

P

Parallel connection of capacitors, 218
Parallel connection of inductors, 224
Parallel connection of resistors, 68
Parallel connections, 30
Parallel resonance, 321
Parallel RLC circuit, 383, 427
Partial fraction expansion, 403
Passive sign convention, 10
Period, 263
Periodic functions, 263
Permeability, 231, 221
Phase angle, 267, 268
Phase voltage, 328
Phasor, 277
Phasor circuit, 277
Phasor diagrams, 314
Phasor method, 274
Photocell, 206
Polar form, 268
Poles, 404

Poles in partial fraction expansion, complex conjugate, 409
Poles in partial fraction expansion, real, distinct, 405
Poles in partial fraction expansion, real, repeated, 407
Poles in partial fraction expansion, 404
Potential difference, 4
Power, apparent VA, 293
Power, average, 292
Power, complex, 293
Power, conservation of, 28, 294
Power, instantaneous, 292
Power, maximum transfer of, 303
Power, reactive, 294
Power, superposition of average, 304
Power, volt-amperes reactive VAR, 294
Power, 10
Power distribution, 37, 327
Power factor, leading, lagging, 300
Power factor, 299
Power factor correction, 302
Printed circuit board, 466
PSPICE, .AC, 333
PSPICE, .DC, 108
PSPICE, .ENDS, 206
PSPICE, .PRINT, 109
PSPICE, .PROBE, 244
PSPICE, .SUBCKT, 206
PSPICE, .TRAN, 242
PSPICE, capacitors and inductors, 241
PSPICE, comments, 109
PSPICE, controlled sources, 111
PSPICE, element specification, 107
PSPICE, frequency response, 335
PSPICE, modeling the impulse source, 453
PSPICE, mutual inductance, 242
PSPICE, node voltages, 106
PSPICE, op amp model, 207
PSPICE, powers of ten, 108
PSPICE, PULSE source specification, 243
PSPICE, PWL source specification, 243
PSPICE, SIN source specification, 243
PSPICE, UIC, 243
PSPICE, 105

Q

Q of bandpass filter, 325
Q of bandreject filter, 325
Q of filters, 325

R

Radian frequency, 261
Radio, AM tuner, 341
RC circuit, 369, 417
Reactance, 278
Reactive power, 294
Rectangular form, 268
Reference node in node-voltage equations, 155
Reflected resistance with ideal transformers, 232
Region of convergence, 391
Reluctance, 231
Residential power distribution, 36
Residues, 458
Resistance, 56
Resistance measurement, 116, 118
Resistor, 55
Resistors, parallel connection, 68
Resistors, series connection, 67
Resonance, 320
Resonant frequency, 321
Response, critically damped, 382, 426
Response, forced, 363, 415, 419
Response, impulse, 435
Response, natural, 363, 415, 419
Response, overdamped, 382, 426
Response, steady state, 365, 415, 419
Response, step, 435
Response, transient, 365, 415, 419
Response, underdamped, 382, 426
Ringing in digital circuits, 465
RC circuit, 369, 417
RL circuit, 362, 413
RLC parallel circuit, 383
RLC series circuit, 373
Rms values of current and voltage, 310
Rms, 52
Roots of characteristic equation, 376

S

S-shift, 389
Saturation in op amps, 195
Second-order circuit, 373, 424
Second-order differential equations, 373
Self inductance, 226
Series connection of capacitors, 219
Series connection of inductors, 224
Series connection of resistors, 67
Series connections, 30
Series resonance, 321
Series RLC circuit, 373, 424, 427
Service entrance panel, 38
Short circuit, 136
Short-circuit current, 148
SI units, 2
Siemens, 56
Sifting property, 392
Single-loop circuits, 59
Single-node-pair circuits, 63
Sinusoid, 261
Sinusoidal response, 266

Slew rate, 194
Source, voltage, 51
Source, current, 53
Source, controlled (dependent), 94
Source impedance, 303
Source modeling, 91
Source resistance, 91, 153
Source transformations, 90
SPICE, 106
Square wave, 264
Stable, 364
Steady-state solution, 365, 415, 419
Steinmetz, Charles Proteus, 274
Step function, 389
Step response, 435
Strain gauge, 205
Stretch-and-bend principle, 34
Summer amplifier, 201
Supernode, 14
Superposition, 136
Susceptance, 279
Switching operation, 361

T

Tellegen's theorem, 29
Thevenin equivalent circuit in first-order RL circuits, 366
Thevenin equivalent circuit in first-order RC circuits, 371
Thevenin equivalent circuit, 143
Three-phase generator, 327
Time constant in first-order RL circuits, 364, 415
Time constant in first-order RC circuits, 370, 419
Time constant, 245, 248, 364, 366, 370, 415, 419, 453, 454
Time invariance, 446
Time shift, 389
Time-domain circuit, 277
Transducer, 204
Transfer function, 237, 316, 436
Transformer, 232
Transformer, ideal, 231
Transformer, step up, step down, 232
Transient solution, 365, 415, 419
Transistor, 95
Transmission line, in digital circuits, 465
Transmission line, in telephone circuits, 342
Transmission line, crosstalk in, 342, 465
Trigonometric formulas, 262
Turns ratio, 232

U

Underdamped response, 377, 382, 426
Undershoot in digital circuits, 465
Unit impulse function, 392
Unit step function, 389
Unity power factor, 300
Unstable, 365

V

Volt, 4
Voltage, 3
Voltage division, 78
Voltage drop, 18
Voltage follower, 203
Voltage rise, 18
Voltmeters, 116

W

Watt, 10
Webers, 220
Wheatstone bridge, 150
Wye connection, 329

Z

Zero-input solution, 414, 419
Zero-state solution, 414, 419
Zeros, 457